"南北极环境综合考察与评估专项"
(CHINARE2015-04-05-03)
极地科技发展战略研究资助

极地科学前沿与热点

顶级期刊论文摘要汇编(1990~2010)

中国科学技术大学出版社

内容简介

本书是“极地国家利益战略评估——极地科技发展战略研究”的一项基础工作，对1990～2010年间发表在*Nature*，*Science*，*PNAS*三大期刊上有关极地科学的650篇论文的摘要进行了分类梳理与翻译，反映了当代国际极地科学的前沿与热点，这为我国的极地科学家、青年研究人员和研究生们提供了一个了解国际极地学术动态的捷径。同时，从对第一作者的国别和所属研究机构以及对不同学科论文发表数量的统计分析，可以了解各个国家极地不同学科的优势研究领域和学科发展动态，这为科学家追踪极地科学前沿、为我国极地科技发展战略的制定提供了重要参考。

本书可以说是一部工具书，它对极地科学相关专业的研究人员、高等院校的教师和学生以及相关政府部门的政策制定者和执行者、基金受理部门的管理和研究人员有重要的参考价值，同时对非极地科学相关学科的研究人员也有一定的参考价值。

图书在版编目(CIP)数据

极地科学前沿与热点：顶级期刊论文摘要汇编：1990～2010/孙立广主编. —合肥：中国科学技术大学出版社，2016.3

ISBN 978-7-312-03742-9

Ⅰ.极…　Ⅱ.孙…　Ⅲ.极地—文集　Ⅳ.P941.6-53

中国版本图书馆CIP数据核字(2015)第175998号

出版　中国科学技术大学出版社
安徽省合肥市金寨路96号，邮编：230026
网址：http://press.ustc.edu.cn

印刷　合肥学苑印务有限公司

发行　中国科学技术大学出版社

经销　全国新华书店

开本　880 mm×1230 mm　1/16

印张　39.25

字数　1208千

版次　2016年3月第1版

印次　2016年3月第1次印刷

定价　200.00元

— 汇编编辑委员会 —

— “极地国家利益战略评估”专题专家委员会 —

（以姓氏笔画为序）

前　言

当今世界，信息网络为我们带来了一个五彩缤纷的世界，与此同时，浩如烟海的文献又让我们在“路”上迷茫。我们在极地古生态、古环境、古气候与大气化学的研究过程中，深感科学文献的重要性，它将引导和启发我们的思路和方向，而海量的文献又让我们无所适从。从学科分类意义上来看，极地科学可能是门类最齐全的科学，几乎涵盖了地球科学和生命科学的所有学科。以地球科学为例，几乎所有的地球科学分支都在极地研究中派上了用场，极地科学是系统地球科学的一部分。地球上除了极地可能没有一块陆地、海洋能容纳这么多国家、这么多门类的科学家聚集在一起探索自然的奥秘。极地是资源的宝库，是发现的摇篮，是国际科学合作的舞台。同时，它也是国家利益和人类利益竞争与平衡的舞台，这就决定了这个舞台的重要性，也就可以非常自然地了解为什么几乎所有有实力的国家都把目光聚焦到了南极和北极，为什么有那么多各个门类的科学家不畏艰险奔向那个冷酷严寒的冰封世界。

我国的极地梦很早就开始了，现在极地研究已成为实现我国“海洋强国梦”的重要组成部分，并得到了国家领导人的高度重视。1984 年，邓小平为南极长城站建站题词：“为人类和平利用南极做出贡献。”习近平主席 2014 年 11 月 18 日在澳大利亚登上了我国的雪龙号极地考察船，他在慰问科考队员时指出：南极科学考察意义重大，是造福人类的崇高事业，中国开展南极科考为人类和平利用南极做出了贡献。他表示愿与国际社会一道，更好地认识南极、保护南极、利用南极。这个讲话是对南极考察队员说的，同样适用于地缘上离我们更近的北极，因为，这是我国国家领导人关心极地、造福人类、和平利用南极的一次宣示。

李克强总理也指出：“极地考察是人类探索地球奥秘的壮举，在我国海洋事业中占有重要地位，对促进可持续发展具有重大意义。”

国家领导人的指示在中国极地考察史上具有里程碑的意义，也为中国的极地科学探索与和平利用南极指出了方向。

我国的极地科学工作者要在和平利用南极上有所作为，首先要在认识南极上有所作为，在这方面要做出更大的贡献。要赶上极地强国在科学认识极地方面所取得的成就，首先应该知道国际极地科学的前沿在哪里，热点在哪里，难点在哪里，也应知道我们所处的位置在哪里，这样我们就可以避免走弯路、走重复的路。同时，在关注不同学科领域的进展中，我们可以打开多个窗户，从不同角度去探索新的方向，寻找新的道路，这也有利于多学科交叉。为此，我们对极地科学按九大研究领域进行了初步划分，即冰川、冰芯古气候，大气科学，海洋古气候、古环境，生物科学，现代海洋科学，地质、测绘，陆地古环境，环境污染，天文、空间物理。对这九个领域在 1990～2010 年中在 *Nature*，*Science*，*PNAS* 三大国际顶级期刊上发表的 650 篇论文摘要分类进行了翻译及分析整理。

从文献的数据统计分析中可以发现，冰川、大气、古气候与古环境及生命科学，不论是在南极还是在北极都是研究的热点；从不同年份文献发表的数据来看，每年的文献发表数量呈波动变化，总体上似乎与国际经济形势的起落一致或略滞后，显然与各国政府对极地研究的投入量有关。

从文献发表的国家来看，美国、英国仍然是亮点论文发表数量的第一梯队，第二梯队是法国、加拿大、德国、瑞士和澳大利亚,中国、日本、俄罗斯等仍是相对落后的,与美英相比还有很大差距。研究水平上的差距是主要的,但是另一方面,由于三大期刊均属于美国或英国,语言方面的劣势和国籍差异也是存在的。因此,数字上的巨大差异并不表示研究水平上有同样大的差异,创造中国科技期刊在国际上的优势地位,是改变这种状态的重要因素。实际上,当我们进一步统计专业学术期刊上各国的论文贡献时,我们发现,上述的巨大差异就减小了,中国的极地论文数量近些年来已经在快速地增长,正在缩小与美英等国家的距离。

重要的不是在于缩小论文数量上的差距,而是在极地科学研究方面,我们能有多少重大的令世人关注的科学发现,极地科学的新发现永远是我们追求的目标。因此我们特别关注三大公认的国际顶级期刊的极地论文,并对这些论文进行了梳理与统计摘编。它将让我们看到极地科学的前沿、热点与高度,尽管我们并不认为这些论文都代表了极地科学的顶级水平,也并不认为在其他综合或专业学术期刊上没有顶级的论文和重大的科学发现。

由于涉及不同的学科领域,译者的专业水平有限,虽然经过美国国立健康研究所(NIH)王玉宏研究员的精细核校,但仍然可能存在误译或译文欠准确的情况。好在,保留了论文摘要原文,读者可以做出自己的判读。

参加本书翻译的有:黄涛、周鑫、何鑫、程文瀚、黄婧、贾楠、刘毅、罗宇涵、晏宏、秦先燕、邵达、邱世灿,臧晶晶进行了后期的编辑工作。

感谢"南北极环境综合考察与评估"专项"极地科技发展战略研究"子专题(CHINARE 2015-04-05-03)[Chinese Polar Environment Comprehensive Investigation & Assessment Programmes (CHINARE 2015-04-05-03)]的资助。

孙立广

2015 年 8 月

目 录

前言 …………………………………………………………………………………………… (*1*)

文献统计概述 ………………………………………………………………………………… (1)

一、冰川、冰芯古气候类文献

1.1 分析概述 ……………………………………………………………………………… (9)

1.2 摘要翻译(南北极) ……………………………………………………………………… (12)

Estimates of Antarctic Precipitation/对南极降雨量的估算 ………………… David H. Bromwich (12)

Physical conditions at the base of a fast moving Antarctic ice stream/一条快速移动的南极冰流底部的物理环境 ……………………………… Engelhardt H, Humphrey N, Kamb B, et al (12)

Recent increase in nitrate concentration of Antarctic snow/近期南极降雪中硝酸盐含量的升高 ………………………………………………………… Mayewski P A, Legrand M R (13)

Changes in the West Antarctic ice-sheet/西南极冰盖的变化 ………… Alley R B, Whillans I M (14)

Satellite-image-derived velocity-field of an Antarctic ice stream/卫星图像获取的一条南极冰流的速度场 …………………………………………… Bindschadler R A, Scambos T A (14)

Rapid disintegration of the Wordie ice shelf in response to atmospheric warming/快速瓦解的Wordie 冰架对大气变暖的响应 ……………………………… Doake C S M, Vaughan D G (15)

Recent variations in Arctic and Antarctic sea-ice covers/近期南北极海冰覆盖的变化 ………………………………………………………………… Gloersen P, Campbell W J (16)

Ice-core record of oceanic emissions of dimethylsulfide during the Last Climate Cycle/末次气候循环期间海洋排放的二甲基硫的冰芯记录 … Legrand M, Fenietsaigne C, Saltzman E S, et al (16)

A collection of diverse micrometeorites recovered from 100 tonnes of Antarctic blue ice/从 100 t 南极蓝冰中获得的不同种类的微陨石 ………… Maurette M, Olinger C, Michellevy M C, et al (17)

Evidence from Antarctic ice cores for recent increases in snow accumulation/南极冰芯中获得的近期雪积累率上升的证据 …………………… Morgan V I, Goodwin I D, Etheridge D M, et al (18)

Antarctic ice volume and contribution to sea-level fall at 20000 yr BP from raised beaches/基于海岸抬升的南极冰量变化及其对 20000 年前海平面下降的研究 ………………………………………… Colhoun E A, Mabin M C G. , Adamson D A, et al (19)

Is the Antarctic ice-sheet growing/南极冰盖是否在增长? ………………………… Jacobs S S (20)

Ice-age atmospheric concentration of nitrous-oxide from an Antarctic ice core/南极冰芯中记录的冰期大气一氧化二氮的浓度 ………………………………… Leuenberger M, Siegenthaler U (20)

Carbon isotope composition of atmospheric CO_2 during the last ice-age from an Antarctic ice core /南极冰芯记录的末次冰盛期大气 CO_2 碳同位素组成 ………………………………………… Leuenberger M, Siegenthaler U, Langway C C (21)

Irregular oscillations of the west Antarctic ice-sheet/西南极冰盖的不规则震荡 ……………………………………………………………………… Macayeal D R (22)

Active volcanism beneath the West Antarctic ice-sheet and implications for ice-sheet stability
/西南极冰盖下的活火山作用及其对冰盖稳定性的影响
…………………………………………………… Blankenship D D,Bell R E,Hodge S M,et al (23)
Nonequilibrium air clathrate hydrates in Antarctic ice:a paleopiezometer for polar ice caps
/南极冰中非平衡空气水包合物:极地冰帽的一种古压力计
…………………………………………………………………… Craig H,Shoji H,Langway C C (24)
Satellite radar interferometry for monitoring ice-sheet motion:application to an Antarctic ice stream
/利用卫星雷达干涉测量法监测冰盖运动:在一条南极冰流上的应用
…………………………………………………… Goldstein R M,Engelhardt H,Kamb B,et al (25)
Extending the Vostok ice-core record of paleoclimate to the penultimate glacial period
/将 Vostok 冰芯古气候记录延伸到倒数第二次冰期
………………………………………………………… Jouzel J,Barkov N I,Barnola J M,et al (26)
Pliocene paleoclimate and east Antarctic ice-sheet history from surficial ash deposits/基于地表
灰沉降的上新世古气候与东南极冰盖历史……… Marchant D R,Swisher C C,Lux D R,et al (26)
The ice record of greenhouse gases/温室气体的冰记录
………………………………………………………… Raynaud D, Jouzel J,Barnola J M,et al (27)
Variations in mercury deposition to Antarctica over the past 34000 years/过去 34000 年南极汞
沉降的变化…………………………………… Vandal G M,Fitzgerald W F,Boutron C F,et al (28)
Climate correlations between Greenland and Antarctica during the past 100000 years
/过去 10 万年中格陵兰和南极的气候关联 ………… Bender M,Sowers T,Dickson M L,et al (28)
Optical-properties of the South-Pole ice at depths between 0.8 and 1 kilometer/南极点深度
0.8～1 km 之间冰的光学特性 ……………… Askebjer P,Barwick S W,Bergstrom L,et al (29)
Kinetics of conversion of air bubbles to air hydrate crystals in Antarctic ice/南极冰中气泡向
空气水合物晶体转换的动力学 ……………………………………………………………… Price P B (30)
Atmospheric gas concentrations over the past century measured in air from firn at the South Pole
/利用南极点粒雪中气体测定的过去一个世纪大气气体浓度
……………………………………………………………… Battle M,Bender M,Sowers T,et al (31)
Climate change during the last deglaciation in Antarctica/南极末次冰消期的气候变化
…………………………………………………… Mayewski P A,Twickler M S,Whitlow S I,et al (32)
Rapid collapse of northern Larsen Ice Shelf,Antarctica/南极北 Larsen 冰架的快速崩塌
…………………………………………………………………………… Rott H,Skvarca P,Nagler T (33)
Recent atmospheric warming and retreat of ice shelves on the Antarctic Peninsula/南极半岛近期
的大气变暖和冰架后退 ………………………………………… Vaughan D G,Doake C S M (33)
Primary production in Antarctic sea ice/南极海冰的初级生产力
…………………………………………………… Arrigo K R,Worthen D L,Lizotte M P,et al (34)
Gases in ice cores/冰芯中的气体 ………………………… Bender M,Sowers T,Brook E (34)
Rapid sea-level rise soon from West Antarctic ice sheet collapse? /西南极冰盖崩溃即将引起快速
海平面上升? ………………………………………………………………………… Bentley C R (35)
Abrupt mid-twentieth-century decline in Antarctic sea-ice extent from whaling records/从捕鲸
记录得出的 20 世纪中期南极海冰范围的突然减退 ………………………… De La Mare W K (36)
Spacecraft offers details of Antarctica/航天器提供了南极洲的细节 ………………… Lawler A (37)

Climate change: Icy message from the Antarctic/气候变化:来自南极的冰冷的信息 …… Murphy E, King J (37)
Predicted reduction in basal melt rates of an Antarctic ice shelf in a warmer climate/预计南极冰架底部融化速率在更暖的气候下将降低 …… Nicholls K W (38)
Influence of subglacial geology on the onset of a West Antarctic ice stream from aerogeophysical observations/基于航空地球物理观测的冰下地质对一条西南极冰流源头的影响 …… Bell R E, Blankenship D D, Finn C A, et al (39)
Changes in the West Antarctic ice sheet since 1963 from declassified satellite photography/基于解密卫星图像获取的 1963 年以来西南极冰盖的变化 …… Bindschadler R, Vornberger P (40)
Breakup and conditions for stability of the northern Larsen Ice Shelf, Antarctica/南极北 Larsen 冰架的分裂及稳定性现状 …… Doake C S M, Corr H F J, Rott H, et al (41)
Asynchrony of Antarctic and Greenland climate change during the last glacial period/末次冰期南极与格陵兰气候变化的异时性 …… Blunier T, Chappellaz J, Schwander J, et al (41)
Global change: signs of past collapse beneath Antarctic ice/全球变化:南极冰下过去崩塌的信号 …… Kerr R A (42)
Global warming and the stability of the West Antarctic Ice Sheet/全球变暖与西南极冰盖的稳定性 …… Oppenheimer M (43)
Perennial Antarctic lake ice: an oasis for life in a polar desert/南极常年湖冰:极地荒漠中的生命绿洲 …… Priscu J C, Fritsen C H, Adams E E, et al (43)
Fast recession of a West Antarctic glacier/一条西南极冰川的快速后退 …… Rignot E J (44)
Pleistocene collapse of the West Antarctic ice sheet/西南极冰盖的更新世崩塌 …… Scherer R P, Aldahan A, Tulaczyk S, et al (45)
Atmospheric CO_2 concentration and millennial-scale climate change during the last glacial period/末次冰期的大气 CO_2浓度与千年尺度气候变化 … Stauffer B, Blunier T, Dallenbach A, et al (45)
Measurements of past ice sheet elevations in interior West Antarctica/西南极内陆过去冰盖高度的测量 …… Ackert R P, Barclay D J, Borns H W, et al (46)
Past and future grounding-line retreat of the West Antarctic Ice Sheet/过去和未来的西南极冰盖接地线的后退 …… Conway H, Hall B L, Denton G H, et al (47)
Ice core records of atmospheric CO_2 around the last three glacial terminations/最后三次冰消期附近大气 CO_2的冰芯记录 …… Fischer H, Wahlen M, Smith J, et al (48)
Tributaries of West Antarctic Ice streams revealed by RADARSAT interferometry/RADARSAT 干涉测量法揭示的西南极冰流的支流 …… Joughin I, Gray L, Bindschadler R, et al (49)
More than 200 meters of lake ice above subglacial lake Vostok, Antarctica/在南极 Vostok 冰下湖之上超过 200 m 的湖冰 …… Jouzel J, Petit J R, Souchez R, et al (49)
Moves are afoot to probe the lake trapped beneath Antarctic ice/探测南极冰下湖的行动在准备中 …… Nadis S (50)
Climate and atmospheric history of the past 420000 years from the Vostok ice core, Antarctica/南极 Vostok 冰芯记录的过去 420000 年的气候与大气历史 …… Petit J R, Jouzel J, Raynaud D, et al (51)
Widespread complex flow in the interior of the Antarctic ice sheet/南极冰盖内部分布广泛的复杂流 …… Bamber J L, Vaughan D G, Joughin I (52)

Millennial-scale instability of the Antarctic ice sheet during the last glaciations/末次冰期南极冰盖千年尺度上的不稳定性 ………………………… Kanfoush S L, Hodell D A, Charles C D, et al (52)
Cenozoic deep-sea temperatures and global ice volumes from Mg/Ca in benthic foraminiferal calcite /得自深海有孔虫碳酸盐 Mg/Ca 记录的新生代深海温度与全球冰量 ………………………… Lear C H, Elderfield H, Wilson P A (53)
The influence of Antarctic sea ice on glacial-interglacial CO_2 variations/南极海冰对冰期-间冰期 CO_2 变化的影响 ………………………… Stephens B B, Keeling R F (54)
Timing of millennial-scale climate change in Antarctica and Greenland during the last glacial period /末次冰期南极和格陵兰千年尺度气候变化的定年 ………………………… Blunier T, Brook E J (54)
Explaining the Weddell Polynya: a large ocean eddy shed at Maud Rise/解释 Weddell 冰间湖:Maud 高地产生的巨大海洋涡流 ………………………… Holland D M (55)
Glaciology: how ice sheets flow/冰川学:冰盖是如何流动的 ………………………… Hulbe C L (56)
Constraints on hydrothermal processes and water exchange in Lake Vostok from helium isotopes /氦同位素揭示的 Vostok 湖热液作用和水交换的约束条件 ………………………… Jean-Baptiste P, Petit J R, Lipenkov V Y, et al (57)
Recent mass balance of polar ice sheets inferred from patterns of global sea-level change /根据全球海平面变化模式推断近期极地冰盖的质量平衡 ………………………… Mitrovica J X, Tamisiea M E, Davis J L, et al (57)
Orbitally induced oscillations in the East Antarctic ice sheet at the Oligocene/Miocene boundary /渐新世/中新世分界处东南极冰盖轨道成因的震荡变化 ………………………… Naish T R, Woolfe K J, Barrett P J, et al (58)
Possible displacement of the climate signal in ancient ice by premelting and anomalous diffusion /古冰中由于预融和不规则扩散而导致气候信号可能的位错 ………………………… Rempel A W, Waddington E D, Wettlaufer J S, et al (60)
Glacial surface temperatures of the southeast Atlantic Ocean/东南大西洋冰期的表面温度 ………………………… Sachs J P, Anderson R F, Lehman S J (60)
Inland thinning of Pine Island Glacier, West Antarctica/西南极松岛冰川的内陆变薄 ………………………… Shepherd A, Wingham D J, Mansley J A D, et al (61)
Physical, chemical and biological processes in Lake Vostok and other Antarctic subglacial lakes /Vostok 湖和其他南极冰下湖的物理、化学和生物过程 ………………………… Siegert M J, Ellis-Evans J C, Tranter M, et al (62)
Climate response to orbital forcing across the Oligocene-Miocene boundary/渐新世-中新世分界处对轨道强迫的气候响应 ………………………… Zachos J C, Shackleton N J, Revenaugh J S, et al (63)
On thickening ice? /不断变厚的冰盖? ………………………… Alley R B (63)
Origin and fate of Lake Vostok water frozen to the base of the East Antarctic ice sheet/冻结在东南极冰盖底部的 Vostok 湖水的来源与去向 ………………………… Bell R E, Studinger M, Tikku A A, et al (64)
Sea-level fingerprinting as a direct test for the source of global meltwater pulse IA/海平面变化作为直接检测全球冰融水量 IA 突变来源的证据 ………………………… Clark P U, Mitrovica J X, Milne G A, et al (65)
Switch of flow direction in an Antarctic ice stream/一条南极冰流的流向转变 ………………………… Conway H, Catania G, Raymond C F, et al (66)

Glacial meltwater dynamics in coastal waters west of the Antarctic peninsula/南极半岛西部近岸水域的冰川融水动力学 …… Dierssen H M, Smith R C, Vernet M (66)
Positive mass balance of the Ross Ice Streams, West Antarctica/西南极 Ross 冰流正向变化的质量平衡 …… Joughin I, Tulaczyk S (67)
Relative timing of deglacial climate events in Antarctica and Greenland/南极和格陵兰冰消期气候事件的相对时间 …… Morgan V, Delmotte M, van Ommen T, et al (68)
Temperature profile for glacial ice at the South Pole: implications for life in a nearby subglacial lake/南极点冰川冰的温度剖面:附近冰下湖生命的推论 …… Price P B, Nagornov O V, Bay R, et al (69)
Ice sheets on the move/冰盖在移动 …… Raymond C F (70)
Rapid bottom melting widespread near Antarctic ice sheet grounding lines/南极冰盖接地线附近普遍的快速底部融化 …… Rignot E, Jacobs S S (71)
An ice sheet remembers/冰盖会记忆 …… Ackert R P (71)
Tidally controlled stick-slip discharge of a West Antarctic ice stream/潮汐控制的一条西南极冰流的黏滑流出 …… Bindschadler R A, King M A, Alley R B, et al (72)
Timing of atmospheric CO_2 and Antarctic temperature changes across termination Ⅲ/在终止Ⅲ期间的大气 CO_2 与南极温度变化的时间 …… Caillon N, Severinghaus J P, Jouzel J, et al (72)
Ice core evidence for Antarctic sea ice decline since the 1950s/20 世纪 50 年代至今南极海冰消退的冰芯证据 …… Curran M A J, van Ommen T D, Morgan V I, et al (73)
Glacier surge after ice shelf collapse/冰架崩塌后的冰川涌动 …… De Angelis H, Skvarca P (74)
Rapid Cenozoic glaciation of Antarctica induced by declining atmospheric CO_2/大气 CO_2 下降导致的新生代南极快速冰川作用 …… De Conto R M, Pollard D (74)
Formation and character of an ancient 19 m ice cover and underlying trapped brine in an “ice-sealed” east Antarctic lake/“冰密封”的东南极湖中 19 m 的覆冰和其下截留的卤水的形成与特征 …… Doran P T, Fritsen C H, McKay C P, et al (75)
Glaciology—Warmer ocean could threaten Antarctic Ice Shelves/冰川学——更暖的海洋可能威胁南极冰架 …… Kaiser J (76)
340000 years centennial-scale marine record of Southern Hemisphere climatic oscillation/340000 年的南半球气候百年尺度震荡的海洋记录 …… Pahnke K, Zahn R, Elderfield H, et al (77)
Larsen ice shelf has progressively thinned/Larsen 冰架已经逐渐变薄 …… Shepherd A, Wingham D, Payne T, et al (78)
Ice core records of atmospheric N_2O covering the last 106000 years/过去 106000 年的大气 N_2O 冰芯记录 …… Sowers T, Alley R B, Jubenville J (78)
Holocene deglaciation of Marie Byrd Land, West Antarctica/西南极 Marie Byrd 地的全新世冰消 …… Stone J O, Balco G A, Sugden D E, et al (79)
Whither Antarctic sea ice? /南极海冰到哪里去? …… Wolff E W (80)
Eight glacial cycles from an Antarctic ice core/一支南极冰芯中的八次冰期旋回 …… Augustin L, Barbante C, Barnes P R F, et al (80)
Bipolar correlation of volcanism with millennial climate change/火山活动与千年尺度气候变化的两极相关性 …… Bay R C, Bramall N, Price P B (81)
Similar meltwater contributions to glacial sea level changes from Antarctic and northern ice sheets/南极和北部冰盖对冰期海平面变化有相似的融水贡献 …… Rohling E J, Marsh R, Wells N C, et al (82)

Ice-sheet and sea-level changes/冰盖和海平面变化
…… Alley R B,Clark P U,Huybrechts P,et al (83)
Retreating glacier fronts on the Antarctic Peninsula over the past half-century/过去半个世纪的南极半岛退却中的冰川前缘 …… Cook A J,Fox A J,Vaughan D G,et al (84)
Snowfall-driven growth in East Antarctic ice sheet mitigates recent sea-level rise/降雪驱动的东南极冰盖增长对近期海平面上升的减缓作用 …… Davis C H,Li Y H,McConnell J R,et al (85)
Stability of the Larsen B ice shelf on the Antarctic Peninsula during the Holocene epoch/全新世时期南极半岛 Larsen B 冰架的稳定性 …… Domack E,Duran D,Leventer A,et al (85)
Antarctic ice puts climate predictions to the test/南极冰使气候预测经受检验 …… Hopkin M (86)
Stable carbon cycle-climate relationship during the late Pleistocene/晚更新世期间稳定的碳循环-气候关系 …… Siegenthaler U,Stocker T F,Monnin E,et al (87)
Atmospheric methane and nitrous oxide of the late Pleistocene from Antarctic ice cores/来自南极冰芯记录的晚更新世大气甲烷和一氧化二氮 …… Spahni R,Chappellaz J,Stocker T F,et al (87)
Eocene bipolar glaciation associated with global carbon cycle changes/与全球碳循环变化相关的始新世两极冰川作用 …… Tripati A,Backman J,Elderfield H,et al (88)
How does the Antarctic ice sheet affect sea level rise? /南极冰盖如何影响海平面上升?
…… Vaughan D G (89)
One-to-one coupling of glacial climate variability in Greenland and Antarctica/格陵兰和南极冰期气候变化的一一耦合 …… Barbante C,Barnola J M,Becagli S,et al (90)
Fortnightly variations in the flow velocity of Rutford Ice Stream,West Antarctica/西南极 Rutford 冰流流速每两周的变化 …… Gudmundsson G H (91)
Paleoclimatic evidence for future ice-sheet instability and rapid sea-level rise/未来冰盖不稳定性和快速海平面上升的古气候证据 …… Overpeck J T,Otto-Bliesner B L,Miller G H,et al (92)
^{10}Be evidence for the Matuyama-Brunhes geomagnetic reversal in the EPICA Dome C ice core /EPICA 冰穹 C 冰芯中 Matuyama-Brunhes 地磁反转的^{10}Be 证据
…… Raisbeck G M,Yiou F,Cattani O,et al (93)
Plio-pleistocene ice volume, Antarctic climate, and the global δ^{18}O record/上一更新世南极冰量、南极气候和全球 δ^{18}O 的记录 …… Raymo M E,Lisiecki L E,Nisancioglu K H (93)
30000 years of cosmic dust in Antarctic ice/南极冰中 30000 年的宇宙尘埃记录
…… Winckler G,Fischer H (94)
Effect of sedimentation on ice-sheet grounding-line stability/沉积物对冰盖接地线稳定性的影响
…… Alley R B,Anandakrishnan S,Dupont T K,et al (95)
No extreme bipolar glaciation during the main Eocene calcite compensation shift/始新世主要碳酸盐补偿深度偏移期间无极端两极冰川作用 …… Edgar K M,Wilson P A,Sexton P F,et al (95)
Orbital and millennial Antarctic climate variability over the past 800000 years/过去 800000 年的轨道与千年尺度南极气候变化 …… Jouzel J,Masson-Delmotte V,Cattani O,et al (97)
Northern Hemisphere forcing of climatic cycles in Antarctica over the past 360000 years/过去 360000 年南极气候旋回的北半球强迫 …… Kawamura K,Parrenin F,Lisiecki L,et al (97)
Deglaciation mysteries/冰消的秘密 …… Keeling R F (99)
Recent sea-level contributions of the Antarctic and Greenland ice sheets/最近南极和格陵兰冰盖对海平面的贡献 …… Shepherd A,Wingham D (99)

Free-drifting icebergs: hot spots of chemical and biological enrichment in the Weddell Sea/自由漂浮的冰山:Weddell 海中化学物质和生物富集的热点
…… Smith K L,Robison B H,Helly J J,et al (100)
Climate change:a matter of firn/气候变化:粒雪的问题 …… Cuffey K M (101)
Thresholds for Cenozoic bipolar glaciation/新生代两极冰川作用的阈值
…… DeConto R M,Pollard D,Wilson P A,et al (101)
Dust-climate couplings over the past 800000 years from the EPICA Dome C ice core/EPICA 冰穹 C 冰芯中过去 800000 年的尘埃-气候耦合 …… Lambert F,Delmonte B,Petit J R,et al (102)
High-resolution carbon dioxide concentration record 650000-800000 years before present /距今 650000～800000 年的高分辨率二氧化碳浓度记录
…… Luthi D,Le Floch M,Bereiter B,et al (103)
Eocene/oligocene ocean de-acidification linked to Antarctic glaciation by sea-level fall/通过海平面下降与南极冰川作用相联系的始新世/渐新世海洋脱酸作用
…… Merico A,Tyrrell T,Wilson P A (104)
Simultaneous teleseismic and geodetic observations of the stick-slip motion of an Antarctic ice stream /同时进行的对一条南极冰流黏滑运动遥测地震学和测地学观测
…… Wiens D A,Anandakrishnan S,Winberry J P,et al (105)
Reassessment of the potential sea-level rise from a collapse of the West Antarctic Ice Sheet /对西南极冰盖崩溃可能引起的海平面上升的重新评估
…… Bamber J L,Riva R E M,Vermeersen B L A,et al (106)
Interhemispheric Atlantic seesaw response during the last deglaciation/末次冰期期间两半球的大西洋跷跷板响应 …… Barker S,Diz P,Vautravers M J,et al (107)
The Last Glacial Maximum/末次冰盛期 …… Clark P U,Dyke A S,Shakun J D,et al (108)
Global change Interglacial and future sea level/全球变化:间冰期和未来的海平面
…… Clark P U, Huybers P (108)
Stable isotope constraints on Holocene carbon cycle changes from an Antarctic ice core/南极冰芯记录的全新世碳循环变化的同位素约束 …… Elsig J,Schmitt J,Leuenberger D,et al (109)
Polar firn air reveals large-scale impact of anthropogenic mercury emissions during the 1970s /极地积雪中的气体揭示了 20 世纪 70 年代人类源汞排放的大范围影响
…… Fain X,Ferrari C P,Dommergue A,et al (110)
Global change West-side story of Antarctic ice/全球变化:南极冰西边的故事
…… Huybrechts P (111)
Climate change early survival of Antarctic ice/气候变化:南极冰的早期保存
…… Lemarchand D (111)
Global cooling during the Eocene-Oligocene climate transition/始新世-渐新世气候转换期间的全球变冷 …… Liu Z H,Pagani M,Zinniker D,et al (112)
Obliquity-paced Pliocene West Antarctic ice sheet oscillations/与地轴倾角变化同步的上新世西南极冰盖震荡 …… Naish T,Powell R,Levy R,et al (113)
The future of ice sheets and sea ice: between reversible retreat and unstoppable loss/冰盖和海冰的未来:可逆转的消退和不可阻挡的损耗之间 …… Notz D (114)
Modelling West Antarctic ice sheet growth and collapse through the past five million years/过去五百万年西南极冰盖增长和崩溃的模拟 …… Pollard D,DeConto R M (115)

Extensive dynamic thinning on the margins of the Greenland and Antarctic ice sheets/格陵兰和南极冰盖边缘的广泛动态变薄 …………… Pritchard H D，Arthern R J，Vaughan D G，et al (116)
Evidence for warmer interglacials in East Antarctic ice cores/南极冰芯中更温暖的间冰期的证据 ………………………………………………………… Sime L C，Wolff E W，Oliver K I C，et al (118)
Warming of the Antarctic ice-sheet surface since the 1957 International Geophysical Year /自1957年国际地球物理年开始的南极冰盖表面变暖 ……………………………………………………… Steig E J，Schneider D P，Rutherford S D，et al (118)
The Gamburtsev mountains and the origin and early evolution of the Antarctic Ice Sheet /Gamburtsev 山脉与南极冰盖的起源和早期演变 …… Sun B，Siegert M J，Mudd S M，et al (119)
The Last Glacial Termination/末次冰期的终止 ……………………………………………… Denton G H，Anderson R F，Toggweiler J R，et al (120)
Abrupt change of Antarctic moisture origin at the end of Termination Ⅱ/终止Ⅱ末端南极水汽来源的突然变化 ……………………………… Masson-Delmotte V，Stenni B，Blunier T，et al (121)

二、大气科学类文献

2.1 分析概述 …………………………………………………………………………………………………… (125)
2.2 大气科学类文献摘要翻译——南极 ………………………………………………………………………… (128)
Wind-driven upwelling in the Southern Ocean and the deglacial rise in atmospheric CO_2/南大洋风驱动上升流与大气 CO_2 在冰消期的上升 ……… Anderson R F，Ali S，Bradtmiller L I，et al (128)
Saturation of the southern ocean CO_2 sink due to recent climate change/最近气候变化导致南大洋 CO_2 汇的饱和 …………… Corinne Le Quéré，Christian Rödenbeck，Erik T Buitenhuis，et al (128)
The sensitivity of polar ozone depletion to proposed geoengineering schemes/极地臭氧损耗对被提议的地球工程计划的敏感性 ……………………………… Tilmes S. Muller R，Salawitch R (129)
Glacial greenhouse-gas fluctuations controlled by ocean circulation changes/海洋环流的变化控制冰川温室气体的波动 ……………………………………………… Schmittner A，Galbraith E D (130)
Climate change：When did the icehouse cometh? /气候变化：冰室什么时候来到？ ……………………………………………………………………………………… Pekar S F (131)
High-resolution carbon dioxide concentration record 650000-800000 years before present/65万至80万年前高分辨率二氧化碳浓度记录 …………… Luthi D，Le Floch M，Bereiter B，et al (131)
Orbital and millennial-scale features of atmospheric CH_4 over the past 800000 years/过去80万年大气中甲烷在轨道尺度和千年尺度的特征 ……………… Loulergue L，Schilt A，Spahni R，et al (132)
Atmospheric CO_2 and climate on millennial time scales during the last glacial period/末次冰期千年时间尺度的大气二氧化碳和气候变化 ……………………………………… Ahn J，Brook E J (133)
Atmospheric science：revisiting ozone depletion/大气科学：再谈臭氧损耗 ……… Von Hobe M (134)
Boundary layer halogens in coastal Antarctica/南极海岸边界层卤素 ……………………………………………… Saiz-Lopez A，Mahajan A S，Salmon R A，et al (135)
Marine radiocarbon evidence for the mechanism of deglacial atmospheric CO_2 rise/冰消期大气 CO_2 浓度上升机制的海洋放射性碳素证据 ……… Marchitto T M，Lehman S J，Ortiz J D，et al (135)
Carbon dioxide release from the North Pacific abyss during the last deglaciation/末次冰消期从北太平洋的深渊中释放的二氧化碳………… Galbraith E D，Jaccard S L，Pedersen T F，et al (136)
The search for signs of recovery of the ozone layer/对臭氧层恢复迹象的搜索 ……………………………………………………… Weatherhead E C，Andersen S B (137)

Significant warming of the Antarctic winter troposphere/南极冬季对流层的明显变暖
…… Turner J, Lachlan-Cope T A, Colwell S, et al (138)
Insignificant change in Antarctic snowfall since the International Geophysical Year /自国际地球物理年来,南极降雪量无显著变化
…… Monaghan A J, Bromwich D H, Fogt R L, et al (139)
Atmospheric science: early peak in Antarctic oscillation index/大气科学:南极涛动指数的早期峰值 …… Jones J M, Widmann M (140)
Interpretation of recent Southern Hemisphere climate change/解读目前南半球气候变化
…… Thompson D W J, Solomon S (140)
Towards robust regional estimates of CO_2 sources and sinks using atmospheric transport models /利用大气传输模型对 CO_2 源与汇进行可靠区域估计
…… Gurney K R, Law R M, Denning A S, et al (141)
Air-snow interactions and atmospheric chemistry/气-雪相互作用与大气化学
…… Domine F, Shepson P B (142)
Atmospheric science—solving the PSC mystery/大气科学——解析极地平流层云之谜
…… Tolbert M A, Toon O B (143)
Tropical tropospheric ozone and biomass burning/热带对流层臭氧和生物质燃烧
…… Thompson A M, Witte J C, Hudson R D, et al (143)
Evidence from the Pacific troposphere for large global sources of oxygenated organic compounds /来自太平洋对流层的含氧有机化合物大规模的全球来源的证据
…… Singh H, Chen Y, Staudt A, et al (144)
Atmospheric CO_2 concentrations over the Last Glacial Termination/末次冰期结束时的大气 CO_2 浓度 …… Monnin E, Indermuhle A, Dallenbach A, et al (145)
Evidence against dust-mediated control of glacial-interglacial changes in atmospheric CO_2/反对冰期-间冰期大气 CO_2 变化为尘埃介导控制的证据 …… Maher B A, Dennis P F (146)
A strong source of methyl chloride to the atmosphere from tropical coastal land/热带沿海地区是大气一氯甲烷的巨大来源 …… Yokouchi Y, Noijiri Y, Barrie L A, et al (147)
Effect of iron supply on Southern Ocean CO_2 uptake and implications for glacial atmospheric CO_2 /铁的供应对南大洋 CO_2 摄取的影响及其对冰消期大气 CO_2 的启示
…… Watson A J, Bakker D C E, Ridgwell A J, et al (148)
Quantifying denitrification and its effect on ozone recovery/反硝化作用量化及其对臭氧层的恢复的影响 …… Tabazadeh A, Santee M L, Danilin M Y, et al (149)
A potent greenhouse gas identified in the atmosphere: SF_5CF_3/在大气中识别的一种强效的温室气体:SF_5CF_3 …… Sturges W T, Wallington T J, Hurley M D, et al (150)
Glacial/interglacial variations in atmospheric carbon dioxide/大气中二氧化碳含量的冰期/间冰期变化
…… Sigman D M, Boyle E A (150)
Satellite mapping of enhanced BrO concentrations in the troposphere/对流层 BrO 含量增大的卫星观测 …… Wagner T, Platt U (151)
Increased polar stratospheric ozone losses and delayed eventual recovery owing to increasing greenhouse-gas concentrations/由于温室气体浓度增加造成的极地平流层臭氧损耗增加和延迟的最终恢复 …… Shindell D T, Rind D, Lonergan P (152)

Interactive effects of ozone depletion and vertical mixing on photosynthesis of Antarctic phytoplankton /臭氧消耗和垂直混合对南极浮游植物光合作用的相互影响 ······ Neale P J, Davis R F, Cullen J J (153)
Midwinter start to Antarctic ozone depletion: evidence from observations and models/仲冬开始的南极臭氧耗损:来自观测和模型的证据 ······ Roscoe H K, Jones A E, Lee A M (154)
Update-Polar clouds and sulfate aerosols/极地云与硫酸盐气溶胶 ······ Tolbert M A (155)
Melting of H_2SO_4 center dot $4H_2O$ particles upon cooling: implications for polar stratospheric clouds /$H_2SO_4 \cdot 4H_2O$ 粒子的融化:极地平流层云的启示 ······ Koop T, Carslaw K S. (155)
Spectroscopic evidence against nitric-acid trihydrate in polar stratospheric clouds/不支持极地平流层云中的 $HNO_3 \cdot 3H_2O$ 的光谱证据 ······ Toon O B, Tolbert M A (156)
Interhemispheric differences in polar stratospheric HNO_3, H_2O, ClO, and O_3 /极地平流层 HNO_3, H_2O, ClO 和 O_3 在两半球间的差异 ······ Santee M L, Read W G, Waters J W, et al (157)
Continued decline of total ozone over Halley, Antarctica, since 1985/自 1985 年以来南极哈雷站上空臭氧总量的持续下降 ······ Jones A E, Shanklin J D (158)
Metastable phases in polar stratospheric aerosols/极地平流层气溶胶的亚稳态 ······ Fox L E, Worsnop D R, Zahniser M S, et al (158)
Antarctic total ozone in 1958/1958 年南极臭氧总量 ······ Newman P A (159)
Vapor-pressures of solid hydrates of nitric-acid: implications for polar stratospheric clouds /极地平流层云的启示:硝酸固体水合物的蒸汽压 ······ Worsnop D R, Fox L E, Zahniser M S, et al (160)
Subtropical stratospheric mixing linked to disturbances in the polar vortices/与极地涡旋扰动相关的亚热带平流层混合 ······ Waugh D W (160)
Stratospheric ClO and ozone from the microwave limb sounder on the upper-atmosphere research satellite /高层大气研究卫星搭载微波声呐观测平流层 ClO 和 O_3 ······ Waters J W, Froidevaux L, Read W G, et al (161)
Stratospheric ozone depletion by $ClONO_2$ photolysis/$ClONO_2$ 光解导致的平流层臭氧损耗 ······ Toumi R, Jones R L, Pyle J A (162)
Ideas flow on Antarctic vortex/南极涡旋引发的思索 ······ Randel W (163)
Physical-chemistry of the $H_2SO_4/HNO_3/H_2O$ system: implications for polar stratospheric clouds /$H_2SO_4/HNO_3/H_2O$ 系统的物理化学过程——极地平流层云的启示 ······ Molina M J, Zhang R, Wooldridge P J, et al (163)
Evidence for heterogeneous reactions in the Antarctic autumn stratosphere/南极秋季平流层中非均相反应的证据 ······ Keys J G, Johnston P V, Blatherwick R D, et al (164)
Decrease in the growth-rates of atmospheric Chlorofluorocarbon-11 and Chlorofluorocarbon-12 /大气 CFC-11 和 CFC-12 生成速率的降低 ······ Elkins J W, Thompson T M, Swanson T H, et al (165)
Measured trends in stratospheric ozone/平流层臭氧观测趋势 ······ Stolarski R, Bojkov R, Bishop L, et al (166)
Ozone depletion—ultraviolet-radiation and phytoplankton biology in Antarctic waters /臭氧损耗——紫外辐射与南极海域的浮游植物 ······ Smith R C, Prezelin B B, Baker K S, et al (167)

Radiative forcing of climate from halocarbon-induced global stratospheric ozone loss /卤代烃引发的全球平流层臭氧损耗与气候辐射强迫 …… Ramaswamy V, Schwarzkopf M D, Shine K P (167)
More rapid polar ozone depletion through the reaction of HOCl with HCl on polar stratospheric clouds /通过极地平流层云中 HOCl 与 HCl 的反应产生的更迅速的臭氧损耗 …… Prather M J (168)
Investigations of the environmental acceptability of fluorocarbon alternatives to chlorofluorocarbons /含氟烃类替代氯氟烃的环境可接受性的调查 …… Mcfarland M (169)
Observation and possible causes of new ozone depletion in Antarctica in 1991/对 1991 年南极新一轮臭氧损耗的观测和可能原因 …… Hofmann D J, Oltmans S J, Harris J M, et. al (170)
Climate forcing by anthropogenic aerosols/来自人为排放气溶胶的气候强迫 …… Charlson R J, Schwartz S E, Hales J M, et al (171)
Evidence for photochemical control of ozone concentrations in unpolluted marine air/对未受污染海洋空气中臭氧浓度的光化学控制的依据 …… Ayers G P, Penkett S A, Gillett R W, et al (172)
Importance of energetic solar protons in ozone depletion/高能太阳质子在臭氧损耗中的作用 …… Stephenson J A E, Scourfield M W J (173)
The dynamics of the stratospheric polar vortex and its relation to springtime ozone depletions /平流层极地涡旋的动力学特征及其与春季臭氧损耗的关系 …… Schoeberl M R, Hartmann D L (173)
Decrease of summer tropospheric ozone concentrations in Antarctica/南极夏季对流层臭氧浓度的减少 …… Schnell R C, Liu S C, Oltmans S J, et al (174)
Wintertime asymmetry of upper tropospheric water between the Northern and Southern Hemispheres /南北半球冬季对流层上部水汽的不对称特征 …… Kelly K K, Tuck A F, Davies T (175)
Reduced Antarctic ozone depletions in a model with hydrocarbon injections/碳氢化合物注入模型中南极臭氧损耗的减少 …… Cicerone R J, Elliott S, Turco R P (176)
Free-radicals within the Antarctic vortex—the role of CFCs in Antarctic ozone loss/南极涡旋中的自由基——氟氯化碳在南极臭氧损耗中的作用 …… Anderson J G, Toohey D W, Brune W H (177)
Spatial variation of ozone depletion rates in the springtime Antarctic Polar vortex/南极春季极涡中臭氧耗损率的空间变异 …… Yung Y L, Allen M, Crisp D, et al (177)
Observational constraints on the global atmospheric CO_2 budget/全球大气二氧化碳排放清单的预测界限 …… Tans P P, Fung I Y, Takahashi T (178)
Progress towards a quantitative understanding of Antarctic ozone depletion/定量理解南极臭氧耗损的进展 …… Solomon S (179)
The CFC-ozone issue—progress on the development of alternatives to CFCs/氟氯烃与臭氧难题——氟氯烃替代品的发展进程 …… Manzer L E (180)
Another deep Antarctic ozone hole/又一个大的南极臭氧空洞 …… Kerr R A (181)
Future changes in stratospheric ozone and the role of heterogeneous chemistry/平流层臭氧的未来变化和非均相化学的作用 …… Brasseur G P, Granier C, Walters S (181)
2.3 大气科学类摘要翻译——北极 …… (183)
Observation of halogen species in the Amundsen Gulf, Arctic, by active long-path differential optical absorption spectroscopy/通过主动长光程差分选择性吸收光谱观测北极阿蒙森湾的卤素 …… Pohler D, Vogel L, Friess U, et al (183)

Decrease in the CO_2 uptake capacity in an ice-free Arctic Ocean Basin/北冰洋无冰海盆区 CO_2 吸收能力的降低 ······ Cai W J, Chen L Q, Chen B S, et al (184)
The sensitivity of polar ozone depletion to proposed geoengineering schemes/极地臭氧亏损对所提出地质工程方案的敏感性 ······ Tilmes S, Muller R, Salawitch R (184)
Tracing the origin and fate of NO_x in the Arctic atmosphere using stable isotopes in nitrate /利用硝酸根中稳定同位素追踪北极大气 NO_x 的源和去向 ······ Morin S, Savarino J, Frey M M, et al (185)
Large tundra methane burst during onset of freezing/冰冻扩张期间大幅度苔原甲烷爆发 ······ Mastepanov M, Sigsgaard C, Dlugokencky E J, et al (186)
Thermokarst lakes as a source of atmospheric CH_4 during the last deglaciation/末次冰消期热溶喀斯特湖是大气甲烷的排放源 ······ Walter K M, Edwards M E, Grosse G, et al (187)
A climatologically significant aerosol longwave indirect effect in the Arctic/北极地区气溶胶长波辐射显著的间接气候效应 ······ Lubin D, Vogelmann A M (188)
Increased Arctic cloud longwave emissivity associated with pollution from mid-latitudes/中纬地区的污染增加了北极云的长波辐射 ······ Garrett T J, Zhao C F (189)
Arctic trends scrutinized as chilly winter destroys ozone/当北极寒冷的冬天破坏臭氧时,人们开始审议北极的趋势 ······ Schiermeier Q (190)
Observational evidence of a change in radiative forcing due to the indirect aerosol effect/气溶胶间接作用下辐射强迫变化的观测证据 ······ Penner J E, Dong X Q, Chen Y (190)
Recent trends in arctic surface, cloud, and radiation properties from space/空间观测的北极地表、云和放射性特征趋势 ······ Wang X J, Key J R (191)
Arctic rockets give glimpse of the atmosphere's top layers/北极火箭窥探大气顶层 ······ Schiermeier Q (192)
Stratospheric memory and skill of extended-range weather forecasts/平流层对更大范围天气的存储和预报能力 ······ Baldwin M P, Stephenson D B, Thompson D W J, et al (192)
Arctic "ozone hole" in a cold volcanic stratosphere/火山爆发影响下的寒冷平流层中的北极"臭氧洞" ······ Tabazadeh A, Drdla K, Schoeberl M R, et al (193)
Role of the stratospheric polar freezing belt in denitrification/平流层极地冰冻带在反硝化过程中的作用 ······ Tabazadeh A, Jensen E J, Toon O B, et al (194)
The role of Br_2 and BrCl in surface ozone destruction at polar sunrise/极地日出时 Br_2 和 BrCl 对表层臭氧的破坏 ······ Foster K L, Plastridge R A, Bottenheim J W, et al (194)
The detection of large HNO_3-containing particles in the winter arctic stratosphere/对北极冬季平流层中含硝酸大颗粒的探测 ······ Fahey D W, Gao R S, Carslaw K S, et al (195)
Extremely large variations of atmospheric ^{14}C concentration during the last glacial period/末次冰期大气 ^{14}C 浓度的极端波动 ······ Beck J W, Richards D A, Edwards R L, et al (196)
Stratospheric harbingers of anomalous weather regimes/反常天气体系在平流层的先兆 ······ Baldwin M P, Dunkerton T J (197)
Quantifying denitrification and its effect on ozone recovery/反硝化及其对臭氧恢复的量化作用 ······ Tabazadeh A, Santee M L, Danilin M Y, et al (198)
Acclimation of ecosystem CO_2 exchange in the Alaskan Arctic in response to decadal climate warming /北极阿拉斯加地区生态系统 CO_2 交换对十年来气候变暖的适应 ······ Oechel W C, Vourlitis G L, Hastings S J, et al (199)

Global warming could be bad news for Arctic ozone layer/全球变暖对北极臭氧层来说可能是坏消息 …… Aldhous P (199)
Contribution of disturbance to increasing seasonal amplitude of atmospheric CO_2/对增加的大气 CO_2 季节性波动的扰动的贡献 …… Zimov S A, Davidov S P, Zimova G M, et al (200)
Arctic ozone loss due to denitrification/反硝化作用造成的北极臭氧流失 …… Waibel A E, Peter T, Carslaw K S, et al (201)
Snowpack production of formaldehyde and its effect on the Arctic troposphere/积雪生成的甲醛及其对北极对流层的影响 …… Sumner A L, Shepson P B (202)
Chemical analysis of polar stratospheric cloud particles/极地平流层云雾粒子的化学分析 …… Schreiner J, Voigt C, Kohlmann A, et al (202)
Evidence for bromine monoxide in the free troposphere during the Arctic polar sunrise/北极日出时自由对流层中 BrO 存在的证据 …… McElroy C T, McLinden C A, McConnell J C (203)
The effect of climate change on ozone depletion through changes in stratospheric water vapour /气候变化产生的平流层水蒸气变化对臭氧亏损的影响 …… Kirk-Davidoff D B, Hintsa E J, Anderson J G, et al (204)
DOAS measurements of tropospheric bromine oxide in mid-latitudes/用 DOAS 手段对中纬度对流层 BrO 的观测 …… Hebestreit K, Stutz J, Rosen D, et al (205)
Relative influences of atmospheric chemistry and transport on Arctic ozone trends/大气化学和传输对北极臭氧趋势的有关影响 …… Chipperfield M P, Jones R L (206)
Satellite mapping of enhanced BrO concentrations in the troposphere/对流层 BrO 浓度增加的卫星成像 …… Wagner T, Platt U (207)
Increased polar stratospheric ozone losses and delayed eventual recovery owing to increasing greenhouse-gas concentrations/日益增加的温室气体浓度造成了极地平流层臭氧层空洞增大并阻碍其恢复 …… Shindell D T, Rind D, Lonergan P (208)
Increased stratospheric ozone depletion due to mountain-induced atmospheric waves/由于山脉地形激发的大气波动加剧了平流层臭氧亏损 …… Carslaw K S, Wirth M, Tsias A, et al (209)
Atmospheric chemistry—a bad winter for Arctic ozone/大气化学——北极臭氧经历严冬 …… Stolarski R (210)
Prolonged stratospheric ozone loss in the 1995-96 Arctic winter/1995～1996 年北极冬季平流层臭氧更长时间的流失 …… Rex M, Harris N R P, von der Gathen P, et al (210)
Severe chemical ozone loss in the Arctic during the winter of 1995-96/1995～1996 年冬季北极发生的严重化学臭氧损耗 …… Muller R, Crutzen P J, Grooss J U, et al (212)
The effect of small-scale inhomogeneities on ozone depletion in the Arctic/北极地区小范围环境的多相性对臭氧耗损的影响 …… Edouard S, Legras B, Lefevre F, et al (213)
Observational evidence for chemical ozone depletion over the Arctic in winter 1991-92 /1991～1992 年期间北极地区化学因素导致的臭氧层耗损的观测证据 …… Vondergathen P, Rex M, Harris N R P, et al (213)
Chemical depletion of ozone in the Arctic-lower stratosphere during winter 1992-93/1992～1993 年冬季北极下平流层臭氧的化学亏损 …… Manney G L, Froidevaux L, Waters J W, et al (215)
Transient nature of CO_2 fertilization in Arctic tundra/北极苔原地区二氧化碳肥化作用的短时性特征 …… Walter C Oechel, Sid Cowles, Nancy Grulke, et al (215)

Vapor-pressures of solid hydrates of nitric-acid—implications for polar stratospheric clouds /固态硝酸水合物的蒸汽压——极地平流层云的指示 ······ Worsnop D R, Fox L E, Zahniser M S, et al (216)
Chlorine chemistry on polar stratospheric cloud particles in the Arctic winter/冬季北极平流层云滴上的氯化学 ······ Webster C R, May R D, Toohey D W, et al (217)
Subtropical stratospheric mixing linked to disturbances in the polar vortices/与极地涡旋扰动相关的亚热带平流层混合 ······ Waugh D W (218)
Stratospheric ClO and ozone from the microwave limb sounder on the upper-atmosphere research satellite /用高层大气研究卫星微波边缘探测器观测平流层 ClO 和臭氧 ······ Waters J W, Froidevaux L, Read W G, et al (219)
Stratospheric ozone depletion by $ClONO_2$ photolysis/$ClONO_2$ 光分解引起的平流层臭氧亏损 ······ Toumi R, Jones R L, Pyle J A (219)
Heterogeneous reaction probabilities, solubilities, and the physical state of cold volcanic aerosols /冷火山冷气溶胶的多相反应概率、溶解性和物态特性 ······ Toon O, Browell E, Gary B, et al (220)
Chemical loss of ozone in the Arctic polar vortex in the winter of 1991-1992/1991～1992 年冬季北极极地涡旋中的臭氧化学亏损 ······ Salawitch R J, Wofsy S C, Gottlieb E W, et al (221)
Stratospheric meteorological conditions in the Arctic polar vortex, 1991 to 1992/1991～1992 年北极极地涡旋的平流层气象条件 ······ Newman P, Lait L R, Schoeberl M, et al (222)
Ozone and aerosol changes during the 1991-1992 airborne Arctic stratospheric expedition /1991～1992 年北极平流层空中观测期间臭氧和气溶胶的变化 ······ Browell E V, Butler C F, Fenn M A, et al (223)
Absence of evidence for greenhouse warming over the arctic ocean in the past 40 years /不存在过去 40 年北冰洋温室变暖的证据 ······ Jonathan D Kahl, Donna J Charlevoix, Nina A Zaftseva, et al (223)
Ozone loss inside the northern polar vortex during the 1991-1992 winter/1991～1992 年冬季北极涡旋内的臭氧损耗 ······ Proffitt M H, Aikin K, Margitan J J, et al (224)
Photochemical bromine production implicated in Arctic boundary-layer ozone depletion/北极边界层臭氧亏损指示的光化学溴的产生 ······ Mcconnell J C, Henderson G S, Barrie L, et al (225)
Surface ozone depletion in Arctic spring sustained by bromine reactions on aerosols/气溶胶上溴反应维持的北极地区春季表面臭氧损耗 ······ Fan S M, Jacob D J (226)
Possibility of an Arctic ozone hole in a doubled-CO_2 climate/CO_2 加倍情况下产生北极臭氧洞的可能性 ······ Austin J, Butchart N, Shine K P (227)
Origin of sulfur in Canadian Arctic haze from isotope measurements/通过同位素方法观测加拿大北极霾中硫的来源 ······ Nriagu J O, Coker R D, Barrie L A (227)
Wintertime asymmetry of upper tropospheric water between the northern and southern hemispheres /南北半球冬季对流层上部水汽的不对称特性 ······ Kelly K K, Tuck A F, Davies T (228)
Evidence from balloon measurements for chemical depletion of stratospheric ozone in the Arctic winter of 1989-90/1989～1990 年北极冬季平流层臭氧化学亏损的气球观测证据 ······ Hofmann D J, Deshler T (229)
The potential for ozone depletion in the Arctic polar stratosphere/北极极地平流层中潜在的臭氧亏损 ······ Brune W H, Anderson J G, Toohey D W, et al (230)

Ozone loss in the Arctic polar vortex inferred from high-altitude aircraft measurements /高海拔飞机观测到的北极极地涡旋中的臭氧流失 …… Proffitt M H, Margitan J J, Kelly K K, et al (231)

Ozone destruction and bromine photochemistry at ground-level in the Arctic spring/北极春季地面臭氧破坏和溴光化学 …… Finlaysonpitts B J, Livingston F E, Berko H N (231)

Observations of denitrification and dehydration in the winter polar stratospheres/极地平流层冬季反硝化和脱水作用的观测 …… Fahey D W, Kelly K K, Kawa S R, et al (232)

三、海洋古气候、古环境类文献

3.1　分析概述 …… (237)

3.2　摘要翻译——南极 …… (239)

Evidence for lower productivity in the Antarctic ocean during the last glaciation/南大洋在末次冰期期间较低海洋生产率的证据 …… Mortlock R A,Charles C D,Froelich P N,et al (239)

Abrupt deep-sea warming, palaeoceanographic changes and benthic extinctions at the end of the Paleocene/深海急剧变暖——古新世末期古海洋变化和深海生物灭绝 …… Kennett J P,Stott L D (240)

Microtektites, microkrystites, and spinels from a late Pliocene asteroid impact in the Southern-Ocean/晚上新世小行星对南大洋冲击所产生的微玻璃陨石、微晶球粒陨石和尖晶石 …… Margolis S V,Claeys P,Kyte F T (240)

Ocean circulation beneath the Ronne ice shelf/南极隆尼冰架下面的海洋环流 …… Nicholls K W,Makinson K,Robinson A V (241)

Forcing mechanisms of the Indian-Ocean monsoon/印度洋季风的驱动机制 …… Clemens S,Prell W,Murray D,et al (242)

Constraints on the age and duration of the last interglacial period and on sea-level variations /末次间冰期的年代、持续时间和对应海平面的参数特征 …… Lambeck K,Nakada M (242)

Isotopic evidence for reduced productivity in the glacial Southern-Ocean/冰期南大洋生产力降低的同位素证据 …… Shemesh A, Macko S A, Charles C D,et al (243)

A model study of the Atlantic thermohaline circulation during the last glacial maximum/末次冰盛期大西洋温盐环流的模拟研究 …… Fichefet T,Hovine S,Duplessy J C (244)

Meltwater input to the Southern-Ocean during the last glacial maximum/末次冰盛期冰融水向南大洋的输入 …… Shemesh A,Burckle L H,Hays J D (244)

Importance of iron for plankton blooms and carbon-dioxide drawdown in the Southern-Ocean /铁对南大洋浮游植物爆发和二氧化碳降低的重要性 …… Debaar H J W,Dejong J T M,Bakker D C E,et al (245)

Increased biological productivity and export production in the glacial Southern-Ocean /冰期南大洋生物生产力的增加与生产力的输出 …… Kumar N,Anderson R F,Mortlock R A,et al (246)

Palaeobotanical evidence for a warm Cretaceous Arctic Ocean/白垩纪北冰洋温暖的古植物学证据 …… Herman A B,Spicer R A (247)

Deglacial changes in ocean circulation from an extended radiocarbon calibration/通过放射性碳的延伸计算了解冰消期大洋环流的变化 …… Hughen K A,Overpeck J T,Lehman S J,et al (247)

A possible 20th-century slowdown of Southern Ocean deep water formation/南大洋深水可能在20世纪停止形成 ………………………… Broecker W S, Sutherland S, Peng T H (248)
Oceanic Cd/P ratio and nutrient utilization in the glacial Southern Ocean/冰期南大洋海洋 Cd/P 比值和营养利用 ………………………………………………… Elderfield H, Rickaby R E M (249)
Is El Nino changing? /厄尔尼诺是否发生了变化? …………… Fedorov A V, Philander S G (250)
Climate impact of late quaternary equatorial Pacific sea surface temperature variations/气候对晚第四纪赤道太平洋海温度变化的影响 ………………………… Lea D W, Pak D K, Spero H J (251)
Old radiocarbon ages in the southwest Pacific Ocean during the last glacial period and deglaciation /末次冰期-冰消期西南太平洋的老碳年龄
……………………………………………………… Sikes E L, Samson C R, Guilderson T P, et al (252)
Proxy evidence for an El Nino-like response to volcanic forcing/厄尔尼诺对火山活动响应的证据
……………………………………………………………… Adams J B, Mann M E, Ammann C M (252)
El Nino/Southern Oscillation and tropical Pacific climate during the last millennium/过去千年热带太平洋气候与 ENSO 的变化………………………… Cobb K M, Charles C D, Cheng H, et al (253)
Magnitude and timing of temperature change in the Indo-Pacific warm pool during deglaciation /印度-太平洋暖池末次间冰期温度变化的幅度和时间 ………… Visser K, Thunell R, Stott L (254)
Meltwater pulse 1A from Antarctica as a trigger of the Bolling-Allerod warm interval /南极洲的冰融水脉动 1A 是布林-阿勒罗德暖期的触发原因之一
……………………………………………………… Weaver A J, Saenko O A, Clark P U, et al (255)
New Zealand maritime glaciation: millennial-scale southern climate change since 3.9 Ma /新西兰沿海冰期作用：自 390 万年前开始的千年尺度的南部气候变化
……………………………………………………………………………… Carter R M, Gammon P (256)
Strong hemispheric coupling of glacial climate through freshwater discharge and ocean circulation /因淡水释放和海洋环流而导致的冰期气候强半球耦合
……………………………………………………… Knutti R, Fluckiger J, Stocker T F, et al (256)
Antarctic timing of surface water changes off Chile and Patagonian ice sheet response/智利和巴塔哥尼亚冰原对南极表面水变化的时间性………… Lamy F, Kaiser J, Ninnemann U, et al (257)
Middle Miocene Southern Ocean cooling and Antarctic cryosphere expansion/中新世中期南大洋的变冷和南极冰冻圈的扩张………………………… Shevenell A E, Kennett J P, Lea D W (258)
Polar ocean stratification in a cold climate/在寒冷气候下极地海洋的分层作用
……………………………………………………………… Sigman D M, Jaccard S L, Haug G H (259)
Rapid stepwise onset of Antarctic glaciation and deeper calcite compensation in the Pacific Ocean /南极冰川作用和太平洋方解石补偿深度加深的快速阶梯式开始
……………………………………………………… Coxall H K, Wilson P A, Palike H, et al (260)
Southern hemisphere water mass conversion linked with North Atlantic climate variability/南半球水体转换与北大西洋气候变化的关系 ………………………………… Pahnke K, Zahn R (261)
Increased productivity in the subantarctic ocean during Heinrich events/在 Heinrich 事件时期亚南极海域生产力的增长 ………………………………………… Sachs J P, Anderson R F (261)
Deep-sea temperature and circulation changes at the Paleocene-Eocene thermal maximum/在古新世-始新世最热期的深海温度和环流变化 ………………………… Tripati A, Elderfield H (262)
Abrupt reversal in ocean overturning during the Palaeocene/Eocene warm period/大洋翻转流在古新世/始新世的温暖期发生突然逆转 ……………………………… Nunes F, Norris R D (263)

The heartbeat of the oligocene climate system/渐新世气候系统的“心跳”(周期性变化)
………………………………………………………… Palike H, Norris R D, Herrle J O, et al (264)
Southern Ocean sea-ice extent, productivity and iron flux over the past eight glacial cycles
/过去 8 个冰期旋回中南大洋的海冰范围、生产力和铁通量
………………………………………………………… Wolff E W, Fischer H, Fundel F, et al (264)
Seasonal characteristics of the Indian Ocean Dipole during the Holocene epoch/全新世期间印度洋偶极现象的季节性特征 ……………………… Abram N J, Gagan M K, Liu Z Y, et al (266)
Absence of cooling in New Zealand and the adjacent ocean during the Younger Dryas chronozone
/新西兰及其周边海域在新仙女木时期没有变冷
………………………………………………… Barrows T T, Lehman S J, Fifield L K, et al (267)
Intense hurricane activity over the past 5000 years controlled by El Nino and the West African monsoon
/过去 5000 年受厄尔尼诺和非洲季风控制的强飓风活动 ……… Donnelly J P, Woodruff J D (267)
The deep ocean during the last interglacial period/末次间冰期的深海
………………………………………………………… Duplessy J C, Roche D M, Kageyama M (268)
Short-circuiting of the overturning circulation in the Antarctic Circumpolar Current/南极绕极流中翻转环流的短路(变短) …………………… Garabato A C N, Stevens D P, Watson A J, et al (269)
Deep ocean impact of a Madden-Julian Oscillation observed by Argo floats Madden-Julian
/海洋实时观测系统观测的季节内震荡的深海效应
……………………………………………… Matthews A J, Singhruck P K, Heywood K J (270)
Millennial-scale trends in west Pacific warm pool hydrology since the Last Glacial Maximum
/末次冰盛期以来西太平洋暖池水文(水汽)千年尺度变化趋势
……………………………………………………… Partin J W, Cobb K M, Adkins J F, et al (271)
Southern hemisphere and deep-sea warming led deglacial atmospheric CO_2 rise and tropical warming
/南半球和深海暖化导致间冰期大气 CO_2 上升和热带变暖
………………………………………………………… Stott L, Timmermann A, Thunell R (272)
Southern Ocean sea-ice extent, productivity and iron flux over the past eight glacial cycles
/在过去八个冰期旋回中南大洋的海冰范围、生产力和铁通量
………………………………………………………… Wolff E W, Fischer H, Fundel F, et al (273)
3.3　摘要翻译——北极 ………………………………………………………………………… (275)
Climate change—the elusive Arctic warming/气候变化——不易确定的北极变暖
………………………………………………………………………………………… Walsh Je (275)
The last deglaciation event in the eastern central Arctic-Ocean/北冰洋东部核心区域的末次冰消期事件 ……………………………………………… Stein R, Nam S I, Schubert C, et al (275)
Arctic environmental change of the last four centuries/过去四个世纪里北极气候环境变化
………………………………………………………… Overpeck J, Hughen K, Hardy D, et al (276)
Decadal variability in the outflow from the Nordic seas to the deep Atlantic Ocean/从诺迪克海到大西洋深水的外流的十年际变化 ………………………………………………… Bacon S (277)
A short circuit in thermohaline circulation: a cause for Northern Hemisphere glaciation?
/温盐环流短路:北半球冰期的原因? ……………………………… Driscoll N W, Haug G H (278)
Abrupt changes in North American climate during early Holocene times/早全新世期间北美快速气候变化 …………………………………………… Hu F S, Slawinski D, Wright H E, et al (278)

Simulating the amplification of orbital forcing by ocean feedbacks in the last glaciation/模拟末次冰期辐射强迫-海洋反馈的放大作用 …………… Khodri M, Leclainche Y, Ramstein G, et al (279)
Regional climate impacts of the Northern Hemisphere annular mode/北半球环流模式对区域气候的影响 ……………………………………………………… Thompson D W J, Wallace J M (280)
The salinity, temperature, and delta ^{18}O of the glacial deep ocean/冰期深海的盐度、温度和氧同位素 ……………………………………………………… Adkins J F, McIntyre K, Schrag D P (281)
The role of the thermohaline circulation in abrupt climate change/温盐环流在快速气候变化中的作用 ……………………………………………………… Clark P U, Pisias N G, Stocker T F, et al (282)
Dynamics of recent climate change in the Arctic/北极最近气候变化的动力学机制 ……………………………………………………… Moritz R E, Bitz C M, Steig E J (283)
Increasing river discharge to the Arctic Ocean/北冰洋河流输入的增加 ……………………………………………… Peterson B J, Holmes R M, McClelland J W, et al (283)
Abrupt climate change/快速气候变化…………… Alley R B, Marotzke J, Nordhaus W D, et al (284)
Asynchronous climate changes in the North Atlantic and Japan during the last termination /北大西洋和日本在末次冰期结束时的气候变化不同步特征 ……………………………………………………… Nakagawa T, Kitagawa H, Yasuda Y, et al (285)
Magnitude and timing of temperature change in the Indo-Pacific warm pool during deglaciation /Indo-Pacific 暖池间冰期期间温度变化的幅度和时间点……… Visser K, Thunell R, Stott L (286)
High temperatures in the Late Cretaceous Arctic Ocean/晚白垩纪北冰洋高温 ……………………………………………………… Jenkyns H C, Forster A, Schouten S, et al (286)
Polar ocean stratification in a cold climate/冷期极地海洋的分层特征 ……………………………………………………… Sigman D M, Jaccard S L, Haug G H (287)
Influence of the Atlantic subpolar gyre on the thermohaline circulation/副极地大西洋回旋对温盐环流的影响……………………………………………… Hatun H, Sando A B, Drange H, et al (288)
Arctic freshwater forcing of the Younger Dryas cold reversal/北冰洋淡水驱动新仙女木冷事件 ……………………………………………………… Tarasov L, Peltier W R (289)
Episodic fresh surface waters in the Eocene Arctic Ocean/始新世北冰洋海表的间插性淡水 ……………………………………………… Brinkhuis H, Schouten S, Collinson M E, et al (290)
The Cenozoic palaeoenvironment of the Arctic Ocean/北冰洋新生代古环境 ……………………………………………………… Moran K, Backman J, Brinkhuis H, et al (291)
Subtropical Arctic ocean temperatures during the Palaeocene/Eocene thermal maximum/始新世温暖期副热带北极海温度 ………………………… Sluijs A, Schouten S, Pagani M, et al (292)

四、生物科学类文献

4.1 分析概述 ……………………………………………………………………… (297)
4.2 生物科学类摘要翻译——南极 ……………………………………………………… (301)
Eco-quandary: what killed the skuas/生态困境:是什么杀死了贼鸥 …………… Barinaga M (301)
Implications of a new acoustic target strength for abundance estimates of Antarctic krill /应用新的声学目标强度方法估算南极磷虾的丰度 ………………………………… Inigo Everson, Jonathan L Watkins, Douglas G Bone, et al (301)

An Antifreeze glycopeptide gene from the Antarctic cod Notothenia-coriiceps-neglecta encodes a polyprotein of high peptide copy number/南极鳕鱼 Notothenia coriiceps neglecta 抗冻糖肽基因编码高拷贝数的聚合蛋白 ……………… Hsiao K C,Cheng C H, Fernandes I E, et al (302)
Pteropod abduction as a chemical defense in a pelagic Antarctic amphipod/诱拐翼足目是远洋南极片脚类动物的一个化学防御措施 …………………… James B Mcclintock,John Janssen (303)
Acoustic estimates of Antarctic krill/南极磷虾的声学估算 ……………………………… Charles H Greene,Timothy K Stanton,Peter H Wie,et al (304)
Krill abundance/磷虾丰度 …………………………………………… Hewitt R P, Demer D A (305)
DNA damage in the Antarctic/南极 DNA 损伤 ………… Karentz S,Cleaver J E,Mitchell D L (305)
Foraging behavior of emperor penguins as a resource detector in winter and summer/冬季和夏季帝企鹅觅食行为可作为一个资源探测器 ……… Ancel A,Kooyman G L,Ponganis P J,et al (306)
Biological weighting function for the inhibition of phytoplankton photosynthesis by ultraviolet-radiation /紫外辐射抑制浮游植物光合作用的生物加权函数 ……… Cullen J J,Neale P J,Lesser M P (307)
Abundant mitochondrial-DNA variation and worldwide population-structure in humpback whales /座头鲸丰富的线粒体 DNA 变异和世界范围内种群结构 ………………………………………………………… Baker C S,Perry A,Bannister J L,et al (308)
Gene organization and primary structure of human hormone-sensitive lipase:possible significance of a sequence homology with a lipase of moraxella Ta144, an Antarctic bacterium /人类激素敏感酯酶的基因组织和基本结构:与南极细菌(摩拉克氏菌属)moraxella Ta144 的序列同源性的可能意义 ……………………………… Langin D,Laurell H,Holst L S,et al (309)
Antarctic benthic diversity/南极底栖生物多样性 …………… Brey T,Klages M,Dahm C,et al (310)
High abundance of archaea in Antarctic marine picoplankton/南极海洋超微型浮游生物古细菌的高丰度 ……………………………………… Delong E F,Wu K Y,Prezelin B B,et al (311)
Genomic remnants of alpha-globin genes in the hemoglobinless Antarctic icefishes/在无血红蛋白的南极银鱼中 α-球蛋白基因染色体组的残留 ……………………………………… Cocca E,Ratnayake Lecamwasam M, Parker S K,et al (311)
Dimethyl sulfide as a foraging cue for Antarctic procellariiformes seabirds/二甲基硫醚作为南极鹱形目海鸟觅食的信号 ………………………………… Nevitt G A,Veit R R,Kareiva P (312)
Pliocene extinction of Antarctic pectinid mollusks/上新世时期南极扇贝科软体动物的灭绝 …………………………………………………… Paul Arthur Berkman,Michael L Prentice (313)
Optical fibres in an Antarctic sponge/南极海绵动物体内的光纤 ……………………………………………… Cattaneo Vietti R,Bavestrello G,Cerrano C,et al (314)
Biological particles over Antarctica/南极上空的生物微粒 ……………………… Marshall W A (315)
Iron stimulation of Antarctic bacteria/铁元素对南极细菌的促进作用 ……………………………………………………… Pakulski J D,Coffin R B,Kelley C A,et al (315)
Bomb signals in old Antarctic brachiopods/南极古老腕足类动物的核弹实验信号 ……………………………………………………………………………… Peck L S,Brey T (316)
Evolution of antifreeze glycoprotein gene from a trypsinogen gene in Antarctic notothenioid fish /南极鱼(notothenioid)胰蛋白酶原基因向抗冻糖蛋白基因的进化转化 ……………………………………………………… Chen L B, De Vries A L, Cheng C H C (316)

Convergent evolution of antifreeze glycoproteins in Antarctic notothenioid fish and Arctic cod /南极鱼(notothenioid)和北极鳕鱼抗冻糖蛋白的协同进化 …… Chen L B, De Vries A L, Cheng C H C (318)

Poultry virus infection in Antarctic penguins/南极企鹅病毒感染 …… Gardner H, Kerry K, Riddle M, et al (319)

Effects of sea-ice extent and krill or salp dominance on the Antarctic food web/海冰覆盖范围和磷虾或樽海鞘的优势地位对南极食物网的影响 …… Loeb V, Siegel V, Holm Hansen O, et al (319)

Solar UVB-induced DNA damage and photoenzymatic DNA repair in Antarctic zooplankton /太阳紫外线引起的南极浮游动物 DNA 损伤和光酶修复 …… Malloy K D, Holman M A, Mitchell D, et al (320)

Variable expression of myoglobin among the hemoglobinless Antarctic icefishes/在无血红蛋白的南极冰鱼中肌红蛋白的变量表达 …… Sidell B D, Vayda M E, Small D J, et al (321)

Biogeochemical controls and feedbacks on ocean primary production/海洋初级生产量的生物地球化学控制和反馈 …… Falkowski P G, Barber R T, Smetacek V (322)

Iron-limited diatom growth and Si∶N uptake ratios in a coastal upwelling regime/沿海岸上升流区铁限制硅藻的生长和硅∶氮吸收比 …… Hutchins D A, Bruland K W (323)

Influence of iron availability on nutrient consumption ratio of diatoms in oceanic waters/海水中铁的可利用性对硅藻营养物质消耗比例的影响 …… Takeda S (324)

Phytoplankton community structure and the drawdown of nutrients and CO_2 in the Southern Ocean /南大洋中浮游植物群落结构和养分及 CO_2 的减少 …… Arrigo K R, Robinson D H, Worthen D L, et al (325)

Hunting behavior of a marine mammal beneath the Antarctic fast ice/南极冰下海洋哺乳动物的狩猎行为 …… Davis R W, Fuiman L A Williams T M, et al (326)

Icy life on a hidden lake/一个隐蔽湖泊的冰冷生活 …… Vincent W F (327)

Importance of stirring in the development of an iron-fertilized phytoplankton bloom /搅拌对于富铁水体里浮游植物暴发的重要性 …… Edward R Abraham, Cliff S Law, Philip W Boyd, et al (327)

A mesoscale phytoplankton bloom in the polar Southern Ocean stimulated by iron fertilization /极地南大洋中由富铁引起的一个中尺度浮游植物暴发现象 …… Philip W Boyd, Andrew J Watson, Cliff S Law, et al (328)

Molecular evidence for genetic mixing of Arctic and Antarctic subpolar populations of planktonic foraminifers/北极和南极亚极地浮游有孔虫种群基因混合的分子证据 …… Darling K F, Wade C M, Stewart I A, et al (329)

Rapid and early export of Phaeocystis Antarctica blooms in the Ross Sea, Antarctica/南极罗斯海中南极棕囊藻大量繁殖和其早期的快速输出 …… Di Tullio G R, Grebmeier J M, Arrigo K R, et al (330)

The animal species-body size distribution of Marion Island/马里恩岛动物物种-体型分布 …… Gaston K J, Chown S L, Mercer R D (331)

Unexpected diversity of small eukaryotes in deep-sea Antarctic plankton/南极深海浮游生物中的小型真核生物出乎意料的多样性 …… Lopez-Garcia P, Rodriguez-Valera F, Pedros-Alio C, et al (332)

Antarctic krill under sea ice: elevated abundance in a narrow band just south of ice edge/南极海冰下的磷虾：在冰缘线南部的狭窄地带数量在上升 …… Andrew S Brierley, Paul G Fernandes, Mark A Brandon, et al (333)
Life in the deep freeze/严寒下的生命 …… Gavaghan H (334)
The crystal structure of a tetrameric hemoglobin in a partial hemichrome state/一个部分高铁血色原状态的血红蛋白四聚体的晶体结构 …… Antonio Riccio, Luigi Vitagliano, Guido di Prisco, et al (334)
Ocean science-Antarctic Sea ice: a habitat for extremophiles/海洋科学-南极海冰：极端生物的一个栖息地 …… Thomas D N, Dieckmann G S (335)
A fly in the biogeographic ointment/在生物地理软质地层中的一只苍蝇 …… Allan C Ashworth, Christian Thompson F (336)
Proteorhodopsin genes are distributed among divergent marine bacterial taxa /Proteorhodopsin 基因分布在多种多样的海洋细菌类群中 …… José R de la Torre, Lynne M Christianson, Oded Béjà, et al (337)
Fish migration: Patagonian toothfish found off Greenland—this catch is evidence of transequatorial migration by a cold-water Antarctic fish/鱼的迁徙：格陵兰岛不远处发现巴塔哥尼亚齿鱼——这个发现是南极冷水鱼类跨赤道迁徙的证据 …… Moller P R, Nielsen J G, Fossen I (338)
Long-term decline in krill stock and increase in salps within the Southern Ocean/南大洋磷虾数量长期下降和樽海鞘数量增加 …… Angus Atkinson, Volker Siegel, Evgeny Pakhomov, et al (339)
Partner-specific odor recognition in an Antarctic seabird/南极海鸟对特定伙伴气味的识别 …… Francesco Bonadonna, Gabrielle A Nevitt (340)
Molecular evidence links cryptic diversification in polar planktonic protists to quaternary climate dynamics/连接神秘的极地浮游原生生物多样化与第四纪气候动力学的分子证据 …… Kate F Darling, Michal Kucera, Carol J Pudsey, et al (341)
Microevolution and mega-icebergs in the Antarctic/微观进化和南极巨型冰山 …… Shepherd L D, Millar C D, Ballard G, et al (342)
Polar ocean ecosystems in a changing world/变化中的极地海洋生态系统 …… Victor Smetacek, Stephen Nicol (343)
Antarctic birds breed later in response to climate change/应对气候变化——南极海鸟繁殖推迟 …… Barbraud C, Weimerskirch H (343)
The Southern Ocean biogeochemical divide/南大洋生物地球化学分界线 …… Marinov I, Gnanadesikan A, Toggweiler J R, et al (344)
Phytoplankton and cloudiness in the Southern Ocean/南大洋浮游植物和云量 …… Nicholas Meskhidze, Athanasios Nenes (345)
Effect of natural iron fertilization on carbon sequestration in the Southern Ocean/自然“铁施肥”对南大洋碳螯合作用的影响 …… Stéphane Blain, Bernard Quéguiner, Leanne Armand, et al (346)
First insights into the biodiversity and biogeography of the Southern Ocean deep sea/对南大洋深海生物多样性和生物地理学的初步研究 …… Angelika Brandt, Andrew J Gooday, Simone N Brandao, et al (347)
The Southern Ocean biological response to Aeolian iron deposition/南大洋对风成铁沉降的生态响应 …… Nicolas Cassar, Michael L Bender, Bruce A Barnett, et al (349)

Antarctic biodiversity/南极生物多样性 …………………………… Convey P,Stevens M I (349)
Isoprene, cloud droplets, and phytoplankton/异戊二烯、云滴和浮游生物 ……… Meskhidze N (350)
Letting the light in on Antarctic ecosystems/让阳光洒向南极生态系统 ……… Odling-Smee L (351)
4.3 摘要翻译——北极 …………………………………………………………………… (352)
A Bangiophyte red alga from the proterozoic of Arctic Canada/加拿大北极地区一种原生代的 Bangiophyte 红藻 …………………………… Butterfield N J,Knoll A H,Sweet. K (352)
Preferential use of organic nitrogen for growth by a nonmycorrhizal Arctic sedge /非菌根北极莎草生长中对有机氮的优先使用 …………………………… F Stuart Chapin Ⅲ,Lori Moilanen,Knut Kielland (352)
Population oscillations of boreal rodents:regulation by mustelid predators leads to chaos /北方鼠害周期受鼬鼠天敌调节而导致混乱 ……… Hanski I,Turchin P,Korpimaeki E,et al (353)
Hypervariable-control-region sequences reveal global population structuring in a long-distance migrant shorebird, the Dunlin (*calidris-alpina*)/在高变控制区域的远距离迁徙的黑腹滨鹬海岸鸟类(*calidris-alpina*)揭示全球数量结构 ………… Wenink P W, Baker A J, Tilanus M G (354)
Acquisition and utilization of transition metal ions by marine organisms/海洋生物体获得和利用过渡金属离子 …………………………… Alison Butler (355)
Sensitivity of boreal forest carbon balance to soil thaw/敏感性寒带森林的碳平衡对土壤解冻的敏感性 …………………………… Goulden M L,Wofsy S C,Harden J W,et al (355)
Evidence for extreme climatic warmth from Late Cretaceous Arctic vertebrates/北极晚白垩世脊椎动物作为极端温暖气候的证据 ………… Tarduno J A,Brinkman D B,Renne P R,et al (356)
Long-distance transport of pollen into the Arctic/花粉长距离输入北极 …………………………… Ian D Campbell,Karen McDonald,Michael D Flannigan,et al (357)
Molecular analysis of plant migration and refugia in the Arctic/北极植物迁移和植物残遗种的分子研究 …………………………… Richard J Abbott, Lisa C Smith, Richard I Milne,et al (357)
A mesoscale phytoplankton bloom in the polar Southern Ocean stimulated by iron fertilization /极地南部海域由于铁营养带来的中尺度浮游植物繁盛 …………………………… Philip W Boyd,Andrew J Watson,Cliff S Law,et al (358)
Molecular evidence for genetic mixing of Arctic and Antarctic subpolar populations of planktonic foraminifers/两极的亚极地地区浮游有孔虫混合种群遗传的分子证据 …………………………… Kate F Darling,Christopher M Wade,Iain A Stewart,et al (359)
Concurrent density dependence and independence in populations of Arctic ground squirrels /北极松鼠种群并发密度依赖和独立性 ……………………… Tim J Karels, Rudy Boonstra (360)
Ecology:bats about the Arctic/生态:北极蝙蝠 …………………………… Moore P D (361)
Migration along orthodromic sun compass routes by Arctic birds/北极鸟类沿着日光罗盘方向迁移 ……………………… Thomas Alerstam,Gudmundur A Gudmundsson,Martin Green,et al (362)
Accelerated regulatory gene evolution in an adaptive radiation/在适应辐射中调控基因的加速进化 ………………………… Marianne Barrier,Robert H Robichaux,Michael D Purugganan (362)
Ornithology—Arctic waders are not capital breeders/鸟类学——北极涉禽类不是资本种畜 …………………………… Klaassen M,Lindström A A,Meltofte H,et al (364)
Colonization of America by Drosophila subobscura:heterotic effect of chromosomal arrangements revealed by the persistence of lethal genes/殖民美国的果蝇致命基因的持久性所揭示的染色体排列的杂种优势效应 …………………………… Mestres F,Balanyà J,Arenas C,et al (364)

Linkage disequilibrium in the human genome/人类基因组连锁不平衡
…… David E Reich, Michele Cargill, Stacey Bolk, et al (365)
Impact of Bt corn pollen on monarch butterfly populations: a risk assessment
/转 Bt 基因玉米授粉对于帝王蝶数量的影响:风险评估
…… Mark K Sears, Richard L Hellmich, Diane E Stanley-Horn, et al (366)
The origin of 15R-prostaglandins in the Caribbean coral *Plexaura homomalla*: molecular cloning and expression of a novel cyclooxygenase/加勒比海珊瑚虫(*Plexaura homomalla*)的 15R 前列腺素的来源:分子克隆和一种新型环氧合酶的表达
…… Karin Valmsen, Ivar Järving, William E Boeglin, et al (367)
Genetic evidence against panmixia in the European eel/遗传证据否定欧洲鳗随机交配
…… Wirth T, Bernatchez L (368)
Robotic observations of dust storm enhancement of carbon biomass in the North Pacific
/对北太平洋沙尘暴中生物碳量增加的遥控观测
…… James K B Bishop, Russ E Davis, Jeffrey T Sherman (369)
Supercool or dehydrate? An experimental analysis of overwintering strategies in small permeable Arctic invertebrates/过度冷却还是脱水?实验分析小型有渗透作用的北极无脊椎动物过冬策略
…… Martin Holmstrup, Mark Bayley, Hans Ramløv (370)
Arctic microorganisms respond more to elevated UV-B radiation than CO_2/北极微生物对 UV-B 增加的反应强于 CO_2 …… David Johnson, Colin D Campbell, John A Lee, et al (371)
Resource-based niches provide a basis for plant species diversity and dominance in Arctic tundra
/北极苔原资源型小环境提供了植物物种多样性和优势的基础
…… Robert B McKane, Loretta C Johnson, Gaius R Shaver, et al (372)
Evidence of hybridity in invasive watermilfoil (Myriophyllum) populations/入侵型水生藻类(狐尾藻属)杂种性证据 …… Michael L Moody, Donald H Les (373)
Stability of forest biodiversity/森林生物多样性的稳定性
…… James S Clark, Jason S McLachlan (374)
Cyclic dynamics in a simple vertebrate predator-prey community/一个简单脊椎动物捕食群落的循环动态…… Olivier Gilg, Ilkka Hanski Benoît Sittler (374)
Natural selection shaped regional mtDNA variation in humans/自然选择造成人类区域性 mtDNA 变异 …… Dan Mishmar, Eduardo Ruiz-Pesini, Pawel Golik, et al (375)
Sequential megafauna collapse in the North Pacific Ocean: an ongoing legacy of industrial whaling?
/北太平洋巨型动物群落的逐渐崩溃:是否由于持续的工业捕鲸?
…… Springer M, Estes J A, van Vliet G B, et al (376)
Molecular evidence links cryptic diversification in polar planktonic protists to quaternary climate dynamics/分子证据表明不可理解的极地浮游原生生物多样化和第四纪气候动态相关
…… Kate F Darling, Michal Kucera, Carol J Pudsey, et al (377)
Prehistoric Inuit whalers affected Arctic freshwater ecosystems/史前因纽特人捕鲸对北极淡水生态系统的影响 …… Marianne S V Douglas, John P Smol, James M Savelle, et al (379)
The Yana RHS site: humans in the Arctic before the Last Glacial Maximum/亚纳 RHS 点:末次冰期前在北极地区的人类 …… Pitulko V V, Nikolsky P A, Girya E Yu, et al (379)
Introduced predators transform subarctic islands from grassland to tundra/引入的捕食者使得亚北极岛屿由草原向苔原区转变 …… Croll D A, Maron J L, Estes J A, et al (380)

Policy strategies to address sustainability of Alaskan boreal forests in response to a directionally changing climate/关于解决阿拉斯加北部森林对于定向气候变化响应的可持续性政策战略 ······ F Stuart Chapin Ⅲ,Amy L Lovecraft,Erika S Zavaleta,et al (381)
High biological species diversity in the Arctic flora/北极植物中生物物种的高多样性 ······ Hanne Hegre Grundt,Siri Kjølner,Liv Borgen,et al (382)
Plant community responses to experimental warming across the tundra biome/苔原生物群落中植物群落对实验性气候变暖的响应 ······ Marilyn D Walker,C Henrik Wahren,Robert D Hollister,et al (383)
Frequent long-distance plant colonization in the changing Arctic/在不断变化的北极植物频繁的长距离迁移 ······ Inger Greve Alsos,Pernille Bronken Eidesen,Dorothee Ehrich,et al (384)
The new face of the Arctic/极地研究:北极新面貌 ······ Quirin Schiermeie (385)
Life in a warming world/变暖世界中的生命 ······ Quirin Schiermeier (386)
Circumpolar synchrony in big river bacterioplankton/极地附近大河流中的浮游细菌的同步性 ······ Byron C Crump,Bruce J Peterson,Peter A Raymond,et al (386)
Ecology of the rare microbial biosphere of the Arctic Ocean/北冰洋罕见微生物圈的生态学 ······ Pierre E Galand,Emilio O Casamayor,David L Kirchman,et al (387)
A constant flux of diverse thermophilic bacteria into the cold Arctic seabed/多样化的嗜热细菌不断流入寒冷的北极海底 ······ Casey Hubert,Alexander Loy,Maren Nickel,et al (388)
A semi-aquatic Arctic mammalian carnivore from the Miocene epoch and origin of Pinnipedia /一种起源自鳍脚亚目的中新世半水生北极食肉类哺乳动物 ······ Natalia Rybczynski,Mary R Dawson,Richard H Tedford (389)
Greenhouse gas mitigation can reduce sea-ice loss and increase polar bear persistence /温室气体减排可以减少海冰的损失并改善北极熊的生存 ······ Steven C Amstrup,Eric T De Weaver,David C Douglas,et al (390)
Essential genes from Arctic bacteria used to construct stable, temperature-sensitive bacterial vaccines/用北极细菌的基础基因构建稳定的温度敏感的细菌疫苗 ······ Barry N Duplantis,Milan Osusky,Crystal L Schmerk,et al (391)
Tracking of Arctic terns Sterna paradisaea reveals longest animal migration/跟踪北极燕鸥揭示最长时间的动物迁徙 ······ Carsten Egevang,Iain J Stenhouse,Richard A Phillips,et al (392)
Complete mitochondrial genome of a Pleistocene jawbone unveils the origin of polar bear /完整的更新世颌骨线粒体基因组揭示了北极熊的起源 ······ Charlotte Lindqvist,Stephan C Schuster,Yazhou Sun,et al (393)
Lower predation risk for migratory birds at high latitudes/高纬度候鸟的低捕食风险 ······ McKinnon L,Smith P A,Nol E,et al (394)
Changes in Arctic vegetation amplify high-latitude warming through the greenhouse effect /北极植被变化放大了温室效应造成的高纬度升温 ······ Abigail L Swann,Inez Y Fung,Samuel Levis,et al (395)

五、现代海洋科学类文献

5.1 分析概述 ······ (399)
5.2 摘要翻译——南极 ······ (401)
Iron in Antarctic waters/南极海水中的铁 ······ Martin J H,Gordon R M,Fitzwater S E (401)

Top predators in the Southern-Ocean—a major leak in the biological carbon pump/南大洋的顶级捕食者——生物碳泵的主要漏洞 …………………… Huntley M E,Lopez M D G,Karl D M (401)
Enhanced particle fluxes in Bay of Bengal induced by injection of fresh-water/淡水输入引发的孟加拉湾海域颗粒物通量增加 ……………… Ittekkot V, Nair R R, Honjo S, et al (402)
Estimates of the effect of Southern-Ocean iron fertilization on atmospheric CO_2 concentrations /关于南大洋铁肥实验对大气中二氧化碳含量影响的估计 ………………………………………………… Joos F, Sarmiento J L,Siegenthaler U (403)
Dynamical limitations on the Antarctic iron fertilization strategy/南极铁肥实验的动力学限制 ………………………………………………………………… Peng T H,Broecker W S (404)
Impact of oceanic sources of biogenic sulphur on sulphate aerosol concentrations at Mawson, Antarctica/南极莫森地区海洋源生物硫对硫酸盐气溶胶浓度的影响 ……………………………………………… Prospero J M,Savoie D L,Saltzman E S,et al (404)
Structure of the upper ocean in the western equatorial Pacific/赤道西太平洋上层海水结构 ……………………………………………………………………… Richards K J,Pollard R T (405)
Foraging behaviour of emperor penguins as a resource detector in winter and summer /帝企鹅的觅食行为——冬夏季节的资源探测器 …………………………………………………… Ancel A,Kooyman G L,Ponganis P J,et al (406)
Warming of the water column in the Southwest Pacific-Ocean/西南太平洋水体变暖 ……………………………………………………………………… Bindoff N L,Church J A (407)
Interhemispheric transport of carbon-dioxide by ocean circulation/二氧化碳通过洋流在南北半球之间传送 ……………………………………………………… Broecker W S,Peng T H (408)
Influence of Southern-Ocean waters on the cadmium phosphate properties of the global ocean /南部大洋海水对全球海洋的镉磷酸盐属性的影响 ………………… Frew R D,Hunter K A (409)
Dissolved organic-carbon in the Atlantic, Southern and Pacific Oceans/大西洋、南大洋和太平洋中的溶解有机碳 ……………………………………………… Martin J H,Fitzwater S E (410)
Flattening of the sea-floor depth age curve as a response to asthenospheric flow/海底深度的年龄曲线的扁平化作为对一个软流圈流动的响应 …………………… Morgan J P,Smith W H F (410)
Eddy momentum flux and its contribution to the Southern-Ocean momentum balance/涡流动量通量及其对南大洋的动量平衡的贡献 ……………… Morrow R,Church J,Coleman R,et al (411)
Evidence for basal marine ice in the Filchner-Ronne ice shelf/在 Filchner-Ronne 冰架上基底海洋冰存在的证据 ………………………………… Oerter H,kipfstuhl J,Determann J,et al (412)
Recent variability in the Southern Oscillation—isotopic results from a Tarawa atoll coral /南方涛动的近期变化——来自塔拉瓦环礁珊瑚的同位素记录 ……………………………………………………………… Cole J E,Fairbanks R G,Shen G T (413)
Deep and bottom water of the Weddell Seas western rim/威德尔海西部边缘的底层水 …………………………………………………… Gordon A L,Huber B A,Hellmer H H,et al (414)
$^{231}Pa/^{230}Th$ ratios in sediments as a proxy for past changes in Southern-Ocean productivity /沉积物中$^{231}Pa/^{230}Th$ 的比率作为过去南大洋生产力变化的指标 …………………………………………………………… Kumar N,Gwiazda R,Anderson R F,et al (415)
Rapid formation of the Shatsky Rise Oceanic plateau inferred from its magnetic anomaly/从磁异常推断 Shatsky Rise 海洋高原的迅速形成 ……………………………… Sager W W,Han H C (416)

Atmospheric carbon-dioxide and the ocean/大气中的二氧化碳和海洋
…… Siegenthaler U, Sarmiento J L (417)
Distributions of phytoplankton blooms in the Southern-Ocean/南大洋大量浮游植物的分布
…… Sullivan C W, Arrigo K R, Mcclain C R, et al (417)
Autumn bloom of Antarctic Pack-Ice algae/南极浮冰藻在秋季爆发性增长
…… Fritsen C H, Lytle V I, Ackley S F, et al (418)
Decade-scale trans-Pacific propagation and warming effects of an El-Nino anomaly/一次厄尔尼诺异常对十年尺度的跨太平洋传播和气候变暖的影响
…… Jacobs G A, Hurlburt H E, Kindle J C, et al (418)
El-Nino on the devils staircase—annual subharmonic steps to chaos/厄尔尼诺困扰阶梯——年度次谐波从有序到混乱 …… Jin F F, Neelin J D, Ghil M (419)
Minimal effects of Uvb-radiation on Antarctic diatoms over the past 20 years/过去 20 年紫外线辐射对南极硅藻的微小影响 …… Mcminn A, Heijnis H, Hodgson D (420)
Rapid climate transitions in a coupled ocean-atmosphere model/海气耦合模型中的快速气候变化
…… Rahmstorf S (421)
El-Nino chaos—overlapping of resonances between the seasonal cycle and the Pacific Ocean-atmosphere oscillator/厄尔尼诺混乱——在季节性周期和太平洋-大气振荡系统之间的响应重叠 …… Tziperman E, Stone L, Cane M A, et al (422)
An Antarctic circumpolar wave in surface pressure, wind, temperature and sea-ice extent/南极环极地表面压力、风、温度和海冰张力的绕极波 …… White W B, Peterson R G (422)
The influence of vegetation-atmosphere-ocean interaction on climate during the mid-Holocene /在全新世中期植被-大气-海洋相互作用对气候的影响
…… Ganopolski A, Kubatzki C, Claussen M, et al (423)
Deep-ocean gradients in the concentration of dissolved organic carbon/深海中的溶解有机碳浓度梯度 …… Hansell D A, Carlson C A (424)
Simulated response of the ocean carbon cycle to anthropogenic climate warming/海洋碳循环对人为造成的气候变暖的模拟响应 …… Sarmiento J L, Hughes T M C, Stouffer R J, et al (425)
A simple predictive model for the structure of the oceanic pycnocline/对海洋密度跃层结构的一个简单的预测模型 …… Gnanadesikan A (426)
Estimation of particulate organic carbon in the ocean from satellite remote sensing/从卫星遥感来估计海洋中的颗粒有机碳 …… Stramski D, Reynolds R A, Kahru M, et al (426)
The role of the Southern Ocean in uptake and storage of anthropogenic carbon dioxide/南大洋在人为源二氧化碳的吸收和储存中的作用 …… Caldeira K, Duffy P B (427)
Emperor penguins and climate change/帝企鹅与气候变化 …… Barbraud C, Weimerskirch H (428)
An abrupt climate event in a coupled ocean-atmosphere simulation without external forcing /在没有外部强迫情况下海洋-大气耦合模拟中一个突发的气候事件
…… Hall A, Stouffer R J (429)
Regulation of oceanic silicon and carbon preservation by temperature control on bacteria/温度控制细菌情况下大洋硅和碳保存的调控 …… Bidle K D, Manganelli M, Azam F (429)
Respiration in the open ocean/开放大洋的呼吸 …… del Giorgio P A, Duarte C M (430)
High mixing rates in the abyssal Southern Ocean/南大洋深海的高混合率
…… Heywood K J, Garabato A C N, Stevens D P (431)

Freshening of the Ross Sea during the late 20th century/在 20 世纪后期罗斯海的海水淡化 …… Jacobs S S, Giulivi C F, Mele P A (432)
The change in oceanic O_2 inventory associated with recent global warming/海洋的氧气储量变化和最近全球变暖之间的联系 …… Keeling R F, Garcia H E (432)
Twentieth century sea level: an enigma/20 世纪的海平面:一个谜 …… Munk W (433)
Detection of human influence on sea-level pressure/人类活动对海平面气压影响的探测 …… Gillett N P, Zwiers F W, Weaver A J. et al (434)
Cool Indonesian throughflow as a consequence of restricted surface layer flow/冷的印度尼西亚贯穿流作为限制表面层流动的结果 …… Gordon A L, Susanto R D, Vranes K (435)
Eocene El Nino: evidence for robust tropical dynamics in the "hothouse"/始新世的厄尔尼诺现象:在"热室"的有力热带动力学证据 …… Huber M, Caballero R (436)
Southern Ocean origin for the resumption of Atlantic thermohaline circulation during deglaciation/在末次冰消期大西洋温盐环流恢复起源于南大洋 …… Knorr G, Lohmann G (437)
Direct observations of North Pacific ventilation: brine rejection in the Okhotsk Sea/北太平洋流场的直接观察:盐水在鄂霍次克海回涌 …… Shcherbina A Y, Talley L D, Rudnick D L (437)
A mesoscale iron enrichment in the western Subarctic Pacific induces a large centric diatom bloom/在亚北极太平洋西部的一个中尺度铁富集引起大范围硅藻爆发 …… Tsuda A, Takeda S, Saito H, et al (438)
Robotic observations of enhanced carbon biomass and export at 55 degrees S during SOFeX/SOFeX 期间在南纬 55°海域机器人观察到的碳生物量的增加和输出 …… Bishop J K B, Wood T J, Davis R E, et al (439)
The effects of iron fertilization on carbon sequestration in the Southern Ocean/在南大洋铁肥对碳汇的影响 …… Buesseler K O, Andrews J E, Pike S M, et al (439)
Southern ocean iron enrichment experiment: carbon cycling in high and low-Si waters/南大洋铁富集实验:高、低硅水域中的碳循环 …… Coale K H, Ohnson K S, Chavez F P, et al (440)
Widespread intense turbulent mixing in the Southern Ocean/在南大洋广泛强烈的湍流混合 …… Garabato A C N, Polzin K L, King B A, et al (441)
Changing concentrations of CO, CH_4, C_5H_8, CH_3Br, CH_3I, and dimethyl sulfide during the southern ocean iron enrichment experiments/在南大洋铁富集实验中, CO、CH_4、C_5H_8、CH_3Br、CH_3I 和二甲基硫的浓度的改变 …… Wingenter O W, Haase K B, Strutton P, et al (441)
Sailing the southern sea/南大洋航行 …… Bohannon J (442)
Mesoscale iron enrichment experiments 1993-2005: synthesis and future directions/1993～2005 年中尺度铁富集实验:综合和未来发展方向 …… Boyd P W, Jickells T, Law C. S, et al (443)
Physical oceanography—super spin in the southern seas/物理海洋学——在南部海域极好的拓展 …… Roemmich D (444)
Ice scour disturbance in Antarctic waters/冰冲入对南极水域的扰动 …… Smale D A, Brown K M, Barnes D K A, et al (444)
Ocean circulation in a warming climate/温暖气候中的海洋循环 …… Toggweiler J R, Russell J (445)
5.3　摘要翻译——北极 …… (446)
Ice flexure forced by internal wave-packets in the Arctic Ocean/北冰洋内波包控制的冰弯曲 …… Czipott P V, Levine M D, Paulson C A, et al (446)

Effects of ice coverage and ice-rafted material on sedimentation in the Fram strait/冰覆盖和冰筏物对弗拉姆海峡沉积物的影响 …………………………………… Hebbeln D, Wefer G (446)
Age of Canada basin deep waters—a way to estimate primary production for the Arctic Ocean /加拿大海盆深海水的年龄——一种估计北极海洋初级生产的方式 …………………………………… Macdonald R W, Carmack E C (447)
Reduction of deep-water formation in the Greenland Sea during the 1980s—evidence from tracer data /20 世纪 80 年代格陵兰海深水形成的减少——来自跟踪物的证据 …………………………………… Schlosser P, Bonisch G, Rhein M, et al (448)
Variability in sea-ice thickness over the North-Pole from 1977 to 1990/1977～1990 年北极海冰厚度的变化 …………………………………… Mclaren A S, Walsh J E, Bourke R H, et al (449)
Bromoform emission from Arctic ice algae/北极冰藻释放的三溴甲烷 …………………………………… Sturges W T, Cota G F, Buckley P T (449)
Absence of evidence for greenhouse warming over the Arctic-Ocean in the past 40 years/过去 40 年在北极海洋的温室效应缺少证据 ……… Kahl J D, Charlevoix D J, Zaitseva N A, et al (450)
Tritium and radiocarbon dating of Canada basin deep waters/加拿大洋盆深水的氚和放射性碳定年 …………………………………… Macdonald R W, Carmack E C, Wallace D W R (451)
Testing the iron hypothesis in ecosystems of the equatorial Pacific-Ocean/赤道太平洋生态系统的铁盐假说 …………………………………… Martin J H, Coale K H, Johnson K S, et al (452)
Global warming and the Arctic/北极与全球变暖 …………………………………… Johannessen O M, Bjorgo E, Miles M W (452)
Awakenings in the Arctic/北极苏醒 …………………………………… Macdonald R (453)
Intense mixing of Antarctic bottom water in the equatorial Atlantic Ocean/南极洋底水在大西洋近赤道海域的强烈混合 …………………………………… Polzin K L, Speer K G, Toole J M, et al (454)
Active cycling of organic carbon in the central Arctic Ocean/北冰洋中心活跃的有机碳循环 …………………………………… Wheeler P A, Gosselin M, Sherr E, et al (454)
Possible predictability in overflow from the Denmark Strait/丹麦海峡溢流可能的可预测性 …………………………………… Dickson B, Meincke J, Vassie I, et al (455)
Simulation of recent northern winter climate trends by greenhouse-gas forcing/温室气体对近期北方地区气候变化趋势驱动的模拟 …………… Shindell D T, Miller R L, Schmidt G A, et al (456)
Mixing and convection in the Greenland Sea from a tracer-release experiment/通过示踪剂释放试验研究格陵兰海域混合传送过程 …………… Watson A J, Messias M J, Fogelqvist E, et al (457)
Climate change: new center gives Japan an Arctic toehold/气候变化:日本立足北极的新中心 …………………………………… Wuethrich B (458)
Isotopic evidence for microbial sulphate reduction in the early Archaean era/早太古代微生物硫酸还原的同位素证据 …………………………………… Shen Y A, Buick R, Canfield D E (459)
Robotic observations of dust storm enhancement of carbon biomass in the North Pacific/尘暴增加北太平洋碳生物量的机器人观测 …………………… Bishop J K B, Davis R E, Sherman J T (460)
Even in the high Arctic, nothing is permanent/在北极,啥都不永久 ……………… Goldman E (461)
A warmer Arctic means change for all/一个温暖的北极意味着一切都会变 …………………………………… Kerr R A (461)
Redistribution of energy available for ocean mixing by long-range propagation of internal waves /在海洋混合过程中的长期范围内波传播的能量再分配能源 …………………… Alford M H (462)

Does the trigger for abrupt climate change reside in the ocean or in the atmosphere? /气候突变的触发在海洋还是在大气? …… Broecker W S (462)
Whales before whaling in the North Atlantic/捕鲸之前北大西洋的鲸 …… Roman J, Palumbi S R (463)
Whither Arctic climate? /北极气候何去何从? …… Shindell D (464)
Degradation of terrigenous dissolved organic carbon in the western Arctic Ocean/北冰洋西部陆源溶解态有机碳的降解 …… Hansell D A, Kadko D, Bates N R (464)
Nitrogen balance and Arctic throughflow/北极流和氮平衡 …… Yamamoto-Kawai M, Carmack E, McLaughlin F (465)
Perspectives on the Arctic's shrinking sea-ice cover/北极退缩海冰覆盖的观点 …… Serreze M C, Holland M M, Stroeve J (466)
Identification of Younger Dryas outburst flood path from Lake Agassiz to the Arctic Ocean /从阿加西斯湖和北冰洋之间"新仙女木"洪水暴发路径的识别 …… Murton J B, Bateman M D, Dallimore S R, et al (466)
Extensive methane venting to the atmosphere from sediments of the East Siberian Arctic Shelf /东西伯利亚北极大陆架沉积物有大量的甲烷排放到大气 …… Shakhova N, Semiletov I, Salyuk A, et al (467)
Decreased frequency of North Atlantic polar lows associated with future climate warming /北大西洋极地低压震荡频率降低与未来气候变暖相关 …… Zahn M, von Storch H (468)

六、地质、测绘类文献

6.1　分析概述 …… (473)
6.2　摘要翻译——南极 …… (475)
Extreme isotopic variations in Heard-Island lavas and the nature of mantle reservoirs /赫德岛熔岩同位素的极端变化与地幔层的性质 …… Barling J, Goldstein S L (475)
Ungrouped iron-meteorites in Antarctica-Origin of anomalously high abundance/南极洲未分类铁陨石数量异常多的原因 …… Wasson J T (476)
Evidence for volcanic-eruption on the southern Juan-De-Fuca ridge between 1981 and 1987 /1981～1987 年间南胡安·德·富卡洋中脊火山喷发的证据 …… Chadwick W W, Embley R W, Fox C G (476)
Geochronological evidence supporting Antarctic deglaciation 3 million years ago/地质年代证据支持 300 万年前南极存在冰消期 …… Barrett P J, Adams C J (477)
Accumulation of suspended barite at mesopelagic depths and export production in the Southern-Ocean/南大洋中层悬浮重晶石的聚集与输出生产力的关系 …… Dehairs F, Baeyens W, Goeyens L (478)
Fresh basalts from the Pacific Antarctic ridge extend the Pacific geochemical province /太平洋-南极洲洋脊的新鲜玄武岩扩大了太平洋地球化学的范围 …… Ferguson E M, Klein E M (479)
Preservation of miocene glacier ice in East Antarctica/东南极中新世时期冰川的保存 …… Sugden D E, Marchant D R, J R N. P. et al (480)
Creation of theta-auroras: the isolation of plasma sheet fragments in the polar-cap /ξ-极光的产生:极地冰冠处等离子层碎片的隔离 …… Newell P T, Meng C I (480)

A large deep freshwater lake beneath the ice of central East Antarctica/东南极中部冰层下有一个大而深的淡水湖 ………………………… Kapitsa A P,Ridley J K,Robin G De Q,et al (481)
A great lake under the ice/冰下的巨大湖泊 ……………… Ellis Evans J C,Wynn Williams D (482)
Nitrification in Antarctic soils/南极土壤中的硝化作用 …………………………………………………… Wilson K,Sprent J I,Hopkins D W (482)
Rifts found as Antarctic ice breaks apart/南极冰盖破裂形成的裂隙 ……………… Rex Dalton (483)
Antarctic tectonics: constraints from an ERS-1 satellite marine gravity field/南极构造地质学:来自 ERS-1 人造卫星海洋重力场的估计 ……………………………… McAdoo D,Laxon S (484)
Evolution of the Pacific-Antarctic Ridge south of the Udintsev fracture zone/Udintsev 断裂带南部太平洋-南极洋中脊的演化 …………………………… Geli L,Bougault H,Aslonian D,et al (485)
The plan to unlock Lake Vostok/解开 Vostok 湖之谜的计划 ………………………… Inman M (485)
Lakes linked beneath Antarctic ice/南极冰盖下的湖泊 ……………………………… Giles J (486)
Rapid discharge connects Antarctic subglacial lakes/水流快速排放的南极冰下湖 ……………………………………………… Wingham D J,Siegert M J,Shepherd A,et al (486)
Measurements of time-variable gravity show mass loss in Antarctica/时变重力测量显示南极地区质量损失 ……………………………………………………………… Velicogna I,Wahr J (487)
Major Australian-Antarctic plate reorganization at Hawaiian-Emperor bend time/夏威夷帝王岛链转折时期澳大利亚-南极大板块的重组 ……………………………………………… Whittaker J M,Muller R D,Leitchenkov G,et al (488)
Elevation changes in Antarctica mainly determined by accumulation variability/南极地区的海拔变化主要由冰雪累积导致 …… Helsen M M,van den Broeke M R,van de Wal R S W,et al (488)
A positive test of east Antarctica-Laurentia juxtaposition within the Rodinia supercontinent/罗迪尼亚超级大陆内部东南极-劳伦古大陆拼合的实证检验 ……………………………………………… Goodge J W,Vervoort J D,Funning C M,et al (489)
Response to comment on “major Australian-Antarctic plate reorganization at Hawaiian-Emperor bend time”/回复对《夏威夷帝王岛链转折时期澳大利亚—南极大板块的重组》一文的质疑 ……………………………………………… Whittaker J M,Muller R D,Leitchenkov G,et al (490)
Comment on “major Australian-Antarctic plate reorganization at Hawaiian-Emperor bend time”/对《夏威夷帝王岛链转折时期澳大利亚-南极大板块的重组》一文的评论 ……………………………………………………………… Tikku A A,Direen N G (491)
6.3 摘要翻译——北极 …………………………………………………………………… (492)
Mid-to Late Pleistocene ice drift in the western Arctic Ocean: evidence for a different circulation in the past/中晚更新世北冰洋西部的冰漂移:过去环流与现在不同的证据 ……………………………………………………………… Bischof J F,Darby D A (492)
Evidence of recent volcanic activity on the ultraslow-spreading Gakkel ridge/Gakkle 超慢速扩张中脊上现代火山活动的证据 …………………… Edwards M H,Kurras G J,Tolstoy M,et al (492)
An ultraslow-spreading class of ocean ridge/一种超慢速扩张类型的洋中脊 ………………………………………………………… Dick H J B,Lin J,Schouten H (493)
Discovery of abundant hydrothermal venting on the ultraslow-spreading Gakkel ridge in the Arctic/北冰洋 Gakkel 超慢速扩张中脊大量热液排放的发现 ……………………………………………… Edmonds H N,Michael P J,Baker E T,et al (494)

Geophysical evidence for reduced melt production on the Arctic ultraslow Gakkel mid-ocean ridge /北极地区超慢速扩张的 Gakkel 洋中脊处熔融物产量减少的地球物理证据 ………………………………………… Jokat W, Ritzmann O, Schrnidt-Aursch M C, et al (495)
Magmatic and amagmatic seafloor generation at the ultraslow-spreading Gakkel ridge, Arctic Ocean /北极扩张速度超慢的 Gakkel 山脊处岩浆和非岩浆型海底的形成 ………………………………………… Michael P J, Langmuir C H, Dick H J B, et al (496)
Ecosystem carbon storage in Arctic tundra reduced by long-term nutrient fertilization /北极苔原地区生态系统碳储量因长期营养性施肥而减少 ………………………………………… Mack M C, Schuur E A G, Bret-Harte M S, et al (497)
Artic Ocean (communications arising): hydrothermal activity on Gakkel Ridge/北冰洋：Gakkel 洋中脊热液活动 ………………………………………… Jean-Baptiste P, Fourre E (498)
Glacial/interglacial changes in subarctic North Pacific stratification/亚北极北太平洋海洋分层的冰期-间冰期变化 ………………………………………… Jaccard S L, Haug G H, Nutman A P (499)
Coupled ^{142}Nd-^{143}Nd isotopic evidence for Hadean mantle dynamics/太古宙地幔动力机制的^{142}Nd/^{143}Nd 同位素证据 ………………………………………… Bennett V C, Brandon A D, Nutman A P (499)
A vestige of Earth's oldest ophiolite/地球上最古老的蛇纹岩的痕迹 ………………………………………… Furnes H, de Wit M, Standigel H, et al (500)
Explosive volcanism on the ultraslow-spreading Gakkel ridge, Arctic Ocean/北冰洋上扩张速率超慢的 Gakkel 洋脊上的爆炸式火山活动 ………………………………………… Sohn R A, Willis C, Humphris S, et al (501)
Ancient, highly heterogeneous mantle beneath Gakkel ridge, Arctic Ocean/北冰洋 Gakkel 洋脊下古老且高度异相的地幔 ………………………………………… Liu C Z, Snow J E, Hellebrand E, et al (502)
Ancient permafrost and a future, warmer Arctic/古冻土与北极未来增暖 ………………………………………… Froese D G, Westgate J A, Reyes A V, et al (503)
Origin of a "Southern Hemisphere" geochemical signature in the Arctic upper mantle/北极上地幔中"南半球"地球化学信号的来源 ………………………………………… Goldstein S L, Soffer G, Langmuir C H, et al (503)

七、陆地古环境类文献

7.1 分析概述 ………………………………………… (507)
7.2 摘要翻译——南极 ………………………………………… (509)
Oxygen supersaturation in ice-covered Antarctic lakes: biological versus physical contributions /南极冰封湖泊过饱和氧：生物过程和物理过程的贡献 ………………………………………… Craig H, Wharton R A Jr, McKay C P (509)
The Antiquity of oxygenic photosynthesis—evidence from stromatolites in sulfate-deficient archean lakes/古老的光合作用——来自太古代缺硫酸盐湖泊中叠藻层的证据 ………………………………………… Roger Buick (509)
Reproductive-biology of the permian glossopteridales and their suggested relationship to flowering plants/二叠纪舌羊齿目的生殖生物学及其与开花植物的指示关系 ………………………………………… Taylor E L, Taylor T N (510)
Cooler estimates of Cretaceous temperatures/更冷的白垩纪气温的估计 ………………………………………… Sellwood B W, Price G D, Valdes P J (511)
Climate variation—cycling around the South Pole/气候变化——南极点附近的循环 ………………………………………… Yuan Xiaojun, Mark A Cane, Douglas G Martinson (512)

Inferring seal populations from lake sediments/从湖泊沉积物恢复海豹数量
………………………………………………………… Dominic A Hodgson,Nadine M Johnston (512)
A 3000-year record of penguin populations/3000 年企鹅数量记录
………………………………………………………… Liguang Sun,Zhouqing Xie,Junlin Zhao (513)
Climate change—devil in the detail/气候变化——难在细节
………………………………………………… Vaughan D G,Marshall G J,Connolley W M,et al (514)
Environmental change and Antarctic seabird populations/环境变化与南极海鸟数量
………………………………………………………… Croxall J P,Trathan P N,Murphy E J (514)
Antarctic climate cooling and terrestrial ecosystem response/南极气候变冷与陆地生态系统的响应
………………………………………………………… Doran P T,Priscu J C,Lyons W B,et al (515)
Rates of evolution in ancient DNA from Adelie penguins/阿德雷企鹅古 DNA 揭示的物种演化速率
………………………………………………… Lambert D M,Ritchie P A,Millar C D,et al (516)
Extreme responses to climate change in Antarctic lakes/南极湖泊对气候变化的极端响应
………………………………………………………… Quayle W C,Peck L S,Peat H,et al (517)
Ecological effects of climate fluctuations/气候变动的生态效应
………………………………………………… Stenseth N C,Mysterud A,Ottersen G,et al (517)
Climate change—recent temperature trends in the Antarctic/气候变化——近来南极温度趋势
………………………………………………… Turner J,King J C,Lachlan-Cope T A,et al (518)
Ecological responses to recent climate change/近来气候变化的生态响应
………………………………………………………… Walther G R,Post E,Convey P,et al (518)
Carbon loss by deciduous trees in a CO_2-rich ancient polar environment/富 CO_2 的古极地环境落叶树的碳流失 ………………………………… Royer D L,Osborne C P,Beerling D J (519)
Microevolution and mega-icebergs in the Antarctic/南极微进化与巨型冰山
………………………………………………… Shepherd L D,Millar C D,Ballard G,et al (520)
Holocene elephant seal distribution implies warmer-than-present climate in the Ross Sea /罗斯海全新世象海豹分布指示的比现代温暖的气候条件
………………………………………………………… Hall B L,Hoelzel A R,Baroni C,et al (521)
Abrupt recent shift in $\delta^{13}C$ and $\delta^{15}N$ values in Adelie penguin eggshell in Antarctica/近来南极阿德雷企鹅蛋壳 ^{13}C、^{15}N 同位素的突变 ………………………… Emslie S D,Patterson W P (521)
7.3 文献摘要——北极 ……………………………………………………………………………… (523)
Arctic lakes and streams as gas conduits to the atmosphere—implications for tundra carbon budgets /北极湖泊河流作为气体通向大气的通道——来自苔原 C 预算的启示
………………………………………………………… Kling G W,Kipphut G W,Miller M C (523)
Rapid response of treeline vegetation and lakes to past climate warming/林木线和湖泊对过去气候变暖的快速响应 …………………… Macdonald G M,Edwards T W D,Moser K A,et al (523)
Recent change of Arctic tundra ecosystems from a net carbon:dioxide sink to a source/北极苔原生态系统近来的变化:从净 CO_2 汇到源 ……… Oechel W C,Hastings S J,Vourlitis G,et al (524)
Holocene dwarf mammoths from Wrangel Island in the siberian Arctic/全新世西伯利亚北高纬弗兰格尔岛矮猛犸象 ………………………………… Vartanyan S L,Garutt V E,Sher A V (525)
Marked post-18th century environmental-change in high-Arctic ecosystems/18 世纪后期北极高纬生态系统显著的环境变化 ………………………… Douglas M S V,Smol J P,Blake W (526)

Paleoindians in Beringia—evidence from Arctic Alaska/白令陆桥古印第安人——来自北极阿拉斯加的证据………………………………………………… Kunz M L, Reanier R E (527)
Transient nature of CO_2 fertilization in Arctic tundra/北极苔原天然 CO_2 施肥的短暂性 …………………………………………………… Oechel W C, Cowles S, Grulke N, et al (527)
Ecosystem size determines food-chain length in lakes/湖泊生态系统大小决定了食物链的长度 ………………………………………………………… Post D M, Pace M L, Hairston N G (528)
Human presence in the European Arctic nearly 40000 years ago/欧洲北极地区 40000 年前人类的出现 ………………………………………………………… Pavlov P, Svendsen J I, Indrelid S (529)
Increasing shrub abundance in the Arctic/北极地区灌木丰度的不断增加 ……………………………………………………………………… Sturm M, Racine C, Tape K (530)
Rapid body size decline in Alaskan Pleistocene horses before extinction/阿拉斯加更新世马灭绝前躯体尺寸的快速减小 ……………………………………………………………… Guthrie R D (531)
Cyclic variation and solar forcing of Holocene climate in the Alaskan subarctic/阿拉斯加亚北极全新世气候的周期变化和太阳辐射强迫………………… Hu F S, Kaufman D, Yoneji S, et al (532)
An Arctic mammal fauna from the Early Pliocene of North America/一种来自于北美早上新世的北极哺乳动物 ………………………………………………… Tedford R H, Harington C R (533)
Correlated terrestrial and marine evidence for global climate changes before mass extinction at the Cretaceous-Paleogene boundary/白垩纪-早第三纪转变时期大灭绝前全球气候变化陆地和海洋的证据间的相互关系……………………………… Wilf P, Johnson K R, Huber B T (533)
Siberian peatlands a net carbon sink and global methane source since the early Holocene /全新世早期以来西伯利亚湿地净碳汇和全球甲烷来源 ……………………………………… Smith L C, MacDonald G M, Velichko A A, et al (534)
Role of land-surface changes in Arctic summer warming/地表变化在北极夏季变暖中的作用 …………………………………………………… Chapin F S, Sturm M, Serreze M C, et al (535)
Disappearing Arctic lakes/消失的北极湖泊……… Smith L C, Sheng Y, MacDonald G M, et al (536)
Climate-driven regime shifts in the biological communities of Arctic lakes/气候驱动北极湖泊生物群落的转变 …………………………………… Smol J P, Wolfe A P, Birks H J B, et al (537)

八、环境污染类文献

8.1 分析概述…………………………………………………………………………………… (541)
8.2 摘要翻译——南极……………………………………………………………………………… (543)
Identification of widespread pollution in the Southern-Hemisphere deduced from satellite analyses /由卫星分析推导识别南半球广泛分布的污染物 …………………………………………………… Fishman J, Fakhruzzaman K, Cros B, et al (543)
Anthropogenic lead in Antarctic sea-water/南极海水中的人为源铅 ………………………………………………………… Flegal A R, Maring H, Niemeyer S (543)
Global influence of the AD1600 eruption of Huaynaputina, Peru/公元 1600 年秘鲁 Huaynaputina 火山喷发的全球性影响 ………………………… Shanaka L de Silva, Gregory A Zielinski (544)
Effects of ship emissions on sulphur cycling and radiative climate forcing over the ocean /船只排放物对海洋硫循环和辐射气候强迫的影响 ………………………………… Kevin Capaldo, James J Corbett, Prasad Kasibhatla, et al (545)

Origins of sulphate in Antarctic dry-valley soils as deduced from anomalous ^{17}O compositions /由^{17}O同位素异常组成来判断南极涸谷地区土壤硫酸盐来源 ······ Huiming Bao, Douglas A Campbell, James G Bockheim, et al (546)

Mass-independent sulfur isotopic compositions in stratospheric volcanic eruptions /火山喷发导致的平流层中硫同位素非质量分馏的组成 ······ Melanie Baroni, Mark H Thiemens, Robert J Delmas, et al (547)

Southern Ocean not so pristine/南大洋并非如此原始 ······ Louise K Blight, David G Ainley (548)

8.3 摘要翻译——北极 ······ (549)

A Scramble for data on Arctic radioactive dumping/对北极放射性物质排放调查的争夺 ······ Marshall E (549)

High-Concentrations of toxaphene in fishes from a Sub-Arctic lake/亚北极地带湖泊中鱼类体内的高浓度毒杀芬 ······ Karen A Kidd, David W Schindler, Derek C G Muir, et al (549)

Accumulation of persistent organochlorine compounds in mountains of western Canada /持久性有机氯化物在加拿大西部山区的聚积 ······ Jules M Blais, David W Schindler, Derek C G Muir, et al (550)

Protactinium-231 and thorium-230 abundances and high scavenging rates in the western Arctic Ocean /西部北冰洋^{231}Pa和^{230}Th的丰度和高清除速率 ······ Henrietta N Edmonds, S Bradley Moran, John A Hoff, et al (551)

Arctic springtime depletion of mercury/春季北极地区汞的消减 ······ Schroeder W H, Anlauf K G, Barrie L A, et al (552)

Eurasian air pollution reaches eastern North America/欧亚大气污染到达了北美东部 ······ Pierre E Biscaye, Francis E Grousset, Anders M Svensson, et al (552)

Atlantic water flow pathways revealed by lead contamination in Arctic basin sediments /北极盆地沉积中铅污染情况揭示了大西洋水流路径 ······ Charles Gobeil, Robie W Macdonald, John N Smith, et al (553)

Adverse effects of acid rain on the distribution of the Wood Thrush Hylocichla mustelina in North America/酸雨对北美一种稀有的黄褐森鸫分布的负作用 ······ Ralph S Hames, Kenneth V Rosenberg, James D Lowe, et al (554)

For precarious populations, pollutants present new perils/处于危险中的民众,污染物带来新危险 ······ Webster P (555)

Roaming polar bears reveal Arctic role of pollutants/迁徙的北极熊揭示北极圈污染物的作用 ······ Marika Willerroider, Munich (556)

Persistent toxic substances—study finds heavy contamination across vast Russian Arctic/研究发现广大的俄罗斯北极地区被持久性有毒物质重度污染 ······ Webster P (557)

Arctic seabirds transport marine-derived contaminants/北极海鸟对海洋源污染物的传输 ······ Jules M Blais, Lynda E Kimpe, Dominique McMahon, et al (558)

Arctic air pollution: origins and impacts/北极大气污染:来源与影响 ······ Kathy S Law, Arolress Stohl (559)

20th-century industrial black carbon emissions altered Arctic climate forcing/20世纪工业黑炭的排放改变了北极地区的气候效应 ······ Joseph R McConnell, Ross Edwards, Gregory L Kok, et al (560)

Counterintuitive carbon-to-nutrient coupling in an Arctic pelagic ecosystem/在北极浮游生态系统营养物质和碳的反常耦合 ………………… Thingstad T F, Bellerby R G J, Bratbak G, et al (560)

九、天文、空间物理类文献

9.1　分析概述 ……………………………………………………………………… (565)

9.2　摘要翻译 ……………………………………………………………………… (567)

The breakup of a meteorite parent body and the delivery of meteorites to Earth/一个陨石母体的解体及到达地球 ……………………………………… Benoit P H, Sears D W G (567)

Dipolar reversal states of the geomagnetic-field and core mantle dynamics/地磁场两极反转和核幔对流 ……………………………………………………… Hoffman K A (567)

Paleomagnetic constraints on the geometry of the geomagnetic-field during/古地磁在地磁场倒转时的几何约束条件 …………………………… Valet J P, Tucholka P, Courtillot V, et al (568)

Longitudinally confined geomagnetic reversal paths from Nondipolar transition fields/来自 Nondipolar 过渡区域的经向封闭地磁倒转 ……………………………… Gubbins D, Coe R S (569)

Persistent patterns in the geomagnetic-field over the past 2.5 Myr/过去 250 万年的地磁场的持续模式 ……………………………………………… Gubbins D, Kelly P (570)

Arctic Ocean gravity field derived from Ers-1 satellite altimetry/卫星测量得到的北冰洋重力场 ……………………………………………………… Laxon S, Mcadoo D (570)

Geophysics of the pitman fracture zone and Pacific-Antarctic plate motions during the cenozoic /新生代皮特曼断裂区及太平洋-南极板块运动的地球物理学 ……………………………… Cande S C, Raymond C A, Stock J, et al (571)

High-altitude observations of the polar wind/高纬观测的极地风 ……………………………… Moore T E, Chappell C R, Chandler M O, et al (572)

A new perspective on the dynamical link between the stratosphere and troposphere/对平流层和对流层动力连接的一个新看法……………… Hartley D E, Villarin J T, Black R X, et al (573)

Isotopic evidence for a solar argon component in the Earths mantle/同位素证据显示地幔中存在来自太阳的氩 ……………………………………………………… Pepin R O (573)

Geomagnetic intensity variations over the past 780 kyr obtained from near-seafloor magnetic anomalies /从近海底磁异常得到的在过去 78 万年的地磁强度变化 ……………………………… Gee J S, Cande S C, Hildebrand J A, et al (574)

Geochemical tracing of Pacific-to-Atlantic upper-mantle flow through the Drake passage /地球化学追踪通过德雷克海峡从太平洋到大西洋的上地幔流 ……………………………… Pearce J A, Leat P T, Barker P F, et al (575)

Repeated and sudden reversals of the dipole field generated by a spherical dynamo action/球形发电机产生的磁极反复而快速的反转……………………… Li J H, Sato T, Kageyama A (576)

Global distribution of neutrons from Mars: results from Mars Odyssey/火星中子的全球性分布：来自火星探测器奥德赛的结果 ……………… Feldman W C, Boynton W V, Tokar R L, et al (577)

Small-scale structure of the geodynamo inferred from Oersted and Magsat satellite data/卫星资料推测的地球发电机的小尺度结构 ………………… Hulot G, Eymin C, Langlais B, et al (577)

Measurement of polarization with the Degree Angular Scale Interferometer/用度角刻度干涉仪测量极化偏振 ……………………………… Leitch E M, Kovac J M, Pryke C, et al (578)

Interplanetary dust from the explosive dispersal of hydratedasteroids by impacts/水合小行星碰撞所产生的星际尘埃 ………………………………… Tomeoka K，Kiriyama K，Nakamura K，et al (579)

Intense equatorial flux spots on the surface of the Earth's core/地心表层赤道位置强烈的通量点 …………………………………………………………………………………… Jackson A (580)

Exceptional astronomical seeing conditions above Dome C in Antarctica/南极冰穹 C 位置卓越的天文观测条件 ………………………………… Lawrence J S，Ashley M C B，Tokovinin A，et al (581)

Removal of meteoric iron on polar mesospheric clouds/极地中层云中陨石铁的去除 ………………………………………………………… Plane J M C，Murray B J，Chu X Z，et al (582)

Heat flux anomalies in Antarctica revealed by satellite magnetic data/卫星磁力数据揭示了在南极洲热通量的异常 ………………………………… Maule C F，Purucker M E，Olsen N，et al (582)

文献统计概述

随着全球气候变化对极地气候影响的日益显著以及持续不断的全球性能源紧张，目前极地资源正在得到国际社会的高度关注，南北极的国际战略地位不断提高。近二十年来我国的极地研究在自然科学领域取得了诸多研究成果，但软科学的研究相对薄弱，表现为研究领域少，研究深度不够。极地软科学是极地自然科学和社会科学的综合交叉学科，目前的软科学研究主要是政治法律、环境保护等介绍性和综合性的论述，研究面和研究深度都有待扩展。本文通过 Web of Science 检索，利用“antarctic or southern oceanor south pol and arctic or north pol”等进行题文检索，并通过对摘要的翻译排除误检，得到 *Nature*，*Science*，*Proceedings of the National Academy of Sciences of the United States of America*（以下简称 *PNAS*）三大期刊上 1990～2010 年发表的关于极地的科学研究论文 650 篇。通过论文摘要翻译，本书展示了极地科学研究各领域的重要成果，并以这 650 篇论文为研究对象，从论文出版年分布、研究领域和第一作者国别及研究机构等方面进行统计分析，指出各国极地科学研究发展趋势，为我国极地科学研究与发展方向提供借鉴。

1. 论文出版年统计与分析

文献量的年度变化是衡量科研成果的重要尺度之一。对 *Nature*，*Science*，*PNAS* 上关于南北极的科学研究论文按出版年份进行数量统计，得出图 1。

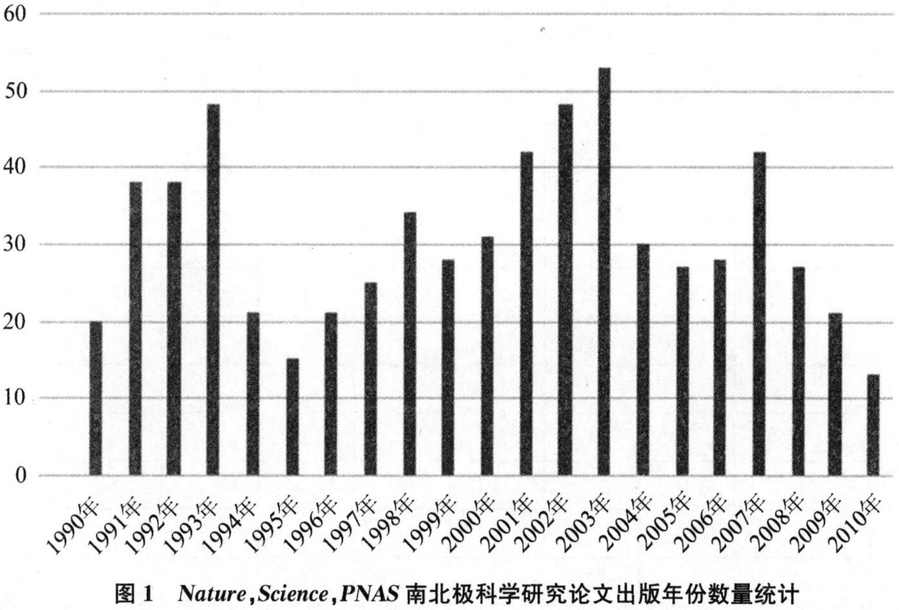

图 1　*Nature*，*Science*，*PNAS* 南北极科学研究论文出版年份数量统计

从图 1 可以看出，各年文献数量呈波浪式发展，2003 年发表文献量最多，有 53 篇；2010 年发表文献量最少，只有 13 篇。2001～2003 年是 1990～2010 年极地研究文献发表的最高峰，平均每年产出文献 47.3 篇；1991～1993 年为次高峰，平均每年产出文献 41.3 篇；1998～2000 年出现一个小高峰，平均每年产出文献 31 篇；2007 年以后极地文献数量呈下降趋势。

2. 研究领域统计与分析

通过对极地科技研究的文献摘要进行解读，按照国际极地文献统计分析方法，结合国际极地热点学科，提取出自然科学类文献，共分类为冰川、冰芯古气候，大气科学，生物科学，现代海洋科学，海洋古气候、古环境，地质、测绘，陆地古环境，环境污染，天文、空间物理九个研究领域，从不同的角度分析各研究领域的特点，具体统计分析如下(表1)。

表1　极地科学研究文献各研究领域文献数量

研究领域	文献数量(篇)	研究领域	文献数量(篇)	研究领域	文献数量(篇)
冰川、冰芯古气候	140	现代海洋科学	89	陆地古环境	37
大气科学	129	海洋古气候、古环境	65	环境污染	22
生物科学	109	地质、测绘	38	天文、空间物理	21

数量最多的是冰川、冰芯古气候类的文献，140篇，约占本书所统计论文总量的21.5%；其次是大气科学类文献，129篇，约占总量的19.8%，接着是生物科学类文献，109篇，约占总量的16.8%。

对九个研究领域文献进行总体分析可以发现：在2001～2003年极地研究文献发表的最高峰阶段，冰川、冰芯古气候方面的研究论文数量呈现高峰期；大气科学和现代海洋科学方面的文献在1991～1993年次高峰这一阶段所占比重较大；海洋古气候、古环境领域从1997年开始到2008年占据优势地位；2007年以后极地各领域文献数量均呈下降趋势；而冰川、冰芯古气候文献量在2009年时显著上升。

3. 极地研究文献第一作者所属国家分布分析

通过对三大顶级期刊上极地相关文献第一作者所属国家(地区、机构)进行分析，前十位的为：美国349篇，英国95篇，加拿大36篇，德国32篇，法国30篇，澳大利亚18篇，瑞士16篇，新西兰12篇，日本8篇，挪威7篇。中国处于第17位，文献数量为3篇。

表2　各研究领域文献第一作者所属国家(地区、机构)分布表

第一作者国家(地区、机构)	冰川、冰芯古气候	大气科学	生物科学	现代海洋科学	海洋古气候、古环境	地质、测绘	陆地古环境	环境污染	天文、空间物理	文献总量
美国	74	86	46	53	33	17	20	10	10	349
英国	25	14	17	9	12	6	7		5	95
加拿大	1	5	11	6	2	1	3	7		36
德国	1	12	7	6	2	3	1			32
法国	14	4	4	1	2	1		2	2	30
澳大利亚	5	1	1	4	3	3			1	18
瑞士	9	3		2	1		1			16
新西兰	2	1	4	2		1	2			12
日本	1	1	1	1	2				2	8
挪威	1		2	1		1	1	1		7
俄罗斯			2		1	1	1	1		6
丹麦			2	2	1				1	6
瑞典	2	1	2							5
荷兰			1		3	1				5
意大利	1		3							4
比利时	1				1	2				4
以色列				1	2					3
中国	1					1	1			3

续表

第一作者国家（地区、机构）	冰川、冰芯古气候	大气科学	生物科学	现代海洋科学	海洋古气候、古环境	地质、测绘	陆地古环境	环境污染	天文、空间物理	文献总量
西班牙			3							3
奥地利	1							1		2
阿根廷	1									1
南非		1								1
芬兰			1							1
爱沙尼亚			1							1
格陵兰			1							1
百慕大生物科技有限公司				1						1

由表 2 可见，本书所统计的文献第一作者共来自 26 个国家（地区、机构）。美国作为极地研究的超级大国，其科研水平远超其他各国，科研论文产出数量占本书编译文献总量的一半以上。其在冰川、冰芯古气候学、大气科学这两个领域产出文献数量最多，说明这两个研究领域是其优势领域，特别是大气科学领域。结合文献发表年代来看，冰川、冰芯古气候学的研究呈波浪式发展，在 1994～1997 年和 2006～2008 年是低谷，在 2009 年达到最高峰；海洋古气候、古环境领域的研究在 1998～2008 年成为美国极地研究的重点领域；大气科学领域在 1990～1993 年是研究的高峰期，这可能与 20 世纪 90 年代初臭氧空洞的发现有密切关系，随后每年均有涉及该领域的研究，但数量已有所减少；2000 年之后，在臭氧的研究上突破点逐渐减少，研究的热点主要转变为导致全球变化的温室气体等方面的研究。

英国极地研究文献数量为 95 篇，文献总量排名仅次于美国，但文献总量尚不足美国的 1/3。其中冰川、冰芯古气候研究，海洋古气候、古环境研究和生物科学研究是英国极地研究的重要领域。

加拿大、德国和法国极地研究文献数量分别为 36 篇、32 篇和 30 篇，仅从本书统计文献产出来看，这三个国家的极地研究水平基本相当，次于美国和英国。加拿大的生物科学研究是其重点研究领域；德国的海洋古气候、古环境和大气科学研究是其重点研究领域；法国的冰川、冰芯古气候和海洋古气候、古环境研究是其重点研究领域。

在汇编涉及的时间内，中国的极地研究文献共 3 篇，涉及冰川、冰芯古气候研究，地质、测绘类研究和陆地古环境研究这三个研究领域，说明中国学者已经逐渐参与进来，但与美英等国的差距还相当大。

4. 极地研究文献第一作者研究机构统计与分析

因文献第一作者所在研究机构涉及较多，我们对文献量达到 5 篇及以上的研究机构进行了统计，详见表 3。

表 3　第一作者所属机构的文献量　　单位：篇（文献量≥5）

序号	极地科技文献第一作者所属研究机构	文献量（篇）
1	British Antarctic Survey, Natural Environment Research Council 英国自然环境研究理事会南极调查局	30
2	National Aeronautics and Space Administration, Goddard Space Flight Center 美国国家航空和航天局戈达德宇宙飞行中心	28
3	Lamont-Doherty Geological Observatory of Columbia University 美国哥伦比亚大学拉蒙特-多尔蒂地质观测站	20

续表

序号	极地科技文献第一作者所属研究机构	文献量(篇)
4	Isotope Laboratory, Scripps Institution of Oceanography, University of California at San Diego 美国加利福尼亚大学圣地亚哥分校斯克里普斯海洋学研究所	19
5	Cnrs, Glaciol & Geophys Environm Lab, Bp 96, F-38402 St Martin Dheres, France 法国国家科学研究院冰川与地球物理环境实验室	14
6	University of California, USA 美国加利福尼亚大学	12
7	Univ Colorado, Dept Geol Sci 美国科罗拉多大学	12
8	University of Bern 瑞士伯尔尼大学	12
9	Harvard University 美国哈佛大学	11
10	Jet Propulsion Laboratory, California Institute of Technology 美国加州理工学院喷气推进实验室	11
11	Woods Hole Oceanographic Institution 美国伍兹霍尔海洋研究所	9
12	Princeton Univ, Atmospher & Ocean Sci Program 美国普林斯顿大学	9
13	Mit, Dept Earth Atmospher & Planetary Sci 美国麻省理工学院	8
14	University of Illinois 美国伊利诺大学	8
15	Noaa, Climate Monitoring & Diagnost, Boulder 美国国家海洋和大气局	8
16	Centre for Polar Observation and Modelling, Scott Polar Research Institute, University of Cambridge 英国剑桥大学斯科特极地研究所极地观测与模拟中心	8
17	Department of Meteorology, Edinburgh University 英国爱丁堡大学	8
18	Pennsylvania State University, University Park 美国宾夕法尼亚州立大学	7
19	Oregon State University 美国俄勒冈州立大学	7
20	University of Alaska 美国阿拉斯加大学	7
21	Alfred Wegener Institute for Polar and Marine Research 德国阿尔弗雷德・魏格纳极地和海洋研究中心	7
22	Geophysics Program, University of Washington 美国华盛顿大学西雅图分校	6

续表

序号	极地科技文献第一作者所属研究机构	文献量(篇)
23	Max Planck Inst Kernphys, Div Atmospher Phys 德国马普研究所	6
24	University of Otago, Dunedin, New Zealand 新西兰奥塔哥大学	6
25	Univ Rhode Isl, Grad Sch Oceanog 美国罗德岛大学	5
26	Geophysical and Polar Research Center, University of Wisconsin, Madison 美国威斯康星大学地球物理和极地研究中心	5
27	Moss Landing Marine Laboratories, California State University 美国加利福尼亚州立大学莫斯兰丁海洋实验室	5
28	Univ Leeds, Sch Chem, Leeds LS2 9JT, W Yorkshire, England 英国利兹大学	5
29	Southampton Oceanog Ctr, Sch Ocean & Earth Sci 英国南安普顿海洋与地球科学学院	5
30	Univ Heidelberg, Inst Environm Phys 德国海德堡大学	5
31	Laboratoire de Modération du Climat et de I'Environnement France 法国原子能委员会气候与环境保护实验室	5

通过对极地科技研究文献产出量 5 篇及以上的第一作者所属机构进行统计得出：美国有 19 家，英国有 5 家，法国有 3 家，德国有 3 家，瑞士有 1 家。英国南极局是文献产出量最多的研究机构，研究领域涉及较广，包括冰川、生物、大气、海洋古气候、陆地环境、现代海洋、地质测绘 7 个领域。

在汇编涉及的时间内，极地科技文献第一作者研究机构统计中先后有 3 家中国极地研究机构，它们是：中国科学技术大学极地环境研究室、中国极地研究中心、中国科学院地质与地球物理研究所岩石圈演化重点实验室，分别在 *Nature*，*Science* 和 *PNAS* 上各发表 1 篇论文。

5. 结论

通过对三大顶级期刊极地科技研究文献的统计分析，编译者得出以下结论：

(1) 通过对文献年代分布进行研究，全部的极地科学研究文献数量随着年代变化呈波浪式变动，出现了几个高峰：2001～2003 年是 1990～2010 年极地研究文献发表的高峰；1991～1993 年为次高峰，在这一阶段，大气科学和现代海洋科学方面的文章所占比重较大；但 2007 年以后极地文献总量呈下降趋势，而反观我国自 2000 年以后，文献数量呈增加趋势，说明我国在极地科学研究方面发展势头良好。

(2) 通过对文献研究领域分布进行分析，冰川、冰芯古气候，大气科学，海洋古气候、古环境，生物科学和现代海洋科学文献量所占比重较大，是极地研究的热点领域。我国应该在这几个领域加强研究投入，扩大我国的极地科研成果影响力，在已有基础上开创特色研究。

(3) 通过对文献第一作者国别和所属机构进行分析，发现美国是极地科技研究产出的超级大国，文献约占总量的 1/2。文献产出量 5 篇及以上的研究机构美国有 19 家，占总量的 2/3。我国应该积极开展国际交流合作，特别是与美、英等的高产出研究机构开展合作，争取有更多高水平的极地科研产出，提升我国在极地舞台上的影响力。

本文研究了 *Nature*，*Science*，*PNAS* 三大顶级期刊的极地研究文献，希望可以从研究产出这一方面推导出世界极地研究的重点和热点领域，为中国的极地研究发展提供参考。

一、冰川、冰芯古气候类文献

1.1 分析概述

1990～2010 年在 *Nature*,*Science*,*Proceedings of the National Academy of Sciences of the United States of America*(以下简称 *PNAS*)共发表冰川、冰芯古气候类相关科学文献 140 篇。其中 *Science* 发表文献 68 篇,*Nature* 发表文献 63 篇,*PNAS* 发表文献 9 篇。由图 1.1 可以看出,*Nature*,*Science* 是发表冰川、冰芯古气候文献的主力刊物,两刊发表数量相当,而 *PNAS* 发表较少。以下分析这 140 篇科学论文的分布情况。

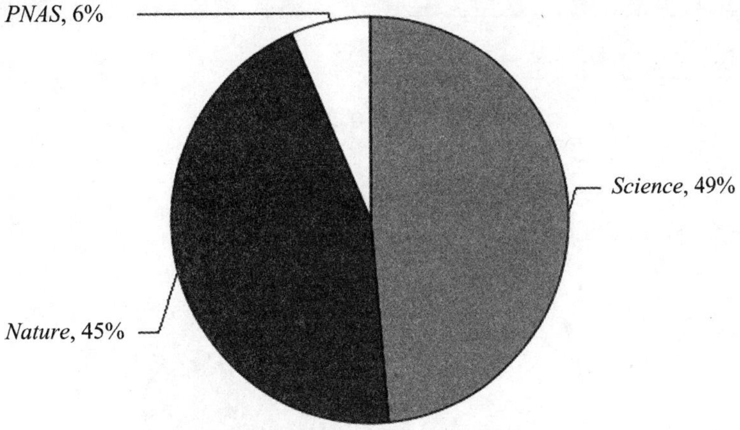

图 1.1 三刊发表科学文献分布比例

从历年文献产出年份分布(图 1.2)可以看出,三刊平均每年发表冰川、冰芯古气候类文献接近 7 篇,其中在 1994 年、1995 年和 2010 年发表文献较少,分别为 1 篇、2 篇和 2 篇,另有 5 年发表文献较多,均达到 10 篇及以上:1998 年发表 10 篇,2001 年发表 11 篇,2002 年发表 10 篇,2003 年发表 13 篇,2009 年发表 16 篇。

除 1994 年只发表 1 篇统观两极的文献外,其余每年均有南极的文献发表,且多数年份不低于 4 篇,可见南极冰川、冰芯古气候研究开展较为充分,已经形成相对完善的研究体系。而北极文献仅有 14 篇,分布在 15 年中,相关科学研究仍有较大增长空间。统观南北两极的文献总计 28 篇,其中 1990～2000 年零星发表 6 篇,2001～2010 年发表了 22 篇,2001、2002 年和 2009 年都发表了 4 篇,相关研究呈现稳定、快速的增长势头。综合看来,冰川、冰芯古气候研究在南极已有一定基础,在北极稳步发展。而在北极相关研究发展的同时,统观两极的相关研究正成为新兴的科学热点。

将所有文献按所关注领域分类归纳可得图 1.3。其中古气候相关文献总计 74 篇,占总数的 53%,且每年均有发表,是冰川、冰芯研究的重点领域。其次是对冰川物理的研究文献总计 30 篇,占总数的 21%。冰川、海冰对气候响应的研究文献 28 篇,占总数的 20%。其他冰川化学、冰下陨石、古地磁等研究文献 8 篇,占总数的 6%。

将历年文献按所关注领域分类可得图 1.4。由图 1.4 可以看出,古气候相关文献呈相对集中出现,这可能与古气候研究载体冰芯的钻取有关。每钻取一支冰芯,均可引起冰芯古气候研究成果的集中出现。冰川对气候响应和冰川物理的研究较为散布。

将 1990～2010 年的文献按第一作者国别归类可得图 1.5。由图 1.5 可以看出,美国是冰川、冰芯古气候研究的主要国家,共发表文献 71 篇,占总数的 51%。英国、法国和瑞士三个欧洲国家共发表文献 47 篇,占总数的 34%。澳大利亚和加拿大分别有 6 篇和 5 篇文献发表,各占总数的 4.3%和 3.6%。其他 9 个国家共发表 11 篇文献,共占总数的 8%。其中我国于 2009 年以第一署名国家在 *Nature* 上发表 1 篇文献,为冰川物理相关研究。

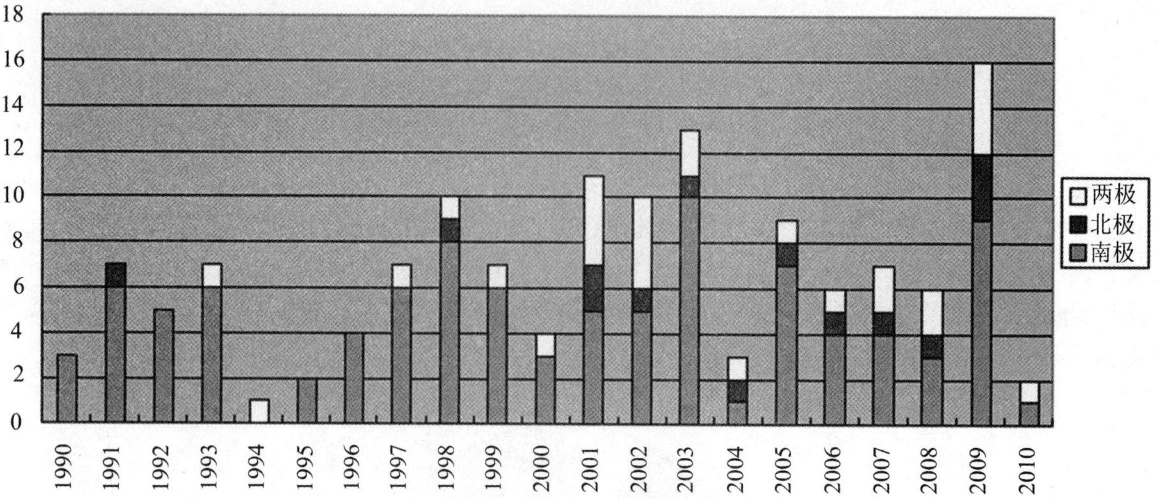

图 1.2　历年科学文献区域分布

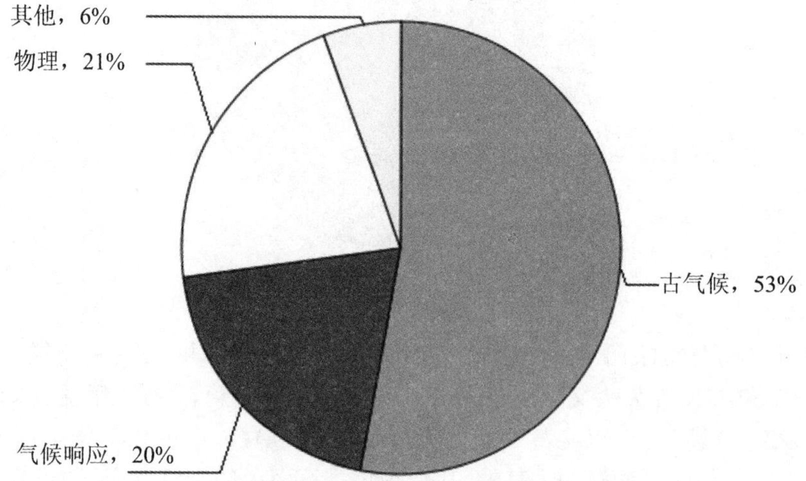

图 1.3　科学文献领域分布

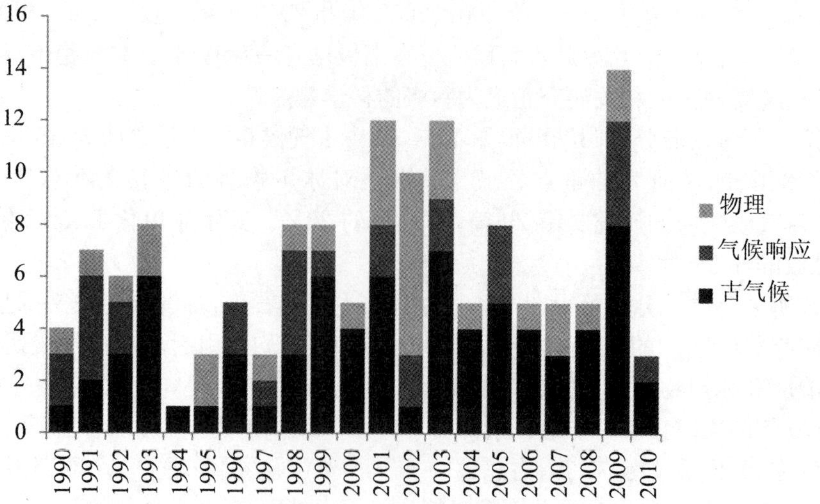

图 1.4　历年科学文献领域分布

由以上初步统计，目前冰川、冰芯古气候类研究重心仍是古气候，但由于无新冰芯的钻取而陷入暂时的低谷。因此中国昆仑站深冰钻项目的实施对该领域具有重大意义，可能引发新的研究热点。冰芯科学研究已有半个世纪的历史，冰芯在古气候与环境变化研究中提供了大量信息。钻取自格陵兰、南极和其他冰川覆盖地区的冰芯涵盖了不同的时间尺度。南极相关领域的研究已有一定基础，对北极的研究正在稳步开展，统观两极的研究正稳步、快速增长，是当前相关科学研究的热点所在。我国在该领域的研究成果已经开始在高水平学术刊物上发表，但与英美等国仍有很大差距。

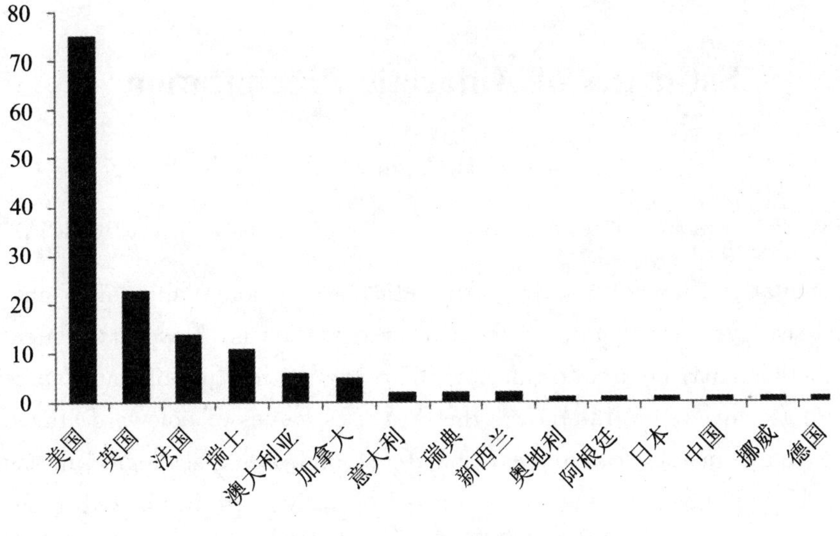

图 1.5 文献第一作者国别分布

1.2 摘要翻译(南北极)

1. David H Bromwich. Estimates of Antarctic Precipitation[J]. Nature,1990,343(6259):627-629.

Estimates of Antarctic Precipitation

David H. Bromwich

(Byrd Polar Research Center, Ohio State University, Columbus, Ohio 43210, USA)

Abstract: Precipitation fluctuations over Antarctica are a potentially important contributor to variations in global sea level. Direct measurement of precipitation is, however, fraught with practical difficulties. Two methods may be used to calculate indirectly the net flux of water (precipitation minus sublimation rate) to the surface of Antarctica: the first uses values of poleward atmospheric moisture transport obtained from climatological studies, and the second uses glaciological measurements of the accumulation rate. Here I show that the two estimates so derived are in marked disagreement for the entire continent but concur for the interior area between 80° S and the pole. I conclude that the large discrepancy near the coast is due to a calculated poleward moisture transport that is smaller than the actual value, as a result of deficiencies in evaluating the effects of cyclones and surface winds at the coast. Improvements in the climatological atmospheric database should therefore make possible reliable estimates of Antarctic precipitation variations.

对南极降雨量的估算

David H. Bromwich

(美国俄亥俄州立大学 Byrd 极地研究中心)

摘要:南极降雨波动对全球海平面变化有潜在的重要贡献。然而技术上直接测量南极降雨量困难重重。有两种方法可以间接计算南极地表的净水汽通量(降水量减去升华量):第一种,利用气候研究得出的向极地地区的大气水汽传输值;第二种,利用冰川测量的积累率。本文按两种方法分别得出估计数值。两组数值在整个南极大陆范围有显著差异,但在南纬 80°以南地区相对一致。本文认为沿岸地区的显著差异是因为缺乏沿岸地区气旋和地面风影响的评估,从而低估了向极地的水汽传输。改进大气气候数据库将有利于对南极降水变化作出可靠的估计。

2. Engelhardt H, Humphrey N, Kamb B, et al. Physical conditions at the base of a fast moving Antarctic ice stream[J]. Science,1990,248(4951):57-59.

Physical conditions at the base of a fast moving Antarctic ice stream

Engelhardt H. , Humphrey N. , Kamb B. , Fahnestock M.

(Division of Geological and Planetary Sciences, California Institute of Technology, Pasadena, CA 91125)

Abstract: Boreholes drilled to the bottom of ice stream B in the West Antarctic Ice Sheet reveal that the base of the ice stream is at the melting point and the basal water pressure is within about 1. 6 bars of the ice overburden pressure. These conditions allow the rapid ice streaming motion to occur by basal sliding or by shear deformation of unconsolidated sediments that underlie the ice in a layer at least 2 meters thick. The mechanics of ice streaming plays a role in the response of the ice sheet to climatic change.

一条快速移动的南极冰流底部的物理环境

Engelhardt H. , Humphrey N. , Kamb B. , Fahnestock M.

（美国加州理工学院地质与行星科学系）

摘要：西南极冰盖冰流B上打到底部的钻孔显示，冰流底部达到融点，且底部水压力未超过1. 6 bar的冰过载压力。这一环境可能导致底部滑动或至少2 m厚的松散沉积层剪切变形，从而引起冰流快速运动。冰流的力学结构在对气候变化的响应中有重要影响。

3. Mayewski P A，Legrand M R. Recent increase in nitrate concentration of Antarctic snow[J]. Nature，1990，346(6281)：258-260.

Recent increase in nitrate concentration of Antarctic snow

Mayewski P. A. , Legrand M. R.

(Glacier Research Group, Institute for the Study of Earth,Oceans and Space, University of New Hampshire, Durham,New Hampshire 03824, USA)

Abstract: Polar ice cores provide a unique record of global climate change. In particular, their records of nitrate concentration can yield new insight into the atmospheric nitrogen cycle, but first it is necessary to understand the processes controlling the spatial distribution of nitrate at the ice-sheet surface, and to define any trends in its temporal distribution. Here we report trends in the nitrate time series deduced from low-accumulation sites such as Dome C and Vostok in Antarctica. These trends must be treated with caution because of the possibility of post-depositional alteration. But the increases in the concentration of the spring maximum in nitrate that we find in the South Pole record for the past few years deserve careful consideration, as they may be a result of denitrification of polar stratospheric clouds in the lower stratosphere and may hence be connected in some way with the Antarctic ozone “hole”.

近期南极降雪中硝酸盐含量的升高

Mayewski P. A. , Legrand M. R.

(美国新罕布什尔大学地球海洋与空间研究所冰川研究组)

摘要:极地冰芯为全球气候变化提供了一个独特的记录。特别是硝酸盐浓度记录可以得出对大气氮循环的新认识,但是首先必须理解控制硝酸盐在冰盖上空间分布的过程及其在时间上变化的趋势。本文研究了南极 Dome C、Vostok 等低积累率地区的硝酸盐时间序列趋势。由于可能的沉积后变化,研究这些变化趋势必须谨慎。但我们在南极点发现的最近几年硝酸盐春季最大值的上升趋势应当引起特别注意,因为这可能是平流层下部的极地平流云反硝化作用的结果,从而可能与南极臭氧层“空洞”有关。

4. Alley R B, Whillans I M. Changes in the West Antarctic ice-sheet[J]. Science, 1991, 254(5034): 959-963.

Changes in the West Antarctic ice-sheet

Alley R. B. , Whillans I. M.

(Earth System Science Center and Department of Geosciences, The Pennsylvania State University, 306 Deike Building, University Park, PA 16802)

Abstract: The portion of the West Antarctic ice sheet that flows into the Ross Sea is thinning in some places and thickening in others. These changes are not caused by any current climatic change, but by the combination of a delayed response to the end of the last global glacial cycle and an internal instability. The near-future impact of the ice sheet on global sea level is largely due to processes internal to the movement of the ice sheet, and not so much to the threat of a possible greenhouse warming. Thus the near-term future of the ice sheet is already determined. However, too little of the ice sheet has been surveyed to predict its overall future behavior.

西南极冰盖的变化

Alley R. B. , Whillans I. M.

(美国宾夕法尼亚州立大学地球系统科学中心与地球科学系)

摘要:西南极冰盖流入 Ross 海的一些区域在变薄,另外一些区域在变厚。这些变化不是由任何当前的气候变化引起的,而是由对末次冰期旋回结束的延迟响应与内在不稳定性引起的。冰盖对全球海平面的近期影响更多是由冰盖内部运动过程引起的,而不是温室效应引起的。因此冰盖的短期未来已经确定。但因为已调查的冰盖过少,冰盖的整体未来趋势难以预测。

5. Bindschadler R A, Scambos T A. Satellite-image-derived velocity-field of an Antarctic ice stream[J]. Science, 1991, 252(5003): 242-246.

Satellite-image-derived velocity-field of an Antarctic ice stream

Bindschadler R. A., Scambos T. A.

(National Aeronautics and Space Administration, Goddard Space Flight Center, Code 971, Greenbelt, MD 20771)

Abstract: The surface velocity of a rapidly moving ice stream has been determined to high accuracy and spatial density with the use of sequential satellite imagery. Variations of ice velocity are spatially related to surface undulations, and transverse velocity variations of up to 30 percent occur. Such large variations negate the concept of plug flow and call into question earlier mass-balance calculations for this and other ice streams where sparse velocity data were used. The coregistration of images with the use of the topographic undulations of the ice stream and the measurement of feature displacement with cross-correlation of image windows provide significant improvements in the use of satellite imagery for ice-flow determination.

卫星图像获取的一条南极冰流的速度场

Bindschadler R. A., Scambos T. A.

(美国国家航空和航天局戈达德宇宙飞行中心)

摘要:本文利用连续卫星成像确定了一条快速运动冰流的高精度、高空间密度的表面速度。冰流速波动和表面起伏在空间上相关,水平速度波动幅度达30%。这样大的波动否定了插入流的观点,并质疑了早期利用稀少数据计算质量平衡时这条冰流和其他冰流的结果。在互相关的图像窗口中利用冰川地形起伏和位移特征测量进行的配准显著改进了卫星图像在冰流测定中的应用。

6. Doake C S M, Vaughan D G. Rapid disintegration of the Wordie ice shelf in response to atmospheric warming[J]. Nature, 1991, 350(6316): 328-330.

Rapid disintegration of the Wordie ice shelf in response to atmospheric warming

Doake C. S. M., Vaughan D. G.

(British Antarctic Survey, Madingley Road, Cambridge CB3 0ET, UK)

Abstract: The breaking up of ice shelves around the Antarctic Peninsula has been cited as a "sign that a dangerous warming is beginning in Antarctica". Here we present satellite images showing the disintegration of the Wordie Ice Shelf, which lies off the west coast of the Antarctic Peninsula. Fracture, either in the form of surface crevasses or rifts extending to the bottom of the ice shelf, has been responsible for iceberg calving and weakening the central region of the ice shelf. These fracture processes, which led to retreat of the ice front, were apparently enhanced by the presence of increased amounts of melt water, resulting from a warming trend recorded in mean annual air temperatures in

Marguerite Bay. If this warming trend continues, other nearby ice shelves on the Antarctic Peninsula may be at risk. But substantial additional warming would be required before similar processes could initiate breakup of the Ross and Filchner-Ronne ice shelves, which help stabilize the West Antarctic ice sheet.

快速瓦解的 Wordie 冰架对大气变暖的响应

Doake C. S. M. ,Vaughan D. G.

(英国南极调查局)

摘要:南极半岛周围冰架的崩塌被认为是“南极变暖开始变得危险的信号”。本文中的卫星图像展示了 Wordie 冰架的解体,Wordie 冰架位于南极半岛西岸以外。无论表面裂缝还是深达冰架底部的断裂,都导致了冰山崩落和冰架中部的削弱。Marguerite 湾记录的年平均气温呈上升趋势,这导致了融水的增加,从而加剧了裂解过程,引起冰架前缘的后退。如果这一变暖趋势持续下去,南极半岛附近的其他冰川也可能会有危险。但 Ross 冰架和 Filchner-Ronne 冰架上相同的过程需要大幅增强的变暖;这些冰架帮助了西南极冰架保持稳定。

7. Gloersen P,Campbell W J. Recent variations in Arctic and Antarctic sea-ice covers[J]. Nature,1991,352(6330):33-36.

Recent variations in Arctic and Antarctic sea-ice covers

Gloersen P. , Campbell W. J.

(Laboratory for Hydrospheric Processes, NASA Goddard Space Flight Center, Code 971, Greenbelt, Maryland 20771, USA)

Abstract: Variations in the extents of sea-ice cover at the poles and the areas of open water enclosed within them were observed every other day during the interval 1978-1987 by a satellite-borne scanning multispectral microwave radiometer. A band-limited regression technique shows that the trends in coverage of the Arctic and Antarctic sea-ice packs are not the same. During these nine years, there are significant decreases in ice extent and open-water areas within the ice cover in the Arctic, whereas in the Antarctic, there are no significant trends.

近期南北极海冰覆盖的变化

Gloersen P. , Campbell W. J.

(美国国家航空和航天局戈达德宇宙飞行中心水圈过程实验室)

摘要:在 1978～1987 年间,卫星搭载的多光谱扫描微波辐射计每隔一天测定一次极区的海冰覆盖范围及其包围的开阔水域的变化。限定频带技术分析显示南北极海冰覆盖变化趋势不同。在这 9 年中,北极海冰覆盖与其中的开阔水域显著减小,而南极没有明显变化趋势。

8. Legrand M, Fenietsaigne C, Saltzman E S, et al. Ice-core record of oceanic emissions of

dimethylsulfide during the Last Climate Cycle[J]. Nature,1991,350(6314):144-146.

Ice-core record of oceanic emissions of dimethylsulfide during the Last Climate Cycle

Legrand M., Fenietsaigne C., Saltzman E. S., Germain C., Barkov N. I., Petrov V. N.

(Lab Glaciol & Geophys Environnement, Bp 96, F-38402 St Martin Dheres, France)

Abstract: The Vostok ice core in Antarctica has provided one of the longest climate records, enabling the stable-isotope, major-ion and gas composition of the atmosphere to be reconstructed over many thousands of years. Here we present depth profiles along this core of methanesulphonate and non-seasalt sulphate (produced by the atmospheric oxidation of dimethylsulphide), which provide the first historical record of biogenic sulphur emissions from the Southern Hemisphere oceans over a complete glacial-interglacial cycle (160 kyr). Those measurements confirm and extend some previous observations made on a very limited data set from the Dome C ice core in Antarctica, which indicated increased oceanic emissions of dimethylsulphide during the later stages of the glacial period, compared with the present day. The observed glacial-interglacial variations in methanesulphonate and non-seasalt sulphate confirm that the ocean-atmosphere sulphur cycle is extremely sensitive to climate change.

末次气候循环期间海洋排放的二甲基硫的冰芯记录

Legrand M., Fenietsaigne C., Saltzman E. S., Germain C., Barkov N. I., Petrov V. N.

(法国冰川与地球物理环境实验室)

摘要:南极 Vostok 冰芯是最长的气候记录之一,可以重建许多千年的稳定同位素、主要离子和大气成分记录。本文得出了甲基磺酸和非海盐硫酸盐(由二甲基硫的大气氧化产生)在这一冰芯中的垂直记录,第一次重建了一个完整冰期—间冰期循环(16 万年)的南半球海洋生物硫排放历史记录。这些记录确认并延伸了之前在南极 Dome C 冰芯有限的数据集中得出的一些结论,表明冰期后段海洋的二甲基硫排放相对于现今更高。冰期-间冰期甲基磺酸和非海盐硫酸盐变化的研究表明,海-气硫循环对气候变化极端敏感。

9. Maurette M, Olinger C, Michellevy M C, et al. A collection of diverse micrometeorites recovered from 100 tonnes of Antarctic blue ice[J]. Nature, 1991, 351(6321):44-47.

A collection of diverse micrometeorites recovered from 100 tonnes of Antarctic blue ice

Maurette M., Olinger C., Michellevy M. C., Kurat G., Pourchet M., Brandstatter F., Bourotdenise M.

(CSNSM, IN2P3-CNRS, Bâtiment 108, 91405 Campus-Orsay, France)

Abstract: Studies of meteorites and interplanetary dust particles (IDPs) have provided constraints on the formation and evolution of the Solar System, and have identified pre-solar interstellar grains. Here we describe a new type of meteoritic material, intermediate in size between meteorites and IDPs. Melting and filtering of approximately 100 tonnes of blue ice near Cap Prudhomme, Antarctica, yielded greater-than-or-equal-to 7500 irregular, friable particles and approximately 1500 melted spherules, approximately 100-mu-m in size, both showing a "chondritic" composition suggestive of an extraterrestrial origin. For the present work, we analysed the composition and texture of 51 irregular particles and 25 spherules. The irregular particles appear to be unmelted, and have similarities with the fine-grained matrix of primitive carbonaceous chondrites, but are extremely diverse in composition. Isotopic analysis of trapped neon confirms an extraterrestrial origin for 16 of 47 irregular particles and 2 of 19 spherules studied, and strongly suggests that they were exposed in space as micrometeoroids. These large Antarctic micrometeorites constitute a new family-or at least a new population-of Solar System objects, in a mass range corresponding to the bulk of extraterrestrial material accreted by the Earth today.

从 100 t 南极蓝冰中获得的不同种类的微陨石

Maurette M., Olinger C., Michellevy M. C., Kurat G., Pourchet M., Brandstatter F., Bourotdenise M.

(法国国家科学研究院核物理与粒子物理研究所)

摘要:对陨石和星际尘埃(IDPs)的研究揭示了太阳系的形成与进化,并识别出了前太阳时期的星际颗粒。本文描述了一种新的陨石物质,尺寸介于陨石和星际尘埃之间。本研究融化并过滤了南极 Prudhomme 冰帽附近的约 100 t 蓝冰,获得了不少于 7500 个不规则易碎微粒和大约 1500 个熔融球粒,尺寸约为 100 μm。两者都呈现指示地外来源的球粒陨石成分。本研究分析了 51 个不规则微粒和 25 个球粒的成分与岩性。不规则颗粒看起来没有熔化过,并与原始碳球粒陨石的填质类似,但成分非常多样。47 个不规则颗粒和 19 个球粒捕集的氖同位素分析证据确认了其中 16 个不规则颗粒及 2 个球粒来自地外空间,并很可能曾作为微流星体暴露在宇宙空间。这些南极的大微陨石建立了太阳系天体的一个新族,或至少一个新种,这一新种对应于大多数附着在地球上的地外物质的质量范围。

10. Morgan V I, Goodwin I D, Etheridge D M, et al. Evidence from Antarctic ice cores for recent increases in snow accumulation[J]. Nature, 1991, 354(6348): 58-60.

Evidence from Antarctic ice cores for recent increases in snow accumulation

Morgan V. I., Goodwin I. D., Etheridge D. M., Wookey C. W.

(Australian Antarctic Division, Channel Highway, Kingston 7050, Australia)

Abstract: Large uncertainties exist in the present knowledge of the mass budget of the Antarctic ice sheet, because of a lack of data on the rates of both ice outflow and snow accumulation. Present estimates indicate that both the outflow and the net accumulation are approximately equal to 2000 km^3 of ice per year (equivalent to about 6 mm of sea level). The temporal variation of accumulation rate is central to determinations of the mass budget, because accumulation can change rapidly in response to short-term climate variations, whereas ice flow varies only on longer timescales. Here we present time series showing changes in the net rate of snow accumulation since 1806 along a 700 km segment of East Antarctica. The accumulation record was derived from the thicknesses of annual layers in ice cores, deduced from seasonal variations in oxygen isotope ratio and in ice-crust stratigraphy. We find a significant increase in the accumulation rate following a minimum around 1960, leading to recent rates that are about 20% above the long-term mean. If this recent increase is widespread, as suggested by shorter-term accumulation data from across a large part of Antarctica, the positive imbalance(5%-25% of the mass input) shown in recent studies of the ice sheet's mass budget may have existed only since the late 1960s. We estimate that this increase in accumulation rate should contribute to a lowering of sea level of 1. 0-1. 2 mm per year.

南极冰芯中获得的近期雪积累率上升的证据

Morgan V. I. ,Goodwin I. D. , Etheridge D. M. , Wookey C. W.

（澳大利亚南极局）

摘要：因为既缺乏冰流速率的数据，又缺乏雪积累率的数据，当前对南极冰盖物质收支的认识还非常不完善。目前估计认为流出和净积累大约均为每年 2000 km^3冰当量（相当于 6 mm 海平面）。因为积累率可能因短期气候波动而快速改变，而冰流只在长时间尺度上变化，因此积累率的时间变化是确定物质收支的关键。本文得出了东南极一条 700 km 断面上自 1806 年以来雪净积累率变化的时间序列。年度积累记录来自年层厚度，年层通过氧同位素季节变化和冰壳地层学推测。我们发现在 1960 年左右出现最小值之后积累率显著上升，导致比长期均值高约 20%的近期积累率。基于整个南极洲大部分的较短时期的积累数据，近期积累率的上升是普遍的；最近冰盖物质收支研究显示的正失调（5%～25%质量收入）可能仅仅从 1960 年代后期开始。我们估计这一积累率的上升可能对海平面下降有每年 1. 0～1. 2 mm的贡献。

11. Colhoun E A,Mabin M C G,Adamson D A,et al. Antarctic ice volume and contribution to sea-level fall at 20000 yr BP from raised beaches[J]. Nature,1992,358(6384):316-319.

Antarctic ice volume and contribution to sea-level fall at 20000 yr BP from raised beaches

Colhoun E. A. , Mabin M. C. G. , Adamson D. A. , Kirk R. M.

(Department of Geography, University of Newcastle, University Drive,Callaghan, Newcastle NSW 2308, Australia)

Abstract: The contribution of the Antarctic ice sheets to global sea-level fall at the Last Glacial Maximum (LGM) depends largely on how the extent and thickness of peripheral ice changed. Model studies suggest that there was widespread thickening (from 500 m to more than 1000 m) of the ice-sheet margins, sufficient to induce a drop in sea level of at least 25 m. Geological evidence, on the other hand, indicates only limited ice expansion and sea-level fall of just 8 m. Here we use recent data on the altitudes and ages of raised beaches from the Ross embayment and East Antarctica to investigate the timing and extent of Antarctic deglaciation. These indicate that the ice margin during the LGM was thinner and less extensive than has been formerly thought, and that its contribution to the drop in sea level was only 0.5-2.5 m. Deglaciation was well advanced by 10 kyr BP and was complete by 6 kyr BP. These findings imply either that sea level during the LGM fell less than present estimates suggest, or that an ice volume considerably greater than currently accepted must have been present in the Northern Hemisphere.

基于海岸抬升的南极冰量变化及其对 20000 年前海平面下降的研究

Colhoun E. A., Mabin M. C. G., Adamson D. A., Kirk R. M.

(澳大利亚纽卡斯尔大学地理系)

摘要:末次冰盛期南极冰盖对全球海平面下降的贡献很大程度上取决于边缘冰的范围与厚度。模型研究认为冰盖边缘普遍较厚(从 500 m 到超过 1000 m),足以引发至少 25 m 的海平面下降。另一方面,地质学证据仅仅显示了有限的冰进和 8 m 的海平面下降。本文利用最近 Ross 湾和东南极海岸抬升的年龄与高度数据,研究南极冰消的时间和范围。研究显示末次冰盛期的冰缘比之前认为的薄且范围小,并且它对海平面下降的贡献仅有 0.5~2.5 m。冰消在 10000 年前充分进展并在 6000 年前完成。这些发现意味着末次冰盛期海平面下降低于当前估计,或者北半球当时冰量比现在认为的多出相当数量。

12. Jacobs S S. Is the Antarctic ice-sheet growing[J]. Nature, 1992, 360(6399): 29-33.

Is the Antarctic ice-sheet growing

Jacobs S. S.

(Lamont-Doherty Geological Observatory of Columbia University, Palisades, New York 10964, USA)

Abstract: A common public perception is that global warming will accelerate the melting of polar ice sheets, causing sea level to rise. A common scientific position is that the volume of grounded Antarctic ice is slowly growing, and will damp future sea-level rise. At present, studies supporting recent shrinkage or growth depend on limited measurements that are subject to high temporal and regional variability, and it is too early to say how the Antarctic ice sheet will behave in a warmer world.

南极冰盖是否在增长?

Jacobs S. S.

(美国哥伦比亚大学拉蒙特-多尔蒂地质观测站)

摘要:公众普遍认为全球变暖会加速极地冰盖融化,导致海平面上升。普遍的科学观点是南极陆基冰量在缓慢增长,可能会抑制未来的海平面上升。当前无论支持冰盖增长还是收缩的研究都是基于有限的测量,而这些测量数据在时间和空间上是多变的。因此南极冰盖在变暖的条件下如何变化,现在下结论还为时过早。

13. Leuenberger M, Siegenthaler U. Ice-age atmospheric concentration of nitrous-oxide from an

Antarctic ice core[J]. Nature, 1992, 360(6403): 449-451.

Ice-age atmospheric concentration of nitrous-oxide from an Antarctic ice core

Leuenberger M., Siegenthaler U.

(Physics Institute, University of Bern, Sidlerstrasse 5, 3012 Bern, Switzerland)

Abstract: Increasing anthropogenic emissions of greenhouse gases are expected to influence the Earth's climate, but the mechanisms for this are not yet fully understood. One way to determine the effect of such gases on climate is to study their atmospheric concentrations during periods of past climate change, such as glacial to interglacial transitions. Previous studies on polar ice cores showed that the concentrations of the greenhouse gases CO_2 and CH_4 were significantly reduced during the last glacial period relative to Holocene values. But no comparable studies have been reported for nitrous oxide (N_2O), which is the next most important greenhouse gas and also affects stratospheric ozone and, potentially, the oxidative capacity of the troposphere. Here we report results from Antarctic ice cores, showing that the atmospheric N_2O concentration was about 30% lower during the Last Glacial Maximum than during the Holocene epoch. Our data also show that present-day N_2O concentrations are unprecedented in the past 45 kyr, and hence provide evidence that recent increases in atmospheric N_2O are of anthropogenic origin.

南极冰芯中记录的冰期大气一氧化二氮的浓度

Leuenberger M., Siegenthaler U.

(瑞士伯尔尼大学物理学院)

摘要：普遍认为人类排放越来越多的温室气体会影响地球气候，但其机制仍未得到充分认识。确定这些气体对气候影响的一个途径是研究过去气候变化时期它们的大气浓度，例如冰期向间冰期的转变。之前对极地冰芯的研究显示，CO_2和CH_4这两种温室气体的浓度在末次冰期相对于全新世有显著降低。但一氧化二氮(N_2O)作为仅次于以上两种的温室气体，同时影响平流层臭氧和可能影响对流层氧化能力，却没有得到相应的研究。本文从南极冰芯中得出末次冰盛期大气N_2O浓度比全新世时期低约30%的结论。得出的数据同时显示当前N_2O浓度在过去45千年中是没有先例的，这一证据表明近期大气N_2O上升是源自人类排放。

14. Leuenberger M, Siegenthaler U, Langway C C. Carbon isotope composition of atmospheric CO_2 during the last ice-age from an Antarctic ice core[J]. Nature, 1992, 357(6378): 488-490.

Carbon isotope composition of atmospheric CO_2 during the last ice-age from an Antarctic ice core

Leuenberger M., Siegenthaler U., Langway C. C.

(Physics Institute, University of Bern, Sidlerstrasse 5, 3012 Bern, Switzerland)

Abstract: Bubbles of ancient air in polar ice cores have revealed that the atmospheric concentration of CO_2 during the Last Glacial Maximum was 180-200 p. p. m. v., substantially lower than the pre-industrial value of about 280 p. p. m. v.. It is generally thought that this reduction in atmospheric CO_2 during glacial time was driven by oceanic processes. The most likely explanations invoke either a decrease in dissolved CO_2 in surface waters because of a more efficient "biological pump" transporting carbon to deep waters, or a higher alkalinity in the glacial ocean as a consequence of changes in carbonate dissolution or sedimentation. Because isotope fractionation during photosynthesis depletes ^{13}C in the organic matter produced, changes in the biological pump would alter the carbon isotope composition of atmospheric CO_2, whereas changes in alkalinity would in themselves have no such effect. Here we report measurements of the carbon isotope content of CO_2 in ice cores from Byrd Station, Antarctica, in an attempt to distinguish between these mechanisms. We find that during the ice age the reduced isotope ratio $\delta^{13}C$ was more negative than pre-industrial values by 0.3 ± 0.2 parts per thousand. Although this result does not allow us to discriminate definitely between the two possible causes of lower glacial atmospheric CO_2, it does indicate that changes in the strength of the biological pump cannot alone have been responsible.

南极冰芯记录的末次冰盛期大气 CO_2 碳同位素组成

Leuenberger M., Siegenthaler U., Langway C. C.

(瑞士伯尔尼大学物理学院)

摘要:极地冰芯的气泡中的远古空气显示末次冰盛期 CO_2 的大气浓度为 180~200 ppmv,显著低于前工业时期 280 ppmv 的值。人们普遍认为冰期大气 CO_2 降低是由海洋过程引起的。最可能的解释或者认为更有效的"生物泵"将碳传送到深层水,导致表层水溶解 CO_2 下降,或者作为碳酸盐溶解与沉淀变化的后果是冰海碱度较高。因为光合作用中的同位素分馏会降低有机产物中的 ^{13}C,生物泵的变化会改变大气 CO_2 的同位素组成,而碱度变化本身没有这样的影响。本文通过南极 Byrd 站冰芯 CO_2 碳同位素的测量,尝试辨别这两种机制。研究发现冰期降低的 $\delta^{13}C$ 同位素比值比前工业时期的值更低 (0.3 ± 0.2)‰。虽然这一结果不能明确辨别冰期大气 CO_2 降低的两种可能原因,但它确实指出生物泵强度的变化不是唯一的原因。

15. Macayeal D R. Irregular oscillations of the west Antarctic ice-sheet[J]. Nature, 1992, 359(6390): 29-32.

Irregular oscillations of the west Antarctic ice-sheet

Macayeal D. R.

(Department of Geophysical Science, University of Chicago, 5734 South Ellis Avenue, Illinois 60637, USA)

Abstract: Model simulations of the West Antarctic ice sheet suggest that sporadic, perhaps chaotic, collapse (complete mobilization) of the ice sheet occurred throughout the past one million years. The irregular behaviour is due to the slow equilibration time of the distribution of basal till, which lubricates ice-sheet motion. This nonlinear response means that predictions of future collapse of the ice sheet in response to global warming must take into account its past history, and in particular whether the present basal till distribution predisposes the ice sheet towards rapid change.

西南极冰盖的不规则震荡

Macayeal D. R.

(美国芝加哥大学地球物理科学系)

摘要:西南极冰盖模型模拟显示,在过去整个一百万年都有零星的或混乱的崩塌(彻底移动)。这一不规则活动是由于底碛分布平衡时间慢,对冰盖移动起到润滑作用。这一非线性响应意味着预报冰盖未来因全球变暖的崩塌必须考虑它过去的历史,特别是当前底碛分布是否会使冰盖快速变化。

16. Blankenship D D, Bell R E, Hodge S M, et al. Active volcanism beneath the West Antarctic ice-sheet and implications for ice-sheet stability[J]. Nature, 1993, 361(6412): 526-529.

Active volcanism beneath the West Antarctic ice-sheet and implications for ice-sheet stability

Blankenship D. D., Bell R. E., Hodge S. M., Brozena J. M., Behrendt J. C., Finn C. A.

(Institute for Geophysics, The University of Texas at Austin, Austin, Texas 78759, USA)

Abstract: It is widely understood that the collapse of the West Antarctic ice sheet (WAIS) would cause a global sea level rise of 6 m, yet there continues to be considerable debate about the detailed response of this ice sheet to climate change. Because its bed is grounded well below sea level, the stability of the WAIS may depend on geologically controlled conditions at the base which are independent of climate. In particular, heat supplied to the base of the ice sheet could increase basal melting and thereby trigger ice streaming, by providing the water for a lubricating basal layer of till on which ice streams are thought to slide. Ice streams act to protect the reservoir of slowly moving inland ice from exposure to oceanic degradation, thus enhancing ice-sheet stability. Here we present aerogeophysical evidence for active volcanism and associated elevated heat flow beneath the WAIS near the critical region where ice streaming begins. If this heat flow is indeed controlling ice-stream formation, then penetration of ocean waters inland of the thin hot crust of the active portion of the West

Antarctic rift system could lead to the disappearance of ice streams, and possibly trigger a collapse of the inland ice reservoir.

西南极冰盖下的活火山作用及其对冰盖稳定性的影响

Blankenship D. D., Bell R. E., Hodge S. M., Brozena J. M., Behrendt J. C., Finn C. A.

(美国德克萨斯大学奥斯汀分校地球物理学院)

摘要:人们普遍认为西南极冰盖(WAIS)的瓦解会导致全球海平面上升 6 m,然而关于这一冰盖对气候变化的响应细节仍存在相当多的争论。因为西南极冰盖在海平面下与基岩彻底接触,所以它的稳定性可能取决于底部受地质控制的条件,与气候无关。尤其是对底部的热量传输可能增加底质融化,融水使底碛层润滑,提供冰流动的基底。冰流保护缓慢移动的内陆冰免于暴露在海洋中融解,增强了冰盖的稳定性。本文提出了西南极冰盖下冰流起源的关键点附近活火山作用和相关高热流的航空物探证据。如果这一热流确实控制了冰流的形成,海水渗透西南极断裂系统活动部分薄热的地壳可能导致冰流消失,或许会引起内陆冰储库的崩溃。

17. Craig H, Shoji H, Langway C C. Nonequilibrium air clathrate hydrates in Antarctic ice: a paleopiezometer for polar ice caps[J]. PNAS, 1993, 90(23): 11416-11418.

Nonequilibrium air clathrate hydrates in Antarctic ice: a paleopiezometer for polar ice caps

Craig H., Shoji H., Langway C. C.

(Isotope Laboratory, Scripps Institution of Oceanography, University of California at San Diego, La Jolla, CA 92093, USA)

Abstract: "Craigite" the mixed-air clathrate hydrate found in polar ice caps below the depth of air-bubble stability, is a clathrate mixed crystal of approximate composition (N_2O_2) • $6H_2O$. Recent observations on the Byrd Station Antarctic core show that the air hydrate is present at a depth of 727 m, well above the predicted depth for the onset of hydrate stability. We propose that the air hydrate occurs some 100 m above the equilibrium phase boundary at Byrd Station because of "piezometry" i. e., that the anomalous depth of hydrate occurrence is a relic of a previous greater equilibrium depth along the flow trajectory, followed by vertical advection of ice through the local phase-boundary depth. Flowline trajectories in the ice based on numerical models show that the required vertical displacement does indeed occur just upstream of Byrd Station. Air-hydrate piezometry can thus be used as a general parameter to study the details of ice flow in polar ice caps and the metastable persistence of the clathrate phase in regions of upwelling blue ice.

南极冰中非平衡空气水包合物：极地冰帽的一种古压力计

Craig H. , Shoji H. , Langway C. C.

(美国加利福尼亚大学圣地亚哥分校斯克里普斯海洋研究所)

摘要："Craigite"是在极地冰帽气泡稳定深度以下发现的混合空气水包合物，它是成分大致为$(N_2O_2)\cdot 6H_2O$的混合包合物晶体。最近对南极Byrd站冰芯的观测显示空气水合物在727 m深度出现，远高于预计的水合物稳定深度。我们认为，空气水合物在Byrd站均相界面以上约100 m出现，因为"压力计"即水合物的异常深度是之前沿着流动轨迹更大平衡深度的遗留现象，以及冰垂直平流通过当地相界面的深度。冰中基于数学模型的流线轨迹显示正是在Byrd站上游发生了必需的垂直位移。因此空气水包合物压力计可以作为一个常规参数研究极地冰帽的冰流细节和蓝冰上升流区域亚稳定持久的包合物相。

18. Goldstein R M, Engelhardt H, Kamb B, et al. Satellite radar interferometry for monitoring ice-sheet motion: application to an Antarctic ice stream[J]. Science, 1993, 262(5139): 1525-1530.

Satellite radar interferometry for monitoring ice-sheet motion: application to an Antarctic ice stream

Goldstein R. M. , Engelhardt H. , Kamb B. , Frolich R. M.

(Jet Propulsion Laboratory, California Institute of Technology, Pasadena, CA 91109)

Abstract: Satellite radar interferometry (SRI) provides a sensitive means of monitoring the flow velocities and grounding-line positions of ice streams, which are indicators of response of the ice sheets to climatic change or internal instability. The detection limit is about 1.5 millimeters for vertical motions and about 4 millimeters for horizontal motions in the radar beam direction. The grounding line, detected by tidal motions where the ice goes afloat, can be mapped at a resolution of approximately 0.5 kilometer. The SRI velocities and grounding line of the Rutford Ice Stream, Antarctica, agree fairly well with earlier ground-based data. The combined use of SRI and other satellite methods is expected to provide data that will enhance the understanding of ice stream mechanics and help make possible the prediction of ice sheet behavior.

利用卫星雷达干涉测量法监测冰盖运动：在一条南极冰流上的应用

Goldstein R. M. , Engelhardt H. , Kamb B. , Frolich R. M.

(美国加州理工学院喷气推进实验室)

摘要：冰流流速和基线位置是冰盖对气候变化或内在不稳定性响应的指示计，卫星雷达干涉测量法(SRI)提供了一种监测它们的灵敏方法。在雷达波方向上垂直分辨率为1.5 mm，水平分辨率为4 mm。基线通过冰漂浮指示的潮汐运动测定，绘图解析度约为0.5 km。南极Rutford冰流的SRI速度和基线与早先基于地面测量的数据相当吻合。SRI 和其他卫星方法的联合应用有望加强对冰流机制的理解,有助于使预测冰盖行为变得可能。

19. Jouzel J,Barkov N I,Barnola J M,et al. Extending the Vostok ice-core record of paleoclimate to the penultimate glacial period[J]. Nature,1993,364(6436):407-412.

Extending the Vostok ice-core record of paleoclimate to the penultimate glacial period

Jouzel J., Barkov N. I., Barnola J. M., Bender M., Chappellaz J., Genthon C., Kotlyakov V. M., Lipenkov V., Lorius C., Petit J. R., Raynaud D., Raisbeck G., Ritz C., Sowers T., Stievenard M., Yiou F., Yiou P.

(Laboratoire de Modération du Climat et de l'Environnement, CEA/DSM CE Saclay, 91191, Gif sur Yvette Cedex, France)

Abstract: The ice-core record of local temperature, dust accumulation and air composition at Vostok station, Antarctica, now extends back to the penultimate glacial period (approximately 140-200 kyr ago) and the end of the preceding interglacial. This yields a new glaciological timescale for the whole record, which is consistent with ocean records. Temperatures at Vostok appear to have been more uniformly cold in the penultimate glacial period than in the most recent one. Concentrations of CO_2 and CH_4 Correlate well with temperature throughout the record.

将 Vostok 冰芯古气候记录延伸到倒数第二次冰期

Jouzel J., Barkov N. I., Barnola J. M., Bender M., Chappellaz J., Genthon C., Kotlyakov V. M., Lipenkov V., Lorius C., Petit J. R., Raynaud D., Raisbeck G., Ritz C., Sowers T., Stievenard M., Yiou F., Yiou P.

(法国原子能委员会气候与环境保护实验室)

摘要:南极 Vostok 站的本地温度、尘埃沉积和空气成分的冰芯记录现在向前延伸到倒数第二次冰期(140 千年~200 千年前)和之前的间冰期末端。这在整个记录上产生了新的冰河学时标。这一时标与大洋记录一致。相对于末次冰期,Vostok 的温度在倒数第二次冰期冷得更均一。CO_2 和 CH_4 的浓度与温度在整个记录中相关良好。

20. Marchant D R, Swisher C C, Lux D R, et al. Pliocene paleoclimate and east Antarctic ice-sheet history from surficial ash deposits[J]. Science,1993,260(5108):667-670.

Pliocene paleoclimate and east Antarctic ice-sheet history from surficial ash deposits

Marchant D. R., Swisher C. C., Lux D. R., West D. P., Denton G. H.

(Institute for Quaternary Studies, University of Maine, Orono, ME 04469)

Abstract: The preservation, age, and stratigraphic relation of an in situ ashfall layer with an underlying desert pavement in Arena Valley, southern Victoria Land, indicate that a cold-desert climate has persisted in Arena Valley during the past 4.3 million years. These data indicate that the present East Antarctic Ice Sheet has endured for this time and that average temperatures during the Pliocene in Arena Valley were no greater than 3 ℃ above present values. One implication is that the collapse of the East Antarctic Ice Sheet due to greenhouse warming is unlikely, even if global atmospheric temperatures rise to levels last experienced during mid-Pliocene times.

基于地表灰沉降的上新世古气候与东南极冰盖历史

Marchant D. R., Swisher C. C., Lux D. R., West D. P., Denton G. H.

(美国缅因大学第四纪研究所)

摘要：在南 Victoria 地的 Arena 谷一处野外底层通道的落灰层的保存、年龄和地层学关系表明，冷干气候在 Arena 谷已经持续了 430 万年。这些数据显示目前的东南极冰盖已经持续了这么长的时间，并且 Arena 谷上新世平均温度不会比现在高 3 ℃以上。这些发现的一个意义是，即使全球大气温度上升到中上新世所经历的水平，东南极冰盖也未必因为温室变暖而瓦解。

21. Raynaud D, Jouzel J, Barnola J M, et al. The ice record of greenhouse gases[J]. Science, 1993, 259 (5097): 926-934.

The ice record of greenhouse gases

Raynaud D., Jouzel J., Barnola J. M., Chappellaz J., Delmas R. J., Lorius C.

(Cnrs, Glaciol & Geophys Environm Lab, Bp 96, F-38402 St Martin Dheres, France)

Abstract: Gases trapped in polar ice provide our most direct record of the changes in greenhouse gas levels during the past 150000 years. The best documented trace-gas records are for CO_2 and CH_4. The measurements corresponding to the industrial period document the recent changes in growth rate. The variability observed over the last 1000 years constrains the possible feedbacks of a climate change on the trace gases under similar conditions as exist today. Changes in the levels of greenhouse gases during the glacial-interglacial cycle overall paralleled, at least at high southern latitudes, changes in temperature; this relation suggests that greenhouse gases play an important role as an amplifier of the initial orbital forcing of Earth's climate and also helps to assess the feedbacks on the biogeochemical cycles in a climate system in which the components are changing at different rates.

温室气体的冰记录

Raynaud D., Jouzel J., Barnola J. M., Chappellaz J., Delmas R. J., Lorius C.

(法国国家科学研究院冰川与地球物理环境实验室)

摘要：极地冰中截留的气体提供了过去 15 万年温室气体水平变化最直接的记录。记载最好的痕量气体记录是 CO_2 和 CH_4。与工业时期对应的测量值证明了最近增长率的变化。最近 1000 年来观测到的变化率限制了类似当今气候条件下对痕量气体可能的气候变化反馈。至少在南半球高纬度,冰期-间冰期旋回中温室气体水平的变化总体上与温度变化一致;这一关系显示温室气体在地球气候中作为最初的轨道强迫的放大器扮演重要角色,也有助于评估在参数以不同变率变化的气候系统中生物地球化学圈的反馈。

22. Vandal G M, Fitzgerald W F, Boutron C F, et al. Variations in mercury deposition to Antarctica over the past 34000 years[J]. Nature, 1993, 362(6421): 621-623.

Variations in mercury deposition to Antarctica over the past 34000 years

Vandal G. M., Fitzgerald W. F., Boutron C. F., Candelone J. P.

(Department of Marine Sciences, University of Connecticut, Avery Point, Groton, Connecticut 06340, USA)

Abstract: Polar ice contains a valuable record of past atmospheric mercury deposition, which can provide information about both the natural biogeochemical cycling of this toxic trace metal and the impact of recent anthropogenic emissions. But existing studies of mercury in polar ice and snow cores suffer from sample contamination and inadequate analytical procedures. Here we report measurements of mercury concentrations spanning the past 34000 years from the Dome C ice core, Antarctica, using the stringent trace-metal clean protocols developed by Patterson and co-workers. Although this record does not extend into the industrial period, it provides an important baseline for future attempts to identify anthropogenic mercury in Antarctic ice and snow. We find that mercury concentrations were strikingly elevated during the last glacial maximum (18000 years ago), when oceanic productivity may have been higher than it is today. As oceanic mercury emission is correlated with productivity, we suggest that this was the principal pre-industrial source of mercury to Antarctica; mercury concentrations in Antarctic ice might therefore serve as a palaeoproductivity indicator for the more distant past.

过去34000年南极汞沉降的变化

Vandal G. M., Fitzgerald W. F., Boutron C. F., Candelone J. P.

(美国康涅狄格大学海洋科学系)

摘要:极地冰保有过去大气汞沉降的有价值的记录,这些记录能提供这一有毒痕量金属自然生物地球化学循环和近期人类排放的影响。但当前极地冰雪中汞的研究受到样品污染和分析手段有限的困扰。本文采用Patterson和他的同事研发的适用于痕量金属研究的严格洁净程序,测定了南极Dome C冰芯中过去34000年汞的浓度。虽然这一记录没有延伸到工业时期,但它为未来识别南极冰雪中人类来源的汞提供了一条重要的基线。我们发现汞浓度在末次冰盛期(18000年前)有惊人的上升,那时海洋生产力可能高于现在。因为海洋汞排放与生产力相关,我们认为这是前工业时代南极汞的主要来源,南极冰中的汞浓度可以作为更远过去的古生产力指示计。

23. Bender M, Sowers T, Dickson M L, et al. Climate correlations between Greenland and Antarctica during the past 100000 years[J]. Nature, 1994, 372(6507): 663-666.

Climate correlations between Greenland and Antarctica during the past 100000 years

Bender M., Sowers T., Dickson M. L., Orchardo J., Grootes P., Mayewski P. A., Meese D. A.

(Univ Rhode Isl, Grad Sch Oceanog, Kingston, Ri 02881 USA)

Abstract: The ice cores recovered from central Greenland by the GRIP and GISP2 projects record 22 interstadial (warm) events during the part of the last glaciation spanning 20-105 kyr before present. The ice core from Vostok, east Antarctica, records nine interstadials during this period. Here we explore links between Greenland and Antarctic climate during the last glaciation using a high-resolution chronology derived by correlating oxygen isotope data for trapped O_2 in the GISP2 and Vostok cores. We find that interstadials occurred in east Antarctica whenever those in Greenland lasted longer than 2000 years. Our results suggest that partial deglaciation and changes in ocean circulation are partly responsible for the climate teleconnection between Greenland and Antarctica. Ice older than 115 kyr in the GISP2 core shows rapid variations in the $\delta^{18}O$ of O_2 that have no counterpart in the Vostok record. The age-depth relationship, and thus the climate record, in this part of the GISP2, core appears to be significantly disturbed.

过去10万年中格陵兰和南极的气候关联

Bender M., Sowers T., Dickson M. L., Orchardo J., Grootes P., Mayewski P. A., Meese D. A.

(美国罗德岛大学海洋研究生院)

摘要:GRIP和GISP2计划在格陵兰中部获得的冰芯记录了末次冰期跨越距今20千年～105千年部分的22个间冰段(暖)事件。东南极的Vostok冰芯记录了这一时期的9个间冰段。本文利用GISP2和Vostok冰芯中捕集的O_2氧同位素数据得出的高分辨年代学序列,探究末次冰期格陵兰和南极气候的关联。我们发现当格陵兰的间冰段持续超过2000年时,它就在东南极出现。我们的结果显示部分冰消和大洋环流变化是格陵兰和南极气候遥相关的部分原因。GISP2冰芯中老于115千年的冰中O_2的$\delta^{18}O$显示出快速变化,而在Vostok冰芯中没有相应的记录。GISP2冰芯中这一部分的年龄-深度的关系看来被显著扰动了,因此气候记录也是如此。

24. Askebjer P, Barwick S W, Bergstrom L, et al. Optical-properties of the South-Pole ice at depths between 0.8 and 1 kilometer[J]. Science, 1995, 267(5201): 1147-1150.

Optical-properties of the South-Pole ice at depths between 0. 8 and 1 kilometer

Askebjer P. , Barwick S. W. , Bergstrom L. , Bouchta A. , Carius S. , Coulthard A. , Engel K. , Erlandsson B. , Goobar A. , Gray L. , Hallgren A. , Halzen F. , Hulth P. O. , Jacobsen J. , Johansson S. , Kandhadai V. , Liubarsky I. , Lowder D. , Miller T. , Mock P. C. , Morse R. , Porrata R. , Price P. B. , Richards A. , Rubinstein H. , Schneider E. , Sun Q. , Tilav S. , Walck G. , Yodh G.

(Stockholm University, Box 6730, S-11385 Stockholm, Sweden)

Abstract: The optical properties of the ice at the geographical South Pole have been investigated at depths between 0. 8 and 1 kilometer. The absorption and scattering lengths of visible light (similar to 515 nanometers) have been measured in situ with the use of the laser calibration setup of the Antarctic Muon and Neutrino Detector Array (AMANDA) neutrino detector. The ice is intrinsically extremely transparent. The measured absorption length is 59±3 meters, comparable with the quality of the ultrapure water used in the Irvine-Michigan-Brookhaven and Kamiokande proton-decay and neutrino experiments and more than twice as long as the best value reported for laboratory ice. Because of a residual density of air bubbles at these depths, the trajectories of photons in the medium are randomized. If the bubbles are assumed to be smooth and spherical, the average distance between collisions at a depth of 1 kilometer is about 25 centimeters. The measured inverse scattering length on bubbles decreases linearly with increasing depth in the volume of ice investigated.

南极点深度 0.8～1 km 之间冰的光学特性

Askebjer P. , Barwick S. W. , Bergstrom L. , Bouchta A. , Carius S. , Coulthard A. , Engel K. , Erlandsson B. , Goobar A. , Gray L. , Hallgren A. , Halzen F. , Hulth P. O. , Jacobsen J. , Johansson S. , Kandhadai V. , Liubarsky I. , Lowder D. , Miller T. , Mock P. C. , Morse R. , Porrata R. , Price P. B. , Richards A. , Rubinstein H. , Schneider E. , Sun Q. , Tilav S. , Walck G. , Yodh G.

(瑞典斯德哥尔摩大学)

摘要：本文研究了地理南极深度在 0. 8～1 km 之间冰的光学特性。在现场利用南极 μ 介子和中微子探测阵列(AMANDA)的中微子探测器测量了可见光(近似于 515 nm)的吸收和散射长度。这里的冰在本质上是非常透明的。测得的吸收长度是 59±3 m，与 Irvine-Michigan-Brookhaven 和 Kamiokande 的质子衰变和中微子实验中使用的超纯水相当，并且比有报道的实验室冰的最好数值高两倍多。光子在介质中的轨迹会因为这些深度的气泡残留密度而随机化。如果假设气泡是平滑球形的，1 km 深度的平均碰撞间距大约是 25 cm。测得的气泡上的反散射长度随研究的冰量深度上升而线性减小。

25. Price P B. Kinetics of conversion of air bubbles to air hydrate crystals in Antarctic ice[J]. Science, 1995,267(5205):1802-1804.

Kinetics of conversion of air bubbles to air hydrate crystals in Antarctic ice

Price P. B.

(Physics Department, University of California, Berkeley, CA 94720, USA)

Abstract: The depth dependence of bubble concentration at pressures above the transition to the air hydrate phase and the optical scattering length due to bubbles in deep ice at the South Pole are modeled with diffusion-growth data from the laboratory, taking into account the dependence of age and temperature on depth in the ice. The model fits the available data on bubbles in cores from Vostok and Byrd and on scattering length in deep ice at the South Pole. It explains why bubbles and air hydrate crystals coexist in deep ice over a range of depths as great as 800 meters and predicts that at depths below similar to 1400 meters the AMANDA neutrino observatory at the South Pole will operate unimpaired by light scattering from bubbles.

南极冰中气泡向空气水合物晶体转换的动力学

Price P. B.

(美国加利福尼亚大学伯克利分校物理系)

摘要:本文基于实验室扩散增长数据,并考虑到冰中年龄和温度对深度的依赖,建立了南极点冰中在高于向空气水合物的转变压力和深部冰中气泡引起的光学散射长度的模型。该模型符合现有的 Vostok 和 Byrd 冰芯数据及南极点深部冰中散射长度。它解释了为何气泡和空气水合物晶体可以在像 800 m 这么大的深度共存,并预言了在南极点低于约 1400 m 深处的 AMANDA 中微子天文台的工作不受气泡的光散射损害。

26. Battle M, Bender M, Sowers T, et al. Atmospheric gas concentrations over the past century measured in air from firn at the South Pole[J]. Nature,1996,383(6597):231-235.

Atmospheric gas concentrations over the past century measured in air from firn at the South Pole

Battle M., Bender M., Sowers T., Tans P. P., Butler J. H., Elkins J. W., Ellis J. T., Conway T., Zhang N., Lang P., Clarke A. D.

(Graduate School of Oceanography, University of Rhode Island, Narragansett, Rhode Island 02882, USA)

Abstract: The extraction and analysis of air from the snowpack (firn) at the South Pole provides atmospheric concentration histories of biogenic greenhouse gases since the beginning of the present century which confirm and expand on those derived from studies of air trapped in ice cores. Furthermore, calculations based on the inferred atmospheric concentrations of oxygen and carbon dioxide indicate that-in contrast to the past few years—the terrestrial biosphere was neither a source nor sink of CO_2 between similar to 1977 and 1985.

利用南极点粒雪中气体测定的过去一个世纪大气气体浓度

Battle M. , Bender M. , Sowers T. , Tans P. P. , Butler J. H. , Elkins J. W. , Ellis J. T. , Conway T. , Zhang N. , Lang P. , Clarke A. D.

(美国罗德岛大学海洋研究生院)

摘要：南极点粒雪中气体的提取与分析提供了从21世纪初开始的生物来源的温室气体大气浓度的历史，确认和扩展了冰芯截留气体的研究结果。此外，基于推测出的氧气和二氧化碳大气浓度的计算显示，与前几年相反，约1977到1985年间陆地生物圈既不是CO_2的源也不是汇。

27. Mayewski P A, Twickler M S, Whitlow S I, et al. Climate change during the last deglaciation in Antarctica[J]. Science, 1996, 272(5268): 1636-1638.

Climate change during the last deglaciation in Antarctica

Mayewski P. A. , Twickler M. S. , Whitlow S. I. , Meeker L. D. , Yang Q. , Thomas J. , Kreutz K. , Grootes P. M. , Morse D. L. , Steig E. J. , Waddington E. D. , Saltzman E. S. , Whung P. Y. , Taylor K. C.

(Department of Earth Sciences, University of New Hampshire, Durham, NH 03824, USA)

Abstract: Greenland ice core records provide clear evidence of rapid changes in climate in a variety of climate indicators. In this work, rapid climate change events in the Northern and Southern hemispheres are compared on the basis of an examination of changes in atmospheric circulation developed from two ice cores. High-resolution glaciochemical series, covering the period 10000 to 16000 years ago, from a central Greenland ice core and a new site in east Antarctica display similar variability. These findings suggest that rapid climate change events occur more frequently in Antarctica than previously demonstrated.

南极末次冰消期的气候变化

Mayewski P. A. , Twickler M. S. , Whitlow S. I. , Meeker L. D. , Yang Q. , Thomas J. , Kreutz K. , Grootes P. M. , Morse D. L. , Steig E. J. , Waddington E. D. , Saltzman E. S. , Whung P. Y. , Taylor K. C.

(美国新罕布什尔大学地球科学系)

摘要：格陵兰冰芯记录提供了气候快速变化的多指示剂的清晰证据。本研究在分析了两支冰芯中得出的大气环流变化的基础上，比较了南北半球的快速气候变化事件。从一支格陵兰中部冰芯和东南极一个新站点得出的覆盖10000年到16000年前的高分辨率记录表现出相似的变化。这些发现显示南极快速气候变化事件出现得比之前显示的更频繁。

28. Rott H, Skvarca P, Nagler T. Rapid collapse of northern Larsen Ice Shelf, Antarctica[J]. Science, 1996, 271(5250): 788-792.

Rapid collapse of northern Larsen Ice Shelf, Antarctica

Rott H., Skvarca P., Nagler T.

(Institut für Meteorologie und Geophysik der Universität Innsbruck, Innrain 52, A-6020 Innsbruck, Austria)

Abstract: In January 1995, 4200 square kilometers of the northern Larsen Ice Shelf, Antarctic Peninsula, broke away. Radar images from the ERS-1 satellite, complemented by field observations, showed that the two northernmost sections of the ice shelf fractured and disintegrated almost completely within a few days. This breakup followed a period of steady retreat that coincided with a regional trend of atmospheric warming. The observations imply that after an ice shelf retreats beyond a critical limit, it may collapse rapidly as a result of perturbated mass balance.

南极北 Larsen 冰架的快速崩塌

Rott H., Skvarca P., Nagler T.

(奥地利因斯布鲁克大学气象与地球物理学院)

摘要:4200 km^2 南极半岛北 Larsen 冰架在 1995 年 1 月崩塌。ERS-1 卫星提供的雷达图像和补充的野外观测显示冰架最北的两个部分在几天内断裂并几乎完全瓦解。这一崩塌紧随一个稳定冰退期,与区域大气变暖趋势相吻合。这些观测显示当冰架后退超过一个临界界限后,它可能因为质量平衡的削弱而快速崩溃。

29. Vaughan D G, Doake C S M. Recent atmospheric warming and retreat of ice shelves on the Antarctic Peninsula[J]. Nature, 1996, 379(6563): 328-331.

Recent atmospheric warming and retreat of ice shelves on the Antarctic Peninsula

Vaughan D. G., Doake C. S. M.

(British Antarctic Survey, Natural Environment Research Council, Madingley Road, Cambridge CB3 0ET, UK)

Abstract: In 1978 Mercer discussed the probable effects of climate warming on the Antarctic Ice Sheet, predicting that one sign of a warming trend in this region would be the retreat of ice shelves on the Antarctic Peninsula. Analyses of 50-year meteorological records have since revealed atmospheric warming on the Antarctic Peninsula, and a number of ice shelves have retreated. Here we present time-series of observations of the areal extent of nine ice shelves on the Antarctic Peninsula, showing that five northerly ones have retreated dramatically in the past fifty years, while those further south show no clear trend. Comparison with air-temperature data shows that the pattern and magnitude of ice-shelf retreat is consistent with the existence of an abrupt thermal limit on ice-shelf viability, the isotherm associated with this limit having been driven south by the atmospheric warming. Ice shelves therefore appear to be sensitive indicators of climate change.

南极半岛近期的大气变暖和冰架后退

Vaughan D. G., Doake C. S. M.

(英国自然环境研究理事会南极调查局)

摘要:1978年Mercer讨论了气候变暖对南极冰盖可能的影响,并预言南极半岛冰架后退将是这一区域变暖趋势的一个信号。对50年气象记录的分析显示了南极半岛的大气变暖和一些冰川已经后退。本文得出了南极半岛区域的9个冰架观测的时间序列,其中北部的5个在过去50年显著后退,然而南部的那些没有显示清晰的趋势。与气温数据的比较显示冰架后退的模式和规模与冰架发育的一个突变热限制一致,与这一限制相关的等温线已经受到大气变暖驱动而南移。由此看来冰架是气候变化的敏感指示器。

30. Arrigo K R, Worthen D L, Lizotte M P, et al. Primary production in Antarctic sea ice[J]. Science, 1997, 276(5311): 394-397.

Primary production in Antarctic sea ice

Arrigo K. R., Worthen D. L., Lizotte M. P., Dixon P., Dieckmann G.

(NASA Oceans and Ice Branch, Goddard Space Flight Center, Code 971.0, Greenbelt, MD 20771, USA)

Abstract: A numerical model shows that in Antarctic sea ice, increased flooding in regions with thick snow cover enhances primary production in the infiltration (surface) layer. Productivity in the freeboard (sea level) layer is also determined by sea ice porosity, which varies with temperature, Spatial and temporal variation in snow thickness and the proportion of first-year ice thus determine regional differences in sea ice primary production. Model results show that of the 40 teragrams of carbon produced annually in the Antarctic ice pack, 75 percent was associated with first-year ice and nearly 50 percent was produced in the Weddell Sea.

南极海冰的初级生产力

Arrigo K. R., Worthen D. L., Lizotte M. P., Dixon P., Dieckmann G.

(美国国家航空和航天局戈达德宇宙飞行中心海洋和冰分局)

摘要:数值模型显示南极海冰中厚雪覆盖区域溢流的增加提高了渗透(表)层的初级生产力。干舷(海平面)层生产力也由海冰随温度变化的多孔性决定。因此雪的厚度和当年冰比例的时空变化决定了海冰初级生产力的区域差异。模拟结果显示南极浮冰每年合成40万亿克碳,其中75%与当年冰有关,并且近50%在Weddell海合成。

31. Bender M, Sowers T, Brook E. Gases in ice cores[J]. PNAS, 1997, 94(16): 8343-8349.

Gases in ice cores

Bender M. , Sowers T. , Brook E.

(Graduate School of Oceanography, University of Rhode Island, Kingston, RI 02881)

Abstract: Air trapped in glacial ice offers a means of reconstructing variations in the concentrations of atmospheric gases over time scales ranging from anthropogenic (last 200 yr) to glacial/interglacial (hundreds of thousands of years). In this paper, we review the glaciological processes by which air is trapped in the ice and discuss processes that fractionate gases in ice cores relative to the contemporaneous atmosphere. We then summarize concentration-time records for CO_2 and CH_4 over the last 200 yr. Finally, we summarize concentration-time records for CO_2 and CH_4 during the last two glacial-interglacial cycles, and their relation to records of global climate change.

冰芯中的气体

Bender M. , Sowers T. , Brook E.

(美国罗德岛大学海洋研究生院)

摘要:冰川冰中捕集的气体提供了重建跨越人类来源尺度(最近 200 年)到冰期/间冰期(数十万年)时间尺度的大气气体浓度的方法。本文综述了冰捕集气体的冰川学过程,并讨论了冰芯与同时期大气相关的气体分馏过程。然后我们总结了最近 200 年 CO_2和 CH_4的浓度-时间记录。最终我们总结了最后两次冰期-间冰期回旋中 CO_2和 CH_4的浓度-时间记录,以及它们同全球气候变化记录的关系。

32. Bentley C R. Rapid sea-level rise soon from West Antarctic ice sheet collapse? [J]. Science,1997,275(5303):1077-1078.

Rapid sea-level rise soon from West Antarctic ice sheet collapse?

Bentley C. R.

(Geophysical and Polar Research Center, University of Wisconsin, Madison, WI 53706-1692, USA)

Abstract:Will worldwide sea level soon rise rapidly because of a shrinkage of the West Antarctic ice sheet (WAIS)? That is a question of widespread interest, great societal import, and considerable controversy. The Intergovernmental Panel on Climate Change (IPCC) Second Scientific Assessment of Climate Change finds that estimating the probability of such an event is not yet possible. Here I give my perspective of that probability.

西南极冰盖崩溃即将引起快速海平面上升?

Bentley C. R.

(美国威斯康星大学地球物理和极地研究中心)

摘要:全世界范围内海平面不久将会因为南极西部冰盖(WAIS)的收缩而迅速上升?这是人们普遍关心的问题,有大量社会因素介入,也有相当大的争议。政府间气候变化专门委员会(IPCC)气候变化第二次科学评估认定,估计此类事件的发生是没有可能的。在这里,我将给出我的观点。

33. De La Mare W K. Abrupt mid-twentieth-century decline in Antarctic sea-ice extent from whaling records[J]. Nature,1997,389(6646):57-60.

Abrupt mid-twentieth-century decline in Antarctic sea-ice extent from whaling records

De La Mare W. K.

(Australian Antarctic Division, Department of the Environment, Sport and Territories, Channel Highway, Kingston, Tasmania 7050, Australia)

Abstract: A decline in Antarctic sea-ice extent is a commonly predicted effect of a warming climate. Direct global estimates of the Antarctic sea-ice cover from satellite observations, only possible since the 1970s have shown no clear trends. Comparisons between satellite observations and ice-edge charts obtained from early ship records suggest that sea-ice extent in the 1970s was less than during the 1930s, an indication supported by limited regional observations. But these observations have been regarded as inconclusive, owing to the limited spatial and temporal scope of the early records. A significant data source has, however, been overlooked. The southern limit of whaling was constrained by sea ice, and since 1931 whaling records have been collected for every whale caught, giving a circumpolar coverage from spring to autumn until 1987. Here, an analysis of these catch records indicates that, averaged over October to April, the Antarctic summer sea-ice edge has moved southwards by 2. 8 degrees of latitude between the mid 1950s and early 1970s. This suggests a decline in the area covered by sea ice of some 25%. This abrupt change poses a challenge to model simulations of recent climate change, and could imply changes in Antarctic deep-water formation and in biological productivity, both important processes affecting atmospheric CO_2 concentrations. Abrupt mid-twentieth-century decline in Antarctic sea-ice extent from whaling records.

从捕鲸记录得出的20世纪中期南极海冰范围的突然减退

De La Mare W. K.

(澳大利亚环境、体育和领土部南极司)

摘要:气候变暖的情况下南极海冰的衰退是普遍的预计。从1970年代开始的卫星观测做出的对南极海冰覆盖的直接全球评估没有显示出明显的趋势。对卫星观测和早期船舶冰缘测绘记录的比较显示海冰范围在1970年代比1930年代小，有限的区域观测也支持这一迹象。但由于早期记录的时空范围限制，这些观测被认为是非定论性的。然而一个显著数据来源被忽视了。鲸类活动范围的南方界限是被海冰限制的，并且从1931年起每头鲸的捕获都有记录，这提供了直到1987年的每年春季到秋季的覆盖环极地的记录。在这里对这些捕获记录的分析显示，从10月到次年4月平均后，在1950年代中期到1970年代早期之间，南极夏季海冰边缘南移了2.8度。这一结论显示海冰覆盖区域减小了约25%。这一突然变化对近期气候变化的模型模拟构成了挑战，并可能显示南极深部水构成和生物生产力的变化，两者都是影响大气CO_2浓度的重要过程。

34. Lawler A. Spacecraft offers details of Antarctica[J]. Science,1997,278(5343):1562.

Spacecraft offers details of Antarctica

Lawler A.

(News)

Abstract: A new Canadian radar satellite is giving polar researchers the first highly detailed map of Antarctica and its ice sheet, revealing unexpected ice flows and a heavily textured surface in areas that were thought to be largely featureless. Researchers say the data offer clues to the topography of the continent hidden under the ice. Comparisons with earlier images could also trace the retreat of ice shelves and glaciers, a possible sign of global warming.

航天器提供了南极洲的细节

Lawler A.

(新闻)

摘要：新的加拿大雷达卫星为极地研究人员描绘出第一幅非常详细的南极冰盖地图，获得了被认为是主要地区的意想不到的冰流和纹理密集的表面。研究人员说，该数据提供了线索来研究大陆冰层下隐藏的地形。与早期的图像比较也可以追踪冰架和冰川、全球变暖可能撤退的标志。

35. Murphy E, King J. Climate change: Icy message from the Antarctic[J]. Nature, 1997, 389(6646): 20-21.

Climate change: Icy message from the Antarctic

Murphy E., King J.

(Marine Life Sciences Division, and John King in the Ice and Climate Division, British Antarctic Survey, NERC, Madingley Road, Cambridge CB3 0ET, UK)

Abstract: Historically, much of the whaling in the Antarctic occurred in the vicinity of the ice edge where the whales congregated to feed on krill. During the austral spring, the whalers tracked the animals as they moved south with the edge of the sea ice, and their records of harvesting position thus provide proxy data for ice extent before monitoring by satellite began. An analysis of those records by de la Mare, published on page 57 of this issue, suggests that there was a large and rapid decrease in Antarctic summertime sea-ice extent between the 1950s and 1970s. The cause of this inferred decrease is unknown. But it will be sadly ironic if the human slaughter of the great whale populations reveals something of the Earth's climate system that may have dramatic implications for our own long-term future on the planet.

气候变化：来自南极的冰冷的信息

Murphy E. , King J.

(英国南极调查局海洋生命科学司)

摘要：从历史上看，南极地区多数捕鲸业发生在邻近的冰缘线处，那里是鲸鱼聚集捕食磷虾的地方。在南半球的春天，捕鲸船跟随动物随着海冰冰缘线向南移动，因而它们记录的捕获位置能够在卫星监控之前提供海冰范围的替代指标数据。De La Mare 在本期 57 页发表了对这些记录的分析，提出在 20 世纪 50 年代至 70 年代之间南极夏季海冰范围有大而快速的减少。该推论的原因仍不知道。如果人类对鲸鱼这一种群的捕杀揭示地球气候系统的某些可能对我们自己在这一星球上的长期未来有着戏剧性的暗示，那将是悲哀的讽刺。

36. Nicholls K W. Predicted reduction in basal melt rates of an Antarctic ice shelf in a warmer climate [J]. Nature, 1997, 388(6641): 460-462.

Predicted reduction in basal melt rates of an Antarctic ice shelf in a warmer climate

Nicholls K. W.

(British Antarctic Survey, Natural Environment Research Council, Cambridge CB3 0ET, UK)

Abstract: Floating ice shelves are vulnerable to climate change at both their upper and lower surfaces. The extent to which the apparently air-temperature-related retreat of some northerly Antarctic Peninsula ice shelves presages the demise of their much larger, more southerly, counterparts is not known, but air-temperature effects are unlikely to be important in the near future. Oceanographic measurements from beneath the most massive of these southerly ice shelves-the Filchner-Ronne Ice Shelf-have confirmed that dense sea water resulting from sea-ice formation north of the ice shelf flows into the sub-ice-shelf cavity. This relatively warm so-called High Salinity Shelf Water (HSSW) is responsible for the net melting at the ice shelf's base. Here I present temperature measurements, from the same sub-ice-shelf cavity, which show a strong seasonality in the inflow of HSSW. This seasonality results from intense wintertime production of sea ice, and I argue that the seasonal springtime warming can be used as an analogue for climate warming. For the present mode of oceanographic circulation, the implication is that warmer winters (a climate warming), leading to lower rates of sea-ice formation, would cause a reduction in the flux of HSSW beneath the ice shelf. The resultant cooling in the sub-ice cavity would lead, in turn, to a reduction in the total melting at the ice shelf's base. A moderate warming of the climate could thus lead to a basal thickening of the Filchner-Ronne Ice Shelf, perhaps increasing its longevity.

预计南极冰架底部融化速率在更暖的气候下将降低

Nicholls K. W.

(英国自然环境研究理事会南极调查局)

摘要:浮动冰架的上下两个表面都易受气候变化影响。一些显然与气温相关的南极半岛北部冰架退缩程度预示更大更南的类似冰架的消亡尚不清楚,但在不久的将来气温的影响似乎并不重要。这些南部冰架中质量最大的——Filchner-Ronne 冰架——之下的海洋学测量已经证实冰架北部海冰形成导致的高密度海水流入冰洞。这些相对温暖的所谓高盐度冰架水(HSSW)是冰架底部净融化的原因。这里我提出了由同一冰洞测量的温度,这一冰洞显示出 HSSW 的强季节性流入。这一季节性源于海冰在冬季的强烈增长,并且我主张季节性的春季变暖可以作为气候变暖的类比。在当前大洋环流状态下,这意味着更暖的冬季(气候变暖),因导致更低的海冰形成率,将导致冰架下 HSSW 通量减少。作为结果的冰洞变冷将反过来导致冰架底部总融解减少。因而气候的适度变暖可能导致 Filchner-Ronne 冰架基部增厚,多半会延长它的寿命。

37. Bell R E,Blankenship D D,Finn C A,et al. Influence of subglacial geology on the onset of a West Antarctic ice stream from aerogeophysical observations[J]. Nature,1998,394(6688):58-62.

Influence of subglacial geology on the onset of a West Antarctic ice stream from aerogeophysical observations

Bell R. E., Blankenship D. D., Finn C. A., Morse D. L., Scambos T. A., Brozena J. M., Hodge S. M.

(Lamont-Doherty Earth Observatory of Columbia University, Palisades, New York 10964, USA)

Abstract: Marine ice-sheet collapse can contribute to rapid sea-level rise Today, the West Antarctic Ice Sheet contains an amount of ice equivalent to approximately six metres of sea-level rise, but most of the ice is in the slowly moving interior reservoir. A relatively small fraction of the ice sheet comprises several rapidly flowing ice streams which drain the ice to the sea. The evolution of this drainage system almost certainly governs the process of ice-sheet collapse. The thick and slow-moving interior ice reservoir is generally fixed to the underlying bedrock while the ice streams glide over lubricated beds at velocities of up to several hundred metres per year. The source of the basal lubricant—a water-saturated till overlain by a water system—may be linked to the underlying geology. The West Antarctic Ice Sheet rests over a geologically complex region characterized by thin crust, high heat flows, active volcanism and sedimentary basins. Here we use aerogeophysical measurements to constrain the geological setting of the onset of an active West Antarctic ice stream. The onset coincides with a sediment-filled basin incised by a steep-sided valley. This observation supports the suggestion that ice-stream dynamics—and therefore the response of the West Antarctice Ice Sheet to changes in climate—are strongly modulated by the underlying geology.

基于航空地球物理观测的冰下地质对一条西南极冰流源头的影响

Bell R. E., Blankenship D. D., Finn C. A., Morse D. L.,
Scambos T. A., Brozena J. M., Hodge S. M.

(美国哥伦比亚大学拉蒙特-多尔蒂地质观测站)

摘要:海洋冰盖崩塌可能导致快速海平面上升。现今西南极冰盖包含了相当于海平面上升 6 m 的冰量,但大多数冰在缓慢移动的内部储库中。相对来说一小部分冰盖形成若干快速流动的冰流流入海中。这一机制几乎肯定控制了冰盖崩塌的进程。缓慢移动的内部厚冰储库总体上固定在下部基岩上,然而冰流在润滑的底部上滑动的速度高达每年数百米。底部润滑剂的来源——水系覆盖的水饱和冰碛物——可能与下部地质有关。西南极冰盖是架在一片地质复杂的区域,特点有薄地壳、高热流、活火山和沉积盆地。这里我们利用航空地球物理测量来测定一条活动西南极冰流源头的地质背景。这一源头与被陡峭谷地切割的、被沉积物填满的盆地一致。观测支持了以下观点:冰流动力学——西南极冰盖对气候变化的响应——受到下部地质的强烈影响。

38. Bindschadler R, Vornberger P. Changes in the West Antarctic ice sheet since 1963 from declassified satellite photography[J]. Science, 1998, 279(5351): 689-692.

Changes in the West Antarctic ice sheet since 1963 from declassified satellite photography

Bindschadler R., Vornberger P.

(Code 971, NASA-Goddard Space Flight Center, Greenbelt, MD 20771, USA)

Abstract: Comparison of declassified satellite photography taken in 1963 with more recent satellite imagery reveals that large changes have occurred in the region where an active ice stream enters the Ross Ice Shelf. Ice stream B has widened by 4 kilometers, at a rate much faster than suggested by models, and has decreased in speed by 50 percent. The ice ridge between ice streams B and C has eroded 14 kilometers. These changes, along with changes in the crevassing around Crary Ice Rise, imply that this region's velocity field shifted during this century.

基于解密卫星图像获取的 1963 年以来西南极冰盖的变化

Bindschadler R., Vornberger P.

(美国国家航空和航天局戈达德宇宙飞行中心)

摘要:解密的 1963 年拍摄的卫星图像与最近的卫星图像比较显示,一条活动冰流汇入 Ross 冰架的区域发生了强烈变化。冰流 B 拓宽了 4 km,拓宽的速率比模型显示得快,并且移动速度下降了 50%。冰流 B 和 C 之间的冰脊受到了 14 km 的侵蚀。这些变化,以及 Crary 海冰隆起周围的裂缝变化,意味着这一区域的速度场在 20 世纪改变了。

39. Doake C S M, Corr H F J, Rott H, et al. Breakup and conditions for stability of the northern Larsen Ice Shelf, Antarctica[J]. Nature, 1998, 391(6669): 778-780.

Breakup and conditions for stability of the northern Larsen Ice Shelf, Antarctica

Doake C. S. M., Corr H. F. J., Rott H., Skvarca P., Young N. W.

(British Antarctic Survey, Madingley Rd, Cambridge CB3 0ET, England)

Abstract: The breakup of ice shelves has been widely regarded as an indicator of climate change, with observations around the Antarctic Peninsula having shown a pattern of gradual retreat, associated with regional atmospheric warming, and increased summer melt and fracturing processes. The rapid collapse of the northernmost section of the Larsen Ice Shelf (Larsen A), over a few days in January 1995, indicated that, after retreat beyond a critical limit, ice shelves may disintegrate rapidly. Here we use a finite-element numerical model that treats ice as a continuum without fracture to examine the breakup history between 1986 and 1997 of the two northern sections of Larsen Ice Shelf (Larsen A and Larsen B), from which we establish stability criteria for ice shelves. Analysis of various ice-shelf configurations reveals characteristic patterns in the strain rates near the ice front which we use to describe the stability of the ice shelf. On Larsen A, only the initial and final ice-front configurations show a stable pattern. Larsen B exhibits a stable pattern, but if the ice front were to retreat by a further few kilometres, it too is likely to enter an irreversible retreat phase.

南极北 Larsen 冰架的分裂及稳定性现状

Doake C. S. M., Corr H. F. J., Rott H., Skvarca P., Young N. W.

(英国南极调查局)

摘要:冰架的分裂被普遍当作气候变化的指示器,环南极半岛的观测显示了与区域大气变暖相关的逐渐退却的格局,以及夏季融化和破裂过程的增加。1995 年 1 月的几天里,Larsen 冰架最北部(Larsen A)的快速崩塌显示,在退却到超过关键界限后,冰架可能快速碎裂。这里我们利用把冰视为无裂缝连续统一体的有限元数字模型分析了 Larsen 冰架的两个北方部分(Larsen A 和 Larsen B)在 1986 和 1997 年间的崩溃历史,由此建立了冰架的稳定性标准。对不同冰架构造的分析显示我们用来描述冰架稳定性的冰缘附近应变率的特有格局。Larsen A 只有最初和最终的冰缘构造展现出稳定的格局。Larsen B 目前展示了稳定的格局,但如果冰缘再退却几千米,它很可能也进入不可逆的退却状态。

40. Blunier T, Chappellaz J, Schwander J, et al. Asynchrony of Antarctic and Greenland climate change during the last glacial period[J]. Nature, 1998, 394(6695): 739-743.

Asynchrony of Antarctic and Greenland climate change during the last glacial period

Blunier T., Chappellaz J., Schwander J., Dallenbach A., Stauffer B., Stocker T. F., Raynaud D., Jouzel J., Clausen H. B., Hammer C. U., Johnsen S. J.

(Climate and Environmental Physics, Physics Institute, University of Bern, Sidlerstrasse 5, CH-3012 Bern, Switzerland)

Abstract: A central issue in climate dynamics is to understand how the Northern and Southern hemispheres are coupled during climate events. The strongest of the fast temperature changes observed in Greenland (so-called Dansgaard-Oeschger events) during the last glaciation have an analogue in the temperature record from Antarctica. A comparison of the global atmospheric concentration of methane as recorded in ice cores from Antarctica and Greenland permits a determination of the phase relationship (in leads or lags) of these temperature variations. Greenland warming events around 36 and 45 kyr before present lag their Antarctic counterpart by more than 1 kyr. On average, Antarctic climate change leads that of Greenland by 1-2.5 kyr over the period 47-23 kyr before present.

末次冰期南极与格陵兰气候变化的异时性

Blunier T., Chappellaz J., Schwander J., Dallenbach A., Stauffer B., Stocker T. F., Raynaud D., Jouzel J., Clausen H. B., Hammer C. U., Johnsen S. J.

(瑞士伯尔尼大学物理学院气候与环境物理系)

摘要:气候动力学的核心问题之一是理解在气候事件中南北半球如何联系。在格陵兰观测到的末次冰期最强的快速温度变化(所谓 Dansgaard-Oeschger 事件)在南极温度记录中有类似信号。通过比较南极和格陵兰冰芯中记录的全球甲烷大气浓度可以确定这些温度变化的相位关系(超前或滞后)。距今 36 千年～45 千年前的格陵兰变暖事件滞后于它们在南极的对应记录超过 1 千年。在距今 47 千年～23 千年的时期,南极气候变化平均比格陵兰超前 1 千年～2.5 千年。

41. Kerr R A. Global change: signs of past collapse beneath Antarctic ice[J]. Science, 1998, 281(5373): 17-19.

Global change: signs of past collapse beneath Antarctic ice

Kerr R. A.

(News)

Abstract: Glaciologists have long been casting a worried eye on the West Antarctic ice sheet (WAIS). Its bed is below sea level, which in theory makes it far less stable than the larger East Antarctic ice sheet. And the western sheet is plenty big. If it melted away in a greenhouse-warmed world, it would raise all the world's oceans by 5 meters. Your favorite beach would be underwater—as would New Orleans, Miami, and Bangkok. Now the worries may deepen. A paper in this issue of

Science (p. 82) confirms suspicions that in the recent geologic past, at a time perhaps not much warmer than today, the WAIS wasted away to a scrap and flooded the world's coasts.

全球变化:南极冰下过去崩塌的信号

Kerr R. A.

(新闻)

摘要:冰川学家长期以来一直忧心忡忡地关注南极西部冰盖(WAIS)。它的基础是海平面,这在理论上使得它远远超过了不太稳定的大东南极冰盖,并且西面足够大。如果它在这个不断升温的温室世界消失了,会导致世界各大洋升高 5 m。你最喜欢的海滩将被海水淹没,像新奥尔良、迈阿密和曼谷。现在的忧虑会更加地深。在这个问题上本期《Science》(第 82 页)的论文证实了怀疑,在最近的地质年代,某一个时间也许比今天没有暖和多少,WAIS 将被毁掉,并淹没全世界的海岸。

42. Oppenheimer M. Global warming and the stability of the West Antarctic Ice Sheet[J]. Nature, 1998,393(6683):325-332.

Global warming and the stability of the West Antarctic Ice Sheet

Oppenheimer M.

(Environmental Defense Fund, 257 Park Avenue South, New York, New York 10010, USA)

Abstract: Of today's great ice sheets, the West Antarctic Ice Sheet poses the most immediate threat of a large sea-level rise, owing to its potential instability. Complete release of its ice to the ocean would raise global mean sea level by four to six metres, causing major coastal flooding worldwide. Humaninduced climate change may play a significant role in controlling the long-term stability of the West Antarctic Ice Sheet and in determining its contribution to sea-level change in the near future.

全球变暖与西南极冰盖的稳定性

Oppenheimer M.

(美国环境防卫基金会)

摘要:在当前的大冰盖中,西南极冰盖由于潜在的不稳定性,对海平面大规模上升具有最直接的威胁。将它的冰完全释放到海洋中会把全球平均海平面提高 4～6 m,引起全世界沿海地区严重的洪水泛滥。人类导致的气候变化可能在控制西南极冰盖的长期稳定性和决定不久的将来它对海平面变化的贡献中扮演重要角色。

43. Priscu J C, Fritsen C H, Adams E E, et al. Perennial Antarctic lake ice: an oasis for life in a polar desert[J]. Science, 1998, 280(5372):2095-2098.

Perennial Antarctic lake ice: an oasis for life in a polar desert

Priscu J. C., Fritsen C. H., Adams E. E., Giovannoni S. J., Paerl H. W., McKay C. P., Doran P. T., Gordon D. A., Lanoil B. D., Pinckney J. L.

(Department of Biological Sciences, Montana State University, Bozeman, MT 59717, USA)

Abstract: The permanent ice covers of Antarctic lakes in the McMurdo Dry Valleys develop liquid water inclusions in response to solar heating of internal aeolian-derived sediments. The ice sediment particles serve as nutrient (inorganic and organic) enriched microzones for the establishment of a physiologically and ecologically complex microbial consortium capable of contemporaneous photosynthesis, nitrogen fixation, and decomposition. The consortium is capable of physically and chemically establishing and modifying a relatively nutrient and organic matter-enriched microbial "oasis" embedded in the lake ice cover.

南极常年湖冰:极地荒漠中的生命绿洲

Priscu J. C., Fritsen C. H., Adams E. E., Giovannoni S. J., Paerl H. W., McKay C. P., Doran P. T., Gordon D. A., Lanoil B. D., Pinckney J. L.

(美国蒙大拿州立大学生物科学系)

摘要:南极 McMurdo 干谷湖泊的永久性覆冰因内在风成沉积被太阳加热而形成液态水包裹。冰沉积颗粒可以作为富含营养(无机和有机)的微区建立生理学和生态学上复杂的微生物聚合体,足以同时进行光合作用、固氮和分解。这样的聚合体足以在物理和化学上建立和改变湖泊冰盖上的富含营养和有机物的微生物"绿洲"。

44. Rignot E J. Fast recession of a West Antarctic glacier[J]. Science, 1998, 281(5376): 549-551.

Fast recession of a West Antarctic glacier

Rignot E. J.

(Jet Propulsion Laboratory, California Institute of Technology, MS 300-235, Pasadena, CA 91109-8099, USA)

Abstract: Satellite radar interferometry observations of Pine Island Glacier, West Antarctica, reveal that the glacier hinge-line position retreated 1.2±0.3 kilometers per year between 1992 and 1996, which in turn implies that the ice thinned by 3.5±0.9 meters per year. The fast recession of Pine Island Glacier, predicted to be a possible trigger for the disintegration of the West Antarctic Ice Sheet, is attributed to enhanced basal melting of the glacier floating tongue by warm ocean waters.

一条西南极冰川的快速后退

Rignot E. J.

（美国加州理工学院喷气推进实验室）

摘要：对西南极松岛冰川的卫星雷达干涉测量法观测显示，冰川脊线位置在1992到1996年间每年退却1.2±0.3 km，同时意味着冰层每年变薄3.5±0.9 m。松岛冰川的快速后退归因于暖海水引起的浮动冰舌底部的加速融化；它可能导致西南极冰盖瓦解。

45. Scherer R P, Aldahan A, Tulaczyk S, et al. Pleistocene collapse of the West Antarctic ice sheet[J]. Science, 1998, 281(5373): 82-85.

Pleistocene collapse of the West Antarctic ice sheet

Scherer R. P., Aldahan A., Tulaczyk S., Possnert G., Engelhardt H., Kamb B.

(Department of Earth Sciences, Uppsala University, Villavägen 16, S-752 36 Uppsala, Sweden)

Abstract: Some glacial sediment samples recovered from beneath the West Antarctic ice sheet at ice stream B contain Quaternary diatoms and up to 10^8 atoms of beryllium-10 per gram. Other samples contain no Quaternary diatoms and only background levels of beryllium-10 (less than 10^6 atoms per gram). The occurrence of young diatoms and high concentrations of beryllium-10 beneath grounded ice indicates that the Ross Embayment was an open marine environment after a late Pleistocene collapse of the marine ice sheet.

西南极冰盖的更新世崩塌

Scherer R. P., Aldahan A., Tulaczyk S., Possnert G., Engelhardt H., Kamb B.

（瑞典乌普萨拉大学地球科学系）

摘要：一些在西南极冰盖下冰流B获得的冰川沉积样品含有第四纪硅藻和每克高达10^8原子数的铍-10。其他样品不含第四纪硅藻，仅含有背景水平的铍-10（低于每克10^6原子数）。在底冰下年轻硅藻和高浓度铍-10的出现显示Ross湾在晚更新世海洋冰盖崩塌后曾经是一片开放海洋环境。

46. Stauffer B, Blunier T, Dallenbach A, et al. Atmospheric CO_2 concentration and millennial-scale climate change during the last glacial period[J]. Nature, 1998, 392(6671): 59-62.

Atmospheric CO_2 concentration and millennial-scale climate change during the last glacial period

Stauffer B. , Blunier T. , Dallenbach A. , Indermuhle A. , Schwander J. , Stocker T. F. , Tschumi J. , Chappellaz J. , Raynaud D. , Hammer C. U. , Clausen H. B.

(Climate and Environmental Physics, Physics Institute, University of Bern, Sidlerstrasse 5, CH-3012 Bern, Switzerland)

Abstract: The analysis of air bubbles trapped in polar ice has permitted the reconstruction of past atmospheric concentrations of CO_2 over various timescales, and revealed that large climate changes over tens of thousands of years are generally accompanied by changes in atmospheric CO_2 concentrations. But the extent to which such covariations occur for fast, millennial-scale climate shifts, such as the Dansgaard-Oeschger events recorded in Greenland ice cores during the last glacial period, is unresolved; CO_2 data from Greenland and Antarctic ice cores have been conflicting in this regard. More recent work suggests that Antarctic ice should provide a more reliable CO_2 record, as the higher dust content of Greenland ice can give rise to artefacts. To compare the rapid climate changes recorded in the Greenland ice with the global trends in atmospheric CO_2 concentrations as recorded in the Antarctic ice, an accurate common timescale is needed. Here we provide such a timescale for the last glacial period using the records of global atmospheric methane concentrations from both Greenland and Antarctic ice. We find that the atmospheric concentration of CO_2 generally varied little with Dansgaard-Oeschger events (<10 parts per million by volume, p. p. m. v.) but varied significantly with Heinrich iceberg-discharge events (similar to 20 p. p. m. v.), especially those starting with a long-lasting Dansgaard-Oeschger event.

末次冰期的大气CO_2浓度与千年尺度气候变化

Stauffer B. , Blunier T. , Dallenbach A. , Indermuhle A. , Schwander J. , Stocker T. F. , Tschumi J. , Chappellaz J. , Raynaud D. , Hammer C. U. , Clausen H. B.

(瑞士伯尔尼大学物理学院气候与环境物理系)

摘要:通过分析极地冰中捕集的气泡可以重建过去不同时间尺度的大气CO_2浓度,并显示了数万年间气候的大变化总体上与大气CO_2浓度的变化同时发生。但这类共变在千年尺度的快速气候变化中的规模尚未得到解决,例如格陵兰冰芯记录的末次冰期 Dansgaard-Oeschger 事件;在这一方面格陵兰和南极冰芯的CO_2数据不一致。最近的研究显示南极冰的CO_2记录更可靠,因为格陵兰冰中更高浓度的尘埃可能引起误差。为了比较格陵兰冰中记录的快速气候变化和南极冰中记录的大气CO_2浓度的全球趋势,我们需要一个准确的共同时间表。这里我们利用格陵兰和南极冰两者中记录的全球大气甲烷浓度,提出了末次冰期的这样一个时间表。我们发现 Dansgaard-Oeschger 事件期间大气CO_2浓度总体上变化不大(小于百万分之 10 体积,ppmv),但在 Heinrich 冰山释放事件期间变化明显(约 20 ppmv),尤其是那些在长期持续的 Dansgaard-Oeschger 事件期间开始的。

47. Ackert R P, Barclay D J, Borns H W, et al. Measurements of past ice sheet elevations in interior West Antarctica[J]. Science, 1999, 286(5438): 276-280.

Measurements of past ice sheet elevations in interior West Antarctica

Ackert R. P., Barclay D. J., Borns H. W., Calkin P. E., Kurz M. D., Fastook J. L., Steig E. J.

(Massachusetts Institute of Technology/Woods Hole Oceanographic Institution Joint Program, MS 25, Clark 419)

Abstract: A Lateral moraine band on Mount Waesche, a volcanic nunatak in Marie Byrd Land, provides estimates of past ice sheet surface elevations in West Antarctica. Helium-3 and chlorine-36 surface exposure ages indicate that the proximal part of the moraine, up to 45 meters above the present ice surface, was deposited about 10000 years ago, substantially later than the maximum ice extent in the Ross Embayment. The upper distal part of the moraine may record multiple earlier ice sheet high stands. A nonequilibrium ice sheet model predicts a delay of several thousand years in maximum ice levels at Mount relative to the maximum ice extent in the Ross Sea. The glacial geologic evidence, coupled with the ice sheet model, indicates that the contribution of the Ross Sea sector of the West Antarctic Ice Sheet to Holocene sea Level rise was only about 3 meters. These results eliminate West Antarctic ice as the principle source of the large meltwater pulse during the early Holocene.

西南极内陆过去冰盖高度的测量

Ackert R. P., Barclay D. J., Borns H. W., Calkin P. E., Kurz M. D., Fastook J. L., Steig E. J.

(美国麻省理工学院/伍兹霍尔海洋研究所联合项目)

摘要:Waesche 山上的一条侧向冰碛带,也即 Marie Byrd 地的一座冰原火山岛峰,提供了对过去西南极冰盖表面高度的估计。氦 3 和氯 36 表面暴露年龄显示,冰碛在目前冰面之上高达 45 m 的最近部分是在约 10000 年前沉积的,大幅迟于 Ross 湾的最大冰范围。冰碛物的上部末梢部分可能记录了多个早期冰盖高位。非平衡态冰盖模型预测了 Waesche 山最大冰水平相对于 Ross 海最大冰范围的数千年滞后。冰川地质学证据与冰盖模型联用显示西南极冰盖 Ross 海部分对全新世海平面上升的贡献仅仅是约 3 m。这些结果将西南极冰盖排除在早全新世巨大融水冲击的主要来源之外。

48. Conway H, Hall B L, Denton G H, et al. Past and future grounding-line retreat of the West Antarctic Ice Sheet[J]. Science, 1999, 286(5438): 280-283.

Past and future grounding-line retreat of the West Antarctic Ice Sheet

Conway H., Hall B. L., Denton G. H., Gades A. M., Waddington E. D.

(Geophysics Program, University of Washington, Seattle, WA 98195, USA)

Abstract: The history of deglaciation of the West Antarctic Ice Sheet (WAIS) gives clues about its future. Southward grounding-line migration was dated past three Locations in the Ross Sea Embayment. Results indicate that most recession occurred during the middle to Late Holocene in the absence of substantial sea Level or climate forcing. Current grounding-line retreat may reflect ongoing ice recession that has been under way since the early Holocene. If so, the WAIS could continue to retreat even in the absence of further external forcing.

过去和未来的西南极冰盖接地线的后退

Conway H., Hall B. L., Denton G. H., Gades A. M., Waddington E. D.

(美国华盛顿大学西雅图分校地球物理项目)

摘要：西南极冰盖(WAIS)冰消的历史给出了它的未来的线索。本文确认了过去 Ross 海湾内三个地点接地线向南的迁移年代。结果显示大部分后退发生在没有充分海平面或气候强迫的中晚全新世。目前的接地线退却可能反映了正在进行的也是从早全新世以来一直在进行的退却。若然，即使没有更多的外部强迫，WAIS 也会继续消退。

49. Fischer H, Wahlen M, Smith J, et al. Ice core records of atmospheric CO_2 around the last three glacial terminations[J]. Science, 1999, 283(5408): 1712-1714.

Ice core records of atmospheric CO_2 around the last three glacial terminations

Fischer H., Wahlen M., Smith J., Mastroianni D., Deck B.

(Scripps Institution of Oceanography, Geosciences Research Division, University of California San Diego, La Jolla, CA 92093-0220, USA)

Abstract: Air trapped in bubbles in polar ice cores constitutes an archive for the reconstruction of the global carbon cycle and the relation between greenhouse gases and climate in the past. High-resolution records from Antarctic ice cores show that carbon dioxide concentrations increased by 80 to 100 parts per million by volume 600±400 years after the warming of the last three deglaciations. Despite strongly decreasing temperatures, high carbon dioxide concentrations can be sustained for thousands of years during glaciations; the size of this phase lag is probably connected to the duration of the preceding warm period, which controls the change in land ice coverage and the buildup of the terrestrial biosphere.

最后三次冰消期附近大气 CO_2 的冰芯记录

Fischer H., Wahlen M., Smith J., Mastroianni D., Deck B.

(美国加利福尼亚大学圣地亚哥分校地球科学研究部斯克里普斯海洋研究所)

摘要：极地冰芯中气泡捕集的空气构成了重建过去全球碳循环和温室气体与气候关系的记录。来自南极冰芯的高分辨率记录显示二氧化碳浓度在最后三次冰消期的变暖后 600±400 年上升了百万分之

80 到 100 体积。冰期期间尽管温度强烈下降，高浓度的二氧化碳可能持续数千年；之前的暖期的持续时间控制了陆冰覆盖和陆地生物圈的建立，这一相位滞后的规模很可能与之相关联。

50. Joughin I, Gray L, Bindschadler R, et al. Tributaries of West Antarctic Ice streams revealed by RADARSAT interferometry[J]. Science, 1999, 286(5438): 283-286.

Tributaries of West Antarctic Ice streams revealed by RADARSAT interferometry

Joughin I., Gray L., Bindschadler R., Price S., Morse D., Hulbe C., Mattar K., Werner C.

(Jet Propulsion Laboratory, California Institute of Technology, M/S 300-235, 4800 Oak Grove Drive, Pasadena, CA 91109, USA)

Abstract: Interferometric RADARSAT data are used to map ice motion in the source areas of four West Antarctic ice streams. The data reveal that tributaries, coincident with subglacial valleys, provide a spatially extensive transition between slow inland flow and rapid ice stream flow and that adjacent ice streams draw from shared source regions. Two tributaries flow into the stagnant ice stream C, creating an extensive region that is thickening at an average rate of 0.49 meters per year. This is one of the largest rates of thickening ever reported in Antarctica.

RADARSAT 干涉测量法揭示的西南极冰流的支流

Joughin I., Gray L., Bindschadler R., Price S., Morse D., Hulbe C., Mattar K., Werner C.

(美国加州理工学院喷气推进实验室)

摘要：本文利用干涉测量的 RADARSAT 数据绘制了四条西南极冰流源区的冰运动地图。数据显示与冰下谷地一致的支流在缓慢内陆流动和快速冰流流动之间有空间上的大范围转变，并且邻近的冰流来自共有的源区。两条支流流入停滞的冰流 C，导致了一片大范围区域平均每年变厚 0.49 m。这是在南极有报告的最大变厚速率之一。

51. Jouzel J, Petit J R, Souchez R, et al. More than 200 meters of lake ice above subglacial Lake Vostok, Antarctica[J]. Science, 1999, 286(5447): 2138-2141.

More than 200 meters of lake ice above subglacial Lake Vostok, Antarctica

Jouzel J., Petit J. R., Souchez R., Barkov N. I., Lipenkov V. Y., Raynaud D., Stievenard M., Vassiliev N. I., Verbeke V., Vimeux F.

(Laboratoire des Sciences du Climat et de l'Environnement (UMR CEA/CNRS 1572), CEA Saclay, 91191 Gif-sur-Yvette Cédex, France)

Abstract: Isotope studies show that the Vostok ice core consists of ice refrozen from Lake Vostok water, from 3539 meters below the surface of the Antarctic ice sheet to its bottom at about 3750 meters. Additional evidence comes from the total gas content, crystal size, and electrical conductivity of the ice. The Vostok site is a likely place for water freezing at the lake-ice interface, because this interface occurs at a higher level here than anywhere else above the lake. Isotopic data suggest that subglacial Lake Vostok is an open system with an efficient circulation of water that was formed during periods that were slightly warmer than those of the past 420000 years. Lake ice recovered by deep drilling is of interest for preliminary investigations of lake chemistry and bedrock properties and for the search for indigenous lake microorganisms. This latter aspect is of potential importance for the exploration of icy planets and moons.

在南极 Vostok 冰下湖之上超过 200 m 的湖冰

Jouzel J., Petit J. R., Souchez R., Barkov N. I., Lipenkov V. Y., Raynaud D., Stievenard M., Vassiliev N. I., Verbeke V., Vimeux F.

(法国原子能委员会气候与环境科学实验室)

摘要:同位素研究显示 Vostok 冰芯从南极冰盖表面 3539 m 以下到约 3750 m 的底部由 Vostok 湖水重新冰冻形成的冰组成。额外的证据来自冰的总气体含量、晶体尺寸和电导率。Vostok 站可能是湖-冰分界面水冻结的地点,因为这一界面在这里比湖上其他任何地方出现的都高。同位素数据显示 Vostok 冰下湖是拥有有效水循环的开放体系,形成于稍暖于过去 420000 年的时期。深钻获得的湖冰对湖的化学和基岩性质的初步研究以及寻找当地湖生微生物研究都有价值。后者对勘探冰覆盖的行星和卫星具有潜在的重要性。

52. Nadis S. Moves are afoot to probe the lake trapped beneath Antarctic ice[J]. Nature, 1999, 401 (6750): 203.

Moves are afoot to probe the lake trapped beneath Antarctic ice

Nadis S.

(News)

Abstract: Lake Vostok, isolated from the rest of the biosphere for at least a million years, could be the subject of an international initiative to plumb its secrets.

Some 80 scientists from 14 countries will meet in Cambridge, England, next week to formulate plans for exploring one of the world's last uncharted natural wonders: Lake Vostok, which is buried beneath four kilometres of ice in East Antarctica.

探测南极冰下湖的行动在准备中

Nadis S.

（新闻）

摘要：沃斯托克湖，与其他的生物圈孤立了至少一万年，这可能是国际首次探索它的秘密的原因。

来自 14 个国家的约 80 名科学家将在英国剑桥于下周举行会议，制订计划，探索世界上最后的未知自然奇观之一：埋在东南极洲 4 km 冰下的沃斯托克湖。

53. Petit J R, Jouzel J, Raynaud D, et al. Climate and atmospheric history of the past 420000 years from the Vostok ice core, Antarctica[J]. Nature, 1999, 399(6735): 429-436.

Climate and atmospheric history of the past 420000 years from the Vostok ice core, Antarctica

Petit J. R., Jouzel J., Raynaud D., Barkov N. I., Barnola J. M., Basile I., Bender M., Chappellaz J., Davis M., Delaygue G., Delmotte M., Kotlyakov V. M., Legrand M., Lipenkov V. Y., Lorius C., Pepin L., Ritz C., Saltzman E., Stievenard M.

(Laboratoire de Glaciologie et Geophysique de l'Environnement, CNRS, BP96, 38402, Saint Martin d'Heres Cedex, France)

Abstract: The recent completion of drilling at Vostok station in East Antarctica has allowed the extension of the ice record of atmospheric composition and climate to the past four glacial-interglacial cycles. The succession of changes through each climate cycle and termination was similar, and atmospheric and climate properties oscillated between stable bounds. Interglacial periods differed in temporal evolution and duration. Atmospheric concentrations of carbon dioxide and methane correlate well with Antarctic air-temperature throughout the record. Present-day atmospheric burdens of these two important greenhouse gases seem to have been unprecedented during the past 420000 years.

南极 Vostok 冰芯记录的过去 420000 年的气候与大气历史

Petit J. R., Jouzel J., Raynaud D., Barkov N. I., Barnola J. M., Basile I., Bender M., Chappellaz J., Davis M., Delaygue G., Delmotte M., Kotlyakov V. M., Legrand M., Lipenkov V. Y., Lorius C., Pepin L., Ritz C., Saltzman E., Stievenard M.

（法国国家科学研究院冰川与地球物理环境实验室）

摘要：最近在东南极 Vostok 站完成的钻孔允许我们把大气成分和气候的冰记录延长到最后四个冰期-间冰期旋回。变化的序列在每个气候旋回和终结都是类似的，并且大气和气候的性质在稳定的范围内震荡。间冰期在时间上演变和持续时间有不同。整个记录中二氧化碳和甲烷的大气浓度与南极气温相关很好。现今大气中这两种重要温室气体的负担在过去 420000 年间已经是空前的了。

54. Bamber J L, Vaughan D G, Joughin I. Widespread complex flow in the interior of the Antarctic ice sheet[J]. Science, 2000, 287(5456): 1248-1250.

Widespread complex flow in the interior of the Antarctic ice sheet

Bamber J. L., Vaughan D. G., Joughin I.

(Bristol Glaciology Centre, School of Geographical Sciences, University of Bristol, University Road, Bristol, BS8 1SS, UK)

Abstract: It has been suggested that as much as 90% of the discharge from the Antarctic Ice Sheet is drained through a small number of fast-moving ice streams and outlet glaciers fed by relatively stable and inactive catchment areas. Here, evidence obtained from balance velocity estimates suggests that each major drainage basin is fed by complex systems of tributaries that penetrate up to 1000 kilometers from the grounding line into the interior of the ice sheet. This finding has important consequences for the modeled or estimated dynamic response time of past and present ice sheets to climate forcing.

南极冰盖内部分布广泛的复杂流

Bamber J. L., Vaughan D. G., Joughin I.

(英国布里斯托尔大学地理科学学院布里斯托尔冰川中心)

摘要:有些研究者持有这样的观点,南极冰盖多达90%的流出量是由少数快速移动的冰流和外流冰川排出的,它们来自相对稳定和非活动的集冰区。这里得自平衡速率估计的证据显示每个主要水系盆地都是由复杂支流系统流入的,支流从接地线向冰盖内部深入到1000 km。这一发现在模拟或估计过去和现在冰盖对气候强迫的动力学响应时间中有重要地位。

55. Kanfoush S L, Hodell D A, Charles C D, et al. Millennial-scale instability of the Antarctic ice sheet during the last glaciations[J]. Science, 2000, 288(5472): 1815-1818.

Millennial-scale instability of the Antarctic ice sheet during the last glaciations

Kanfoush S. L., Hodell D. A., Charles C. D., Guilderson T. P., Mortyn P. G., Ninnemann U. S.

(Department of Geological Sciences, University of Florida, Gainesville, FL 32611, USA)

Abstract: Records of ice-rafted detritus (IRD) concentration in deep-sea cores from the southeast Atlantic Ocean reveal millennial-scale pulses of IRD delivery between 20000 and 74000 years ago. Prominent IRD Layers correlate across the Polar Frontal Zone, suggesting episodes of Antarctic Ice Sheet instability. Carbon isotopes ($\delta^{13}C$) of benthic foraminifers, a proxy of deep water circulation, reveal that South Atlantic IRD events coincided with strong increases in North Atlantic Deep Water (NADW) production and inferred warming (interstadials) in the high-latitude North Atlantic. Sea level rise or increased NADW production associated with strong interstadials may have resulted in destabilization of grounded ice shelves and possible surging in the Weddell Sea region of West Antarctica.

末次冰期南极冰盖千年尺度上的不稳定性

Kanfoush S. L., Hodell D. A., Charles C. D., Guilderson T. P., Mortyn P. G., Ninnemann U. S.

(美国佛罗里达大学地质科学系)

摘要:来自东南大西洋深海岩芯中冰筏残屑(IRD)集合体记录显示了在20000年到74000年前的千年尺度IRD输送脉冲。显著的IRD层在整个极锋锋区都相关,揭示了南极冰盖的不稳定事件。

深海环流的代用性指标深海有孔虫的碳同位素($\delta^{13}C$)显示,南大西洋IRD事件与北大西洋深水(NADW)产量的强烈增加和由此推断的北大西洋高纬的变暖(间冰段)一致。与强间冰段相关的海平面上升或NADW产量的增加可能造成接地冰架的扰动与西南极Weddell海区域可能的扰动。

56. Lear C H, Elderfield H, Wilson P A. Cenozoic deep-sea temperatures and global ice volumes from Mg/Ca in benthic foraminiferal calcite[J]. Science, 2000, 287(5451): 269-272.

Cenozoic deep-sea temperatures and global ice volumes from Mg/Ca in benthic foraminiferal calcite

Lear C. H., Elderfield H., Wilson P. A.

(Department of Earth Sciences, University of Cambridge, Downing Street, Cambridge CB2 3EQ, UK)

Abstract: A deep-sea temperature record for the past 50 million years has been produced from the magnesium/calcium ratio (Mg/Ca) in benthic foraminiferal calcite. The record is strikingly similar in form to the corresponding benthic oxygen isotope ($\delta^{18}O$) record and defines an overall cooling of about 12 degrees C in the deep oceans with four main cooling periods. Used in conjunction with the benthic $\delta^{18}O$ record, the magnesium temperature record indicates that the first major accumulation of Antarctic ice occurred rapidly in the earliest Oligocene (34 million years ago) and was not accompanied by a decrease in deep-sea temperatures.

得自深海有孔虫碳酸盐Mg/Ca记录的新生代深海温度与全球冰量

Lear C. H., Elderfield H., Wilson P. A.

(英国剑桥大学地球科学学系)

摘要:本文利用深海有孔虫碳酸盐的镁钙比(Mg/Ca)得出了过去5000万年以来的深海温度记录。这一记录与相应的深海氧同位素($\delta^{18}O$)记录显著相似,并确认了4个主要冷期中深海大约12 ℃的全面变冷。与深海$\delta^{18}O$记录联合使用,镁温度记录显示南极冰的第一次主要积累在渐新世初期(3400万年前)快速发生，并且没有相伴的深海温度下降。

57. Stephens B B, Keeling R F. The influence of Antarctic sea ice on glacial-interglacial CO_2 variations [J]. Nature, 2000, 404(6774): 171-174.

The influence of Antarctic sea ice on glacial-interglacial CO_2 variations

Stephens B. B., Keeling R. F.

(Scripps Institution of Oceanography, University of California, San Diego, La Jolla, California 92093±0244, USA)

Abstract: Ice-core measurements indicate that atmospheric CO_2 concentrations during glacial periods were consistently about 80 parts per million lower than during interglacial periods. Previous explanations for this observation have typically had difficulty accounting for either the estimated glacial O_2 concentrations in the deep sea, $^{13}C/^{12}C$ ratios in Antarctic surface waters, or the depth of calcite saturation; also lacking is an explanation for the strong link between atmospheric CO_2 and Antarctic air temperature. There is growing evidence that the amount of deep water upwelling at low latitudes is significantly overestimated in most ocean general circulation models and simpler box models previously used to investigate this problem. Here we use a box model with deep water upwelling confined to south of 55°S to investigate the glacial-interglacial linkages between Antarctic air temperature and atmospheric CO_2 variations. We suggest that low glacial atmospheric CO_2 levels might result from reduced deep water ventilation associated with either year-round Antarctic sea-ice coverage, or wintertime coverage combined with ice-induced stratification during the summer. The model presented here reproduces 67 parts per million of the observed glacial-interglacial CO_2 difference, as a result of reduced air-sea gas exchange in the Antarctic region, and is generally consistent with the additional observational constraints.

南极海冰对冰期-间冰期 CO_2变化的影响

Stephens B. B., Keeling R. F.

(美国加利福尼亚大学圣地亚哥分校斯克里普斯海洋研究所)

摘要：冰芯测量显示冰期大气 CO_2浓度相对于间冰期一贯低 80%左右。之前对这一观测资料的解释在解决冰期深海 O_2浓度或南极表层水$^{13}C/^{12}C$ 比值或碳酸盐饱和深度中有困难；对大气 CO_2与南极气温之间的密切联系也缺乏解释。越来越多的证据表明低纬深水上升流量在之前研究这一问题使用的大多数大洋全面环流模型和更简单的箱式模型中被过高估计。这里我们利用把深水上升流限制在 55°S 以南的箱式模型来研究南极气温和大气 CO_2 变化的联系。我们提出冰期低的大气 CO_2水平可能起因于深水透气性减弱，这与全年南极海冰覆盖或冬季覆盖与夏季冰引起的分层相关。作为南极地区空气-海洋气体交换减少的结果，我们的模型重现了观察到的冰期-间冰期 CO_2差异中的 67%，并大体上与另外的观测约束条件一致。

58. Blunier T, Brook E J. Timing of millennial-scale climate change in Antarctica and Greenland during the last glacial period[J]. Science, 2001, 291(5501): 109-112.

Timing of millennial-scale climate change in Antarctica and Greenland during the last glacial period

Blunier T., Brook E. J.

(Department of Geology and Program in Environmental Science, Washington State University, 14204 Northeast Salmon Creek Avenue, Vancouver, WA 98686, USA)

Abstract: A precise relative chronology for Greenland and West Antarctic paleotemperature is extended to 90000 years ago, based on correlation of atmospheric methane records from the Greenland Ice Sheet Project 2 and Byrd ice cores. Over this period, the onset of seven major millennial-scale warmings in Antarctica preceded the onset of Greenland warmings by 1500 to 3000 years. In general, Antarctic temperatures increased gradually while Greenland temperatures were decreasing or constant, and the termination of Antarctic warming was apparently coincident with the onset of rapid warming in Greenland. This pattern provides further evidence for the operation of a "bipolar see-saw" in air temperatures and an oceanic teleconnection between the hemispheres on millennial time scales.

末次冰期南极和格陵兰千年尺度气候变化的定年

Blunier T., Brook E. J.

(美国华盛顿州立大学地质和环境科学规划系)

摘要:基于格陵兰冰盖计划 2 和 Byrd 冰芯大气甲烷记录的相互关系,我们将格陵兰和西南极古温度精确的相对年表延伸到 90000 年前。在这一时期内,南极 7 个千年尺度增温的开端先于格陵兰的升温开端 1500~3000 年。总体上在格陵兰温度下降或持平时南极温度逐渐上升,南极升温的终止显然与格陵兰快速升温的开端一致。这一模式为气温的"两极跷跷板"的运转方式以及千年尺度上两半球间海洋的遥相关提供了进一步的证据。

59. Holland D M. Explaining the Weddell Polynya:a large ocean eddy shed at Maud Rise[J]. Science, 2001,292(5522):1697-1700.

Explaining the Weddell Polynya:a large ocean eddy shed at Maud Rise

Holland D. M.

(Center for Atmosphere-Ocean Science, Courant Institute of Mathematical Sciences and Faculty of Arts and Science, New York University, New York, NY 10012, USA)

Abstract: Satellite observations have shown the occasional occurrence of a large opening in the sea-ice cover of the Weddell Sea, Antarctica, a phenomenon known as the Weddell Polynya. The transient appearance, position, size, and shape of the polynya is explained here by a mechanism by which modest variations in the large-scale oceanic flow past the Maud Rise seamount cause a horizontal cyclonic eddy to be shed from its northeast flank. The shed eddy transmits a divergent Ekman stress into the sea ice, leading to a crescent-shaped opening in the pack. Atmospheric thermodynamical interaction further enhances the opening by inducing oceanic convection. A sea-ice-ocean computer model simulation vividly demonstrates how this mechanism fully accounts for the characteristics that mark Weddell Polynya events.

解释 Weddell 冰间湖：Maud 高地产生的巨大海洋涡流

Holland D. M.

(美国纽约大学 Courant 数学科学研究所和艺术与科学学院大气-海洋科学中心)

摘要：卫星观测已经显示了南极 Weddell 海海冰覆盖中不时出现的巨大开口，这一现象被称为 Weddell 冰间湖。这里我们提出一个机制解释冰间湖的短暂出现、位置、规模和形状，这一机制中流经 Maud 高地海岭的大尺度洋流的适度变化引起它的东北侧产生水平气旋涡流。该涡流向海冰传送发散的 Ekman 胁强，引起大块浮冰中的新月形开口。大气热力学交互作用通过诱导海洋对流进一步增强了开口。海冰-海洋计算机模型生动描述了这一机制并充分解释了标志 Weddell 冰间湖事件的特征。

60. Hulbe C L. Glaciology：how ice sheets flow[J]. Science，2001，294(5550)：2300-2301.

Glaciology：how ice sheets flow

Hulbe C. L.

(Department of Geology, Portland State University, Portland, OR 97207, USA)

Abstract: Melt waters at the bases of glaciers and ice sheets play an important role in their flow behavior. In her perspective, Hulbe highlights the report by Fahnestock et al., who have identified high basal melting rates below a part of the Greenland Ice Sheet as the cause for fast ice sheet flow. They attribute the observation to higher than expected geothermal heat flow. But melt waters are not uniformly distributed underneath the ice. Many regions are frozen to the bedrock. Determining the conditions under which basal melting occurs is important for understanding ice sheet histories and predicting their future behaviors.

冰川学：冰盖是如何流动的

Hulbe C. L.

(美国波特兰州立大学地质系)

摘要：融水在冰川和冰盖的流动行为中发挥着重要作用。在 Hulbe 看来，她突出了法内斯托克等人的观点，他们已经确定格陵兰冰盖快速流动的原因是其底部融化较快。他们将此归因于他们实际观察到的比预期高的地热流量。但融水不是均匀的冰下分布，许多地区是被冻结的基石。确定冰盖底部融化发生的条件是了解冰盖历史和预测其未来的行为的重要研究。

61. Jean-Baptiste P, Petit J R, Lipenkov V Y, et al. Constraints on hydrothermal processes and water exchange in Lake Vostok from helium isotopes[J]. Nature, 2001, 411(6836): 460-462.

Constraints on hydrothermal processes and water exchange in Lake Vostok from helium isotopes

Jean-Baptiste P., Petit J. R., Lipenkov V. Y., Raynaud D., Barkov N. I.

(Laboratoire des Sciences du Climat et de l'Environnement, CEA/CNRS, Centre d'études de Saclay, 91191 Gif sur Yvette, France)

Abstract: Lake Vostok, the largest subglacial lake in Antarctica, is covered by the East Antarctic ice sheet, which varies in thickness between 3750 and 4100 m. At a depth of 3539 m in the drill hole at Vostok station, sharp changes in stable isotopes and the gas content of the ice delineate the boundary between glacier ice and ice accreted through refreezing of lake water. Unlike most gases, helium can be incorporated into the crystal structure of ice during freezing, making helium isotopes in the accreted ice a valuable source of information on lake environment. Here we present helium isotope measurements from the deep section of the Vostok ice core that encompasses the boundary between the glacier ice and accreted ice, showing that the accreted ice is enriched by a helium source with a radiogenic isotope signature typical of an old continental province. This result rules out any significant hydrothermal energy input into the lake from high-enthalpy mantle processes, which would be expected to produce a much higher $^{3}He/^{4}He$ ratio. Based on the average helium flux for continental areas, the helium budget of the lake leads to a renewal time of the lake of the order of 5000 years.

氦同位素揭示的 Vostok 湖热液作用和水交换的约束条件

Jean-Baptiste P., Petit J. R., Lipenkov V. Y., Raynaud D., Barkov N. I.

(法国原子能委员会/国家科学研究院气候与环境科学实验室)

摘要：南极最大的冰下湖 Vostok 湖被东南极冰盖所覆盖，湖上冰盖厚度在 3750～4100 m 之间变化。Vostok 站钻孔深度 3539 m 处冰的稳定同位素和气体含量的突变描绘出冰川冰和通过湖水重新冰冻附着的冰之间的界面。与其他气体不同，氦可以在冻结过程中一体化进入晶体结构，使得附着冰中的氦同位素成为湖环境信息的有价值的来源。这里我们介绍了 Vostok 冰芯包含冰川冰和附着冰界面的深部氦同位素测量，结果显示附着冰富集了带有典型古老大陆区域的放射性同位素信号的氦源。这一结果排除了湖的任何来自高焓地幔过程的热液能量输入，因为这一过程预计将产生高得多的$^{3}He/^{4}He$ 比值。基于大陆地区的平均氦通量，湖的氦收支得出的湖更新时间是 5000 年级别的。

62. Mitrovica J X, Tamisiea M E, Davis J L, et al. Recent mass balance of polar ice sheets inferred from patterns of global sea-level change[J]. Nature, 2001, 409(6823): 1026-1029.

Recent mass balance of polar ice sheets inferred from patterns of global sea-level change

Mitrovica J. X., Tamisiea M. E., Davis J. L., Milne G. A.

(Department of Physics, University of Toronto, 60 St George Street, Toronto, M5S 1A7, Canada)

Abstract: Global sea level is an indicator of climate change, as it is sensitive to both thermal expansion of the oceans and a reduction of land-based glaciers. Global sea-level rise has been estimated by correcting observations from tide gauges for glacial isostatic adjustment—the continuing sea-level response due to melting of Late Pleistocene ice—and by computing the global mean of these residual trends. In such analyses, spatial patterns of sealevel rise are assumed to be signals that will average out over geographically distributed tide-gauge data. But a long history of modelling studies has demonstrated that non-uniform—that is, non-eustatic—sea-level redistributions can be produced by variations in the volume of the polar ice sheets. Here we present numerical predictions of gravitationally consistent patterns of sea-level change following variations in either the Antarctic or Greenland ice sheets or the melting of a suite of small mountain glaciers. These predictions are characterized by geometrically distinct patterns that reconcile spatial variations in previously published sea-level records. Under the-albeit coarse—assumption of a globally uniform thermal expansion of the oceans, our approach suggests melting of the Greenland ice complex over the last century equivalent to similar to 0.6 mm • yr^{-1} of sea-level rise.

根据全球海平面变化模式推断近期极地冰盖的质量平衡

Mitrovica J. X., Tamisiea M. E., Davis J. L., Milne G. A.

(加拿大多伦多大学物理系)

摘要：全球海平面是气候变化的指示器，因为它对海洋热膨胀和地基冰川的减少都敏感。研究者已经通过验潮站对冰川均衡调整的正确观测估计了全球海平面上升——海平面对晚更新世冰融化的持续响应——也计算了这些残留趋势的全球平均。在这些分析中，海平面上升的空间模式被假定为将把验潮站数据的地理学分布平均掉的信号。但是模型研究的长期历史已经证明是非均匀的——也就是非海平面升降的——海平面重新分布可能由极地冰盖体积变化引起。这里我们提出在南极或格陵兰冰盖变化或者一系列小山地冰川融化后的海平面变化的引力一致模式的数字预测。这些预测由调和了先前发表的海平面空间变化的独特的几何模式描述。在这一关于海洋的全球统一热膨胀的假设下——虽然依然粗糙——我们的提议显示整个20世纪格陵兰冰联合体的融化相当于约0.6 mm • yr^{-1}的海平面上升。

63. Naish T R, Woolfe K J, Barrett P J, et al. Orbitally induced oscillations in the East Antarctic ice sheet at the Oligocene/Miocene boundary[J]. Nature, 2001, 413(6857): 719-723.

Orbitally induced oscillations in the East Antarctic ice sheet at the Oligocene/Miocene boundary

Naish T. R. , Woolfe K. J. , Barrett P. J. , Wilson G. S. , Atkins C. , Bohaty S. M. , Bucker C. J. , Claps M. , Davey F. J. , Dunbar G. B. , Dunn A. G. , Fielding C. R. , Florindo F. , Hannah M. J. , Harwood D. M. , Henrys S. A. , Krissek L. A. , Lavelle M. , van der Meer J. , McIntosh W. C. , Niessen F. , Passchier S. , Powell R. D. , Roberts A. P. , Sagnotti L. , Scherer R. P. , Strong C. P. , Talarico F. , Verosub K. L. , Villa G. , Watkins D. K. , Webb P. N. , Wonik T.

(Institute of Geological and Nuclear Sciences, PO Box 30368, LowerHutt, New Zealand)

Abstract: Between 34 and 15 million years (Myr) ago, when planetary temperatures were 3-4 ℃ warmer than at present and atmospheric CO_2 concentrations were twice as high as today, the Antarctic ice sheets may have been unstable. Oxygen isotope records from deep-sea sediment cores suggest that during this time fluctuations in global temperatures and high-latitude continental ice volumes were influenced by orbital cycles. But it has hitherto not been possible to calibrate the inferred changes in ice volume with direct evidence for oscillations of the Antarctic ice sheets. Here we present sediment data from shallow marine cores in the western Ross Sea that exhibit well dated cyclic variations, and which link the extent of the East Antarctic ice sheet directly to orbital cycles during the Oligocene/Miocene transition (24. 1-23. 7 Myr ago). Three rapidly deposited glaci-marine sequences are constrained to a period of less than 450 kyr by our age model, suggesting that orbital influences at the frequencies of obliquity (40 kyr) and eccentricity (125 kyr) controlled the oscillations of the ice margin at that time. An erosional hiatus covering 250 kyr provides direct evidence for a major episode of global cooling and ice-sheet expansion about 23. 7 Myr ago, which had previously been inferred from oxygen isotope data (Mil event).

渐新世/中新世分界处东南极冰盖轨道成因的震荡变化

Naish T. R. , Woolfe K. J. , Barrett P. J. , Wilson G. S. , Atkins C. , Bohaty S. M. , Bucker C. J. , Claps M. , Davey F. J. , Dunbar G. B. , Dunn A. G. , Fielding C. R. , Florindo F. , Hannah M. J. , Harwood D. M. , Henrys S. A. , Krissek L. A. , Lavelle M. , van der Meer J. , McIntosh W. C. , Niessen F. , Passchier S. , Powell R. D. , Roberts A. P. , Sagnotti L. , Scherer R. P. , Strong C. P. , Talarico F. , Verosub K. L. , Villa G. , Watkins D. K. , Webb P. N. , Wonik T.

(新西兰地质和核科学研究所)

摘要:在 3400 万年和 1500 万年(Myr)前,当地球温度比现在暖 3～4 ℃且大气 CO_2 浓度是今天的两倍时,南极冰盖可能是不稳定的。深海沉积柱的氧同位素记录显示这段时间全球温度的波动和高纬大陆冰量受到轨道循环的影响。但迄今为止仍未能用南极冰盖震荡的直接证据校准推断的冰量变化。这里我们提出了来自 Ross 海西部的浅海沉积柱数据。该数据展示了良好定年的循环变化,并与东南极冰盖范围和渐新世/中新世分界处(24. 1～23. 7 Myr 前)的轨道循环直接相关。通过我们的年龄模型,3 个快速沉积冰海序列被限制在少于 450 千年的时间内,显示来自地轴倾角(40 千年)和偏心率(125 千年)周期的轨道影响控制了当时的冰缘震荡。一条覆盖250千年的侵蚀裂缝提供了约23.7 Myr前全球变冷和冰盖扩张的主要时间段的直接证据，这也是之前从氧同位素数据中推断出的(Mil事件)。

64. Rempel A W, Waddington E D, Wettlaufer J S, et al. Possible displacement of the climate signal in ancient ice by premelting and anomalous diffusion[J]. Nature, 2001, 411(6837): 568-571.

Possible displacement of the climate signal in ancient ice by premelting and anomalous diffusion

Rempel A. W., Waddington E. D., Wettlaufer J. S., Worster M. G.

(Applied Physics Laboratory, University of Washington, Box 355640, Seattle, Washington 98105, USA)

Abstract: The best high-resolution records of climate over the past few hundred millennia are derived from ice cores retrieved from Greenland and Antarctica. The interpretation of these records relies on the assumption that the trace constituents used as proxies for past climate have undergone only modest post-depositional migration. Many of the constituents are soluble impurities found principally in unfrozen liquid that separates the grain boundaries in ice sheets. This phase behaviour, termed premelting, is characteristic of polycrystalline material. Here we show that premelting influences compositional diffusion in a manner that causes the advection of impurity anomalies towards warmer regions while maintaining their spatial integrity. Notwithstanding chemical reactions that might fix certain species against this prevailing transport, we find that—under conditions that resemble those encountered in the Eemian interglacial ice of central Greenland (from about 125000 to 115000 years ago)—impurity fluctuations may be separated from ice of the same age by as much as 50 cm. This distance is comparable to the ice thickness of the contested sudden cooling events in Eemian ice from the GRIP core.

古冰中由于预融和不规则扩散而导致气候信号可能的位错

Rempel A. W., Waddington E. D., Wettlaufer J. S., Worster M. G.

(美国华盛顿大学西雅图分校应用物理实验室)

摘要：过去数十万年间最好的高分辨率气候记录是得自格陵兰和南极采集的冰芯。对这些记录的解释要依靠过去气候代用性指标的痕量成分仅仅经历了有限的沉积后迁移这一假设。很多成分是主要存在于分散在冰盖颗粒边界中的未冻结液体中的可溶性杂质。这一被称为预融化的相行为是多晶物质的特征。这里我们展示了预融化通过引起杂质畸形物向更暖区域的水平对流并同时保持它们的空间完整性的手段来影响成分扩散。尽管化学反应可能在这一普遍传输中固定某些种类的成分，我们发现——在像格陵兰中部Eemian冰(从约125000年到115000年前)中遇到的条件下——杂质波动可能与同年龄的冰分离多达50 cm。这个距离相当于有争议的GRIP冰芯Eemian冰中突然变冷事件的冰厚度。

65. Sachs J P, Anderson R F, Lehman S J. Glacial surface temperatures of the southeast Atlantic Ocean [J]. Science, 2001, 293(5537): 2077-2079.

Glacial surface temperatures of the southeast Atlantic Ocean

Sachs J. P., Anderson R. F., Lehman S. J.

(Department of Earth, Atmospheric and Planetary Sciences, Massachusetts Institute of Technology, 77 Massachusetts Avenue, Room E34-254, Cambridge, MA 02139, USA)

Abstract: A detailed record of sea surface temperature from sediments of the Cape Basin in the subtropical South Atlantic indicates a previously undocumented progression of marine climate change between 41 and 18 thousand years before the present (ky B. P.), during the last glacial period. Whereas marine records typically indicate a long-term cooling into the Last Glacial Maximum (around 21 ky B. P.) consistent with gradually increasing global ice volume, the Cape Basin record documents an interval of substantial temperate ocean warming from 41 to 25 ky B. P. The pattern is similar to that expected in response to changes in insolation owing to variations in Earth's tilt.

东南大西洋冰期的表面温度

Sachs J. P., Anderson R. F., Lehman S. J.

(美国麻省理工学院地球大气行星与科学系)

摘要:得自亚热带南大西洋的 Cape 海盆沉积柱的海表面温度详细记录显示末次冰期中距今 41 千年和 18 千年(ky B. P.)之间有以前未发现的海洋气候变化过程。然而典型的海洋记录显示末次冰盛期(21 ky B. P. 左右)是长期变冷同时全球冰量逐渐上升,而 Cape 海盆记录了从 41～25 ky B. P. 的一段实质性的温和海洋变暖。这一模式相似于期望的对地轴倾角变化引起的日射能量变化的响应。

66. Shepherd A, Wingham D J, Mansley J A D, et al. Inland thinning of Pine Island Glacier, West Antarctica[J]. Science, 2001, 291(5505): 862-864.

Inland thinning of Pine Island Glacier, West Antarctica

Shepherd A., Wingham D. J., Mansley J. A. D., Corr H. F. J.

(Centre for Polar Observation & Modelling, Department of Space and Climate Physics, University College London, 17 Gordon Street, London, WC1H0AH, UK)

Abstract: The Pine island Glacier (PIG) transports 69 cubic kilometers of ice each year from similar to 10% of the West Antarctic Ice Sheet (WAIS). It is possible that a retreat of the PIG may accelerate ice discharge from the WAIS interior. Satellite altimetry and interferometry show that the grounded PIG thinned by up to 1.6 meters per year between 1992 and 1999, affecting 150 kilometers of the inland glacier. The thinning cannot be explained by short-term variability in accumulation and must result from glacier dynamics.

西南极松岛冰川的内陆变薄

Shepherd A., Wingham D. J., Mansley J. A. D., Corr H. F. J.

(英国伦敦大学空间和气候物理系极地观测与模拟中心)

摘要:松岛冰川(PIG)每年运送69立方千米的冰,相当于西南极冰盖(WAIS)的10%。PIG的退却可能加速WAIS内部冰的流失。卫星高度测量与干涉测量显示1992~1999年间的PIG每年变薄高达1.6米,影响了150千米的内陆冰川。这一变薄不能用积累率的短期变化解释,应当是冰川动力学的结果。

67. Siegert M J, Ellis-Evans J C, Tranter M, et al. Physical, chemical and biological processes in Lake Vostok and other Antarctic subglacial lakes[J]. Nature, 2001, 414(6864): 603-609.

Physical, chemical and biological processes in Lake Vostok and other Antarctic subglacial lakes

Siegert M. J., Ellis-Evans J. C., Tranter M., Mayer C., Petit J. R., Salamatin A., Priscu J. C.

(Bristol Glaciology Centre, School of Geographical Sciences, University of Bristol, Bristol BS8 1SS, UK)

Abstract: Over 70 lakes have now been identified beneath the Antarctic ice sheet. Although water from none of the lakes has been sampled directly, analysis of lake ice frozen (accreted) to the underside of the ice sheet above Lake Vostok, the largest of these lakes, has allowed inferences to be made on lake water chemistry and has revealed small quantities of microbes. These findings suggest that Lake Vostok is an extreme, yet viable, environment for life. All subglacial lakes are subject to high pressure (similar to 350 atmospheres), low temperatures (about −3 ℃) and permanent darkness. Any microbes present must therefore use chemical sources to power biological processes. Importantly, dissolved oxygen is available at least at the lake surface, from equilibration with air hydrates released from melting basal glacier ice. Microbes found in Lake Vostok's accreted ice are relatively modern, but the probability of ancient lake-floor sediments leads to a possibility of a very old biota at the base of subglacial lakes.

Vostok湖和其他南极冰下湖的物理、化学和生物过程

Siegert M. J., Ellis-Evans J. C., Tranter M., Mayer C., Petit J. R., Salamatin A., Priscu J. C.

(英国布里斯托尔大学地理科学学院布里斯托尔冰川中心)

摘要:在南极冰盖下已经识别出超过70个湖泊。虽然还没有从湖中直接采得水样,对冻结(附着)在Vostok湖上方冰盖下侧的湖冰的分析使得推测湖水化学成分成为可能,并显示了少量微生物。这些发现显示Vostok湖对生命来说是极端的但仍能存活的环境。所有冰下湖都受到高压(约350个大气压)、

低温(约－3 ℃)和永久黑暗的影响。因此任何出现的微生物都必须用化学能源来为生物过程提供能量。重要的是,溶解氧至少在湖表面是可通过与融化的冰川底部冰中的空气水合物平衡得到的。在 Vostok 湖附着冰中发现的微生物相对现代,但古湖底沉积物存在的可能性导致了冰下湖底部存在非常古老的生物群的可能性。

68. Zachos J C, Shackleton N J, Revenaugh J S, et al. Climate response to orbital forcing across the Oligocene-Miocene boundary[J]. Science, 2001, 292(5515): 274-278.

Climate response to orbital forcing across the Oligocene-Miocene boundary

Zachos J. C., Shackleton N. J., Revenaugh J. S., Palike H., Flower B. P.

(Earth Sciences Department, Center for Dynamics and Evolution of the Land-Sea Interface, University of California, Santa Cruz, CA 95064, USA)

Abstract: Spectral analyses of an uninterrupted 5.5-million-year (My)-long chronology of late Oligocene-early Miocene climate and ocean carbon chemistry from two deep-sea cores recovered in the western equatorial Atlantic reveal variance concentrated at all Milankovitch frequencies. Exceptional spectral power in climate is recorded at the 406-thousand-year (ky) period eccentricity band over a 3.4-million-year period [20 to 23.4 My ago (Ma)] as well as in the 125 ky and 95 ky bands over a 1.3-million-year period (21.7 to 23.0 Ma) of suspected Low greenhouse gas Levels. Moreover, a major transient glaciation at the epoch boundary (similar to 23 Ma), Mi-1, corresponds with a rare orbital congruence involving obliquity and eccentricity. The anomaly, which consists of low-amplitude variance in obliquity (a node) and a minimum in eccentricity, results in an extended period (similar to 200 ky) of low seasonality orbits favorable to ice-sheet expansion on Antarctica.

渐新世-中新世分界处对轨道强迫的气候响应

Zachos J. C., Shackleton N. J., Revenaugh J. S., Palike H., Flower B. P.

(美国加利福尼亚大学圣克鲁斯分校地球科学系海陆界面动力学与演化中心)

摘要:对两根采自西赤道大西洋的晚渐新世-早中新世年龄长达 5.5 百万年(My)深海沉积柱中气候和海洋碳化学的光谱分析显示变化集中在所有米兰科维奇频率上。异常的光谱强度在一个 3.4 百万年的时期[20 到 23.4 百万年前(Ma)]记录了 406 千年(ky)的偏心率周期,还有一个 1.3 百万年时期(21.7～23 Ma)的令人怀疑的低温室气体水平的 125 ky 和 95 ky 周期。此外,世代界限处(约 23 Ma)的一个主要短暂冰川期 Mi-1 对应一种罕见的地轴倾角和偏心率的轨道一致性。这一由地轴倾角低幅变化(轨道交点)和偏心率最小值构成的异常导致有利南极冰盖扩张的长期(约 200 ky)的低季节性轨道。

69. Alley R B. On thickening ice? [J]. Science, 2002, 295(5554): 451-452.

On thickening ice?

Alley R. B.

(EMS Environment Institute and the Department of Geosciences, Pennsylvania State University, University Park, PA 16802, USA)

Abstract: The large ice sheets in Greenland and Antarctica play an important role in global climate. Much attention has focused on the West Antarctic Ice Sheet. In his perspective, Alley highlights the report by Joughin et al., who show that in the region of the West Antarctic drainage into the Ross Sea, on average, the ice sheet is thickening slowly. Earlier studies indicated a net thinning in this region. The results may indicate the stabilization or readvance of the ice sheet, but the system is complex and existing data are insufficient to identify a trend.

不断变厚的冰盖?

Alley R. B.

(美国宾夕法尼亚州立大学 EMS 环境研究所与地球科学系)

摘要:格陵兰和南极的大范围冰盖在地球气候中起着重要作用,人们对西南极冰盖给予很高的关注。Alley 在文中强调了 Joughin 等人的报道,展示了在西南极流入罗斯海流域,平均来说冰盖正在缓慢变厚,而早期研究显示该区域是净变薄。该结果可能表明冰盖的稳定化或再前进,但该系统是复杂的且现有数据不足以确定这一趋势。

70. Bell R E, Studinger M, Tikku A A, et al. Origin and fate of Lake Vostok water frozen to the base of the East Antarctic ice sheet[J]. Nature, 2002, 416(6878): 307-310.

Origin and fate of Lake Vostok water frozen to the base of the East Antarctic ice sheet

Bell R. E., Studinger M., Tikku A. A., Clarke G. K. C., Gutner M. M., Meertens C.

(Lamont-Doherty Earth Observatory, Columbia University, Palisades, New York 10964, USA)

Abstract: The subglacial Lake Vostok may be a unique reservoir of genetic material and it may contain organisms with distinct adaptations, but it has yet to be explored directly. The lake and the overlying ice sheet are closely linked, as the ice-sheet thickness drives the lake circulation, while melting and freezing at the ice-sheet base will control the flux of water, biota and sediment through the lake. Here we present a reconstruction of the ice flow trajectories for the Vostok core site, using ice-penetrating radar data and Global Positioning System (GPS) measurements of surface ice velocity. We find that the ice sheet has a significant along-lake flow component, persistent since the Last Glacial Maximum. The rates at which ice is frozen (accreted) to the base of the ice sheet are greatest at the shorelines, and the accreted ice layer is subsequently transported out of the lake. Using these new flow field and velocity measurements, we estimate the time for ice to traverse Lake Vostok to be 16000-20000 years. We infer that most Vostok ice analysed to date was accreted to the ice sheet close to the western shoreline, and is therefore not representative of open lake conditions. From the amount of accreted lake water we estimate to be exported along the southern shoreline, the lake water residence time is about 13300 years.

冻结在东南极冰盖底部的 Vostok 湖水的来源与去向

Bell R. E., Studinger M., Tikku A. A., Clarke G. K. C., Gutner M. M., Meertens C.

(美国哥伦比亚大学拉蒙特-多尔蒂地质观测站)

摘要:Vostok 冰下湖可能是独特的遗传物质储库并可能保有具备独特适应性的生物体,但它还没有被直接探测过。湖和上覆的冰盖联系紧密,因为冰盖厚度驱动了湖的循环,同时冰盖底部的融解和冻结控制了湖的水通量、生物群和沉积物。这里我们利用冰雷达数据和表面冰流速的全球定位系统(GPS)测量重建了 Vostok 冰芯处的冰流轨迹。我们发现冰盖有一个重要的沿湖流分量从末次冰盛期持续至今。湖岸线处冰冻结(附着)在冰盖底部的速率最大,并且附着的冰层随后被输送到湖外。应用这些新的流场和速度测量,我们估计冰穿过 Vostok 湖的时间是 16000~20000 年。我们推断迄今为止分析的大多数 Vostok 冰是附着在西湖岸线附近的冰盖上的,因此不能代表典型的开放湖泊环境。根据我们估计的南部湖岸线沿线输出的附着湖水量,湖水的停留时间约为 13300 年。

71. Clark P U, Mitrovica J X, Milne G A, et al. Sea-level fingerprinting as a direct test for the source of global meltwater pulse IA[J]. Science, 2002, 295(5564): 2438-2441.

Sea-level fingerprinting as a direct test for the source of global meltwater pulse IA

Clark P. U., Mitrovica J. X., Milne G. A., Tamisiea M. E.

(Department of Geosciences, Oregon State University, Corvallis, OR 97331, USA)

Abstract: The ice reservoir that served as the source for the meltwater pulse IA remains enigmatic and controversial. We show that each of the melting scenarios that have been proposed for the event produces a distinct variation, or fingerprint, in the global distribution of meltwater. We compare sea-level fingerprints associated with various melting scenarios to existing sea-level records from Barbados and the Sunda Shelf and conclude that the southern Laurentide Ice Sheet could not have been the sole source of the meltwater pulse, whereas a substantial contribution from the Antarctic Ice Sheet is consistent with these records.

海平面变化作为直接检测全球冰融水量 IA 突变来源的证据

Clark P. U., Mitrovica J. X., Milne G. A., Tamisiea M. E.

(美国俄勒冈州立大学地球科学系)

摘要:作为融水脉冲 IA 来源的冰储库依然难以确认并存在争议。我们指出为这一事件提出的每个融化的情况都会在全球融水分布中产生独特的变化，也就是指纹。我们把与多种融化场景相关的海平面指纹与现有的Barbados和Sunda大陆架的海平面记录作了对比，推断Laurentide冰架南部不可能是融水脉动的唯一来源，然而南极冰盖对其有显著贡献的观点与这些记录吻合。

72. Conway H, Catania G, Raymond C F, et al. Switch of flow direction in an Antarctic ice stream[J]. Nature, 2002, 419(6906): 465-467.

Switch of flow direction in an Antarctic ice stream

Conway H., Catania G., Raymond C. F., Gades A. M., Scambos T. A., Engelhardt H.

(Earth and Space Sciences, University of Washington, Seattle, Washington 98195, USA)

Abstract: Fast-flowing ice streams transport ice from the interior of West Antarctica to the ocean, and fluctuations in their activity control the mass balance of the ice sheet. The mass balance of the Ross Sea sector of the West Antarctic ice sheet is now positive—that is, it is growing—mainly because one of the ice streams (ice stream C) slowed down about 150 years ago. Here we present evidence from both surface measurements and remote sensing that demonstrates the highly dynamic nature of the Ross drainage system. We show that the flow in an area that once discharged into ice stream C has changed direction, now draining into the Whillans ice stream (formerly ice stream B). This switch in flow direction is a result of continuing thinning of the Whillans ice stream and recent thickening of ice stream C. Further abrupt reorganization of the activity and configuration of the ice streams over short timescales is to be expected in the future as the surface topography of the ice sheet responds to the combined effects of internal dynamics and long-term climate change. We suggest that caution is needed when using observations of short-term mass changes to draw conclusions about the large-scale mass balance of the ice sheet.

一条南极冰流的流向转变

Conway H., Catania G., Raymond C. F., Gades A. M., Scambos T. A., Engelhardt H.

(美国华盛顿大学西雅图分校地球与空间科学系)

摘要：快速流动的冰流把冰从西南极内部输送到海洋，并且它们的活动性波动控制了冰盖的质量平衡。西南极冰盖的Ross海地区目前处于正的质量平衡——也就是说，它在增长——主要因为其中一条冰流(冰流C)约150年前变慢了。这里我们提供了地表测量和遥感两方面的证据，都能证明Ross水系的高动态性。我们指出某一曾经流入冰流C的区域已经改变方向，现在流入Whillans冰流(以前的冰流B)。这一流向的转变是Whillans冰流的持续变薄和冰流C最近变厚的结果。未来短时期内冰流活动性和构造的进一步突然重组是可能的，因为冰盖的表面地形要对内部动力学和长期气候变化的联合影响作出响应。我们建议当利用短期质量变化观测为冰盖大尺度质量平衡下结论时需要谨慎。

73. Dierssen H M, Smith R C, Vernet M. Glacial meltwater dynamics in coastal waters west of the Antarctic peninsula[J]. PNAS, 2002, 99(4): 1790-1795.

Glacial meltwater dynamics in coastal waters west of the Antarctic peninsula

Dierssen H. M., Smith R. C., Vernet M.

(Moss Landing Marine Laboratories, California State University, Moss Landing, CA 95039)

Abstract: The annual advance and retreat of sea ice has been considered a major physical determinant of spatial and temporal changes in the structure of the Antarctic coastal marine ecosystem. However, the role of glacial meltwater on the hydrography of the Antarctic Peninsula ecosystem has been largely ignored, and the resulting biological effects have only been considered within a few kilometers from shore. Through several lines of evidence collected in conjunction with the Palmer Station Long-Term Ecological Research Project, we show that the freshening and warming of the coastal surface water over the summer months is influenced not solely by sea ice melt, as suggested by the literature, but largely by the influx of glacial meltwater. Moreover, the seasonal variability in the amount and extent of the glacial meltwater plume plays a critical role in the functioning of the biota by influencing the physical dynamics of the water (e. g., water column stratification, nearshore turbidity). From nearly a decade of observations (1991-1999), the presence of surface meltwater is correlated not only to phytoplankton blooms nearshore, but spatially over 100 km off shore. The amount of meltwater will also have important secondary effects on the ecosystem by influencing the timing of sea ice formation. Because air temperatures are statistically increasing along the Antarctic Peninsula region, the presence of glacial meltwater is likely to become more prevalent in these surface waters and continue to play an ever-increasing role in driving this fragile ecosystem.

南极半岛西部近岸水域的冰川融水动力学

Dierssen H. M., Smith R. C., Vernet M.

(美国加利福尼亚州立大学莫斯兰丁海洋实验室)

摘要:年度的海冰进退被认为是南极沿海海洋生态系统构造的时间和空间改变的主要自然决定因素。然而,冰川融水在南极半岛生态系统的水温地理学中的角色很大程度上被忽视了,并且由其导致的生物学影响被认为仅限于距海岸数千米以内。通过在与 Palmer 站长期生态研究计划合作中收集的若干条证据,我们指出沿海表面水在夏季月份中的淡化和变暖不是如同在文献中提出的受到海冰融化的单独影响,而是很大程度上受到冰川融水流入的影响。冰川融水水团的数量和范围的季节性变化通过影响水的物理动力学(例如水柱层化、近岸水域浊度)在生物区机能中起到关键作用。通过近十年的观测(1991～1999),表面融水的出现不仅与近岸水域浮游植物暴发相关,而且空间上影响到离岸 100 km 以上。融水量通过影响海冰形成的时间也将对生态系统有重要的次生影响。因为环南极半岛地区统计学上气温在上升,冰川融水的出现可能在这些表面水中变得更普遍,并在驱动这一脆弱的生态系统中继续发挥日益重要的作用。

74. Joughin I, Tulaczyk S. Positive mass balance of the Ross Ice Streams, West Antarctica[J]. Science, 2002,295(5554):476-480.

Positive mass balance of the Ross Ice Streams, West Antarctica

Joughin I., Tulaczyk S.

(Jet Propulsion Laboratory, California Institute of Technology, Mailstop 300-235, 4800 Oak Grove Drive, Pasadena, CA 91109, USA)

Abstract: We have used ice-flow velocity measurements from synthetic aperture radar to reassess the mass balance of the Ross Ice Streams, West Antarctica. We find strong evidence for ice-sheet growth (+26.8 gigatons per year), in contrast to earlier estimates indicating a mass deficit (−20.9 gigatons per year). Average thickening is equal to similar to 25% of the accumulation rate, with most of this growth occurring on Ice Stream C. Whillans Ice Stream, which was thought to have a significantly negative mass balance, is close to balance, reflecting its continuing slowdown. The overall positive mass balance may signal an end to the Holocene retreat of these ice streams.

西南极 Ross 冰流正向变化的质量平衡

Joughin I., Tulaczyk S.

(美国加州理工学院喷气推进实验室)

摘要:我们利用合成孔径雷达的冰流速度测量来重新评估西南极 Ross 冰流的质量平衡。与早先估计的质量赤字(每年-2.09×10^{11}吨)相反,我们发现了冰盖增长(每年$+2.68\times10^{11}$吨)的强有力的证据。平均增厚约为积累率的 25%,这一增长多数发生在冰流 C。曾被认为有显著负物质平衡的 Whillans 冰流现在接近平衡,反映了它的持续减速。总体的正物质平衡可能是这些冰流全新世退却结束的信号。

75. Morgan V, Delmotte M, van Ommen T, et al. Relative timing of deglacial climate events in Antarctica and Greenland[J]. Science, 2002, 297(5588): 1862-1864.

Relative timing of deglacial climate events in Antarctica and Greenland

Morgan V., Delmotte M., van Ommen T., Jouzel J., Chappellaz J., Woon S., Masson-Delmotte V., Raynaud D.

(Antarctic Cooperative Research Centre and Australian Antarctic Division, GPO Box 252-80, Hobart, Tasmania, Australia)

Abstract: The last deglaciation was marked by large, hemispheric, millennial-scale climate variations: the Bolling-Allerod and Younger Dryas periods in the north, and the Antarctic Cold Reversal in the south. A chronology from the high-accumulation Law Dome East Antarctic ice core constrains the relative timing of these two events and provides strong evidence that the cooling at the start of the Antarctic Cold Reversal did not follow the abrupt warming during the northern Bolling transition around 14500 years ago. This result suggests that southern changes are not a direct response to abrupt changes in North Atlantic thermohaline circulation, as is assumed in the conventional picture of a hemispheric temperature seesa.

南极和格陵兰冰消期气候事件的相对时间

Morgan V., Delmotte M., van Ommen T., Jouzel J., Chappellaz J., Woon S., Masson-Delmotte V., Raynaud D.

(澳大利亚南极合作研究中心和澳大利亚南极局)

摘要:末次冰消伴随着显著的千年尺度半球大气候变化:北方的 Bolling-Allerod 和新仙女木时期,以及南方的南极冷反转。高积累率的东南极冰穹 Law 冰芯的年表限定了这两个事件的相对时间,并提供了强有力的证据,表明南极冷反转开端的变冷不是跟随着大约 14500 年前北方的 Bolling 转变的突然变暖。这一结果显示南部的变化并不是对北大西洋温盐环流突然变化的直接响应,与半球温度跷跷板常规描述所假定的不同。

76. Price P B, Nagornov O V, Bay R, et al. Temperature profile for glacial ice at the South Pole: implications for life in a nearby subglacial lake[J]. PNAS, 2002, 99(12): 7844-7847.

Temperature profile for glacial ice at the South Pole: implications for life in a nearby subglacial lake

Price P. B., Nagornov O. V., Bay R., Chirkin D., He Y. D., Miocinovic P., Richards A., Woschnagg K., Koci B., Zagorodnov V.

(Physics Department, University of California, Berkeley, CA 94720)

Abstract: Airborne radar has detected approximate to 100 lakes under the Antarctic ice cap, the largest of which is Lake Vostok. International planning is underway to search in Lake Vostok for microbial life that may have evolved in isolation from surface life for millions of years. It is thought, however, that the lakes may be hydraulically interconnected. If so, unsterile drilling would contaminate not just one but many of them. Here we report measurements of temperature vs. depth down to 2345 m in ice at the South Pole, within 10 km from a subglacial lake seen by airborne radar profiling. We infer a temperature at the 2810 m deep base of the South Pole ice and at the lake of −9 ℃, which is 7 ℃ below the pressure-induced melting temperature of freshwater ice. To produce the strong radar signal, the frozen lake must consist of a mix of sediment and ice in a flat bed, formed before permanent Antarctic glaciation. it may, like Siberian and Antarctic permafrost, be rich in microbial life. Because of its hydraulic isolation, proximity to South Pole Station infrastructure, and analog to a Martian polar cap, it is an ideal place to test a sterile drill before risking contamination of Lake Vostok. From the semiempirical expression for strain rate vs. shear stress, we estimate shear vs. depth and show that the IceCube neutrino observatory will be able to map the three-dimensional ice-flow field within a larger volume (0.5 km^3) and at lower temperatures (−20 ℃ to −35 ℃) than has heretofore been possible.

南极点冰川冰的温度剖面:附近冰下湖生命的推论

Price P. B., Nagornov O. V., Bay R., Chirkin D., He Y. D., Miocinovic P., Richards A., Woschnagg K., Koci B., Zagorodnov V.

(美国加利福尼亚大学伯克利分校物理系)

摘要:机载雷达探测到了南极冰帽下约100个湖泊,其中最大的是Vostok湖。调查Vostok湖中可能与表面生物隔绝进化了数百万年的微生物生命的国际计划正在进行中。然而有人认为这些湖泊可能是水力学上相互连接的。若如此,未消毒的钻探可能不仅污染其中一个,而会污染很多个湖泊。这里我们报告了南极点冰下2345 m温度相对于深度的测量结果,距离机载雷达剖面上发现的一个冰下湖不到10 km。我们推断南极点冰底部2810 m深和湖中的温度为−9 ℃,比压力导致的淡水冰融化温度低7 ℃。冻结的湖泊必须由在南极永久冰盖之前形成的沉积物和冰的混合物在平坦的底部上组成,才能产生强烈的雷达信号。它可能像西伯利亚和南极永久冻土带一样富含微生物生命。因为水力隔绝,接近南极点站设施以及与火星极地冰帽类似,它是在冒污染Vostok湖的风险之前消毒钻探的理想测试地点。通过应变率对剪应力的半经验公式,我们估计了切边对深度的关系,并指出冰块中微子天文台能够比迄今为止可能的描绘更大体积(0.5 km^3)和更低温度(−20 ℃到−35 ℃)的三维冰流场。

77. Raymond C F. Ice sheets on the move[J]. Science, 2002, 298(5601): 2147-2148.

Ice sheets on the move

Raymond C. F.

(Department of Earth and Space Sciences, University of Washington, Seattle, WA 98195, USA)

Abstract: Are today's polar ice sheets in danger of collapse? Even with climate warming, increases in snowfall should compensate for additional melting of the ice sheets, but ice flow speeds can change abruptly by orders of magnitude. In his perspective, Raymond reviews recent advances toward understanding the dynamics of ice sheets. Spatial connections in basal lubrication help to explain how speed changes can spread over large areas from localized triggers. But it remains unclear how quickly flow acceleration can be initiated at low elevation and spread deep into the interiors of ice sheets.

冰盖在移动

Raymond C. F.

(美国华盛顿大学西雅图分校地球与空间科学系)

摘要:今天的极地冰盖有崩塌的危险么?即使随着气候变暖,降雪的增长会补偿额外的冰盖融化,但冰流速能仍突然改变了几个数量级。Raymond在文中回顾了最近对冰盖机制理解的进展。基部润滑作用的空间联系能帮助解释速度变化怎样由局部触发到大面积传播。但快速流体加速如何在低海拔发起且传播深入冰盖内部依然未知。

78. Rignot E,Jacobs S S. Rapid bottom melting widespread near Antarctic ice sheet grounding lines[J]. Science,2002,296(5575):2020-2023.

Rapid bottom melting widespread near Antarctic ice sheet grounding lines

Rignot E., Jacobs S. S.

(Jet Propulsion Laboratory, California Institute of Technology, Mail Stop 300-235, Pasadena, CA 91109-8099, USA)

Abstract: As continental ice from Antarctica reaches the grounding line and begins to float, its underside melts into the ocean. Results obtained with satellite radar interferometry reveal that bottom melt rates experienced by large outlet glaciers near their grounding lines are far higher than generally assumed. The melting rate is positively correlated with thermal forcing, increasing by 1 meter per year for each 0.1 ℃ rise in ocean temperature. Where deep water has direct access to grounding lines, glaciers and ice shelves are vulnerable to ongoing increases in ocean temperature.

南极冰盖接地线附近普遍的快速底部融化

Rignot E., Jacobs S. S.

(美国加州理工学院喷气推进实验室等)

摘要:当南极大陆冰川到达接地线并开始漂浮时,它的下侧开始融化到海里。卫星雷达干涉测量得到的结果显示大的冰川入海口经受的底部融化速率远高于通常认为的。融化速率与热强迫正相关,海洋温度每上升0.1℃,融化速率每年提高1 m。在深部水直接接触接地线的区域,冰川和冰架容易受到正在上升的海洋温度的影响。

79. Ackert R P. An ice sheet remembers[J]. Science,2003,299(5603):57-58.

An ice sheet remembers

Ackert R. P.

(Department of Marine Chemistry and Geochemistry, Woods Hole Oceanographic Institution, Woods Hole, MA 02543, USA)

Abstract: Are we witnessing the early stages of the collapse of the West Antarctic Ice Sheet? In his Perspective, Ackert explains that to answer this question, we must study the behavior of the ice sheet not just in recent decades but over thousands of years. He highlights the report by Stone et al., who document slow but steady thinning of the ice sheet over the last 9300 years, punctuated by more rapid thinning as the ice margin approaches. The record is consistent with earlier suggestions that the ice sheet is still responding to changes in climate and sea level that occurred thousands of years ago, making it difficult to predict its response to present climate warming.

冰盖会记忆

Ackert R. P.

(美国伍兹霍尔海洋研究所海洋化学与地球化学部)

摘要：我们正在见证西南极冰盖崩塌的初期阶段么？Ackert 在文中解释了这一问题的答案。我们必须研究最近几十年甚至超过千年冰盖的行为。他强调了 Stone 等人报道的冰盖在过去 9300 年间在慢速但稳定地变薄，夹杂着例如冰缘接触之类的更快速变薄的过程。该记录与早期的观点一致即冰盖仍然响应着几千年前就发生的海平面和气候变化，使得预测冰盖对现代气候变暖的响应变得困难。

80. Bindschadler R A, King M A, Alley R B, et al. Tidally controlled stick-slip discharge of a West Antarctic ice stream[J]. Science, 2003, 301(5636): 1087-1089.

Tidally controlled stick-slip discharge of a West Antarctic ice stream

Bindschadler R. A., King M. A., Alley R. B., Anandakrishnan S., Padman L.

(Ocean and Ice Branch, NASA Goddard Space Flight Center, Greenbelt, MD 20771, USA)

Abstract: A major West Antarctic ice stream discharges by sudden and brief periods of very rapid motion paced by oceanic tidal oscillations of about 1 meter. Acceleration to speeds greater than 1 meter per hour and deceleration back to a stationary state occur in minutes or less. Slip propagates at approximately 88 meters per second, suggestive of a shear wave traveling within the subglacial till. A model of an episodically slipping friction-locked fault reproduces the observed quasi-periodic event timing, demonstrating an ice stream's ability to change speed rapidly and its extreme sensitivity to subglacial conditions and variations in sea level.

潮汐控制的一条西南极冰流的黏滑流出

Bindschadler R. A., King M. A., Alley R. B., Anandakrishnan S., Padman L.

(美国国家航空和航天局戈达德宇宙飞行中心海洋与冰分局)

摘要：一条主要的西南极冰流突发而短暂地流出，这一移动是受到海洋潮汐震荡(大约 1 m)控制的。加速到超过 1 m/h 的速度，然后在几分钟或者更短时间内减速到静止状态。滑动以约 88 m/s 的速度传播，显示有横波在底碛中传播。短暂滑动的摩擦锁定断层模型重现了观察到的准周期事件时序，显示了冰流快速改变速度的能力和它对冰下条件和海平面变化的极端敏感性。

81. Caillon N, Severinghaus J P, Jouzel J, et al. Timing of atmospheric CO_2 and Antarctic temperature changes across termination Ⅲ[J]. Science, 2003, 299(5613): 1728-1731.

Timing of atmospheric CO_2 and Antarctic temperature changes across termination Ⅲ

Caillon N., Severinghaus J. P., Jouzel J., Barnola J. M., Kang J. C., Lipenkov V. Y.

(Institute Pierre Simon Laplace/Laboratoire des Sciences du Climat et de l'Environnement, Commissariat à l'Energie Atomique/CNRS, L'Orme des Merisiers, CEA Saclay, 91191, Gif sur Yvette, France)

Abstract: The analysis of air bubbles from ice cores has yielded a precise record of atmospheric greenhouse gas concentrations, but the timing of changes in these gases with respect to temperature is not accurately known because of uncertainty in the gas age-ice age difference. We have measured the isotopic composition of argon in air bubbles in the Vostok core during Termination Ⅲ (similar to 240000 years before the present). This record most likely reflects the temperature and accumulation change, although the mechanism remains unclear. The sequence of events during Termination Ⅲ suggests that the CO_2 increase tagged Antarctic deglacial warming by 800 +/−200 years and preceded the Northern Hemisphere deglaciation.

在终止Ⅲ期间的大气 CO_2 与南极温度变化的时间

Caillon N., Severinghaus J. P., Jouzel J., Barnola J. M., Kang J. C., Lipenkov V. Y.

(法国原子能委员会,国家科学研究院气候与环境科学实验室,皮埃尔西蒙拉普拉斯研究所)

摘要:对冰芯中气泡的分析已经取得大气温室气体浓度的精确记录,但因为气体年龄-冰年龄差异的不确定性,这些气体随温度变化的时间还没有准确的认识。我们测量了 Vostok 冰芯中在终止Ⅲ(距今约 240000 年前)期间气泡的氩同位素组成。虽然机制仍不明确,但这一记录最可能反映了温度和积累率变化。终止Ⅲ期间的事件序列显示 CO_2 上升以 800±200 年滞后于南极冰消变暖并超前于北半球冰消。

82. Curran M A J, Van Ommen T D, Morgan V I, et al. Ice core evidence for Antarctic sea ice decline since the 1950s[J]. Science, 2003, 302(5648): 1203-1206.

Ice core evidence for Antarctic sea ice decline since the 1950s

Curran M. A. J., Van Ommen T. D., Morgan V. I., Phillips K. L., Palmer A. S.

(Department of the Environment and Heritage, Australian Antarctic Division, and Antarctic Climate and Ecosystem Cooperative Research Centre, Private Bag 80, Hobart, Tasmania 7001, Australia)

Abstract: The instrumental record of Antarctic sea ice in recent decades does not reveal a clear signature of warming despite observational evidence from coastal Antarctica. Here we report a significant correlation ($P<0.002$) between methanesulphonic acid (MSA) concentrations from a Law Dome ice core and 22 years of satellite-derived sea ice extent (SIE) for the 80°E to 140°E sector. Applying this instrumental calibration to longer term MSA data (1841 to 1995 A. D.) suggests that there has been a 20% decline in SIE since about 1950. The decline is not uniform, showing large cyclical variations, with periods of about 11 years, that confuse trend detection over the relatively short satellite era.

20世纪50年代至今南极海冰消退的冰芯证据

Curran M. A. J., Van Ommen T. D., Morgan V. I., Phillips K. L., Palmer A. S.

(澳大利亚环境遗产部，澳大利亚南极局，南极气候和生态系统合作研究中心)

摘要：尽管有南极洲沿海的证据，最近几十年南极海冰的器测记录没有提供变暖的清晰信号。这里我们报告了一支冰穹Law冰芯中甲基磺酸(MSA)浓度和卫星获得的22年80°E到140°E地区海冰范围(SIE)之间的显著相关($P<0.002$)。将这一仪器校准应用到更长期的MSA数据(1841～1995年)显示从1950年左右至今SIE衰退了20%。这一衰退并不均匀，显示了大的周期性变化，周期约为11年，这使得在相对较短的卫星年代发现趋势变得不清楚。

83. De Angelis H, Skvarca P. Glacier surge after ice shelf collapse[J]. Science, 2003, 299(5612): 1560-1562.

Glacier surge after ice shelf collapse

De Angelis H., Skvarca P.

(Instituto Antártico Argentino, Cerrito 1248, C1010AAZ Buenos Aires, Argentina)

Abstract: The possibility that the West Antarctic Ice Sheet will collapse as a consequence of ice shelf disintegration has been debated for many years. This matter is of concern because such an event would imply a sudden increase in sea level. Evidence is presented here showing drastic dynamic perturbations on former tributary glaciers that fed sections of the Larsen Ice Shelf on the Antarctic Peninsula before its collapse in 1995. Satellite images and airborne surveys allowed unambiguous identification of active surging phases of Boydell, Sjogren, Edgeworth, Bombardier, and Drygalski glaciers. This discovery calls for a reconsideration of former hypotheses about the stabilizing role of ice shelves.

冰架崩塌后的冰川涌动

De Angelis H., Skvarca P.

(阿根廷南极研究所)

摘要：作为冰架瓦解的结果，西南极冰盖崩溃的可能性已经争论多年。这一事件引起关注是因为它意味着海平面的突然上升。这里提供了1995年南极半岛Larsen冰架崩塌前注入它的一部分以前的支流冰川极端动力学扰动的证据。卫星图像和机载观测使得Boydell、Sjogren、Edgeworth、Bombardier和Drygalski冰川活动涌动状态的明确鉴定。这一发现表明需重新审议之前关于冰架稳定作用的假说。

84. De Conto R M, Pollard D. Rapid Cenozoic glaciation of Antarctica induced by declining atmospheric

CO_2[J]. Nature,2003,421(6920):245-249.

Rapid Cenozoic glaciation of Antarctica induced by declining atmospheric CO_2

De Conto R. M. , Pollard D.

(Department of Geosciences, University of Massachusetts, Amherst, Massachusetts 01003, USA)

Abstract: The sudden, widespread glaciation of Antarctica and the associated shift towards colder temperatures at the Eocene/Oligocene boundary (similar to 34 million years ago) is one of the most fundamental reorganizations of global climate known in the geologic record. The glaciation of Antarctica has hitherto been thought to result from the tectonic opening of Southern Ocean gateways, which enabled the formation of the Antarctic Circumpolar Current and the subsequent thermal isolation of the Antarctic continent. Here we simulate the glacial inception and early growth of the East Antarctic Ice Sheet using a general circulation model with coupled components for atmosphere, ocean, ice sheet and sediment, and which incorporates palaeogeography, greenhouse gas, changing orbital parameters, and varying ocean heat transport. In our model, declining Cenozoic CO_2 first leads to the formation of small, highly dynamic ice caps on high Antarctic plateaux. At a later time, a CO_2 threshold is crossed, initiating ice-sheet height/mass-balance feedbacks that cause the ice caps to expand rapidly with large orbital variations, eventually coalescing into a continental-scale East Antarctic Ice Sheet. According to our simulation the opening of Southern Ocean gateways plays a secondary role in this transition, relative to CO_2 concentration.

大气 CO_2 下降导致的新生代南极快速冰川作用

De Conto R. M. , Pollard D.

(美国马萨诸塞大学地球科学系)

摘要:在始新世/渐新世分界处(约 3400 万年前)南极突然广泛的冰川作用和相关的向更冷温度的转变是地质记录中全球气候最根本的重组之一。迄今为止南极冰川作用被认为是南大洋门户构造开放的结果,这一结果使得南极绕极流和南极大陆的热力学隔离成为可能。这里我们利用全面环流模型模拟了冰期起始和东南极冰盖的早期发育,这一模型带有大气、海洋、冰盖和沉积物的耦合组件并结合了古地理、温室气体、变化的轨道参量和变化的海洋热传输。在我们的模型中,新生代 CO_2 下降首先导致南极高原高海拔地区高动态性的小冰帽的形成。在稍后的时间里,CO_2 超过了阈值,触发了冰盖高度/质量平衡反馈,引起冰帽在大的轨道变化的同时快速扩张,最终合并为大陆尺度的东南极冰盖。依据我们的模拟,相对于 CO_2 浓度,南大洋门户开放在这个转变中起次要作用。

85. Doran P T, Fritsen C H, McKay C P, et al. Formation and character of an ancient 19 m ice cover and underlying trapped brine in an "ice-sealed" east Antarctic lake[J]. PNAS, 2003, 100(1): 26-31.

Formation and character of an ancient 19 m ice cover and underlying trapped brine in an "ice-sealed" east Antarctic lake

Doran P. T., Fritsen C. H., McKay C. P., Priscu J. C., Adams E. E.

(Department of Earth and Environmental Sciences, University of Illinois, 845 West Taylor Street, MS 186, Chicago, IL 60607)

Abstract: Lake Vida, one of the largest lakes in the McMurdo Dry Valleys of Antarctica, was previously believed to be shallow (<10 m) and frozen to its bed year-round. New ice-core analysis and temperature data show that beneath 19 m of ice is a water column composed of a NaCl brine with a salinity seven times that of seawater that remains liquid below −10 ℃. The ice cover thickens at both its base and surface, sealing concentrated brine beneath. The ice cover is stabilized by a negative feedback between ice growth and the freezing-point depression of the brine. The ice cover contains frozen microbial mats throughout that are viable after thawing and has a history that extends to at least 2800 ^{14}C years B. P., suggesting that the brine has been isolated from the atmosphere for as long. To our knowledge, Lake Vida has the thickest subaerial lake ice cover recorded and may represent a previously undiscovered end-member lacustrine ecosystem on Earth.

"冰密封"的东南极湖中 19 m 的覆冰和其下截留的卤水的形成与特征

Doran P. T., Fritsen C. H., McKay C. P., Priscu J. C., Adams E. E.

(美国伊利诺斯大学地球与环境科学系)

摘要：南极 McMurdo 干谷中最大的湖泊之一 Vida 湖曾被认为是终年连底冻的浅湖(<10 m)。新的冰芯分析和温度数据显示在 19 m 的冰下是 NaCl 盐水构成的水体，其盐度七倍于在−10 ℃以下仍保持液态的海水。覆冰在底部和表面都变厚，将浓缩的盐水密封在其下。覆冰因冰增长和盐水的冰点降低之间的负反馈而保持稳定。覆冰到处都含有冻结的菌垫，在融化后能萌发生存并有延伸至距今至少 2800 年的 ^{14}C 历史，显示盐水已经与大气隔离了这么久。据我们所知，Vida 湖拥有有记录的最厚的陆上湖覆冰，可能呈现了一个之前地球上未发现的湖泊生态系统端元。

86. Kaiser J. Glaciology: Warmer ocean could threaten Antarctic Ice Shelves[J]. Science, 2003, 302 (5646): 759-759.

Glaciology—Warmer ocean could threaten Antarctic Ice Shelves

Kaiser J.

(News)

Abstract: When two chunks of ice the size of a small country broke off the Antarctic Peninsula's Larsen Ice Shelf in 1995 and 2002, experts scrambled to figure out how it had happened. The pat answer, global warming, was too simple: some parts of Antarctica are cooling, and the soaring air temperatures along the peninsula that seemed to have triggered the collapse have not yet been convincingly linked to a worldwide pattern. But as a model for what could happen elsewhere in Antarctica as temperatures rise, it was crucial to understand the Larsen Ice Shelf's demise.

冰川学——更暖的海洋可能威胁南极冰架

Kaiser J.

(新闻)

摘要:当两块大如一个小国家的冰盖分别在1995年和2002年从南极半岛拉森冰架崩离时,专家们争论着这是如何发生的。过去的答案是全球变暖,这过于简单了,因为南极洲某些地方正在变凉;沿着半岛的气温猛增似乎引发了崩塌,仍未能令人信服地与全球模式联系起来。但正如一个研究在南极其他地方可能发生类似温度上升事件的模型,其对于理解拉森冰盖消亡是至关重要的。

87. Pahnke K, Zahn R, Elderfield H, et al. 340000 years centennial-scale marine record of Southern Hemisphere climatic oscillation[J]. Science, 2003, 301(5635): 948-952.

340000 years centennial-scale marine record of Southern Hemisphere climatic oscillation

Pahnke K., Zahn R., Elderfield H., Schulz M.

(School of Earth, Ocean and Planetary Sciences, Cardiff University, Park Place, Cardiff CF10 3YE, UK)

Abstract: In order to investigate rapid climatic changes at mid-southern latitudes, we have developed centennial-scale paleoceanographic records from the southwest Pacific that enable detailed comparison with Antarctic ice core records. These records suggest close coupling of mid-southern latitudes with Antarctic climate during deglacial and interglacial periods. Glacial sections display higher variability than is seen in Antarctic ice cores, which implies climatic decoupling between mid and high southern latitudes due to enhanced circum-Antarctic circulation. Structural and temporal similarity with the Greenland ice core record is evident in glacial sections and suggests a degree of interhemispheric synchroneity not predicted from bipolar ice core correlations.

340000年的南半球气候百年尺度震荡的海洋记录

Pahnke K., Zahn R., Elderfield H., Schulz M.

(英国加的夫大学地球海洋与行星科学学院)

摘要:为了研究南半球中纬的快速气候变化,我们在西南太平洋开展了百年尺度的古海洋学记录研究,使其与南极冰芯记录的详细比较成为可能。这些记录显示了冰期和间冰期南半球中纬和南极气候的紧密耦合。冰期部分显示了比南极冰芯中看出的更高的变化性,意味着由于绕南极流增强而退耦的南半球中纬和高纬气候。冰期部分与格陵兰冰芯记录在结构上和时间上相似性明显,显示了在两极冰芯相关中没有被预言的一定程度上的两半球同步性。

88. Shepherd A, Wingham D, Payne T, et al. Larsen ice shelf has progressively thinned[J]. Science, 2003, 302(5646): 856-859.

Larsen ice shelf has progressively thinned

Shepherd A., Wingham D., Payne T., Skvarca P.

(Centre for Polar Observation and Modelling, Scott Polar Research Institute, University of Cambridge, Cambridge CB2 1ER, UK)

Abstract: The retreat and collapse of Antarctic Peninsula ice shelves in tandem with a regional atmospheric warming has fueled speculation as to how these events may be related. Satellite radar altimeter measurements show that between 1992 and 2001 the Larsen Ice Shelf lowered by up to 0.27 ± 0.11 meters per year. The lowering is explained by increased summer melt-water and the loss of basal ice through melting. Enhanced ocean-driven melting may provide a simple link between regional climate warming and the successive disintegration of sections of the Larsen Ice Shelf.

Larsen冰架已经逐渐变薄

Shepherd A., Wingham D., Payne T., Skvarca P.

(英国剑桥大学斯科特极地研究所极地观测与模拟中心)

摘要:南极半岛冰架的依次退却和崩溃连带区域大气变暖激起了这些事件可能如何联系的思考。卫星雷达高度计测量显示1992～2001年间Larsen冰架每年降低了最多0.27±0.11 m。这一降低用夏季融水增长和融化导致的底部冰损耗来解释。海洋驱动的融化增强可能解释区域气候变暖和Larsen冰盖各部分连续瓦解之间的联系。

89. Sowers T, Alley R B, Jubenville J. Ice core records of atmospheric N_2O covering the last 106000 years[J]. Science, 2003, 301(5635): 945-948.

Ice core records of atmospheric N_2O covering the last 106000 years

Sowers T., Alley R. B., Jubenville J.

(Department of Geosciences and the EMS Environment Institute, Pennsylvania State University, University Park, PA 16802, USA)

Abstract: Paleoatmospheric records of trace-gas concentrations recovered from ice cores provide important sources of information on many biogeochemical cycles involving carbon, nitrogen, and oxygen. Here, we present a 106000 year record of atmospheric nitrous oxide (N_2O) along with corresponding isotopic records spanning the last 30000 years, which together suggest minimal changes in the ratio of marine to terrestrial N_2O production. During the last glacial termination, both marine and oceanic N_2O emissions increased by (40±8)%. We speculate that our records do not support those hypotheses that invoke enhanced export production to explain low carbon dioxide values during glacial periods.

过去 106000 年的大气 N_2O 冰芯记录

Sowers T., Alley R. B., Jubenville J.

(美国宾夕法尼亚州立大学地球科学系与 EMS 环境研究所)

摘要:从冰芯中恢复的古大气痕量气体浓度的记录提供了很多有关生物地球化学循环的信息来源,包括碳、氮和氧循环。这里我们提出了 106000 年的大气一氧化二氮(N_2O)记录和最近 30000 年相关的同位素记录,两者联系起来显示海洋和陆地产生 N_2O 的极小的变化。在末次冰期结束时海和洋的 N_2O 排放都增加了(40±8)%。我们推测我们的记录不支持那些借助输出产量增加来解释冰期二氧化碳低值的假说。

90. Stone J O, Balco G A, Sugden D E, et al. Holocene deglaciation of Marie Byrd Land, West Antarctica [J]. Science, 2003, 299(5603): 99-102.

Holocene deglaciation of Marie Byrd Land, West Antarctica

Stone J. O., Balco G. A., Sugden D. E., Caffee M. W., Sass L. C., Cowdery S. G., Siddoway C.

(Quaternary Research Center and Department of Earth and Space Sciences, Box 351360, University of Washington, Seattle, WA 98195-1360, USA)

Abstract: Surface exposure ages of glacial deposits in the Ford Ranges of western Marie Byrd Land indicate continuous thinning of the West Antarctic Ice Sheet by more than 700 meters near the coast throughout the past 10000 years. Deglaciation lagged the disappearance of ice sheets in the Northern Hemisphere by thousands of years and may still be under way. These results provide further evidence that parts of the West Antarctic Ice Sheet are on a long-term trajectory of decline. West Antarctic melting contributed water to the oceans in the late Holocene and may continue to do so in the future.

西南极 Marie Byrd 地的全新世冰消

Stone J. O., Balco G. A., Sugden D. E., Caffee M. W., Sass L. C., Cowdery S. G., Siddoway C.

(美国华盛顿大学西雅图分校第四纪研究中心与地球和空间科学系)

摘要:Marie Byrd 地西部的 Ford 山脉冰川沉积物的表面暴露年龄显示了最近 10000 年海岸附近的西南极冰盖连续变薄了超过 700 m。冰消作用滞后于北半球冰盖消失数千年并可能仍在进行中。这一结果提供了部分西南极冰盖处于长期衰退轨迹的进一步证据。西南极融化在全新世晚期向海洋贡献水,并可能在将来继续这样做。

91. Wolff E W. Whither Antarctic sea ice? [J]. Science,2003,302(5648):1164.

Whither Antarctic sea ice?

Wolff E. W.

(British Antarctic Survey, High Cross, Madingley Road, Cambridge CB3 0ET, UK)

Abstract: To an observer from space, the growth and retreat of the white ring of sea ice around Antarctica would be one of the clearest signs that Earth's climate varies on seasonal and longer time scales. However, the record of satellite observations is too short to assess whether there have been climate-related trends in sea ice extent.

南极海冰到哪里去?

Wolff E. W.

(英国南极调查局)

摘要:对于一个太空中的观察者,环绕南极洲海冰的白色圆环的消长是季节和更长时间尺度地球气候变化的最清晰标志之一。然而,卫星观测记录对评价是否有气候相关的海冰范围变化趋势来说就太短了。

92. Augustin L,Barbante C,Barnes P R F,et al. Eight glacial cycles from an Antarctic ice core[J]. Nature,2004,429(6992):623-628.

Eight glacial cycles from an Antarctic ice core

Augustin L., Barbante C., Barnes P. R. F., Barnola J. M., Bigler M., Castellano E., Cattani O., Chappellaz J., DahlJensen D., Delmonte B., Dreyfus G., Durand G., Falourd S., Fischer H., Fluckiger J., Hansson M. E., Huybrechts P., Jugie R., Johnsen S. J., Jouzel J., Kaufmann P., Kipfstuhl J., Lambert F., Lipenkov V. Y., Littot G. V. C., Longinelli A., Lorrain R., Maggi V., Masson-Delmotte V., Miller H., Mulvaney R., Oerlemans J., Oerter H., Orombelli G., Parrenin F., Peel D. A., Petit J. R., Raynaud D., Ritz C., Ruth U., Schwander J., Siegenthaler U., Souchez R., Stauffer B., Steffensen J. P., Stenni B., Stocker T. F., Tabacco I. E., Udisti R., van de Wal R. S. W., van den Broeke M., Weiss J., Wilhelms F., Winther J. G., Wolff E. W., Zucchelli M., EPICA Community Members

(Laboratoire de Glaciologie et Géophysique de l'Environnement (CNRS), BP 96, 38402 St Martin d'Hères Cedex, France)

Abstract: The Antarctic Vostok ice core provided compelling evidence of the nature of climate, and of climate feedbacks, over the past 420000 years. Marine records suggest that the amplitude of climate variability was smaller before that time, but such records are often poorly resolved. Moreover, it is not possible to infer the abundance of greenhouse gases in the atmosphere from marine records. Here we report the recovery of a deep ice core from Dome C, Antarctica, that provides a climate record for the past 740000 years. For the four most recent glacial cycles, the data agree well with the record from Vostok. The earlier period, between 740000 and 430000 years ago, was characterized by less pronounced warmth in interglacial periods in Antarctica, but a higher proportion of each cycle was spent in the warm mode. The transition from glacial to interglacial conditions about 430000 years ago (Termination Ⅴ) resembles the transition into the present interglacial period in terms of the magnitude of change in temperatures and greenhouse gases, but there are significant differences in the patterns of change. The interglacial stage following Termination Ⅴ was exceptionally long 28000 years compared to, for example, the 12000 years recorded so far in the present interglacial period. Given the similarities between this earlier warm period and today, our results may imply that without human intervention, a climate similar to the present one would extend well into the future.

一支南极冰芯中的八次冰期旋回

Augustin L., Barbante C., Barnes P. R. F., Barnola J. M., Bigler M., Castellano E., Cattani O., Chappellaz J., DahlJensen D., Delmonte B., Dreyfus G., Durand G., Falourd S., Fischer H., Fluckiger J., Hansson M. E., Huybrechts P., Jugie R., Johnsen S. J., Jouzel J., Kaufmann P., Kipfstuhl J., Lambert F., Lipenkov V. Y., Littot G. V. C., Longinelli A., Lorrain R., Maggi V., Masson-Delmotte V., Miller H., Mulvaney R., Oerlemans J., Oerter H., Orombelli G., Parrenin F., Peel D. A., Petit J. R., Raynaud D., Ritz C., Ruth U., Schwander J., Siegenthaler U., Souchez R., Stauffer B., Steffensen J. P., Stenni B., Stocker T. F., Tabacco I. E., Udisti R., van de Wal R. S. W., van den Broeke M., Weiss J., Wilhelms F., Winther J. G., Wolff E. W., Zucchelli M., EPICA Community Members

(法国国家科学研究院冰川与地球物理环境实验室)

摘要:南极 Vostok 冰芯提供了过去 420000 年气候的性质与反馈的令人信服的证据。海洋记录显示在那一时间之前气候变化幅度较小,但这种记录常常分辨率很差。此外,海洋记录中不可能推断出大气温室气体的丰度。这里我们报告了南极 Dome C 一支深冰芯中恢复的过去 740000 年的气候记录。关于最近的四个冰期旋回的数据与 Vostok 记录的数据吻合得很好。在 740000 年前与 430000 年前之间的更早时期,南极间冰期的特点是不太明显的温暖,但每个旋回有更长的时间处于暖期。约 430000 年前(终止Ⅴ)从冰期到间冰期转变的条件在温度和温室气体变化幅度上类似于向当前间冰期的转变,但在变化模式上有显著差异。在终止Ⅴ之后的 28000 年的间冰期阶段与其他间冰期相比较来说格外长,例如当前间冰期迄今记录到的 12000 年。鉴于这个更早暖期与当今之间的相似性,我们的结果可能意味着在没有人类干预的情况下,类似当前的气候会充分延伸到未来。

93. Bay R C, Bramall N, Price P B. Bipolar correlation of volcanism with millennial climate change[J]. PNAS, 2004, 101(17): 6341-6345.

Bipolar correlation of volcanism with millennial climate change

Bay R. C., Bramall N., Price P. B.

(Physics Department, University of California, Berkeley, CA 94720)

Abstract: Analyzing data from our optical dust logger, we find that volcanic ash layers from the Siple Dome (Antarctica) borehole are simultaneous (with >99% rejection of the null hypothesis) with the onset of millennium-timescale cooling recorded at Greenland Ice Sheet Project 2 (GISP2; Greenland). These data are the best evidence yet for a causal connection between volcanism and millennial climate change and lead to possibilities of a direct causal relationship. Evidence has been accumulating for decades that volcanic eruptions can perturb climate and possibly affect it on long timescales and that volcanism may respond to climate change. If rapid climate change can induce volcanism, this result could be further evidence of a southern-lead North-South climate asynchrony. Alternatively, a volcanic-forcing viewpoint is of particular interest because of the high correlation and relative timing of the events, and it may involve a scenario in which volcanic ash and sulfate abruptly increase the soluble iron in large surface areas of the nutrient-limited Southern Ocean, stimulate growth of phytoplankton, which enhance volcanic effects on planetary albedo and the global carbon cycle, and trigger northern millennial cooling. Large global temperature swings could be limited by feedback within the volcano-climate system.

火山活动与千年尺度气候变化的两极相关性

Bay R. C., Bramall N., Price P. B.

(美国加利福尼亚大学伯克利分校物理系)

摘要:通过分析我们的光学尘埃记录仪的数据,我们发现来自Siple冰穹(南极)的火山灰层与格陵兰冰盖计划2(GISP2,格陵兰)记录的千年时间尺度变冷的开端具有同时性(>99%拒绝零假设)。这些数据是火山活动与千年尺度气候变化之间因果关系迄今最好的证据,并引出了直接因果关系的可能性。火山爆发能引起气候扰动并可能在长时间尺度上影响它,并且火山活动可能响应气候变化,相关的证据已经积累了几十年。如果快速气候变化能诱发火山活动,这一结果可能成为南方超前的南北气候异步的进一步证据。另一方面,因为这些事件的高相关性和相对时间,火山强迫的观点受到特别关注,并且它可能包括火山灰和硫酸盐增加贫营养的南大洋的巨大表面区域的溶解铁,刺激浮游植物的生长,加强火山活动对行星反照率和全球碳循环的影响,并触发北部千年尺度的变冷。在火山-气候系统的反馈下全球温度的大幅度波动可能被限制。

94. Rohling E J, Marsh R, Wells N C, et al. Similar meltwater contributions to glacial sea level changes from Antarctic and northern ice sheets[J]. Nature, 2004, 430(7003): 1016-1021.

Similar meltwater contributions to glacial sea level changes from Antarctic and northern ice sheets

Rohling E. J. , Marsh R. , Wells N. C. , Siddall M. , Edwards N. R.

(Southampton Oceanography Centre, Southampton SO14 3ZH, UK)

Abstract: The period between 75000 and 20000 years ago was characterized by high variability in climate and sea level. Southern Ocean records of ice-rafted debris suggest a significant contribution to the sea level changes from melt water of Antarctic origin, in addition to likely contributions from northern ice sheets, but the relative volumes of melt water from northern and southern sources have yet to be established. Here we simulate the first-order impact of a range of relative meltwater releases from the two polar regions on the distribution of marine oxygen isotopes, using an intermediate complexity model. By comparing our simulations with oxygen isotope data from sediment cores, we infer that the contributions from Antarctica and the northern ice sheets to the documented sea level rises between 65000 and 35000 years ago were approximately equal, each accounting for a rise of about 15 m. The reductions in Antarctic ice volume implied by our analysis are comparable to that inferred previously for the Antarctic contribution to meltwater pulse 1A, which occurred about 14200 years ago, during the last deglaciation.

南极和北部冰盖对冰期海平面变化有相似的融水贡献

Rohling E. J. , Marsh R. , Wells N. C. , Siddall M. , Edwards N. R.

(英国南安普顿海洋中心)

摘要:距今 75000 年和 20000 年前之间的时期的特点是气候和海平面的高变化性。除了来自北部冰盖的可能贡献,南大洋的冰筏残屑记录显示了南极来源的融水对海平面变化的显著贡献,但来自北部和南部来源的融水的相对体积尚未建立。这里我们利用中等复杂度的模型模拟了来自两个极地地区的相对融水排放在一定范围内对海洋氧同位素分布的一阶影响。通过对比我们的模拟和来自沉积柱的氧同位素数据,我们推断南极和北部冰盖对 65000 年前和 35000 年前之间有记录的海平面上升的贡献大致相当,各贡献了大约 15 m 的上升。包含在我们分析中的南极冰量缩减与之前推断的出现在末次冰期期间约 14200 年前的融水脉动 1A 的南极贡献相当。

95. Alley R B, Clark P U, Huybrechts P, et al. Ice-sheet and sea-level changes[J]. Science, 2005, 310 (5747): 456-460.

Ice-sheet and sea-level changes

Alley R. B. , Clark P. U. , Huybrechts P. , Joughin I.

(Department of Geosciences and Earth and Environmental Systems Institute, Pennsylvania State University, Deike Building, University Park, PA 16802, USA)

Abstract: Future sea-level rise is an important issue related to the continuing build up of atmospheric greenhouse gas concentrations. The Greenland and Antarctic ice sheets, with the potential to raise sea level similar to 70 meters if completely melted, dominate uncertainties in projected sea-level change. Freshwater fluxes from these ice sheets also may affect oceanic circulation, contributing to climate change. Observational and modeling advances have reduced many uncertainties related to ice-sheet behavior, but recently detected, rapid ice-marginal changes contributing to sea-level rise may indicate greater ice-sheet sensitivity to warming than previously considered.

冰盖和海平面变化

Alley R. B. , Clark P. U. , Huybrechts P. , Joughin I.

（美国宾夕法尼亚州立大学地球科学系与地球和环境系统研究所）

摘要：未来的海平面上升是与持续增加的大气温室气体浓度相关的重要问题。格陵兰和南极冰盖如果完全融化将有提高海平面约 70 m 的潜在可能，它们主宰了预测海平面变化的不确定性。来自这些冰盖的淡水冲击也可能影响洋流，从而对气候变化作出贡献。观测和建模的进步已经降低了许多与冰盖行为相关的不确定性，但最近发现的对海平面上升有贡献的快速冰缘变化可能意味着冰盖有对变暖比以前认为的更强烈的敏感性。

96. Cook A J, Fox A J, Vaughan D G, et al. Retreating glacier fronts on the Antarctic Peninsula over the past half-century[J]. Science, 2005, 308(5721): 541-544.

Retreating glacier fronts on the Antarctic Peninsula over the past half-century

Cook A. J. , Fox A. J. , Vaughan D. G. , Ferrigno J. G.

（British Antarctic Survey, Natural Environment Research Council, Madingley Road, Cambridge CB3 0ET, UK）

Abstract: The continued retreat of ice shelves on the Antarctic Peninsula has been widely attributed to recent atmospheric warming, but there is little published work describing changes in glacier margin positions. We present trends in 244 marine glacier fronts on the peninsula and associated islands over the past 61 years. Of these glaciers, 87% have retreated and a clear boundary between mean advance and retreat has migrated progressively southward. The pattern is broadly compatible with retreat driven by atmospheric warming, but the rapidity of the migration suggests that this may not be the sole driver of glacier retreat in this region.

过去半个世纪的南极半岛退却中的冰川前缘

Cook A. J. , Fox A. J. , Vaughan D. G. , Ferrigno J. G.

（英国自然环境研究理事会南极调查局）

摘要：南极半岛持续的冰盖退却被广泛认为是近期大气变暖的结果，但很少有关于冰川边缘位置的工作发表。我们提供了这一半岛上过去 61 年 244 个海洋冰川前缘和相关岛屿的变化趋势。这些冰川中

87%发生了退却，并且平均前进和退却的清晰分界线日益南移。这一模式大致与大气变暖驱动的退却兼容，但冰缘移动的急速显示这可能不是该区域冰川退却的单一驱动力。

97. Davis C H, Li Y H, McConnell J R, et al. Snowfall-driven growth in East Antarctic ice sheet mitigates recent sea-level rise[J]. Science, 2005, 308(5730): 1898-1901.

Snowfall-driven growth in East Antarctic ice sheet mitigates recent sea-level rise

Davis C. H., Li Y. H., McConnell J. R., Frey M. M., Hanna E.

(Department of Electrical and Computer Engineering, University of Missouri-Columbia, Columbia, MO 65211, USA)

Abstract: Satellite radar altimetry measurements indicate that the East Antarctic ice-sheet interior north of 81.6°S increased in mass by 45±7 billion metric tons per year from 1992 to 2003. Comparisons with contemporaneous meteorological model snowfall estimates suggest that the gain in mass was associated with increased precipitation. A gain of this magnitude is enough to slow sea-level rise by 0.12±0.02 millimeters per year.

降雪驱动的东南极冰盖增长对近期海平面上升的减缓作用

Davis C. H., Li Y. H., McConnell J. R., Frey M. M., Hanna E.

(美国密苏里-哥伦比亚大学电气和计算机工程系)

摘要：卫星雷达高度测量显示东南极冰盖内部 81.6°S 以北在 1992 到 2003 年间质量每年增长了 $(45\pm7)\times10^{10}$ 吨。与气象模型的同时期降雪估计对比显示质量的增加与降水增强相关。这一程度的增加足以减缓每年 0.12±0.02 mm 的海平面上升。

98. Domack E, Duran D, Leventer A, et al. Stability of the Larsen B ice shelf on the Antarctic Peninsula during the Holocene epoch[J]. Nature, 2005, 436(7051): 681-685.

Stability of the Larsen B ice shelf on the Antarctic Peninsula during the Holocene epoch

Domack E., Duran D., Leventer A., Ishman S., Doane S., McCallum S., Amblas D., Ring J., Gilbert R., Prentice M.

(Department of Geosciences, Hamilton College, Clinton, New York 13323, USA)

Abstract: The stability of the Antarctic ice shelves in a warming climate has long been discussed, and the recent collapse of a significant part, over 12500 km^2 in area, of the Larsen ice shelf off the Antarctic Peninsula has led to a refocus toward the implications of ice shelf decay for the stability of Antarctica's grounded ice. Some smaller Antarctic ice shelves have undergone periodic growth and decay over the past 11000 yr, but these ice shelves are at the climatic limit of ice shelf viability and are therefore expected to respond rapidly to natural climate variability at century to millennial scales. Here we use records of diatoms, detrital material and geochemical parameters from six marine sediment cores in the vicinity of the Larsen ice shelf to demonstrate that the recent collapse of the Larsen B ice shelf is unprecedented during the Holocene. We infer from our oxygen isotope measurements in planktonic foraminifera that the Larsen B ice shelf has been thinning throughout the Holocene, and we suggest that the recent prolonged period of warming in the Antarctic Peninsula region, in combination with the long-term thinning, has led to collapse of the ice shelf.

全新世时期南极半岛 Larsen B 冰架的稳定性

Domack E., Duran D., Leventer A., Ishman S., Doane S., McCallum S., Amblas D., Ring J., Gilbert R., Prentice M.

(美国汉密尔顿学院地球科学系)

摘要：在变暖气候下的南极冰架的稳定性早有论述，并且最近南极半岛附近 Larsen 冰架一块面积超过 12500 km^2 的重要部分的崩塌重新引起了关于冰架衰退对南极陆缘冰稳定性的意义。一些较小的南极冰架在过去 11000 年内经历了周期性的增长和衰退，但这些冰架处于冰架耐久性在气候上的极限，并因此可以预期它们对百年和千年尺度上自然气候变化的快速响应。这里我们利用来自 Larsen 冰架附近六支海洋沉积柱的硅藻、岩屑物质和地球化学参数记录来证明 Larsen B 冰架最近的崩塌在全新世期间是没有先例的。我们从我们的浮游有孔虫氧同位素测量中推断 Larsen 冰架在整个全新世始终在变薄，并且我们认为南极半岛地区最近长时间的变暖结合长期变薄导致了冰架的崩塌。

99. Hopkin M. Antarctic ice puts climate predictions to the test[J]. Nature, 2005, 438(7068): 536-537.

Antarctic ice puts climate predictions to the test

Hopkin M.

(News)

Abstract: Frozen record of the past reveals models' shortcomings. A record of greenhouse gases spanning the past 650000 years made headlines around the globe last week. The painstaking work proves that levels of carbon dioxide and methane in the atmosphere today massively outstrip those of the pre-industrial era. But it also reveals how little we understand about the way in which these gases influence global climate.

南极冰使气候预测经受检验

Hopkin M.

(新闻)

摘要：过去的冰冻记录揭示了模型的不足。一个横跨过去 65 万年的温室气体记录在上周成为了全球的头条。这一辛苦的工作证明了今日大气中二氧化碳和甲烷的水平大大超过了前工业时代，但其也反映出我们对这些气体如何影响全球气候了解甚少。

100. Siegenthaler U, Stocker T F, Monnin E, et al. Stable carbon cycle-climate relationship during the late Pleistocene[J]. Science, 2005, 310(5752): 1313-1317.

Stable carbon cycle-climate relationship during the late Pleistocene

Siegenthaler U., Stocker T. F., Monnin E., Luthi D., Schwander J., Stauffer B., Raynaud D., Barnola J. M., Fischer H., Masson-Delmotte V., Jouzel J.

(Climate and Environmental Physics, Physics Institute, University of Bern, Sidlerstrasse 5, CH-3012 Bern, Switzerland)

Abstract: A record of atmospheric carbon dioxide (CO_2) concentrations measured on the EPICA (European Project for Ice Coring in Antarctica) Dome Concordia ice core extends the Vostok CO_2 record back to 650000 years before the present (yr B. P.). Before 430000 yr B. P., partial pressure of atmospheric CO_2 lies within the range of 260 and 180 parts per million by volume. This range is almost 30% smaller than that of the last four glacial cycles; however, the apparent sensitivity between deuterium and CO_2 remains stable throughout the six glacial cycles, suggesting that the relationship between CO_2 and Antarctic climate remained rather constant over this interval.

晚更新世期间稳定的碳循环-气候关系

Siegenthaler U., Stocker T. F., Monnin E., Luthi D., Schwander J., Stauffer B., Raynaud D., Barnola J. M., Fischer H., Masson-Delmotte V., Jouzel J.

(瑞士伯尔尼大学物理学院气候与环境物理系)

摘要:得自 EPICA(欧洲南极冰芯计划)Concordia 冰穹冰芯的大气二氧化碳(CO_2)浓度测量记录将 Vostok 的 CO_2 记录延伸到距今 650000 年前(yr B. P.)。在 430000 yr B. P.,大气 CO_2 分压处于百万分之 260 体积与百万分之 180 体积之间。这一范围比最近四个冰期旋回中的几乎小 30%;然而氘和 CO_2 之间的表观灵敏度在六个冰期旋回中始终保持稳定,意味着在这一时期内 CO_2 和南极气候之间的关系保持了相当的稳定性。

101. Spahni R, Chappellaz J, Stocker T F, et al. Atmospheric methane and nitrous oxide of the late Pleistocene from Antarctic ice cores[J]. Science, 2005, 310(5752): 1317-1321.

Atmospheric methane and nitrous oxide of the late Pleistocene from Antarctic ice cores

Spahni R., Chappellaz J., Stocker T. F., Loulergue L., Hausammann G., Kawamura K., Fluckiger J., Schwander J., Raynaud D., Masson-Delmotte V., Jouzel J.

(Climate and Environmental Physics, Physics Institute, University of Bern, Sidlerstrasse 5, CH-3012 Bern, Switzerland)

Abstract: The European Project for Ice Coring in Antarctica Dome C ice core enables us to extend existing records of atmospheric methane (CH_4) and nitrous oxide (N_2O) back to 650000 years before the present. A combined record of CH_4 measured along the Dome C and the Vostok ice cores demonstrates, within the resolution of our measurements, that preindustrial concentrations over Antarctica have not exceeded 773±15 ppbv (parts per billion by volume) during the past 650000 years. Before 420000 years ago, when interglacials were cooler, maximum CH_4 concentrations were only about 600 ppbv, similar to lower Holocene values. In contrast, the N_2O record shows maximum concentrations of 278±7 ppbv, slightly higher than early Holocene values.

来自南极冰芯记录的晚更新世大气甲烷和一氧化二氮

Spahni R., Chappellaz J., Stocker T. F., Loulergue L., Hausammann G., Kawamura K., Fluckiger J., Schwander J., Raynaud D., Masson-Delmotte V., Jouzel J.

(瑞士伯尔尼大学物理学院气候与环境物理系)

摘要:南极冰穹C的欧洲南极冰芯计划使我们能将现有大气甲烷(CH_4)和一氧化二氮(N_2O)记录向前延伸到距今650000年。沿冰穹C和Vostok冰芯测量的CH_4联合记录证明,在我们测量的分辨率以内,过去650000年间工业化前南极的浓度没有超过773±15 ppbv(每十亿分之体积)。距今420000年以前间冰期较冷的时期,CH_4的最大浓度约仅为600 ppbv,与全新世的较低值相近。相反,N_2O记录显示了278±7 ppbv的最大浓度,比早全新世的值略高。

102. Tripati A, Backman J, Elderfield H, et al. Eocene bipolar glaciation associated with global carbon cycle changes[J]. Nature, 2005, 436(7049): 341-346.

Eocene bipolar glaciation associated with global carbon cycle changes

Tripati A., Backman J., Elderfield H., Ferretti P.

(Department of Earth Sciences, University of Cambridge, Downing Street, Cambridge CB2 3EQ, UK)

Abstract: The transition from the extreme global warmth of the early Eocene "greenhouse" climate similar to 55 million years ago to the present glaciated state is one of the most prominent changes in Earth's climatic evolution. It is widely accepted that large ice sheets first appeared on Antarctica similar to 34 million years ago, coincident with decreasing atmospheric carbon dioxide concentrations and a deepening of the calcite compensation depth in the world's oceans, and that glaciation in the Northern Hemisphere began much later, between 10 and 6 million years ago. Here we present records of sediment and foraminiferal geochemistry covering the greenhouse-icehouse climate transition. We report evidence for synchronous deepening and subsequent oscillations in the calcite compensation depth in the tropical Pacific and South Atlantic oceans from similar to 42 million years ago, with a permanent deepening 34 million years ago. The most prominent variations in the calcite compensation depth coincide with changes in seawater oxygen isotope ratios of up to 1.5 per mil, suggesting a lowering of global sea level through significant storage of ice in both hemispheres by at least 100 to 125 metres. Variations in benthic carbon isotope ratios of up to similar to 1.4 per mil occurred at the same time, indicating large changes in carbon cycling. We suggest that the greenhouse-icehouse transition was closely coupled to the evolution of atmospheric carbon dioxide, and that negative carbon cycle feedbacks may have prevented the permanent establishment of large ice sheets earlier than 34 million years ago.

与全球碳循环变化相关的始新世两极冰川作用

Tripati A., Backman J., Elderfield H., Ferretti P.

(英国剑桥大学地球科学系)

摘要:从约5500万年前的早始新世极端全球暖化的"温室"气候到当前冰川作用状态的转变是地球气候演变中最显著的变化之一。大冰盖首先在约34百万年前在南极出现,同时大气二氧化碳浓度下降并且全球海洋方解石补偿深度加深,北半球冰川作用在1000万年和600万年前之间开始,相对晚得多,这一认识被广泛接受。这里我们提供了覆盖温室-冰室气候转换的沉积物和有孔虫地球化学记录。我们报告了自约4200万年前起的热带太平洋和南大西洋方解石补偿深度的同时加深和后续的震荡,以及3400万年前的永久加深。方解石补偿深度最显著的变化与海水氧同位素比率高达千分之1.5的变化一致,显示通过在两个半球大量保有冰全球海平面降低了至少100～125 m。深海碳同位素比率高达约千分之1.4的变化在同一时间出现,意味着碳循环的大变化。我们推测温室-冰室转变与大气二氧化碳演变紧密耦合,并且碳循环的负反馈可能阻止了早于3400万年前的大冰盖的永久建立。

103. Vaughan D G. How does the Antarctic ice sheet affect sea level rise? [J]. Science, 2005, 308 (5730): 1877-1878.

How does the Antarctic ice sheet affect sea level rise?

Vaughan D. G.

(British Antarctic Survey, Natural Environment Research Council, Madingley Road, Cambridge CB3 0ET, UK)

Abstract: Global sea levels are predicted to rise as a result of global warming, but many contributions to this sea level rise are poorly understood. The contribution of the Antarctic ice sheet is particularly uncertain. In his perspective, Vaughan highlights the report by Davis et al., who have used satellite data to compile an 11 year record of surface elevation change in Antarctica. The resulting maps show that the ice sheet is thickening in some areas and thinning in others. Vaughan cautions that the complex patterns of change in Antarctica preclude prediction of which effect will dominate in the future.

南极冰盖如何影响海平面上升?

Vaughan D. G.

(英国自然环境研究理事会南极调查局)

摘要:全球海平面上升预计是由全球变暖导致的,但许多关于海平面上升的投稿知之甚少。南极冰盖的贡献尤其不确定。戴维斯从他的角度来看,突出显示了沃恩的报告,他们利用卫星数据分析了南极洲11年来表面高程变化的记录。生成的地图显示,冰盖在一些地区增厚,在另一些地区变薄。沃恩警告说,在南极洲复杂模式的变化,排除法将在未来预测占据主导地位。

104. Barbante C, Barnola J M, Becagli S, et al. One-to-one coupling of glacial climate variability in Greenland and Antarctica[J]. Nature, 2006, 444(7116): 195-198.

One-to-one coupling of glacial climate variability in Greenland and Antarctica

Barbante C., Barnola J. M., Becagli S., Beer J., Bigler M., Boutron C., Blunier T., Castellano E., Cattani O., Chappellaz J., Dahl-Jensen D., Debret M., Delmonte B., Dick D., Falourd S., Faria S., Federer U., Fischer H., Freitag J., Frenzel A., Fritzsche D., Fundel F., Gabrielli P., Gaspari V., Gersonde R., Graf W., Grigoriev D., Hamann I., Hansson M., Hoffmann G., Hutterli M. A., Huybrechts P., Isaksson E., Johnsen S., Jouzel J., Kaczmarska M., Karlin T., Kaufmann P., Kipfstuhl S., Kohno M., Lambert F., Lambrecht A., Lambrecht A., Landais A., Lawer G., Leuenberger M., Littot G., Loulergue L., Luthi D., Maggi V., Marino F., Masson-Delmotte V., Meyer H., Miller H., Mulvaney R., Narcisi B., Oerlemans J., Oerter H., Parrenin F., Petit J. R., Raisbeck G., Raynaud D., Rothlisberger R., Ruth U., Rybak O., Severi M., Schmitt J., Schwander J., Siegenthaler U., Siggaard-Andersen M. L., Spahni R., Steffensen J. P., Stenni B., Stocker T. F., Tison J. L., Traversi R., Udisti R., Valero-Delgado F., van den Broeke M. R., van de Wal R. S. W., Wagenbach D., Wegner A., Weiler K., Wilhelms F., Winther J. G., Wolff E., EPICA Community Members

(Department of Environmental Sciences, University Ca' Foscari of Venice, Italy)

Abstract: Precise knowledge of the phase relationship between climate changes in the two hemispheres is a key for understanding the Earth's climate dynamics. For the last glacial period, ice core studies have revealed strong coupling of the largest millennial-scale warm events in Antarctica with the longest Dansgaard-Oeschger events in Greenland through the Atlantic meridional overturning circulation. It has been unclear, however, whether the shorter Dansgaard-Oeschger events have counterparts in the shorter and less prominent Antarctic temperature variations, and whether these events are linked by the same mechanism. Here we present a glacial climate record derived from an ice core from Dronning Maud Land, Antarctica, which represents South Atlantic climate at a resolution comparable with the Greenland ice core records. After methane synchronization with an ice core from North Greenland, the oxygen isotope record from the Dronning Maud Land ice core shows a one-to-one coupling between all Antarctic warm events and Greenland Dansgaard-Oeschger events by the bipolar seesaw. The amplitude of the Antarctic warm events is found to be linearly dependent on the duration of the concurrent stadial in the North, suggesting that they all result from a similar reduction in the meridional overturning circulation.

格陵兰和南极冰期气候变化的一一耦合

Barbante C. , Barnola J. M. , Becagli S. , Beer J. , Bigler M. , Boutron C. , Blunier T. , Castellano E. , Cattani O. , Chappellaz J. , Dahl-Jensen D. , Debret M. , Delmonte B. , Dick D. , Falourd S. , Faria S. , Federer U. , Fischer H. , Freitag J. , Frenzel A. , Fritzsche D. , Fundel F. , Gabrielli P. , Gaspari V. , Gersonde R. , Graf W. , Grigoriev D. , Hamann I. , Hansson M. , Hoffmann G. , Hutterli M. A. , Huybrechts P. , Isaksson E. , Johnsen S. , Jouzel J. , Kaczmarska M. , Karlin T. , Kaufmann P. , Kipfstuhl S. , Kohno M. , Lambert F. , Lambrecht A. , Lambrecht A. , Landais A. , Lawer G. , Leuenberger M. , Littot G. , Loulergue L. , Luthi D. , Maggi V. , Marino F. , Masson-Delmotte V. , Meyer H. , Miller H. , Mulvaney R. , Narcisi B. , Oerlemans J. , Oerter H. , Parrenin F. , Petit J. R. , Raisbeck G. , Raynaud D. , Rothlisberger R. , Ruth U. , Rybak O. , Severi M. , Schmitt J. , Schwander J. , Siegenthaler U. , Siggaard-Andersen M. L. , Spahni R. , Steffensen J. P. , Stenni B. , Stocker T. F. , Tison J. L. , Traversi R. , Udisti R. , Valero-Delgado F. , van den Broeke M. R. , van de Wal R. S. W. , Wagenbach D. , Wegner A. , Weiler K. , Wilhelms F. , Winther J. G. , Wolff E. , EPICA Community Members

(意大利威尼斯大学环境科学系)

摘要：对两个半球间气候变化相位关系的精确认识是理解地球气候动力学的关键。对末次冰期的冰芯研究揭示了南极千年尺度上最大的暖事件与格陵兰最长的 Dansgaard-Oeschger 事件通过大西洋南欧翻转流的强耦合。然而较短的 Dansgaard-Oeschger 事件在南极是否有对应的较短和较不显著的温度变化，以及这些事件是否由相同的机制连接在一起，还一直不清楚。这里我们提供了得自南极 Dronning Maud 地一支冰芯的冰期气候记录，在与格陵兰冰芯记录相当的分辨率下描述了南大西洋的气候。在与一支北格陵兰冰芯的甲烷同步之后，Dronning Maud 地冰芯的氧同位素记录显示了所有南极暖事件和格陵兰 Dansgaard-Oeschger 事件通过两极跷跷板效应的一一耦合。我们发现南极暖事件的幅度与北部并发的冰退阶段的持续时间线性相关，意味着它们都是南欧翻转流的类似减弱导致的。

105. Gudmundsson G H. Fortnightly variations in the flow velocity of Rutford Ice Stream, West Antarctica[J]. Nature,2006,444(7122):1063-1064.

Fortnightly variations in the flow velocity of Rutford Ice Stream, West Antarctica

Gudmundsson G. H.

(British Antarctic Survey, High Cross, Madingley Road, Cambridge CB3 0ET, UK)

Abstract: Most of the ice lost from the Antarctic ice sheet passes through a few fast-flowing and highly dynamic ice streams. Quantifying temporal variations in flow in these ice streams, and understanding their causes, is a prerequisite for estimating the potential contribution of the Antarctic ice sheet to global sea-level change. Here I show that surface velocities on a major West Antarctic Ice Stream, Rutford Ice Stream, vary periodically by about 20 percent every two weeks as a result of tidal forcing. Tidally induced motion on ice streams has previously been thought to be limited to diurnal or even shorter-term variations. The existence of strong fortnightly variations in flow demonstrates the potential pitfalls of using repeated velocity measurements over intervals of days to infer long-term change.

西南极 Rutford 冰流流速每两周的变化

Gudmundsson G. H.

(英国南极调查局)

摘要:南极冰盖损耗的大多数冰是通过少数快速流动的高度动态的冰流完成的。定量这些冰流在时间上流动的变化并理解其原因是估计南极冰盖对全球海平面上升的潜在贡献的先决条件。这里我展示了作为潮汐强迫的结果,西南极一条主要冰流 Rutford 冰流的表面速度每两周变化约百分之 20。潮汐引起的冰流运动曾被认为限于每日甚至更短期的变化。流动中两周的强变化证明利用以数天间隔重复的速度测量推断长期变化中潜在的缺陷。

106. Overpeck J T, Otto-Bliesner B L, Miller G H, et al. Paleoclimatic evidence for future ice-sheet instability and rapid sea-level rise[J]. Science, 2006, 311(5768): 1747-1750.

Paleoclimatic evidence for future ice-sheet instability and rapid sea-level rise

Overpeck J. T., Otto-Bliesner B. L., Miller G. H., Muhs D. R., Alley R. B., Kiehl J. T.

(Institute for the Study of Planet Earth, Department of Geosciences, and Department of Atmospheric Sciences, University of Arizona, Tucson, AZ 85721, USA)

Abstract: Sea-level rise from melting of polar ice sheets is one of the largest potential threats of future climate change. Polar warming by the year 2100 may reach levels similar to those of 130000 to 127000 years ago that were associated with sea levels several meters above modern levels; both the Greenland Ice Sheet and portions of the Antarctic Ice Sheet may be vulnerable. The record of past ice-sheet melting indicates that the rate of future melting and related sea-level rise could be faster than widely thought.

未来冰盖不稳定性和快速海平面上升的古气候证据

Overpeck J. T., Otto-Bliesner B. L., Miller G. H., Muhs D. R., Alley R. B., Kiehl J. T.

(美国亚利桑那大学行星地球学院地球科学系和大气科学系)

摘要:极地冰盖引起的海平面上升是未来气候变化最大的潜在威胁之一。到 2100 年的极地变暖也许会达到类似于 130000 到 127000 年前的水平,与海平面高于现代水平数米相关;格陵兰冰盖和部分南极冰盖可能受到影响。过去冰盖融化的记录指出未来融化的速率和相关的海平面上升可能比普遍认为的要快。

107. Raisbeck G M, Yiou F, Cattani O, et al. ^{10}Be evidence for the Matuyama-Brunhes geomagnetic reversal in the EPICA Dome C ice core[J]. Nature, 2006, 444(7115): 82-84.

^{10}Be evidence for the Matuyama-Brunhes geomagnetic reversal in the EPICA Dome C ice core

Raisbeck G. M., Yiou F., Cattani O., Jouzel J.

(Centre de Spectrométrie Nucléaire et de Spectrométrie de Masse, IN2P3-CNRS-Université de Paris-Sud, Bât. 108, 91405 Orsay, France)

Abstract: An ice core drilled at Dome C, Antarctica, is the oldest ice core so far retrieved. On the basis of ice flow modelling and a comparison between the deuterium signal in the ice with climate records from marine sediment cores, the ice at a depth of 3190 m in the Dome C core is believed to have been deposited around 800000 years ago, offering a rare opportunity to study climatic and environmental conditions over this time period. However, an independent determination of this age is important because the deuterium profile below a depth of 3190 m depth does not show the expected correlation with the marine record. Here we present evidence for enhanced ^{10}Be deposition in the ice at 3160-3170 m, which we interpret as a result of the low dipole field strength during the Matuyama-Brunhes geomagnetic reversal, which occurred about 780000 years ago. If correct, this provides a crucial tie point between ice cores, marine cores and a radiometric timescale.

EPICA 冰穹 C 冰芯中 Matuyama-Brunhes 地磁反转的^{10}Be 证据

Raisbeck G. M., Yiou F., Cattani O., Jouzel J.

(法国国家科学研究院-巴黎大学核光谱和质谱中心)

摘要：钻取自南极冰穹 C 的一支冰芯是迄今为止获得的最古老的冰芯。在冰流模型和冰中氘信号与海洋沉积柱中气候变化对比的基础上，冰穹 C 冰芯 3190 m 深度的冰被认为是在约 800000 年前沉积的，提供了研究这段时间内气候和环境条件的宝贵机会。然而因为 3190 m 深度以下的氘剖面没有展示出与海洋记录期望中的相关性，对这一年龄的独立测定是重要的。这里我们提出了在 3160～3170 m 冰中^{10}Be沉降增强的证据，我们解释为出现在约 780000 年前的 Matuyama-Brunhes 地磁反转期间低偶极子磁场强度的结果。这如果正确，将提供冰芯、海洋沉积柱和辐射测量的时间标尺之间关键的连接点。

108. Raymo M E, Lisiecki L E, Nisancioglu K H. Plio-pleistocene ice volume, Antarctic climate, and the global $\delta^{18}O$ record[J]. Science, 2006, 313(5786): 492-495.

Plio-pleistocene ice volume, Antarctic climate, and the global δ¹⁸O record

Raymo M. E., Lisiecki L. E., Nisancioglu K. H.

(Palaeoclimates, Bjerknes Center for Climate Research, Allegaten 55, Bergen 5007, Norway)

Abstract: We propose that from similar to 3 to 1 million years ago, ice volume changes occurred in both the Northern and Southern Hemispheres, each controlled by local summer insolation. Because Earth's orbital precession is out of phase between hemispheres, 23000 years changes in ice volume in each hemisphere cancel out in globally integrated proxies such as ocean $\delta^{18}O$ or sea level, leaving the in-phase obliquity (41000 years) component of insolation to dominate those records. Only a modest ice mass change in Antarctica is required to effectively cancel out a much larger northern ice volume signal. At the mid-Pleistocene transition, we propose that marine-based ice sheet margins replaced terrestrial ice margins around the perimeter of East Antarctica, resulting in a shift to in-phase behavior of northern and southern ice sheets as well as the strengthening of 23000 years cyclicity in the marine $\delta^{18}O$ record.

上一更新世南极冰量、南极气候和全球 δ¹⁸O 的记录

Raymo M. E., Lisiecki L. E., Nisancioglu K. H.

(挪威 Bjerknes 古气候和气候研究中心)

摘要:我们主张从约 300 万年到 100 万年前冰量变化在南北两半球都有出现,均受到当地夏季日照强度的控制。由于地球轨道岁差在半球之间的异相,每个半球 23000 年的冰量变化在全球综合记录如海洋 $\delta^{18}O$ 或海平面记录中抵消,因而这些 $\delta^{18}O$ 记录主要受同相的地轴倾角(41000 年)日照组分控制。仅仅有限的南极冰质量变化就能有效抵消大得多的北部冰量信号。在上一更新世过渡期,我们认为海基冰盖边缘取代了环东南极的陆地冰缘,导致南北冰盖行为转换为同相以及海洋 $\delta^{18}O$ 记录中 23000 年周期的增强。

109. Winckler G, Fischer H. 30000 years of cosmic dust in Antarctic ice[J]. Science, 2006, 313 (5786):491.

30000 years of cosmic dust in Antarctic ice

Winckler G., Fischer H.

(Lamont-Doherty Earth Observatory (L-DEO), Earth Institute at Columbia University, Palisades, NY 10964, USA)

Abstract: Polar ice provides an archive for the influx of cosmic dust. Here, we present a high-resolution, glacial-to-interglacial record of cosmic dust using helium isotope analysis of the European Project for Ice Coring in Antarctica (EPICA) ice core drilled in Dronning Maud Land. We obtained a relatively constant ^{3}He flux over the past 30000 years. This finding excludes ^{3}He as a pacemaker of late

Pleistocene glacial cycles. Rather, it supports ^3He as a constant flux parameter in paleoclimatic studies. A last glacial-to-Holocene shift of the ^4He/non-sea salt Ca^{2+} ratio appears to indicate a glacial-to-interglacial change in the terrestrial dust source.

南极冰中30000年的宇宙尘埃记录

Winckler G., Fischer H.

(美国哥伦比亚大学地球研究所拉蒙特-多尔蒂地质观测站)

摘要:极冰成为宇宙尘埃流的一份研究档案。在毛德皇后地的欧洲南极冰芯钻探计划中,我们提出了一个高分辨率的、冰期-间冰期宇宙尘埃的氦同位素分析南极冰芯气候记录。我们获得一个在过去的30000年中相对恒定的通量,这一发现不包括作为更新世冰期旋回起搏器的^3He。相反,它支持^3He作为古气候研究的一个恒定的流量参数。末次冰盛期至全新世^4He/非海盐Ca^{2+}的比例转变似乎表明冰川在地面尘源的冰期-间冰期的变化。

110. Alley R B, Anandakrishnan S, Dupont T K, et al. Effect of sedimentation on ice-sheet grounding-line stability[J]. Science, 2007, 315(5820): 1838-1841.

Effect of sedimentation on ice-sheet grounding-line stability

Alley R. B., Anandakrishnan S., Dupont T. K., Parizek B. R., Pollard D.

(Department of Geosciences and Earth and Environmental Systems Institute, Pennsylvania State University, University Park, PA 16802, USA)

Abstract: Sedimentation filling space beneath ice shelves helps to stabilize ice sheets against grounding-line retreat in response to a rise in relative sea level of at least several meters. Recent Antarctic changes thus cannot be attributed to sea-level rise, strengthening earlier interpretations that warming has driven ice-sheet mass loss. Large sea-level rise, such as the approximate to 100-meter rise at the end of the last ice age, may overwhelm the stabilizing feedback from sedimentation, but smaller sea-level changes are unlikely to have synchronized the behavior of ice sheets in the past.

沉积物对冰盖接地线稳定性的影响

Alley R. B., Anandakrishnan S., Dupont T. K., Parizek B. R., Pollard D.

(美国宾夕法尼亚州立大学地球科学系与地球和环境系统研究所)

摘要:沉积物对冰架下空间的填充在对相对海平面上升至少数米的响应中阻止接地线的退却并帮助冰架稳定。因此最近的南极变化不能归因于海平面上升,巩固了更早的变暖驱动了冰盖质量损耗的解释。大的海平面上升,例如在末次冰期结束时大约100 m的上升,可能压制了来自沉积物的稳定化反馈,但更小的海平面变化未必能同步过去的冰盖行为。

111. Edgar K M, Wilson P A, Sexton P F, et al. No extreme bipolar glaciation during the main Eocene

calcite compensation shift[J]. Nature, 2007, 448(7156): 908-911.

No extreme bipolar glaciation during the main Eocene calcite compensation shift

Edgar K. M., Wilson P. A., Sexton P. F., Suganuma Y.

(National Oceanography Centre, School of Ocean and Earth Science, European Way, Southampton, SO14 3ZH, UK)

Abstract: Major ice sheets were permanently established on Antarctica approximately 34 million years ago, close to the Eocene/Oligocene boundary, at the same time as a permanent deepening of the calcite compensation depth in the world's oceans. Until recently, it was thought that Northern Hemisphere glaciation began much later, between 11 and 5 million years ago. This view has been challenged, however, by records of ice rafting at high northern latitudes during the Eocene epoch and by estimates of global ice volume that exceed the storage capacity of Antarctica at the same time as a temporary deepening of the calcite compensation depth 41. 6 million years ago. Here we test the hypothesis that large ice sheets were present in both hemispheres 41. 6 million years ago using marine sediment records of oxygen and carbon isotope values and of calcium carbonate content from the equatorial Atlantic Ocean. These records allow, at most, an ice budget that can easily be accommodated on Antarctica, indicating that large ice sheets were not present in the Northern Hemisphere. The records also reveal a brief interval shortly before the temporary deepening of the calcite compensation depth during which the calcite compensation depth shoaled, ocean temperatures increased and carbon isotope values decreased in the equatorial Atlantic. The nature of these changes around 41. 6 million years ago implies common links, in terms of carbon cycling, with events at the Eocene/Oligocene boundary and with the "hyperthermals" of the Early Eocene climate optimum. Our findings help to resolve the apparent discrepancy between the geological records of Northern Hemisphere glaciation and model results that indicate that the threshold for continental glaciation was crossed earlier in the Southern Hemisphere than in the Northern Hemisphere.

始新世主要碳酸盐补偿深度偏移期间无极端两极冰川作用

Edgar K. M., Wilson P. A., Sexton P. F., Suganuma Y.

(英国国家海洋中心海洋与地球科学学院)

摘要：主要冰盖在大约3400万年前在南极永久建立，接近始新世/渐新世分界线，与全球海洋方解石补偿深度的永久加深同时。直到最近研究者还认为北半球冰川作用开始于1100到500万年前，要晚得多。然而这一观点受到了两个记录的挑战，分别是始新世期间北半球高纬冰筏记录和与4160万年前方解石补偿深度暂时加深同时的超过南极大陆存储容量的全球冰量估计。这里我们利用氧和碳同位素的海洋沉积记录和赤道大西洋的碳酸钙比例记录来检验4160万年前两个半球均出现大冰盖的假说。这些记录最多允许的冰保有量可以被南极大陆轻松容纳，意味着当时北半球并没有出现大冰盖。这些记录也揭示了方解石补偿深度暂时加深之前不久的一段短暂时期，其间方解石补偿深度变浅，海洋温度升高并且赤道大西洋的碳同位素值下降。大约4160万年前这些变化意味着在碳循环方面始新世/渐新世分界线处的事件和早始新世"极高热的"气候适宜期的共同联系。我们的发现帮助解决了北半球冰川作用的地质记录与指出南半球大陆性冰川作用超前于北半球的模型结果之间的明显矛盾。

112. Jouzel J, Masson-Delmotte V, Cattani O, et al. Orbital and millennial Antarctic climate variability over the past 800000 years[J]. Science, 2007, 317(5839): 793-796.

Orbital and millennial Antarctic climate variability over the past 800000 years

Jouzel J., Masson-Delmotte V., Cattani O., Dreyfus G., Falourd S., Hoffmann G., Minster B., Nouet J., Barnola J. M., Chappellaz J., Fischer H., Gallet J. C., Johnsen S., Leuenberger M., Loulergue L., Luethi D., Oerter H., Parrenin F., Raisbeck G., Raynaud D., Schilt A., Schwander J., Selmo E., Souchez R., Spahni R., Stauffer B., Steffensen J. P., Stenni B., Stocker T. F., Tison J. L., Werner M., Wolff E. W.

(Laboratoire des Sciences du Climat et l'Environnement/Institut Pierre Simon Laplace, CEA-CNRS-Université de Versailles Saint-Quentin en Yvelines, CE Saclay, 91191, Gif-sur-Yvette, France)

Abstract: A high-resolution deuterium profile is now available along the entire European Project for Ice Coring in Antarctica Dome C ice core, extending this climate record back to marine isotope stage 20.2, similar to 800000 years ago. Experiments performed with an atmospheric general circulation model including water isotopes support its temperature interpretation. We assessed the general correspondence between Dansgaard-Oeschger events and their smoothed Antarctic counterparts for this Dome C record, which reveals the presence of such features with similar amplitudes during previous glacial periods. We suggest that the interplay between obliquity and precession accounts for the variable intensity of interglacial periods in ice core records.

过去800000年的轨道与千年尺度南极气候变化

Jouzel J., Masson-Delmotte V., Cattani O., Dreyfus G., Falourd S., Hoffmann G., Minster B., Nouet J., Barnola J. M., Chappellaz J., Fischer H., Gallet J. C., Johnsen S., Leuenberger M., Loulergue L., Luethi D., Oerter H., Parrenin F., Raisbeck G., Raynaud D., Schilt A., Schwander J., Selmo E., Souchez R., Spahni R., Stauffer B., Steffensen J. P., Stenni B., Stocker T. F., Tison J. L., Werner M., Wolff E. W.

(法国原子能委员会—国家科学研究院—凡尔赛大学圣昆廷分校 Pierre Simon Laplace 研究所气候与环境科学实验室)

摘要:现在已经得到沿欧洲南极冰芯计划冰穹C冰芯的高分辨率氘剖面,将气候记录向后延伸到约80万年前的海洋氧同位素阶段20.2。利用包含水同位素大气综合环流模型进行的实验支持了它的温度解释。我们评估了Dansgaard-Oeschger事件与冰穹C这一记录中它们平滑的南极对应记录之间的总体对应性,显示了具有前一次冰期期间类似幅度的这样特征的出现。我们认为地轴倾角与岁差之间的相互影响是冰芯记录的间冰期期间强度变化的原因。

113. Kawamura K, Parrenin F, Lisiecki L, et al. Northern Hemisphere forcing of climatic cycles in Antarctica over the past 360000 years[J]. Nature, 2007, 448(7156): 912-914.

Northern Hemisphere forcing of climatic cycles in Antarctica over the past 360000 years

Kawamura K., Parrenin F., Lisiecki L., Uemura R., Vimeux F., Severinghaus J. P., Hutterli M. A., Nakazawa T., Aoki S., Jouzel J., Raymo M. E., Matsumoto K., Nakata H., Motoyama H., Fujita S., Goto-Azuma K., Fujii Y., Watanabe O.

(Center for Atmospheric and Oceanic Studies, Graduate School of Science, Tohoku University, Sendai 980-8578, Japan)

Abstract: The Milankovitch theory of climate change proposes that glacial interglacial cycles are driven by changes in summer insolation at high northern latitudes. The timing of climate change in the Southern Hemisphere at glacial-interglacial transitions (which are known as terminations) relative to variations in summer insolation in the Northern Hemisphere is an important test of this hypothesis. So far, it has only been possible to apply this test to the most recent termination, because the dating uncertainty associated with older terminations is too large to allow phase relationships to be determined. Here we present a new chronology of Antarctic climate change over the past 360000 years that is based on the ratio of oxygen to nitrogen molecules in air trapped in the Dome Fuji and Vostok ice cores. This ratio is a proxy for local summer insolation, and thus allows the chronology to be constructed by orbital tuning without the need to assume a lag between a climate record and an orbital parameter. The accuracy of the chronology allows us to examine the phase relationships between climate records from the ice cores and changes in insolation. Our results indicate that orbital-scale Antarctic climate change lags Northern Hemisphere insolation by a few millennia, and that the increases in Antarctic temperature and atmospheric carbon dioxide concentration during the last four terminations occurred within the rising phase of Northern Hemisphere summer insolation. These results support the Milankovitch theory that Northern Hemisphere summer insolation triggered the last four deglaciations.

过去 360000 年南极气候旋回的北半球强迫

Kawamura K., Parrenin F., Lisiecki L., Uemura R., Vimeux F., Severinghaus J. P., Hutterli M. A., Nakazawa T., Aoki S., Jouzel J., Raymo M. E., Matsumoto K., Nakata H., Motoyama H., Fujita S., Goto-Azuma K., Fujii Y., Watanabe O.

(日本东北大学科学研究生院大气和海洋研究中心)

摘要:气候变化的米兰科维奇理论认为冰期-间冰期旋回受到北半球高纬夏季日照的驱动。与北半球夏季日照相关的冰期-间冰期转换(被称为终止)时的南半球气候变化的时间是对这一假说的重要检验。迄今为止仅能在最近一次终止应用这一检验,因为与更早终止相关的定年不确定性太大而不能确定相位关系。这里我们提出了基于 Fuji 冰穹和 Vostok 冰芯中捕集气泡的氧氮分子比率的过去 360000 年南极气候变化的新年表。这一比率是当地日照的代用性指标,因此可以无需假设气候记录与轨道参数之间的滞后而通过轨道调制建立年表。这一年表的精确性允许我们检验冰芯中气候记录与日照变化之间的相位关系。我们的结果表明轨道尺度的南极气候变化滞后于北半球日照变化数千年,并且过去四个终止时南极气温和大气二氧化碳浓度的上升出现在北半球日照的上升相位中。这些结果支持了关于北半球夏季日照触发了最近四个冰消期的米兰科维奇理论。

114. Keeling R F. Deglaciation mysteries[J]. Science,2007,316(5830):1440-1441.

Deglaciation mysteries

Keeling R. F.

(Scripps Institution of Oceanography, University of California, San Diego, CA 92093, USA)

Abstract: Results from a sediment core provide insights into ocean circulation changes during the last deglaciation.

冰消的秘密

Keeling R. F.

(美国加利福尼亚大学圣地亚哥分校斯克里普斯海洋研究所)

摘要:沉积物岩心分析结果提供了深入了解海洋环流在末次冰消期的变化的机会。

115. Shepherd A,Wingham D. Recent sea-level contributions of the Antarctic and Greenland ice sheets [J]. Science,2007,315(5818):1529-1532.

Recent sea-level contributions of the Antarctic and Greenland ice sheets

Shepherd A., Wingham D.

(Centre for Polar Observation and Modelling, School of Geosciences, University of Edinburgh, EH8 9XP, UK)

Abstract: After a century of polar exploration, the past decade of satellite measurements has painted an altogether new picture of how Earth's ice sheets are changing. As global temperatures have risen, so have rates of snowfall, ice melting, and glacier flow. Although the balance between these opposing processes has varied considerably on a regional scale, data show that Antarctica and Greenland are each losing mass overall. Our best estimate of their combined imbalance is about 125 gigatons per year of ice, enough to raise sea level by 0. 35 millimeters per year. This is only a modest contribution to the present rate of sea-level rise of 3. 0 millimeters per year. However, much of the loss from Antarctica and Greenland is the result of the flow of ice to the ocean from ice streams and glaciers, which has accelerated over the past decade. In both continents, there are suspected triggers for the accelerated ice discharge—surface and ocean warming, respectively—and, over the course of the 21st century, these processes could rapidly counteract the snowfall gains predicted by present coupled climate models.

最近南极和格陵兰冰盖对海平面的贡献

Shepherd A., Wingham D.

(英国爱丁堡大学地球科学学院极地观测与模拟中心)

摘要：在一个世纪的极地探测之后，过去十年的卫星测量为地球冰盖如何变化描绘了全新的图像。随着全球温度上升，降雪率、冰融化率和冰川流速均增加。虽然这些相互制约过程间的平衡在区域尺度上发生了相当的变化，数据显示南极和格陵兰整体上均受到质量损耗。我们对它们联合不平衡的最好估计是大约每年 1.25×10^{12} 吨冰，足以每年提高海平面 0.35 mm。这仅仅是对当前每年 3.0 mm 的海平面上升速率的有限贡献。然而南极和格陵兰的很多损耗是冰通过冰流和冰川向海里流动的结果，这在过去十年加快了。在两块大陆都有猜测的冰排放触发因素——分别是表面和海洋变暖——并且在 21 世纪的进程中，这些过程可以快速抵消当前耦合气候模型预测的降雪收益。

116. Smith K L, Robison B H, Helly J J, et al. Free-drifting icebergs: hot spots of chemical and biological enrichment in the Weddell Sea[J]. Science, 2007, 317(5837): 478-482.

Free-drifting icebergs: hot spots of chemical and biological enrichment in the Weddell Sea

Smith K. L., Robison B. H., Helly J. J., Kaufmann R. S., Ruhl H. A., Shaw T. J., Twining B. S., Vernet M.

(Monterey Bay Aquarium Research Institute, 7700 Sandholdt Road, Moss Landing, CA 95039, USA)

Abstract: The proliferation of icebergs from Antarctica over the past decade has raised questions about their potential impact on the surrounding pelagic ecosystem. Two free-drifting icebergs, 0.1 and 30.8 square kilometers in aerial surface area, and the surrounding waters were sampled in the northwest Weddell Sea during austral spring 2005. There was substantial enrichment of terrigenous material, and there were high concentrations of chlorophyll, krill, and seabirds surrounding each iceberg, extending out to a radial distance of similar to 3.7 kilometers. Extrapolating these results to all icebergs in the same size range, with the use of iceberg population estimates from satellite surveys, indicates that they similarly affect 39% of the surface ocean in this region. These results suggest that free-drifting icebergs can substantially affect the pelagic ecosystem of the Southern Ocean and can serve as areas of enhanced production and sequestration of organic carbon to the deep sea.

自由漂浮的冰山：Weddell 海中化学物质和生物富集的热点

Smith K. L., Robison B. H., Helly J. J., Kaufmann R. S., Ruhl H. A., Shaw T. J., Twining B. S., Vernet M.

(美国蒙特利湾海洋馆研究所)

摘要：过去十年来自南极的冰山扩散提出了关于它们对周围远洋生态系统的潜在影响的问题。我们在2005年的南半球春季在西北Weddell海两座暴露表面积分别为0.1 km^3 和30.8 km^3 的自由漂浮冰山及周边水体采集了样品。该处有陆源物质的大量富集，并且每个冰山周围都有高浓度的叶绿素、磷虾和海鸟，辐射到约3.7 km距离的半径。将这一结果外推到在这一尺寸范围的所有冰山，并利用卫星观测得出的冰山数量估计，可推论它们影响了这一地区约39%的海面。这些结果意味着自由漂浮的冰山可以极大影响南大洋的远洋生态系统，并可作为强生产力区域并将有机碳封存到深海。

117. Cuffey K M. Climate change: a matter of firn[J]. Science, 2008, 320(5883): 1596-1597.

Climate change: a matter of firn

Cuffey K. M.

(The university of California, Berkeley, department of geography)

Abstract: Estimating ice sheet mass changes from elevation surveys requires adjustments for snow density variations at the ice sheet surface.

气候变化：粒雪的问题

Cuffey K. M.

（美国加利福尼亚大学伯克利分校地理系）

摘要：预测冰盖质量变化的高程测量需要随冰盖表层雪密度变化调整。

118. DeConto R M, Pollard D, Wilson P A, et al. Thresholds for Cenozoic bipolar glaciation[J]. Nature, 2008, 455(7213): 652-656.

Thresholds for Cenozoic bipolar glaciation

DeConto R. M., Pollard D., Wilson P. A., Palike H., Lear C. H., Pagani M.

(Department of Geosciences, University of Massachusetts, Amherst, Massachusetts 01003, USA)

Abstract: The long-standing view of Earth's Cenozoic glacial history calls for the first continental-scale glaciation of Antarctica in the earliest Oligocene epoch (similar to 33.6 million years ago), followed by the onset of northern-hemispheric glacial cycles in the late Pliocene epoch, about 31 million years later. The pivotal early Oligocene event is characterized by a rapid shift of 1.5 parts per thousand in deep-sea benthic oxygen-isotope values (Oi-1) within a few hundred thousand years, reflecting a combination of terrestrial ice growth and deep-sea cooling. The apparent absence of contemporaneous cooling in deep-sea Mg/Ca records, however, has been argued to reflect the growth of more ice than can be accommodated on Antarctica; this, combined with new evidence of continental cooling and ice-rafted debris in the Northern Hemisphere during this period, raises the possibility that Oi-1 represents a precursory bipolar glaciation. Here we test this hypothesis using an isotope-capable global climate/ice-sheet model that accommodates both the long-term decline of Cenozoic atmospheric CO_2 levels and the effects of orbital forcing. We show that the CO_2 threshold below which glaciation occurs in the Northern Hemisphere (similar to 280 p. p. m. v.) is much lower than that for Antarctica (similar to 750 p. p. m. v.). Therefore, the growth of ice sheets in the Northern Hemisphere immediately following Antarctic glaciation would have required rapid CO_2 drawdown within the Oi-1 time-frame, to levels lower than those estimated by geochemical proxies and carbon-cycle models. Instead of bipolar glaciation, we find that Oi-1 is best explained by Antarctic glaciation alone, combined with deep-sea cooling of up to 4 ℃ and Antarctic ice that is less isotopically depleted (−30‰ to −35‰) than previously suggested. Proxy CO_2 estimates remain above our model's northern-hemispheric glaciation threshold of similar to 280 p. p. m. v. until similar to 25 Myr ago, but have been near or below that level ever since. This implies that episodic northern-hemispheric ice sheets have been possible some 20 million years earlier than currently assumed (although still much later than Oi-1) and could explain some of the variability in Miocene sea-level records.

新生代两极冰川作用的阈值

DeConto R. M. , Pollard D. , Wilson P. A. , Palike H. , Lear C. H. , Pagani M.

(美国马萨诸塞州大学阿默斯特学院地球科学系)

摘要：长期存在的关于新生代冰川历史的观点认为南极最初的大陆尺度的冰川作用发生在渐新世初期(约3360万年前)，接着是3100万年后上新世晚期北半球冰川旋回的开端。关键的早渐新世事件的特征是深海底的氧同位素值在数十万年内快速改变了千分之1.5(Oi-1)，反映出同时进行的陆地冰增长和深海变冷。然而深海Mg/Ca记录中同时期变冷的显然缺失被辩称是反映了增长的冰量不能被南极大陆容纳；这和这一时期北半球大陆变冷和冰筏残屑的新证据结合起来，提出了Oi-1代表两极冰川作用前兆的可能性。这里我们利用调和了新生代大气CO_2水平长期下降和轨道强迫的影响两者并容纳同位素参数的全球气候/冰盖模型来检验这一假说。我们指出北半球发生冰川作用的CO_2阈值(约280 ppmv)比南极的相似阈值(约750 ppmv)低得多。因此紧随着南极冰川作用立即发生的北半球冰盖增长会要求在Oi-1时限内CO_2快速降低到比那些地球化学代用性指标和碳循环模型估计的更低水平。与两极冰川作用相反，我们发现Oi-1最好仅用南极冰川作用解释，并与最多达4℃的深海变冷和比之前认为的更少同位素亏损(−30‰到−35‰)的南极冰结合起来解释。代用性指标的CO_2估计直到约25百万年前都保持在我们模型的约280 ppmv的北半球冰川作用阈值之上，但从那时起已经接近或低于那一水平。这意味着间歇式的北半球冰盖可能的出现比现在认为的早大约20百万年(尽管仍比Oi-1晚得多)，并可以解释部分中新世海平面记录的变化。

119. Lambert F, Delmonte B, Petit J R, et al. Dust-climate couplings over the past 800000 years from the EPICA Dome C ice core[J]. Nature, 2008, 452(7187): 616-619.

Dust-climate couplings over the past 800000 years from the EPICA Dome C ice core

Lambert F. , Delmonte B. , Petit J. R. , Bigler M. , Kaufmann P. R. , Hutterli M. A. , Stocker T. F. , Ruth U. , Steffensen J. P. , Maggi V.

(Climate and Environmental Physics, Physics Institute, University of Bern, Sidlerstrasse 5, 3012 Bern, Switzerland)

Abstract: Dust can affect the radiative balance of the atmosphere by absorbing or reflecting incoming solar radiation; it can also be a source of micronutrients, such as iron, to the ocean. It has been suggested that production, transport and deposition of dust is influenced by climatic changes on glacial-interglacial timescales. Here we present a high-resolution record of aeolian dust from the EPICA Dome C ice core in East Antarctica, which provides an undisturbed climate sequence over the past eight climatic cycles. We find that there is a significant correlation between dust flux and temperature records during glacial periods that is absent during interglacial periods. Our data suggest that dust flux is increasingly correlated with Antarctic temperature as the climate becomes colder. We interpret this as progressive coupling of the climates of Antarctic and lower latitudes. Limited changes in glacial-interglacial atmospheric transport time suggest that the sources and lifetime of dust are the main factors controlling the high glacial dust input. We propose that the observed similar to 25-fold increase in glacial dust flux over all eight glacial periods can be attributed to a strengthening of South American dust sources, together with a longer lifetime for atmospheric dust particles in the upper troposphere resulting from a reduced hydrological cycle during the ice ages.

EPICA 冰穹 C 冰芯中过去 800000 年的尘埃-气候耦合

Lambert F., Delmonte B., Petit J. R., Bigler M., Kaufmann P. R., Hutterli M. A., Stocker T. F., Ruth U., Steffensen J. P., Maggi V.

(瑞士伯尔尼大学物理学院气候与环境物理系)

摘要:尘埃能通过吸收和反射入射的太阳辐射来影响大气的辐射平衡;它也可以作为海洋的微量营养元素如铁的来源。研究者认为尘埃的产生、输送和沉降受到冰期-间冰期尺度的气候变化的影响。这里我们提出了东南极 EPICA 冰穹 C 冰芯风成尘埃的记录,提供了未受干扰的过去八个气候旋回的气候序列。我们发现在冰期有尘埃通量和温度记录之间的显著相关,而间冰期没有这一关系。我们的数据显示气候变得越冷尘埃通量与南极温度的相关越高。我们将其解释为南极与更低纬度气候的逐步耦合。冰期-间冰期大气传输时间的有限变化说明尘埃的来源和寿命是控制高冰川尘埃输入的主要因素。我们提出在所有八个冰期观测到的冰川尘埃通量约 25 倍的上升可以归因于南美尘埃来源的增强和冰期期间水文循环减弱导致的对流层上部大气尘埃颗粒更长的寿命。

120. Luthi D, Le Floch M, Bereiter B, et al. High-resolution carbon dioxide concentration record 650000-800000 years before present[J]. Nature, 2008, 453(7193): 379-382.

High-resolution carbon dioxide concentration record 650000-800000 years before present

Luthi D., Le Floch M., Bereiter B., Blunier T., Barnola J. M., Siegenthaler U., Raynaud D., Jouzel J., Fischer H., Kawamura K., Stocker T. F.

(Climate and Environmental Physics, Physics Institute, University of Bern, Sidlerstrasse 5, CH-3012 Bern, Switzerland)

Abstract: Changes in past atmospheric carbon dioxide concentrations can be determined by measuring the composition of air trapped in ice cores from Antarctica. So far, the Antarctic Vostok and EPICA Dome C ice cores have provided a composite record of atmospheric carbon dioxide levels over the past 650000 years. Here we present results of the lowest 200 m of the Dome C ice core, extending the record of atmospheric carbon dioxide concentration by two complete glacial cycles to 800000 yr before present. From previously published data and the present work, we find that atmospheric carbon dioxide is strongly correlated with Antarctic temperature throughout eight glacial cycles but with significantly lower concentrations between 650000 and 750000 yr before present. Carbon dioxide levels are below 180 parts per million by volume (p. p. m. v.) for a period of 3000 yr during Marine Isotope Stage 16, possibly reflecting more pronounced oceanic carbon storage. We report the lowest carbon dioxide concentration measured in an ice core, which extends the pre-industrial range of carbon dioxide concentrations during the late Quaternary by about 10 p. p. m. v. to 172-300 p. p. m. v.

距今 650000～800000 年的高分辨率二氧化碳浓度记录

Luthi D., Le Floch M., Bereiter B., Blunier T., Barnola J. M., Siegenthaler U., Raynaud D., Jouzel J., Fischer H., Kawamura K., Stocker T. F.

(瑞士伯尔尼大学物理学院气候与环境物理系)

摘要:过去大气二氧化碳浓度的变化可以通过测量南极冰芯中捕集的空气成分来确定。到目前为止,南极 Vostok 和 EPICA 冰穹 C 冰芯提供了过去 650000 年大气二氧化碳水平的综合记录。这里我们提出了冰穹 C 冰芯底部 200 m 的结果,将大气二氧化碳浓度的记录延伸了两个完整的冰期旋回到距今 800000 年。通过之前发表的数据和目前的工作,我们发现贯穿八个冰期旋回大气二氧化碳与南极温度都有很强的相关性,但在距今 650000 和 700000 年之间浓度显著降低。在海洋同位素阶段 16 期间二氧化碳水平在 3000 年间低于百万分之 180 体积(ppmv),可能反映了更显著的海洋碳存储。我们报告了冰芯中测量到的最低的二氧化碳浓度,将晚第四纪前工业化时代二氧化碳浓度范围扩展了约 10 ppmv,扩展到 172～300 ppmv。

121. Merico A, Tyrrell T, Wilson P A. Eocene/oligocene ocean de-acidification linked to Antarctic glaciation by sea-level fall[J]. Nature, 2008, 452(7190): 979-982.

Eocene/oligocene ocean de-acidification linked to Antarctic glaciation by sea-level fall

Merico A., Tyrrell T., Wilson P. A.

(National Oceanography Centre, Southampton, European Way, Southampton SO14 3ZH, UK)

Abstract: One of the most dramatic perturbations to the Earth system during the past 100 million years was the rapid onset of Antarctic glaciation near the Eocene/Oligocene epoch boundary (similar to 34 million years ago). This climate transition was accompanied by a deepening of the calcite compensation depth—the ocean depth at which the rate of calcium carbonate input from surface waters equals the rate of dissolution. Changes in the global carbon cycle, rather than changes in continental configuration, have recently been proposed as the most likely root cause of Antarctic glaciation, but the mechanism linking glaciation to the deepening of calcite compensation depth remains unclear. Here we use a global biogeochemical box model to test competing hypotheses put forward to explain the Eocene/Oligocene transition. We find that, of the candidate hypotheses, only shelf to deep sea carbonate partitioning is capable of explaining the observed changes in both carbon isotope composition and calcium carbonate accumulation at the sea floor. In our simulations, glacioeustatic sea-level fall associated with the growth of Antarctic ice sheets permanently reduces global calcium carbonate accumulation on the continental shelves, leading to an increase in pelagic burial via permanent deepening of the calcite compensation depth. At the same time, fresh limestones are exposed to erosion, thus temporarily increasing global river inputs of dissolved carbonate and increasing seawater $\delta^{13}C$. Our work sheds new light on the mechanisms linking glaciation and ocean acidity change across arguably the most important climate transition of the Cenozoic era.

通过海平面下降与南极冰川作用相联系的始新世/渐新世海洋脱酸作用

Merico A., Tyrrell T., Wilson P. A.

(英国国家海洋中心)

摘要：过去 1×10^8 年对地球系统最剧烈的变化之一是始新世/渐新世分界线附近(约 3.4×10^7 年前)的南极冰川作用的快速开始。这一气候转变是与方解石补偿深度加深相伴发生的——方解石补偿深度是来自表层水的碳酸钙输入与溶解速率平衡的海洋深度。最近有研究者提出全球碳循环的变化而不是大陆构造的变化最可能是南极冰川作用的根本原因，但是将冰川作用联系到方解石补偿深度加深的机制仍不清楚。这里我们利用全球生物地球化学的箱式模型来检验为解释始新世/渐新世转变提出的相互竞争的假说。我们发现在这些候选的假说中仅有陆架到深海的碳酸盐分区能够解释观测到的碳同位素组成变化与海底碳酸钙积累。在我们的模拟中与南极冰盖增长相关的冰期海平面下降永久降低了全球大陆架上的碳酸钙积累，导致通过方解石补偿深度加深来实现的碳酸钙远洋埋藏。在同一时间新鲜的石灰石暴露在侵蚀下，因此暂时增加了全球溶解碳酸盐的河流输入并提高了海水的 $\delta^{13}C$。我们的研究揭示了新生代有争议的最重要气候转变中联系冰川作用和海洋酸度变化的机制。

122. Wiens D A, Anandakrishnan S, Winberry J P, et al. Simultaneous teleseismic and geodetic observations of the stick-slip motion of an Antarctic ice stream[J]. Nature, 2008, 453(7196): 770-774.

Simultaneous teleseismic and geodetic observations of the stick-slip motion of an Antarctic ice stream

Wiens D. A., Anandakrishnan S., Winberry J. P., King M. A.

(Department of Earth and Planetary Sciences, Washington University, St Louis, Missouri 63130, USA)

Abstract: Long-period seismic sources associated with glacier motion have been recently discovered, and an increase in ice flow over the past decade has been suggested on the basis of secular changes in such measurements. Their significance, however, remains uncertain, as a relationship to ice flow has not been confirmed by direct observation. Here we combine long-period surface-wave observations with simultaneous Global Positioning System measurements of ice displacement to study the tidally modulated stick-slip motion of the Whillans Ice Stream in West Antarctica. The seismic origin time corresponds to slip nucleation at a region of the bed of the Whillans Ice Stream that is likely stronger than in surrounding regions and, thus, acts like an "asperity" in traditional fault models. In addition to the initial pulse, two seismic arrivals occurring 10-23 minutes later represent stopping phases as the slip terminates at the ice stream edge and the grounding line. Seismic amplitude and average rupture velocity are correlated with tidal amplitude for the different slip events during the spring-to-neap tidal cycle. Although the total seismic moment calculated from ice rigidity, slip displacement, and rupture area is equivalent to an earthquake of moment magnitude seven (M_w 7), seismic amplitudes are modest (M_s 3.6-4.2), owing to the source duration of 20-30 minutes. Seismic radiation from ice movement is proportional to the derivative of the moment rate function at periods of 25-100 seconds and very long-period radiation is not detected, owing to the source geometry. Long-period seismic waves are thus useful for detecting and studying sudden ice movements but are insensitive to the total amount of slip.

同时进行的对一条南极冰流黏滑运动遥测地震学和测地学观测

Wiens D. A., Anandakrishnan S., Winberry J. P., King M. A.

(美国华盛顿大学圣路易斯分校地球与行星科学系)

摘要:研究者最近发现了与冰川运动相关的长周期地震源,并且在这种测量的长期变化基础上提出了过去十年冰流动的增加。然而因为与冰流动的关系尚未被直接观测确认,它们的意义仍不确定。这里我们结合了长期表面波观测和同时期的全球定位系统对冰位移的测量来研究西南极 Whillans 冰流被潮汐调制的黏滑运动。地震的起源时间对应 Whillans 冰流基部的某个区域的滑动成核时间;该区域的滑动成核可能比周围区域更强,作为传统断层模型中的“关键步”。除了最初的脉冲,当滑动在冰流边缘和接地线停止时两次在 10～23 分钟后的代表地震停止阶段出现。地震振幅和平均断裂速率在大小潮循环期间不同的滑动事件与潮汐幅度相关。虽然通过冰硬度、滑动位移和断裂面积计算的总地震时间与第七矩震级(M_w 7)相等,但由于震源持续时间为 20～30 分钟,地震振幅是有限的(M_s 3.6～4.2)。来自冰运动的地震波辐射与为期 25～100 秒的矩率函数的导数成常比并且因为源的几何结构未发现很长周期的辐射,因此长周期地震波在探测和研究冰的突然移动中有用但对滑动总量不敏感。

123. Bamber J L, Riva R E M, Vermeersen B L A, et al. Reassessment of the potential sea-level rise from a collapse of the West Antarctic Ice Sheet[J]. Science, 2009, 324(5929): 901-903.

Reassessment of the potential sea-level rise from a collapse of the West Antarctic Ice Sheet

Bamber J. L., Riva R. E. M., Vermeersen B. L. A., Le Brocq A. M.

(Bristol Glaciology Centre, School of Geographical Sciences, University of Bristol, Bristol BS8 1SS, UK)

Abstract: Theory has suggested that the West Antarctic Ice Sheet may be inherently unstable. Recent observations lend weight to this hypothesis. We reassess the potential contribution to eustatic and regional sea level from a rapid collapse of the ice sheet and find that previous assessments have substantially overestimated its likely primary contribution. We obtain a value for the global, eustatic sea-level rise contribution of about 3.3 meters, with important regional variations. The maximum increase is concentrated along the Pacific and Atlantic seaboard of the United States, where the value is about 25% greater than the global mean, even for the case of a partial collapse.

对西南极冰盖崩溃可能引起的海平面上升的重新评估

Bamber J. L. , Riva R. E. M. , Vermeersen B. L. A. , Le Brocq A. M.

(英国布里斯托尔大学地理科学学院 Bristol 冰川中心)

摘要:理论研究显示西南极冰盖可能具有内在的不稳定性。最近的观测支持了这一假说。我们重新评估了快速冰盖崩塌对全球和区域海平面变化的潜在贡献,发现之前的评估大幅高估了它可能的初级贡献。我们得到了对全球海平面上升 3.3 m 的贡献值和显著区域变化。最大的上升集中在美国的太平洋和大西洋沿岸,这些地区的上升值比全球平均高约 25%,即使部分崩塌也是如此。

124. Barker S, Diz P, Vautravers M J, et al. Interhemispheric Atlantic seesaw response during the last deglaciation[J]. Nature, 2009, 457(7233): 1097-1102.

Interhemispheric Atlantic seesaw response during the last deglaciation

Barker S. , Diz P. , Vautravers M. J. , Pike J. , Knorr G. , Hall I. R. , Broecker W. S.

(School of Earth and Ocean Sciences, Cardiff University, Cardiff CF10 3YE, UK)

Abstract: The asynchronous relationship between millennial-scale temperature changes over Greenland and Antarctica during the last glacial period has led to the notion of a bipolar seesaw which acts to redistribute heat depending on the state of meridional overturning circulation within the Atlantic Ocean. Here we present new records from the South Atlantic that show rapid changes during the last deglaciation that were instantaneous (within dating uncertainty) and of opposite sign to those observed in the North Atlantic. Our results demonstrate a direct link between the abrupt changes associated with variations in the Atlantic meridional overturning circulation and the more gradual adjustments characteristic of the Southern Ocean. These results emphasize the importance of the Southern Ocean for the development and transmission of millennial-scale climate variability and highlight its role in deglacial climate change and the associated rise in atmospheric carbon dioxide.

末次冰期期间两半球的大西洋跷跷板响应

Barker S. , Diz P. , Vautravers M. J. , Pike J. , Knorr G. , Hall I. R. , Broecker W. S.

(英国加的夫大学地球与海洋科学学院)

摘要:末次冰期期间,南极和格陵兰之间的千年尺度温度变化的异步关系引发了两极跷跷板的概念,它对热的再分配取决于大西洋内的南欧翻转流的状态。这里我们提出了来自南大西洋的新记录,显示了末次冰期期间瞬时(在定年不确定性以内)的快速变化,与北大西洋观测到的信号相反。我们的结果证明了与大西洋南欧翻转流相关的突然变化和南大洋更渐进的调节特征之间的直接联系。这些结果强调了南大洋在千年尺度气候变化的发展和传播中的重要性,并突出了它在冰消气候变化和相关的大气二氧化碳上升中的角色。

125. Clark P U,Dyke A S,Shakun J D,et al. The Last Glacial Maximum[J]. Science,2009,325(5941):710-714.

The Last Glacial Maximum

Clark P. U., Dyke A. S., Shakun J. D., Carlson A. E., Clark J., Wohlfarth B., Mitrovica J. X., Hostetler S. W., McCabe A. M.

(Department of Geosciences, Oregon State University, Corvallis, OR 97331, USA)

Abstract: We used 5704 ^{14}C, ^{10}Be, and ^{3}He ages that span the interval from 10000 to 50000 years ago (10 to 50 ka) to constrain the timing of the Last Glacial Maximum (LGM) in terms of global ice-sheet and mountain-glacier extent. Growth of the ice sheets to their maximum positions occurred between 33.0 and 26.5 ka in response to climate forcing from decreases in northern summer insolation, tropical Pacific sea surface temperatures, and atmospheric CO_2. Nearly all ice sheets were at their LGM positions from 26.5 ka to 19 to 20 ka, corresponding to minima in these forcings. The onset of Northern Hemisphere deglaciation 19 to 20 ka was induced by an increase in northern summer insolation, providing the source for an abrupt rise in sea level. The onset of deglaciation of the West Antarctic Ice Sheet occurred between 14 and 15 ka, consistent with evidence that this was the primary source for an abrupt rise in sea level similar to 14.5 ka.

末次冰盛期

Clark P. U., Dyke A. S., Shakun J. D., Carlson A. E., Clark J., Wohlfarth B., Mitrovica J. X., Hostetler S. W., McCabe A. M.

(美国俄勒冈州立大学地球科学系)

摘要:我们利用5704个跨度从10000到50000年(10~50 ka)^{14}C、^{10}Be和^{3}He数据来限定以全球冰盖和山地冰川的扩张为标志的末次冰盛期的时间。作为对北部夏季日照、热带太平洋海表面温度和大气CO_2下降的响应,冰盖在33.0 ka和26.5 ka之间增长到最大位置。从26.5 ka起到19~20 ka止几乎所有冰盖都位于LGM的最大位置,对应这些强迫的最低值。19~20 ka北半球冰消作用的开端是由北部夏季日照增加引起的,提供了海平面突然上升的来源。西南极冰盖冰消作用的开端出现在14~15 ka之间,与它是约14.5 ka海平面快速上升的主要来源的证据一致。

126. Clark P U, Huybers P. Global change Interglacial and future sea level[J]. Nature, 2009, 462(7275):856-857.

Global change Interglacial and future sea level

Clark P. U., Huybers P.

(Department of Geosciences, Oregon State University, Corvallis, Oregon 97331, USA)

Abstract: A merger of data and modelling using a probabilistic approach indicates that sea level was much higher during the last interglacial than it is now, providing telling clues about future ice-sheet responses to warming.

全球变化:间冰期和未来的海平面

Clark P. U. , Huybers P.

(美国俄勒冈州立大学地球科学系)

摘要:数据和概率方法合并建模表明,海平面在末次间冰期要比现在高得多,证明这是未来冰盖应对气候变暖有说服力的因素。

127. Elsig J, Schmitt J, Leuenberger D, et al. Stable isotope constraints on Holocene carbon cycle changes from an Antarctic ice core[J]. Nature, 2009, 461(7263): 507-510.

Stable isotope constraints on Holocene carbon cycle changes from an Antarctic ice core

Elsig J. , Schmitt J. , Leuenberger D. , Schneider R. , Eyer M. , Leuenberger M. , Joos F. , Fischer H. , Stocker T. F.

(Climate and Environmental Physics, Physics Institute, University of Bern, Sidlerstrasse 5, CH-3012 Bern, Switzerland)

Abstract: Reconstructions of atmospheric CO_2 concentrations based on Antarctic ice cores reveal significant changes during the Holocene epoch, but the processes responsible for these changes in CO_2 concentrations have not been unambiguously identified. Distinct characteristics in the carbon isotope signatures of the major carbon reservoirs (ocean, biosphere, sediments and atmosphere) constrain variations in the CO_2 fluxes between those reservoirs. Here we present a highly resolved atmospheric $\delta^{13}C$ record for the past 11000 years from measurements on atmospheric CO_2 trapped in an Antarctic ice core. From mass-balance inverse model calculations performed with a simplified carbon cycle model, we show that the decrease in atmospheric CO_2 of about 5 parts per million by volume (p. p. m. v.). The increase in $\delta^{13}C$ of about 0. 25‰ during the early Holocene is most probably the result of a combination of carbon uptake of about 290 gigatonnes of carbon by the land biosphere and carbon release from the ocean in response to carbonate compensation of the terrestrial uptake during the termination of the last ice age. The 20 p. p. m. v. increase of atmospheric CO_2 and the small decrease in $\delta^{13}C$ of about 0. 05‰ during the later Holocene can mostly be explained by contributions from carbonate compensation of earlier land-biosphere uptake and coral reef formation, with only a minor contribution from a small decrease of the land-biosphere carbon inventory.

南极冰芯记录的全新世碳循环变化的同位素约束

Elsig J., Schmitt J., Leuenberger D., Schneider R., Eyer M., Leuenberger M., Joos F., Fischer H., Stocker T. F.

(瑞士伯尔尼大学物理学院气候与环境物理系)

摘要:基于南极冰芯的大气 CO_2 浓度重建揭示了全新世时期的显著变化,但这些 CO_2 浓度变化相应的过程尚未明确确定。主要碳储库(海洋、生物圈、沉积物和大气)中碳同位素信号的明显特征决定了这些储库之间的 CO_2 通量变化。这里我们通过测定一支南极冰芯中捕集的大气 CO_2 得出了过去 11000 年的高分辨率大气 $\delta^{13}C$ 记录。通过基于简化碳循环模型进行的质量平衡反转模型计算,我们指出了大气 CO_2 约百万分之 5 体积(ppmv)的下降。早全新世时期 $\delta^{13}C$ 约 0.25‰的上升最可能是陆地生物圈的约 2.9×10^{12} 吨碳吸收和作为对末次冰期结束时陆地碳吸收的碳酸盐补偿响应的海洋碳释放联合的结果。晚全新世大气 CO_2 的 20 ppmv 的上升和 $\delta^{13}C$ 的 0.05‰的小幅下降可以主要通过来自对早先的陆地生物圈吸收和珊瑚礁形成的碳酸盐补偿的贡献来解释,加上来自陆地生物圈碳存储的小幅下降的次要贡献。

128. Fain X, Ferrari C P, Dommergue A, et al. Polar firn air reveals large-scale impact of anthropogenic mercury emissions during the 1970s[J]. PNAS, 2009, 106(38): 16114-16119.

Polar firn air reveals large-scale impact of anthropogenic mercury emissions during the 1970s

Fain X., Ferrari C. P., Dommergue A., Albert M. R., Battle M., Severinghaus J., Arnaud L., Barnola J. M., Cairns W., Barbante C., Boutron C.

(Laboratoire de Glaciologie et Géophysique de l'Environnement (Unité Mixte de Recherche 5183 Centre National de la Recherche Scientifique/Université Joseph Fourier), 54 Rue Molière, B. P. 96, 38402 St. Martin d'Hères Cedex, France)

Abstract: Mercury (Hg) is an extremely toxic pollutant, and its biogeochemical cycle has been perturbed by anthropogenic emissions during recent centuries. In the atmosphere, gaseous elemental mercury (GEM; Hg^0) is the predominant form of mercury (up to 95%). Here we report the evolution of atmospheric levels of GEM in mid-to high-northern latitudes inferred from the interstitial air of firn (perennial snowpack) at Summit, Greenland. GEM concentrations increased rapidly after World War Ⅱ from approximate to 1.5 ng • m^{-3} reaching a maximum of approximate to 3 ng • m^{-3} around 1970 and decreased until stabilizing at approximate to 1.7 ng • m^{-3} around 1995. This reconstruction reproduces real-time measurements available from the Arctic since 1995 and exhibits the same general trend observed in Europe since 1990. Anthropogenic emissions caused a two-fold rise in boreal atmospheric GEM concentrations before the 1970s, which likely contributed to higher deposition of mercury in both industrialized and remotes areas. Once deposited, this toxin becomes available for methylation and, subsequently, the contamination of ecosystems. Implementation of air pollution regulations, however, enabled a large-scale decline in atmospheric mercury levels during the 1980s. The results shown here suggest that potential increases in emissions in the coming decades could have a similar large-scale impact on atmospheric Hg levels.

极地积雪中的气体揭示了20世纪70年代人类源汞排放的大范围影响

Fain X., Ferrari C. P., Dommergue A., Albert M. R., Battle M., Severinghaus J., Arnaud L., Barnola J. M., Cairns W., Barbante C., Boutron C.

(法国冰川与地球物理环境实验室(国家科学研究院5183联合研究组/约瑟夫傅立叶大学))

摘要:汞(Hg)是一种极其有毒的污染物,并且在最近几个世纪它的生物地球化学循环受到了人类源排放的扰动。在大气中,气态元素汞(GEM;Hg^0)是汞的主要形态(高达95%)。这里我们报告了通过格陵兰Summit积雪(终年积雪带)孔隙中空气推断的中高纬度GEM大气水平的演化。第二次世界大战后GEM浓度从约1.5 ng · m^{-3}提高到1970年左右的3 ng · m^{-3},并在1995年左右稳定在1.7 ng · m^{-3}之前持续下降。这一重建再现了从1995年起的北极实时记录,并展现出与1990年起在欧洲的观测相同的总体趋势。人类源排放引起了1970年代前北部大气GEM浓度两倍的上升,可能对工业化地区和偏远地区更高的汞沉降都有贡献。一旦沉降下来,这一毒素可以被甲基化,随后可以污染生态系统。然而空气污染规章的执行使得1980年代大气汞浓度出现大幅下降。这里显示的结果意味着在未来数十年潜在的排放上升会对大气Hg浓度产生类似的大幅影响。

129. Huybrechts P. Global change West-side story of Antarctic ice[J]. Nature, 2009, 458(7236): 295-296.

Global change West-side story of Antarctic ice

Huybrechts P.

(Earth System Sciences Group and the Department of Geography, Vrije Universiteit Brussel, Pleinlaan 2, B-1050 Brussels, Belgium)

Abstract: During the past five million years, the West Antarctic ice sheet has waxed and waned in size. A two-pronged reconstruction of that history provides clues to the ice sheet's future behaviour.

全球变化:南极冰西边的故事

Huybrechts P.

(比利时布鲁塞尔自由大学地球系统科学组和地理系)

摘要:在过去的500万年中,南极西部冰盖大小有跌宕起伏的变化。这为冰盖历史的重建和冰盖的未来行为两方面的研究提供了线索。

130. Lemarchand D. Climate change early survival of Antarctic ice[J]. Nature, 2009, 461(7267): 1065-1066.

Climate change early survival of Antarctic ice

Lemarchand D.

(Laboratoire d'Hydrologie et de Geochimie de Strasbourg, UMR 7517 CNRS, EOST/UdS, 67084 Strasbourg Cedex, France)

Abstract: Analyses of boron isotopes in ancient marine carbonate sediments provide an enlightening perspective on the links between carbon dioxide and ice-cap cover at a climatically momentous time in Earth's history.

气候变化：南极冰的早期保存

Lemarchand D.

(法国国家科学研究院 7517 联合研究单位斯特拉斯堡水文地球化学实验室)

摘要：利用古代海相碳酸盐沉积物硼同位素的分析，大气中二氧化碳含量及冰盖覆盖之间的联系为地球历史重要时刻的气候变化提供了启发性的观点。

131. Liu Z H, Pagani M, Zinniker D, et al. Global cooling during the Eocene-Oligocene climate transition [J]. Science, 2009, 323(5918): 1187-1190.

Global cooling during the Eocene-Oligocene climate transition

Liu Z. H., Pagani M., Zinniker D., DeConto R., Huber M., Brinkhuis H., Shah S. R., Leckie R. M., Pearson A.

(Department of Geology and Geophysics, Yale University, New Haven, CT 06520, USA)

Abstract: About 34 million years ago, Earth's climate shifted from a relatively ice-free world to one with glacial conditions on Antarctica characterized by substantial ice sheets. How Earth's temperature changed during this climate transition remains poorly understood, and evidence for Northern Hemisphere polar ice is controversial. Here, we report proxy records of sea surface temperatures from multiple ocean localities and show that the high-latitude temperature decrease was substantial and heterogeneous. High-latitude (45 degrees to 70 degrees in both hemispheres) temperatures before the climate transition were similar to 20 ℃ and cooled an average of similar to 5 ℃. Our results, combined with ocean and ice-sheet model simulations and benthic oxygen isotope records, indicate that Northern Hemisphere glaciation was not required to accommodate the magnitude of continental ice growth during this time.

始新世-渐新世气候转换期间的全球变冷

Liu Z. H., Pagani M., Zinniker D., DeConto R., Huber M., Brinkhuis H., Shah S. R., Leckie R. M., Pearson A.

(美国耶鲁大学地质和地球物理系)

摘要:约 3400 万年前,地球气候从相对无冰的世界转变到南极具有永久冰盖特征的冰川条件。人们对这一转变期间地球温度如何变化仍知之甚少,并且北半球极地冰的证据还有争议。这里我们报告了多个海洋位置的海表面温度代用性指标的记录并指出高纬温度下降是实质性的和不均匀的。气候转变前高纬(两半球 45 度到 70 度)约为 20 ℃并平均变冷了约 5 ℃。我们的结论与海洋和冰盖模型以及深海氧同位素记录结合起来,显示要容纳大陆冰在此期间的增长幅度,北半球冰期不是必需的。

132. Naish T, Powell R, Levy R, et al. Obliquity-paced Pliocene West Antarctic ice sheet oscillations [J]. Nature, 2009, 458(7236): 322-328.

Obliquity-paced Pliocene West Antarctic ice sheet oscillations

Naish T., Powell R., Levy R., Wilson G., Scherer R., Talarico F., Krissek L., Niessen F., Pompilio M., Wilson T., Carter L., DeConto R., Huybers P., McKay R., Pollard D., Ross J., Winter D., Barrett P., Browne G., Cody R., Cowan E., Crampton J., Dunbar G., Dunbar N., Florindo F., Gebhardt C., Graham I., Hannah M., Hansaraj D., Harwood D., Helling D., Henrys S., Hinnov L., Kuhn G., Kyle P., Laufer A., Maffioli P., Magens D., Mandernack K., McIntosh W., Millan C., Morin R., Ohneiser C., Paulsen T., Persico D., Raine I., Reed J., Riesselman C., Sagnotti L., Schmitt D., Sjunneskog C., Strong P., Taviani M., Vogel S., Wilch T., Williams T.

(Antarctic Research Centre, Victoria University of Wellington, Kelburn Parade, PO Box 600, Wellington 6012, New Zealand)

Abstract: Thirty years after oxygen isotope records from microfossils deposited in ocean sediments confirmed the hypothesis that variations in the Earth's orbital geometry control the ice ages, fundamental questions remain over the response of the Antarctic ice sheets to orbital cycles. Furthermore, an understanding of the behaviour of the marine-based West Antarctic ice sheet (WAIS) during the "warmer-than-present" early-Pliocene epoch (similar to 5-3 Myr ago) is needed to better constrain the possible range of ice-sheet behaviour in the context of future global warming. Here we present a marine glacial record from the upper 600 m of the AND-1B sediment core recovered from beneath the northwest part of the Ross ice shelf by the ANDRILL programme and demonstrate well-dated, similar to 40 kyr cyclic variations in ice-sheet extent linked to cycles in insolation influenced by changes in the Earth's axial tilt (obliquity) during the Pliocene. Our data provide direct evidence for orbitally induced oscillations in the WAIS, which periodically collapsed, resulting in a switch from grounded ice, or ice shelves, to open waters in the Ross embayment when planetary temperatures were up to similar to 3 ℃ warmer than today and atmospheric CO_2 concentration was as high as similar to 400 p. p. m. v.. The evidence is consistent with a new ice-sheet/ice-shelf model that simulates fluctuations in Antarctic ice volume of up to +7 m in equivalent sea level associated with the loss of the WAIS and up to +3 m in equivalent sea level from the East Antarctic ice sheet, in response to ocean-induced melting paced by obliquity. During interglacial times, diatomaceous sediments indicate high surface-water productivity, minimal summer sea ice and air temperatures above freezing, suggesting an additional influence of surface melt under conditions of elevated CO_2.

与地轴倾角变化同步的上新世西南极冰盖震荡

Naish T., Powell R., Levy R., Wilson G., Scherer R., Talarico F., Krissek L., Niessen F., Pompilio M., Wilson T., Carter L., DeConto R., Huybers P., McKay R., Pollard D., Ross J., Winter D., Barrett P., Browne G., Cody R., Cowan E., Crampton J., Dunbar G., Dunbar N., Florindo F., Gebhardt C., Graham I., Hannah M., Hansaraj D., Harwood D., Helling D., Henrys S., Hinnov L., Kuhn G., Kyle P., Laufer A., Maffioli P., Magens D., Mandernack K., McIntosh W., Millan C., Morin R., Ohneiser C., Paulsen T., Persico D., Raine I., Reed J., Riesselman C., Sagnotti L., Schmitt D., Sjunneskog C., Strong P., Taviani M., Vogel S., Wilch T., Williams T.

（新西兰惠灵顿维多利亚大学南极研究中心）

摘要：沉积在海洋沉积物中的微化石氧同位素记录证实了地球轨道的几何变化控制了冰期的假说三十年后，南极冰盖对轨道循环的响应仍是基本的问题。此外，需要对“比现在更暖”的早上新世时期（5～3 Myr 前）海基的西南极冰盖（WAIS）行为的理解来限定在未来全球变暖背景下冰盖行为的可能范围。这里我们提出了在 ANDRILL 计划中采自 Ross 冰架西北部下的 AND-1B 沉积柱表层 600 m 的海洋冰川记录，并证明了上新世中冰盖范围准确定年的约 40 ky 周期性变化，这一周期性变化与受到地轴倾角（倾斜度）影响的日照循环相联系。我们的数据提供了轨道诱导的 WAIS 震荡的直接证据，当地球温度比今天暖到约 3 ℃且大气 CO_2 浓度高达 400 ppmv 时它周期性地崩塌并引起从陆缘冰或冰架向 Ross 湾开放水域的转换。这一证据与新的冰盖/冰架模型一致，这一模型模拟了作为对地轴倾角调制的海洋引起的融化中与 WAIS 损耗相关的多达相当于海平面+7 m 和来自西南极冰盖多达相当于+3 m 的南极冰量波动。间冰期期间硅藻沉积揭示了高的表面水生产力，最少的夏季海冰和高于冰点的气温，意味着在 CO_2 升高的条件下额外的表面融化影响。

133. Notz D. The future of ice sheets and sea ice: between reversible retreat and unstoppable loss[J]. PNAS, 2009, 106(49): 20590-20595.

The future of ice sheets and sea ice: between reversible retreat and unstoppable loss

Notz D.

(Max planck Institute for Meteorology, 20146 Hamburg, Germany)

Abstract: We discuss the existence of cryospheric “tipping points” in the Earth’s climate system.

Such critical thresholds have been suggested to exist for the disappearance of Arctic sea ice and the retreat of ice sheets: once these ice masses have shrunk below an anticipated critical extent, the ice-albedo feedback might lead to the irreversible and unstoppable loss of the remaining ice. We here give an overview of our current understanding of such threshold behavior. By using conceptual arguments, we review the recent findings that such a tipping point probably does not exist for the loss of Arctic summer sea ice. Hence, in a cooler climate, sea ice could recover rapidly from the loss it has experienced in recent years. In addition, we discuss why this recent rapid retreat of Arctic summer sea ice might largely be a consequence of a slow shift in ice-thickness distribution, which will lead to strongly increased year-to-year variability of the Arctic summer sea-ice extent. This variability will render seasonal forecasts of the Arctic summer sea-ice extent increasingly difficult. We also discuss why, in contrast to Arctic summer sea ice, a tipping point is more likely to exist for the loss of the Greenland ice sheet and the West Antarctic ice sheet.

冰盖和海冰的未来:可逆转的消退和不可阻挡的损耗之间

Notz D.

(德国马克斯普朗克气象研究所)

摘要:我们讨论了地球气候系统中冰冻圈"临界点"的存在。这一关键阈值被认为在北极海冰消失和冰盖退却中存在:一旦这些冰体收缩到预期的关键范围以下,冰-反照率反馈可能导致剩余冰的不可逆转和不可阻挡的损耗。我们在这里概述对这一阈值行为的理解。通过理论论证,我们回顾了最近关于这一临界点在北极夏季海冰损耗中不存在的发现。因此在更冷的气候下海冰可以快速从它近年经历的损耗中恢复。此外,我们讨论了为何北极夏季海冰近期快速退却很大可能是冰厚度分布缓慢转换的结果,这一转换会导致强烈增加的北极夏季海冰范围的年际变化。这一变化将使北极夏季海冰范围的季节预测更加困难。我们也讨论了为何在格陵兰冰盖和西南极冰盖的损耗中与北极夏季海冰相反,更可能存在临界点。

134. Pollard D,DeConto R M. Modelling West Antarctic ice sheet growth and collapse through the past five million years[J]. Nature,2009,458(7236):329-332.

Modelling West Antarctic ice sheet growth and collapse through the past five million years

Pollard D., DeConto R. M.

(Earth and Environmental Systems Institute, Pennsylvania State University, University Park, Pennsylvania 16802, USA)

Abstract: The West Antarctic ice sheet (WAIS), with ice volume equivalent to similar to 5 m of sea level, has long been considered capable of past and future catastrophic collapse. Today, the ice sheet is fringed by vulnerable floating ice shelves that buttress the fast flow of inland ice streams. Grounding lines are several hundred metres below sea level and the bed deepens upstream, raising the prospect of runaway retreat. Projections of future WAIS behaviour have been hampered by limited understanding of past variations and their underlying forcing mechanisms. Its variation since the Last Glacial Maximum, with grounding lines advancing to the continental-shelf edges around similar to 15 kyr ago before retreating to near-modern locations by similar to 3 kyr ago. Prior collapses during the warmth of the early Pliocene epoch and some Pleistocene interglacials have been suggested indirectly from records of sea level and deep-sea-core isotopes, and by the discovery of open-ocean diatoms in subglacial sediments. Until now, however, little direct evidence of such behaviour has been available. Here we use a combined ice sheet/ice shelf model capable of high-resolution nesting with a new treatment of grounding-line dynamics and ice-shelf buttressing 5 to simulate Antarctic ice sheet variations over the past five million years. Modelled WAIS variations range from full glacial extents with grounding lines near the continental shelf break, intermediate states similar to modern, and brief but dramatic retreats, leaving only small, isolated ice caps on West Antarctic islands. Transitions between glacial, intermediate and collapsed states are relatively rapid, taking one to several thousand years. Our simulation is in good agreement with a new sediment record (ANDRILL AND-1B) recovered from the western Ross Sea, indicating a long-term trend from more frequently collapsed to more glaciated states, dominant 40-kyr cyclicity in the Pliocene, and major retreats at marine isotope stage 31 (similar to 1.07 Myr ago) and other super-interglacials.

过去五百万年西南极冰盖增长和崩溃的模拟

Pollard D., DeConto R. M.

(美国宾夕法尼亚州立大学地球和环境系统学院)

摘要：拥有相当于约 5 m 海平面冰量的西南极冰盖(WAIS)长期以来被认为在过去和未来可能遭遇灾难性崩溃。今天这一冰盖的边缘已经碎裂为支持内陆冰流快速流动的脆弱浮动冰架。接地线低于海平面数百米并且基部在上游加深，提出了失控退却的前景。对 WAIS 未来行为的预测受到了对过去变化和内在强迫机制的有限认识的妨碍。最著名的是自末次冰盛期至今它的变化，接地线在约 3 kyr 前退却近现代的位置前，在约 15 kyr 前前进到大陆架边缘。海平面和深海沉积柱同位素记录与冰川下沉积物中开放大洋硅藻的发现，间接显示了早上新世时期温暖期间和部分更新世间冰期的先前的崩塌。然而至今这些行为几乎没有可用的直接证据。这里我们利用对接地线动力学和冰架支撑作了新的处理的、高分辨率的联合冰盖/冰架模型来模拟过去五百万年南极冰盖的变化。模拟的 WAIS 变化范围从接地线在大陆架断层附近的冰川完全扩张，到类似于现代的中间状态，再到短暂但剧烈的退却后仅仅在西南极诸岛上留下孤立的小冰帽。冰期、中间态和崩溃后状态间的转变相对快速，用时一到数千年。我们的模拟与得自 Ross 海西部的新沉积物记录(ANDRILL AND-1B)吻合良好，揭示了从更频繁的崩塌到更多的冰期状态的长期趋势，上新世支配性的 40 kyr 周期性和海洋氧同位素阶段 31(约 1.07 Myr 前)以及其他超级间冰期时的显著退却。

135. Pritchard H D, Arthern R J, Vaughan D G, et al. Extensive dynamic thinning on the margins of the Greenland and Antarctic ice sheets[J]. Nature, 2009, 461(7266): 971-975.

Extensive dynamic thinning on the margins of the Greenland and Antarctic ice sheets

Pritchard H. D., Arthern R. J., Vaughan D. G., Edwards L. A.

(British Antarctic Survey, Natural Environment Research Council, Madingley Road, Cambridge CB3 0ET, UK)

Abstract: Many glaciers along the margins of the Greenland and Antarctic ice sheets are accelerating and, for this reason, contribute increasingly to global sea-level rise. Globally, ice losses contribute similar to 1.8 mm • yr^{-1}, but this could increase if the retreat of ice shelves and tidewater glaciers further enhances the loss of grounded ice or initiates the large-scale collapse of vulnerable parts of the ice sheets. Ice loss as a result of accelerated flow, known as dynamic thinning, is so poorly understood that its potential contribution to sea level over the twenty-first century remains unpredictable. Thinning on the ice-sheet scale has been monitored by using repeat satellite altimetry observations to track small changes in surface elevation, but previous sensors could not resolve most fast-flowing coastal glaciers. Here we report the use of high-resolution ICESat (Ice, Cloud and land Elevation Satellite) laser altimetry to map change along the entire grounded margins of the Greenland and Antarctic ice sheets. To isolate the dynamic signal, we compare rates of elevation change from both fast-flowing and slow-flowing ice with those expected from surface mass-balance fluctuations. We find that dynamic thinning of glaciers now reaches all latitudes in Greenland, has intensified on key Antarctic grounding lines, has endured for decades after ice-shelf collapse, penetrates far into the interior of each ice sheet and is spreading as ice shelves thin by ocean-driven melt. In Greenland, glaciers flowing faster than 100 m • yr^{-1} thinned at an average rate of 0.84 m • yr^{-1}, and in the Amundsen Sea embayment of Antarctica, thinning exceeded 9.0 m • yr^{-1} for some glaciers. Our results show that the most profound changes in the ice sheets currently result from glacier dynamics at ocean margins.

格陵兰和南极冰盖边缘的广泛动态变薄

Pritchard H. D., Arthern R. J., Vaughan D. G., Edwards L. A.

(英国自然环境研究理事会南极调查局)

摘要:沿着格陵兰和南极冰盖边缘的许多冰川正在加速流动,并且因为这个原因对全球海平面上升的贡献日益增加。全球冰损耗贡献了约 1.8 mm • yr^{-1},但如果冰架的退却和受潮水影响的冰川增加了陆缘冰的损耗或冰盖脆弱部分开始大范围崩塌,海平面可能会上升。作为加速流动结果冰损耗被称为动态变薄尚知之甚少,因此它在 21 世纪对海平面的潜在贡献仍不可预测。冰盖尺度的变薄已经反复通过卫星测高观测跟踪表面高程的小变化来监测,但之前的传感器不能分辨多数快速流动的沿岸冰川。这里我们报告了应用高分辨率 ICESat(冰、云和陆地高程卫星)激光测高法描绘沿着格陵兰和南极冰盖整个接地边缘的变化。为分辨动态信号,我们将快速流动和慢速流动的冰两者的高程变化率与表面质量平衡波动预测的值作了比较。我们发现冰川的动态变薄已经延伸到格陵兰的所有地区,强化了关键的南极接地线,在冰架崩塌后持续了数十年,远远深入到每个冰盖内部并在由海洋驱动的冰架变薄的同时传播开来。格陵兰流速超过 100 m • yr^{-1} 的冰川以平均 0.84 m • yr^{-1} 的速率变薄,而南极 Amundsen 海湾部分冰川变薄超过 9.0 m • yr^{-1}。我们的结果显示冰盖最深刻的变化目前是海洋边缘的冰盖动力学的结果。

136. Sime L C, Wolff E W, Oliver K I C, et al. Evidence for warmer interglacials in East Antarctic ice cores[J]. Nature, 2009, 462(7271): 342-345.

Evidence for warmer interglacials in East Antarctic ice cores

Sime L. C., Wolff E. W., Oliver K. I. C., Tindall J. C.

(British Antarctic Survey, Cambridge CB3 0ET, UK)

Abstract: Stable isotope ratios of oxygen and hydrogen in the Antarctic ice core record have revolutionized our understanding of Pleistocene climate variations and have allowed reconstructions of Antarctic temperature over the past 800000 years (800 kyr). The relationship between the D/H ratio of mean annual precipitation and mean annual surface air temperature is said to be uniform ±10% over East Antarctica and constant with time ±20%. In the absence of strong independent temperature proxy evidence allowing us to calibrate individual ice cores, prior general circulation model (GCM) studies have supported the assumption of constant uniform conversion for climates cooler than that of the present day. Here we analyse the three available 340 kyr East Antarctic ice core records alongside input from GCM modelling. We show that for warmer interglacial periods the relationship between temperature and the isotopic signature varies among ice core sites, and that therefore the conversions must be nonlinear for at least some sites. Model results indicate that the isotopic composition of East Antarctic ice is less sensitive to temperature changes during warmer climates. We conclude that previous temperature estimates from interglacial climates are likely to be too low. The available evidence is consistent with a peak Antarctic interglacial temperature that was at least 6 K higher than that of the present day—approximately double the widely quoted 3±1.5 K.

南极冰芯中更温暖的间冰期的证据

Sime L. C., Wolff E. W., Oliver K. I. C., Tindall J. C.

(英国南极调查局)

摘要:南极冰芯记录中氧和氢的稳定同位素比率彻底改变了我们对更新世气候变化的理解,并且使重建南极过去 800000 年(800 kyr)的温度成为可能。东南极 D/H 比率和年平均降水及年平均表面气温的关系被认为是均衡的;在东南极其空间变化幅度为±10%;其时间变化幅度为±20%。因为缺少允许我们校正个别冰芯的有力的独立替代性指标证据,之前的综合环流模型(GCM)研究支持了比当今更冷的气候的恒定均匀转换的假设。这里我们与 GCM 建模输入同时分析了三支现有的 340 kyr 东南极冰芯记录。我们指出在更暖的间冰期温度和同位素信号之间的关系在冰芯站点之间不同,因此至少在部分站点转换必须是非线性的。模型结果显示东南极冰的同位素组成在更暖的气候下对温度变化更不敏感。我们推断之前通过间冰期气候估计的温度可能太低。现有的证据与比现今高至少 6 K 的南极间冰期温度峰值一致——约为普遍认为的 3±1.5 K 的两倍。

137. Steig E J, Schneider D P, Rutherford S D, et al. Warming of the Antarctic ice-sheet surface since the 1957 International Geophysical Year[J]. Nature, 2009, 457(7228): 459-462.

Warming of the Antarctic ice-sheet surface since the 1957 International Geophysical Year

Steig E. J., Schneider D. P., Rutherford S. D., Mann M. E., Comiso J. C., Shindell D. T.

(Department of Earth and Space Sciences and Quaternary Research Center, University of Washington, Seattle, Washington 98195, USA)

Abstract: Assessments of Antarctic temperature change have emphasized the contrast between strong warming of the Antarctic Peninsula and slight cooling of the Antarctic continental interior in recent decades. This pattern of temperature change has been attributed to the increased strength of the circumpolar westerlies, largely in response to changes in stratospheric ozone. This picture, however, is substantially incomplete owing to the sparseness and short duration of the observations. Here we show that significant warming extends well beyond the Antarctic Peninsula to cover most of West Antarctica, an area of warming much larger than previously reported. West Antarctic warming exceeds 0.1 ℃ per decade over the past 50 years, and is strongest in winter and spring. Although this is partly offset by autumn cooling in East Antarctica, the continent-wide average near-surface temperature trend is positive. Simulations using a general circulation model reproduce the essential features of the spatial pattern and the long-term trend, and we suggest that neither can be attributed directly to increases in the strength of the westerlies. Instead, regional changes in atmospheric circulation and associated changes in sea surface temperature and sea ice are required to explain the enhanced warming in West Antarctica.

自 1957 年国际地球物理年开始的南极冰盖表面变暖

Steig E. J., Schneider D. P., Rutherford S. D., Mann M. E., Comiso J. C., Shindell D. T.

(美国华盛顿大学西雅图分校地球与空间科学系和第四纪研究中心)

摘要:对南极温度变化的评估强调了过去数十年南极半岛的强烈变暖和南极大陆内部轻微变冷之间的对比。这一温度变化模式被归因于绕极西风带强度的增加,主要作为平流层臭氧变化的响应。然而这一图景因为观测的稀少和持续时间较短而极不完整。这里我们展示了显著的变暖超过南极半岛,并覆盖了大部分西南极,比之前报告的变暖区域大得多。过去 50 年西南极变暖超过每十年 0.1 ℃,并在冬季和春季最强。虽然这被东南极的秋季变冷部分抵消,但全大陆的平均近地表温度趋势是正的。利用综合环流模型的模拟再现了空间模式和长期趋势的基本特征,我们认为两者都不能被直接归因于西风带强度的上升。相反,大气环流的区域变化和相关的海平面温度与海冰变化是解释西南极变暖所必需的。

138. Sun B, Siegert M J, Mudd S M, et al. The Gamburtsev mountains and the origin and early evolution of the Antarctic Ice Sheet[J]. Nature, 2009, 459(7247): 690-693.

The Gamburtsev mountains and the origin and early evolution of the Antarctic Ice Sheet

Sun B., Siegert M. J., Mudd S. M., Sugden D., Fujita S., Cui X. B., Jiang Y. Y., Tang X. Y., Li Y. S.

(Polar Research Institute of China, 451 Jinqiao Road, Pudong, Shanghai, 200136, China)

Abstract: Ice-sheet development in Antarctica was a result of significant and rapid global climate change about 34 million years ago. Ice-sheet and climate modelling suggest reductions in atmospheric carbon dioxide (less than three times the pre-industrial level of 280 parts per million by volume) that, in conjunction with the development of the Antarctic Circumpolar Current, led to cooling and glaciation paced by changes in Earth's orbit. Based on the present subglacial topography, numerical models point to ice-sheet genesis on mountain massifs of Antarctica, including the Gamburtsev mountains at Dome A, the centre of the present ice sheet. Our lack of knowledge of the present-day topography of the Gamburtsev mountains means, however, that the nature of early glaciation and subsequent development of a continental-sized ice sheet are uncertain. Here we present radar information about the base of the ice at Dome A, revealing classic Alpine topography with pre-existing river valleys overdeepened by valley glaciers formed when the mean summer surface temperature was around 3 ℃. This landscape is likely to have developed during the initial phases of Antarctic glaciation. According to Antarctic climate history (estimated from offshore sediment records) the Gamburtsev mountains are probably older than 34 million years and were the main centre for ice-sheet growth. Moreover, the landscape has most probably been preserved beneath the present ice sheet for around 14 million years.

Gamburtsev 山脉与南极冰盖的起源和早期演变

Sun B., Siegert M. J., Mudd S. M., Sugden D., Fujita S., Cui X. B., Jiang Y. Y., Tang X. Y., Li Y. S.

(中国极地研究中心)

摘要：南极的冰盖发展是约 3400 万年前显著而快速的全球气候变化的结果。冰盖与气候模拟显示大气二氧化碳的减少(低于前工业化水平百万分之 280 体积的三倍)与南极绕极流的发展联合导致了地球轨道变化调制的变冷和冰川作用。基于当前的冰下地形学，数字模型指出冰盖起源于南极山体，包括当前冰盖中心冰穹 A 的 Gamburtsev 山脉。然而我们对现今 Gamburtsev 山脉地形学认识的缺乏意味着早期冰川作用和随后的大陆尺度冰盖发展的特性是不确定的。这里我们提供了关于冰穹 A 冰底部的雷达信息，显示了典型的带有之前存在的河谷的高山地形，河谷受到过形成于平均夏季表面温度约为 3 ℃ 时的山谷冰川的切削。这一地形很可能是在南极冰川作用的初始阶段发展出来的。根据南极气候历史(由海洋沉积记录估计的)Gamburtsev 山脉或许老于 3400 万年，并曾是冰盖增长的主要中心。此外，这一地形很可能在目前冰盖下被保存了约 1400 万年。

139. Denton G H, Anderson R F, Toggweiler J R, et al. The Last Glacial Termination[J]. Science, 2010, 328(5986): 1652-1656.

The Last Glacial Termination

Denton G. H. , Anderson R. F. , Toggweiler J. R. , Edwards R. L. , Schaefer J. M. , Putnam A. E.

(Department of Earth Sciences and Climate Change Institute, Bryand Global Sciences Center, University of Maine, Orono, ME 04469, USA)

Abstract: A major puzzle of paleoclimatology is why, after a long interval of cooling climate, each late Quaternary ice age ended with a relatively short warming leg called a termination. We here offer a comprehensive hypothesis of how Earth emerged from the last global ice age. A prerequisite was the growth of very large Northern Hemisphere ice sheets, whose subsequent collapse created stadial conditions that disrupted global patterns of ocean and atmospheric circulation. The Southern Hemisphere westerlies shifted poleward during each northern stadial, producing pulses of ocean upwelling and warming that together accounted for much of the termination in the Southern Ocean and Antarctica. Rising atmospheric CO_2 during southern upwelling pulses augmented warming during the last termination in both polar hemispheres.

末次冰期的终止

Denton G. H. , Anderson R. F. , Toggweiler J. R. , Edwards R. L. , Schaefer J. M. , Putnam A. E.

(美国缅因大学 Bryand 全球科学中心地球科学系和气候变化学院)

摘要:古气候学的一个主要难题是为什么在长期的变冷气候后每个晚第四纪冰期都在被称为终止的相对短的变暖期内结束。我们在这里提出了一个关于地球如何从末次冰期中摆脱出来的全面假说。增长到非常大的北半球冰盖是一个先决条件,它随后的崩溃建立的冰退阶段的环境破坏了海洋大气环流的全球模式。在每个北部的冰退阶段南半球西风带都向极地移动,产生海洋上升流和变暖的脉冲很大程度上共同导致了南大洋和南极洲的终止。南部上升流脉冲期间增加的大气 CO_2 增大了两极半球末次终止期间的变暖。

140. Masson-Delmotte V, Stenni B, Blunier T, et al. Abrupt change of Antarctic moisture origin at the end of Termination Ⅱ[J]. PNAS, 2010, 107(27): 12091-12094.

Abrupt change of Antarctic moisture origin at the end of Termination Ⅱ

Masson-Delmotte V., Stenni B., Blunier T., Cattani O., Chappellaz J., Cheng H., Dreyfus G., Edwards R. L., Falourd S., Govin A., Kawamura K., Johnsen S. J., Jouzel J., Landais A., Lemieux-Dudon B., Lourantou A., Marshall G., Minster B., Mudelsee M., Pol K., Rothlisberger R., Selmo E., Waelbroeck C.

(Laplace/Commissariat *à* l'Energie Atomique—Centre National de la Recherche Scientifique—Université Versailles St-Quentin Unité Mixte de Recherche 8212), L'Orme des Merisiers, CEA Saclay, 91191 Gif-sur-Yvette cédex, France)

Abstract: The deuterium excess of polar ice cores documents past changes in evaporation conditions and moisture origin. New data obtained from the European Project for Ice Coring in Antarctica Dome C East Antarctic ice core provide new insights on the sequence of events involved in Termination Ⅱ, the transition between the penultimate glacial and interglacial periods. This termination is marked by a north-south seesaw behavior, with first a slow methane concentration rise associated with a strong Antarctic temperature warming and a slow deuterium excess rise. This first step is followed by an abrupt north Atlantic warming, an abrupt resumption of the East Asian summer monsoon, a sharp methane rise, and a CO_2 overshoot, which coincide within dating uncertainties with the end of Antarctic optimum. Here, we show that this second phase is marked by a very sharp Dome C centennial deuterium excess rise, revealing abrupt reorganization of atmospheric circulation in the southern Indian Ocean sector.

终止Ⅱ末端南极水汽来源的突然变化

Masson-Delmotte V., Stenni B., Blunier T., Cattani O., Chappellaz J., Cheng H., Dreyfus G., Edwards R. L., Falourd S., Govin A., Kawamura K., Johnsen S. J., Jouzel J., Landais A., Lemieux-Dudon B., Lourantou A., Marshall G., Minster B., Mudelsee M., Pol K., Rothlisberger R., Selmo E., Waelbroeck C.

(法国气候与环境科学实验室(皮埃尔西蒙拉普拉斯研究所/原子能委员会—国家科学研究中心—凡尔赛大学圣昆廷分校 8212 联合研究组))

摘要:极地冰芯中的氘过剩记录了过去蒸发条件和水汽来源的变化。欧洲南极冰芯计划从东南极冰穹C冰芯中获得的新数据提供了对包括在倒数第二次冰期和间冰期转换,也即终止Ⅱ中事件顺序的新认识。这一终止的特点是南北跷跷板行为,首先是与强烈的南极气温变暖和缓慢氘过剩上升相关的缓慢甲烷浓度上升。随后是北大西洋的突然变暖、东亚夏季风的突然恢复、甲烷急剧上升以及二氧化碳过量,这些与南极适宜期末端定年不确定性相一致。这里我们指出这一第二相位的特点是冰穹C百年尺度氘过剩的非常急剧的上升,揭示了印度洋南部大气环流的突然改组。

二、大气科学类文献

2.1 分析概述

据统计，1990～2010 年 *Nature*，*Science*，*PNAS* 刊物上共有“大气科学类”文献 129 篇，其中科学论文 112 篇，杂志通讯、关于科学论文的评论和回应 17 篇。科学论文中包括南极方面的文献 38 篇，北极方面的文献 42 篇，南北极及其他方面的文献 32 篇。杂志通讯和文章评论未计算在下面的统计中。

1. 按年代将文章数量进行统计

(1) 1990～2010 年在 *Nature*，*Science*，*PNAS* 刊物上平均每年发表大气科学类文章 5.33 篇，数量在整体上有下降的趋势。其中，1990～1993 年为发表高峰期，平均每年发表 10.25 篇，2001 年和 2008 年也均在 9 篇以上，2005 年没有文章发表。如图 2.1 所示。

(2) 从研究领域来看，南极和北极领域的研究成果相当，只有在 1999 年主要以北极方面的研究为主。

由此可见，南北极地区的大气科学研究早在 20 世纪 90 年代初就已经非常热门，并呈现起伏式的发展，北半球由于对人类影响更大而得到更多的关注。

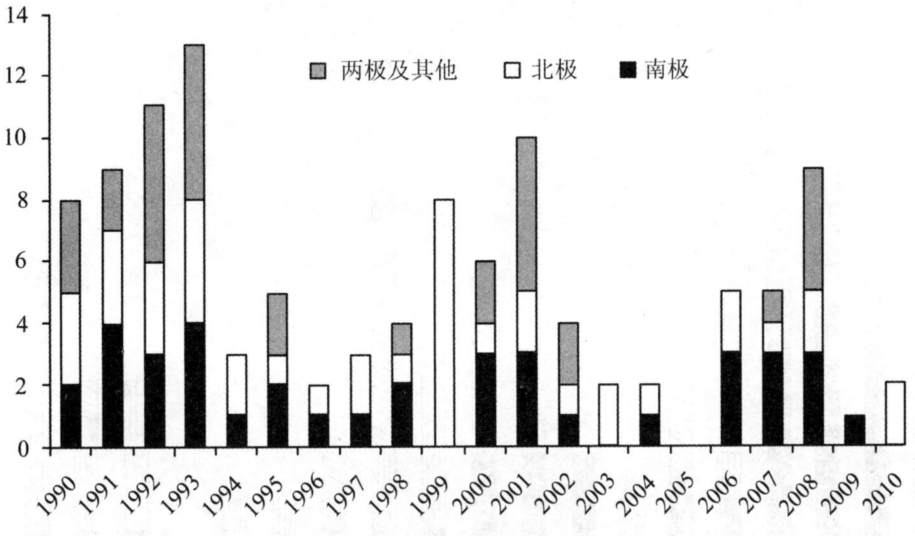

图 2.1 文章数量随年代的变化

2. 将文章研究的对象进行统计

(1) 南北极大气科学研究的对象主要集中在臭氧空洞、温室气体、极地平流层化学、卤素化学、气溶胶等方面。由各个研究对象所占的比例来看(图 2.2)，自臭氧空洞在南北极被发现之后，关于臭氧、极地平流层化学和卤素化学的研究占南北极大气科学研究的绝大多数。当然，在全球变暖的大背景下，二氧化碳、气溶胶、甲烷等温室气体的研究也占有非常重要的位置。

(2) 由各类文章随时间的分布(图 2.3)可以看出，20 世纪对南北极的研究主要以臭氧为主，臭氧空洞的发现引起了科学界的强烈关注，同时围绕臭氧空洞发生的环境——极地平流层和臭氧损耗发生的原因——卤素化学和极地平流层化学进行了相关研究；2000 年之后，在臭氧的研究上突破点逐渐减少，研究的热点主要转变为导致全球变化的温室气体的研究和南北极涛动、对流层大气化学等方面的研究。

由此可见，臭氧空洞早在 20 世纪 90 年代就已成为人们关注的焦点，二十多年来科学家一直致力于研究导致臭氧空洞的原因，并希望能通过努力抑制臭氧空洞的发展。目前，全球气候的变化再一次敲响了人类的警钟，科学家同样希望通过科学研究引发人们的关注和思考。

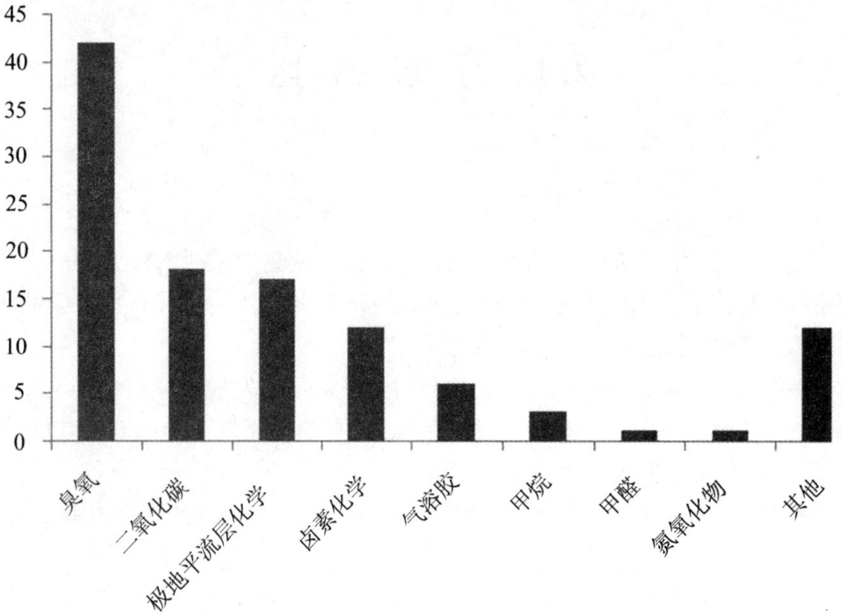

图 2.2　各研究对象文章数量分布

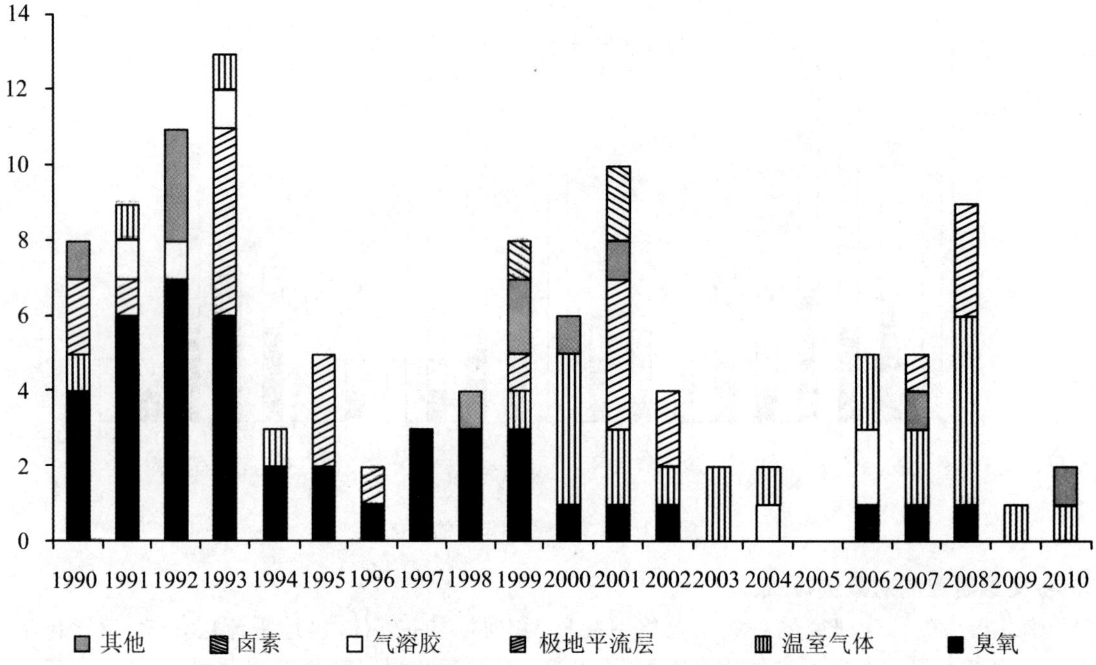

图 2.3　各类文章随时间的分布

3. 文章第一作者署名单位所在国家的统计

将文章数量按第一作者署名单位所在国家进行排列(图 2.4)，我们发现这 20 年在大气科学领域发表的文章数最多的国家是美国，高达 71 篇，为总数量的 63.4%，其次的 5 个国家依次为英国 12 篇、德国 11 篇、加拿大 5 篇、法国 4 篇、瑞士 3 篇。日本、新西兰、澳大利亚、南非、瑞典和俄罗斯各为 1 篇。中国只有在 2010 年 *Science* 上有一篇研究北冰洋 CO_2 的文章中排列第二单位，其他并无中国科研单位的身影出现。

由以上初步统计可知，目前极地大气科学研究发展的重心已由臭氧空洞向全球气候变化在南北极的响应研究转变。研究的区域也并不局限于南极或北极部分地区，而逐渐向全球的大尺度循环发展。未来，与全球气候相关的大气化学过程和南北极气候变化对人类的影响可能会成为极地大气科学研究的增长点。中国在极地大气科学的高水平研究与美国等国仍有相当大的差距。

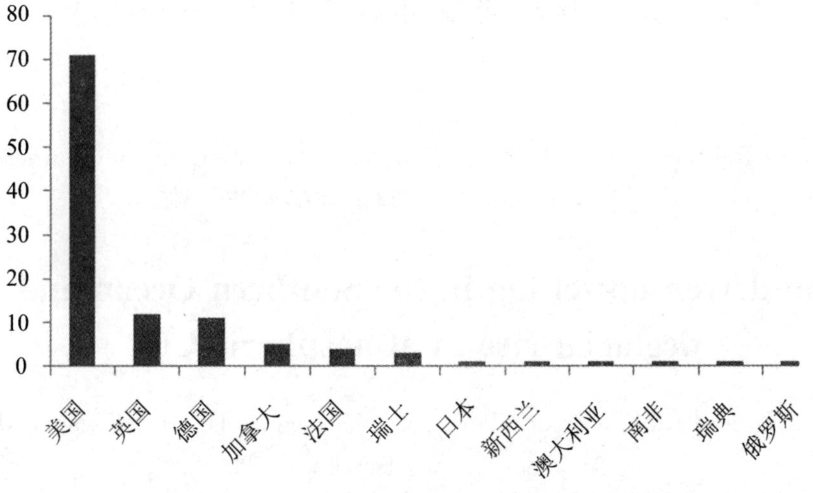

图 2.4　第一作者署名单位所在国家发表的文章数量

2.2　大气科学类文献摘要翻译——南极

1. Anderson R F, Ali S, Bradtmiller L I, et al. Wind-driven upwelling in the Southern Ocean and the deglacial rise in atmospheric CO_2[J]. Science, 2009, 323(5920): 1443-1448.

Wind-driven upwelling in the Southern Ocean and the deglacial rise in atmospheric CO_2

Anderson R. F., Ali S., Bradtmiller L. I., Nielsen S. H. H., Fleisher M. Q., Anderson B. E., Burckle L. H.

(Columbia Univ, Lamont-Doherty Earth Observ, POB 1000, Palisades, NY 10964 USA)

Abstract: Wind-driven upwelling in the ocean around Antarctica helps regulate the exchange of carbon dioxide (CO_2) between the deep sea and the atmosphere, as well as the supply of dissolved silicon to the euphotic zone of the Southern Ocean. Diatom productivity south of the Antarctic Polar Front and the subsequent burial of biogenic opal in underlying sediments are limited by this silicon supply. We show that opal burial rates, and thus upwelling, were enhanced during the termination of the last ice age in each sector of the Southern Ocean. In the record with the greatest temporal resolution, we find evidence for two intervals of enhanced upwelling concurrent with the two intervals of rising atmospheric CO_2 during deglaciation. These results directly link increased ventilation of deep water to the deglacial rise in atmospheric CO_2.

南大洋风驱动上升流与大气 CO_2 在冰消期的上升

Anderson R. F., Ali S., Bradtmiller L. I., Nielsen S. H. H., Fleisher M. Q., Anderson B. E., Burckle L. H.

(美国哥伦比亚大学)

摘要：在南极附近大洋的风驱动上升流帮助调节深海和大气之间的二氧化碳(CO_2)交换，以及南大洋透光层中溶解硅的供应。南极极锋之南的硅藻生产力，以及在随后的底层沉积物中生物蛋白石的掩埋都受到这种硅补给的制约。本研究表明，蛋白石掩埋率以及上升流于最后一个冰期结束时在南大洋的各个区域均增强。在最大的时间分辨率的记录中，我们找到两个区间上升流的增强与末次冰消期大气中二氧化碳不断上涨同时进行的证据。这些结果直接将深水换气的增加与大气二氧化碳在冰消期的上升联系起来。

2. Corinne Le Quéré, Christian Rödenbeck, Erik T Buitenhuis, et al. Saturation of the southern ocean CO_2 sink due to recent climate change[J]. Science, 2007, 316(5832): 1735-1738.

Saturation of the southern ocean CO_2 sink due to recent climate change

Corinne Le Quéré, Christian Rödenbeck, Erik T. Buitenhuis, Thomas J. Conway, Ray Langenfelds, Antony Gomez, Casper Labuschagne, Michel Ramonet, Takakiyo Nakazawa, Nicolas Metzl, Nathan P. Gillett, Martin Heimann

(School of Environmental Sciences, University of East Anglia, Norwich NR4 7TJ, UK)

Abstract: Based on observed atmospheric carbon dioxide (CO_2) concentration and an inverse method, we estimate that the Southern Ocean sink of CO_2 has weakened between 1981 and 2004 by 0.08 petagrams of carbon per year per decade relative to the trend expected from the large increase in atmospheric CO_2. We attribute this weakening to the observed increase in Southern Ocean winds resulting from human activities, which is projected to continue in the future. Consequences include a reduction of the efficiency of the Southern Ocean sink of CO_2 in the short term (about 25 years) and possibly a higher level of stabilization of atmospheric CO_2 on a multicentury time scale.

最近气候变化导致南大洋 CO_2 汇的饱和

Corinne Le Quéré, Christian Rödenbeck, Erik T. Buitenhuis, Thomas J. Conway, Ray Langenfelds, Antony Gomez, Casper Labuschagne, Michel Ramonet, Takakiyo Nakazawa, Nicolas Metzl, Nathan P. Gillett, Martin Heimann

(英国东安格利亚大学)

摘要:基于观察到的大气二氧化碳(CO_2)浓度和一个反推的方法,我们估计南大洋二氧化碳汇在1981年至2004年,相对于大气中大量增加的二氧化碳趋势所预期的,每十年减弱了0.08 pg碳每年。我们认为这种削弱归因于由人类活动的增加导致南大洋海风的增强,并预计该趋势将在未来继续。后果包括短期内(约25年)南大洋的二氧化碳汇的效率降低,以及在几百年的时间尺度大气 CO_2 将稳定在一个较高的水平。

3. Tilmes S, Muller R, Salawitch R. The sensitivity of polar ozone depletion to proposed geoengineering schemes[J]. Science, 2008, 320(5880): 1201-1204.

The sensitivity of polar ozone depletion to proposed geoengineering schemes

Tilmes S., Muller R., Salawitch R.

(Natl Ctr Atmospher Res, POB 3000, Boulder, CO 80307 USA)

Abstract: The large burden of sulfate aerosols injected into the stratosphere by the eruption of Mount Pinatubo in 1991 cooled Earth and enhanced the destruction of polar ozone in the subsequent few years. The continuous injection of sulfur into the stratosphere has been suggested as a "geoengineering" scheme to counteract global warming. We use an empirical relationship between ozone depletion and chlorine activation to estimate how this approach might influence polar ozone. An injection of sulfur large enough to compensate for surface warming caused by the doubling of atmospheric CO_2 would strongly increase the extent of Arctic ozone depletion during the present century for cold winters and would cause a considerable delay, between 30 and 70 years, in the expected recovery of the Antarctic ozone hole.

极地臭氧损耗对被提议的地球工程计划的敏感性

Tilmes S. , Muller R. , Salawitch R.

(美国国家大气研究中心)

摘要:1991 年皮纳图博火山喷发产生的大量硫酸盐气溶胶注入平流层,冷却了地球并在随后的几年里增强了对极地臭氧的破坏。有人建议一个"地球工程"计划,不断将二氧化硫注入平流层以对付全球变暖。我们使用臭氧消耗与氯活化经验关系来估计这种做法可能会影响极地臭氧。如果注入足够多的硫来弥补由大气中二氧化碳增加一倍引起的地球表面变暖,将会大大增加本世纪的北极的冷冬臭氧损耗程度,并会对南极臭氧洞的恢复造成 30～70 年的延误。

4. Schmittner A, Galbraith E D. Glacial greenhouse-gas fluctuations controlled by ocean circulation changes[J]. Nature, 2008, 456(7220): 373-376.

Glacial greenhouse-gas fluctuations controlled by ocean circulation changes

Schmittner A. , Galbraith E. D.

(Oregon State Univ, Coll Ocean & Atmospher Sci, Corvallis, OR 97331 USA)

Abstract: Earth's climate and the concentrations of the atmospheric greenhouse gases carbon dioxide (CO_2) and nitrous oxide (N_2O) varied strongly on millennial timescales during past glacial periods. Large and rapid warming events in Greenland and the North Atlantic were followed by more gradual cooling, and are highly correlated with fluctuations of N_2O as recorded in ice cores. Antarctic temperature variations, on the other hand, were smaller and more gradual, showed warming during the Greenland cold phase and cooling while the North Atlantic was warm, and were highly correlated with fluctuations in CO_2. Abrupt changes in the Atlantic meridional overturning circulation (AMOC) have often been invoked to explain the physical characteristics of these Dansgaard-Oeschger climate oscillations, but the mechanisms for the greenhouse-gas variations and their linkage to the AMOC have remained unclear. Here we present simulations with a coupled model of glacial climate and biogeochemical cycles, forced only with changes in the AMOC. The model simultaneously reproduces characteristic features of the Dansgaard-Oeschger temperature, as well as CO_2 and N_2O fluctuations. Despite significant changes in the land carbon inventory, CO_2 variations on millennial timescales are dominated by slow changes in the deep ocean inventory of biologically sequestered carbon and are correlated with Antarctic temperature and Southern Ocean stratification. In contrast, N_2O co-varies more rapidly with Greenland temperatures owing to fast adjustments of the thermocline oxygen budget.

These results suggest that ocean circulation changes were the primary mechanism that drove glacial CO_2 and N_2O fluctuations on millennial timescales.

海洋环流的变化控制冰川温室气体的波动

Schmittner A. , Galbraith E. D.

(美国俄勒冈州立大学)

摘要:在过去的冰河时期,地球的气候和大气中的温室气体二氧化碳(CO_2)与氧化亚氮(N_2O)的浓度在千年时间尺度变化强烈。在格陵兰岛和北大西洋,大范围和迅速的升温常伴随着更加平缓的降温,这一变化与冰芯记录中 N_2O 的波动高度相关。另一方面,南极的温度变化更小更缓慢,在格陵兰岛的冷期南极却是变暖的,同时北大西洋也变温暖,并与二氧化碳的波动高度相关。大西洋经向翻转环流(AMOC)突然变化常常被援引来解释这些物理特性的 DO 气候振荡,但对温室气体变化及其与 AMOC 联系的机制仍然不清楚。本文阐述一个只受 AMOC 驱动的冰期气候和生物地球化学循环耦合模型。该模型同时再现 DO 温度的典型特征,以及 CO_2 和 N_2O 的波动。尽管陆地碳储库有着显著变化,二氧化碳在千年时间尺度的变化主要是与在深海的生物固碳储库缓慢变化有关,并与南极和南大洋温度分层相关。与此相反,由于温跃层氧含量的快速响应,氧化亚氮和格陵兰温度的变化更为迅速。这些结果表明,海洋环流的变化是驱动冰川 CO_2 和 N_2O 在千年时间尺度上波动的主要机制。

5. Pekar S F. Climate change:when did the icehouse cometh? [J]. Nature,2008,455(7213):602-603.

Climate change:when did the icehouse cometh?

Pekar S. F.

(CUNY Queens Coll, Sch Earth & Environm Sci, 65-30 Kissena Blvd, Flushing, NY 11367 USA)

Abstract: The concentration of atmospheric carbon dioxide decreased between 45 million and 25 million years ago, a trend accompanied by glaciation at the poles. Modelling results suggest when and where the ice closed in.

气候变化:冰室什么时候来到?

Pekar S. F.

(美国纽约城市大学)

摘要:4500 万～2500 万年前,大气二氧化碳浓度下降,相伴而来的是两极的冰川构造的形成。模拟结果表明在何时何地冰川逼近。

6. Luthi D, Le Floch M, Bereiter B, et al. High-resolution carbon dioxide concentration record 650000-800000 years before present[J]. Nature,2008,453(7193):379-382.

High-resolution carbon dioxide concentration record 650000-800000 years before present

Luthi D., Le Floch M., Bereiter B., Blunier T., Barnola J. M., Siegenthaler U., Raynaud D., Jouzel J., Fischer H., Kawamura K., Stocker T. F.

(Univ Bern, Inst Phys, Sidlerstr 5, CH-3012 Bern, Switzerland)

Abstract: Changes in past atmospheric carbon dioxide concentrations can be determined by measuring the composition of air trapped in ice cores from Antarctica. So far, the Antarctic Vostok and EPICA Dome C ice cores have provided a composite record of atmospheric carbon dioxide levels over the past 650000 years. Here we present results of the lowest 200 m of the Dome C ice core, extending the record of atmospheric carbon dioxide concentration by two complete glacial cycles to 800000 yr before present. From previously published data and the present work, we find that atmospheric carbon dioxide is strongly correlated with Antarctic temperature throughout eight glacial cycles but with significantly lower concentrations between 650000 and 750000 yr before present. Carbon dioxide levels are below 180 parts per million by volume (p. p. m. v.) for a period of 3000 yr during Marine Isotope Stage 16, possibly reflecting more pronounced oceanic carbon storage. We report the lowest carbon dioxide concentration measured in an ice core, which extends the pre-industrial range of carbon dioxide concentrations during the late Quaternary by about 10 p. p. m. v. to 172-300 p. p. m. v..

65 万至 80 万年前高分辨率二氧化碳浓度记录

Luthi D., Le Floch M., Bereiter B., Blunier T., Barnola J. M., Siegenthaler U., Raynaud D., Jouzel J., Fischer H., Kawamura K., Stocker T. F.

(瑞士伯尔尼大学)

摘要:过去大气中二氧化碳浓度的变化可通过被滞留在南极冰芯中的气体成分测得。到目前为止,南极东方站冰芯和欧洲南极冰芯计划在冰穹C的冰芯都提供了在过去65万年的大气二氧化碳水平的综合记录。本文阐述了冰穹C冰芯最下部200米的结果,将大气二氧化碳浓度延长两个完整的冰期旋回到80万年前。从先前公布的数据和目前的工作,我们发现,在整个八个冰期旋回大气中二氧化碳与南极气温强烈相关,但至今65万至75万年前浓度显著降低。在对应于海洋同位素阶段16期的3000年,二氧化碳含量低于180 ppmv,这可能反映了更加显著的大洋碳储存。本文报告了所含二氧化碳浓度最低的冰芯,从而将晚第四纪工业化前的二氧化碳浓度范围扩展10 ppmv至172～300 ppmv。

7. Loulergue L, Schilt A, Spahni R, et al. Orbital and millennial-scale features of atmospheric CH_4 over the past 800000 years[J]. Nature, 2008, 453(7193): 383-386.

Orbital and millennial-scale features of atmospheric CH_4 over the past 800000 years

Loulergue L., Schilt A., Spahni R., Masson-Delmotte V., Blunier T., Lemieux B., Barnola J. M., Raynaud D., Stocker T. F., Chappellaz J.

(Univ Grenoble, CNRS, Lab Glaciol & Geophys Environm, 54 Rue Moliere, F-38402 St Martin Dheres, France)

Abstract: Atmospheric methane is an important greenhouse gas and a sensitive indicator of climate change and millennial-scale temperature variability. Its concentrations over the past 650000 years have varied between 350 and 800 parts per 109 by volume (p. p. b. v.) during glacial and interglacial periods, respectively. In comparison, present-day methane levels of 1770 p. p. b. v. have been reported. Insights into the external forcing factors and internal feedbacks controlling atmospheric methane are essential for predicting the methane budget in a warmer world. Here we present a detailed atmospheric methane record from the EPICA Dome C ice core that extends the history of this greenhouse gas to 800000 yr before present. The average time resolution of the new data is similar to 380 yr and permits the identification of orbital and millennial-scale features. Spectral analyses indicate that the long-term variability in atmospheric methane levels is dominated by 100000 yr glacial-interglacial cycles up to, 400000 yr ago with an increasing contribution of the precessional component during the four more recent climatic cycles. We suggest that changes in the strength of tropical methane sources and sinks (wetlands, atmospheric oxidation), possibly influenced by changes in monsoon systems and the position of the intertropical convergence zone, controlled the atmospheric methane budget.

过去 80 万年大气中甲烷在轨道尺度和千年尺度的特征

Loulergue L., Schilt A., Spahni R., Masson-Delmotte V., Blunier T., Lemieux B., Barnola J. M., Raynaud D., Stocker T. F., Chappellaz J.

(法国格勒诺布尔大学)

摘要:大气中的甲烷是一种重要的温室气体,是气候变化和千年尺度的温度变化的敏感指示剂。过去 65 万年来它的浓度在冰期和间冰期分别为 350～800 ppbv。相比之下,当今甲烷浓度为 1770 ppbv。理解外力因素和内部反馈控制大气甲烷对预测全球变暖下的甲烷含量至关重要。本文提交了从欧洲南极冰芯计划冰穹 C 冰芯的大气甲烷的详细记录,将这种温室气体的历史扩展到 80 万年前。新数据的平均时间分辨率大约是 380 年,可以鉴别在轨道尺度和千年尺度的特征。光谱分析表明,大气中甲烷水平变化主要由 10 万年的冰期和间冰期循环所主导。从 40 万年前,在最近的四个气候周期,岁差成分的贡献越来越多。我们认为热带甲烷源和汇的变化(湿地、大气氧化),可能受季风系统的变化和热带辐合带的位置强度变化所影响,并控制了大气中甲烷的含量。

8. Ahn J, Brook E J. Atmospheric CO_2 and climate on millennial time scales during the last glacial period[J]. Science, 2008, 322(5898): 83-85.

Atmospheric CO_2 and climate on millennial time scales during the last glacial period

Ahn J. , Brook E. J.

(Oregon State Univ, Dept Geosci, Corvallis, OR 97331 USA)

Abstract: Reconstructions of ancient atmospheric carbon dioxide (CO_2) variations help us better understand how the global carbon cycle and climate are linked. We compared CO_2 variations on millennial time scales between 20000 and 90000 years ago with an Antarctic temperature proxy and records of abrupt climate change in the Northern Hemisphere. CO_2 concentration and Antarctic temperature were positively correlated over millennial-scale climate cycles, implying a strong connection to Southern Ocean processes. Evidence from marine sediment proxies indicates that CO_2 concentration rose most rapidly when North Atlantic Deep Water shoaled and stratification in the Southern Ocean was reduced. These increases in CO_2 concentration occurred during stadial (cold) periods in the Northern Hemisphere, several thousand years before abrupt warming events in Greenland.

末次冰期千年时间尺度的大气二氧化碳和气候变化

Ahn J. , Brook E. J.

(美国俄勒冈州立大学)

摘要:古代大气二氧化碳(CO_2)变化的重建有助于我们更好地了解全球碳循环和气候是如何联系的。我们比较了距今20000～90000年间千年尺度的二氧化碳变化和南极温度的替代性指标与北半球气候突变的记录。千年尺度的气候周期上,CO_2浓度和南极温度呈正相关,这意味着与南大洋过程的一个强大的连接。海洋沉积物替代性指标证据表明,二氧化碳浓度迅速上升时,北大西洋水深变浅,南大洋分层减少。二氧化碳浓度的增加发生在北半球冰期,在格陵兰突然变暖事件数千年以前。

9. Von Hobe M. Atmospheric science: revisiting ozone depletion[J]. Science, 2007, 318(5858): 1878-1879.

Atmospheric science: revisiting ozone depletion

Von Hobe M.

(Forschungszentrum Julich, Inst Chem & Dynam Geosphere ICG Stratosphere 1, Postfach 1913, D-52425 Julich, Germany)

Abstract: In 1985, Farman et al discovered a substantial thinning of the stratospheric ozone layer over Antarctica in spring. This "ozone hole" took the atmospheric research community by surprise because it could not be explained by any catalytic cycles known to remove ozone in the stratosphere. Today, the consensus is that the chemical processes responsible for the formation of this "ozone hole" are reasonably well understood. New laboratory data published recently by Pope et al call this consensus into question, but the results must be treated with caution.

大气科学:再谈臭氧损耗

Von Hobe M.

(德国尤利希研究中心)

摘要:1985 年,法曼等人在春季南极上空发现了平流层臭氧显著稀薄化。"臭氧洞"震惊了大气研究界,因为它不能用任何消除平流层臭氧的催化过程解释。如今,人们对于"臭氧洞"形成的化学机理已经达成了广泛的共识。Pope 等人最新发表的实验数据对这样的共识提出了质疑,但结果要谨慎对待。

10. Saiz-Lopez A, Mahajan A S, Salmon R A, et al. Boundary layer halogens in coastal Antarctica[J]. Science, 2007, 317(5836): 348-351.

Boundary layer halogens in coastal Antarctica

Saiz-Lopez A., Mahajan A. S., Salmon R. A., Bauguitte S. J. B., Jones A. E., Roscoe H. K., Plane J. M. C.

(Univ Leeds, Sch Chem, Leeds LS2 9JT, W Yorkshire, England)

Abstract: Halogens influence the oxidizing capacity of Earth's troposphere, and iodine oxides form ultrafine aerosols, which may have an impact on climate. We report year-round measurements of boundary layer iodine oxide and bromine oxide at the near-coastal site of Halley Station, Antarctica. Surprisingly, both species are present throughout the sunlit period and exhibit similar seasonal cycles and concentrations. The springtime peak of iodine oxide (20 parts per trillion) is the highest concentration recorded anywhere in the atmosphere. These levels of halogens cause substantial ozone depletion, as well as the rapid oxidation of dimethyl sulfide and mercury in the Antarctic boundary layer.

南极海岸边界层卤素

Saiz-Lopez A., Mahajan A. S., Salmon R. A., Bauguitte S. J. B., Jones A. E., Roscoe H. K., Plane J. M. C.

(英国利兹大学)

摘要:卤素会影响地球对流层大气的氧化能力。碘的氧化物会形成极细小的气溶胶,对气候产生影响。本文报道南极哈雷站附近海岸检测到的大气边界层 IO_x 和 BrO_x 的年际变化值。两种气体令人惊奇地在整个极昼期间一直存在,并展现出相似的季节循环和含量变化。IO_x 的春季峰值(20 ppt)是全球大气中的最高含量,这么高浓度的卤素导致大量臭氧损耗,以及二甲基硫化物和 Hg 在南极边界层的迅速氧化。

11. Marchitto T M, Lehman S J, Ortiz J D, et al. Marine radiocarbon evidence for the mechanism of deglacial atmospheric CO_2 rise[J]. Science, 2007, 316(5830): 1456-1459.

Marine radiocarbon evidence for the mechanism of deglacial atmospheric CO_2 rise

Marchitto T. M., Lehman S. J., Ortiz J. D., Fluckiger J., van Geen A.

(Univ Colorado, Dept Geol Sci, Boulder, CO 80309 USA)

Abstract: We reconstructed the radiocarbon activity of intermediate waters in the eastern North Pacific over the past 38000 years. Radiocarbon activity paralleled that of the atmosphere, except during deglaciation, when intermediate-water values fell by more than 300 per mil. Such a large decrease requires a deglacial injection of very old waters from a deep-ocean carbon reservoir that was previously well isolated from the atmosphere. The timing of intermediate-water radiocarbon depletion closely matches that of atmospheric carbon dioxide rise and effectively traces the redistribution of carbon from the deep ocean to the atmosphere during deglaciation.

冰消期大气 CO_2 浓度上升机制的海洋放射性碳素证据

Marchitto T. M., Lehman S. J., Ortiz J. D., Fluckiger J., van Geen A.

(美国科罗拉多大学)

摘要:我们重建了过去38000年在北太平洋东部中层海域的放射性碳活动。除去在冰消期放射性碳与大气活动同步变化,中层水的放射性碳值下降超过0.03%。如此大的下降需要大量来自深海大洋碳储库的非常古老的冰融水注入。中层水放射性碳消耗的时间与大气二氧化碳的增加时间密切相关,有效地追溯了冰消期从深海向大气的碳的再分配过程。

12. Galbraith E D, Jaccard S L, Pedersen T F, et al. Carbon dioxide release from the North Pacific abyss during the last deglaciation[J]. Nature,2007,449(7164):890-893.

Carbon dioxide release from the North Pacific abyss during the last deglaciation

Galbraith E. D., Jaccard S. L., Pedersen T. F., Sigman D. M., Haug G. H., Cook M., Southon J. R., Francois R.

(Univ British Columbia, Dept Earth & Ocean Sci, Vancouver, BC V6T 1Z4, Canada)

Abstract: Atmospheric carbon dioxide concentrations were significantly lower during glacial periods than during intervening interglacial periods, but the mechanisms responsible for this difference remain uncertain. Many recent explanations call on greater carbon storage in a poorly ventilated deep ocean during glacial periods, but direct evidence regarding the ventilation and respired carbon content of the glacial deep ocean is sparse and often equivocal. Here we present sedimentary geochemical records from sites spanning the deep subarctic Pacific that-together with previously published results show that a poorly ventilated water mass containing a high concentration of respired carbon dioxide occupied the North Pacific abyss during the Last Glacial Maximum. Despite an inferred increase in deep Southern Ocean ventilation during the first step of the deglaciation (18000-15000 years ago), we find no evidence for improved ventilation in the abyssal subarctic Pacific until a rapid transition similar to 14600 years ago: this change was accompanied by an acceleration of export production from the surface waters above but only a small increase in atmospheric carbon dioxide concentration. We speculate that these changes were mechanistically linked to a roughly coeval increase in deep water formation in the North Atlantic, which flushed respired carbon dioxide from northern abyssal waters, but also increased the supply of nutrients to the upper ocean, leading to greater carbon dioxide sequestration at mid-depths and stalling the rise of atmospheric carbon dioxide concentrations. Our findings are qualitatively consistent with hypotheses invoking a deglacial flushing of respired carbon dioxide from an isolated, deep ocean reservoir, but suggest that the reservoir may have been released in stages, as vigorous deep water ventilation switched between North Atlantic and Southern Ocean source regions.

末次冰消期从北太平洋的深渊中释放的二氧化碳

Galbraith E. D., Jaccard S. L., Pedersen T. F., Sigman D. M., Haug G. H., Cook M., Southon J. R., Francois R.

(加拿大不列颠哥伦比亚大学)

摘要：大气二氧化碳浓度在冰期比在间冰期显著降低，但这种差别的机制仍然不确定。最近许多解释认为，在冰期更多的碳会储存在通风不充分的深海，但直接证据很少，而且常常模棱两可。本研究提交的跨越亚北极太平洋深海的沉积地球化学记录与先前公布的结果显示，末次盛冰期期间，北太平洋深的通风不良的水体含有高浓度呼吸作用产生的二氧化碳。尽管在间冰期的第一阶段(18000～15000 年前)南大洋深处通风增强，但是在 1.46 万年以前的迅速过渡过程前，我们没有发现在亚北极太平洋深海存在通风增强的证据：这种变化伴随着地表水外溢的加速过程，但大气二氧化碳浓度只有很少的增加。我们推测，这些变化与在同时代北大西洋深海水的形成是相关的，深海水不仅增加了北部深海水域冲入海中的二氧化碳，也增加了供应海洋上层的营养源，导致更多的二氧化碳封存在中层及大气二氧化碳浓度的降低。我们的研究结果定性与从一个孤立的深海储库的二氧化碳冲入的假设相一致，但也表明，当深海水通风从北大西洋转向南大洋源区时，储库分阶段性释放。

13. Weatherhead E C, Andersen S B. The search for signs of recovery of the ozone layer[J]. Nature, 2006, 441(7089): 39-45.

The search for signs of recovery of the ozone layer

Weatherhead E. C., Andersen S. B.

(Univ Colorado, Cooperat Inst Res Environm Sci, Campus Box 216, Boulder, CO 80307 USA)

Abstract: Evidence of mid-latitude ozone depletion and proof that the Antarctic ozone hole was caused by humans spurred policy makers from the late 1980s onwards to ratify the Montreal Protocol and subsequent treaties, legislating for reduced production of ozone-depleting substances. The case of anthropogenic ozone loss has often been cited since as a success story of international agreements in the regulation of environmental pollution. Although recent data suggests that total column ozone abundances have at least not decreased over the past eight years for most of the world, it is still uncertain whether this improvement is actually attributable to the observed decline in the amount of ozone-depleting substances in the Earth's atmosphere. The high natural variability in ozone abundances, due in part to the solar cycle as well as changes in transport and temperature, could override the relatively small changes expected from the recent decrease in ozone-depleting substances. Whatever the benefits of the Montreal agreement, recovery of ozone is likely to occur in a different atmospheric environment, with changes expected in atmospheric transport, temperature and important trace gases. It is therefore unlikely that ozone will stabilize at levels observed before 1980, when a decline in ozone concentrations was first observed.

对臭氧层恢复迹象的搜索

Weatherhead E. C., Andersen S. B.

(美国科罗拉多大学)

摘要:中纬度地区臭氧消耗的证据和南极上空的臭氧洞是由人为因素造成的这一事实促进了决策者自 20 世纪 80 年代末批准了减少臭氧消耗物质的蒙特利尔议定书和随后的条约。这一事件经常被引用作为国际协定调控环境污染的成功故事。虽然最近的数据表明,在世界大多数地方总臭氧柱含量在过去的 8 年至少没有下降,但这是否是由于地球大气中臭氧消耗物质的数量下降使这种情况得到改善仍是未知数。由太阳活动周期,以及在传输和由于温度变化导致的臭氧丰度的自然波动,可能大大掩盖了臭氧消耗物质减少所导致相对较小的变化。无论蒙特利尔协议带来什么好处,臭氧复苏很可能在不同的大气环境中发生,并伴随着大气传输、温度和重要的微量气体含量的变化。因此,当首次观察到臭氧浓度下降时,臭氧含量就不可能稳定在 1980 年以前的水平。

14. Turner J, Lachlan-Cope T A, Colwell S, et al. Significant warming of the Antarctic winter troposphere[J]. Science, 2006, 311(5769): 1914-1917.

Significant warming of the Antarctic winter troposphere

Turner J., Lachlan-Cope T. A., Colwell S., Marshall G. J., Connolley W. M.

(British Antarctic Survey, NERC, Cambridge CB3 0ET, England)

Abstract: We report an undocumented major warming of the Antarctic winter troposphere that is larger than any previously identified regional tropospheric warming on Earth. This result has come to light through an analysis of recently digitized and rigorously quality controlled Antarctic radiosonde observations. The data show that regional midtropospheric temperatures have increased at a statistically significant rate of 0.5° to 0.7°Celsius per decade over the past 30 years. Analysis of the time series of radiosonde temperatures indicates that the data are temporally homogeneous. The available data do not allow us to unambiguously assign a cause to the tropospheric warming at this stage.

南极冬季对流层的明显变暖

Turner J., Lachlan-Cope T. A., Colwell S., Marshall G. J., Connolley W. M.

(英国南极调查局)

摘要:本文报告了南极冬季对流层变暖,比地球上任何先前确定对流层变暖的区域都大。这一结果最近因严格的质量控制数字化的南极无线电探空仪的观测分析被发现。这些数据表明,在过去30年中对流层的局部温度已经升高了每十年0.5~0.7℃。无线电探空仪的温度和时间序列分析表明,该数据在时间上均匀。现有的数据还不足以让我们完全肯定这个阶段对流层变暖的原因。

15. Monaghan A J, Bromwich D H, Fogt R L, et al. Insignificant change in Antarctic snowfall since the International Geophysical Year[J]. Science, 2006, 313(5788): 827-831.

Insignificant change in Antarctic snowfall since the International Geophysical Year

Monaghan A. J., Bromwich D. H., Fogt R. L., Wang S. H., Mayewski P. A., Dixon D. A., Ekaykin A., Frezzotti M., Goodwin I., Isaksson E., Kaspari S. D., Morgan V. I., Oerter H., Van Ommen T. D., Van der Veen C. J., Wen J. H.

(Ohio State Univ, Byrd Polar Res Ctr, Polar Meteorol Grp, Columbus, OH 43210 USA)

Abstract: Antarctic snowfall exhibits substantial variability over a range of time scales, with consequent impacts on global sea level and the mass balance of the ice sheets. To assess how snowfall has affected the thickness of the ice sheets in Antarctica and to provide an extended perspective, we derived a 50-year time series of snowfall accumulation over the continent by combining model simulations and observations primarily from ice cores. There has been no statistically significant change in snowfall since the 1950s, indicating that Antarctic precipitation is not mitigating global sea level rise as expected, despite recent winter warming of the overlying atmosphere.

自国际地球物理年来,南极降雪量无显著变化

Monaghan A. J., Bromwich D. H., Fogt R. L., Wang S. H., Mayewski P. A., Dixon D. A., Ekaykin A., Frezzotti M., Goodwin I., Isaksson E., Kaspari S. D., Morgan V. I., Oerter H., Van Ommen T. D., Van der Veen C. J., Wen J. H.

(美国俄亥俄州立大学)

摘要:南极降雪显示了在不同时间尺度范围内的较大幅度变化,从而影响了全球海平面和冰盖物质平衡。为了评估降雪已影响到南极冰盖的厚度,并提供一个更广阔的视角,我们主要从冰芯模型模拟和冰核数据得出过去50年在南极大陆的降雪积累时间序列。自20世纪50年代以来,降雪量还没有显著变化,这表明尽管近年冬季上覆大气变暖,但南极降水并没有减缓全球海平面的上升。

16. Jones J M, Widmann M. Atmospheric science: early peak in Antarctic oscillation index[J]. Nature, 2004, 432(7015): 290-291.

Atmospheric science: early peak in Antarctic oscillation index

Jones J. M., Widmann M.

(GKSS Forschungszentrum Geesthacht GmbH, Inst Coastal Res, D-21502 Geesthacht, Germany)

Abstract: The principal extratropical atmospheric circulation mode in the Southern Hemisphere, the Antarctic oscillation (or Southern Hemisphere annular mode), represents fluctuations in the strength of the circumpolar vortex and has shown a trend towards a positive index in austral summer in recent decades, which has been linked to stratospheric ozone depletion and to increased atmospheric greenhouse-gas concentrations. Here we reconstruct the austral summer (December-January) Antarctic oscillation index from sea-level pressure measurements over the twentieth century and find that large positive values, and positive trends of a similar magnitude to those of past decades, also occurred around 1960, and that strong negative trends occurred afterwards. This positive Antarctic oscillation index and large positive trend during a period before ozone-depleting chemicals were released into the atmosphere and before marked anthropogenic warming, together with the later negative trend, indicate that natural forcing factors or internal mechanisms in the climate system must also strongly influence the state of the Antarctic oscillation.

大气科学：南极涛动指数的早期峰值

Jones J. M., Widmann M.

(德国基斯塔公司研究中心)

摘要：南半球主要的高纬大气环流模式：南极涛动(或南半球环模式)，代表了绕极涡旋强度的波动，并在最近数十年来在南半球夏季有朝着正向指数变化的趋势，这与平流层臭氧损耗和大气中的温室气体浓度的增加相关。本文由20世纪海平面气压的测量重建了南半球夏季(12月到次年1月)南极涛动指数，发现过去几十年有较大的正值和一个类似规模的正向趋势出现，以及大约1960年发生的正值和这之后发生强烈的负面趋势。这种南极涛动指数正向趋势及其在消耗臭氧化学物质被释放到大气中之前以及人为造成的显著变暖之前的正向趋势连同后来的消极趋势表明，气候系统的自然因素或者内部机制也强烈影响到南极涛动的状态。

17. Thompson D W J, Solomon S. Interpretation of recent Southern Hemisphere climate change[J]. Science, 2002, 296(5569): 895-899.

Interpretation of recent Southern Hemisphere climate change

Thompson D. W. J., Solomon S.

(Colorado State Univ, Dept Atmospher Sci, Foothills Campus, Ft Collins, CO 80523 USA)

Abstract: Climate variability in the high-latitude Southern Hemisphere (SH) is dominated by the SH annular mode, a large-scale pattern of variability characterized by fluctuations in the strength of the circumpolar vortex. We present evidence that recent trends in the SH tropospheric circulation can be interpreted as a bias toward the high-index polarity of this pattern, with stronger westerly flow encircling the polar cap. It is argued that the largest and most significant tropospheric trends can be traced to recent trends in the lower stratospheric polar vortex, which are due largely to photochemical ozone losses. During the summer-fall season, the trend toward stronger circumpolar flow has contributed substantially to the observed warming over the Antarctic Peninsula and Patagonia and to the cooling over eastern Antarctica and the Antarctic plateau.

解读目前南半球气候变化

Thompson D. W. J., Solomon S.

(美国科罗拉多州立大学)

摘要:南半球高纬度气候变化由南半球(SH)环形模式占主导地位,这是一种以极地涡旋强度波动为主要特征的大型变化模式。本文提交一些证据,表明南半球对流层环流最近趋势,可解释为这种模式转向高指数极性的结果,伴随较强的西风流包围极冠。一般认为最大和最重要的对流层的趋势可以追溯到最近较低的平流层极涡的趋势,平流层极涡主要与光化学臭氧损耗相关。在夏秋季,更强的极地流动的趋势显著促进了南极半岛和巴塔哥尼亚观测到的变暖与东南极和南极高原的变冷。

18. Gurney K R, Law R M, Denning A S, et al. Towards robust regional estimates of CO_2 sources and sinks using atmospheric transport models[J]. Nature, 2002, 415(6872): 626-630.

Towards robust regional estimates of CO_2 sources and sinks using atmospheric transport models

Gurney K. R., Law R. M., Denning A. S., Rayner P. J., Baker D., Bousquet P., Bruhwiler L., Chen Y. H., Ciais P., Fan S., Fung I. Y., Gloor M., Heimann M., Higuchi K., John J., Maki T., Maksyutov S., Masarie K., Peylin P., Prather M., Pak B. C., Randerson J., Sarmiento J., Taguchi S., Takahashi T., Yuen C. W.

(Colorado State Univ, Dept Atmospher Sci, Ft Collins, CO 80523 USA)

Abstract: Information about regional carbon sources and sinks can be derived from variations in observed atmospheric CO_2 concentrations via inverse modelling with atmospheric tracer transport models. A consensus has not yet been reached regarding the size and distribution of regional carbon fluxes obtained using this approach, partly owing to the use of several different atmospheric transport models. Here we report estimates of surface-atmosphere CO_2 fluxes from an intercomparison of atmospheric CO_2 inversion models (the TransCom 3 project), which includes 16 transport models and model variants. We find an uptake of CO_2 in the southern extratropical ocean less than that estimated from ocean measurements, a result that is not sensitive to transport models or methodological approaches. We also find a northern land carbon sink that is distributed relatively evenly among the continents of the Northern Hemisphere, but these results show some sensitivity to transport differences among models, especially in how they respond to seasonal terrestrial exchange of CO_2. Overall, carbon fluxes integrated over latitudinal zones are strongly constrained by observations in the middle to high latitudes. Further significant constraints to our understanding of regional carbon fluxes will therefore require improvements in transport models and expansion of the CO_2 observation network within the tropics.

利用大气传输模型对 CO_2 源与汇进行可靠区域估计

Gurney K. R., Law R. M., Denning A. S., Rayner P. J., Baker D., Bousquet P., Bruhwiler L., Chen Y. H., Ciais P., Fan S., Fung I. Y., Gloor M., Heimann M., Higuchi K., John J., Maki T., Maksyutov S., Masarie K., Peylin P., Prather M., Pak B. C., Randerson J., Sarmiento J., Taguchi S., Takahashi T., Yuen C. W.

(美国科罗拉多州立大学)

摘要:关于区域碳的源和汇的信息可以通过对观测到的大气二氧化碳浓度的变化用大气示踪传输模型反演得到。通过这种方法观测到的区域碳通量的规模和分布尚没有达到共识,一部分原因是由于几个不同的大气输送模式的使用。本文用一个大气二氧化碳反演模型(TransCom 3 项目),其中包括 16 种不同传输模式和模式的变异来估算地面-大气二氧化碳通量。我们发现在南部温带海洋二氧化碳的吸收比海洋测量的少,这个结果对传输模型或方法不敏感。我们还发现一个北半球大陆碳汇分布相对均匀,但这些结果对传输模型之间的差异较灵敏,特别是在如何应对陆地的二氧化碳季节性交换方面。总体而言,纬度区域的综合碳通量受中高纬度地区的观测强烈制约。这些对于区域碳通量的严重制约需要我们进一步改善传输模型并扩展对热带地区 CO_2 的观测网络。

19. Domine F, Shepson P B. Air-snow interactions and atmospheric chemistry[J]. Science, 2002, 297 (5586): 1506-1510.

Air-snow interactions and atmospheric chemistry

Domine F., Shepson P. B.

(CNRS, Lab Glaciol & Geophys Enviornm, BP 96, 54 Rue Moliere, F-38042 Grenoble, France)

Abstract: The presence of snow greatly perturbs the composition of near-surface polar air, and the higher concentrations of hydroxyl radicals (OH) observed result in greater oxidative capacity of the lower atmosphere. Emissions of nitrogen oxides, nitrous acid, light aldehydes, acetone, and molecular halogens have also been detected. Photolysis of nitrate ions contained in the snow appears to play an important role in creating these perturbations. OH formed in the snowpack can oxidize organic matter and halide ions in the snow, producing carbonyl compounds and halogens that are released to the atmosphere or incorporated into snow crystals. These reactions modify the composition of the snow, of the interstitial air, and of the overlying atmosphere. Reconstructing the composition of past atmospheres from ice-core analyses may therefore require complex corrections and modeling for reactive species.

气-雪相互作用与大气化学

Domine F. , Shepson P. B.

(法国科学研究中心)

摘要:雪的存在极大地扰乱了近地表极地空气的成分,观察到的高浓度氢氧自由基(OH)导致低层大气氧化能力更强。氮氧化物、亚硝酸、光醛、丙酮和卤素分子的排放也被检测到。雪中的硝酸根离子光解似乎在创造这些扰动中起了重要的作用。在积雪中形成的OH可以氧化雪中的有机物和卤素离子,产生羰基化合物和卤素释放到大气中或包裹在冰晶中。这些反应改变了雪、间隙空气和上覆大气的组成。因此通过冰芯分析重建过去大气成分可能需要复杂的修正和多种反应组分的模拟。

20. Tolbert M A, Toon O B. Atmospheric science: solving the PSC mystery[J]. Science, 2001, 292 (5514): 61-63.

Atmospheric science—solving the PSC mystery

Tolbert M. A. , Toon O. B.

(Univ Colorado, Boulder, CO 80309 US)

Abstract: Until the discovery of the ozone hole in the mid-1980s, polar stratospheric clouds (PSCs) were considered beautiful curiosities. In their perspective, Tolbert and Toon describe the insights gained since then into how PSCs form and how they affect the destruction of stratospheric ozone.

大气科学——解析极地平流层云之谜

Tolbert M. A. , Toon O. B.

(美国科罗拉多大学)

摘要:直到在20世纪80年代中期发现臭氧空洞,极地平流层云都被认为是美丽的奇观。本文中,Tolbert和Toon描述了自从发现臭氧空洞后对极地平流层云的见解,即PSC是如何形成的,以及它们如何影响到平流层臭氧的破坏。

21. Thompson A M, Witte J C, Hudson R D, et al. Tropical tropospheric ozone and biomass burning [J]. Science, 2001, 291(5511): 2128-2132.

Tropical tropospheric ozone and biomass burning

Thompson A. M., Witte J. C., Hudson R. D., Guo H., Herman J. R., Fujiwara M.

(NASA, Goddard Space Flight Ctr, Code 916, Greenbelt, MD 20771 USA)

Abstract: New methods for retrieving tropospheric ozone column depth and absorbing aerosol (smoke and dust) from the Earth Probe-Total Ozone Mapping Spectrometer (EP/TOMS) are used to follow pollution and to determine interannual variability and trends. During intense fires over Indonesia (August to November 1997), ozone plumes, decoupled from the smoke below, extended as far as India. This ozone overlay a regional ozone increase triggered by atmospheric responses to the El Nino and Indian Ocean Dipole. Tropospheric ozone and smoke aerosol measurements from the Nimbus 7 TOMS instrument show El Nino signals but no tropospheric ozone trend in the 1980s. Offsets between smoke and ozone seasonal maxima point to multiple factors determining tropical tropospheric ozone variability.

热带对流层臭氧和生物质燃烧

Thompson A. M., Witte J. C., Hudson R. D., Guo H., Herman J. R., Fujiwara M.

(美国国家航空和航天局)

摘要:利用地球探测-臭氧总量测绘光谱仪(EP/TOMS)反演对流层臭氧柱深度和吸收气溶胶(烟雾和粉尘)的新方法被用来追踪污染并确定年际变化和趋势。在印尼(1997 年 8 月至 11 月)的剧烈火灾期间,臭氧柱与下面的烟雾脱离,最远蔓延至印度。这个区域与厄尔尼诺和印度洋偶极子对大气的响应引发的区域重叠。从雨云 7 TOMS 仪器观测到的对流层臭氧和烟雾气溶胶显示 20 世纪 80 年代有厄尔尼诺信号,但没有对流层臭氧的趋势。烟雾和臭氧的季节最大值点之间的弥补,指明热带对流层臭氧变化由多个因素决定。

22. Singh H, Chen Y, Staudt A, et al. Evidence from the Pacific troposphere for large global sources of oxygenated organic compounds[J]. Nature, 2001, 410(6832): 1078-1081.

Evidence from the Pacific troposphere for large global sources of oxygenated organic compounds

Singh H., Chen Y., Staudt A., Jacob D., Blake D., Heikes B., Snow J.

(NASA, Ames Res Ctr, Moffett Field, CA 94035 USA)

Abstract: The presence of oxygenated organic compounds in the troposphere strongly influences key atmospheric processes. Such oxygenated species are, for example, carriers of reactive nitrogen and are easily photolysed, producing free radicals and so influence the oxidizing capacity and the ozone-forming potential of the atmosphere and may also contribute significantly to the organic component of aerosols. But knowledge of the distribution and sources of oxygenated organic compounds, especially in the Southern Hemisphere, is limited. Here we characterize the tropospheric composition of oxygenated organic species, using data from a recent airborne survey conducted over the tropical Pacific Ocean (30 degrees N to 30 degrees S). Measurements of a dozen oxygenated chemicals (carbonyls, alcohols, organic nitrates, organic pernitrates and peroxides), along with several C-2-C-8 hydrocarbons, reveal that abundances of oxygenated species are extremely high, and collectively, oxygenated species are nearly five times more abundant than non-methane hydrocarbons in the Southern Hemisphere. Current atmospheric models are unable to correctly simulate these findings, suggesting that large, diffuse, and hitherto-unknown sources of oxygenated organic compounds must therefore exist. Although the origin of these sources is still unclear, we suggest that oxygenated species could be formed via the oxidation of hydrocarbons in the atmosphere, the photochemical degradation of organic matter in the oceans, and direct emissions from terrestrial vegetation.

来自太平洋对流层的含氧有机化合物大规模的全球来源的证据

Singh H., Chen Y., Staudt A., Jacob D., Blake D., Heikes B., Snow J.

(美国国家航空和航天局)

摘要:在对流层中的有机含氧化合物的存在强烈影响着关键大气过程。举个例子,这种含氧化合物是活性氮的载体,很容易光解生成自由基从而影响大气的氧化能力和臭氧形成潜力,也可能极大地增加了气溶胶的有机组成部分。但是对于含氧有机化合物的分布和来源的知识,特别在南半球,是有限的。本文采用最近在热带太平洋(北纬 30 度至南纬 30 度)进行的航测数据,描述了含氧有机物的对流层组成。十几种含氧化合物(羰基、醇类、有机硝酸盐、有机过氧硝酸盐和过氧化物),以及一些 C-2-C-8 碳氢化合物的测量反映出含氧化合物的丰度非常高;在南半球含氧化合物之和是非甲烷碳氢化合物丰度的近五倍以上。目前存在的大气模型尚不能正确模拟这些研究结果,表明必定存在大量弥漫性并迄今未知的含氧有机化合物的来源。虽然这些来源的起源尚不清楚,我们认为含氧化合物有可能是通过大气碳氢化合物的氧化、海洋有机物光解及陆地植被的直接排放形成的。

23. Monnin E, Indermuhle A, Dallenbach A, et al. Atmospheric CO_2 concentrations over the Last Glacial Termination[J]. Science, 2001, 291(5501): 112-114.

Atmospheric CO_2 concentrations over the Last Glacial Termination

Monnin E., Indermuhle A., Dallenbach A., Fluckiger J., Stauffer B., Stocker T. F., Raynaud D., Barnola J. M.

(Univ Bern, Inst Phys, Sidlerstr 5, CH-3012 Bern, Switzerland)

Abstract: A record of atmospheric carbon dioxide (CO_2) concentration during the transition from the Last Glacial Maximum to the Holocene, obtained from the Dome Concordia, Antarctica, ice core, reveals that an increase of 76 parts per million by volume occurred over a period of 6000 years in four clearly distinguishable intervals. The close correlation between CO_2 concentration and Antarctic temperature indicates that the Southern Ocean played an important role in causing the CO_2 increase. However, the similarity of changes in CO_2 concentration and variations of atmospheric methane concentration suggests that processes in the tropics and in the Northern Hemisphere, where the main sources for methane are located, also had substantial effects on atmospheric CO_2 concentrations.

末次冰期结束时的大气 CO_2 浓度

Monnin E., Indermuhle A., Dallenbach A., Fluckiger J., Stauffer B., Stocker T. F., Raynaud D., Barnola J. M.

（瑞士伯尔尼大学）

摘要：南极冰穹C冰芯获得的末次盛冰期向全新世过渡时期大气中二氧化碳（CO_2）浓度的记录显示，在四个清晰可辨的时间间隔包含的6000多年里，CO_2 增加了76 ppmv。二氧化碳浓度和南极温度的密切相关表明，南大洋发挥了造成二氧化碳增加的重要作用。然而，二氧化碳浓度和大气甲烷浓度的相似性变化表明，热带与北半球——主要的甲烷源所在地，也对大气中二氧化碳浓度有重大影响。

24. Maher B A, Dennis P F. Evidence against dust-mediated control of glacial-interglacial changes in atmospheric CO_2[J]. Nature, 2001, 411(6834): 176-180.

Evidence against dust-mediated control of glacial-interglacial changes in atmospheric CO_2

Maher B. A., Dennis P. F.

(Univ E Anglia, Ctr Environm Magnetism & Palaeomagnetism, Norwich NR4 7TJ, Norfolk, England)

Abstract: The low concentration of atmospheric CO_2 inferred to have been present during glacial periods is thought to have been partly caused by an increased supply of iron-bearing dust to the ocean surface. This is supported by a recent model that attributes half of the CO_2 reduction during past glacial stages to iron-stimulated uptake of CO_2 by phytoplankton in the Southern Ocean. But atmospheric dust fluxes to the Southern Ocean, even in glacial periods, are thought to be relatively low and therefore it has been proposed that Southern Ocean productivity might be influenced by iron deposited elsewhere—for example, in the Northern Hemisphere—which is then transported south via ocean circulation (similar to the distal supply of iron to the equatorial Pacific Ocean). Here we examine the timing of dust fluxes to the North Atlantic Ocean, in relation to climate records from the Vostok ice core in Antarctica around the time of the penultimate deglaciation (about 130 kyr ago). Two main dust peaks occurred 155 kyr and 130 kyr ago, but neither was associated with the CO_2 rise recorded in the Vostok ice core. This mismatch, together with the low dust flux supplied to the Southern Ocean, suggests that dust-mediated iron fertilization of the Southern Ocean did not significantly influence atmospheric CO_2 at the termination of the penultimate glaciation.

反对冰期-间冰期大气 CO_2 变化为尘埃介导控制的证据

Maher B. A. , Dennis P. F.

（英国东安格利亚大学）

摘要：大气二氧化碳在冰期的低浓度被认为部分是由洋面含铁的尘埃造成的。最近一个模型支持这一解释，这一模型将在末次冰期二氧化碳减少的一半都归因于南大洋浮游植物由于铁的输入而引发的对 CO_2 的吸收。但是，南大洋的大气粉尘通量，即使在冰期，也被认为是比较低的。因此人们又提出南大洋生产力可能是由沉积在其他地方的铁(例如北半球)所影响的，它们通过洋流向南输送(类似赤道太平洋铁的远端供应)。本文考察了北大西洋中粉尘通量的时间，与南极沃斯托克冰芯的大约倒数第二次冰消期时(约 13 万年前)的气候记录相比较。两个主要尘峰发生在 15.5 万年和 13 万年前，但与沃斯托克冰芯记录的二氧化碳上升均无关。这种不匹配加上向南大洋供给的粉尘低通量表明，粉尘介导的南大洋富铁并没有显著影响倒数第二次冰期大气中二氧化碳的含量。

25. Yokouchi Y, Noijiri Y, Barrie L A, et al. A strong source of methyl chloride to the atmosphere from tropical coastal land[J]. Nature, 2000, 403(6767): 295-298.

A strong source of methyl chloride to the atmosphere from tropical coastal land

Yokouchi Y. , Noijiri Y. , Barrie L. A. , Toom-Sauntry D. , Machida T. , Inuzuka Y. , Akimoto H. , Li H. J. , Fujinuma Y. , Aoki S.

(Natl Inst Environm Studies, 16-2 Onogawa, Tsukuba, Ibaraki 3050053, Japan)

Abstract: Methyl chloride (CH_3Cl), the most abundant halocarbon in the atmosphere, has received much attention as a natural source of chlorine atoms in the stratosphere. The annual global flux of CH_3Cl has been estimated to be around 3.5 Tg on the grounds that this must balance the loss through reaction with OH radicals (which gives a lifetime for atmospheric CH_3Cl of 1.5 yr). The most likely main source of methyl chloride has been thought to be oceanic emission, with biomass burning the second largest source. But recent seawater measurements indicate that oceanic fluxes cannot account for more than 12% of the estimated global flux of CH_3Cl, raising the question of where the remainder comes from. Here we report evidence of significant CH_3Cl emission from warm coastal land, particularly from tropical islands. This conclusion is based on a global monitoring study and spot measurements, which show enhancement of atmospheric CH_3Cl in the tropics, a close correlation between CH_3Cl concentrations and those of biogenic compounds emitted by terrestrial plants, and OH-linked seasonality of CH_3Cl concentrations in middle and high latitudes. A strong, equatorially located source of this nature would explain why the distribution of CH_3Cl is uniform between the Northern and Southern hemispheres, despite their differences in ocean and land area.

热带沿海地区是大气一氯甲烷的巨大来源

Yokouchi Y., Noijiri Y., Barrie L. A., Toom-Sauntry D., Machida T., Inuzuka Y., Akimoto H., Li H. J., Fujinuma Y., Aoki S.

(日本国家环境研究中心)

摘要:一氯甲烷(CH_3Cl),大气中最丰富的卤代烃,已经作为平流层中氯原子的自然源受到广泛的关注。据估计每年全球 CH_3Cl 通量约 3.5 Tg,因为它必须平衡与 OH 自由基反应的损失(这就给出了大气中 CH_3Cl 的寿命约 1.5 年)。一氯甲烷的主要来源最有可能是海洋,其次是生物质燃烧。但最近的海水测量表明,海洋通量不可能超过全球 CH_3Cl 通量的 12%,这就提出了一个问题,其余的一氯甲烷来自哪里。本文阐述了 CH_3Cl 从温暖的沿海地区排放,特别是来自热带岛屿的重要证据。这一结论基于全球监测研究和现场测量,结果显示大气 CH_3Cl 在热带地区增加,CH_3Cl 浓度和陆生植物所释放的生源化合物和在中高纬度与 OH 有关的 CH_3Cl 季节性升高密切相关。一个强大的位于赤道的自然来源可以解释尽管存在海陆分布上的差异,CH_3Cl 仍均匀分布在南北半球之间。

26. Watson A J, Bakker D C E, Ridgwell A J, et al. Effect of iron supply on Southern Ocean CO_2 uptake and implications for glacial atmospheric CO_2[J]. Nature, 2000, 407(6805): 730-733.

Effect of iron supply on Southern Ocean CO_2 uptake and implications for glacial atmospheric CO_2

Watson A. J., Bakker D. C. E., Ridgwell A. J., Boyd P. W., Law C. S.

(Univ E Anglia, Sch Environm Sci, Norwich NR4 7TJ, Norfolk, England)

Abstract: Photosynthesis by marine phytoplankton in the Southern Ocean, and the associated uptake of carbon, is thought to be currently limited by the availability of iron. One implication of this limitation is that a larger iron supply to the region in glacial times could have stimulated algal photosynthesis, leading to lower concentrations of atmospheric CO_2. Similarly, it has been proposed that artificial iron fertilization of the oceans might increase future carbon sequestration. Here we report data from a whole-ecosystem test of the iron-limitation hypothesis in the Southern Ocean, which show that surface uptake of atmospheric CO_2 and uptake ratios of silica to carbon by phytoplankton were strongly influenced by nanomolar increases of iron concentration. We use these results to inform a model of global carbon and ocean nutrients, forced with atmospheric iron fluxes to the region derived from the Vostok ice-core dust record. During glacial periods, predicted magnitudes and timings of atmospheric CO_2 changes match ice-core records well. At glacial terminations, the model suggests that forcing of Southern Ocean biota by iron caused the initial similar to 40 p. p. m. of glacial-interglacial CO_2 change, but other mechanisms must have accounted for the remaining 40 p. p. m. increase. The experiment also confirms that modest sequestration of atmospheric CO_2 by artificial additions of iron to the Southern Ocean is in principle possible, although the period and geographical extent over which sequestration would be effective remain poorly known.

铁的供应对南大洋 CO_2 摄取的影响及其对冰消期大气 CO_2 的启示

Watson A. J., Bakker D. C. E., Ridgwell A. J., Boyd P. W., Law C. S.

(英国东安格利亚大学)

摘要:南大洋海洋浮游植物的光合作用及相关的碳吸收,一般被认为受控于铁的可用性。这个限制的一个含义是,在冰期一个更大的铁的供给可能刺激藻类光合作用,导致大气中的二氧化碳浓度较低。同样人们也已经提出,海洋人造铁施肥可能会增加未来的碳封存。本研究列出了铁限制假说在南大洋的整个生态系统的测试报告的数据,阐明了大气中二氧化碳的表面吸收和浮游植物 Si/C 的吸收率受铁含量纳摩尔增加的强烈影响。我们利用这些结果构建全球碳和海洋营养素的模型,并用南极 Vostok 冰芯粉尘记录对该地区的大气中铁通量加以约束。在冰期,大气中二氧化碳变化的预测的幅度和冰芯记录相匹配。在冰期结束期,模型则表明南大洋生物群造成了冰期-间冰期的大气中二氧化碳 40 ppm 的变化,但其他机制要解释余下的 40 ppm 的增加。实验还证实,人工补充铁对南大洋大气中二氧化碳的封存在理论上是可能的,但对有效碳的封存所需的时间和地域范围仍然知之甚少。

27. Tabazadeh A, Santee M L, Danilin M Y, et al. Quantifying denitrification and its effect on ozone recovery[J]. Science, 2000, 288(5470): 1407-1411.

Quantifying denitrification and its effect on ozone recovery

Tabazadeh A., Santee M. L., Danilin M. Y., Pumphrey H. C., Newman P. A., Hamill P. J., Mergenthaler J. L.

(NASA, Ames Res Ctr, MS 245-4, Moffett Field, CA 94035 USA)

Abstract: Upper Atmosphere Research Satellite observations indicate that extensive denitrification without significant dehydration currently occurs only in the Antarctic during mid to late June. The fact that denitrification occurs in a relatively warm month in the Antarctic raises concern about the likelihood of its occurrence and associated effects on ozone recovery in a colder and possibly more humid future Arctic lower stratosphere. Polar stratospheric cloud lifetimes required for Arctic denitrification to occur in the future are presented and contrasted against the current Antarctic cloud lifetimes. Model calculations show that widespread severe denitrification could enhance future Arctic ozone loss by up to 30%.

反硝化作用量化及其对臭氧层的恢复的影响

Tabazadeh A., Santee M. L., Danilin M. Y., Pumphrey H. C., Newman P. A., Hamill P. J., Mergenthaler J. L.

(美国国家航空和航天局)

摘要:高层大气研究卫星观测表明,目前仅在南极的 6 月中下旬发生无明显脱水的大量反硝化作用。反硝化作用发生在南极相对温暖的月份的事实引起其发生的关联效应和在较冷及可能比较潮湿的未来北极平流层下部臭氧层恢复影响的关注。本文还阐明了北极在未来发生的反硝化作用所需的极地平流层云的寿命,并与当前的南极平流层云的寿命进行了对比。模型计算表明,普遍严重的反硝化作用可提高未来北极臭氧损耗高达30%。

28. Sturges W T, Wallington T J, Hurley M D, et al. A potent greenhouse gas identified in the atmosphere: SF_5CF_3[J]. Science, 2000, 289(5479): 611-613.

A potent greenhouse gas identified in the atmosphere: SF_5CF_3

Sturges W. T., Wallington T. J., Hurley M. D., Shine K. P., Sihra K., Engel A., Oram D. E., Penkett S. A., Mulvaney R., Brenninkmeijer C. A. M.

(Univ E Anglia, Sch Environm Sci, Norwich NR4 7TJ, Norfolk, England)

Abstract: We detected a compound previously unreported in the atmosphere, trifluoromethyl sulfur pentafluoride (SF_5CF_3). Measurements of its infrared absorption cross section show SF_5CF_3 to have a radiative forcing of 0.57 watt per square meter per parts per billion. This is the largest radiative forcing, on a per molecule basis, of any gas found in the atmosphere to date. Antarctic firn measurements show it to have grown from near zero in the late 1960s to about 0.12 part per trillion in 1999. It is presently growing by about 0.008 part per trillion per year, or 6% per year. Stratospheric profiles of SF_5CF_3 suggest that it is long-lived in the atmosphere (on the order of 1000 years).

在大气中识别的一种强效的温室气体: SF_5CF_3

Sturges W. T., Wallington T. J., Hurley M. D., Shine K. P., Sihra K., Engel A., Oram D. E., Penkett S. A., Mulvaney R., Brenninkmeijer C. A. M.

(英国东安格利亚大学)

摘要:我们检测出大气中一种以前未报道的化合物,五氟化三氟甲基硫(SF_5CF_3)。其红外吸收截面的测量显示 SF_5CF_3 的辐射强迫为每平方米每十亿分之一浓度0.57瓦。这是迄今在大气中发现的分子基础上的最大的辐射强迫。南极粒雪的测量结果表明,从20世纪60年代中后期的近零增长到1999年的约0.12 ppt。目前增长约0.008 ppt每年,或每年6%。SF_5CF_3 平流层廓线表明,它是大气中的长寿气体(千年量级)。

29. Sigman D M, Boyle E A. Glacial/interglacial variations in atmospheric carbon dioxide[J]. Nature, 2000, 407(6806): 859-869.

Glacial/interglacial variations in atmospheric carbon dioxide

Sigman D. M., Boyle E. A.

(Princeton Univ, Dept Geosci, Guyot Hall, Princeton, NJ 08544 USA)

Abstract: Twenty years ago, measurements on ice cores showed that the concentration of carbon dioxide in the atmosphere was lower during ice ages than it is today. As yet, there is no broadly accepted explanation for this difference. Current investigations focus on the ocean's "biological pump", the sequestration of carbon in the ocean interior by the rain of organic carbon out of the surface ocean, and its effect on the burial of calcium carbonate in marine sediments. Some researchers surmise that the whole-ocean reservoir of algal nutrients was larger during glacial times, strengthening the biological pump at low latitudes, where these nutrients are currently limiting. Others propose that the biological pump was more efficient during glacial times because of more complete utilization of nutrients at high latitudes, where much of the nutrient supply currently goes unused. We present a version of the latter hypothesis that focuses on the open ocean surrounding Antarctica, involving both the biology and physics of that region.

大气中二氧化碳含量的冰期/间冰期变化

Sigman D. M. , Boyle E. A. .

（美国普林斯顿大学）

摘要：二十年前，对冰芯的测量结果表明，冰期大气中的二氧化碳浓度比现在低。到目前为止，对这种差异还没有被广泛接受的解释。目前的调查重点在于海洋“生物泵”、海洋表面的有机碳通过雨的作用在海洋内部封存和其对海洋沉积物中碳酸钙封存的影响。一些研究人员推测，整个海洋藻类营养储库在冰期更大，加强了低纬度地区生物泵的作用，而通常这些地区的营养素是有限的。另一些人则认为在冰期生物泵更有效是因为高纬度地区营养利用更充分，而通常这些营养供给是闲置无用的。我们对后一种假设进行深入讨论，专注于开放南极洲周围的海洋，涉及该地区的生物和物理。

30. Wagner T, Platt U. Satellite mapping of enhanced BrO concentrations in the troposphere[J]. Nature,1998,395(6701):486-490.

Satellite mapping of enhanced BrO concentrations in the troposphere

Wagner T. , Platt U.

(Univ Heidelberg, Inst Umweltphys, INF 366, D-69120 Heidelberg, Germany)

Abstract: Reactive bromine species contribute significantly to the destruction of ozone in the polar stratosphere. Reactive halogen compounds can have a strong effect not only on the chemistry of the stratosphere but also on that of the underlying troposphere. For example, severe ozone depletion events that are less persistent than those in the stratosphere occur in the Arctic and Antarctic boundary layer during springtime and are also associated with enhanced BrO abundances. Observations of BrO and ClO, which is less important at ground level during these ozone depletion events have revealed halogen oxide mixing ratios of up to 30 parts per trillion-sufficient to destroy within one to two days the 30-40 parts per billion of ozone typically present in the boundary layer. The catalytic mechanism leading to so-called "tropospheric ozone holes" is well established, but the origin of the increased BrO concentrations and the spatial and temporal extent of these events remains poorly understood. Here we present satellite observations showing that tropospheric air masses enriched in BrO are always situated close to sea ice and typically extend over areas of about 300-2000 km. The BrO abundances remain enhanced for periods of 1 to 3 days. These observations support the suggestion that autocatalytic release of bromine from sea salt gives rise to significant BrO formation which, in turn, initiates ozone depletion in the polar troposphere.

对流层 BrO 含量增大的卫星观测

Wagner T., Platt U.

(德国海德堡大学)

摘要：活性 Br 化物对极地平流层的臭氧损耗有重要的作用。活性卤素化合物不仅对平流层化学有很大影响，对其下层的对流层也有影响。比如，南北极边界层比平流层在春季存在更为严重的臭氧损耗以及与之相关的 BrO 含量增加。在这期间，地面 BrO 的观测显示卤素氧化物的混合比高达 30 ppt，足够在一两天内消耗 30～40 ppb 臭氧，尤其在边界层。这种导致的所谓臭氧空洞催化机制已经被证实，但是 BrO 增加的源头和臭氧洞的时空分布还未完全明晰。我们通过卫星数据发现对流层富含 BrO 的气团始终存在于海冰之上，在 300～2000 km 的区域内存在。BrO 的丰度在 1～3 天内上升，这也表明海盐自催化释放 Br 使 BrO 含量明显上升，也解释了极区对流层臭氧的减少。

31. Shindell D T, Rind D, Lonergan P. Increased polar stratospheric ozone losses and delayed eventual recovery owing to increasing greenhouse-gas concentrations[J]. Nature, 1998, 395(6701): 486-490.

Increased polar stratospheric ozone losses and delayed eventual recovery owing to increasing greenhouse-gas concentrations

Shindell D. T., Rind D., Lonergan P.

(NASA, Goddard Inst Space Studies, 2880 Broadway, New York, NY 10025 USA)

Abstract: The chemical reactions responsible for stratospheric ozone depletion are extremely sensitive to temperature. Greenhouse gases warm the Earth's surface but cool the stratosphere radiatively and therefore affect ozone depletion. Here we investigate the interplay between projected future emissions of greenhouse gases and levels of ozone-depleting halogen species using a global climate model that incorporates simplified ozone-depletion chemistry. Temperature and wind changes induced by the increasing greenhouse-gas concentrations alter planetary-wave propagation in our model, reducing the frequency of sudden stratospheric warmings in the Northern Hemisphere. This results in a more stable Arctic polar vortex, with significantly colder temperatures in the lower stratosphere and concomitantly increased ozone depletion. Increased concentrations of greenhouse gases might therefore be at least partly responsible for the very large Arctic ozone losses observed in recent winters. Arctic losses reach a maximum in the decade 2010 to 2019 in our model, roughly a decade after the maximum in stratospheric chlorine abundance. The mean losses are about the same as those over the Antarctic during the early 1990s, with geographically localized losses of up to two-thirds of the Arctic ozone column in the worst years. The severity and the duration of the Antarctic ozone hole are also predicted to increase because of greenhouse-gas-induced stratospheric cooling over the coming decades.

由于温室气体浓度增加造成的极地平流层臭氧损耗增加和延迟的最终恢复

Shindell D. T. , Rind D. , Lonergan P.

(美国国家航空和航天局)

摘要:平流层臭氧损耗的化学反应对温度极其敏感。温室气体温暖了地球的表面,但冷却了平流层辐射,从而影响臭氧耗损。本文使用全球气候模型,采用简化的化学臭氧消耗,探讨了消耗臭氧的温室气体和卤素含量与预计的未来温室气体排放之间的相互作用。在我们的模型中,不断增加的温室气体浓度诱导的温度和风速的变化改变了行星波传播,减少了北半球平流层突然变暖的频率。这导致了一个更加稳定的北极极涡,伴随较低的平流层的气温显著变冷,臭氧耗损随之增加。因此温室气体浓度的增加可能至少部分地导致了最近几个冬季非常大的北极地区的臭氧损耗。在我们的模型中,北极臭氧损耗于2010年到2019年的十年间达到最高,大约是在平流层氯的丰度达到最大的十年后。与20世纪90年代初南极上空的平均损耗基本相同,最差的年份里达到北极地区的臭氧柱损失的三分之二。模型也预测了未来几十年由于增加的温室气体引起的平流层降温导致南极臭氧洞的严重性和持续时间。

32. Neale P J, Davis R F, Cullen J J. Interactive effects of ozone depletion and vertical mixing on photosynthesis of Antarctic phytoplankton[J]. Nature, 1998, 392(6676): 585-589.

Interactive effects of ozone depletion and vertical mixing on photosynthesis of Antarctic phytoplankton

Neale P. J. , Davis R. F. , Cullen J. J.

(Dalhousie Univ, Dept Oceanog, Ctr Environm Observat Technol & Res, Halifax, NS B3H 4J1, Canada)

Abstract: Photosynthesis of Antarctic phytoplankton is inhibited by ambient ultraviolet (UV) radiation during incubations, and the inhibition is worse in regions beneath the Antarctic ozone "hole". But to evaluate such effects, experimental results on, and existing models of, photosynthesis cannot be extrapolated directly to the conditions of the open waters of the Antarctic because vertical mixing of phytoplankton alters UV exposure and has significant effects on the integrated inhibition through the water column. Here we present a model of UV-influenced photosynthesis in the presence of vertical mixing, which we constrain with comprehensive measurements from the Weddell-Scotia Confluence during the austral spring of 1993. Our calculations of photosynthesis integrated through the water column (denoted PT) show that photosynthesis is strongly inhibited by near-surface UV radiation, this inhibition can be either enhanced or decreased by vertical mixing, depending on the depth of the mixed layer. Predicted inhibition is most severe when mixing is rapid, extending to the lower part of the photic zone. Our analysis reveals that an abrupt 50% reduction in stratospheric ozone could, in the worst case, lower P-T by as much as 8.5%. However, stronger influences on inhibition can come from realistic changes in vertical mixing (maximum effect on P-T of about ±37%), measured differences in the sensitivity of phytoplankton to UV radiation (±46%) and cloudiness (±15%).

臭氧消耗和垂直混合对南极浮游植物光合作用的相互影响

Neale P. J., Davis R. F., Cullen J. J.

(加拿大达尔豪斯大学)

摘要:南极浮游植物的光合作用在繁殖期被环境紫外线(UV)辐射抑制,在南极臭氧洞下方的地区这种抑制作用更强。但为了评估这种影响,对光合作用的实验结果和现有模型不具直接推广到南极开放水域的条件,因为浮游植物的垂直混合改变了紫外辐射并对整个水柱的综合抑制作用有重要影响。本文提出了有垂直混合条件的紫外辐射影响光合作用模型,并以南半球1993年春季Weddell-Scotia洋流的全面测量为控制。对穿过水柱的光合作用(记作PT)的计算表明,光合作用受近地表的紫外辐射强烈抑制,这种抑制作用根据混合层深度,可以增强或减少垂直混合。当混合迅速时,预测的抑制作用最严重,可延伸到下部的透光区。我们的分析显示,在最坏的情况下,平流层臭氧突然减少50%,PT降低多达8.5%。不过,更强的抑制影响来自现实的垂直混合的变化(PT最大的效果约±37%)、浮游植物的紫外线辐射(±46%)和浊度(±15%)的敏感性的差异。

33. Roscoe H K, Jones A E, Lee A M. Midwinter start to Antarctic ozone depletion: evidence from observations and models[J]. Science, 1997, 278(5335): 93-96.

Midwinter start to Antarctic ozone depletion: evidence from observations and models

Roscoe H. K., Jones A. E., Lee A. M.

(British Antarctic Survey, Natural Environment Research Council, High Cross, Madingley Road, Cambridge CB3 0ET, UK)

Abstract: Measurements of total ozone at Faraday, Antarctica (65 degrees S), by a visible spectrometer show a winter maximum. This new observation is consistent with the descent of air within the polar vortex during early winter, together with ozone depletion starting in midwinter. Chemical depletion at these latitudes in midwinter is suggested by existing satellite observations of enhanced chlorine monoxide above 100 hectapascals and by reduced ozone in sonde profiles. New three-dimensional model calculations for 1994 confirm that chemical ozone depletion started in June at the sunlit vortex edge and became substantial by late July. This would not have been observed by most previous techniques, which either could not operate in winter or were closer to the Pole.

仲冬开始的南极臭氧耗损:来自观测和模型的证据

Roscoe H. K., Jones A. E., Lee A. M.

(英国南极调查局)

摘要:在南极洲(南纬65度)法拉第站,可见光谱仪观测到冬季臭氧总量的最大值。这项新发现连同仲冬开始的臭氧消耗与初冬极地涡旋内的空气是相关的。现有的卫星观测显示100 kPa以上氧化氯升高,探空廓线显示臭氧减少,指示了仲冬该纬度的化学损耗。对1994年新的三维模型计算表明,化学臭氧耗损始于6月阳光照射的极涡边缘，在7月下旬实质性开始。大多数以前的技术，不能在冬季或不能在极地运行，不会观测到这种变化。

34. Tolbert M A. Update-Polar clouds and sulfate aerosols[J]. Science，1996，272(5268)：1597.

Update-Polar clouds and sulfate aerosols

Tolbert M. A.

(Department of Chemistry and Biochemistry and CIRES, University of Colorado, Boulder, Colorado 80309, USA)

Abstract: Polar stratospheric clouds are known to play an important role in ozone destruction over the north and south poles. One puzzle is how the clouds can continuously form throughout the winter. In her perspective, Tolbert describes results by Koop and Carslaw that reveal a new pathway for cloud regeneration.

极地云与硫酸盐气溶胶

Tolbert M. A.

(美国科罗拉多大学)

摘要：极地平流层云对南北极的臭氧破坏发挥了重要作用。一个难题是云是如何在整个冬季不断形成的。从Tolbert的角度来看，她解释了Koop和Carslaw的结果，揭示了一个云再生的新途径。

35. Koop T, Carslaw K S. Melting of H_2SO_4 center dot $4H_2O$ particles upon cooling：implications for polar stratospheric clouds[J]. Science，1996，272(5268)：1638-1641.

Melting of H_2SO_4 center dot $4H_2O$ particles upon cooling: implications for polar stratospheric clouds

Koop T., Carslaw K. S.

(Max Planck Institute for Chemistry, Postfach 3060, 55020 Mainz, Germany)

Abstract: Polar stratospheric clouds (PSCs) are important for the chemical activation of chlorine compounds and subsequent ozone depletion. Solid PSCs can form on sulfuric acid tetrahydrate (SAT) ($H_2SO_4 \cdot 4H_2O$) nuclei, but recent laboratory experiments have shown that PSC nucleation on SAT is strongly hindered. A PSC formation mechanism is proposed in which SAT particles melt upon cooling in the presence of HNO_3 to form liquid HNO_3-H_2SO_4-H_2O droplets 2 to 3 kelvin above the ice frost point. This mechanism offers a PSC formation temperature that is defined by the ambient conditions and sets a temperature limit below which PSCs should form.

$H_2SO_4 \cdot 4H_2O$ 粒子的融化:极地平流层云的启示

Koop T. , Carslaw K. S.

(德国马普化学所)

摘要:极地平流层云(PSCs)对氯化物的化学活化和随后的臭氧消耗有重要的作用。固态的 PSCs 可以在硫酸水合物(SAT)($H_2SO_4 \cdot 4H_2O$)成核,但最近的实验室研究表明,PSC 在 SAT 上成核受到强烈的阻碍。文章提出了 PSC 的形成机制,在硝酸的存在下,在比霜点高 2～3 开尔文的温度下,SAT 颗粒融化后冷却成液态的硝酸-硫酸水合物的水滴。该机制提供了一个 PSC 的形成温度,该温度由环境条件及设置的 PSC 应该形成的温度下限定义。

36. Toon O B, Tolbert M A. Spectroscopic evidence against nitric-acid trihydrate in polar stratospheric clouds[J]. Nature,1995,375(6528):218-221.

Spectroscopic evidence against nitric-acid trihydrate in polar stratospheric clouds

Toon O. B. , Tolbert M. A.

(NASA Ames Research Center, Moffett Field, California 94035, USA)

Abstract: Heterogeneous reactions on polar stratospheric clouds (PSCs) play a key role in the photochemical mechanism thought to be responsible for ozone depletion in the Antarctic and the Arctic. Reactions on PSC particles activate chlorine to forms that are capable of photochemical ozone destruction, and sequester nitrogen oxides (NO_x) that would otherwise deactivate the chlorine. Although the heterogeneous chemistry is now well established, the composition of the clouds themselves is uncertain. It is commonly thought that they are composed of nitric acid trihydrate, although observations have left this question unresolved. Here we reanalyse infrared spectra of type Ⅰ PSCs obtained in Antarctica in September 1987, using recently measured optical constants of the various compounds that might be present in PSCs. We find that these PSCs were not composed of nitric acid trihydrate but instead had a more complex composition, perhaps that of a ternary solution. Because cloud formation is sensitive to their composition, this finding will alter our understanding of the locations and conditions in which PSCs form. In addition, the extent of ozone loss depends on the ability of the PSCs to remove NO_x permanently through sedimentation. The sedimentation rates depend on PSC particle size which in turn is controlled by the composition and formation mechanism.

不支持极地平流层云中的 $HNO_3 \cdot 3H_2O$ 的光谱证据

Toon O. B. , Tolbert M. A.

(美国国家航空和航天局)

摘要:极地平流层云(PSCs)的多相反应在被认为是南北极臭氧耗损主要原因的光化学机制中发挥

了关键作用。在 PSC 粒子上的反应将氯激活为具有光化学反应破坏臭氧的形式，并隔绝了能够降低氯活性的氮氧化物(NO_x)。虽然多相化学反应已经建立，但云本身的组成是不确定的，人们普遍认为是由硝酸的三水合物组成的，虽然在观测中这个问题还悬而未决。本文使用最近观测的可能在 PSC 中出现的多种化合物的光学常量，重新分析了 1987 年 9 月在南极获得的 Ⅰ 型 PSC 的红外光谱。我们发现，这些 PSC 不是由硝酸的三水合物组成的，而是一个更复杂的结构，也许是一个三元的构成。由于云的形成对其成分十分敏感，这一发现将改变我们对 PSC 形成的位置和条件的理解。此外，臭氧损耗的程度取决于 PSCs 通过沉淀永久去除氮氧化物隔绝 NO_x 的能力。沉淀的速率取决于 PSC 的颗粒大小，而这又受组成及形成机制的控制。

37. Santee M L, Read W G, Waters J W, et al. Interhemispheric differences in polar stratospheric HNO_3, H_2O, ClO, and O_3[J]. Science, 1995, 267(5199): 849-852.

Interhemispheric differences in polar stratospheric HNO_3, H_2O, ClO, and O_3

Santee M. L., Read W. G., Waters J. W., Froidevaux L., Manney G. L., Flower D. A., Jarnot R. F., Harwood R. S., Peckham G. E.

(Mail Stop 183701, Jet Propulsion Laboratory, California Institute of Technology, Pasadena, CA 91109, USA)

Abstract: Simultaneous global measurements of nitric acid (HNO_3), water (H_2O), chlorine monoxide (ClO), and ozone (O_3) in the stratosphere have been obtained over complete annual cycles in both hemispheres by the Microwave Limb Sounder on the Upper Atmosphere Research Satellite. A sizeable decrease in gas-phase HNO_3 was evident in the lower stratospheric vortex over Antarctica by early June 1992, followed by a significant reduction in gas-phase H_2O after mid-July. By mid-August, near the time of peak ClO, abundances of gas-phase HNO_3 and H_2O were extremely low. The concentrations of HNO_3 and H_2O over Antarctica remained depressed into November, well after temperatures in the lower stratosphere had risen above the evaporation threshold for polar stratospheric clouds, implying that denitrification and dehydration had occurred. No large decreases in either gas-phase HNO_3 or H_2O were observed in the 1992-1993 Arctic winter vortex. Although ClO was enhanced over the Arctic as it was over the Antarctic, Arctic O_3 depletion was substantially smaller than that over Antarctica. A major factor currently limiting the formation of an Arctic ozone "hole" is the lack of denitrification in the northern polar vortex, but future cooling of the lower stratosphere could lead to more intense denitrification and consequently larger losses of Arctic ozone.

极地平流层 HNO_3, H_2O, ClO 和 O_3 在两半球间的差异

Santee M. L., Read W. G., Waters J. W., Froidevaux L., Manney G. L., Flower D. A., Jarnot R. F., Harwood R. S., Peckham G. E.

(美国加州理工学院空气动力实验室)

摘要：高层大气研究卫星上的微波探测器的全球同步测量已取得较完整的两个半球平流层硝酸(HNO_3)、水(H_2O)、氧化氯(ClO)和臭氧(O_3)的年周期数据。气相硝酸的显著减少在 1992 年 6 月初在南极下部平流层中被观测到，随之是 7 月中旬以后气相 H_2O 的显著减少。截至 8 月中旬，在 ClO 达到峰

值的时候,气相 HNO_3 和 H_2O 的含量极低。进入 11 月,在较低的平流层温度已经升高至极地平流层云的蒸发限之后,南极上空的 HNO_3 和 H_2O 的浓度仍然很低,这意味着反硝化和脱水作用发生了。在 1992~1993 年的北极冬季极涡,气相硝酸或水都没有大的减少。虽然同南极一样,北极上空的 ClO 升高,但北极 O_3 的损耗大大小于南极。一个限制北极地区的臭氧洞形成的主要因素是北半球极地涡旋缺乏反硝化作用,但未来平流层下部的冷却可能会导致更加激烈的反硝化作用和北极地区的臭氧损耗。

38. Jones A E, Shanklin J D. Continued decline of total ozone over Halley, Antarctica, since 1985[J]. Nature, 1995, 376(6539): 409-411.

Continued decline of total ozone over Halley, Antarctica, since 1985

Jones A. E., Shanklin J. D.

(British Antarctic Survey, Natural Environment Research Council, High Cross, Madingley Road, Cambridge CBS OET, UK)

Abstract: In 1985, Farman et al announced that a dramatic reduction in total ozone was occurring in the atmosphere over Halley, Antarctica, during the polar spring. Analysis of satellite data revealed that this ozone depletion was an Antarctic-wide phenomenon. Combined theoretical, observational and laboratory work has shown that chlorine radicals derived from the photolysis of chlorofluorocarbons were the dominant cause of the ozone loss. Ten years later, we review here the status of the "ozone hole" based on the continued total-ozone measurements at Halley. The springtime "ozone hole" continues to deepen, with both the October mean and minimum total ozone persistently decreasing. The ozone loss extends into January and February, so that significant increases in ultraviolet-B radiation can be expected at the surface over Antarctica during the summer. A signal of ozone loss is now apparent in the spring and summer temperature records, with recent temperatures at the 100-mbar level consistently close to, or colder than, the historical (1957-1972) minima for the period October to January. These low temperatures may well enable the maintenance of springtime ozone-loss mechanisms until later in the year.

自 1985 年以来南极哈雷站上空臭氧总量的持续下降

Jones A. E., Shanklin J. D.

(英国南极调查局)

摘要:1985 年,法曼等人宣布在南极的哈雷站发现春季大气中的臭氧总量急剧减少。对卫星数据的分析显示,这种臭氧耗损是南极的普遍现象。结合理论、观测和实验室的工作表明,氯氟烃光解产生的氯自由基是臭氧损耗的主导原因。10 年后,我们在哈雷站持续观测臭氧总量的基础上回顾"臭氧洞"的状况。春季"臭氧洞"不断深化,10 月平均和最低的臭氧总量均持续下降。臭氧损耗延伸到 1 月和 2 月,可以预计在夏季南极地表的 UV-B 辐射会明显增加。现在看来,臭氧损耗的信号与春季和夏季的气温记录明显相关,最近的温度在 100 毫巴水平,持续接近或比历史最低的(1957~1972 年)10 月至 1 月期间更冷。这低温可能直到在今年晚些时候仍然维持春季臭氧损耗机制。

39. Fox L E, Worsnop D R, Zahniser M S, et al. Metastable phases in polar stratospheric aerosols[J].

Science,1995,267(5196):351-355.

Metastable phases in polar stratospheric aerosols

Fox L. E., Worsnop D. R., Zahniser M. S., Wofsy S. C.

(Division of Applied Sciences, Harvard University, 29 Oxford Street, Cambridge, MA 02138, USA)

Abstract: Phase changes in stratospheric aerosols were studied by cooling a droplet of sulfuric acid (H_2SO_4) in the presence of nitric acid (HNO_3) and water vapor. A sequence of solid phases was observed to form that followed Ostwald's rule for phase nucleation. For stratospheric partial pressures at temperatures between 193 and 195 kelvin, a metastable ternary H_2SO_4-HNO_3 hydrate, $H_2SO_4 \cdot HNO_3 \cdot 5H(_2)O$, formed in coexistence with binary $H_2SO_4 \cdot kH(_2)O$ hydrates ($k=2$, 3, and 4) and then transformed to nitric acid dihydrate, $HNO_3 \cdot 2H(_2)O$, within a few hours. Metastable $HNO_3 \cdot 2H(_2)O$ always formed before stable nitric acid trihydrate, $HNO_3 \cdot 3H(_2)O$, under stratospheric conditions and persisted for long periods. The formation of metastable phases provides a mechanism for differential particle growth and sedimentation of HNO_3 from the polar winter stratosphere.

极地平流层气溶胶的亚稳态

Fox L. E., Worsnop D. R., Zahniser M. S., Wofsy S. C.

(美国哈佛大学)

摘要:通过在硝酸(HNO_3)和水蒸气存在下冷却硫酸(H_2SO_4)液滴对平流层气溶胶的相位变化进行研究。固相序列依奥斯特瓦尔德定律阶段成核。平流层的部分压力在 193 K 和 195 K 之间,亚稳态的三元硫酸-硝酸水合物 $H_2SO_4 \cdot HNO_3 \cdot 5H(_2)O$,在 $H_2SO_4 \cdot kH(_2)O$ 水合物($k=2,3,4$)共同作用下形成,然后在几个小时内转化为二水硝酸 $HNO_3 \cdot 2H(_2)O$。在平流层的条件下,亚稳态 $HNO_3 \cdot 2H(_2)O$ 总是在稳定的硝酸三水态 $HNO_3 \cdot 3H(_2)O$ 之前形成,并持续很长一段时间。亚稳相的形成提供了一个极地冬季平流层硝酸不同粒子生长和沉淀的机制。

40. Newman P A. Antarctic total ozone in 1958[J]. Science,1994,264(5158):543-546.

Antarctic total ozone in 1958

Newman P. A.

(Nasa, Goddard Space Flight Ctr, Atmospher Chem & Dynam Branch, Code 916, Greenbelt, Md 20771, USA)

Abstract: The Antarctic ozone hole results from catalytic destruction of ozone by chlorine radicals. The hole develops in August, reaches its full depth in early October, and is gone by early December of each year. Extremely low total ozone measurements were made at the Antarctic Dumont d'Urville station in 1958. These measurements were derived from spectrographic plates of the blue sky, the moon, and two stars. These Dumont plate data are inconsistent with 1958 Dobson spectrophotometer ozone measurements, inconsistent with present-day Antarctic observations, and inconsistent with meteorological and theoretical information. There is no credible evidence for an ozone hole in 1958.

1958 年南极臭氧总量

Newman P. A.

(美国国家航空和航天局戈达德宇宙飞行中心)

摘要:南极臭氧洞是氯自由基催化破坏臭氧的结果。臭氧洞在每年 8 月形成发展,在 10 月初达到最大,在 12 月上旬结束。1958 年南极迪蒙·迪尔维尔站观测到非常低的臭氧总量。这些测量由蓝天、月亮和两颗星的光谱平板分析得到。这些杜蒙特平板数据与 1958 年多布森分光光度计臭氧测量数据不一致,不符合现今的南极观测,并与气象和理论信息不一致。因此没有可信的证据证明 1958 年的臭氧洞。

41. Worsnop D R, Fox L E, Zahniser M S, et al. Vapor-pressures of solid hydrates of nitric-acid: implications for polar stratospheric clouds[J]. Science, 1993, 259(5091): 71-74.

Vapor-pressures of solid hydrates of nitric-acid: implications for polar stratospheric clouds

Worsnop D. R., Fox L. E., Zahniser M. S., Wofsy S. C.

(Aerodyne Res Inc, Ctr Chem & Environm Phys, Billerica, Ma 01821)

Abstract: Thermodynamic data are presented for hydrates of nitric acid: $HNO_3 \cdot H_2O$, $HNO_3 \cdot 2H_2O$, $HNO_3 \cdot 3H_2O$, and a higher hydrate. Laboratory data indicate that nucleation and persistence of metastable $HNO_3 \cdot 2H_2O$ may be favored in polar stratospheric clouds over the slightly more stable $HNO_3 \cdot 3H_2O$. Atmospheric observations indicate that some polar stratospheric clouds may be composed of $HNO_3 \cdot 2H_2O$ and $HNO_3 \cdot 3H_2O$. Vapor transfer from $HNO_3 \cdot 2H_2O$ to $HNO_3 \cdot 3H_2O$ could be a key step in the sedimentation of HNO_3, which plays an important role in the depletion of polar ozone.

极地平流层云的启示:硝酸固体水合物的蒸汽压

Worsnop D. R., Fox L. E., Zahniser M. S., Wofsy S. C.

(美国重航空器研究中心)

摘要:本文列出了硝酸水合物 $HNO_3 \cdot H_2O$、$HNO_3 \cdot 2H_2O$、$HNO_3 \cdot 3H_2O$ 和更高的水合物热力学数据。实验数据表明,极地平流层云中亚稳 $HNO_3 \cdot 2H_2O$ 的成核和持久性可能会比稍微稳定的 $HNO_3 \cdot 3H_2O$ 更受青睐。大气观测表明,一些极地平流层云由 $HNO_3 \cdot 2H_2O$ 和 $HNO_3 \cdot 3H_2O$ 组成。蒸汽从 $HNO_3 \cdot 2H_2O$ 转移到 $HNO_3 \cdot 3H_2O$ 是硝酸沉淀的关键一步,它在极地臭氧损耗过程中起着重要作用。

42. Waugh D W. Subtropical stratospheric mixing linked to disturbances in the polar vortices[J]. Nature, 1993, 365(6446): 535-537.

Subtropical stratospheric mixing linked to disturbances in the polar vortices

Waugh D. W.

(Mit, Ctr Meteorol & Phys Oceanog, Cambridge, Ma 02139, USA)

Abstract: Randel et al have observed tongues of stratospheric air stretching from the tropics into middle latitudes, and conclude that such events may be responsible for transporting significant amounts of stratospheric air across the tropical-mid-latitude barrier. Here I examine the movements of air parcels during these events using high-resolution contour-trajectory calculations. My calculations suggest that the tongues of tropical air are associated with disturbances of the stratospheric polar vortices. The edge of the disturbed polar vortex reaches low latitudes, and draws a long tongue of tropical air around the vortex into middle latitudes. This process occurs in the winter of both hemispheres, although the edge of the larger Antarctic polar vortex reaches farther toward the Equator, and draws up material from lower latitudes, than its Arctic counterpart.

与极地涡旋扰动相关的亚热带平流层混合

Waugh D. W.

(美国麻省理工学院)

摘要:兰德尔等人观察到从热带延伸到中纬度地区的平流层气团,并总结出这样的事件可能会对运送大量的平流层空气穿过整个热带-中纬度屏障产生重要影响。本文利用高分辨率的廓线轨迹计算对这个事件中的大气运动进行研究。计算结果表明,热带空气舌与平流层极涡的干扰一致。扰乱的极涡边缘到达低纬度地区,热带空气在中纬度地区形成长长的气舌。这一过程发生在两个半球的冬天,较大的南极极涡的边缘比北极极涡向赤道到达得更远,并从低纬度地区携带物质。

43. Waters J W, Froidevaux L, Read W G, et al. Stratospheric ClO and ozone from the microwave limb sounder on the upper-atmosphere research satellite[J]. Nature,1993,362(6421):597-602.

Stratospheric ClO and ozone from the microwave limb sounder on the upper-atmosphere research satellite

Waters J. W., Froidevaux L., Read W. G., Manney G. L., Elson L. S., Flower D. A., Jarnot R. F., Harwood R. S.

(Jet Prop Lab, Pasadena, Ca 91109 USA)

Abstract: Concentrations of atmospheric ozone and of ClO (the predominant form of reactive chlorine responsible for stratospheric ozone depletion) are reported for both the Arctic and Antarctic winters of the past 18 months. Chlorine in the lower stratosphere was almost completely converted to chemically reactive forms in both the northern and southern polar winter vortices. This occurred in the south long before the development of the Antarctic ozone hole, suggesting that ozone loss can be masked by influx of ozone-rich air.

高层大气研究卫星搭载微波声呐观测平流层 ClO 和 O_3

Waters J. W., Froidevaux L., Read W. G., Manney G. L., Elson L. S., Flower D. A., Jarnot R. F., Harwood R. S.

(美国加州理工学院喷气推进实验室)

摘要:本文对过去 18 个月的北极和南极的冬季大气臭氧和 ClO 的浓度(负责平流层臭氧耗损的活性氯的主要形式)进行报告。在北极和南极的冬季涡旋,平流层下部的氯几乎完全转化为化学反应的形式。这在南极臭氧洞形成之前很长一段时间就已经产生,表明臭氧损耗可能被丰富的臭氧涌入所掩盖。

44. Toumi R, Jones R L, Pyle J A. Stratospheric ozone depletion by $ClONO_2$ photolysis[J]. Nature, 1993,365(6441):37-39.

Stratospheric ozone depletion by $ClONO_2$ photolysis

Toumi R., Jones R. L., Pyle J. A.

(Univ Cambridge, Ctr Atmospher Sci, Dept Chem, Lensfield Rd, Cambridge Cb2 1ew, England)

Abstract: Springtime ozone depletion over Antarctica is thought to be due to catalytic cycles involving chlorine monoxide, which is formed as a result of reactions on the surface of polar stratospheric clouds (PSCs). When the PSCs evaporate, ClO in the polar air can react with NO_2 to form the reservoir species $ClONO_2$. High concentrations of $ClONO_2$ can also be found at lower latitudes because of direct transport of polar air or mixing of ClO and NO_2 at the edges of the polar vortex. $ClONO_2$ can take part in an ozone-depleting catalytic cycle, but the significance of this cycle has not been clear. Here we present model simulations of ozone concentrations from March to May both within the Arctic vortex and at a mid-latitude Northern Hemisphere site. We find increasing ozone loss from March to May. The $ClONO_2$ cycle seems to be responsible for a significant proportion of the simulated ozone loss. An important aspect of this cycle is that it is not as limited as the other chlorine cycles to the timing and location of PSCs; it may therefore play an important role in ozone depletion at warm middle latitudes.

$ClONO_2$ 光解导致的平流层臭氧损耗

Toumi R., Jones R. L., Pyle J. A.

(英国剑桥大学)

摘要:春天,南极上空的臭氧损耗被认为是由于氧化氯的催化循环在极地平流层云(PSCs)表面反应的结果。当 PSCs 蒸发时,极地空气 ClO 可以与 NO_2 反应形成储库 $ClONO_2$。高浓度的 $ClONO_2$ 也在低纬度地区被发现,由于极地空气直接运输或 ClO 和 NO_2 在极涡的边缘混合,$ClONO_2$ 可以作为消耗臭氧催化循环的一部分,但这个周期的意义尚未明确。本文列举了目前的模型模拟北极涡旋和在北半球中纬度站点从 3 月至 5 月的臭氧浓度,我们发现从 3 月至 5 月臭氧的损耗越来越多。$ClONO_2$ 的循环似乎是模拟臭氧损耗的一个重要部分,这个循环的一个重要方面是,它并不像其他含氯化合物一样,受 PSCs 的时间和地点的限制,因此它可能会在温暖的中纬度地区的臭氧消耗中发挥重要作用。

45. Randel W. Ideas flow on Antarctic vortex[J]. Nature,1993,364(6433):105-106.

Ideas flow on Antarctic vortex

Randel W.

(National Center for Atmospheric Research, Boulder, Colorado 80307, USA)

Abstract: Ozone destruction within the vortex that forms over the Antarctic during polar winter gives us the all too familiar "ozone hole". But if large volumes of air are swept through the vortex, ozone-depleting reactions at the poles could also be affecting mid-latitudes, where most of the world's population lives. New results from the Upper Atmosphere Research Satellite (UARS), presented at a conference in late May, have rekindled arguments over just how fast material does flow through the stratospheric vortex.

南极涡旋引发的思索

Randel W.

(美国国家大气研究中心)

摘要:南极冬季涡旋中的臭氧损耗形成了我们所熟知的臭氧空洞。但是如果大量的空气席卷了涡旋,那么极地的臭氧损耗还将影响中纬度地区——世界上大部分人类居住的地方。在 5 月下旬,来自 UARS 的最近研究结果重新引发了穿过平流层涡旋的气流有多快的讨论。

46. Molina M J,Zhang R,Wooldridge P J,et al. Physical-chemistry of the $H_2SO_4/HNO_3/H_2O$ system: implications for polar stratospheric clouds[J]. Science,1993,261(5127):1418-1423.

Physical-chemistry of the $H_2SO_4/HNO_3/H_2O$ system: implications for polar stratospheric clouds

Molina M. J., Zhang R., Wooldridge P. J., Mcmahon J. R., Kim J. E., Chang H. Y., Beyer K. D.

(Mit, Dept Earth Atmospher & Planetary Sci, Cambridge, Ma 02139)

Abstract: Polar stratospheric clouds (PSCs) play a key role in stratospheric ozone depletion. Surface-catalyzed reactions on PSC particles generate chlorine compounds that photolyze readily to yield chlorine radicals, which in turn destroy ozone very efficiently. The most prevalent PSCs form at temperatures several degrees above the ice frost point and are believed to consist of HNO_3 hydrates; however, their formation mechanism is unclear. Results of laboratory experiments are presented which indicate that the background stratospheric H_2SO_4/H_2O aerosols provide an essential link in this mechanism: these liquid aerosols absorb significant amounts of HNO_3 vapor, leading most likely to the crystallization of nitric acid trihydrate (NAT). The frozen particles then grow to form PSCs by condensation of additional amounts of HNO_3 and H_2O vapor. Furthermore, reaction probability measurements reveal that the chlorine radical precursors are formed readily at polar stratospheric temperatures not just on NAT and ice crystals, but also on liquid H_2SO_4 solutions and on solid H_2SO_4 hydrates. These results imply that the chlorine activation efficiency of the aerosol particles increases rapidly as the temperature approaches the ice frost point regardless of the phase or composition of the particles.

$H_2SO_4/HNO_3/H_2O$ 系统的物理化学过程——极地平流层云的启示

Molina M. J., Zhang R., Wooldridge P. J., Mcmahon J. R., Kim J. E., Chang H. Y., Beyer K. D.

(美国麻省理工学院)

摘要:极地平流层云(PSCs)在平流层臭氧耗损中发挥了关键作用。PSC 颗粒上的表面催化反应生成易光解氯自由基的氯化合物,这反过来又非常有效地破坏臭氧层。最常见的 PSC 在冰霜冻点几摄氏度以上的温度形成,被认为是由硝酸水合物构成的,但其形成机制尚不清楚。实验结果表明,背景平流层的 H_2SO_4/H_2O 气溶胶在这个机制中提供了一个必不可少的环节:这些液体气溶胶吸收大量的硝酸蒸汽,最有可能导致硝酸三水合物(NAT)的结晶。然后冷冻颗粒通过 HNO_3 和水蒸气的凝结增长形成 PSCs。此外,反应概率的测量表明,在极地平流层温度下,氯自由基的前体,不仅在 NAT 和冰晶上,而且在液体 H_2SO_4 和固体硫酸水合物上,也容易形成。这些结果意味着随着温度接近冰的霜冻点,无论颗粒的组成或相怎样,气溶胶粒子的氯活化效率都迅速增加。

47. Keys J G, Johnston P V, Blatherwick R D, et al. Evidence for heterogeneous reactions in the Antarctic autumn stratosphere[J]. Nature, 1993, 361(6407): 49-51.

Evidence for heterogeneous reactions in the Antarctic autumn stratosphere

Keys J. G., Johnston P. V., Blatherwick R. D., Murcray F. J.

(Natl Inst Water & Atmospher Res, Lauder, New Zealand Univ Denver, Dept Phys, Denver, Co 80208)

Abstract: Reactive chlorine compounds are known to cause ozone depletion in the Antarctic stratosphere, but they can be bound into an inactive form through reactions with nitrogen dioxide. In the spring, NO_2 can be converted to a long-lived reservoir species, HNO_3, on the surface of polar stratospheric clouds. This removes NO_2 from the stratosphere and allows chlorine-catalysed ozone destruction to proceed. It has been suggested that similar reactions may take place on background sulphate aerosols in the Antarctic stratosphere, but as yet there has been no unambiguous evidence for these reactions in the absence of polar stratospheric clouds (although there have been observations of ozone loss attributed to volcanic aerosols). Here we present measurements of Antarctic stratospheric NO_2 and HNO_3 concentrations taken in 1991. Our results demonstrate that reactive nitrogen was converted to HNO_3 in autumn, before temperatures were low enough for polar stratospheric clouds to form. We conclude that heterogeneous chemistry on background aerosols was responsible for this conversion, which brought with it the potential for additional ozone loss in the autumn.

南极秋季平流层中非均相反应的证据

Keys J. G., Johnston P. V., Blatherwick R. D., Murcray F. J.

(新西兰国家水与大气研究中心)

摘要：活性氯化合物是已知的导致南极平流层臭氧耗损的物质，但它们可以通过与二氧化氮反应形成一个非活性形式。在春季，二氧化氮可在极地平流层云表面形成长寿命储库形式——硝酸。这个过程从平流层中移除二氧化氮并允许氯催化进行臭氧破坏。有人提出类似的反应可能会在南极平流层背景硫酸盐气溶胶下发生，但在极地平流层云的情况下这些反应尚未有明确的证据（虽然已经有火山气溶胶对臭氧损耗贡献的观测）。本文观测了 1991 年南极平流层二氧化氮和硝酸浓度。我们的研究结果表明，秋季温度足够低到极地平流层云的形成之前，活性氮转化为硝酸。我们的结论是本底气溶胶的非均相化学过程对这个转换起重要作用，带来秋季额外的臭氧损耗的潜力。

48. Elkins J W, Thompson T M, Swanson T H, et al. Decrease in the growth-rates of atmospheric Chlorofluorocarbon-11 and Chlorofluorocarbon-12[J]. Nature, 1993, 364(6440): 780-783.

Decrease in the growth-rates of atmospheric Chlorofluorocarbon-11 and Chlorofluorocarbon-12

Elkins J. W., Thompson T. M., Swanson T. H., Butler J. H., Hall B. D., Cummings S. O., Fisher D. A., Raffo A. G.

(Noaa, Climate Monitoring & Diagnost, Boulder, Co 80303)

Abstract: The discovery of the Antarctic ozone hole in 1985 led to international efforts to reduce emissions of ozone-destroying chlorofluorocarbons. These efforts culminated in the Montreal Protocol and its subsequent amendments, which called for the elimination of CFC production by 1996. Here we focus on CFC-11 (CCl_3F) and CFC-12 (CCl_2F_2), which are used for refrigeration, air conditioning and the production of aerosols and foams, and which together make up about half of the total abundance of stratospheric organic chlorine. We report a significant recent decrease in the atmospheric growth rates of these two species, based on measurements spanning the past 15 years and latitudes ranging from 83-degrees-N to 90-degrees-S. This is consistent with CFC-producers' own estimates of reduced emissions. If the atmospheric growth rates of these two species continue to slow in line with predicted changes in industrial emissions, global atmospheric mixing ratios will reach a maximum before the turn of the century, and then begin to decline.

大气 CFC-11 和 CFC-12 生成速率的降低

Elkins J. W., Thompson T. M., Swanson T. H., Butler J. H., Hall B. D., Cummings S. O., Fisher D. A., Raffo A. G.

(美国国家海洋和大气局)

摘要:1985 年的南极臭氧洞的发现导致国际性的致力于减少破坏臭氧层的含氯氟烃排放。这些努力使《蒙特利尔议定书》及其后的修正案呼吁到 1996 年前消除氟氯化碳的生产。本文重点对用于制冷、空调、气溶胶和泡沫生产的 CFC-11(CCl_3F)和 CFC-12(CCl_2F_2),以及它们一起组成了约一半的平流层中的有机氯的丰度进行研究。根据覆盖范围从 83 °N 到 90 °S,在过去 15 年测量结果表明这两个物种的大气增长率显著下降。这与氟氯化碳生产者估计的减少的排放一致。如果这两个物种的大气增长率继续按照工业排放量预测的变化放缓,全球大气中的混合比例将在世纪之交前达到最大值,然后开始下降。

49. Stolarski R, Bojkov R, Bishop L, et al. Measured trends in stratospheric ozone[J]. Science, 1992, 256(5055):342-349.

Measured trends in stratospheric ozone

Stolarski R., Bojkov R., Bishop L., Zerefos C., Staehelin J., Zawodny J.

(Nasa, Goddard Space Flight Ctr, Greenbelt, Md 20771)

Abstract: Recent findings, based on both ground-based and satellite measurements, have established that there has been an apparent downward trend in the total column amount of ozone over mid-latitude areas of the Northern Hemisphere in all seasons. Measurements of the altitude profile of the change in the ozone concentration have established that decreases are taking place in the lower stratosphere in the region of highest ozone concentration. Analysis of updated ozone records, through March of 1991, including 29 stations in the former Soviet Union, and analysis of independently calibrated satellite data records from the Total Ozone Mapping Spectrometer and Stratospheric Aerosol and Gas Experiment instruments confirm many of the findings originally derived from the Dobson record concerning northern mid-latitude changes in ozone. The data from many instruments now provide a fairly consistent picture of the change that has occurred in stratospheric ozone levels.

平流层臭氧观测趋势

Stolarski R., Bojkov R., Bishop L., Zerefos C., Staehelin J., Zawodny J.

(美国国家航天和宇航局)

摘要:基于地面和卫星测量,最近的研究结果表明北半球中纬度地区的各个季节臭氧的柱总量有明显下降趋势。测量臭氧浓度的高度变化廓线已确立在平流层下部的臭氧浓度最高的地区臭氧减少。根据 1991 年 3 月期间,包括苏联的 29 站更新的臭氧记录,以及从臭氧总量测绘光谱仪及平流层气溶胶和气体实验仪器独立校准的卫星数据记录分析确认了许多来自 dobson 记录的关于臭氧在北半球中纬度地区变化的发现。现在许多仪器的数据提供了一个相当一致的平流层臭氧变化的图片。

50. Smith R C, Prezelin B B, Baker K S, et al. Ozone depletion: ultraviolet-radiation and phytoplankton biology in Antarctic waters[J]. Science,1992,255(5047):952-959.

Ozone depletion—ultraviolet-radiation and phytoplankton biology in Antarctic waters

Smith R. C., Prezelin B. B., Baker K. S., Bidigare R. R., Boucher N. P., Coley T., Karentz D., Macintyre S., Matlick H. A., Menzies D., Ondrusek M., Wan Z., Waters K. J.

(Univ Calif Santa Barbara, Dept Geog, Santa Barbara, Ca 93106)

Abstract: The springtime stratospheric ozone layer over the Antarctic is thinning by as much as 50 percent, resulting in increased midultraviolet (UVB) radiation reaching the surface of the Southern Ocean. There is concern that phytoplankton communities confined to near-surface waters of the marginal ice zone will be harmed by increased UVB irradiance penetrating the ocean surface, thereby altering the dynamics of Antarctic marine ecosystems. Results from a 6-week cruise (Icecolors) in the marginal ice zone of the Bellingshausen Sea in austral spring of 1990 indicated that as the O_3 layer thinned: (ⅰ) sea surface and depth-dependent ratios of UVB irradiance (280 to 320 nanometers) to total irradiance (280 to 700 nanometers) increased and (ⅱ) UVB inhibition of photosynthesis increased. These and other Icecolors findings suggest that O_3-dependent shifts of in-water spectral irradiances alter the balance of spectrally dependent phytoplankton processes, including photoinhibition, photo-reactivation, photoprotection, and photosynthesis. A minimum 6 to 12 percent reduction in primary production associated with O_3 depletion was estimated for the duration of the cruise.

臭氧损耗——紫外辐射与南极海域的浮游植物

Smith R. C., Prezelin B. B., Baker K. S., Bidigare R. R., Boucher N. P., Coley T., Karentz D., Macintyre S., Matlick H. A., Menzies D., Ondrusek M., Wan Z., Waters K. J.

(美国加州大学圣巴巴拉分校)

摘要:春季南极上空的平流层臭氧层变薄 50%,导致到达南大洋表面的 UVB 辐射增强。有人担心冰区边缘的近地表水域浮游植物群落将受到穿透海洋表面增强的 UVB 辐射伤害,从而改变南极海洋生态系统动力学。从 1990 年别林斯高晋海的边际冰区 6 周巡航的结果表明,随着南半球春天的臭氧层变薄:① 海表面和与深度有关的紫外线辐射(280～320 nm)比总辐射(280～700 nm)的比率增大;② UVB 对光合作用的抑制增加。这些和其他航线结果表明,依赖臭氧转变的水中的光谱辐照度改变光谱依赖浮游植物过程的平衡,包括光抑制、光再活化、光保护和光合作用。在巡航期间,据估计最低有 6%～12% 的初级产品生产的减少与臭氧耗损有关。

51. Ramaswamy V, Schwarzkopf M D, Shine K P. Radiative forcing of climate from halocarbon-

induced global stratospheric ozone loss[J]. Nature,1992,355(6363):810-812.

Radiative forcing of climate from halocarbon-induced global stratospheric ozone loss

Ramaswamy V., Schwarzkopf M. D., Shine K. P.

(Princeton Univ, Atmospher & Ocean Sci Program, Princeton, Nj 08542)

Abstract: Observations from satellite and ground-based instruments indicate that between 1979 and 1990 there have been statistically significant losses of ozone in the lower stratosphere of the middle to high latitudes in both hemispheres. Here we determine the radiative forcing of the surface-troposphere system due to the observed decadal ozone losses, and compare it with that due to the increased concentrations of the other main radiatively active gases (CO_2, CH_4, N_2O and chlorofluorocarbons) over the same time period. Our results indicate that a significant negative radiative forcing results from ozone losses in middle to high latitudes, in contrast to the positive forcing at all latitudes caused by the CFCs and other gases. As the anthropogenic emissions of CFCs and other halocarbons are thought to be largely responsible for the observed ozone depletions, our results suggest that the net decadal contribution of CFCs to the greenhouse climate forcing is substantially less than previously estimated.

卤代烃引发的全球平流层臭氧损耗与气候辐射强迫

Ramaswamy V., Schwarzkopf M. D., Shine K. P.

(美国普林斯顿大学)

摘要:基于卫星和地面仪器的观测结果表明,1979~1990年间在两个半球中高纬度的平流层下部已有显著的臭氧损耗。本文根据十年间观测到的由于臭氧损耗造成的对流层表面系统的辐射强迫,并与在同一时间由于其他主要辐射活性气体(二氧化碳、甲烷、氧化亚氮和氟氯烃)浓度增加所导致的辐射强迫进行比较。我们的研究结果表明,与氟氯化碳和其他气体造成的所有纬度的正向强迫相比,中高纬度臭氧亏损导致显著的负辐射强迫。人为排放的氟氯化碳和其他卤烃的量被认为是所观察到的臭氧消耗的主要原因,我们的研究结果表明,十年的氟氯化碳对温室气候净强迫的贡献大大低于以前的估计。

52. Prather M J. More rapid polar ozone depletion through the reaction of HOCl with HCl on polar stratospheric clouds[J]. Nature,1992,355(6360):534-537.

More rapid polar ozone depletion through the reaction of HOCl with HCl on polar stratospheric clouds

Prather M. J.

(Nasa, Goddard Inst Space Studies, New York, Ny 10025)

Abstract: The direct reaction of HOCl with HCl, known to occur in liquid water and on glass surfaces, has now been measured on surfaces similar to polar stratospheric clouds and is shown here to play a critical part in polar ozone loss. Two keys to understanding the chemistry of the Antarctic ozone hole are, one, the recognition that reactions on polar stratospheric clouds transform HCl into more reactive species denoted by $ClO_{(x)}$ and, two, the discovery of the ClO-dimer (Cl_2O_2) mechanism that rapidly catalyses destruction of O_3. Observations of high levels of OClO and ClO in the springtime Antarctic stratosphere confirm that most of the available chlorine is in the form of $ClO_{(x)}$. But current photochemical models have difficulty converting HCl to $ClO_{(x)}$ rapidly enough in early spring to account fully for the observations. Here I show, using a chemical model, that the direct reaction of HOCl with HCl provides the missing mechanism. As alternative sources of nitrogen-containing oxidants, such as N_2O_5 and $ClONO_2$, have been converted in the late autumn to inactive HNO_3 by known reactions on the sulphate-layer aerosols, the reaction of HOCl with HCl on polar stratospheric clouds becomes the most important pathway for releasing that stratospheric chlorine which goes into polar night as HCl.

通过极地平流层云中 HOCl 与 HCl 的反应产生的更迅速的臭氧损耗

Prather M. J.

(美国国家航空和航天局)

摘要:次氯酸与盐酸发生在液态水和玻璃表面的直接反应现在已经在类似极地平流层云的表面上测定,并显示在极地臭氧损耗过程中发挥关键作用。了解南极臭氧洞化学有两个关键点,其一,在极地平流层云上发生 HCl 转化为活性更高的 $ClO_{(x)}$的反应;其二,能够迅速催化破坏臭氧的 ClO 二聚体(Cl_2O_2)机制。在春季南极平流层中,观测到的高浓度 OClO 与 ClO 证实,氯最有可能存在的形式是 $ClO_{(x)}$。但是,目前的光化学模型在初春 HCl 快速转换为 $ClO_{(x)}$的模拟上仍有困难,不能充分解释观测数据。本文使用化学模型,次氯酸与盐酸直接反应提供了所缺少的机制。像含氮氧化剂的可替代源,如 N_2O_5 和 $ClONO_2$,在深秋通过硫酸盐层气溶胶上进行的已知反应转换为不活跃的硝酸一样,在极地平流层云上次氯酸与盐酸的反应成为最重要的以 HCl 形式在夜晚释放平流层氯的途径。

53. Mcfarland M. Investigations of the environmental acceptability of fluorocarbon alternatives to chlorofluorocarbons[J]. PNAS, 1992, 89(3): 807-811.

Investigations of the environmental acceptability of fluorocarbon alternatives to chlorofluorocarbons

Mcfarland M.

(Dupont Co, Fluorochem, B-13230, Wilmington, De 19898)

Abstract: Chlorofluorocarbons (CFCs) are currently used in systems for preservation of perishable foods and medical supplies, increasing worker productivity and consumer comfort, conserving energy and increasing product reliability. As use of CFCs is phased out due to concerns of ozone depletion, a variety of new chemicals and technologies will be needed to serve these needs. In choosing alternatives, industry must balance concerns over safety and environmental acceptability and still meet the performance characteristics of the current CFC-based products. About 60% of projected CFC demand will either be eliminated by improved conservation practices or will be satisfied by nonfluorocarbon alternatives. With current technology, the only viable alternatives meeting the safety, performance, and environmental requirements for the remaining 40% of demand are fluorocarbons, hydrochlorofluorocarbons (HCFCs), and hydrofluorocarbons (HFCs). HCFCs and HFCs possess many of the desirable properties of the CFCs, but because of the, hydrogen, they react with hydroxyl in the lower atmosphere. This results in shorter atmospheric lifetimes compared to CFCs and reduces their potential to contribute to stratospheric ozone depletion or global warming; HFCs do not contain chlorine and have no potential to destroy ozone. This paper provides an overview of challenges faced by industry, regulators, and society in general in continuing to meet societal needs and consumer demands while reducing risk to the environment without compromising consumer or worker safety.

含氟烃类替代氯氟烃的环境可接受性的调查

Mcfarland M.

(美国杜邦公司)

摘要:氯氟烃(CFCs)一般用于保存易腐食品和医疗用品,提高工人的生产力和消费舒适度,节约能源和提高产品的可靠性等。由于对臭氧的损耗,氟氯化碳的使用正逐步淘汰,需要各种新的化学品和技术以满足这些需要。在选择替代品时,必须考虑对安全和环境的可接受度,同时满足当前使用 CFC 产品的性能特点。约 60%的氟氯化碳的预计需求将被改进的清洁手段淘汰,或由不含氟碳的替代品满足。以现有技术,能够同时满足安全、性能和环保要求的其余 40%的需求的唯一可行的替代品是碳氟化合物、水和氟氯烃(HCFCs)以及水和氟化烃(HFCs)。HCFCs 和 HFCs 化合物具有氟氯化碳的许多优良特性,但是因为含氢离子,它们在低层大气中与羟基发生反应。这将导致相比氟氯化碳,它们在大气中寿命较短,从而减少了它们在平流层的臭氧耗损或全球变暖上的贡献;氢氟碳化合物不含有氯,也没有破坏臭氧的潜力。本文概述了生产部门、监管机构和社会都面临的挑战,在不断满足社会的需求和消费者的需求,而不损害消费者或工人的安全下,同时降低对环境的风险的概况。

54. Hofmann D J, Oltmans S J, Harris J M, et. al. Observation and possible causes of new ozone depletion in Antarctica in 1991[J]. Nature,1992,359(6393):283-287.

Observation and possible causes of new ozone depletion in Antarctica in 1991

Hofmann D. J. , Oltmans S. J. , Harris J. M. , Solomon S. , Deshler T. , Johnson B. J.

(Noaa, Climate Monitoring & Diagnost Lab, Boulder, Co 80303)

Abstract: Local ozone reductions approaching 50% in magnitude were observed during the Antarctic spring in the 11-13 and 25-30 km altitude regions over South Pole and McMurdo Stations in 1991. These reductions, at altitudes where depletion has not been observed previously, resulted in a late September total ozone column 10%-15% lower than previous years. The added depletion in the lower stratosphere was observed to coincide with penetration into the polar vortex of highly enhanced concentrations of aerosol particles from volcanic activity in 1991.

对 1991 年南极新一轮臭氧损耗的观测和可能原因

Hofmann D. J. , Oltmans S. J. , Harris J. M. , Solomon S. , Deshler T. , Johnson B. J.

（美国国家海洋和大气局）

摘要：1991 年春季的南极点和麦克默多站地区在海拔 11～13 km 和 25～30 km 观测到本地的臭氧减少的幅度接近 50%。在臭氧损耗尚未观测到的高度，这些减少导致 9 月下旬的臭氧总量较往年低 10%～15%。此外，平流层下部臭氧的减少与观察到的渗透到极地涡旋内的 1991 年火山活动的气溶胶粒子浓度大大增强吻合。

55. Charlson R J, Schwartz S E, Hales J M, et al. Climate forcing by anthropogenic aerosols[J]. Science,1992,255(5043):423-430.

Climate forcing by anthropogenic aerosols

Charlson R. J. , Schwartz S. E. , Hales J. M. , Cess R. D. , Coakley J. A. , Hansen J. E. , Hofmann D. J.

(Univ Washington, Dept Atmospher Sci, Seattle, Wa 98195 USA)

Abstract: Although long considered to be of marginal importance to global climate change, tropospheric aerosol contributes substantially to radiative forcing, and anthropogenic sulfate aerosol in particular has imposed a major perturbation to this forcing. Both the direct scattering of short-wavelength solar radiation and the modification of the shortwave reflective properties of clouds by sulfate aerosol particles increase planetary albedo, thereby exerting a cooling influence on the planet. Current climate forcing due to anthropogenic sulfate is estimated to be −1 to −2 watts per square meter, globally averaged. This perturbation is comparable in magnitude to current anthropogenic greenhouse gas forcing but opposite in sign. Thus, the aerosol forcing has likely offset global greenhouse warming to a substantial degree. However, differences in geographical and seasonal distributions of these forcings preclude any simple compensation. Aerosol effects must be taken into account in evaluating anthropogenic influences on past, current, and projected future climate and in formulating policy regarding controls on emission of greenhouse gases and sulfur dioxide. Resolution of such policy issues requires integrated research on the magnitude and geographical distribution of aerosol climate forcing and on the controlling chemical and physical processes.

来自人为排放气溶胶的气候强迫

Charlson R. J. , Schwartz S. E. , Hales J. M. , Cess R. D. , Coakley J. A. , Hansen J. E. , Hofmann D. J.

（美国华盛顿大学）

摘要：虽然对流层气溶胶长期被认为对全球气候变化只有边缘的重要性，但它对辐射强迫有很大的贡献,人为硫酸盐气溶胶更加剧了这一强迫。短波太阳辐射的直接散射和硫酸盐气溶胶粒子对云的短波反射性能的改变增加了行星反照率,从而使这个星球变冷。人为排放硫酸盐造成的当前的气候强迫变化估计为全球平均每平方米−1～−2 W。这种扰动的幅度,与目前人为排放的温室气体造成的强迫强度相当,但符号相反。因此,气溶胶强迫有可能在很大程度上抵消全球温室效应。然而,这些强迫在地域和季节分布的差异不能用任何简单的过程来弥补。在评价过去、现在和预测未来的气候和人为影响,在制定关于控制温室气体和二氧化硫排放的政策时,必须考虑气溶胶的影响。这样的政策问题的解决需要气溶胶强迫气候的严重程度和地理分布以及控制化学和物理过程的综合研究。

56. Ayers G P, Penkett S A, Gillett R W, et al. Evidence for photochemical control of ozone concentrations in unpolluted marine air[J]. Nature, 1992, 360(6403): 446-449.

Evidence for photochemical control of ozone concentrations in unpolluted marine air

Ayers G. P., Penkett S. A., Gillett R. W., Bandy B., Galbally I. E., Meyer C. P., Elsworth C. M., Bentley S. T., Forgan B. W.

(Csiro, Div Atmospher Res, Private Bag, Mordialloc, Vic 3195, Australia)

Abstract: Ozone in the troposphere is an important greenhouse gas, and a key participant in the oxidation of other trace species, but the mechanisms for its formation and destruction are not fully understood. In polluted regions of the Northern Hemisphere, seasonal increases in ozone concentration have been observed; such changes could arise from photochemical reactions, but they could also involve transport from the ozone-rich stratosphere. In remote, unpolluted (low-$NO_{(x)}$) regions, photochemical theory predicts net destruction of ozone. Here we present observations of a large summer minimum in ozone concentration in the unpolluted marine boundary layer of the Southern Hemisphere. Our results show a clear link between ozone loss and hydrogen peroxide production in the region, demonstrating that in situ photochemistry, rather than transport, is the major cause of the seasonal ozone cycle in the boundary layer. These findings emphasize the role of photochemical processes in the lower atmosphere, and may suggest only a limited role for transport in other, more polluted regions.

对未受污染海洋空气中臭氧浓度的光化学控制的依据

Ayers G. P., Penkett S. A., Gillett R. W., Bandy B., Galbally I. E., Meyer C. P., Elsworth C. M., Bentley S. T., Forgan B. W.

(澳大利亚联邦科工组织)

摘要:对流层中的臭氧是一种重要的温室气体,是其他痕量物种氧化的关键参与者,但它的形成和破坏机制尚不完全清楚。在北半球的污染地区,已观测到臭氧浓度的季节性增加,这种变化可能由光化学反应引起,但也可能涉及从富含臭氧的平流层的传送。在偏远的未受污染(低 NO_x)地区,光化学理论预计了臭氧的净损耗。本文报道目前在南半球未受污染的海洋边界层中臭氧浓度在夏季的低浓度的观测。我们的结果显示,臭氧损耗和该地区的过氧化氢之间有明确的联系,这表明原位光化学比远距离传输更加可能是边界层的季节性臭氧循环的重要原因。这些发现强调了低层大气中的光化学过程的作用,并可能指示出在其他污染的地区,远距离传输的作用也十分有限。

57. Stephenson J A E, Scourfield M W J. Importance of energetic solar protons in ozone depletion[J]. Nature, 1991, 352(6331): 137-139.

Importance of energetic solar protons in ozone depletion

Stephenson J. A. E., Scourfield M. W. J.

(Univ Natal, Space Phys Res Inst, Durban 4001, South Africa)

Abstract: Chlorine-catalysed depletion of the stratospheric ozone layer has commanded considerable attention since 1985, when Farman et al observed a decrease of 50% in the total column ozone over Antarctica in the austral spring. Here we examine the depletion of stratospheric ozone caused by the reaction of ozone with nitric oxide generated by energetic solar protons, associated with solar flares. During large solar flares in March 1989, satellite observations indicated that total column ozone was depleted by approximately 9% over approximately 20% of the total area between the South Pole and latitude 70-degrees-S. Chlorine-catalysed ozone depletion takes place over a much larger area, but our results indicate that the influence of solar protons on atmospheric ozone concentrations should not be ignored.

高能太阳质子在臭氧损耗中的作用

Stephenson J. A. E., Scourfield M. W. J.

(南非纳塔尔大学)

摘要：自 1985 年法曼等人观测到南极上空的臭氧总量在南半球春季减少 50%后，氯催化造成的平流层臭氧损耗已受到人们的重视。本文研究了高能太阳质子连同太阳耀斑产生的一氧化氮与臭氧反应所造成的平流层臭氧损耗。卫星观测表明，在 1989 年 3 月大的太阳耀斑期间，南极至南纬 70 度之间总面积的 20%地区，臭氧柱总量亏损约 9%。氯催化臭氧的损耗发生在一个更大的区域，但我们的结果表明，太阳质子对大气臭氧浓度的影响不应该被忽视。

58. Schoeberl M R, Hartmann D L. The dynamics of the stratospheric polar vortex and its relation to springtime ozone depletions[J]. Science, 1991, 251(4989): 46-52.

The dynamics of the stratospheric polar vortex and its relation to springtime ozone depletions

Schoeberl M. R., Hartmann D. L.

(Nasa, Goddard Space Flight Ctr, Greenbelt, Md 20771)

Abstract: Dramatic springtime depletions of ozone in polar regions require that polar stratospheric air has a high degree of dynamical isolation and extremely cold temperatures necessary for the formation of polar stratospheric clouds. Both of these conditions are produced within the stratospheric winter polar vortex. Recent aircraft missions have provided new information about the structure of polar vortices during winter and their relation to polar ozone depletions. The aircraft data show that gradients of potential vorticity and the concentration of conservative trace species are large at the transition from mid-latitude to polar air. The presence of such sharp gradients at the boundary of polar air implies that the inward mixing of heat and constituents is strongly inhibited and that the perturbed polar stratospheric chemistry associated with the ozone hole is isolated from the rest of the stratosphere until the vortex breaks up in late spring. The overall size of the polar vortex thus limits the maximum areal coverage of the annual polar ozone depletions. Because it appears that this limit has not been reached for the Antarctic depletions, the possibility of future increases in the size of the Antarctic ozone hole is left open. In the Northern Hemisphere, the smaller vortex and the more restricted region of cold temperatures suggest that this region has a smaller theoretical maximum for column ozone depletion, about 40 percent of the currently observed change in the Antarctic ozone column in spring.

平流层极地涡旋的动力学特征及其与春季臭氧损耗的关系

Schoeberl M. R. , Hartmann D. L.

(美国国家航空和航天局)

摘要:在极地地区春季戏剧性的臭氧耗损需要极地平流层空气高度的动力隔离和极地平流层云形成所必需的极为寒冷的气温。这些条件都在冬季平流层极涡中产生。最近,一次飞行任务提供了有关冬季极涡结构的新信息及其与极地臭氧耗损的关系。这架飞机的数据显示,位势涡度和保守的痕量物质的浓度梯度在中纬度向极地转换的过程中较大。极地空气的边界如此尖锐的梯度存在意味着外来的热量和成分混合强烈抑制,扰动的极地平流层化学与臭氧洞的关联直到晚春旋涡爆发都与平流层的其余部分隔离。因此,极涡的总体规模限制了年际极地臭氧消耗的最大覆盖面。因为它似乎尚未达到南极臭氧损耗的限度,未来的南极臭氧洞大小增加的可能性是开放的。在北半球,在较小的旋涡和寒冷的气温更受限制的地区,有一个较小的臭氧耗损理论最大值,大约为南极春季臭氧柱目前观察到的变化的40%。

59. Schnell R C, Liu S C, Oltmans S J, et al. Decrease of summer tropospheric ozone concentrations in Antarctica[J]. Nature, 1991, 351(6329): 726-729.

Decrease of summer tropospheric ozone concentrations in Antarctica

Schnell R. C. , Liu S. C. , Oltmans S. J. , Stone R. S. , Hofmann D. J. , Dutton E. G. , Deshler T. , Sturges W. T. , Harder J. W. , Sewell S. D. , Trainer M. , Harris J. M.

(Univ Colorado, Noaa, Cooperat Inst Res Environm Sci, Boulder, Co 80309)

Abstract: As an oxidant and a precursor for other highly reactive oxidants, ozone plays an important role in tropospheric photochemistry. In the upper troposphere, ozone absorbs infrared radiation and is thus an effective greenhouse gas. Here we show that surface ozone concentrations at the South Pole in the austral summer decreased by 17% over the period 1976-1990. Over the same period, solar irradiance at the South Pole in January and February decreased by 7% as a result of a 25% increase in cloudiness. We suggest that the trend in the summer ozone concentrations is caused by enhanced photochemical destruction of ozone in the lower troposphere caused by the increased penetration of ultraviolet radiation associated with stratospheric ozone depletion, coupled with enhanced transport of ozone-poor marine air from lower latitudes to the South Pole.

南极夏季对流层臭氧浓度的减少

Schnell R. C., Liu S. C., Oltmans S. J., Stone R. S., Hofmann D. J., Dutton E. G., Deshler T., Sturges W. T., Harder J. W., Sewell S. D., Trainer M., Harris J. M.

(美国科罗拉多大学)

摘要:作为氧化剂和其他高活性氧化剂的前体,臭氧在对流层的光化学反应中起着重要作用。在对流层上部,臭氧吸收红外辐射,因此是一种有效的温室气体。本文中,我们指出1976~1990年间在南半球的夏季南极地面臭氧浓度下降17%。同一时期的1月和2月由于云增多25%使南极的太阳辐照度下降了7%。我们指出夏季臭氧浓度的趋势是由下对流层臭氧损耗的光化学反应增强造成的,而下对流层臭氧的损耗又是与平流层臭氧损耗相联系的紫外辐射渗透的增加,加上贫臭氧的海洋大气从低纬度地区向南极的传输增强所造成的。

60. Kelly K K, Tuck A F, Davies T. Wintertime asymmetry of upper tropospheric water between the Northern and Southern Hemispheres[J]. Nature, 1991, 353(6341): 244-247.

Wintertime asymmetry of upper tropospheric water between the Northern and Southern Hemispheres

Kelly K. K., Tuck A. F., Davies T.

(Noaa, Aeron Lab, 325 Broadway, Boulder, Co 80303)

Abstract: Water vapour is an important greenhouse gas and yet its abundance in the upper troposphere is poorly known. Upper-tropospheric water vapour is particularly important despite its low mixing ratios, because it has large effects on the flux of infrared radiation near the tropopause. In addition, the distribution and supply of water vapour are central to cloud formation; the effects of cloud on the Earth's radiation budget are in turn central to understanding the climate response to increasing atmospheric concentrations of greenhouse gases. From airborne measurements of total water (vapour plus ice crystal) during the winters of 1987 in the Southern Hemisphere and of 1988-1989 in the Northern Hemisphere, we find that the upper troposphere in middle, subpolar and high latitudes is a factor of 2-4 drier during austral winter than during boreal winter. As the lower-latitude air moves towards the pole in austral winter, it is forced to cool to lower temperatures than in the north-more of the water vapour therefore condenses to form ice crystals, which then precipitate, thereby removing moisture from the air mass. Clearly, climate models must be able to reproduce this asymmetry if their predictions are to be credible. We also note that the asymmetry in water vapour implies an asymmetry in the production rate of the hydroxyl radical, and hence in the tropospheric chemistry of each hemisphere, for example in the rate of methane loss.

南北半球冬季对流层上部水汽的不对称特征

Kelly K. K. , Tuck A. F. , Davies T.

(美国国家海洋和大气局)

摘要：水蒸气是一种非常重要的温室气体，但是关于其在对流层上部的赋存状态我们了解得还非常少。虽然上对流层水蒸气混合比很低，但由于其对对流层顶部红外辐射通量的巨大作用而显得非常重要。除此以外，水蒸气的分布和供给是形成云的关键，而云在地球辐射平衡方面的影响是理解气候对大气温室气体浓度增加响应的关键。从南半球1987年和北半球1988～1989年冬季对总水(水蒸气和冰晶加和)的机载观测中，我们发现中纬度、亚极地和高纬度地区的上对流层是造成南方冬季比北方冬季干旱2～4倍的一个因素。由于在南方的冬季低纬度气团向极地迁移，使得较北方的水蒸气温度降得更低，形成冰晶，进而沉降，并转移走气团中的水汽。显然，气候模型如果想要预测结果可信则必须能够模拟这种不对称性。我们也注意到水蒸气的不对称性同时意味着羟基自由基沉降速率的不对称性，并进而影响两个半球的对流层化学(如甲烷损耗率)。

61. Cicerone R J, Elliott S, Turco R P. Reduced Antarctic ozone depletions in a model with hydrocarbon injections[J]. Science, 1991, 254(5035): 1191-1194.

Reduced Antarctic ozone depletions in a model with hydrocarbon injections

Cicerone R. J. , Elliott S. , Turco R. P.

(Univ Calif Irvine, Dept Geosci, Irvine, Ca 92717)

Abstract: Motivated by increased losses of Antarctic stratospheric ozone and by improved understanding of the mechanism, a concept is suggested for action to arrest this ozone loss: injecting the alkanes ethane or propane (E or P) into the Antarctic stratosphere. A numerical model of chemical processes was used to explore the concept. The model results suggest that annual injections of about 50000 tons of E or P could suppress ozone loss, but there are some scenarios where smaller E or P injections could increase ozone depletion. Further, key uncertainties must be resolved, including initial concentrations of nitrogen-oxide species in austral spring, and several poorly defined physical and chemical processes must be quantified. There would also be major difficulties in delivering and distributing the needed alkanes.

碳氢化合物注入模型中南极臭氧损耗的减少

Cicerone R. J. , Elliott S. , Turco R. P.

(美国加州大学)

摘要：出于南极平流层臭氧损耗增加及对该机制更好地理解，人们建议采取行动，阻止臭氧损耗：向南极平流层注入正构烷烃乙烷或丙烷(E或P)。一个化学过程的数值模型被用来探索这个概念。模型结果表明，每年约 50000 吨 E 或 P 的注射可以抑制臭氧损耗，但也可能发生少量的 E 或 P 注射增加臭氧消耗情况。此外，一些关键的不确定性必须解决，包括在南半球春季的氮氧化物的初始浓度与几个定义不清的物理和化学过程必须量化。此外，运输和分配所需的烷烃也有很大困难。

62. Anderson J G, Toohey D W, Brune W H. Free-radicals within the Antarctic vortex: the role of CFCs in Antarctic ozone loss[J]. Science, 1991, 251(4989): 39-46.

Free-radicals within the Antarctic vortex—the role of CFCs in Antarctic ozone loss

Anderson J. G., Toohey D. W., Brune W. H.

(Harvard Univ, Dept Chem, Cambridge, Ma 02138)

Abstract: How strong is the case linking global release of chlorofluorocarbons to episodic disappearance of ozone from the Antarctic stratosphere each austral spring? Three lines of evidence defining a link are (ⅰ) observed containment in the vortex of ClO concentrations two orders of magnitude greater than normal levels; (ⅱ) in situ observations obtained during ten high-altitude aircraft flights into the vortex as the ozone hole was forming that show a decrease in ozone concentrations as ClO concentrations increased; and (ⅲ) a comparison between observed ozone loss rates and those predicted with the use of absolute concentrations of ClO and BrO, the rate-limiting radicals in an array of proposed catalytic cycles. Recent advances in our understanding of the kinetics, photochemistry, and structural details of key intermediates in these catalytic cycles as well as an improved absolute calibration for ClO and BrO concentrations at the temperatures and pressures encountered in the lower antarctic stratosphere have been essential for defining the link.

南极涡旋中的自由基——氟氯化碳在南极臭氧损耗中的作用

Anderson J. G., Toohey D. W., Brune W. H.

（美国哈佛大学）

摘要：南半球的每个春天，南极平流层中间歇性的臭氧损耗与氟氯化碳的全球释放有多大的相关性？有三条证据支持这一联系：① 涡旋中观察到的 ClO 浓度超过正常水平的两个数量级；② 10 个高海拔的飞机航线上原位观测表明在涡旋中当臭氧洞形成时，臭氧浓度减少而 ClO 浓度增加；③ 观测到的和用绝对浓度 ClO 和 BrO(一系列催化循环中的限制速率的自由基粒子)预测的臭氧损耗率之间的比较。根据我们对动力学、光化学的了解和在这些催化循环的关键中间体的结构细节以及改进的 ClO 和 BrO 浓度在南极平流层下部中遇到的温度和压力的绝对校准最新进展的了解基本可以确定这种联系。

63. Yung Y L, Allen M, Crisp D, et al. Spatial variation of ozone depletion rates in the springtime Antarctic Polar vortex[J]. Science, 1990, 248(4956): 721-724.

Spatial variation of ozone depletion rates in the springtime Antarctic Polar vortex

Yung Y. L., Allen M., Crisp D., Zurek R. W., Sander S. P.

(Caltech, Div Geol & Planetary Sci, Pasadena, Ca 91125)

Abstract: An area-mapping technique, designed to filter out synoptic perturbations of the Antarctic polar vortex such as distortion or displacement away from the pole, was applied to the Nimbus-7 TOMS (Total Ozone Mapping Spectrometer) data. This procedure reveals the detailed morphology of the temporal evolution of column O_3. The results for the austral spring of 1987 suggest the existence of a relatively stable collar region enclosing an interior that is undergoing large variations. There is tentative evidence for quasi-periodic (15 to 20 days) O_3 fluctuations in the collar and for upwelling of tropospheric air in late spring. A simplified photochemical model of O_3 loss and the temporal evolution of the area-mapped polar O_3 are used to constrain the chlorine monoxide (ClO) concentrations in the springtime Antarctic vortex. The concentrations required to account for the observed loss of O_3 are higher than those previously reported by Anderson et al but are comparable to their recently revised values. However, the O_3 loss rates could be larger than deduced here because of underestimates of total O_3 by TOMS near the terminator. This uncertainty, together with the uncertainties associated with measurements acquired during the Airborne Antarctic Ozone Experiment, suggests that in early spring, closer to the vortex center, there may be even larger ClO concentrations than have yet been detected.

南极春季极涡中臭氧耗损率的空间变异

Yung Y. L., Allen M., Crisp D., Zurek R. W., Sander S. P.

(美国加州理工学院)

摘要：一种旨在筛选出如变形或距极位移的南极极涡天气扰动的面积测绘技术在云雨 7-TOMS(臭氧总量测绘光谱仪)上应用。该过程揭示了臭氧柱的时空演变的详细形态。1987 年南半球春天的研究结果表明存在一个相对稳定的包围内部的环形区域正在发生巨大变化。在春末，有初步的证据表明 O_3 在环形区域和大气对流层上升流波动的准周期(15～20 天)。简化的光化学臭氧损耗模型和极地臭氧区域映射模型的时间演化用来约束春季南极旋涡中氧化氯(ClO)的浓度。解释观察到的臭氧损耗所需的浓度高于以前由安德森等人报道的浓度，但其最近的修正值是相当的。然而，因为 TOMS 对臭氧总量的低估，臭氧损耗率可能会比这里推断的高。这种不确定性，连同在南极臭氧机载实验获得的测量相关的不确定性，表明在早春，更接近涡旋的中心，有可能会有更高的 ClO 浓度尚未发现。

64. Tans P P, Fung I Y, Takahashi T. Observational constraints on the global atmospheric CO_2 budget[J]. Science, 1990, 247(4949): 1431-1438.

Observational constraints on the global atmospheric CO_2 budget

Tans P. P., Fung I. Y., Takahashi T.

(Univ Colorado, Noaa, Cooperat Inst Res Environm Sci, Campus Box 216, Boulder, Co 80309)

Abstract: Observed atmospheric concentrations of CO_2 and data on the partial pressures of CO_2 in surface ocean waters are combined to identify globally significant sources and sinks of CO_2. The atmospheric data are compared with boundary layer concentrations calculated with the transport fields generated by a general circulation model (GCM) for specified source-sink distributions. In the model the observed north-south atmospheric concentration gradient can be maintained only if sinks for CO_2 are greater in the Northern than in the Southern Hemisphere. The observed differences between the partial pressure of CO_2 in the surface waters of the Northern Hemisphere and the atmosphere are too small for the oceans to be the major sink of fossil fuel CO_2. Therefore, a large amount of the CO_2 is apparently absorbed on the continents by terrestrial ecosystems.

全球大气二氧化碳排放清单的预测界限

Tans P. P., Fung I. Y., Takahashi T.

(美国科罗拉多大学)

摘要:观测到的大气 CO_2 浓度和海水中的 CO_2 分压相结合用以确定全球 CO_2 源和汇。大气数据用来与特定源汇分布的循环传输模型计算得到的边界层浓度相比较。在模型中,除非 CO_2 的汇在北半球比南半球更强,才能使得观测到的由北向南分布的大气浓度梯度维持。在北半球水表面和大气中的 CO_2 分压的差异非常小以至于海洋不能成为化石燃料燃烧 CO_2 的主要汇。因此,大量的 CO_2 显然是被陆地生态系统所吸收的。

65. Solomon S. Progress towards a quantitative understanding of Antarctic ozone depletion[J]. Nature, 1990,347(6291):347-354.

Progress towards a quantitative understanding of Antarctic ozone depletion

Solomon S.

(Noaa, Aeron Lab, Boulder, Co 80303)

Abstract: The possibility that the stratospheric ozone layer could be depleted by half at certain latitudes and seasons would have been deemed a preposterous and alarmist suggestion in the early 1980s. A decade later, the statement is acknowledged as proved beyond reasonable scientific doubt. Observations of the composition of the Antarctic stratosphere have established that the chemistry of this region is highly unusual because of its extreme cold temperatures, leading to a greatly enhanced susceptibility to chlorine-catalysed ozone depletion.

定量理解南极臭氧耗损的进展

Solomon S.

(美国国家海洋和大气局)

摘要:在20世纪80年代初,平流层臭氧在一定的纬度和季节损耗一半的可能性已被视为一个荒唐的和危言耸听的建议。十年后,这一陈述被证实超出了合理的科学疑问而被承认。南极平流层组成的观测确定是由于极端寒冷的气温,该地区的化学成分极不寻常,导致氯催化臭氧消耗的可能性大大增强。

66. Manzer L E. The CFC-ozone issue: progress on the development of alternatives to CFCs[J]. Science,1990,249(4964):31-35.

The CFC-ozone issue—progress on the development of alternatives to CFCs

Manzer L. E.

(Dupont Co, Dept Cent Res & Dev, Expt Stn, Wilmington, De 19880)

Abstract: Chlorofluorocarbons (CFCs) are now believed to be major contributors to the seasonal ozone depletion over the Antarctic continent. However, because they are so important to many aspects of Modern society, it would be irresponsible to immediately cease their production. The identification of suitable substitutes is difficult when issues such as toxicity, flammability, cost, environmental impact, and physical properties are considered. Several candidates that meet these criteria have been selected by the industry and significant research and development programs are under way to commercialize them. Unlike the simple, fully halogenated CFCs, which can only be made in the single step, there are many potentially viable routes to the alternatives, but these will require significant improvements in catalysis. Many other important issues such as materials compatibility, energy efficiency, the needs of developing countries, and the product life cycle of the alternatives need to be resolved before a timely transition to substitutes can be accomplished.

氟氯烃与臭氧难题——氟氯烃替代品的发展进程

Manzer L. E.

(美国杜邦公司)

摘要:氯氟烃(CFC)被认为是南极大陆季节性臭氧耗损的主要贡献者。然而,因为它们在现代社会的许多方面是如此重要,立即停止其生产是不负责任的。当毒性、易燃性、成本、环境影响、物理性质、化学性质等问题都需要考虑时,物色合适的替代品是很难的。一些符合这些条件的候选者已经通过商业化的方式被行业和重大的研究和发展计划选定。完全卤化的氟氯化碳的生产简单,只需一步,还有许多潜在可行路线的替代品却没那么简单,它们都需要对催化进行显著改善。在及时过渡到替代品前还必须解决其他许多重要的问题，如材料的相容性、能源利用效率、发展中国家的需求和替代品的产品生命周期等。

67. Kerr R A. Another deep Antarctic ozone hole[J]. Science, 1990, 250(4979): 370-370.

Another deep Antarctic ozone hole

Kerr R. A.

(News)

Abstract: Again in 1990, drastic depletion of stratospheric ozone over the South Pole has been measured, in August 140 Dobson units, far below the 220 Dobson units typically seen over Antarctica. This extensive destruction of ozone is determined to be brought about by sunshine acting in combination with the chlorine released from chlorofluorohydrocarbons (CFCs) by icy stratospheric clouds. It is concluded that CFC concentrations have now reached a level that will almost totally destroy the ozone in the lower stratosphere in most years.

又一个大的南极臭氧空洞

Kerr R. A.

(新闻)

摘要：同样是在 1990 年 8 月，已观测到南极平流层臭氧急剧下降至 140 DU，远低于南极上空通常的 220 DU。这么大量的臭氧损耗被确定为是由阳光激活冰冷的平流层云中的氟氯烃所释放的氯而带来的广泛破坏臭氧。结论是在大多数年份氟氯烃已经达到的浓度，将几乎完全摧毁较低平流层的臭氧。

68. Brasseur G P, Granier C, Walters S. Future changes in stratospheric ozone and the role of heterogeneous chemistry[J]. Nature, 1990, 348(6302): 626-628.

Future changes in stratospheric ozone and the role of heterogeneous chemistry

Brasseur G. P., Granier C., Walters S.

(Natl Ctr Atmospher Res, Div Atmospher Chem, Pob 3000, Boulder, Co 80307, USA)

Abstract: Heterogeneous chemical reactions on the surfaces of solid or liquid particles present in the lower stratosphere may be an important influence on the levels of ozone depletion resulting from increased concentrations of anthropogenic chlorine compounds in the atmosphere. Such processes, occurring on ice particles in polar stratospheric clouds, have been invoked to explain the formation of the springtime ozone hole over Antarctica. Sulphuric acid aerosols injected into the atmosphere following a volcanic eruption may also provide sites for heterogeneous chemistry leading to ozone depletion. Here we present model calculations that assess the importance of heterogeneous processes in future ozone depletion, assuming that trace-gas concentrations follow the protocol agreed at the recent international convention in London. We find that, even if this protocol is adhered to, reactions on the surface of sulphuric acid aerosol particles could produce significant ozone depletion into the beginning of the next century, especially if a major volcanic eruption (similar to the El Chichón eruption of 19824) takes place.

平流层臭氧的未来变化和非均相化学的作用

Brasseur G. P. , Granier C. , Walters S.

(美国国家大气研究中心)

摘要:较低平流层中的固体或液体颗粒表面的多相化学反应可能会对由大气中人为氯化合物浓度增加导致的臭氧损耗的水平有重要影响。这一发生在极地平流层云冰粒上的过程,已用来解释南极上空春季臭氧洞的形成。火山喷发后,硫酸气溶胶注入到大气中也提供了导致臭氧损耗的非均相化学反应。在本文中,我们假设痕量气体的浓度遵守最近在伦敦举行的国际公约协议,用模型计算评估非均相过程对未来臭氧损耗的重要性。我们发现,即使坚持这个协议,到22世纪初,硫酸气溶胶粒子表面上的反应仍可能会产生显著臭氧耗损,尤其是如果一个大规模的火山喷发(类似El Chichón 19824喷发)的发生。

2.3 大气科学类摘要翻译——北极

1. Pohler D, Vogel L, Friess U, et al. Observation of halogen species in the Amundsen Gulf, Arctic, by active long-path differential optical absorption spectroscopy[J]. PNAS,2010,107(15):6582-6587.

Observation of halogen species in the Amundsen Gulf, Arctic, by active long-path differential optical absorption spectroscopy

Pohler D., Vogel L., Friess U., Platt U.

(Institute of Environmental Physics, University of Heidelberg, Im Neuenheimer Feld 229, 69120 Heidelberg, Germany)

Abstract: In the polar tropospheric boundary layer, reactive halogen species (RHS) are responsible for ozone depletion as well as the oxidation of elemental mercury and dimethyl sulphide. After polar sunrise, air masses enriched in reactive bromine cover areas of several million square kilometers. Still, the source and release mechanisms of halogens are not completely understood. We report measurements of halogen oxides performed in the Amundsen Gulf, Arctic, during spring 2008. Active long-path differential optical absorption spectroscopy (LP-DOAS) measurements were set up offshore, several kilometers from the coast, directly on the sea ice, which was never done before. High bromine oxide concentrations were detected frequently during sunlight hours with a characteristic daily cycle showing morning and evening maxima and a minimum at noon. The, so far, highest observed average mixing ratio in the polar boundary layer of 41 pmol/mol (equal to pptv) was detected. Only short sea ice contact is required to release high amounts of bromine. An observed linear decrease of maximum bromine oxide levels with ambient temperature during sunlight, between −24 degrees C and −15 degrees C, provides indications on the conditions required for the emission of RHS. In addition, the data indicate the presence of reactive chlorine in the Arctic boundary layer. In contrast to Antarctica, iodine oxide was not detected above a detection limit of 0.3 pmol/mol.

通过主动长光程差分选择性吸收光谱观测北极阿蒙森湾的卤素

Pohler D., Vogel L., Friess U., Platt U.

(德国海德堡大学)

摘要:在极地对流层边界层,活性卤素粒子是造成臭氧亏损、原子汞和二甲基硫氧化的原因。极夜之后,富含活性卤素的气团分布在数百万平方公里的范围内。然而,卤素的源和释放机制并未完全被人们所了解。本文报道了2008年春季在北极阿蒙森湾对卤素氧化物的观测。利用主动长光程差分吸收光谱在离海岸几公里的海冰面上进行直接观测,这是过去从未有过的。白天频繁观测到高浓度的BrO,并且BrO的浓度有明显的日循环特征:早晚浓度最高,中午浓度最低。此外,我们还观测到了迄今为止极地边界层最高的平均混合比(41 pptv)。很短的海冰接触即可释放大量的溴。观测结果显示,在日光下,温度在−24 ℃至−15 ℃间观测到最大BrO浓度的线性减少,表明了活性卤素粒子释放的条件。此外,结果表明在北极边界层有活性氯的存在。与南极不同的是,IO没有观测到超过检出限(0.3 pptv)的结果。

2. Cai W J, Chen L Q, Chen B S, et al. Decrease in the CO_2 uptake capacity in an ice-free Arctic Ocean Basin[J]. Science, 2010, 329(5991): 556-559.

Decrease in the CO_2 uptake capacity in an ice-free Arctic Ocean Basin

Cai W. J., Chen L. Q., Chen B. S., Gao Z. Y., Lee S. H., Chen J. F., Pierrot D., Sullivan K., Wang Y. C., Hu X. P., Huang W. J., Zhang Y. H., Xu S. Q., Murata A., Grebmeier J. M., Jones E. P., Zhang H. S.

(Department of Marine Sciences, University of Georgia, Athens, GA 30602, USA)

Abstract: It has been predicted that the Arctic Ocean will sequester much greater amounts of carbon dioxide (CO_2) from the atmosphere as a result of sea ice melt and increasing primary productivity. However, this prediction was made on the basis of observations from either highly productive ocean margins or ice-covered basins before the recent major ice retreat. We report here a high-resolution survey of sea-surface CO_2 concentration across the Canada Basin, showing a great increase relative to earlier observations. Rapid CO_2 invasion from the atmosphere and low biological CO_2 drawdown are the main causes for the higher CO_2, which also acts as a barrier to further CO_2 invasion. Contrary to the current view, we predict that the Arctic Ocean basin will not become a large atmospheric CO_2 sink under ice-free conditions.

北冰洋无冰海盆区 CO_2 吸收能力的降低

Cai W. J., Chen L. Q., Chen B. S., Gao Z. Y., Lee S. H., Chen J. F., Pierrot D., Sullivan K., Wang Y. C., Hu X. P., Huang W. J., Zhang Y. H., Xu S. Q., Murata A., Grebmeier J. M., Jones E. P., Zhang H. S.

(美国乔治亚大学)

摘要:过去有预测认为由于海冰融化和海洋初级生产力的上升,北冰洋能够从大气中吸收更多的CO_2。然而,这一预测是建立在两个观察结果至少其一基础上的:目前大冰退之前海洋边缘或冰覆盖盆地的高生产力。我们报道了一组横跨加拿大海盆的海表面 CO_2 浓度的高分辨观测结果,相对过去的结果显示出了明显的增长。大气中 CO_2 的快速增长和减弱的 CO_2 生物吸收是造成高 CO_2 的主要原因,同样也是高 CO_2 进一步入侵的一个障碍。与此相反,我们预言北冰洋海盆在无冰的条件下不会成为一个大的大气 CO_2 汇。

3. Tilmes S, Muller R, Salawitch R. The sensitivity of polar ozone depletion to proposed geoengineering schemes[J]. Science, 2008, 320(5880): 1201-1204.

The sensitivity of polar ozone depletion to proposed geoengineering schemes

Tilmes S. , Muller R. , Salawitch R.

(Natl Ctr Atmospher Res, Div Atmospher Chem, Pob 3000, Boulder, Co 80307, USA)

Abstract: The large burden of sulfate aerosols injected into the stratosphere by the eruption of Mount Pinatubo in 1991 cooled Earth and enhanced the destruction of polar ozone in the subsequent few years. The continuous injection of sulfur into the stratosphere has been suggested as a "geoengineering" scheme to counteract global warming. We use an empirical relationship between ozone depletion and chlorine activation to estimate how this approach might influence polar ozone. An injection of sulfur large enough to compensate for surface warming caused by the doubling of atmospheric CO_2 would strongly increase the extent of Arctic ozone depletion during the present century for cold winters and would cause a considerable delay, between 30 and 70 years, in the expected recovery of the Antarctic ozone hole.

极地臭氧亏损对所提出地质工程方案的敏感性

Tilmes S. , Muller R. , Salawitch R.

(美国国家大气研究中心)

摘要:1991 年 Pinatubo 火山爆发,向平流层注入大量硫酸盐气溶胶,在随后几年使全球降温并进一步破坏了极地臭氧。向平流层中持续注入硫,被认为是对全球变暖起中和作用的一种地质工程方案。我们根据臭氧亏损和氯的活化作用的经验方程估算这一过程会怎样影响极地臭氧。当注入的硫多到足够补偿大气 CO_2 加倍产生的表面温度升高作用时,也会造成本世纪冬季北极臭氧亏损面积的大幅度扩展,南极臭氧洞的恢复估算会延迟 30～70 年。

4. Morin S, Savarino J, Frey M M, et al. Tracing the origin and fate of NO_x in the Arctic atmosphere using stable isotopes in nitrate[J]. Science, 2008, 322(5902): 730-732.

Tracing the origin and fate of NO_x in the Arctic atmosphere using stable isotopes in nitrate

Morin S. , Savarino J. , Frey M. M. , Yan N. , Bekki S. , Bottenheim J. W. , Martins J. M. F.

(Inst Natl Sci, CNRS, Paris, France)

Abstract: Atmospheric nitrogen oxides ($NO_x = NO + NO_2$) play a pivotal role in the cycling of reactive nitrogen (ultimately deposited as nitrate) and the oxidative capacity of the atmosphere. Combined measurements of nitrogen and oxygen stable isotope ratios of nitrate collected in the Arctic atmosphere were used to infer the origin and fate of NO_x and nitrate on a seasonal basis. In spring, photochemically driven emissions of reactive nitrogen from the snowpack into the atmosphere make local oxidation of NO_x by bromine oxide the major contributor to the nitrate budget. The comprehensive isotopic composition of nitrate provides strong constraints on the relative importance of the key atmospheric oxidants in the present atmosphere, with the potential for extension into the past using ice cores.

利用硝酸根中稳定同位素追踪北极大气 NO_x 的源和去向

Morin S., Savarino J., Frey M. M., Yan N., Bekki S., Bottenheim J. W., Martins J. M. F.

(法国科学研究中心)

摘要：大气氮氧化物(NO_x=NO+NO_2)在活性氮(最终转化为硝态氮)循环和大气氧化能力中扮演重要角色。本文对采自北极大气中的硝酸根的氮、氧稳定同位素比进行测量来追踪 NO_x 和硝酸根季节上的来源和去向。在春季，光化学驱动下活性氮从积雪中释放到大气，从而使氧化溴造成的局部 NO_x 的氧化成为硝酸根的主要贡献。硝酸盐的同位素组成为当前大气中关键大气氧化剂的相对重要性提供了强约束，对于利用冰芯研究过去变化也有潜在意义。

5. Mastepanov M, Sigsgaard C, Dlugokencky E J, et al. Large tundra methane burst during onset of freezing[J]. Nature, 2008, 456(7222): 628-631.

Large tundra methane burst during onset of freezing

Mastepanov M., Sigsgaard C., Dlugokencky E. J., Houweling S., Strom L., Tamstorf M. P., Christensen T. R.

(Lund Univ, Geobiosphere Sci Ctr, Solvegatan 12, S-22362 Lund, Sweden)

Abstract: Terrestrial wetland emissions are the largest single source of the greenhouse gas methane. Northern high-latitude wetlands contribute significantly to the overall methane emissions from wetlands, but the relative source distribution between tropical and high-latitude wetlands remains uncertain. As a result, not all the observed spatial and seasonal patterns of atmospheric methane concentrations can be satisfactorily explained, particularly for high northern latitudes. For example, a late-autumn shoulder is consistently observed in the seasonal cycles of atmospheric methane at high-latitude sites, but the sources responsible for these increased methane concentrations remain uncertain. Here we report a data set that extends hourly methane flux measurements from a high Arctic setting into the late autumn and early winter, during the onset of soil freezing. We find that emissions fall to a low steady level after the growing season but then increase significantly during the freeze-in period. The integral of emissions during the freeze-in period is approximately equal to the amount of methane emitted during the entire summer season. Three-dimensional atmospheric chemistry and transport model simulations of global atmospheric methane concentrations indicate that the observed early winter emission burst improves the agreement between the simulated seasonal cycle and atmospheric data from latitudes north of 60 degrees N. Our findings suggest that permafrost-associated freeze-in bursts of methane emissions from tundra regions could be an important and so far unrecognized component of the seasonal distribution of methane emissions from high latitudes.

冰冻扩张期间大幅度苔原甲烷爆发

Mastepanov M. , Sigsgaard C. , Dlugokencky E. J. , Houweling S. , Strom L. , Tamstorf M. P. , Christensen T. R.

（瑞典兰德大学）

摘要：陆地湿地排放是甲烷温室气体的最大单一来源。北半球高纬度湿地在湿地对甲烷排放中有显著贡献，但是热带和高纬度地区湿地相对源的分布还不明确。因此，并非所有大气甲烷浓度的空间和季节分布模式都能得到合理解释，尤其是北半球高纬度地区。例如，在高纬度地区大气甲烷季节性循环的观测中一直可以发现秋季末期的升高现象，但是产生这些升高的甲烷源还不清楚。本文我们对北极高纬度的一个点的土壤冻结期间（秋季末期至冬季早期）进行了每小时甲烷通量的观测。我们发现，在生长季节结束之后排放量降到了一个很低但很稳定的水平，但是在冻结期间有显著上升。冰冻期间的排放通量与整个夏季的甲烷排放量近似相等。对全球大气甲烷浓度进行的三维大气化学和传输模型模拟结果显示，观测到的冬季初期大规模排放进一步加强了北纬 60 度大气数据与模拟的季节性循环的良好一致性。我们的研究结果表明永久冻结带与冰冻带苔原地区甲烷大规模排放在高纬度甲烷排放的季节性分布中起到重要的作用，但尚未受到关注。

6. Walter K M, Edwards M E, Grosse G, et al. Thermokarst lakes as a source of atmospheric CH_4 during the last deglaciation[J]. Science, 2007, 318(5850): 633-636.

Thermokarst lakes as a source of atmospheric CH_4 during the last deglaciation

Walter K. M. , Edwards M. E. , Grosse G. , Zimov S. A. , Chapin F. S.

(Univ Alaska, Water & Environm Res Ctr, Fairbanks, AK 99775 USA
Russian Acad Sci, Pacific InstGeog, NE Sci Stn, Moscow 117901, Russia)

Abstract: Polar ice-core records suggest that an arctic or boreal source was responsible for more than 30% of the large increase in global atmospheric methane (CH_4) concentration during deglacial climate warming; however, specific sources of that CH_4 are still debated. Here we present an estimate of past CH_4 flux during deglaciation from bubbling from thermokarst (thaw) lakes. Based on high rates of CH_4 bubbling from contemporary arctic thermokarst lakes, high CH_4 production potentials of organic matter from Pleistocene-aged frozen sediments, and estimates of the changing extent of these deposits as thermokarst lakes developed during deglaciation, we find that CH_4 bubbling from newly forming thermokarst lakes comprised 33% to 87% of the high-latitude increase in atmospheric methane concentration and, in turn, contributed to the climate warming at the Pleistocene-Holocene transition.

末次冰消期热溶喀斯特湖是大气甲烷的排放源

Walter K. M., Edwards M. E., Grosse G., Zimov S. A., Chapin F. S.

(美国阿拉斯加大学;俄罗斯科学院)

摘要:极地冰芯记录表明间冰期气候温暖时期全球大气甲烷浓度大幅度升高30%归因于北极或寒带源;然而,甲烷的具体来源仍然存在争议。在本文中,我们估算了冰消期热溶喀斯特湖气泡所产生的甲烷气体通量。基于当代北极热溶喀斯特湖甲烷气泡的高生成速度,更新世冰冻沉积中的有机质有产生大量甲烷的潜力,以及冰消期时所发育的热溶喀斯特湖沉积物改变范围的估算,我们发现高纬度地区大气甲烷浓度的增加33%~87%由新形成的热溶喀斯特湖的甲烷冒泡所造成,并进而对更新世-全新世过渡期间的气候变暖做出贡献。

7. Lubin D, Vogelmann A M. A climatologically significant aerosol longwave indirect effect in the Arctic[J]. Nature, 2006, 439(7075): 453-456.

A climatologically significant aerosol longwave indirect effect in the Arctic

Lubin D., Vogelmann A. M.

(Univ Calif San Diego, Scripps Inst Oceanog, La Jolla, CA 92093 USA)

Abstract: The warming of Arctic climate and decreases in sea ice thickness and extent observed over recent decades are believed to result from increased direct greenhouse gas forcing, changes in atmospheric dynamics having anthropogenic origin, and important positive reinforcements including ice-albedo and cloud-radiation feedbacks. The importance of cloud-radiation interactions is being investigated through advanced instrumentation deployed in the high Arctic since 1997. These studies have established that clouds, via the dominance of longwave radiation, exert a net warming on the Arctic climate system throughout most of the year, except briefly during the summer. The Arctic region also experiences significant periodic influxes of anthropogenic aerosols, which originate from the industrial regions in lower latitudes. Here we use multisensor radiometric data to show that enhanced aerosol concentrations alter the microphysical properties of Arctic clouds, in a process known as the "first indirect" effect. Under frequently occurring cloud types we find that this leads to an increase of an average 3.4 watts per square metre in the surface longwave fluxes. This is comparable to a warming effect from established greenhouse gases and implies that the observed longwave enhancement is climatologically significant.

北极地区气溶胶长波辐射显著的间接气候效应

Lubin D., Vogelmann A. M.

(美国加利福尼亚大学圣地亚哥分校)

摘要:近几十年来北极气候逐渐变暖,海冰的厚度和范围也在下降,普遍认为这一现象是逐渐增长的温室气体强迫、人类活动造成的大气动力学的改变，以及冰面和云反照的正反馈等因素作用的结果。自1997年起，研究者开始在北极高纬度地区通过先进仪器研究云辐射交互作用的重要性。这些研究明确了云通过长波辐射的优势使得一年中大部分时间(除短暂的夏季以外)北极气候系统产生净变暖。北极地区也经历了来自低纬度工业地区的人为气溶胶定期涌入。本研究中，我们使用多传感的辐射测量数据，结果显示增加的气溶胶浓度通过第一间接影响过程改变了北极云的微物理特性。在经常出现的云的类型中，我们发现这将导致每平方米3.4 W的表面长波通量的增加。与温室气体所造成的气候变暖效应相比而言，这意味着所观测到的长波增强在气候变化上也有显著作用。

8. Garrett T J, Zhao C F. Increased Arctic cloud longwave emissivity associated with pollution from mid-latitudes[J]. Nature, 2006, 440(7085): 787-789.

Increased Arctic cloud longwave emissivity associated with pollution from mid-latitudes

Garrett T. J., Zhao C. F.

(Univ Utah, Dept Meteorol, Salt Lake City, UT 84112 USA)

Abstract: There is consensus among climate models that Arctic climate is particularly sensitive to anthropogenic greenhouse gases and that, over the next century, Arctic surface temperatures are projected to rise at a rate about twice the global mean. The response of Arctic surface temperatures to greenhouse gas thermal emission is modified by Northern Hemisphere synoptic meteorology and local radiative processes. Aerosols may play a contributing factor through changes to cloud radiative properties. Here we evaluate a previously suggested contribution of anthropogenic aerosols to cloud emission and surface temperatures in the Arctic. Using four years of ground-based aerosol and radiation measurements obtained near Barrow, Alaska, we show that, where thin water clouds and pollution are coincident, there is an increase in cloud longwave emissivity resulting from elevated haze levels. This results in an estimated surface warming under cloudy skies of between 3.3 and 5.2 $W \cdot m^{-2}$ or 1 and 1.6 degrees C. Arctic climate is closely tied to cloud longwave emission, but feedback mechanisms in the system are complex and the actual climate response to the described sensitivity remains to be evaluated.

中纬地区的污染增加了北极云的长波辐射

Garrett T. J., Zhao C. F.

(美国犹他大学)

摘要：对气候模型的研究一般认为北极气候对人类排放的温室气体异常敏感，在下个世纪中北极表面温度的上升速度是全球平均值的两倍。北半球天气气象和当地辐射过程，修正了北极地表温度对温室气体热辐射的响应。气溶胶可能对云辐射特征的变化产生贡献。本文中，我们评估了过去人为气溶胶对北极云的产生和地表温度的贡献。通过在阿拉斯加Barrow附近对气溶胶和辐射所作的四年地表观测，结果显示当薄水云层和污染的地区一致时，由于阴霾水平的上升造成云长波辐射的升高。这导致有云地区表面温度上升3.3～5.2 W/m^2，相当于1～1.6 ℃。北极气候与云长波辐射紧密联系在一起，但是系统的反馈机制很复杂，实际气候响应的灵敏度仍有待进一步评估。

9. Schiermeier Q. Arctic trends scrutinized as chilly winter destroys ozone[J]. Nature, 2005, 435 (7038): 6-6.

Arctic trends scrutinized as chilly winter destroys ozone

Schiermeier Q.

(News)

Abstract: In an effort to determine the cause of the hole in the ozone layer, researchers believe that pollutants such as chlorofluorocarbons and greenhouse gases are to blame. Both global warming and the rising level of pollutants in the atmosphere are leading to changes in Arctic temperatures.

当北极寒冷的冬天破坏臭氧时，人们开始审议北极的趋势

Schiermeier Q.

(新闻)

摘要：为了确定臭氧层空洞的原因，研究人员认为，像氟氯化碳这样的污染物和温室气体都是造成臭氧空洞的原因。无论是全球变暖还是大气中的污染物含量的不断升高，都将导致北极气温的变化。

10. Penner J E, Dong X Q, Chen Y. Observational evidence of a change in radiative forcing due to the indirect aerosol effect[J]. Nature, 2004, 427(6971): 231-234.

Observational evidence of a change in radiative forcing due to the indirect aerosol effect

Penner J. E., Dong X. Q., Chen Y.

(Univ Michigan, Dept Atmospher Ocean & Space Sci, Ann Arbor, MI 48109 USA)

Abstract: Anthropogenic aerosols enhance cloud reflectivity by increasing the number concentration of cloud droplets, leading to a cooling effect on climate known as the indirect aerosol effect. Observational support for this effect is based mainly on evidence that aerosol number concentrations are connected with droplet concentrations, but it has been difficult to determine the impact of these indirect effects on radiative forcing. Here we provide observational evidence for a substantial alteration of radiative fluxes due to the indirect aerosol effect. We examine the effect of aerosols on cloud optical properties using measurements of aerosol and cloud properties at two North American sites that span polluted and clean conditions-a continental site in Oklahoma with high aerosol concentrations, and an Arctic site in Alaska with low aerosol concentrations. We determine the cloud optical depth required to fit the observed shortwave downward surface radiation. We then use a cloud parcel model to simulate the cloud optical depth from observed aerosol properties due to the indirect aerosol effect. From the good agreement between the simulated indirect aerosol effect and observed surface radiation, we conclude that the indirect aerosol effect has a significant influence on radiative fluxes.

气溶胶间接作用下辐射强迫变化的观测证据

Penner J. E. , Dong X. Q. , Chen Y.

(美国密歇根大学)

摘要:人为产生的气溶胶通过增加云滴的计数浓度增强了云的反射率,从而导致气候变冷,即间接气溶胶效应。支持这一效应的观测结果主要是基于气溶胶的计数浓度与云滴浓度相关,但是明确这些间接影响对辐射强迫的作用还很困难。在本文中,我们为间接气溶胶效应对辐射通量所产生的显著变化提供了观测证据。我们通过对北美两个气溶胶和云光学特征的观测点(一个高气溶胶浓度,位于俄克拉荷马州监测污染和洁净大气环境;一个低气溶胶浓度,位于北极阿拉斯加)检测了气溶胶对云的光学特征的影响。我们确定需要符合观测的向下表面的短波辐射的云光学厚度。之后我们采用云包裹模型,根据间接气溶胶效应通过观测到的气溶胶性质来模拟云的光学厚度。从模拟气溶胶的间接影响和实际观测的表面辐射有很好的一致性来看,我们认为气溶胶的间接影响对辐射通量起显著影响。

11. Wang X J, Key J R. Recent trends in arctic surface, cloud, and radiation properties from space[J]. Science,2003,299(5613):1725-1728.

Recent trends in arctic surface, cloud, and radiation properties from space

Wang X. J. , Key J. R.

(Univ Wisconsin, Cooperat Inst Meteorol Satellite Studies, 1225 W Dayton St, Madison, WI 53706 USA)

Abstract: Trends in satellite-derived cloud and surface properties for 1982 to 1999 show that the Arctic has warmed and become cloudier in spring and summer but has cooled and become less cloudy in winter. The increase in spring cloud amount radiatively balances changes in surface temperature and albedo, but during summer, fall, and winter, cloud forcing has tended toward increased cooling. This implies that, if seasonal cloud amounts were not changing, surface warming would be even greater than that observed. Strong correlations with the Arctic Oscillation indicate that the rise in surface temperature and changes in cloud amount are related to large-scale circulation rather than to local processes.

空间观测的北极地表、云和放射性特征趋势

Wang X. J. , Key J. R.

(美国威斯康星大学)

摘要:卫星观测得到的 1982～1999 年云层和地表特征趋势显示,北极在春夏季更加温暖、多云,而在冬季则变得少云和寒冷。春季云量的增加改变了表面温度和反照率的辐射平衡,但在夏、秋、冬季,云的辐射强迫趋向气候更加变冷。这表明如果季节性云量没有发生变化,地表温度升高甚至会比已经观测到的更加强烈。与北极涛动的高相关性表明,地表温度的升高和云量的改变与大尺度循环而非本地过程

相关。

12. Schiermeier Q. Arctic rockets give glimpse of the atmosphere's top layers[J]. Nature, 2003, 424(6946): 243-243.

Arctic rockets give glimpse of the atmosphere's top layers

Schiermeier Q.

(News)

Abstract: A series of rocket-borne atmospheric reconnaissance missions in the Arctic Circle is yielding potentially valuable data on the little-known thermodynamics of the middle and upper atmosphere.

北极火箭窥探大气顶层

Schiermeier Q.

(新闻)

摘要：一系列火箭搭载的北极圈的大气侦察为鲜为人知的中层和上层大气热力学研究提供有价值的数据。

13. Baldwin M P, Stephenson D B, Thompson D W J, et al. Stratospheric memory and skill of extended-range weather forecasts[J]. Science, 2003, 301(5633): 636-640.

Stratospheric memory and skill of extended-range weather forecasts

Baldwin M. P., Stephenson D. B., Thompson D. W. J., Dunkerton T. J., Charlton A. J., O'Neill A.

(NW Res Associates Inc, 14508 NE 20th St, Bellevue, WA 98007 USA)

Abstract: We use an empirical statistical model to demonstrate significant skill in making extended-range forecasts of the monthly-mean Arctic Oscillation (AO). Forecast skill derives from persistent circulation anomalies in the lowermost stratosphere and is greatest during boreal winter. A comparison to the Southern Hemisphere provides evidence that both the time scale and predictability of the AO depend on the presence of persistent circulation anomalies just above the tropopause. These circulation anomalies most likely affect the troposphere through changes to waves in the upper troposphere, which induce surface pressure changes that correspond to the AO.

平流层对更大范围天气的存储和预报能力

Baldwin M. P. , Stephenson D. B. , Thompson D. W. J. , Dunkerton T. J. ,
Charlton A. J. , O'Neill A.

(美国西北研究协会)

摘要:我们采用经验统计模型演示了北极涛动月平均扩展范围预测的能力。预测能力来自平流层最底层持久性环流异常,并在北方冬季最强。与南半球的对比表明北极涛动的时间尺度和可预测性依赖于对流层顶持久性环流异常。这些环流异常很可能通过改变对流层上部的波影响对流层;这些波能够诱导与北极涛动相关的表面压力的改变。

14. Tabazadeh A, Drdla K, Schoeberl M R, et al. Arctic "ozone hole" in a cold volcanic stratosphere [J]. PNAS,2002,99(5):2609-2612.

Arctic "ozone hole" in a cold volcanic stratosphere

Tabazadeh A. , Drdla K. , Schoeberl M. R. , Hamill P. , Toon O. B.

(NASA, Ames Res Ctr, Div Earth Sci, MS-245-4, Moffett Field, CA 94035 USA
San Jose State Univ, Dept Phys, San Jose, CA 95192 USA)

Abstract: Optical depth records indicate that volcanic aerosols from major eruptions often produce clouds that have greater surface area than typical Arctic polar stratospheric clouds (PSCs). A trajectory cloud-chemistry model is used to study how volcanic aerosols could affect springtime Arctic ozone loss processes, such as chlorine activation and denitrification, in a cold winter within the current range of natural variability. Several studies indicate that severe denitrification can increase Arctic ozone loss by up to 30%. We show large PSC particles that cause denitrification in a nonvolcanic stratosphere cannot efficiently form in a volcanic environment. However, volcanic aerosols, when present at low altitudes, where Arctic PSCs cannot form, can extend the vertical range of chemical ozone loss in the lower stratosphere. Chemical processing on volcanic aerosols over a 10 km altitude range could increase the current levels of springtime column ozone loss by up to 70% independent of denitrification. Climate models predict that the lower stratosphere is cooling as a result of greenhouse gas built-up in the troposphere. The magnitude of column ozone loss calculated here for the 1999-2000 Arctic winter, in an assumed volcanic state, is similar to that projected for a colder future nonvolcanic stratosphere in the 2010 decade.

火山爆发影响下的寒冷平流层中的北极"臭氧洞"

Tabazadeh A. , Drdla K. , Schoeberl M. R. , Hamill P. , Toon O. B.

(美国国家航空和航天局)

摘要:光学厚度记录显示,主要的火山爆发引起的火山气溶胶常常会产生要远大于典型的北极极地平流层云(PSCs)的云层的表面积。云化学轨迹模型被用来研究冬季火山气溶胶在当前的范围和变化下如何影响春季北极臭氧流失过程,比如氯的活化作用和脱氮作用。一些研究显示脱氮作用可以使北极臭氧损失量上升30%。本文指出,能够引起非火山平流层地区脱氮作用的大的平流层云粒子不能在火山环境中有效地形成。然而,当火山气溶胶出现在不能生成北极平流层云的低海拔地区时,可以将低平流层地区化学臭氧亏损的垂直距离扩大。在超过10 km海拔范围的火山气溶胶上发生的化学过程可以使目前春季臭氧柱浓度的损耗独立于脱氮作用之外增长70%。气候模型预测,对流层温室气体的增加使得平流层底部正在变冷。本文计算的北极1999~2000冬季在假定处于火山影响下的臭氧柱浓度的损耗量级与至2010年的未来10年间非火山影响下平流层的变冷效果相似。

15. Tabazadeh A, Jensen E J, Toon O B, et al. Role of the stratospheric polar freezing belt in denitrification[J]. Science, 2001, 291(5513): 2591-2594.

Role of the stratospheric polar freezing belt in denitrification

Tabazadeh A., Jensen E. J., Toon O. B., Drdla K., Schoeberl M. R.

(Univ Colorado, Program Atmospher & Ocean Sci, Atmospher & Space Phys Lab, Boulder, CO 80309 USA)

Abstract: Homogeneous freezing of nitric acid hydrate particles can produce a polar freezing belt in either hemisphere that can cause denitrification. Computed denitrification profiles for one Antarctic and two Arctic cold winters are presented. The vertical range over which denitrification occurs is normally quite deep in the Antarctic but limited in the Arctic. A 4 kelvin decrease in the temperature of the Arctic stratosphere due to anthropogenic and/or natural effects can trigger the occurrence of widespread severe denitrification. Ozone loss is amplified in a denitrified stratosphere, so the effects of falling temperatures in promoting denitrification must be considered in assessment studies of ozone recovery trends.

平流层极地冰冻带在反硝化过程中的作用

Tabazadeh A., Jensen E. J., Toon O. B., Drdla K., Schoeberl M. R.

(美国科罗拉多大学)

摘要:硝酸水合物颗粒的均匀冻结可以在南北半球均形成极地冻结带,从而引起脱氮作用。本文计算了一个南极冬季和两个北极冬季的脱氮作用模式。其中脱氮作用发生的垂直范围在南极一般很深,而在北极则范围很有限。由于人为和(或)自然因素的影响,北极平流层会产生4开尔文的温度降低,从而引发分布广泛的脱氮作用。在脱氮后的平流层中,臭氧损耗被放大,因此在考虑评估臭氧恢复趋势的时候,必需注意到温度降低促发的脱氮作用的影响。

16. Foster K L, Plastridge R A, Bottenheim J W, et al. The role of Br_2 and BrCl in surface ozone destruction at polar sunrise[J]. Science, 2001, 291(5503): 471-474.

The role of Br_2 and BrCl in surface ozone destruction at polar sunrise

Foster K. L., Plastridge R. A., Bottenheim J. W., Shepson P. B., Finlayson-Pitts B. J., Spicer C. W.

(Battelle Mem Inst, 505 King Ave, Columbus, OH 43201 USA)

Abstract: Bromine atoms are believed to play a central role in the depletion of surface-level ozone in the Arctic at polar sunrise. Br_2, BrCl, and HOBr have been hypothesized as bromine atom precursors, and there is evidence for chlorine atom precursors as well, but these species have not been measured directly. We report here measurements of Br_2, BrCl, and Cl_2 made using atmospheric pressure chemical ionization-mass spectrometry at Alert, Nunavut, Canada. In addition to Br_2 at mixing ratios up to 25 parts per trillion, BrCl was found at levels as high as 35 parts per trillion. Molecular chlorine was not observed, implying that BrCl is the dominant source of chlorine atoms during polar sunrise, consistent with recent modeling studies. Similar formation of bromine compounds and tropospheric ozone destruction may also occur at mid-latitudes but may not be as apparent owing to more efficient mixing in the boundary layer.

极地日出时 Br_2 和 BrCl 对表层臭氧的破坏

Foster K. L., Plastridge R. A., Bottenheim J. W., Shepson P. B., Finlayson-Pitts B. J., Spicer C. W.

(美国巴特尔纪念研究院)

摘要:一般认为在北极日出时溴原子在近地面臭氧的亏损中起主要作用。有假设认为 Br_2、BrCl 和 HOBr 是溴原子的前体,氯原子的前体也有同样证据,但是这些粒子没有被直接测量到。我们在加拿大 Alert 地区通过大气压化学电离质谱测量了 Br_2、BrCl 和 Cl_2。除 Br_2 的混合比达到 25 ppt 以外,BrCl 也高达 35 ppt。没有观测到分子氯,显示极地日出期间 BrCl 是氯原子的主要源,这也与最近的模型研究结果一致。溴化物类似的形成过程和对流层臭氧亏损同样可能发生在中纬度,但由于边界层的混合效率较高,可能没有高纬度一样常见。

17. Fahey D W, Gao R S, Carslaw K S, et al. The detection of large HNO_3-containing particles in the winter arctic stratosphere[J]. Science, 2001, 291(5506): 1026-1031.

The detection of large HNO_3-containing particles in the winter arctic stratosphere

Fahey D. W., Gao R. S., Carslaw K. S., Kettleborough J., Popp P. J., Northway M. J., Holecek J. C., Ciciora S. C., McLaughlin R. J., Thompson T. L., Winkler R. H., Baumgardner D. G., Gandrud B., Wennberg P. O., Dhaniyala S., McKinney K., Peter T., Salawitch R. J., Bui T. P., Elkins J. W., Webster C. R., Atlas E. L., Jost H., Wilson J. C., Herman R. L., Kleinbohl A., von Konig M.

(NOAA, Aeron Lab, Boulder, CO 80305 USA)

Abstract: Large particles containing nitric acid (HNO_3) were observed in the 1999/2000 Arctic winter stratosphere. These in situ observations were made over a large altitude range (16 to 21 kilometers) and horizontal extent (1800 kilometers) on several airborne sampling flights during a period of several weeks. With diameters of 10 to 20 micrometers, these sedimenting particles have significant potential to denitrify the lower stratosphere. A microphysical model of nitric acid trihydrate particles is able to simulate the growth and sedimentation of these large sites in the lower stratosphere, but the nucleation process is not yet known. Accurate modeling of the formation of these Large particles is essential for understanding Arctic denitrification and predicting future Arctic ozone abundances.

对北极冬季平流层中含硝酸大颗粒的探测

Fahey D. W., Gao R. S., Carslaw K. S., Kettleborough J., Popp P. J., Northway M. J., Holecek J. C., Ciciora S. C., McLaughlin R. J., Thompson T. L., Winkler R. H., Baumgardner D. G., Gandrud B., Wennberg P. O., Dhaniyala S., McKinney K., Peter T., Salawitch R. J., Bui T. P., Elkins J. W., Webster C. R., Atlas E. L., Jost H., Wilson J. C., Herman R. L., Kleinbohl A., von Konig M.

(美国国家海洋和大气局)

摘要:1999～2000 年北极冬季平流层中我们观测到了含有硝酸的大颗粒,这些原位观测结果是通过在高达 16～21 km 的高度范围以及 1800 km 的水平距离内的几周之内一系列的空中采样获得的。这些沉积物颗粒的直径在 10～20 μm,很有可能会使下层平流层进行脱氮作用。硝酸三水合物微粒的微物理模型可以很好地模拟平流层下层中大颗粒的增长和沉积,但是对于聚核作用还不了解。这些大颗粒形成的准确模型对于理解北极地区的脱氮作用和预测未来臭氧丰度都意义重大。

18. Beck J W, Richards D A, Edwards R L, et al. Extremely large variations of atmospheric ^{14}C concentration during the last glacial period[J]. Science,2001,292(5526):2453-2458.

Extremely large variations of atmospheric ^{14}C concentration during the last glacial period

Beck J. W., Richards D. A., Edwards R. L., Silverman B. W., Smart P. L., Donahue D. J., Hererra-Osterheld S., Burr G. S., Calsoyas L., Jull A. J. T., Biddulph D.

(Univ Arizona, Dept Phys, NSF, Arizona Accelerator Mass Spectrometry Facil, Tucson, AZ 85721 USA)

Abstract: A long record of atmospheric ^{14}C concentration, from 45 to 11 thousand years ago (ka), was obtained from a stalagmite with thermal-ionization mass-spectrometric Th-230 and accelerator mass-spectrometric ^{14}C measurements. This record reveals highly elevated Delta ^{14}C between 45 and 33 ka, portions of which may correlate with peaks in cosmogenic ^{36}Cl and ^{10}Be isotopes observed in polar ice cores. Superimposed on this broad peak of Delta ^{14}C are several rapid excursions, the largest of which occurs between 44.3 and 43.3 ka. Between 26 and 11 ka, atmospheric Delta ^{14}C decreased from similar to 700 to similar to 100 per mil, modulated by numerous minor excursions. Carbon cycle models suggest that the major features of this record cannot be produced with solar or terrestrial magnetic field modulation alone but also require substantial fluctuations in the carbon cycle.

末次冰期大气^{14}C浓度的极端波动

Beck J. W., Richards D. A., Edwards R. L., Silverman B. W., Smart P. L., Donahue D. J., Hererra-Osterheld S., Burr G. S., Calsoyas L., Jull A. J. T., Biddulph D.

(美国亚利桑那大学)

摘要:通过热电离质谱 Th230 和加速器质谱^{14}C的测量方法分析石笋获得了 45～11 ka 以来大气^{14}C的长时间记录。结果显示,45～33 ka 期间,$\delta^{14}C$有显著提高,一部分可能与极地冰芯中观测到的来自宇宙射线的^{36}Cl和^{10}Be的峰相关。在$\delta^{14}C$这一较宽的峰中规律地分布着几个快速背离的过程,其中最大的一次发生在 44.3～43.3 ka 期间。在 26～11 ka 期间,大气$\delta^{14}C$从接近 700/ml 降至约 100/ml,中间伴随许多小波动。碳循环模式显示此记录的主要特征并非由太阳活动或者陆地磁场活动单独产生,它仍然需要在碳循环中的波动。

19. Baldwin M P, Dunkerton T J. Stratospheric harbingers of anomalous weather regimes[J]. Science, 2001,294(5542):581-584.

Stratospheric harbingers of anomalous weather regimes

Baldwin M. P., Dunkerton T. J.

(NW Res Associates, 14508 NE 20th St, Bellevue, WA 98007 USA)

Abstract: Observations show that large variations in the strength of the stratospheric circulation, appearing first above similar to 50 kilometers, descend to the lowermost stratosphere and are followed by anomalous tropospheric weather regimes. During the 60 days after the onset of these events, average surface pressure maps resemble closely the Arctic Oscillation pattern. These stratospheric events also precede shifts in the probability distributions of extreme values of the Arctic and North Atlantic Oscillations, the location of storm tracks, and the local likelihood of mid-latitude storms. Our observations suggest that these stratospheric harbingers may be used as a predictor of tropospheric weather regimes.

反常天气体系在平流层的先兆

Baldwin M. P., Dunkerton T. J.

(美国西北研究协会)

摘要：观测表明平流层环流强度的巨大变化，首先出现在约 50 km 以上的位置，转而降到平流层最底层，并紧接着出现对流层异常天气模式。在这些事件发生后的 60 天内，平均表面压力地图与北极涛动模式很相似。这些平流层事件也先于北极和北大西洋涛动，风暴路径的位置和当地中纬度风暴的可能性。我们的观测显示这些平流层信号可以用来预测对流层天气模式。

20. Tabazadeh A, Santee M L, Danilin M Y, et al. Quantifying denitrification and its effect on ozone recovery[J]. Science, 2000, 288(5470): 1407-1411.

Quantifying denitrification and its effect on ozone recovery

Tabazadeh A., Santee M. L., Danilin M. Y., Pumphrey H. C., Newman P. A., Hamill P. J., Mergenthaler J. L.

(NASA Ames Research Center, MS 245-4, Moffett Field, CA 94035-1000, USA)

Abstract: Upper Atmosphere Research Satellite observations indicate that extensive denitrification without significant dehydration currently occurs only in the Antarctic during mid to late June. The fact that denitrification occurs in a relatively warm month in the Antarctic raises concern about the likelihood of its occurrence and associated effects on ozone recovery in a colder and possibly more humid future Arctic lower stratosphere. Polar stratospheric cloud lifetimes required for Arctic denitrification to occur in the future are presented and contrasted against the current Antarctic cloud lifetimes. Model calculations show that widespread severe denitrification could enhance future Arctic ozone loss by up to 30%.

反硝化及其对臭氧恢复的量化作用

Tabazadeh A., Santee M. L., Danilin M. Y., Pumphrey H. C., Newman P. A., Hamill P. J., Mergenthaler J. L.

(美国国家航空和航天局)

摘要：高层大气研究卫星观测结果显示，没有明显脱水作用伴随的广泛的脱氮作用通常仅在南极 6 月中下旬发生。脱氮作用发生在南极相对温暖月份的事实增加了对其发生可能性的关注，以及在未来更加寒冷，同时可能更加潮湿的北极平流层底部臭氧恢复过程中的影响。本文指出了未来依赖北极脱氮作用的极地平流层云的生存周期并与当前南极云的生存周期进行对比。模型计算显示广泛发生的脱氮作用会使未来北极臭氧流失增加达30%。

21. Oechel W C, Vourlitis G L, Hastings S J, et al. Acclimation of ecosystem CO_2 exchange in the Alaskan Arctic in response to decadal climate warming[J]. Nature, 2000, 406(6799): 978-981.

Acclimation of ecosystem CO_2 exchange in the Alaskan Arctic in response to decadal climate warming

Oechel W. C., Vourlitis G. L., Hastings S. J., Zulueta R. C., Hinzman L., Kane D.

(Global Change Research Group, Department of Biology, San Diego State University, San Diego, California 92182-4614, USA)

Abstract: Long-term sequestration of carbon in Alaskan Arctic tundra ecosystems was reversed by warming and drying of the climate in the early 1980s, resulting in substantial losses of terrestrial carbon. But recent measurements suggest that continued warming and drying has resulted in diminished CO_2 efflux, and in some cases, summer CO_2 sink activity. Here we compile summer CO_2 flux data for two Arctic ecosystems from 1960 to the end of 1998. The results show that a return to summer sink activity has come during the warmest and driest period observed over the past four decades, and indicates a previously undemonstrated capacity for ecosystems to metabolically adjust to long-term (decadal or longer) changes in climate. The mechanisms involved are likely to include changes in nutrient cycling, physiological acclimation, and population and community reorganization. Nevertheless, despite the observed acclimation, the Arctic ecosystems studied are still annual net sources of CO_2 to the atmosphere of at least 40 g · C · m^{-2} · yr^{-1}, due to winter release of CO_2, implying that further climate change may still exacerbate CO_2 emissions from Arctic ecosystems.

北极阿拉斯加地区生态系统 CO_2 交换对十年来气候变暖的适应

Oechel W. C., Vourlitis G. L., Hastings S. J., Zulueta R. C., Hinzman L., Kane D.

(美国圣地亚哥州立大学)

摘要：20世纪80年代早期，由于气候的温暖和干燥，造成了阿拉斯加北极苔原生态系统长期封存的碳的排放，从而引起了陆源碳的大量流失。但是，最近结果显示持续的温暖和干旱导致了一些情况下 CO_2 排放的减少，以及夏季 CO_2 汇的行为。本文中我们汇编了北极生态系统中1960～1998年夏季 CO_2 通量的数据。结果显示在过去40年间的温暖干燥时期观测到了夏季汇行为的反弹，并指示了过去尚未证实的生态系统通过新陈代谢以适应气候长期(十年或更长)变化的能力。相关的机制可能包括养分循环的变化、生理适应和数量与群落的重组。然而，尽管观测到了适应现象，由于冬季 CO_2 的释放，北极生态系统仍然是大气 CO_2 的一个净源(每年至少40 g · C/m^2)，这意味着北极生态系统的 CO_2 排放会使得气候进一步恶化。

22. Aldhous P. Global warming could be bad news for Arctic ozone layer[J]. Nature, 2000, 404(6778):

531.

Global warming could be bad news for Arctic ozone layer

Aldhous P.

(News)

Abstract: Environmental scientists have struggled for years to dispel a popular misconception that global warming and the thinning ozone layer are one and the same. But researchers who have just completed the largest ever survey of Arctic stratospheric ozone now warn that the two processes are, indeed, connected.

全球变暖对北极臭氧层来说可能是坏消息

Aldhous P.

(新闻)

摘要:为了消除人们对全球气候变暖等同于臭氧层变薄的普遍误解,环境科学家已经为此奋斗了多年,但刚刚完成了有史以来最大的北极平流层臭氧调查的研究人员警告,这两个过程实际上是相互关联的。

23. Zimov S A, Davidov S P, Zimova G M, et al. Contribution of disturbance to increasing seasonal amplitude of atmospheric CO_2[J]. Science,1999,284(5422):1973-1976.

Contribution of disturbance to increasing seasonal amplitude of atmospheric CO_2

Zimov S. A., Davidov S. P., Zimova G. M., Davidova A. I., Chapin F. S., Chapin M. C., Reynolds J. F.

(Univ Alaska, Inst Arctic Biol, Fairbanks, AK 99775 USA)

Abstract: Recent increases in the seasonal amplitude of atmospheric carbon dioxide (CO_2) at high latitudes suggest a widespread biospheric response to high-latitude warming. The seasonal amplitude of net ecosystem carbon exchange by northern Siberian ecosystems is shown to be greater in disturbed than undisturbed sites, due to increased summer influx and increased winter efflux. Increased disturbance could therefore contribute significantly to the amplified seasonal cycle of atmospheric carbon dioxide at high latitudes. Warm temperatures reduced summer carbon influx, suggesting that high-latitude warming, if it occurred, would be unlikely to increase seasonal amplitude of carbon exchange.

对增加的大气 CO_2 季节性波动的扰动的贡献

Zimov S. A., Davidov S. P., Zimova G. M., Davidova A. I., Chapin F. S., Chapin M. C., Reynolds J. F.

(美国阿拉斯加大学)

摘要:最近高纬度地区大气 CO_2 季节性振幅的增加指示了生物圈对高纬度气候变暖的广泛响应。由于夏季流入通量和冬季流出通量的增加,西伯利亚北部生态系统引起的生态系统碳净交换的季节性振幅在扰动的地点较未被扰动的地点更大。因此增加的扰动能够对高纬度地区大气 CO_2 放大的季节循环产生显著贡献。温度的升高降低了夏季碳的流入通量,表明如果发生高纬度地区气候变暖,可能不会造成碳交换季节振幅的增大。

24. Waibel A E, Peter T, Carslaw K S, et al. Arctic ozone loss due to denitrification[J]. Science, 1999,283(5410):2064-2069.

Arctic ozone loss due to denitrification

Waibel A. E., Peter T., Carslaw K. S., Oelhaf H., Wetzel G., Crutzen P. J., Poschl U., Tsias A., Reimer E., Fischer H.

(ETH Zurich, HPP Honggerberg, Inst Atmospher Sci, CH-8093 Zurich, Switzerland)

Abstract: Measurements from the winter of 1994-1995 indicating removal of total reactive nitrogen from the Arctic stratosphere by particle sedimentation were used to constrain a microphysical model. The model suggests that denitrification is caused predominantly by nitric acid trihydrate particles in small number densities. The denitrification is shown to increase Arctic ozone loss substantially. Sensitivity studies indicate that the Arctic stratosphere is currently at a threshold of denitrification. This implies that future stratospheric cooling, induced by an increase in the anthropogenic carbon dioxide burden, is likely to enhance denitrification and to delay until late in the next century the return of Arctic stratospheric ozone to preindustrial values.

反硝化作用造成的北极臭氧流失

Waibel A. E., Peter T., Carslaw K. S., Oelhaf H., Wetzel G., Crutzen P. J., Poschl U., Tsias A., Reimer E., Fischer H.

(瑞士苏黎世理工学院)

摘要:1994～1995 年冬季的观测显示,粒子沉降可造成总活性氮从北极平流层被除去,这一观测被用来约束微观物理模型。模型显示脱氮作用主要是由小密度的三水合硝酸粒子所引发的,脱氮作用显著增强了北极臭氧的损失,敏感性研究表明北极平流层脱氮作用当前处于一个阀值点。这一结果表明,由增加的人为源 CO_2 负担所导致的未来平流层变冷,有可能会增强脱氮作用,并使北极平流层臭氧恢复至工业革命前水平的时间推迟至下世纪晚期。

25. Sumner A L, Shepson P B. Snowpack production of formaldehyde and its effect on the Arctic troposphere[J]. Nature, 1999, 398(6724): 230-233.

Snowpack production of formaldehyde and its effect on the Arctic troposphere

Sumner A. L., Shepson P. B.

(Purdue Univ, Dept Chem, W Lafayette, IN 47907 USA)

Abstract: The oxidative capacity of the atmosphere determines the lifetime and ultimate fate of atmospheric trace species. It is controlled by the presence of highly reactive radicals, particularly OH formed as a result of ozone photolysis. The dramatic depletion of ozone in Arctic surface air during polar sunrise therefore offers an opportunity to improve our understanding of the processes controlling ozone abundance and hence the oxidative capacity of the atmosphere. Ozone destruction is catalysed by bromine atoms' and is terminated once bromine reacts with formaldehyde to form relatively inert hydrogen bromide, but neither the activation of bromine nor the contribution of formaldehyde are fully understood. Particularly troubling is the failure of current models to simulate the high formaldehyde concentrations in Arctic surface air. Here we report measurements in Arctic snow and near-surface air, which suggest that photochemical production at the air-snow interface accounts for the discrepancy between observed and predicted formaldehyde concentrations. The strength of this source is comparable to that of the dominant formaldehyde source in the free troposphere (the reaction between OH and methane) and implies that formaldehyde photolysis can be a dominant source of oxidizing free radicals in the lower polar troposphere. We expect that formaldehyde will also affect photochemistry at the snow surface to facilitate the release of bromine into the lower troposphere—the initial step in Arctic tropospheric ozone depletion.

积雪生成的甲醛及其对北极对流层的影响

Sumner A. L., Shepson P. B.

(美国普渡大学)

摘要：大气的氧化能力决定了大气中痕量物质的寿命和最终命运。而这一能力受到了高活性分子，特别是臭氧光分解所产生的 OH 自由基的控制。因此极地日出期间北极近地面大气臭氧的快速损失为我们进一步了解控制臭氧丰度的过程以及大气的氧化能力提供了机会。臭氧的损失由溴原子催化引起，一旦溴与甲醛反应生成相对惰性的 HBr 则终止。但是不论是溴的活化作用还是甲醛的贡献机制都没有完全研究清楚。尤其是目前的模型无法模拟北极表层空气中的高甲醛浓度。本文中，我们报道了对北极雪面和近地面空气的观测，结果表明大气-雪界面中的光化学产物能够解释甲醛的观测浓度与预测浓度之间的矛盾。这一来源的强度相当于自由对流层中起主要作用的甲醛源(OH 和甲烷间的反应)，并意味着甲醛光分解是极地对流层底部氧化性自由基的主要来源。我们估计甲醛也能影响雪表面的光化学以促进溴向对流层底部的释放，进而引发北极对流层臭氧的亏损。

26. Schreiner J, Voigt C, Kohlmann A, et al. Chemical analysis of polar stratospheric cloud particles [J]. Science, 1999, 283(5404): 968-970.

Chemical analysis of polar stratospheric cloud particles

Schreiner J., Voigt C., Kohlmann A., Amold F., Mauersberger K., Larsen N.

(Max Planck Inst Kernphys, Div Atmospher Phys, POB 103 980, D-69029 Heidelberg, Germany)

Abstract: A balloon-borne gondola carrying a particle analysis system, a backscatter sonde, and pressure and temperature sensors was launched from Kiruna, Sweden, on 25 January 1998. Measurements within polar stratospheric cloud layers inside the Arctic polar vortex show a close correlation between Large backscatter ratios and enhanced particle-related water and nitric acid signals at low temperatures. Periodic structures in the data indicate the presence of lee waves. The H_2O/HNO_3 molar ratios are consistently found to be above 10 at atmospheric temperatures between 189 and 192 kelvin. Such high ratios indicate ternary solution particles of H_2O, HNO_3, and H_2SO_4 rather than the presence of solid hydrates.

极地平流层云雾粒子的化学分析

Schreiner J., Voigt C., Kohlmann A., Amold F., Mauersberger K., Larsen N.

(德国马普研究所)

摘要:1998 年 1 月 25 日,在瑞典基律纳地区使用气球搭载吊舱携带粒子分析系统、后向散射探测仪和压力温度传感器进行观测。北极极地涡旋内极地平流层云的观测显示,在低温情况下高的后向散射比和增强的与粒子相关的水和硝酸信号之间有强相关性。数据中的周期结构指示了下风波的存在。当大气温度介于 189～192 开尔文时,H_2O/HNO_3 摩尔比保持在 10 以上。如此高的比值指示 H_2O、HNO_3 和 H_2SO_4 三元溶液微粒的存在,而不是固态氢氧化物。

27. McElroy C T, McLinden C A, McConnell J C. Evidence for bromine monoxide in the free troposphere during the Arctic polar sunrise[J]. Nature,1999,397(6717):338-341.

Evidence for bromine monoxide in the free troposphere during the Arctic polar sunrise

McElroy C. T., McLinden C. A., McConnell J. C.

(Environm Canada, Downsview, ON M3H 5T4, Canada)

Abstract: During the Arctic polar springtime, dramatic ozone losses occur not only in the stratosphere but also in the underlying troposphere. These tropospheric ozone loss events have been observed over large areas in the planetary boundary layer (PBL) throughout the Arctic. They are associated with enhanced concentrations of halogen species and are probably caused by catalytic reactions involving bromine monoxide (BrO) and perhaps also chlorine monooxide (ClO). The origin of the BrO, the principle species driving the ozone destruction, is thought to be the autocatalytic release of bromine from sea salt accumulated on the Arctic snow pack, followed by photolytic and heterogeneous reactions which produce and recycle the oxide. Satellite observations have shown the horizontal and temporal extent of large BrO enhancements in the Arctic troposphere, but the vertical distribution of the BrO has remained uncertain. Here we report BrO observations obtained from a high-altitude aircraft that suggest the presence of significant amounts of BrO not only in the PBL but also in the free troposphere above it. We believe that the BrO is transported from the PBL into the free troposphere through convection over large Arctic ice leads (openings in the pack ice). The convective transport also lifts ice crystals and water droplets well above the PBL, thus providing surfaces for heterogeneous reactions that can recycle BrO from less-reactive forms and thereby maintain its ability to affect the chemistry of the free troposphere.

北极日出时自由对流层中 BrO 存在的证据

McElroy C. T., McLinden C. A., McConnell J. C.

(加拿大环境部)

摘要：北极极地春季期间，臭氧的大量流失不仅发生在平流层，对流层臭氧损耗事件也在北极边界层大面积观测到。这与卤素浓度的增加相关，可能是由于 BrO(也可能包括 ClO)所引起的催化反应而造成的。一般认为 BrO——引起臭氧流失的基本粒子的来源，除了光解和导致氧化物循环的多相反应以外，还有来自北极积雪上积累的海盐中溴的自催化过程。卫星观测显示，北极对流层中 BrO 在水平范围和时间尺度上有大量增加，但是在垂直方向上的分布尚不明确。本文中，我们通过高海拔飞机观测得到了 BrO 的变化，结果显示 BrO 高含量的出现不仅在地球边界层，还有其上的自由对流层。我们相信在北极海冰前缘的对流中 BrO 从地球边界层传输到自由对流层。这一对流同样将冰晶和水滴抬升于边界层之上，并为多相反应提供界面，从而使 BrO 从弱活性状态再循环，并继续影响对流层的化学。

28. Kirk-Davidoff D B, Hintsa E J, Anderson J G, et al. The effect of climate change on ozone depletion through changes in stratospheric water vapour[J]. Nature, 1999, 402(6760): 399-401.

The effect of climate change on ozone depletion through changes in stratospheric water vapour

Kirk-Davidoff D. B., Hintsa E. J., Anderson J. G., Keith D. W.

(Harvard Univ, Dept Earth & Planetary Sci, 20 Oxford St, Cambridge, MA 02138 USA)

Abstract: Several studies have predicted substantial increases in Arctic ozone depletion due to the stratospheric cooling induced by increasing atmospheric CO_2 concentrations. But climate change may additionally influence Arctic ozone depletion through changes in the water vapour cycle. Here we investigate this possibility by combining predictions of tropical tropopause temperatures from a general circulation model with results from a one-dimensional radiative convective model, recent progress in understanding the stratospheric water vapour budget, modelling of heterogeneous reaction rates and the results of a general circulation model on the radiative effect of increased water vapour. Whereas most of the stratosphere will cool as greenhouse-gas concentrations increase, the tropical tropopause may become warmer, resulting in an increase of the mean saturation mixing ratio of water vapour and hence an increased transport of water vapour from the troposphere to the stratosphere. Stratospheric water vapour concentration in the polar regions determines both the critical temperature below which heterogeneous reactions on cold aerosols become important (the mechanism driving enhanced ozone depletion) and the temperature of the Arctic vortex itself. Our results indicate that ozone loss in the later winter and spring Arctic vortex depends critically on water vapour variations which are forced by sea surface temperature changes in the tropics. This potentially important effect has not been taken into account in previous scenarios of Arctic ozone loss under climate change conditions.

气候变化产生的平流层水蒸气变化对臭氧亏损的影响

Kirk-Davidoff D. B., Hintsa E. J., Anderson J. G., Keith D. W.

(美国哈佛大学)

摘要：大气 CO_2 浓度增加导致了平流层变冷，基于此许多研究预测了北极臭氧损耗的大幅度增加。但是气候变化可能会通过水汽循环的改变影响北极臭氧的损耗。综合大气环流模型中得到的热带对流层顶部温度、一维空间辐射对流模型的结果，最近在解释平流层水汽平衡、模拟异构反应的速率和大气环流模型结果在水汽增加上的辐射影响的进展，我们研究了气候变化通过水汽循环而影响臭氧损耗的可能性。然而，由于温室气体浓度的增加，大部分平流层会变冷，热带对流层顶部会变得更温暖，导致水汽的平均饱和混合比上升，并进而引起水汽从对流层到平流层传输的增加。极地地区平流层水汽的浓度决定了在冷气溶胶上发生重要的多相反应（驱动臭氧亏损增强的机制）的最高临界温度和北极涡旋自身的温度。我们的结果表明，晚冬和春季北极涡旋中的臭氧流失严重依赖于受热带海表面温度的改变所驱动的水汽的变化。而这潜在的重要影响当前还没有被考虑作为气候变化条件下北极臭氧损耗的一个环节。

29. Hebestreit K, Stutz J, Rosen D, et al. DOAS measurements of tropospheric bromine oxide in mid-latitudes[J]. Science, 1999, 283(5398): 55-57.

DOAS measurements of tropospheric bromine oxide in mid-latitudes

Hebestreit K., Stutz J., Rosen D., Matveiv V., Peleg M., Luria M., Platt U.

(Univ Heidelberg, Inst Umweltphys, Neuenheimer Feld 366, D-69120 Heidelberg, Germany)

Abstract: Episodes of elevated bromine oxide (BrO) concentration are known to occur at high latitudes in the Arctic boundary layer and to lead to catalytic destruction of ozone at those latitudes; these events have not been observed at lower latitudes. With the use of differential optical absorption spectroscopy (DOAS), locally high BrO concentrations were observed at mid-latitudes at the Dead Sea, Israel, during spring 1997. Mixing ratios peaked daily at around 80 parts per trillion around noon and were correlated with low boundary-layer ozone mixing ratios.

用 DOAS 手段对中纬度对流层 BrO 的观测

Hebestreit K., Stutz J., Rosen D., Matveiv V., Peleg M., Luria M., Platt U.

(德国海德堡大学)

摘要:普遍认为在高纬度地区北极边界层氧化溴的浓度有升高现象,从而导致了在这些纬度臭氧的催化破坏,而这些事件没有在较低纬度地区被观测到。1997 年春季在以色列死海的中纬度地区使用差分光学吸收光谱(DOAS)观测到了当地较高的 BrO 浓度。每天中午时出现 80 ppt 的混合比峰值,与低边界层臭氧混合比相一致。

30. Chipperfield M P, Jones R L. Relative influences of atmospheric chemistry and transport on Arctic ozone trends[J]. Nature, 1999, 400(6744): 551-554.

Relative influences of atmospheric chemistry and transport on Arctic ozone trends

Chipperfield M. P., Jones R. L.

(Univ Leeds, Ctr Environm, Leeds LS2 9JT, W Yorkshire, England)

Abstract: The reduction in the amount of ozone in the atmospheric column over the Arctic region, observed during the 1990s, resembles the onset of the Antarctic ozone "hole" in the mid-1980s, but the two polar regions differ significantly with respect to the relative contributions of chemistry and atmospheric dynamics to the ozone abundance. In the strong, cold Antarctic vortex, rapid springtime chemical ozone loss occurs throughout a large region of the lower stratosphere, whereas in the Arctic, although chemical ozone depletion has been observed, the vortex is generally much smaller, weaker and more variable. Here we report a model-based analysis of the relative importance of dynamics and chemistry in causing the Arctic ozone trend in the 1990s, using a state-of-the-art three-dimensional stratospheric chemistry-transport model. North of 63 degrees N we find that, on average, dynamical variations dominate the interannual variability, with little evidence for a trend towards more wintertime chemical depletion. However, increases in the burden of atmospheric halogens since the early 1970s are responsible for a large (14%) reduction in the average March column ozone, but this effect is mostly caused by increased destruction throughout the year rather than by halogen chemistry associated with wintertime polar statospheric clouds. Any influence of climate change on future average Arctic ozone amounts may thus be dominated by possible circulation changes, rather than by changes in chemical loss.

大气化学和传输对北极臭氧趋势的有关影响

Chipperfield M. P., Jones R. L.

(英国利兹大学)

摘要:20 世纪 90 年代观测到的北极地区臭氧大气柱含量的减少与 20 世纪 80 年代中期发生的南极

臭氧洞相似,但是两处极地地区在大气动力学和化学对臭氧浓度的贡献方面非常不一致。在极端寒冷的南极涡旋中,平流层下部的较大范围发生春季臭氧快速损耗,而在北极虽然观测到了化学臭氧亏损,涡旋一般小很多,更弱且变数更多。本研究中,我们采用前沿的三维平流层化学传输模型分析了动力学和化学在北极臭氧趋势中的相对重要性。我们发现,一般而言在北纬63°,年际变化主要受动力学变化所支配,并且没有证据显示冬季化学亏损的趋向。然而从20世纪70年代初以来的大气卤素负担的增加,造成3月平均臭氧柱含量大幅度减少(14%),但是这种影响主要是由于全年损耗的增加所造成的,而非与冬季极地平流层云相关的卤素化学。气候变化对未来北极臭氧平均浓度的影响可能是由环流变化所造成的,而不是化学损耗的变化。

31. Wagner T, Platt U. Satellite mapping of enhanced BrO concentrations in the troposphere[J]. Nature,1998,395(6701):486-490.

Satellite mapping of enhanced BrO concentrations in the troposphere

Wagner T. , Platt U.

(Institut für Umweltphysik, University of Heidelberg, INF 366, 69120 Heidelberg, Germany)

Abstract: Reactive bromine species contribute significantly to the destruction of ozone in the polar stratosphere. Reactive halogen compounds can have a strong effect not only on the chemistry of the stratosphere but also on that of the underlying troposphere. For example, severe ozone depletion events that are less persistent than those in the stratosphere occur in the Arctic and Antarctic boundary layer during springtime and are also associated with enhanced BrO abundances. Observations of BrO (and ClO, which is less important) at ground level during these ozone depletion events have revealed halogen oxide mixing ratios of up to 30 parts per trillion-sufficient to destroy within one to two days the 30-40 parts per billion of ozone typically present in the boundary layer. The catalytic mechanism leading to so-called "tropospheric ozone holes" is well established, but the origin of the increased BrO concentrations and the spatial and temporal extent of these events remains poorly understood. Here we present satellite observations showing that tropospheric air masses enriched in BrO are always situated close to sea ice and typically extend over areas of about 300-2000 km. The BrO abundances remain enhanced for periods of 1 to 3 days. These observations support the suggestion that autocatalytic release of bromine from sea salt gives rise to significant BrO formation which, in turn, initiates ozone depletion in the polar troposphere.

对流层 BrO 浓度增加的卫星成像

Wagner T. , Platt U.

(德国海德堡大学)

摘要:活性溴对极地平流层中臭氧的损耗产生重大影响。活性卤素化合物不仅对平流层化学而且对对流层底层都会产生重大影响。例如,发生在春季北极和南极边界层的严重臭氧损耗事件没有平流层中的那样持久,但也都与BrO的增加有关。对臭氧亏损事件期间地面BrO(以及次重要的ClO)的观测显示,混合比高达30 ppt的卤素氧化物足够在1~2天内使边界层中的臭氧损失30~40 ppb。该催化机理引发了所谓的"对流层臭氧洞",但是对增加的BrO源和这些事件的时空尺度仍然知之甚少。本文中,我们通过卫星观测显示,富集BrO的对流层气团总是位于海冰附近,并且通常延展到300～2000 km。BrO的含量保持1～3天的增长。这些观测结果能够支持溴可以从海盐中经自催化释放从而生成引起极地对流层臭氧损耗的BrO形成的观点。

32. Shindell D T, Rind D, Lonergan P. Increased polar stratospheric ozone losses and delayed eventual recovery owing to increasing greenhouse-gas concentrations[J]. Nature, 1998, 395(6701): 486-490.

Increased polar stratospheric ozone losses and delayed eventual recovery owing to increasing greenhouse-gas concentrations

Shindell D. T., Rind D., Lonergan P.

(NASA Goddard Institute for Space Studies and Center for Climate Systems Research, Columbia University, 2880 Broadway, New York, New York 10025, USA)

Abstract: The chemical reactions responsible for stratospheric ozone depletion are extremely sensitive to temperature. Greenhouse gases warm the Earth's surface but cool the stratosphere radiatively and therefore affect ozone depletion. Here we investigate the interplay between projected future emissions of greenhouse gases and levels of ozone-depleting halogen species using a global climate model that incorporates simplified ozone-depletion chemistry. Temperature and wind changes induced by the increasing greenhouse-gas concentrations alter planetary-wave propagation in our model, reducing the frequency of sudden stratospheric warmings in the Northern Hemisphere. This results in a more stable Arctic polar vortex, with significantly colder temperatures in the lower stratosphere and concomitantly increased ozone depletion. Increased concentrations of greenhouse gases might therefore be at least partly responsible for the very large Arctic ozone losses observed in recent winters. Arctic losses reach a maximum in the decade 2010 to 2019 in our model, roughly a decade after the maximum in stratospheric chlorine abundance. The mean losses are about the same as those over the Antarctic during the early 1990s, with geographically localized losses of up to two-thirds of the Arctic ozone column in the worst years. The severity and the duration of the Antarctic ozone hole are also predicted to increase because of greenhouse-gas-induced stratospheric cooling over thecoming decades.

日益增加的温室气体浓度造成了极地平流层臭氧层空洞增大并阻碍其恢复

Shindell D. T., Rind D., Lonergan P.

(美国国家航空和航天局)

摘要:导致平流层臭氧耗损的化学反应对温度极其敏感。温室气体造成地球表面的温度升高,但却导致平流层大气的温度相对降低,并且由此影响了平流层的臭氧层耗损。本文中,我们利用一个综合了简化的臭氧层消耗化学反应的全球气候模型来研究未来温室气体的排放量和臭氧消耗卤族元素的水平之间的相互作用关系。增加的温室气体浓度导致的温度和风的变化改变了行星波的传播,减少了北半球平流层突然变暖事件发生的频率。这导致北极涡流系统变得更加稳定,并且低平流层变得极为寒冷,从而导致臭氧损耗的速度增加。因此,温室气体浓度的增加至少在一定程度上导致了近些年来冬季北极十分大的臭氧层空洞。利用我们的模型进行模拟的结果显示，北极臭氧层空洞在2010～2019年这10年间将达到最大，比平流层中氯的含量达到最大值的时间晚了将近10年。臭氧层的平均耗损量大约和20世纪90年代南极地区的相当。对于最严重的年份，臭氧层空洞在地理位置上占了北极臭氧层的至少2/3的面积。我们也预测了南极臭氧层空洞的严重程度和持续时间，结果表明温室气体在未来几十年里引发的平流层降温会导致南极臭氧空洞的增加。

33. Carslaw K S, Wirth M, Tsias A, et al. Increased stratospheric ozone depletion due to mountain-induced atmospheric waves[J]. Nature, 1998, 391(6668): 675-678.

Increased stratospheric ozone depletion due to mountain-induced atmospheric waves

Carslaw K. S., Wirth M., Tsias A., Luo B. P., Dörnbrack A., Leutbecher M., Volkert H., Renger W., Bacmeister J. T., Reimer E., Peter T.

(Max Planck Inst Chem, Postfach 3060, D-6500 Mainz, Germany)

Abstract: Chemical reactions on polar stratospheric cloud (PSC) particles are responsible for the production of reactive chlorine species (chlorine "activation") which cause ozone destruction. Gas-phase deactivation of these chlorine species can take several weeks in the Arctic winter stratosphere, so that ozone destruction can be sustained even in air parcels that encounter PSCs only intermittently. Chlorine activation during a PSC encounter proceeds much faster at low temperatures when cloud particle surface area and heterogeneous reaction rates are higher. Although mountain-induced atmospheric gravity waves are known to cause local reductions in stratospheric temperature of as much as 10-15 K, and are often associated with mesoscale PSCs, their effect on chlorine activation and ozone depletion has not been considered. Here we describe aircraft observations of mountain-wave-induced mesoscale PSCs in which temperatures were 12 K lower than expected synoptically, Model calculations show that despite their localized nature, these PSCs can cause almost complete conversion of inactive chlorine species to ozone-destroying forms in air flowing through the clouds. Using a global mountain-wave model, we identify regions where mountain waves can develop, and show that they can cause frequent chlorine activation of air in the Arctic stratosphere. Such mesoscale processes offer a possible explanation for the underprediction of reactive chlorine concentrations and ozone depletion rates calculated by three-dimensional models of the Arctic stratosphere.

由于山脉地形激发的大气波动加剧了平流层臭氧亏损

Carslaw K. S., Wirth M., Tsias A., Luo B. P., Dörnbrack A., Leutbecher M., Volkert H., Renger W., Bacmeister J. T., Reimer E., Peter T.

(德国马普研究所)

摘要：极地平流层云雾粒子上的化学反应是产生能够引起臭氧损耗的活性氯的原因。这些活性氯在北极冬季平流层能够保持数周的气相惰性，因此即使是间歇遇到极地平流层云的空气粒子，它们依旧能够破坏臭氧。在云雾粒子表面积更大、多相反应速率更高的情况下，在低温时，氯遇到极地平流层云会有很高的活化速度。尽管已经知道山脉地形激发的大气重力波会引起平流层温度的区域性降低(10～

15 K)且常常与中尺度极地平流层云相关,但是对氯的活化作用和臭氧亏损方面的影响人们还没有加以考虑。本文中,我们对山脉地形波激发的中尺度极地平流层云(温度比一般预期低 12 K)进行了飞行器观测。模型计算显示不论局部情况如何,这些极地平流层云都能使惰性氯在随空气通过这些云时几乎全部转换成破坏臭氧的形态。通过全球山脉地形波模型,我们识别出了可以产生山脉波的地区,并显示这些地区可以导致北极平流层地区大气中频繁的氯活化作用。这种中尺度作用为北极平流层活性氯浓度的三维模型计算和臭氧亏损速率的预测偏低提供了一种可能的解释。

34. Stolarski R. Atmospheric chemistry:a bad winter for Arctic ozone[J]. Nature,1997,389(6653):788-789.

Atmospheric chemistry—a bad winter for Arctic ozone

Stolarski R.

(News and Views)

Abstract:Ozone loss in the Antarctic, resulting in a seasonal "ozone hole", is a familiar problem. But what of the Arctic? Why are the Northern polar regions less susceptible to ozone loss and could that situation change? Two papers (by Müller et al in last week's Nature and Rex et al on page 835 of this issue) now report large ozone loss during the winter and early spring of 1995-96 in the Arctic. This comes after considerable losses over the previous few winters. What has been peculiar about the past few winters is that the Arctic stratosphere has stayed cold for longer than average, and this seems to have led to low ozone concentrations.

大气化学——北极臭氧经历严冬

Stolarski R.

(新闻观点)

摘要:在南极由季节性的"臭氧洞"导致的臭氧损耗是一个常见的问题。但是北极的臭氧是怎样的呢?为什么北极地区少有臭氧损耗发生,这一形势会有变化吗?这次有两篇论文(Müller 等在上周的《自然》杂志和 Rex 等在本期第 835 页)报告在 1995～1996 年的冬季和早春发生了北极臭氧损耗,前几个冬天也有相当大的臭氧损耗。与过去几年的冬天不同的是,北极平流层比以往寒冷的时间更长,这似乎是导致臭氧浓度低的原因。

35. Rex M, Harris N R P, von der Gathen P, et al. Prolonged stratospheric ozone loss in the 1995-96 Arctic winter[J]. Nature,1997,389(6653):835-838.

Prolonged stratospheric ozone loss in the 1995-96 Arctic winter

Rex M., Harris N. R. P., von der Gathen P., Lehmann R., Braathen G. O., Reimer E., Beck A., Chipperfield M. P., Alfier R., Allaart M., Oconnor F., Dier H., Dorokhov V., Fast H., Gil M., Kyro E., Litynska Z., Mikkelsen I. S., Molyneux M. G., Nakane H., Notholt J., Rummukainen M., Viatte P., Wenger J.

(Alfred Wegener Institute for Polar and Marine Research, PO Box 60 01 49, D-14401 Potsdam, Germany)

Abstract: It is well established that extensive depletion of ozone, initiated by heterogenous reactions on polar stratospheric clouds (PSCs) can occur in both the Arctic and Antarctic lower stratosphere. Moreover, it has been shown that ozone loss rates in the Arctic region in recent years reached values comparable to those over the Antarctic. But until now the accumulated ozone losses over the Arctic have been the smaller, mainly because the period of Arctic ozone loss has not-unlike over the Antarctic-persisted well into springtime. Here we report the occurrence-during the unusually cold 1995-96 Arctic winter of the highest recorded chemical ozone loss over the Arctic region. Two new kinds of behaviour were observed. First, ozone loss at some altitudes was observed long after the last exposure to PSCs. This continued loss appears to be due to a removal of the nitrogen species that slow down chemical ozone depletion. Second, in another altitude range ozone loss rates decreased while PSCs were still present, apparently because of an early transformation of the ozone-destroying chlorine species into less active chlorinenitrate. The balance between these two counteracting mechanisms is probably a fine one, determined by small differences in wintertime stratospheric temperatures. If the apparent cooling trend in the Arctic stratosphere is real, more dramatic ozone losses may occur in the future.

1995～1996 年北极冬季平流层臭氧更长时间的流失

Rex M., Harris N. R. P., von der Gathen P., Lehmann R., Braathen G. O., Reimer E., Beck A., Chipperfield M. P., Alfier R., Allaart M., Oconnor F., Dier H., Dorokhov V., Fast H., Gil M., Kyro E., Litynska Z., Mikkelsen I. S., Molyneux M. G., Nakane H., Notholt J., Rummukainen M., Viatte P., Wenger J.

(德国魏格纳极地和海洋研究中心)

摘要:众所周知,由极地平流层云中多相反应引发的大规模的臭氧层损耗可以在北极和南极低平流层大气中发生。除此之外,已经有研究表明,近些年来北极地区臭氧层损耗的速率已经和南极地区相当。但是直到现在,北极上空中的臭氧层消耗的累计量仍然十分小,主要是由于北极臭氧层损耗的时间不像南极地区那样持续到第二年的春季。本文中,我们研究了 1995～1996 年北极冬季异常冷的事件中,北极地区记录到的最大的臭氧层空洞事件。我们观察到了两个新的现象:第一,在最后一次暴露在平流层云之后,一些高度上的臭氧层空洞持续了较长时间。这种持续的臭氧层空洞明显可以减慢臭氧层消耗反应的氮粒子的去除。第二,在其他高度尺度臭氧层损耗的速率有所降低而极地平流云仍然存在。这显然是由于前期消耗臭氧的氯转化成活性低的硝酸氯。这两个反作用机制的平衡可能是由冬季平流层温度的微小差异来决定的。如果北极平流层有明显变冷的趋势是真实的,未来发生臭氧损耗的可能性会更大。

36. Muller R, Crutzen P J, Grooss J U, et al. Severe chemical ozone loss in the Arctic during the winter of 1995-96[J]. Nature,1997,389(6652):709-712.

Severe chemical ozone loss in the Arctic during the winter of 1995-96

Muller R., Crutzen P. J., Grooss J. U., Bruhl C., Russell J. M., Gernandt H., McKenna D. S., Tuck A. F.

(Forschungszentrum Jülich, Institute for Stratospheric Chemistry (ICG-1), 52425 Jülich, Germany)

Abstract: Severe stratospheric ozone depletion is the result of perturbations of chlorine chemistry owing to the presence of polar stratospheric clouds (PSCs) during periods of Limited exchange of air between the polar vortex and midlatitudes and partial exposure of the vortex to sunlight. These conditions are consistently encountered over Antarctica during the austral spring. In the Arctic, extensive PSC formation occurs only during the coldest winters, when temperatures fall as low as those regularly found in the Antarctic. Moreover, ozone levels in late winter and early spring are significantly higher than in the corresponding austral season, and usually strongly perturbed by atmospheric dynamics. For these reasons, chemical ozone loss in the Arctic is difficult to quantify. Here we use the correlation between CH_4 and O_3 in the Arctic polar vortex to discriminate between changes in ozone concentration due to chemical and dynamical effects. Our results indicate that 120-160 Dobson units (DU) of ozone were chemically destroyed between January and March 1996—a loss greater than observed in Antarctica in 1985, when the "ozone hole" was first reported. This loss outweighs the expected increase in total ozone over the same period through dynamical effects, leading to an observed net decrease of about 50 DU. This ozone loss arises through the simultaneous occurrence of extremely low Arctic stratospheric temperatures and large stratospheric chlorine loadings. Comparable depletion is likely to recur because stratospheric cooling and elevated chlorine concentrations are expected to persist for several decades.

1995～1996年冬季北极发生的严重化学臭氧损耗

Muller R., Crutzen P. J., Grooss J. U., Bruhl C., Russell J. M., Gernandt H., McKenna D. S., Tuck A. F.

(德国尤里希研究中心平流层化学研究所)

摘要：严重的平流层臭氧损耗可能是在极地环流和中纬度地区之间有限的大气交换以及环流部分暴露于阳光期间在极地平流层云的存在下氯的化学扰动造成的。这些情况在南半球春季的南极地区是基本一致的。在北极地区，大量的平流层云只有在最冷的冬季时期，当温度下降到南极地区的平常水平时才形成。在这些时期，温度可以下降到南极地区的平常水平。除此之外，臭氧浓度在冬季和早春季节明显高于南半球相应季节，并且经常受到大气动力学的强烈扰动。由于这些原因，北极地区化学臭氧空洞难以被定量化。本文中，我们利用北极环流中甲烷和臭氧之间的相关关系来区分化学原因和大气动力学原因导致的臭氧浓度变化。我们的结果显示在1996年1月至3月期间，120～160 DU的臭氧被损耗，这比1985年第一次报道在南极观测到的臭氧层空洞时所损耗的要高。这些损耗的量比同一时期预测的通过动力学因素增加的浓度值要高，表明观察到一个净的50 DU的浓度减少。这种臭氧的损耗在极低的北极平流层温度和高平流层氯浓度时发生。类似的损耗可能会重现，因为平流层变冷和氯浓度升高的趋势可能会持续数十年。

37. Edouard S, Legras B, Lefevre F, et al. The effect of small-scale inhomogeneities on ozone depletion in the Arctic[J]. Nature, 1996, 384(6608): 444-447.

The effect of small-scale inhomogeneities on ozone depletion in the Arctic

Edouard S., Legras B., Lefevre F., Eymard R.

(Laboratoire de Météorologie Dynamique du CNRS, Ecole Normale Supérieure, 24 Rue Lhomond, 75231 Paris Cedex 05, France)

Abstract: The chemical processes involved in the depletion of polar stratospheric ozone are now fairly well understood. But the effect of small-scale stirring and mixing of the chemical species involved can be misrepresented in three dimensional chemical-transport models because of their coarse resolution. Because of the nonlinearities in the chemical rate laws, especially those involving chlorine in the main catalytic cycle, these effects can be important-particularly in the Arctic, where the polar vortex is less uniform and less isolated from surrounding air than in the Antarctic. Here we use a very-high-resolution model with simplified ozone-depletion chemistry to show that the depletion is sensitive to small-scale inhomogeneities in the distribution of reactant species. Under the conditions of the winter of 1994-95 the effect is large enough to account for the observed discrepancies of about 40% between modelled and observed ozone depletion in the Arctic environment.

北极地区小范围环境的多相性对臭氧耗损的影响

Edouard S., Legras B., Lefevre F., Eymard R.

(法国科学研究中心)

摘要:极地平流层臭氧耗损所涉及的化学过程现在已经比较清楚了。但是,小尺度的化学物质的扰动和混合的影响可能会被三维的化学传输模型错误预测,因为这些模型的分辨率不高。由于化学反应速率规律的非线性特征,尤其是那些在主要催化循环中涉及的氯的反应,这些因素产生的影响可能是重要的,尤其是在北极这个相对于南极地区而言极地环流和周围大气不是很一致和独立的地区。本文中,我们利用一个高分辨率的模型以及简化的臭氧损耗的化学过程,说明臭氧的损耗过程对于反应物的小尺度分布的不均匀敏感。在1994～1995年冬季,这种影响足以导致模型模拟和实际观测的北极地区臭氧耗损的结果达到大约40%的差异。

38. Vondergathen P, Rex M, Harris N R P, et al. Observational evidence for chemical ozone depletion over the Arctic in winter 1991-92[J]. Nature, 1995, 375(6527): 131-134.

Observational evidence for chemical ozone depletion over the Arctic in winter 1991-92

Vondergathen P. , Rex M. , Harris N. R. P. , Lucic D. , Knudsen B. M. , Braathen G. O. , Debacker H. , Fabian R. , Fast H. , Gil M. , Kyro E. , Mikkelsen I. S. , Rummukainen M. , Stahelin J. , Varotsos C.

(Alfred Wegener Institute for Polar and Marine Research, PO Box 60 01 49, D-14401 Potsdam, Germany)

Abstract: Long-term depletion of ozone has been observed since the early 1980s in the Antarctic polar vortex, and more recently at midlatitudes in both hemispheres, with most of the ozone loss occurring in the lower stratosphere. Insufficient measurements of ozone exist, however, to determine decadal trends in ozone concentration in the Arctic winter. Several studies of ozone concentrations in the Arctic vortex have inferred that chemical ozone loss has occurred; but because natural variations in ozone concentration at any given location can be large, deducing long-term trends from time series is fraught with difficulties. The approaches used previously have often been indirect, typically relying on relationships between ozone and long-lived tracers. Most recently Manney et al used such an approach, based on satellite measurements, to conclude that the observed ozone decrease of about 20% in the lower stratosphere in February and March 1993 was caused by chemical, rather than dynamical, processes. Here we report the results of a new approach to calculate chemical ozone destruction rates that allows us to compare ozone concentrations in specific air parcels at different times, thus avoiding the need to make assumptions about ozone/tracer ratios. For the Arctic vortex of the 1991-92 winter we find that, at 20 km altitude, chemical ozone loss occurred only between early January and mid February and that the loss is proportional to the exposure to sunlight. The timing and magnitude are broadly consistent with existing understanding of photochemical ozone-depletion processes.

1991～1992 年期间北极地区化学因素导致的臭氧层耗损的观测证据

Vondergathen P. , Rex M. , Harris N. R. P. , Lucic D. , Knudsen B. M. , Braathen G. O. , Debacker H. , Fabian R. , Fast H. , Gil M. , Kyro E. , Mikkelsen I. S. , Rummukainen M. , Stahelin J. , Varotsos C.

（德国魏格纳极地和海洋研究中心）

摘要：20 世纪 80 年代早期，臭氧的长期损耗现象首先在南极极地涡旋内被观测到，最近在南北半球的中纬度地区也发现同样的现象，大部分的臭氧耗损过程发生在较低的平流层。然而，关于臭氧的相关测量对于确定北极冬季期间臭氧浓度的十年趋势而言依然不足。在北极涡旋中所做的一些关于臭氧浓度的研究说明化学因素导致的臭氧损耗过程已经发生；但是，由于任意给定位置上臭氧浓度的自然变化会非常大，所以从时间序列中推断长期的变化趋势充满了困难。以往用的方法经常是间接的，通常依赖臭氧和长时间尺度的示踪物质之间的关系。最近，Manney 等人利用一种依赖卫星测量的方法观测到 1993 年 2 月和 3 月时低平流层中约 20%的臭氧层减少源于化学过程，而非动力学过程。本文中，我们报道一种计算化学破坏臭氧层的速率的新方法，从而使得我们能比较一个气团不同时间的臭氧浓度，避免了做关于臭氧浓度和示踪物质比值的假设。对于 1991～1992 年的北极涡旋，我们发现，在 20 km 高度上，化学过程导致的臭氧损耗只在1月和2月中旬之间发生，并且损耗量与暴露在太阳光中的时间成正比。臭氧层空洞的时间和程度与当前关于臭氧损耗的光化学反应过程的理解相一致。

39. Manney G L, Froidevaux L, Waters J W, et al. Chemical depletion of ozone in the Arctic-lower stratosphere during winter 1992-93[J]. Nature, 1994, 370(6489): 429-434.

Chemical depletion of ozone in the Arctic-lower stratosphere during winter 1992-93

Manney G. L., Froidevaux L., Waters J. W., Zurek R. W., Read W. G., Elson L. S., Kumer J. B., Mergenthaler J. L., Roche A. E., Oneill A., Harwood R. S., Mackenzie I., Swinbank R.

(Caltech, Jet Prop Lab, Pasadena, Ca 91109)

Abstract: Satellite observations of ozone and chlorine monoxide concentrations during winter 1992-1993 show that in February 1993 chlorine in the lower stratosphere was mostly in chemically reactive forms. Decreases in stratospheric ozone concentration during February and early March 1993 are consistent with chemical destruction by this reactive chlorine. Comparison with changes in the distribution of long-lived chemical and dynamical tracers shows that the observed decrease cannot have been caused solely by dynamical processes.

1992～1993年冬季北极下平流层臭氧的化学亏损

Manney G. L., Froidevaux L., Waters J. W., Zurek R. W., Read W. G., Elson L. S., Kumer J. B., Mergenthaler J. L., Roche A. E., Oneill A., Harwood R. S., Mackenzie I., Swinbank R.

（美国加州理工大学喷气推进实验室）

摘要：对1992～1993冬季臭氧和ClO浓度的卫星观测结果显示，1993年2月下平流层的氯主要处于化学活性状态。而1993年2月至3月上旬平流层臭氧浓度的降低也与这些活性氯的化学破坏作用相一致。与动力学示踪和其他化学物质的分布对比发现，观测到的降低不是由动力学过程所引起的。

40. Walter C Oechel, Sid Cowles, Nancy Grulke, et al. Transient nature of CO_2 fertilization in Arctic tundra[J]. Nature, 1994, 371: 500-503.

Transient nature of CO_2 fertilization in Arctic tundra

Walter C. Oechel, Sid Cowles, Nancy Grulke, Steven J. Hastings, Bill Lawrence, Tom Prudhomme, George Riechers, Boyd Strain, David Tissue, George Vourlitis

(Global Change Research Group, and Systems Ecology Research Group, San Diego State University, San Diego, California 92182, USA)

Abstract: There has been much debate about the effect of increased atmospheric CO_2 concentrations on plant net primary production and on net ecosystem CO_2 flux. Apparently conflicting experimental findings could be the result of differences in genetic potential and resource availability, different experimental conditions and the fact that many studies have focused on individual components of the system rather than the whole ecosystem. Here we present results of an in situ experiment on the response of an intact native ecosystem to elevated CO_2. An undisturbed patch of tussock tundra at Toolik Lake, Alaska, was enclosed in greenhouses in which the CO_2 level, moisture and temperature could be controlled, and was subjected to ambient (340 p. p. m.) and elevated (680 p. p. m.) levels of CO_2 and temperature (+4 ℃). Air humidity, precipitation and soil water table were maintained at ambient control levels. For a doubled CO_2 level alone, complete homeostasis of the CO_2 flux was re-established within three years, whereas the regions exposed to a combination of higher temperatures and doubled CO_2 showed persistent fertilization effect on net ecosystem carbon sequestration over this time. This difference may be due to enhanced sink activity from the direct effects of higher temperatures on growth and to indirect effects from enhanced nutrient supply caused by increased mineralization. These results indicate that the responses of native ecosystems to elevated CO_2 may not always be positive, and are unlikely to be straightforward. Clearly, CO_2 fertilization effects must always be considered in the context of genetic limitation, resource availability and other such factors.

北极苔原地区二氧化碳肥化作用的短时性特征

Walter C. Oechel, Sid Cowles, Nancy Grulke, Steven J. Hastings,
Bill Lawrence, Tom Prudhomme, George Riechers, Boyd Strain,
David Tissue, George Vourlitis

(美国圣迭戈州立大学)

摘要：关于大气中二氧化碳浓度的增长对植物净初级生产力和生态系统净二氧化碳通量的影响一直存在着许多争论。各种实验结果彼此之间存在的显著争议很可能是由多个方面导致的，如植物的基因潜力、资源可利用性、不同的实验条件等，同时很多研究局限于生态系统的个别组分而非整个体系。我们进行了一项原位实验以测试一个完整的天然生态系统对二氧化碳浓度升高的响应。我们选取阿拉斯加Toolik湖附近的一小块未经扰动的苔原草丛，将其封闭在二氧化碳浓度、湿度、温度均可控制的温室之中，控制 CO_2 和温度条件与周围环境相同(CO_2 浓度为 340 ppm)和升高(CO_2 浓度为 680 ppm)，温度设为 4 ℃。空气湿度、降水量、土壤水位都保持在周围环境的控制水平。在仅二氧化碳浓度加倍的情况下，二氧化碳通量完整的内稳态重建耗时三年，而暴露在高温和加倍二氧化碳浓度环境下的区域在这段时间里，在净生态系统固碳方面显示出持久的肥化作用。这些差异可能是由于受高温直接影响的渗透活动的加强，同时也受矿化作用加强导致的营养供给增加的间接影响。这些研究结果都表明，天然生态系统对二氧化碳浓度增加的响应未必是正面的，也并非那么简单。无疑，二氧化碳肥化作用应该在遗传限度、资源可利用性和其他影响因子的大背景下被重新考虑。

41. Worsnop D R, Fox L E, Zahniser M S, et al. Vapor-pressures of solid hydrates of nitric-acid: implications for polar stratospheric clouds[J]. Science, 1993, 259(5091): 71-74.

Vapor-pressures of solid hydrates of nitric-acid—implications for polar stratospheric clouds

Worsnop D. R. , Fox L. E. , Zahniser M. S. , Wofsy S. C.

(Center for Chemical and Environmental Physics, Aerodyne Research, Inc. , Billerica, MA 01821)

Abstract: Thermodynamic data are presented for hydrates of nitric acid: $HNO_3 \cdot H_2O$, $HNO_3 \cdot 2H_2O$, $HNO_3 \cdot 3H_2O$, and a higher hydrate. Laboratory data indicate that nucleation, and persistence of metastable $HNO_3 \cdot 2H_2O$ may be favored in polar stratospheric clouds over the slightly more stable $HNO_3 \cdot 3H_2O$. Atmospheric observations indicate that some polar stratospheric clouds may be composed of $HNO_3 \cdot 2H_2O$ and $HNO_3 \cdot 3H_2O$. Vapor transfer from $HNO_3 \cdot 2H_2O$ to $HNO_3 \cdot 3H_2O$ could be a key step in the sedimentation of HNO_3, which plays an important role in the depletion of polar ozone.

固态硝酸水合物的蒸汽压——极地平流层云的指示

Worsnop D. R. , Fox L. E. , Zahniser M. S. , Wofsy S. C.

(美国重航空器研究所)

摘要:本文列举了硝酸水合物($HNO_3 \cdot H_2O$,$HNO_3 \cdot 2H_2O$,$HNO_3 \cdot 3H_2O$和更多水化物)的热力学数据。实验室数据显示极地平流层中亚稳态的$HNO_3 \cdot 2H_2O$的晶核形成和维持比更加稳定的$HNO_3 \cdot 3H_2O$更有利。大气观测显示,一些极地平流层云可能由$HNO_3 \cdot 2H_2O$和$HNO_3 \cdot 3H_2O$组成,从$HNO_3 \cdot 2H_2O$到$HNO_3 \cdot 3H_2O$的蒸汽转换是HNO_3沉降的关键环节,而HNO_3的沉降在极地臭氧损耗中扮演重要角色。

42. Webster C R, May R D, Toohey D W, et al. Chlorine chemistry on polar stratospheric cloud particles in the Arctic winter[J]. Science, 1993, 261(5125): 1130-1134.

Chlorine chemistry on polar stratospheric cloud particles in the Arctic winter

Webster C. R. , May R. D. , Toohey D. W. , Avallone L. M. , Anderson J. G. , Newman P. , Lait L. , Schoeberl M. R. , Elkins J. W. , Chan K. R.

(Jet Propulsion Laboratory, California Institute of Technology, MS 183-401, 4800 Oak Grove Drive, Pasadena, California 91109)

Abstract: Simultaneous in situ measurements of hydrochloric acid (HCl) and chlorine monoxide (ClO) in the Arctic winter vortex showed large HCl losses, of up to 1 part per billion by volume (ppbv), which were correlated with high ClO levels of up to 1. 4 ppbv. Air parcel trajectory analysis identified that this conversion of inorganic chlorine occurred at air temperatures of less than 196+/−4 kelvin. High ClO was always accompanied by loss of HCl mixing ratios equal to $1/2(ClO+2Cl_2O_2)$.

These data indicate that the heterogeneous reaction $HCl + ClONO_2 \rightarrow Cl_2 + HNO_3$ on particles of polar stratospheric clouds establishes the chlorine partitioning, which, contrary to earlier notions, begins with an excess of $ClONO_2$, not HCl.

冬季北极平流层云滴上的氯化学

Webster C. R., May R. D., Toohey D. W., Avallone L. M., Anderson J. G., Newman P., Lait L., Schoeberl M. R., Elkins J. W., Chan K. R.

(美国加州理工学院喷气推进实验室)

摘要：对北极冬季涡旋中盐酸(HCl)和一氧化氯(ClO)同时进行的原位观测显示，高达 1 ppbv 的盐酸损耗与高达 1.4 ppbv 的 ClO 水平有关。对气团轨迹分析显示这种无机氯的转化发生在空气温度低于 196±—4 开尔文时。而高 ClO 总是伴随 HCl 混合比 $1/2(ClO + 2Cl_2O_2)$ 的损耗。这些数据表明极地平流层云粒子上的多相反应 $HCl + ClONO_2 \rightarrow Cl_2 + HNO_3$ 确定了氯的分部，而并非早期的观点所认为的起始于 $ClONO_2$(而非 HCl)的过剩。

43. Waugh D W. Subtropical stratospheric mixing linked to disturbances in the polar vortices[J]. Nature, 1993, 365(6446): 535-537.

Subtropical stratospheric mixing linked to disturbances in the polar vortices

Waugh D. W.

(Center for Meteorology and Physical Oceanography, Massachusetts Institute of Technology, Cambridge, Massachusetts 02139, USA)

Abstract: Randel et al have observed tongues of stratospheric air stretching from the tropics into middle latitudes, and conclude that such events may be responsible for transporting significant amounts of stratospheric air across the tropical-mid-latitude barrier. Here I examine the movements of air parcels during these events using high-resolution contour-trajectory calculations. My calculations suggest that the tongues of tropical air are associated with disturbances of the stratospheric polar vortices. The edge of the disturbed polar vortex reaches low latitudes, and draws a long tongue of tropical air around the vortex into middle latitudes. This process occurs in the winter of both hemispheres, although the edge of the larger Antarctic polar vortex reaches farther toward the Equator, and draws up material from lower latitudes, than its Arctic counterpart.

与极地涡旋扰动相关的亚热带平流层混合

Waugh D. W.

(美国麻省理工学院)

摘要：Randel 等人观测到平流层从热带向中纬度地区延展的气舌，并认为这类事件可能是平流层气体越过热带-中纬度屏障进行传输的原因。本文中，我采用高分辨廓线-轨迹计算调查了这些事件期间气团的运动。我的计算表明热带气舌与平流层极地涡旋的扰动有关。扰动的极地涡旋的边缘抵达至低纬度地区，并将热带气舌拖至中纬度地区。这一过程在南北半球的冬季发生，但相较而言，南极极地涡旋更大，边缘向赤道延展，与北极相比能够从更低的纬度带走物质。

44. Waters J W, Froidevaux L, Read W G, et al. Stratospheric ClO and ozone from the microwave limb sounder on the upper-atmosphere research satellite[J]. Nature, 1993, 362(6421): 597-602.

Stratospheric ClO and ozone from the microwave limb sounder on the upper-atmosphere research satellite

Waters J. W., Froidevaux L., Read W. G., Manney G. L., Elson L. S., Flower D. A., Jarnot R. F., Harwood R. S.

(Jet Propulsion Laboratory, California Institute of Technology, Pasadena, CA 91109 USA)

Abstract: Concentrations of atmospheric ozone and of ClO (the predominant form of reactive chlorine responsible for stratospheric ozone depletion) are reported for both the Arctic and Antarctic winters of the past 18 months. Chlorine in the lower stratosphere was almost completely converted to chemically reactive forms in both the northern and southern polar winter vortices. This occurred in the south long before the development of the Antarctic ozone hole, suggesting that ozone loss can be masked by influx of ozone-rich air.

用高层大气研究卫星微波边缘探测器观测平流层 ClO 和臭氧

Waters J. W., Froidevaux L., Read W. G., Manney G. L., Elson L. S., Flower D. A., Jarnot R. F., Harwood R. S.

(美国加州理工学院喷气推进实验室)

摘要：本研究指出过去 18 个月北极和南极冬季大气臭氧和 ClO(造成平流层臭氧损耗的活性氯的主要形态)的浓度。在南北半球极地冬季涡旋中低平流层中的氯几乎全部转化为化学活性态，这在南极臭氧洞发展很久以前就在南半球发生了，表明臭氧损耗会被富集臭氧的气团涌入所掩盖。

45. Toumi R, Jones R L, Pyle J A. Stratospheric ozone depletion by $ClONO_2$ photolysis[J]. Nature, 1993, 365(6441): 37-39.

Stratospheric ozone depletion by $ClONO_2$ photolysis

Toumi R., Jones R. L., Pyle J. A.

(University of Cambridge, UK)

Abstract: Springtime ozone depletion over Antarctica is thought to be due to catalytic cycles involving chlorine monoxide, which is formed as a result of reactions on the surface of polar stratospheric clouds (PSCs). When the PSCs evaporate, ClO in the polar air can react with NO_2 to form the reservoir species $ClONO_2$. High concentrations of $ClONO_2$ can also be found at lower latitudes because of direct transport of polar air or mixing of ClO and NO_2 at the edges of the polar vortex. $ClONO_2$ can take part in an ozone-depleting catalytic cycle, but the significance of this cycle has not been clear. Here we present model simulations of ozone concentrations from March to May both within the Arctic vortex and at a mid-latitude Northern Hemisphere site. We find increasing ozone loss from March to May. The $ClONO_2$ cycle seems to be responsible for a significant proportion of the simulated ozone loss. An important aspect of this cycle is that it is not as limited as the other chlorine cycles to the timing and location of PSCs; it may therefore play an important role in ozone depletion at warm middle latitudes.

$ClONO_2$ 光分解引起的平流层臭氧亏损

Toumi R. , Jones R. L. , Pyle J. A.

(英国剑桥大学)

摘要：目前认为南极洲春季臭氧亏损是由ClO(极地平流层表面反应的产物)催化的循环过程所引起的。当极地平流层云消失时，极地大气中的ClO会与NO_2反应生成储存物$ClONO_2$。由于极地大气向低纬度地区的直接传输或ClO和NO_2在极地涡旋边缘的混合，低纬度地区也发现了高浓度的$ClONO_2$。$ClONO_2$会参与臭氧亏损的循环，但是这一循环的重要性还不明确。我们对北极涡旋和北半球中纬度一个地区3～5月的臭氧浓度进行了模型模拟，发现从3～5月臭氧亏损增加。而$ClONO_2$的循环似乎是导致模拟臭氧亏损的重要原因。这一循环的一个重要方面在于并不像其他氯循环一样受限于极地平流层云出现的时间和空间位置，因此对温暖的中纬度地区的臭氧亏损可能起重要作用。

46. Toon O, Browell E, Gary B, et al. Heterogeneous reaction probabilities, solubilities, and the physical state of cold volcanic aerosols[J]. Science, 1993, 261(5125): 1136-1140.

Heterogeneous reaction probabilities, solubilities, and the physical state of cold volcanic aerosols

Toon O. , Browell E. , Gary B. , Lait L. , Livingston J. , Newman P. , Pueschel R. , Russell P. , Schoeberl M. , Toon G. , Traub W. , Valero F. P. J. , Selkirk H. , Jordan J.

(Nasa, Ames Res Ctr, Moffett Field, Ca 94035)

Abstract: On 19 January 1992, heterogeneous loss of HNO_3, $ClNO_3$, and HCl was observed in part of the Mount Pinatubo volcanic cloud that had cooled as a result of forced ascent. Portions of the volcanic cloud froze near 191 kelvin. The reaction probability of $ClNO_3$ and the solubility of HNO_3 were close to laboratory measurements on liquid sulfuric acid. The magnitude of the observed loss of HCl suggests that it underwent a heterogeneous reaction. Such reactions could lead to substantial loss of HCl on background sulfuric acid particles and so be important for polar ozone loss.

冷火山冷气溶胶的多相反应概率、溶解性和物态特性

Toon O. , Browell E. , Gary B. , Lait L. , Livingston J. , Newman P. , Pueschel R. , Russell P. , Schoeberl M. , Toon G. , Traub W. , Valero F. P. J. , Selkirk H. , Jordan J.

(美国国家航空和航天局)

摘要:1992 年 1 月 19 日,在部分 Pinatubo 火山云(由于被动上升而已经冷却)中观测到了 HNO_3、$ClNO_3$ 和 HCl 的非均相损耗,部分火山云降至约 191 K。$ClNO_3$ 的反应概率和 HNO_3 的溶解度与在实验室中观测到的液态硫酸的结果一致。观测到的 HCl 流失的量级表明盐酸经历了多相反应,这种反应能够导致背景硫酸颗粒上的 HCl 流失,因而对极地臭氧流失很重要。

47. Salawitch R J, Wofsy S C, Gottlieb E W, et al. Chemical loss of ozone in the Arctic polar vortex in the winter of 1991-1992[J]. Science, 1993, 261(5125): 1146-1149.

Chemical loss of ozone in the Arctic polar vortex in the winter of 1991-1992

Salawitch R. J. , Wofsy S. C. , Gottlieb E. W. , Lait L. R. , Newman P. A. , Schoeberl M. R. , Loewenstein M. , Podolske J. R. , Strahan S. E. , Proffitt M. H. , Webster C. R. , May R. D. , Fahey D. W. , Baumgardner D. , Dye J. E. , Wilson J. C. , Kelly K. K. , Elkins J. W. , Chan K. R. , Anderson J. G.

(Harvard Univ, Div Appl Sci, Cambridge, Ma 02138)

Abstract: In situ measurements of chlorine monoxide, bromine monoxide, and ozone are extrapolated globally, with the use of meteorological tracers, to infer the loss rates for ozone in the Arctic lower stratosphere during the Airborne Arctic Stratospheric Expedition Ⅱ (AASE Ⅱ) in the winter of 1991-1992. The analysis indicates removal of 15 to 20 percent of ambient ozone because of elevated concentrations of chlorine monoxide and bromine monoxide. Observations during AASE Ⅱ define rates of removal of chlorine monoxide attributable to reaction with nitrogen dioxide (produced by photolysis of nitric acid) and to production of hydrochloric acid. Ozone loss ceased in March as concentrations of chlorine monoxide declined. Ozone losses could approach 50 percent if regeneration of nitrogen dioxide were inhibited by irreversible removal of nitrogen oxides (denitrification), as presently observed in the Antarctic, or without denitrification if inorganic chlorine concentrations were to double.

1991～1992 年冬季北极极地涡旋中的臭氧化学亏损

Salawitch R. J.，Wofsy S. C.，Gottlieb E. W.，Lait L. R.，Newman P. A.，Schoeberl M. R.，Loewenstein M.，Podolske J. R.，Strahan S. E.，Proffitt M. H.，Webster C. R.，May R. D.，Fahey D. W.，Baumgardner D.，Dye J. E.，Wilson J. C.，Kelly K. K.，Elkins J. W.，Chan K. R.，Anderson J. G.

（美国哈佛大学）

摘要：1991～1992 年冬季机载北极平流层考察期间（AASE Ⅱ），通过使用气相追踪剂，对 ClO、BrO 和 O_3 的原位观测已经推广至全球范围用于推断北极平流层底部的臭氧损耗速率。分析显示由于 ClO 和 BrO 浓度的升高造成了环境中臭氧 15%～20%的损耗。AASE Ⅱ观测表明 ClO 的损耗速率可归因于与 NO_2（硝酸光分解的产物）的反应和 HCl。3 月时由于 ClO 浓度的下降造成了臭氧损耗的终止。正如目前在南极所观测到的，如果 NO_2 的再生由于 NO（脱氮作用）的不可逆损耗受到了抑制，或者如果无机氯浓度加倍而没有发生脱氮作用，臭氧损耗会达到 50%。

48. Newman P，Lait L R，Schoeberl M，et al. Stratospheric meteorological conditions in the Arctic polar vortex，1991 to 1992[J]. Science，1993，261(5125)：1143-1146.

Stratospheric meteorological conditions in the Arctic polar vortex，1991 to 1992

Newman P.，Lait L. R.，Schoeberl M.，Nash E. R.，Kelly K.，Fahey D. W.，Nagatani R.，Toohey D.，Avallone L.，Anderson J.

（Nasa，Goddard Space Flight Ctr，Greenbelt，Md 20771）

Abstract：Stratospheric meteorological conditions during the Airborne Arctic Stratospheric Expedition Ⅱ（AASE Ⅱ）presented excellent observational opportunities from Bangor，Maine，because the polar vortex was located over southeastern Canada for significant periods during the 1991-1992 winter. Temperature analyses showed that nitric acid trihydrates（NAT temperatures below 195 K）should have formed over small regions in early December. The temperatures in the polar vortex warmed beyond NAT temperatures by late January（earlier than normal）. Perturbed chemistry was found to be associated with these cold temperatures.

1991～1992 年北极极地涡旋的平流层气象条件

Newman P.，Lait L. R.，Schoeberl M.，Nash E. R.，Kelly K.，Fahey D. W.，Nagatani R.，Toohey D.，Avallone L.，Anderson J.

（美国国家航空和航天局）

摘要：由于 1991～1992 年冬季机载北极平流层考察期间（AASE Ⅱ）极地涡旋位于加拿大东南部，同温层的气象条件为缅因州 Bangor 的观测提供了很好的机会。温度分析显示三水合硝酸（NAT 温度低于

195 K)应该在 12 月上旬在小区域内形成，极地涡旋内的温度到 1 月下旬(较平常早)远高于 NAT 温度。扰乱的化学反应与低温的形成有关。

49. Browell E V, Butler C F, Fenn M A, et al. Ozone and aerosol changes during the 1991-1992 airborne Arctic stratospheric expedition[J]. Science, 1993, 261(5125): 1155-1158.

Ozone and aerosol changes during the 1991-1992 airborne Arctic stratospheric expedition

Browell E. V., Butler C. F., Fenn M. A., Grant W. B., Ismail S., Schoeberl M. R., Toon O. B., Loewenstein M., Podolske J. R.

(Nasa, Langley Res Ctr, Div Atmospher Sci, Hampton, Va 23681)

Abstract: Stratospheric ozone and aerosol distributions were measured across the wintertime Arctic vortex from January to March 1992 with an airborne lidar system as part of the 1992 Airborne Arctic Stratospheric Expedition (AASE Ⅱ). Aerosols from the Mount Pinatubo eruption were found outside and inside the vortex with distinctly different distributions that clearly identified the dynamics of the vortex. Changes in aerosols inside the vortex indicated advection of air from outside to inside the vortex below 16 kilometers. No polar stratospheric clouds were observed and no evidence was found for frozen volcanic aerosols inside the vortex. Between January and March, ozone depletion was observed inside the vortex from 14 to 20 kilometers with a maximum average loss of about 23 percent near 18 kilometers.

1991～1992 年北极平流层空中观测期间臭氧和气溶胶的变化

Browell E. V., Butler C. F., Fenn M. A., Grant W. B., Ismail S., Schoeberl M. R., Toon O. B., Loewenstein M., Podolske J. R.

(美国国家航空和航天局)

摘要：作为 1992 年北极平流层机载观测项目(AASE Ⅱ)的一部分，1992 年 1～3 月对冬季北极涡旋中平流层臭氧和气溶胶的分布进行了观测。在涡旋内外观测到的来自 Pinatubo 火山爆发的气溶胶表现出了显著不同的分布，清晰地指示了涡旋动力学。涡旋内气溶胶的改变显示在 16 km 之下气团从外向内的水平对流。在涡旋内部没有观测到极地平流层云，也没有冻结的火山气溶胶的证据。在 1～3 月期间，涡旋内部 14～20 km 处观测到了臭氧亏损，在约 18 km 处达到最大平均亏损约 23%。

50. Jonathan D Kahl, Donna J Charlevoix, Nina A Zaftseva, et al. Absence of evidence for greenhouse warming over the arctic ocean in the past 40 years[J]. Nature, 1993, 361: 335-337.

Absence of evidence for greenhouse warming over the arctic ocean in the past 40 years

Jonathan D. Kahl, Donna J. Charlevoix, Nina A. Zaftseva,
Russell C. Schnell, Mark C. Serreze

(Department of Geosciences, University of Wisconsin-Milwaukee, PO Box 413, Milwaukee, Wisconsin 53201, USA)

Abstract: Atmospheric general circulation models predict enhanced greenhouse warming at high latitudes owing to positive feedbacks between air temperature, ice extent and surface albedo. Previous analyses of Arctic temperature trends have been restricted to land-based measurements on the periphery of the Arctic Ocean. Here we present temperatures measured in the lower troposphere over the Arctic Ocean during the period 1950-90. We have analysed more than 27000 temperature profiles, measured by radiosonde at Russian drifting ice stations and by dropsonde from US "Ptarmigan" weather reconnaissance aircraft, for trends as a function of season and altitude. Most of the trends are not statistically significant. In particular, we do not observe the large surface warming trends predicted by models; indeed, we detect significant surface cooling trends over the western Arctic Ocean during winter and autumn. This discrepancy suggests that present climate models do not adequately incorporate the physical processes that affect the polar regions.

不存在过去40年北冰洋温室变暖的证据

Jonathan D. Kahl, Donna J. Charlevoix, Nina A. Zaftseva,
Russell C. Schnell, Mark C. Serreze

(美国维斯康辛大学密尔沃基分校)

摘要：根据气温、冰盖范围和表面反照率的正反馈效应，大气环流模型预测高纬度地区的变暖效应会增强。之前对北极温度趋势的观测局限于北冰洋边界的对地基观测。本文中，我们报道了1950～1990年期间北冰洋对流层下部所观测的温度。我们使用俄罗斯浮冰观测站的无线电控空仪，以及美国“雷鸟”天气探测航天器的下投式探空仪，共获得超过27000个温度廓线，并分析温度与季节和海拔的趋势。大多数趋向在统计学上意义不大。特别是我们没有观测到模型中所预测到的表面温度大幅度升高的趋势；事实上，我们发现北冰洋西部地区在东秋季有明显的表面温度降低的趋势。这一矛盾显示，目前的气候模型不能充分解释极地地区的物理过程。

51. Proffitt M H, Aikin K, Margitan J J, et al. Ozone loss inside the northern polar vortex during the 1991-1992 winter[J]. Science, 1993, 261(5125): 1150-1154.

Ozone loss inside the northern polar vortex during the 1991-1992 winter

Proffitt M. H., Aikin K., Margitan J. J., Loewenstein M., Podolske J. R., Weaver A., Chan K. R., Fast H., Elkins J. W.

(Noaa, Aeron Lab, 325 Broadway, Boulder, Co 80303;
Univ Colorado, Noaa, Cooperat Inst Res Environm Sci, Boulder, Co 80309)

Abstract: Measurements made in the outer ring of the northern polar vortex from October 1991 through March 1992 reveal an altitude-dependent change in ozone, with a decrease at the bottom of the vortex and a substantial increase at the highest altitudes accessible to measurement. The increase is the result of ozone-rich air entering the vortex, and the decrease reflects ozone loss accumulated after the descent of the air through high concentrations of reactive chlorine. The depleted air that is released out of the bottom of the vortex is sufficient to significantly reduce column ozone at mid-latitudes.

1991～1992 年冬季北极涡旋内的臭氧损耗

Proffitt M. H., Aikin K., Margitan J. J., Loewenstein M., Podolske J. R., Weaver A., Chan K. R., Fast H., Elkins J. W.

(美国国家海洋和大气局)

摘要:从 1991 年 10 月至 1992 年 3 月在北极涡旋外圈的测量揭示了臭氧的变化对高度的依赖,臭氧在底部的旋涡减少,在海拔最高的测量中大幅增加。增加的原因是富含臭氧的空气进入旋涡,而减少则反映了积聚的空气通过高浓度的活性氯后臭氧的损失。空气涡旋底部释放出的气体足够使中纬度地区臭氧柱显著减少。

52. Mcconnell J C, Henderson G S, Barrie L, et al. Photochemical bromine production implicated in Arctic boundary-layer ozone depletion[J]. Nature,1992,355(6356):150-152.

Photochemical bromine production implicated in Arctic boundary-layer ozone depletion

Mcconnell J. C., Henderson G. S., Barrie L., Bottenheim J., Niki H., Langford C. H., Templeton E. M. J.

(York Univ, Dept Earth & Atmospher, 4700 Keele St, N York M3j 1p3, Ontario, Canada)

Abstract: Recent measurements in the Arctic have revealed episodic destruction of boundary-layer ozone from 30-40 parts per 109 by volume (p. p. b. v.) to undetectable levels on a timescale of less than a day, during periods when the boundary layer is very stable. The ozone destruction begins at polar sunrise, continues for the months of March and April, and is strongly associated with levels of filterable bromine which are much greater than during the rest of the year. Here we suggest that sea-salt Brreaches high concentrations in the snow pack during the long polar night, and is evolved into the atmosphere as Br_2 at polar sunrise. Ordinarily, gas-phase photochemistry would convert Br_2 to HBr or brominated organic compounds with consequently little destruction of boundary-layer ozone. In view of several laboratory experiments, and by analogy with the marine boundary layer, we propose that the HBr and brominated organic compounds will be scavenged by the ambient aerosols and ice crystals, and that these heterogeneous reactions release Br_2 back to the atmosphere. We argue that this cycling of bromine between the aerosol and the gas phase should maintain sufficiently high levels of Br atoms and BrO radicals to destroy ozone, in agreement with observations.

北极边界层臭氧亏损指示的光化学溴的产生

Mcconnell J. C., Henderson G. S., Barrie L., Bottenheim J., Niki H., Langford C. H., Templeton E. M. J.

(加拿大约克大学)

摘要：最近在北极的观测显示边界层臭氧存在间歇性损耗的现象，在边界层非常稳定期间，能够在不足一天的时间内从 30～40 ppbv 降至检出限以下。臭氧损耗开始于极地日出时，在 3、4 月持续，这与比剩下月份要高的可滤性溴有很强的联系。我们认为在漫长的极夜期间，海盐 Br 在积雪中达到较高的浓度，并在极地日出时以 Br_2 的形态释放到大气中。通常情况下，气相光化学会将 Br_2 转换成 HBr 或者有机溴化物，并因此导致少量的边界层臭氧损耗。鉴于许多实验室实验并根据海洋边界层类推，我们认为 HBr 和有机溴化物会被周围的气溶胶及冰晶清除，而这些多相反应会向大气释放回 Br_2。这种气溶胶和气相之间的溴循环应该可以维持足够高水平的溴原子和 BrO，并像观测到的结果一样对臭氧产生破坏。

53. Fan S M, Jacob D J. Surface ozone depletion in Arctic spring sustained by bromine reactions on aerosols[J]. Nature, 1992, 359(6395): 522-524.

Surface ozone depletion in Arctic spring sustained by bromine reactions on aerosols

Fan S. M., Jacob D. J.

(Harvard Univ, Div Appl Sci, Pierce Hall, 29 Oxford St, Cambridge, Ma 02138)

Abstract: Near-total depletion of the ozone in surface air is often observed in the Arctic spring, coincident with high atmospheric concentrations of inorganic bromine. Barrie et al suggested that the ozone depletion was due to a catalytic cycle involving the radicals Br and BrO; however, these species are rapidly converted to the nonradical species HBr, HOBr and $BrNO_3$, quenching ozone loss. McConnell et al proposed that cycling of inorganic bromine between aerosols and the gas phase could maintain sufficiently high levels of Br and BrO to destroy ozone, but they did not specify a mechanism for aerosol-phase production of active bromine species. Here we propose such a mechanism, based on known aqueous-phase chemistry, which rapidly converts HBr, HOBr and $BrNO_3$ back to Br and BrO radicals. This mechanism should be particularly efficient in the presence of the high concentrations of sulphuric acid aerosols observed during ozone depletion events.

气溶胶上溴反应维持的北极地区春季表面臭氧损耗

Fan S. M. , Jacob D. J.

(美国哈佛大学)

摘要:在北极地区春季观测到地面大气臭氧经常几乎全部损耗,同时还发现无机溴化物的大气浓度异常偏高。Barrie 等人认为臭氧枯竭是由于 Br 和 BrO 自由基参与的一个催化循环导致的,然而这些物质都会迅速地转化成 HBr、$BrNO_3$ 和 HOBr 这些非自由基,进而抑制臭氧的损失。McConnell 提出,无机溴在气溶胶相和气相之间的循环可以充分维持 Br 和 BrO 高浓度以损耗臭氧,但是他们并没有阐明气溶胶-气相产生活化溴化物的机制。本文中我们根据已知的水相化学原理,提出了可以迅速将 HBr、$BrNO_3$ 和 HOBr 转化为 Br 和 BrO 自由基的一个机制。特别是在臭氧空洞事件中观测到高浓度硫酸气溶胶的时候,这一机制会特别有效。

54. Austin J, Butchart N, Shine K P. Possibility of an Arctic ozone hole in a doubled-CO_2 climate[J]. Nature,1992,360(6401):221-225.

Possibility of an Arctic ozone hole in a doubled-CO_2 climate

Austin J. , Butchart N. , Shine K. P.

(Meteorol Off, London Rd, Bracknell Rb12 2sz, Berks, England)

Abstract: Increased atmospheric carbon dioxide concentrations are expected to cause cooling of the lower stratosphere. This could enhance the formation of polar stratospheric clouds, which convert potential ozone-depleting species to their active forms. In an idealized three-dimensional numerical simulation of the Northern Hemisphere winter stratosphere, doubling the CO_2 concentration leads to the formation of an Arctic ozone hole comparable to that observed over Antarctica, with nearly 100% local depletion of lower-stratospheric ozone.

CO_2 加倍情况下产生北极臭氧洞的可能性

Austin J. , Butchart N. , Shine K. P.

(英国气象办公室)

摘要:目前认为大气 CO_2 浓度的增加会引起下平流层的变冷,从而促进极地平流层云的形成,并引起臭氧亏损的物质处于活性状态。在北半球冬季平流层理想化三维数值模型中,CO_2 浓度的加倍会导致可以与观测到的南极臭氧洞相当的北极臭氧洞的形成,会导致当地下平流层臭氧 100%的损耗。

55. Nriagu J O, Coker R D, Barrie L A. Origin of sulfur in Canadian Arctic haze from isotope measurements[J]. Nature,1991,349(6305):142-145.

Origin of sulfur in Canadian Arctic haze from isotope measurements

Nriagu J. O. , Coker R. D. , Barrie L. A.

(Natl Water Res Inst Branch, Box 5050, Burlington L7r 4a6, Ontario, Canada)

Abstract: Since the mid-1950s, there has been a marked increase in levels of air pollution in the Arctic region. This is apparent from the pervasive haze that has been detected over large areas of the Northern Hemisphere, which can cover up to 9% of the Earth's surface. The haze is most pronounced during January to April, and is characterized by a reduction in visibility and anomalously high levels of organic and inorganic compounds of the type found in polluted environments. Most of the present knowledge about the origin of the chemical constituents of the haze comes from studies of elemental ratios and trajectory analyses. Here we report the results of an analysis of the isotopic composition of sulphur in the Arctic haze. We find that most the sulphur in the haze comes from Europe rather than from more local anthropogenic or biogenic sources, indicating that this form of air pollution becomes distributed globally.

通过同位素方法观测加拿大北极霾中硫的来源

Nriagu J. O. , Coker R. D. , Barrie L. A.

(加拿大国家水研究所)

摘要:从20世纪50年代中期,北极地区空气污染水平就有了明显上升。这显然与北半球广泛检测到的阴霾有关,阴霾面积巨大,可以达到地表面积的9%。在1~4月间阴霾最明显,能见度降低,在污染区出现异常高的有机和无机化合物。目前对阴霾化学要素的了解多基于元素比和轨迹分析的研究,本文中我们对北极阴霾中硫的同位素组成进行了分析。我们发现阴霾中的硫大部分来自于欧洲而不是当地人为或生物源,这一结果表明空气污染的形成具有全球性。

56. Kelly K K, Tuck A F, Davies T. Wintertime asymmetry of upper tropospheric water between the northern and southern hemispheres[J]. Nature, 1991, 353(6341): 244-247.

Wintertime asymmetry of upper tropospheric water between the northern and southern hemispheres

Kelly K. K. , Tuck A. F. , Davies T.

(National Oceanic and Atmospheric Administration Aeronomy Laboratory, 325 Broadway, Boulder, Colorado 80303-3328 USA)

Abstract: Water vapour is an important greenhouse gas and yet its abundance in the upper troposphere is poorly known. Upper-tropospheric water vapour is particularly important despite its low mixing ratios, because it has large effects on the flux of infrared radiation near the tropopause. In addition, the distribution and supply of water vapour are central to cloud formation; the effects of cloud on the Earth's radiation budget are in turn central to understanding the climate response to increasing atmospheric concentrations of greenhouse gases. From airborne measurements of total water (vapour plus ice crystal) during the winters of 1987 in the Southern Hemisphere and of 1988-89 in the Northern Hemisphere, we find that the upper troposphere in middle, subpolar and high latitudes is a factor of 2-4 drier during austral winter than during boreal winter. As the lower-latitude air moves towards the pole in austral winter, it is forced to cool to lower temperatures than in the north-more of the water vapour therefore condenses to form ice crystals, which then precipitate, thereby removing moisture from the air mass. Clearly, climate models must be able to reproduce this asymmetry if their predictions are to be credible. We also note that the asymmetry in water vapour implies an asymmetry in the production rate of the hydroxyl radical, and hence in the tropospheric chemistry of each hemisphere, for example in the rate of methane loss.

南北半球冬季对流层上部水汽的不对称特性

Kelly K. K. , Tuck A. F. , Davies T.

(美国国家海洋和大气局)

摘要：水蒸气是一种非常重要的温室气体，但是关于其在对流层上部的赋存状态我们了解得还非常少。虽然上对流层水蒸气混合比很低，但由于其对对流层顶部红外辐射通量的巨大作用而显得非常重要。除此以外，水蒸气的分布和供给是形成云的关键，而云在地球辐射平衡方面的影响是理解气候对大气温室气体浓度增加的反馈作用的关键之一。从南半球 1987 年和北半球 1988～1989 年冬季对总水(水蒸气和冰晶加和)的机载观测中，我们发现中纬度、亚极地和高纬度地区的上对流层造成南半球冬季比北半球冬季干旱 2～4 倍。由于在南半球的冬季低纬度气团向极地迁移，使得较北半球的水蒸气温度降得更低，形成冰晶，进而沉降，并转移走气团中的水汽。显然，气候模型如果想要预测结果可信则必须能够模拟这种不对称性。我们也注意到水蒸气的不对称性同时意味着羟基自由基沉降速率的不对称性，并进而影响两个半球的对流层化学(如甲烷损耗率)。

57. Hofmann D J, Deshler T. Evidence from balloon measurements for chemical depletion of stratospheric ozone in the Arctic winter of 1989-90[J]. Nature, 1991, 349(6307): 300-305.

Evidence from balloon measurements for chemical depletion of stratospheric ozone in the Arctic winter of 1989-90

Hofmann D. J. , Deshler T.

(Univ Wyoming, Dept Phys & Astron, Laramie, Wy 82071)

Abstract: Measurements of the vertical ozone profile with balloon-borne sensors in the Arctic during January and February 1990 indicate repeated minima in the 22 km region. A general negative correlation of the 22 km ozone mixing ratio with time spent by the air parcel in regions cold enough for the formation of polar stratospheric clouds, together with the presence of sunlight on a portion of the air-parcel trajectory, and an inability to explain the ozone minimum by conventional dynamical processes, suggest that this feature is related to chemical ozone depletion.

1989～1990 年北极冬季平流层臭氧化学亏损的气球观测证据

Hofmann D. J.，Deshler T.

(美国怀俄明大学)

摘要：1990 年 1～2 月球载传感器测量了臭氧垂直廓线，结果显示 22 km 区域出现极小值。22 km 处的臭氧混合比和传感器在温度低到可以形成极地平流层云的地区所处时间呈反相关。这一结果与阳光在部分传感器轨迹中的出现以及利用传统动力学过程无法解释臭氧低值共同表明，22 km 的特征与化学臭氧亏损相关。

58. Brune W H，Anderson J G，Toohey D W，et al. The potential for ozone depletion in the Arctic polar stratosphere[J]. Science，1991，252(5010)：1260-1266.

The potential for ozone depletion in the Arctic polar stratosphere

Brune W. H.，Anderson J. G.，Toohey D. W.，Fahey D. W.，Kawa S. R.，Jones R. L.，Mckenna D. S.，Poole L. R.

(Harvard Univ，Dept Chem，Cambridge，Ma 02138)

Abstract：The nature of the Arctic polar stratosphere is observed to be similar in many respects to that of the Antarctic polar stratosphere，where an ozone hole has been identified. Most of the available chlorine (HCl and $ClONO_2$) was converted by reactions on polar stratospheric clouds to reactive ClO and Cl_2O_2 throughout the Arctic polar vortex before midwinter. Reactive nitrogen was converted to HNO_3，and some，with spatial inhomogeneity，fell out of the stratosphere. These chemical changes ensured characteristic ozone losses of 10% to 15% at altitudes inside the polar vortex where polar stratospheric clouds had occurred. These local losses can translate into 5% to 8% losses in the vertical column abundance of ozone. As the amount of stratospheric chlorine inevitably increases by 50% over the next two decades，ozone losses recognizable as an ozone hole may well appear.

北极极地平流层中潜在的臭氧亏损

Brune W. H.，Anderson J. G.，Toohey D. W.，Fahey D. W.，Kawa S. R.，Jones R. L.，Mckenna D. S.，Poole L. R.

(美国哈佛大学)

摘要：从观测来看，北极极地平流层与检测到的臭氧洞南极极地平流层在很多方面相似。大多数有效氯(HCl 和 $ClONO_2$)会在冬季中期以前的北极极地涡旋中经极地平流层云上的反应转化成活性的 ClO 和 Cl_2O_2。活性氮转化成 HNO_3，其中一些由于空间的不均一性，从平流层中消失。这些化学改变保证在产生极地平流层云的极地涡旋高度内部的臭氧损失为 10%～15%。这些损耗转换成臭氧垂直柱浓度相当于 5%～8%。考虑到平流层氯在未来 20 年后至少会上升 50%，臭氧洞将很可能会出现。

59. Proffitt M H, Margitan J J, Kelly K K, et al. Ozone loss in the Arctic polar vortex inferred from high-altitude aircraft measurements[J]. Nature, 1990, 347(6288): 31-36.

Ozone loss in the Arctic polar vortex inferred from high-altitude aircraft measurements

Proffitt M. H., Margitan J. J., Kelly K. K., Loewenstein M., Podolske J. R., Chan K. R.

(Noaa, Aeron Lab, 325 Broadway, Boulder, Co 80303, USA)

Abstract: The Arctic polar vortex in winter is known to be chemically primed for ozone depletion, yet it does not exhibit the large seasonal ozone decrease that characterizes its southern counterpart. This difference may be due in part to a net flux of ozone-rich air through the Arctic vortex, which can mask ozone loss. But by using a chemically con-served tracer as a reference, significant ozone loss can be identified. This loss is found to be correlated with high levels of chlorine monoxide, suggesting that much of the decrease in ozone is caused by anthropogenic emissions of chlorofluorocarbons.

高海拔飞机观测到的北极极地涡旋中的臭氧流失

Proffitt M. H., Margitan J. J., Kelly K. K., Loewenstein M., Podolske J. R., Chan K. R.

(美国国家海洋和大气局)

摘要：冬季北极极地涡旋为臭氧损耗提供了化学基础，但并没有像南半球一样造成巨大的季节性臭氧亏损。这一差异可能部分取决于北极涡旋富含臭氧气团的净通量，掩盖了臭氧的流失。但是通过化学示踪剂作为参照物，可以将显著的臭氧损耗识别出来。这一损耗被发现与高水平的 ClO 相关，表明臭氧的损失大多是由于人类来源的氯氟烃的排放引起的。

60. Finlaysonpitts B J, Livingston F E, Berko H N. Ozone destruction and bromine photochemistry at ground-level in the Arctic spring[J]. Nature, 1990, 343(6259): 622-625.

Ozone destruction and bromine photochemistry at ground-level in the Arctic spring

Finlaysonpitts B. J., Livingston F. E., Berko H. N.

(Calif State Univ Fullerton, Dept Chem & Biochem, Fullerton, Ca 92634)

Abstract: Recent Arctic field studies in the spring at Alert, Canada (82.5°N, 62.3°W), show that ground-level ozone concentrations fell from 30-40 parts per billion (p. p. b.) to almost 0 p. p. b. in a period of time ranging from a few hours to a few days. Simultaneously, the concentration of brominated species collected on cellulose filters increases dramatically, with 50% or more of this bromine attributed to gaseous compounds such as HBr. The dramatic change in surface ozone and increase in filterable bromine seem to be due to advection of air below the inversion layer off the polar ice cap; in this air mass, surface ozone has been depleted and filterable bromine formed from photochemical reactions involving bromine compounds. Here we report laboratory experiments and computer kinetic modelling studies which indicate that the photochemically active bromine compound may be nitryl bromide ($BrNO_2$), formed in the reaction of N_2O_5 with NaBr found in sea-salt particles. $BrNO_2$ will photolyse rapidly at sunrise, forming ozone-destroying bromine atoms. We propose key field experiments to test this hypothesis.

北极春季地面臭氧破坏和溴光化学

Finlaysonpitts B. J., Livingston F. E., Berko H. N.

（美国加州州立大学富尔顿分校）

摘要：最近对加拿大 Alert 地区（82.5°N，62.3°W）的春季北极现场观测显示，地面臭氧浓度在数小时到数天的时间内从 30～40 ppb 降至约 0 ppb。同时，纤维素膜上收集到的 Br 浓度发生了显著增长，而这一部分 Br 约 50%或更多归因于如 HBr 等气态化合物。臭氧的显著性改变和可滤性溴的增加可能受极地冰盖逆温层下空气水平对流的影响。在这一气团中，由于溴化物的光化学反应导致表面臭氧的亏损和可滤性溴的生成。本文中，室内实验和计算机动力学模型的研究结果显示，引起反应的光化学活性溴化物可能是 $BrNO_2$，它是 N_2O_5 和海洋中的 NaBr 反应形成的。$BrNO_2$ 在日出时迅速光分解，形成破坏臭氧的溴原子。我们计划对这一假设的关键部分进行实验检验。

61. Fahey D W, Kelly K K, Kawa S R, et al. Observations of denitrification and dehydration in the winter polar stratospheres[J]. Nature, 1990, 344: 321-324.

Observations of denitrification and dehydration in the winter polar stratospheres

Fahey D. W., Kelly K. K., Kawa S. R., Tuck A. F., Loewenstein M., Chan K. R., Heidt L. E.

(Aeronomy Laboratory, National Oceanic and Atmospheric Administration, Boulder, Colorado 80303, USA)

Abstract: Water vapour and reactive nitrogen species are removed from the polar stratosphere in both hemispheres in the winter and early spring, probably by the sedimentation of aerosol particles containing water and nitric acid, which co-condense at low stratospheric temperatures. In the Antarctic in 1987, intense dehydration invariably accompanied intense denitrification. However, from the marked difference between water vapour concentrations in the two hemispheres for similar concentrations of reactive nitrogen, we deduce that in the Antarctic some dehydration may have occurred without denitrification. In the Arctic in 1989, despite higher temperatures than in the Antarctic, intense denitrification occurred but without intense dehydration. These results provide important constraints for the uncertain microphysics controlling the growth and sedimentation of aerosol particles effecting the removal. We argue that the Arctic denitrification can be explained by the selective growth and sedimentation of aerosol particles rich in nitric acid. Because reactive nitrogen species moderate the destruction of ozone by chlorine-catalysed reactions, by sequestering chlorine in reservoir species such as $ClONO_2$, the possibility of the removal of reactive nitrogen without dehydration should be allowed for in attempts to model ozone depletion in the Arctic. Indeed, denitrification along with elevated concentrations of reactive chlorine1 observed in 1989 indicate that the Arctic was chemically primed for ozone destruction without an extended period of temperatures below the frost point, as is characteristic of the Antarctic.

极地平流层冬季反硝化和脱水作用的观测

Fahey D. W., Kelly K. K., Kawa S. R., Tuck A. F., Loewenstein M., Chan K. R., Heidt L. E.

(美国国家海洋和大气局)

摘要：可能是由于含有水和硝酸的气溶胶粒子(源于平流层的低温浓缩)的沉降，造成了南北半球冬季和早春水汽与活性氮分子从极地平流层迁出。1987 年在南极，强烈的脱水作用总是伴随着强烈的脱氮作用。然而，从两个半球活性氮相近浓度中显著的水汽浓度的差异，我们认为在南极脱水作用发生时并没有发生脱氮作用。尽管比南极的温度要高，在北极 1989 年还是发生了强烈的脱氮作用(没有伴随强烈的脱水作用)。这些结果为控制去除作用的气溶胶粒子的产生和沉降的微观物理学提供了重要的约束。富含硝酸的气溶胶粒子的选择性生长和沉降可以解释北极脱氮作用。由于活性氮通过氯催化反应与氯结合生成 $ClONO_2$ 及其他化合物，从而缓和了臭氧的破坏过程，因此在尝试模拟北极臭氧亏损时，应该考虑独立于脱水作用的活性氮被去除的可能性。事实上，1989 年观测到的去氮作用以及活性氯浓度的升高表明，北极化学作用可破坏臭氧，并不是如南极一样需要长期低于霜点的温度。

三、海洋古气候、古环境类文献

3.1 分析概述

自1990年以来*Nature*和*Science*刊物上共有“海洋古气候、古环境类”文献65篇，其中南极方面文献43篇，北极方面文献22篇，南北极均包括的文献4篇。

1. 发表年代特征

将文章数量分区域并按照年代进行分类发现，平均来看两刊物上每年发表北极古气候、古环境类文章在1篇左右；其中1990～1996年较少，总共只有两篇；2001～2006年出现发表高峰，平均每年约发表3篇；2007、2008两年无论文发表。如图3.1所示。

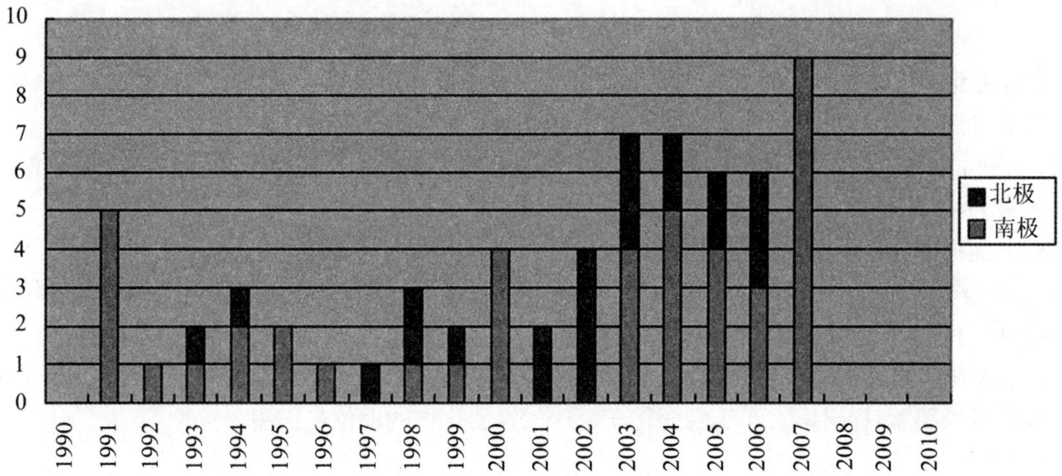

图3.1 文章数量随年代的变化

在两刊物上发表的关于南极海洋古气候、古环境的文章每年约3篇；1991年论文发表5篇，成为1990～1999年间的一次高峰，但随后论文量均较少，1997年甚至没有南极海洋古气候、古环境的文章，1997年之后，发表论文数量有所增加，但2001、2002两年中还是没有南极相关的文章，随后2007年发表南极相关论文9篇，形成一个高峰。

从两极的对比来看，每年发表的南极海洋古气候、古环境论文明显多于北极，约为后者的2倍。此外，*Nature*和*Science*上发表的南极文章的高峰期也要比北极早，有关南极海洋古气候、古环境的文章在1997年之后迅速增加，而北极方面在2001年之后才出现较大幅度的增加。值得注意的是，两刊在经历发表高峰期后，发表数量在最近均出现了下降的趋势。

2. 发表国别特征

对文章作者署名单位进行分析(图3.2)发现，署名单位为美国的作者参与发表了65篇文章中的37篇，占总数的57%。其次为英国，参与发表了其中的15篇文章，占总数的23%。其后还有德国、法国和瑞士的作者参与发表了5篇以上的文章。近20年来署名单位为中国的作者共参与了2篇两极海洋古气候、古环境类的文章，其中第一署名单位的为1篇。说明在这一研究领域，中国学者已经逐渐参与进来，但与英美等国的差距还相当大。

3. 发表文章研究领域特征

对发表文章的内容进行分析，发现在*Nature*和*Science*上发表的极地海洋古气候、古环境论文多集中在以下几个研究方向。

其中北极的研究主要集中在两个方向：

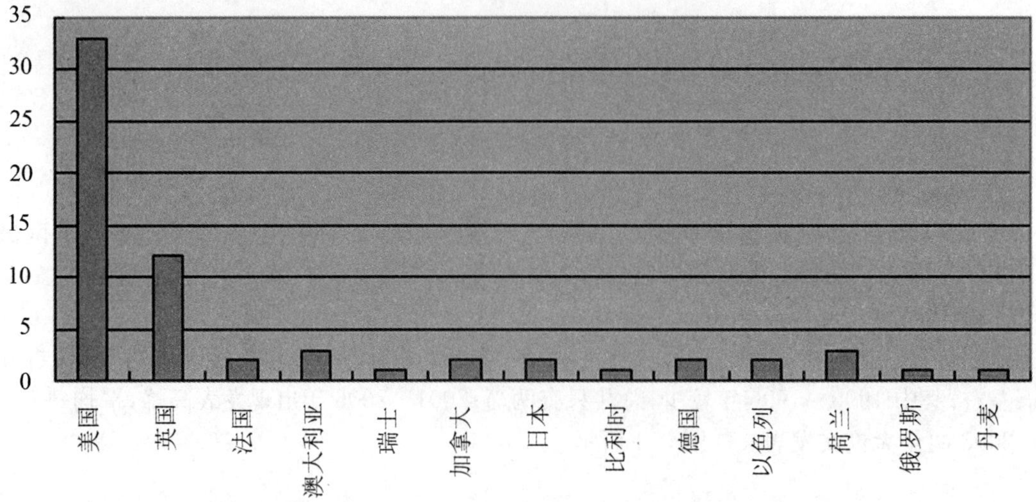

图 3.2 第一署名单位所在国家发表的文章数量

(1) 有关北大西洋与气候变化的研究。这部分研究主要利用亚极地北大西洋区域的 ODP 钻孔进行古气候研究,探讨北大西洋温盐环流的变化及其对全球气候变化的影响。

(2) 北极古温度变化。这部分研究主要利用北极一些陆架区钻孔来研究地质历史时期北极的温度、海冰状况等,并将其与当前的北极海冰融化结合起来。

南极方面的研究也主要集中在两个方向:

(1) 南大洋古生产力变化。这部分研究主要利用南大洋的 ODP 钻孔研究历史时期南大洋生产力的变化,这其中有很大一部分文章是有关南大洋古生产力与南大洋表层铁元素含量之间关系的。

(2) 南大洋气候变化。这部分内容主要利用海洋深钻,研究历史时期南大洋古温度、上升流强度、温跃层深度等气候要素,探讨南大洋气候变化在全球气候变化中的地位与作用。

3.2 摘要翻译——南极

1. Mortlock R A, Charles C D, Froelich P N, et al. Evidence for lower productivity in the Antarctic ocean during the last glaciation[J]. Nature, 1991, 351(6323): 220-223.

Evidence for lower productivity in the Antarctic ocean during the last glaciation

Mortlock R. A., Charles C. D., Froelich P. N., Zibello M. A., Saltzman J., Hays J. D., Burckle L. H.

(Columbia University, New York, United States)

Abstract: Both increased biological productivity and more efficient uptake of upwelled nutrients in high-latitude oceans have been proposed as mechanisms responsible for the glacial reduction in atmospheric concentrations of carbon dioxide deduced from ice-core measurements. These glacial models invoke more efficient "biological pumping" of carbon into the deep sea by increasing the uptake of "excess" biolimiting nutrients in the Antarctic surface ocean or by reorganizing chemical circulation patterns within the ocean. Here we challenge this conventional view with new evidence from tracers of palaeoproductivity preserved in Antarctic sediments. Records of the accumulation rates of diatom shells, the ratio of germanium to silicon in diatomaceous opal and the carbon isotope ratio in foraminiferal carbonate all suggest lower glacial productivity and less efficient uptake of nutrients. Although alternative interpretations are possible, our results support previous studies that indicate lower glacial productivity in the Southern Ocean and raise new questions about the role of ocean productivity in models of the causes (or remedies) for changes in atmospheric concentrations of carbon dioxide.

南大洋在末次冰期期间较低海洋生产率的证据

Mortlock R. A., Charles C. D., Froelich P. N., Zibello M. A., Saltzman J., Hays J. D., Burckle L. H.

(美国哥伦比亚大学)

摘要:高纬度海洋生物生产力的增加和上升流带来的营养物质的吸收被认为是冰期大气二氧化碳浓度降低的机制。这种模型认为冰期南极海洋增加对生物增长重要的营养物的吸收和重组化学环流导致生物泵作用加强,将碳泵入深海。这里,我们根据南极沉积物中古生产力的跟踪物证据对这一理论提出了挑战。生物硅壳体的累积速率,蛋白石的锗硅比和有孔虫碳酸盐碳同位素的记录全部显示出在冰期的时候,南极附近的生产力较低,且营养元素吸收低效。虽然其他解释是可能的,但是我们的结论支持了前人一些关于南大洋冰期低生产力的研究,同时对模型认为的冰期海洋生产力增加导致大气二氧化碳浓度下降的模式提出了质疑。

2. Kennett J P, Stott L D. Abrupt deep-sea warming, palaeoceanographic changes and benthic extinctions at the end of the Paleocene[J]. Nature,1991,353(6341):225-229.

Abrupt deep-sea warming, palaeoceanographic changes and benthic extinctions at the end of the Paleocene

Kennett J. P. , Stott L. D.

(University of California, Santa Barbara, Santa Barbara, California, United States)

Abstract: A remarkable oxygen and carbon isotope excursion occurred in Antarctic waters near the end of the Palaeocene (approximately 57. 33 Myr ago), indicating rapid global warming and oceanographic changes that caused one of the largest deep-sea benthic extinctions of the past 90 million years. In contrast, the oceanic plankton were largely unaffected, implying a decoupling of the deep and shallow ecosystems. The data suggest that for a few thousand years, ocean circulation underwent fundamental changes producing a transient state that, although brief, had long-term effects on environmental and biotic evolution.

深海急剧变暖——古新世末期古海洋变化和深海生物灭绝

Kennett J. P. , Stott L. D.

(美国加利福尼亚大学)

摘要:在古新世末期(大约5733万年前),南极附近海水的氧碳同位素出现了一次偏离,说明快速的全球变暖和海洋变化导致了一次过去9000万年最大的深海生物灭绝事件。相反,该时段的海洋浮游生物却并没有受到明显的影响,说明深海和浅海生态系统在这一时间段里的变化呈现出不相关特征。我们的数据说明,在几千年的时间里面,海洋环流发生了重大的变化,虽然时间短暂,但是对环境和生命演化产生了长期的影响。

3. Margolis S V, Claeys P, Kyte F T. Microtektites, microkrystites, and spinels from a late Pliocene asteroid impact in the Southern-Ocean[J]. Science,1991,251(5001):1594-1597.

Microtektites, microkrystites, and spinels from a late Pliocene asteroid impact in the Southern-Ocean

Margolis S. V. , Claeys P. , Kyte F. T.

(The United States, the University of California, Davis)

Abstract: The properties of glassy spherules found in sedimentary deposits of a late Pliocene asteroid impact into the southeast Pacific are similar to those of both microtektites and microkrystites. These spherules probably formed from molten silicate droplets that condensed from an impact-generated vapor cloud. The spherules contain inclusions of magnesioferrite spinels similar to those in spherules found at the Cretaceous-Tertiary boundary, indicating that both sets of spherules are impact debris formed under similar physical and chemical conditions.

晚上新世小行星对南大洋冲击所产生的微玻璃陨石、微晶球粒陨石和尖晶石

Margolis S. V., Claeys P., Kyte F. T.

(美国加州大学戴维斯分校)

摘要:在南太平洋沉积物中发现的晚上新世小行星冲击留下的玻璃状球体与微玻璃陨石和微球粒陨石都很相似。这些小球体很有可能产生于小行星碰撞产生的蒸汽云中的熔化硅酸盐。这些小球体里面还有镁铁矿-尖晶石的包裹体,与白垩纪-第三纪发现的玻璃球体类似。说明两者产生于类似的物理化学条件下。

4. Nicholls K W, Makinson K, Robinson A V. Ocean circulation beneath the Ronne ice shelf[J]. Nature, 1991, 354(6350): 221-223.

Ocean circulation beneath the Ronne ice shelf

Nicholls K. W., Makinson K., Robinson A. V.

(British Antarctic Survey, NERC, Cambridge CB3 0ET, England)

Abstract: The intimate thermal contact between the base of Antarctic ice shelves and the underlying ocean enables changes in climate to have a rapid impact on the outflow of ice from the interior of Antarctica. Furthermore, water modified by passage under ice shelves, particularly in the Weddell Sea, is believed to be an important constituent of Antarctic Bottom Water—a water mass that can be observed as far north as 50-degrees-N in the deep oceans. Antarctic Bottom Water is both cold and oxygen-rich, and plays an important part in the cooling and ventilation of the world's oceans. Because of the difficulty in gaining access, the oceanographic regime beneath ice shelves is very poorly sampled. By successfully drilling through the ice, however, we were able to obtain oceanographic data from beneath the largest Antarctic ice shelf, the Ronne-Filchner ice shelf in the southern Weddell Sea. We find that our data agree well with the predictions of a relatively simple oceanographic plume model of sub-ice-shelf circulation. This model can therefore be used with some confidence to investigate the links between climate changes, ice-shelf melting and bottom-water production.

南极隆尼冰架下面的海洋环流

Nicholls K. W., Makinson K., Robinson A. V.

(英国南极调查局)

摘要:南极冰架下面和海洋下面存在的密切热接触使得气候变化有可能对南极冰盖内部的冰外流产生快速的影响。此外,冰架下面的水运动,尤其是威德尔海的水运动,被认为是南极底部水的一个重要组

分——南极底部水能够在北纬50度深海观测到。南极底部水不仅冷而且富氧，在全球海洋的冷却和交换方面起到了重要作用。由于开槽评估比较困难，对于冰盖下面的海流模式目前还很不清楚。通过成功的钻透冰架，我们能够获得南极大冰架下面的海洋数据，如威德尔海的隆尼-菲尔切南冰架。我们发现，我们的数据与副冰架环流模式预测的结果很一致。因此这一模式可以被用来研究气候变化、冰盖融化以及底部水的产生之间的关联。

5. Clemens S, Prell W, Murray D, et al. Forcing mechanisms of the Indian-Ocean monsoon[J]. Nature, 1991, 353(6346): 720-725.

Forcing mechanisms of the Indian-Ocean monsoon

Clemens S., Prell W., Murray D., Shimmield G., Weedon G.

(Brown University, Department of Geological Sciences)

Abstract: Sediments in the Arabian Sea provide biological, biogeochemical and lithogenic evidence of past changes in the Indian Ocean summer monsoon winds. For the past 350000 years, this system has been externally forced by cyclical changes in solar radiation, and internally phase-locked to the transport of latent heat from the southern subtropical Indian Ocean to the Tibetan Plateau. In contrast to the results of general circulation models, these geological data suggest that the climate change associated with variability in global ice volume is not a primary factor in determining the strength and timing of the monsoon winds.

印度洋季风的驱动机制

Clemens S., Prell W., Murray D., Shimmield G., Weedon G.

(美国布朗大学)

摘要：来自阿拉伯海的沉积物提供了印度季风的历史变化的生物学证据、生物地球化学证据和岩性学证据。在过去35万年里，印度季风在外受太阳辐射的影响，在内受从南部副热带印度洋到青藏高原的热量输送的影响。与大气环流模型的结果不同的是，这些地质学证据表明全球冰量变化相关的气候变化并不是季风强度和变化时间的主要影响因素。

6. Lambeck K, Nakada M. Constraints on the age and duration of the last interglacial period and on sea-level variations[J]. Nature, 1992, 357(6374): 125-128.

Constraints on the age and duration of the last interglacial period and on sea-level variations

Lambeck K., Nakada M.

(Research School of Earth Sciences, The Australian National University, GPO Box 4, Canberra, Australian Capital Territory 2601, Australian)

Abstract: The relation between height and age of shorelines formed during the last interglacial period, as revealed by coral reefs, cannot be related directly to changes in ocean volume because of the effect of isostatic uplift in response to changes in ice-sheet loading. Sea-level changes at sites near the melting ice sheet, such as Bermuda and the Caribbean islands, differ from those along the Australian margin. Modelling of these differences constrains the times of onset and termination of the last interglacial, which are at variance with those deduced from oxygen-isotope studies of deep-sea cores.

末次间冰期的年代、持续时间和对应海平面的参数特征

Lambeck K., Nakada M.

(澳大利亚国立大学地球科学研究学院)

摘要:珊瑚礁反映出来的末次间冰期期间的海岸高度和年代之间的关联,不能直接归因于海洋量的变化,这其中还涉及冰盖融化过程中的均衡抬升作用。冰架融化附近的海平面变化,比如加勒比海的百慕大,与澳大利亚边缘就不一样。模拟显示这些差异影响的末次间冰期开始和结束的时间,与深海柱得到的海洋氧同位素阶段不一致。

7. Shemesh A, Macko S A, Charles C D, et al. Isotopic evidence for reduced productivity in the glacial Southern-Ocean[J]. Science, 1993, 262(5132): 407-410.

Isotopic evidence for reduced productivity in the glacial Southern-Ocean

Shemesh A., Macko S. A., Charles C. D., Rau G. H.

(Department of Environment Sciences and Energy Research, Weizmann Institute of Science Rehcvct 76100 Israel)

Abstract: Records of carbon and nitrogen isotopes in biogenic silica and carbon isotopes in planktonic foraminifera from deep-sea sediment cores from the Southern Ocean reveal that the primary production during the last glacial maximum was lower than Holocene productivity. These observations conflict with the hypothesis that the low atmospheric carbon dioxide concentrations were introduced by an increase in the efficiency of the high-latitude biological pump. Instead, different oceanic sectors may have had high glacial productivity, or alternative mechanisms that do not involve the biological pump must be considered as the primary cause of the low glacial atmospheric carbon dioxide concentrations.

冰期南大洋生产力降低的同位素证据

Shemesh A., Macko S. A., Charles C. D., Rau G. H.

(以色列魏茨曼科学研究所)

摘要:南大洋深海沉积柱生物硅酸盐碳氮同位素记录和浮游有孔虫碳同位素记录显示,末次冰期最盛期的海洋初级生产力要低于全新世的生产力。这个结果与冰期海洋生物泵作用强导致大气二氧化碳浓度降低的假设不相符合。这一结果有如下两个推断:第一,不同的海洋区域可能有不同的冰期海洋生产力;第二,非生物泵作用的机制可能是冰期二氧化碳浓度低的主要原因。

8. Fichefet T,Hovine S,Duplessy J C. A model study of the Atlantic thermohaline circulation during the last glacial maximum[J]. Nature,1994,372(6503):252-255.

A model study of the Atlantic thermohaline circulation during the last glacial maximum

Fichefet T., Hovine S., Duplessy J. C.

(Institut d'Astronomie et de Géophysique G. Lemaître, Université Catholique de Louvain, B-1348 Louvain-la-Neuve, Belgium)

Abstract: Stable isotope measurements in deep-sea sediment cores have indicated that the Atlantic thermohaline circulation experienced significant changes during the last glacial maximum: the North Atlantic Deep Water (NADW) was shallower than today and the Antarctic Bottom Water (AABW) penetrated much farther north. Numerical ocean models have, so far, been unable to simulate these circulation changes realistically. Here we show that a zonally averaged, three-basin ocean model, driven by glacial boundary conditions, reproduces the main trends of the geochemically constrained glacial Atlantic circulation. In addition, we provide quantitative estimates of the meridional water transport during glacial times. Our results suggest that the glacial production of AABW was slightly higher than at present, whereas that of NADW was reduced by similar to 40%, resulting in an intermediate circulation cell which closed within the Atlantic basin. We also show that the strength of the Atlantic conveyor belt strongly depends on the surface density contrast between the high latitudes of the Northern and Southern hemispheres.

末次冰盛期大西洋温盐环流的模拟研究

Fichefet T., Hovine S., Duplessy J. C.

(比利时法语鲁汶天主教大学)

摘要:深海沉积柱稳定同位素结果显示末次冰盛期大西洋深海环流发生了重大的变化:北大西洋深水变浅,南极底部水能够到达更北的区域。海洋模拟研究至今不能模拟这种变化。这里我们报道一个纬向平均-三维海洋模型,可以实现冰期大西洋环变化的主要趋势。此外,我们还定量估计了冰期的经向水传输。我们的结果表明冰期南极底部水比现代要高,而北大西洋深水则减弱了40%,这种结果导致形成了一个大西洋的封闭环流圈。我们的结果还说明大西洋传送带的强度受南北半球高纬度表层水密度差的影响。

9. Shemesh A,Burckle L H,Hays J D. Meltwater input to the Southern-Ocean during the last glacial maximum[J]. Science,1994,266(5190):1542-1544.

Meltwater input to the Southern-Ocean during the last glacial maximum

Shemesh A., Burckle L. H., Hays J. D.

(Weizmann Institute of Science, Israel)

Abstract: Three records of oxygen isotopes in biogenic silica from deep-sea sediment cores from the

Atlantic and Indian sectors of the Southern Ocean reveal the presence of isotopically depleted diatomaceous opal in sediment from the last glacial maximum. This depletion is attributed to the presence of lids of meltwater that mixed with surface water along certain trajectories in the Southern Ocean. An increase in the drainage from Antarctica or extensive northward transport of icebergs are among the main mechanisms that could have produced the increase in meltwater input to the glacial Southern Ocean. Similar isotopic trends were observed in older climatic cycles at the same cores.

末次冰盛期冰融水向南大洋的输入

Shemesh A. , Burckle L. H. , Hays J. D.

(以色列魏茨曼科学研究所)

摘要:我们在南大洋的大西洋和印度洋部分采集了三根沉积柱,测试了生物成因二氧化硅的氧同位素比值,结果显示在末次冰盛期的时候硅藻蛋白石存在氧同位素亏损。这种氧同位素偏负现象主要归因于冰盖融水参入了南大洋表层水流。南极冰融水排放的增加以及冰山向北的扩展可能是末次冰期南大洋冰融水增加的主要原因。在这沉积物中相似的同位素趋势在更早的气候循环中也有存在。

10. Debaar H J W, Dejong J T M, Bakker D C E, et al. Importance of iron for plankton blooms and carbon-dioxide drawdown in the Southern-Ocean[J]. Nature, 1995, 373(6513): 412-415.

Importance of iron for plankton blooms and carbon-dioxide drawdown in the Southern-Ocean

Debaar H. J. W. , Dejong J. T. M. , Bakker D. C. E. , Loscher B. M. , Veth C. , Bathmann U. , Smetacek V.

(Royal Netherlands Institute for Sea Research, Department of Marine Chemistry and Geology, P. O Box 59, NL-1790 AB Den Burg, The Netherlands)

Abstract: The iron hypothesis—the suggestion that iron is a limiting nutrient for plankton productivity and consequent CO_2 drawdown—has been tested by small-scale experiments in incubation bottles in the subarctic Pacific and Southern Oceans, and by a recent large-scale experiment in the equatorial Pacific Ocean. Here we test the idea by looking at natural levels of productivity in regions of the Southern Ocean with differing iron abundance. In the southerly branch of the Antarctic circumpolar current (ACC), upwelling of deep waters supplies sufficient iron to the surface to sustain moderate primary production but not to permit blooms to develop. In contrast, within the fast-flowing, iron-rich jet of the polar front (PF), spring blooms produced phytoplankton biomass an order of magnitude greater than that in southern ACC waters, leading to CO_2 undersaturation. The plankton-rich PF waters were sharply delineated from adjacent iron-poor waters, indicating that iron availability was the critical factor in allowing blooms to occur.

铁对南大洋浮游植物爆发和二氧化碳降低的重要性

Debaar H. J. W., Dejong J. T. M., Bakker D. C. E., Loscher B. M., Veth C., Bathmann U., Smetacek V.

(荷兰皇家海洋研究所)

摘要:铁假说——即铁是海洋浮游生物爆发和二氧化碳降低的限制性因素的假说已经在亚北极和南大洋进行的小尺度的孵化瓶中被测试,在赤道太平洋目前也在进行着一些大规模的测试。这里我们利用南大洋的自然状况对这个假说进行检验,在南极绕极流的南方分支,上升流的作用使得这一区域铁离子可以支持中等丰度的浮游生物,而在南极极锋交汇区,铁离子的含量比较丰富,可以支持丰富的浮游生物,春季浮游生物爆发生产比南极绕极流的南方水域高一个量级的生物体,导致 CO_2 低饱和。从南极极锋交汇区往外铁离子含量迅速降低,相应的浮游生物也迅速减少,表明可利用的铁离子是浮游生物爆发的决定性因素。

11. Kumar N, Anderson R F, Mortlock R A, et al. Increased biological productivity and export production in the glacial Southern-Ocean[J]. Nature, 1995, 378(6558): 675-680.

Increased biological productivity and export production in the glacial Southern-Ocean

Kumar N., Anderson R. F., Mortlock R. A., Froelich P. N., Kubik P., Dittrichhannen B., Suter M.

(Columbia Univ, Lamont-Doherty Earth Observ, POB 1000, Palisades, NY 10964 USA)

Abstract: A range of complementary radionuclide proxies in sediments of the southernmost Atlantic Ocean over the past 140000 years indicate that glacial periods were characterized by greatly increased fluxes of biogenic detritus out of surface waters. This increase in export production, which may have contributed to lower concentrations of carbon dioxide in the glacial atmosphere, was accompanied by more than a fivefold increase in accumulation of lithogenic iron transported by winds from Patagonian deserts. These observations support the hypothesis that the iron limitation of today's Southern Ocean productivity was relieved in glacial periods by a greater supply of iron from wind-blown dust.

冰期南大洋生物生产力的增加与生产力的输出

Kumar N., Anderson R. F., Mortlock R. A., Froelich P. N., Kubik P., Dittrichhannen B., Suter M.

(美国哥伦比亚大学拉蒙特-多尔蒂地质观测站)

摘要:一系列来自南大洋大西洋部分的互补的沉积物放射性替代性指标显示,在过去的 140000 年里,来自于南大洋表层水的生物碎屑在冰期大幅度增加。这种生产力的增加可能是冰期大气二氧化碳减少的一个原因,同时 Patagonian 沙漠的铁增加了五倍。这一结论支持了目前南大洋生产力低是铁限制的说法，同时也支持了冰期南大洋生产力增加是由来自风尘物质的铁所导致的这一说法。

12. Herman A B, Spicer R A. Palaeobotanical evidence for a warm Cretaceous Arctic Ocean[J]. Nature, 1996, 380(6572): 330-333.

Palaeobotanical evidence for a warm Cretaceous Arctic Ocean

Herman A. B. , Spicer R. A.

(Russian Academy of Sciences)

Abstract: The Cretaceous period was a time of global warmth. MidCretaceous equatorial temperatures were similar to today's, but the equator-to-pole temperature gradient is the subject of some controversy. Although it is unlikely that the poles were ice-frees, fossil evidence indicates that near-polar temperatures were much higher than they are today. Little is known, moreover, about oceanic poleward heat transport, and this makes it hard to model the Cretaceous climate or to evaluate the extent to which it provides an analogue for a "greenhouse" world warmed by increased atmospheric CO_2 alone. Here we use relationships between leaf physiognomy (such as shape and size) and modern climate to determine Cretaceous climate conditions in the Arctic region from fossil leaves. We find that the Arctic Ocean was relatively warm, remaining above 0 ℃ even during the winter months. This implies that there was significant poleward heat transport during all seasons.

白垩纪北冰洋温暖的古植物学证据

Herman A. B. , Spicer R. A.

（俄罗斯科学院地质所）

摘要：白垩纪是地球历史上较为温暖的时期。中白垩纪赤道的温度跟现在类似，但是从尺度到极地的温度梯度目前还存在一些争论。虽然极地不太可能是无冰的，但是化石证据显示在近极地的区域温度要比现代高得多。此外，对于白垩纪海洋向极的热传输也知道得很少，这就使得模拟白垩纪气候和评估温室气体效应变得困难。这里我们从化石植物叶，利用植物相貌和现代气候的关系来推测白垩纪的北极气候。我们发现北冰洋是相对温暖的，在冬季都有大约 0 ℃，这暗示着在所有季节都可能存在显著的向极热传输。

13. Hughen K A, Overpeck J T, Lehman S J, et al. Deglacial changes in ocean circulation from an extended radiocarbon calibration[J]. Nature, 1998, 391(6662): 65-68.

Deglacial changes in ocean circulation from an extended radiocarbon calibration

Hughen K. A., Overpeck J. T., Lehman S. J., Kashgarian M., Southon J., Peterson L. C., Alley R., Sigman D. M.

(INSTAAR and Department of Geological Sciences, University of Colorado, Boulder, Colorado 80309, USA)

Abstract: Temporal variations in the atmospheric concentration of radiocarbon sometimes result in radiocarbon-based age-estimates of biogenic material that do not agree with true calendar age. This problem is particularly severe beyond the limit of the high-resolution radiocarbon calibration based on tree-ring data, which stretches back only to about 11.8 kyr before present, near the termination of the Younger Dryas cold period if a wide range of palaeoclimate records are to be exploited for better understanding the rates and patterns of environmental change during the last deglaciation, extending the well-calibrated radiocarbon timescale back further in time is crucial. Several studies attempting such an extension, using uranium/thorium-dated corals and laminae counts in varved sediments, show conflicting results. Here we use radiocarbon data from varved sediments in the Cariaco basin, in the southern Caribbean Sea, to construct an accurate and continuous radiocarbon calibration (for the period 9 to 14.5 kyr sp), nearly 3000 years beyond the tree-ring-based calibration. A simple model compared to the calculated atmospheric radiocarbon concentration and palaeoclimate data from the same sediment core suggests that North Atlantic Deep Water formation shut down during the Younger Dryas period, but was gradually replaced by an alternative mode of convection, possibly via the formation of North Atlantic Intermediate Water.

通过放射性碳的延伸计算了解冰消期大洋环流的变化

Hughen K. A., Overpeck J. T., Lehman S. J., Kashgarian M., Southon J., Peterson L. C., Alley R., Sigman D. M.

(美国科罗拉多大学地质系)

摘要：大气中放射性碳随时间的变化有时会导致生物材料放射性定年的不准确。这个问题在11.8 kyr之前显得尤为突出，因为在11.8 kyr之后我们可以用树轮数据进行校正，11.8 kyr接近新仙女木事件冷期的中止时间。这给理解末次冰消期的环境变化带来困难，因此迫切需要将放射性碳定年校正延伸。一些研究，包括铀铅珊瑚定年、Cariaco盆地纹层沉积物的纹层计数都尝试做这个工作，但是结果还存在很大争议。这里，我们通过对Cariaco盆地的纹层沉积物的放射性碳数据研究，建立精确且连续的大约比树轮校正外扩3000年的放射性碳校正。进一步的分析显示北大西洋深水形成在新仙女木时期突然停止，取而代之的是一种对流模式，有可能是形成了北大西洋中层水。

14. Broecker W S, Sutherland S, Peng T H. A possible 20th-century slowdown of Southern Ocean deep water formation[J]. Science, 1999, 286(5442): 1132-1135.

A possible 20th-century slowdown of Southern Ocean deep water formation

Broecker W. S. , Sutherland S. , Peng T. H.

(Lamont-Doherty Earth Observatory of Columbia University, Palisades, NY 10964, USA)

Abstract: Chlorofluorocarbon-11 inventories for the deep Southern Ocean appear to confirm physical oceanographic and geochemical studies in the Southern Ocean, which suggest that no more than 5×10^6 cubic meters per second of ventilated deep water is currently being produced. This result conflicts with conclusions based on the distributions of the carbon-14/carbon ratio and a quasi-conservative property, PO_4, in the deep sea, which seem to require an ;average of about 15×10^6 cubic meters per second of Southern Ocean deep ventilation over about the past 800 years. A major reduction in Southern Ocean deep water production during the 20th century (from high rates during the Little Ice Age) may explain this apparent discordance. If this is true, a seesawing of deep water production between the northern Atlantic and Southern oceans may lie at the heart of the 1500-year ice-rafting cycle.

南大洋深水可能在20世纪停止形成

Broecker W. S. , Sutherland S. , Peng T. H.

(美国哥伦比亚大学拉蒙特-多尔蒂地质观测站)

摘要:南大洋深海的物理氯氟碳-11的库存看来支持海洋和地球化学研究结果,显示当前深水形成不超过 5×10^6 立方米每秒。这一结果与基于 $^{14}C/C$ 比值和 PO_4 准保守性质的深海分布不一致,这些分布表明在过去800年南大洋深水形成速度为平均 15×10^6 立方米每秒。这个不一致可用南大洋深水形成的减少来解释,该减少可能主要源于20世纪的一次减少(从小冰期的高峰期)。如果是这样的话,南大洋与北大西洋深水的秋千效应可能与1500年周期的冰筏循环有关。

15. Elderfield H, Rickaby R E M. Oceanic Cd/P ratio and nutrient utilization in the glacial Southern Ocean[J]. Nature, 2000, 405(6784): 305-310.

Oceanic Cd/P ratio and nutrient utilization in the glacial Southern Ocean

Elderfield H. , Rickaby R. E. M.

(Department of Earth Sciences, University of Cambridge, Downing Street, Cambridge CB2 3EQ, UK)

Abstract: During glacial periods, low atmospheric carbon dioxide concentration has been associated with increased oceanic carbon uptake, particularly in the southern oceans. The mechanism involved remains unclear. Because ocean productivity is strongly influenced by nutrient levels, palaeo-oceanographic proxies have been applied to investigate nutrient utilization in surface water across glacial transitions. Here we show that present-day cadmium and phosphorus concentrations in the global oceans can be explained by a chemical fractionation during particle formation, whereby uptake of cadmium occurs in preference to uptake of phosphorus. This allows the reconstruction of past surface water phosphate concentrations from the cadmium/calcium ratio of planktonic foraminifera. Results from the Last Glacial Maximum show similar phosphate utilization in the subantarctic to that of today, but much smaller utilization in the polar Southern Ocean, in a model that is consistent with the expansion of glacial sea ice and which can reconcile all proxy records of polar nutrient utilization. By restricting communication between the ocean and atmosphere, sea ice expansion also provides a mechanism for reduced CO_2 release by the Southern Ocean and lower glacial atmospheric CO_2.

冰期南大洋海洋 Cd/P 比值和营养利用

Elderfield H. , Rickaby R. E. M.

(英国剑桥大学地球科学系)

摘要:在末次冰期,大气二氧化碳浓度降低与海洋,尤其南大洋碳利用增加同时发生,但是机制目前还不清楚。因为海洋生产力主要受营养水平的影响,人们用古海洋指示剂来研究末次冰期期间海洋表层海水营养利用。这里我们证明粒子形成过程中 Cd 和 P 的分馏能够解释全球海洋 Cd 和 P 浓度的分布,Cd 的吸收要优先于 P 的吸收。这种关系使得我们可以通过浮游有孔虫中的 Cd/Ca 比来估算海洋表层的 P 的水平。从末次冰盛期的结果显示副北极地区磷的利用跟当前差不多,但是在南大洋的利用要小得多;这一结果与冰期海冰扩展一致,并可解释所有极地营养利用指示剂记录。通过限制海洋和大气的交流,海冰扩张也为冰期大气二氧化碳偏低提供了解释。

16. Fedorov A V,Philander S G. Is El Nino changing? [J]. Science,2000,288(5473):1997-2002.

Is El Nino changing?

Fedorov A. V. , Philander S. G.

(Department of Geosciences, Princeton University, Princeton, NJ 08544, USA)

Abstract: Recent advances in observational and theoretical studies of El Nino have shed Light on controversies concerning the possible effect of global warming on this phenomenon over the past few decades and in the future. El Nino is now understood to be one phase of a natural mode of oscillation—La Nina is the complementary phase—that results from unstable interactions between the tropical Pacific Ocean and the atmosphere. Random disturbances maintain this neutrally stable mode, whose properties depend on the background (time-averaged) climate state. Apparent changes in the properties of El Nino could reflect the importance of random disturbances, but they could also be a consequence of decadal variations of the background state. The possibility that global warming is affecting those variations cannot be excluded.

厄尔尼诺是否发生了变化?

Fedorov A. V., Philander S. G.

(美国普林斯顿大学地质系)

摘要:最近的关于厄尔尼诺的理论和观测研究澄清了一些对全球变暖对厄尔尼诺的过去和未来影响的争议。厄尔尼诺和拉尼娜现在被理解为是一种热带太平洋大气海洋的不稳定相互作用的自然现象。这种随机震荡维持中性稳定态,其性质取决于气候背景状态。厄尔尼诺性质变化可反映随机震荡的重要性,也可能是背景状态变化的结果。全球变暖引起这些变化的可能性不能被排除。

17. Lea D W, Pak D K, Spero H J. Climate impact of late quaternary equatorial Pacific sea surface temperature variations[J]. Science, 2000, 289(5485): 1719-1724.

Climate impact of late quaternary equatorial Pacific sea surface temperature variations

Lea D. W., Pak D. K., Spero H. J.

(University of California, Santa Barbara, California, United States)

Abstract: Magnesium/calcium data from planktonic foraminifera in equatorial Pacific sediment cores demonstrate that tropical Pacific sea surface temperatures (SSTs) were 2.8 degrees ±0.7 degrees C colder than the present at the last glacial maximum. Glacial-interglacial temperature differences as great as 5 degrees C are observed over the last 450 thousand years. Changes in SST coincide with changes in Antarctic air temperature and precede changes in continental ice volume by about 3 thousand years, suggesting that tropical cooling played a major role in driving ice-age climate. Comparison of SST estimates from eastern and western sites indicates that the equatorial Pacific zonal SST gradient was similar or somewhat Larger during glacial episodes. Extraction of a salinity proxy from the magnesium/calcium and oxygen isotope data indicates that transport of water vapor into the western Pacific was enhanced during glacial episodes.

气候对晚第四纪赤道太平洋海温度变化的影响

Lea D. W., Pak D. K., Spero H. J.

(美国加利福尼亚大学圣巴巴拉分校)

摘要:浮游有孔虫的 Mg/Ca 比值数据恢复的海表面温度变化显示在末次冰盛期赤道太平洋温度要比现在低 2.8±0.7 ℃。过去 45 万年冰期间冰期的温度变化可大到 5 ℃。海表面温度的变化与南极大气温度一致,但超前于大陆冰量变化 3000 年,这表明热带的变冷是冰期到来的一个重要驱动因素。东西太平洋海表面温度的对比显示赤道太平洋纬向温度梯度在冰期时与现代类似或偏大。从镁钙比和氧同位素的盐度恢复结果显示在冰期的时候向西太平洋输送的水汽增加。

18. Sikes E L, Samson C R, Guilderson T P, et al. Old radiocarbon ages in the southwest Pacific Ocean during the last glacial period and deglaciation[J]. Nature, 2000, 405(6786): 555-559.

Old radiocarbon ages in the southwest Pacific Ocean during the last glacial period and deglaciation

Sikes E. L., Samson C. R., Guilderson T. P., Howard W. R.

(School of Environmental and Marine Sciences and Department of Geology, University of Auckland, New Zealand)

Abstract: Marine radiocarbon (C-14) dates are widely used for dating oceanic events and as tracers of ocean circulation, essential components for understanding ocean-climate interactions. Past ocean ventilation rates have been determined by the difference between radiocarbon ages of deep-water and surface-water reservoirs, but the apparent age of surface waters (currently similar to 400 years in the tropics and similar to 1200 years in Antarctic waters) might not be constant through time, as has been assumed in radiocarbon chronologies and palaeoclimate studies. Here we present independent estimates of surface-water and deep-water reservoir ages in the New Zealand region since the last glacial period, using volcanic ejecta (tephras) deposited in both marine and terrestrial sediments as stratigraphic markers. Compared to present-day values, surface-reservoir ages from 11900 C-14 years ago were twice as large (800 years) and during glacial times were five times as large (2000 years), contradicting the assumption of constant surface age. Furthermore, the ages of glacial deepwater reservoirs were much older (3000-5000 years). The increase in surface-to-deep water age differences in the glacial Southern Ocean suggests that there was decreased ocean ventilation during this period.

末次冰期-冰消期西南太平洋的老碳年龄

Sikes E. L., Samson C. R., Guilderson T. P., Howard W. R.

(新西兰奥克兰大学环境和海洋科学与地质学系)

摘要:海洋放射性^{14}C年龄被广泛用于界定海洋事件、追踪海洋环流、评估海洋气候变化。通过测定深层水储库浅深层水储库的碳年龄差,可以确定过去海洋风流通速度,但是表层水的年龄(热带400年,南极1200年)可能并不是像在碳定年和古气候研究所假设的那样不变。这里我们用火山在海洋和陆地的沉积物,独立地评估新西兰附近末次冰期的表层水和深水储库年龄。与现代^{14}C储库相比较,11900前的表层储库年龄是两倍大(800年),冰期的表层储库年龄是五倍大(2000年),这与表层储库年龄不变的假设是冲突的。进一步地,冰期深层水储库要明显偏老,一般是3000～5000年。这一在南大洋表层水-深层水年龄差异在冰期的增加表明海洋风流通减弱。

19. Adams J B, Mann M E, Ammann C M. Proxy evidence for an El Nino-like response to volcanic forcing[J]. Nature, 2003, 426(6964): 274-278.

Proxy evidence for an El Nino-like response to volcanic forcing

Adams J. B., Mann M. E., Ammann C. M.

(Department of Environmental Sciences, University of Virginia, Clark Hall, Charlottesville, Virginia 22903, USA)

Abstract: Past studies have suggested a statistical connection between explosive volcanic eruptions and subsequent El Nino climate events. This connection, however, has remained controversial. Here we present support for a response of the El Nino/Southern Oscillation (ENSO) phenomenon to forcing from explosive volcanism by using two different palaeoclimate reconstructions of El Nino activity and two independent, proxy-based chronologies of explosive volcanic activity from AD 1649 to the present. We demonstrate a significant, multi-year, El Nino-like response to explosive tropical volcanic forcing over the past several centuries. The results imply roughly a doubling of the probability of an El Nino event occurring in the winter following a volcanic eruption. Our empirical findings shed light on how the tropical Pacific ocean-atmosphere system may respond to exogenous (both natural and anthropogenic) radiative forcing.

厄尔尼诺对火山活动响应的证据

Adams J. B., Mann M. E., Ammann C. M.

(美国弗吉尼亚大学环境科学系)

摘要:过去的研究显示火山活动与厄尔尼诺事件存在统计学上的联系,但是这一结论仍然存在争议。这里我们通过自 1649 年来两个不同的厄尔尼诺古气候重建、两个独立的火山重建记录来支持火山活动对厄尔尼诺变化的影响。我们的结果证明在过去几个世纪,厄尔尼诺事件显著受到火山活动影响。我们的结果暗示厄尔尼诺活动在火山喷发后的冬季可能加倍。这一发现证明了热带海气系统对外部辐射的响应。

20. Cobb K M, Charles C D, Cheng H, et al. El Nino/Southern Oscillation and tropical Pacific climate during the last millennium[J]. Nature, 2003, 424(6946): 271-276.

El Nino/Southern Oscillation and tropical Pacific climate during the last millennium

Cobb K. M., Charles C. D., Cheng H., Edwards R. L.

(Scripps Institution of Oceanography, University of California-San Diego, La Jolla, California 92093, USA)

Abstract: Any assessment of future climate change requires knowledge of the full range of natural variability in the El Nino/Southern Oscillation (ENSO) phenomenon. Here we splice together fossil-coral oxygen isotopic records from Palmyra Island in the tropical Pacific Ocean to provide 30-150-year windows of tropical Pacific climate variability within the last 1100 years. The records indicate mean climate conditions in the central tropical Pacific ranging from relatively cool and dry during the tenth century to increasingly warmer and wetter climate in the twentieth century. But the corals also document a broad range of ENSO behaviour that correlates poorly with these estimates of mean climate. The most intense ENSO activity within the reconstruction occurred during the mid-seventeenth century. Taken together, the coral data imply that the majority of ENSO variability over the last millennium may have arisen from dynamics internal to the ENSO system itself.

过去千年热带太平洋气候与ENSO的变化

Cobb K. M., Charles C. D., Cheng H., Edwards R. L.

(美国加州大学圣地亚哥分校斯克里普斯海洋研究所)

摘要:了解过去ENSO的变化对于评估未来气候变化有着重要意义,这里我们通过采自中太平洋Palmyra岛的珊瑚来重建了过去1100年的在30～150年尺度上的热带太平洋气候变化。结果显示中太平洋在10世纪的时候气候均态相对冷干,而在20世纪的时候暖湿。珊瑚同样也记录了ENSO的循环变化,但是与气候均态不一致。结果显示ENSO在17世纪中期最强。总的来看,大多数ENSO活动在过去千年来自于内在的动力。

21. Visser K, Thunell R, Stott L. Magnitude and timing of temperature change in the Indo-Pacific warm pool during deglaciation[J]. Nature, 2003, 421(6919): 152-155.

Magnitude and timing of temperature change in the Indo-Pacific warm pool during deglaciation

Visser K., Thunell R., Stott L.

(Department of Geological Sciences, University of South Carolina, Columbia, South Carolina 29205, USA)

Abstract: Ocean-atmosphere interactions in the tropical Pacific region have a strong influence on global heat and water vapour transport and thus constitute an important component of the climate system. Changes in sea surface temperatures and convection in the tropical Indo-Pacific region are thought to be responsible for the interannual to decadal climate variability observed in extratropical regions, but the role of the tropics in climate changes on millennial and orbital timescales is less clear. Here we analyse oxygen isotopes and Mg/Ca ratios of foraminiferal shells from the Makassar strait in the heart of the Indo-Pacific warm pool, to obtain synchronous estimates of sea surface temperatures and ice volume. We find that sea surface temperatures increased by 3. 5-4. 0 degrees C during the last two glacial-interglacial transitions, synchronous with the global increase in atmospheric CO_2 and Antarctic warming, but the temperature increase occurred 2000-3000 years before the Northern Hemisphere ice sheets melted. Our observations suggest that the tropical Pacific region plays an important role in driving glacial-interglacial cycles, possibly through a system similar to how El Nino/Southern Oscillation regulates the poleward flux of heat and water vapour.

印度-太平洋暖池末次间冰期温度变化的幅度和时间

Visser K., Thunell R., Stott L.

（美国南卡罗莱纳大学哥伦比亚分校地质科学系）

摘要：热带太平洋地区海气交感作用对全球水热输送有重要影响，因此是气候系统的一个重要分量。印度-太平洋暖池区域的温度和对流作用被认为与年际、年代际温带气候变化联系在一起，但是热带在千年和轨道尺度上对气候变化的影响目前还不清楚。这里，我们分析采自印度尼西亚的海洋沉积物浮游有孔虫氧同位素和Mg/Ca，探讨轨道尺度的海温变化和冰量变化。我们发现海温在过去两个冰期-间冰期转换中增加了3.5～4 ℃，与南极变暖以及大气二氧化碳升高同步，但温度增加超前于北半球冰盖的融化2000～3000年。我们的结果发现热带在冰期-间冰期的转换中发挥了重要的作用，也许作用过程类似于现代ENSO对中高纬度气候的影响。

22. Weaver A J, Saenko O A, Clark P U, et al. Meltwater pulse 1A from Antarctica as a trigger of the Bolling-Allerod warm interval[J]. Science, 2003, 299(5613): 1709-1713.

Meltwater pulse 1A from Antarctica as a trigger of the Bolling-Allerod warm interval

Weaver A. J., Saenko O. A., Clark P. U., Mitrovica J. X.

(University of Victoria in Canada)

Abstract: Meltwater pulse 1A (mwp-1A) was a prominent feature of the last deglaciation, which led to a sea-level rise of similar to 20 meters in less than 500 years. Concurrent with mwp-1A was the onset of the Bolling-Allerod interstadial event (14600 years before the present), which marked the termination of the last glacial period. Previous studies have been unable to reconcile a warm Northern Hemisphere with mwp-1A originating from the Laurentide or Fennoscandian ice sheets. With the use of a climate model of intermediate complexity, we demonstrate that with mwp-1A originating from the Antarctic Ice Sheet, consistent with recent sea-level fingerprinting inferences, the strength of North Atlantic Deep Water (NADW) formation increases, thereby warming the North Atlantic region and providing an explanation for the onset of the Boiling-Allerod warm interval. The established mode of active NADW formation is then able to respond to subsequent freshwater forcing from the Laurentide and Fennoscandian ice sheets, setting the stage for the Younger Dryas cold period.

南极洲的冰融水脉动1A是布林-阿勒罗德暖期的触发原因之一

Weaver A. J., Saenko O. A., Clark P. U., Mitrovica J. X.

（加拿大维多利亚大学）

摘要：冰融水脉动1A(MWP-1A)是末次冰消期一个突出特点，它导致了在不到500年时间中海平面上升约20 m。与MWP-1A并发的Bolling-Allerod(布林-阿勒罗德)间冰期事件(距今14600年)标志着末次冰期的终止。以往的研究已经无法调和温暖的北半球和来自 Laurentide 或 Fennoscandian 冰盖的 MWP-1A。我们使用一个中等复杂的气候模型证明从南极冰盖起源的 MWP-1A 符合最近的海平面指纹推论，北大西洋深层水(NADW)形成强度的增加，从而使北大西洋地区变暖，并提供了一个 Boiling-Allerod 暖期开始的解释。活跃的 NADW 形成的既定模式能够应对来自 Laurentide 和 Fennoscandian 冰盖的淡水驱动力，并为新仙女木冷期做好准备。

23. Carter R M, Gammon P. New Zealand maritime glaciation: millennial-scale southern climate change since 3.9 Ma[J]. Science, 2004, 304(5677): 1659-1662.

New Zealand maritime glaciation: millennial-scale southern climate change since 3.9 Ma

Carter R. M., Gammon P.

(School of Earth and Environmental Sciences, University of Adelaide, Tarndarnya, South Australia, Australia)

Abstract: Ocean Drilling Program Site 1119 is ideally located to intercept discharges of sediment from the mid-latitude glaciers of the New Zealand Southern Alps. The natural gamma ray signal from the site's sediment core contains a history of the South Island mountain ice cap since 3.9 million years ago (Ma). The younger record, to 0.37 Ma, resembles the climatic history of Antarctica as manifested by the Vostok ice core. Beyond, and back to the late Pliocene, the record may serve as a proxy for both mid-latitude and Antarctic polar plateau air temperature. The gamma ray signal, which is atmospheric, also resembles the ocean climate history represented by oxygen isotope time series.

新西兰沿海冰期作用：自 390 万年前开始的千年尺度的南部气候变化

Carter R. M., Gammon P.

(澳大利亚阿德莱德大学地球与环境科学学院)

摘要：大洋钻探计划 1119 点非常理想地拦截了来自新西兰南阿尔卑斯山的中纬度地区的冰川的沉积物。该点的沉积岩芯的自然伽马射线信号包含了南岛山区 390 万年前以来的冰帽的历史。到 37 万年的更年轻的纪录与沃斯托克冰芯展示的南极洲的气候历史类似。更进一步，到上新世晚期，该记录可能作为中纬度和南极高原的空气温度的替代性指标。展示大气的伽马射线信号也类似于氧同位素时间序列所代表的海洋气候的历史。

24. Knutti R, Fluckiger J, Stocker T F, et al. Strong hemispheric coupling of glacial climate through freshwater discharge and ocean circulation[J]. Nature, 2004, 430(7002): 851-856.

Strong hemispheric coupling of glacial climate through freshwater discharge and ocean circulation

Knutti R., Fluckiger J., Stocker T. F., Timmermann A.

(University of Bern, Sidlerstrasse 5, 3012 Bern, Switzerland)

Abstract: The climate of the last glacial period was extremely variable, characterized by abrupt warming events in the Northern Hemisphere, accompanied by slower temperature changes in Antarctica and variations of global sea level. It is generally accepted that this millennial-scale climate variability was caused by abrupt changes in the ocean thermohaline circulation. Here we use a coupled ocean-atmosphere-sea ice model to show that freshwater discharge into the North Atlantic Ocean, in addition to a reduction of the thermohaline circulation, has a direct effect on Southern Ocean temperature. The related anomalous oceanic southward heat transport arises from a zonal density gradient in the subtropical North Atlantic caused by a fast wave-adjustment process. We present an extended and quantitative bipolar seesaw concept that explains the timing and amplitude of Greenland and Antarctic temperature changes, the slow changes in Antarctic temperature and its similarity to sea level, as well as a possible time lag of sea level with respect to Antarctic temperature during Marine Isotope Stage 3.

因淡水释放和海洋环流而导致的冰期气候强半球耦合

Knutti R., Fluckiger J., Stocker T. F., Timmermann A.

(瑞士伯尔尼大学)

摘要:末次冰期的气候是一个变化极大、以北半球突然变暖为特征的事件,伴随着南极洲缓慢的温度变化和全球海平面的变化。普遍接受的观点是这千年尺度的气候变化是由海洋温盐环流的突然变化造成的。这里我们用一个耦合的海洋-大气-海冰模型来显示,北大西洋的淡水注入,和温盐环流减弱一样,也对南大洋的温度产生直接影响。相关异常的海洋向南热量输送来自一个快波调整过程中所造成的在北大西洋亚热带纬向的密度梯度。我们提出了一个长期和定量的双极跷跷板的概念来解释格陵兰岛和南极的温度变化的时间和幅度、南极温度和海平面的缓慢变化以及在海洋同位素3期关于南极温度的海平面的时间滞后。

25. Lamy F, Kaiser J, Ninnemann U, et al. Antarctic timing of surface water changes off Chile and Patagonian ice sheet response[J]. Science, 2004, 304(5679): 1959-1962.

Antarctic timing of surface water changes off Chile and Patagonian ice sheet response

Lamy F., Kaiser J., Ninnemann U., Hebbeln D., Arz H. W., Stoner J.

(Geo Forschungs Zentrum-Potsdan, Telegrafenberg, 14473 Potsdam, Germany; DFG Research Center Ocean Margins, Klagenfurter Strasse, 28359 Bremen, Germany)

Abstract: Marine sediments from the Chilean continental margin are used to infer millennial-scale changes in southeast Pacific surface ocean water properties and Patagonian ice sheet extent since the last glacial period. Our data show a clear "Antarctic" timing of sea surface temperature changes, which appear systematically linked to meridional displacements in sea ice, westerly winds, and the circumpolar current system. Proxy data for ice sheet changes show a similar pattern as oceanographic variations offshore, but reveal a variable glacier-response time of up to similar to 1000 years, which may explain some of the current discrepancies among terrestrial records in southern South America.

智利和巴塔哥尼亚冰原对南极表面水变化的时间性

Lamy F., Kaiser J., Ninnemann U., Hebbeln D., Arz H. W., Stoner J.

(德国地理协会;DFG 海洋研究中心)

摘要:智利大陆边缘的海洋沉积物被用来推断末次冰期以来千年尺度上的东南太平洋海洋表面水的性质和巴塔哥尼亚冰原范围的变化。我们的数据显示表层海水温度明显的"南极"时间性,其与海冰经向位移、西风和极地环流系统关联。冰盖变化的替代性指标变化与近海海洋变化一致,但揭示冰川响应时间长达约 1000 年,这也许可以解释一些在南美洲南部陆地记录之间的当前差异。

26. Shevenell A E, Kennett J P, Lea D W. Middle Miocene Southern Ocean cooling and Antarctic cryosphere expansion[J]. Science, 2004, 305(5691): 1766-1770.

Middle Miocene Southern Ocean cooling and Antarctic cryosphere expansion

Shevenell A. E., Kennett J. P., Lea D. W.

(Department of Geological Sciences and Marine Science Institute, University of California, Santa Barbara, CA 93106-9630, USA)

Abstract: Magnesium/calcium data from Southern Ocean planktonic foraminifera demonstrate that high-latitude (similar to 55 degrees S) southwest Pacific sea surface temperatures (SSTs) cooled 6 degrees to 7 degrees C during the middle Miocene climate transition (14.2 to 13.8 million years ago). Stepwise surface cooling is paced by eccentricity forcing and precedes Antarctic cryosphere expansion by similar to 60 thousand years, suggesting the involvement of additional feedbacks during this interval of inferred low-atmospheric partial pressure of CO_2 (pCO_2). Comparing SSTs and global carbon cycling proxies challenges the notion that episodic pCO_2 drawdown drove this major Cenozoic climate transition. SST, salinity, and ice-volume trends suggest instead that orbitally paced ocean circulation changes altered meridional heat/vapor transport, triggering ice growth and global cooling.

中新世中期南大洋的变冷和南极冰冻圈的扩张

Shevenell A. E., Kennett J. P., Lea D. W.

(美国加州大学圣巴巴拉分校地质和海洋科学研究所)

摘要:南大洋浮游有孔虫的镁/钙数据表明,高纬度地区(约南纬 55 度)西南太平洋的海表面温度

(SST)在中新世中期的气候转型期(1420～1380 万年前)降低了 6～7 ℃。表面逐步冷却受到离心率驱动的控制并领先于南极冰雪圈的扩大约 6 万年，暗示在大气二氧化碳低分压的间隔期内额外反馈的影响。比较海表面温度和全球碳循环的概念对阵发性二氧化碳分压下降导致了主要的新生代气候过渡的概念提出了挑战。海表面温度、盐度和冰量趋势表明，是轨道节奏(周期)的海洋环流变化改变了南部热/水汽输送，引发冰增长和全球变冷。

27. Sigman D M, Jaccard S L, Haug G H. Polar ocean stratification in a cold climate[J]. Nature, 2004, 428(6978): 59-63.

Polar ocean stratification in a cold climate

Sigman D. M., Jaccard S. L., Haug G. H.

(Department of Geosciences, Princeton University, Princeton, New Jersey 08544, USA)

Abstract: The low-latitude ocean is strongly stratified by the warmth of its surface water. As a result, the great volume of the deep ocean has easiest access to the atmosphere through the polar surface ocean. In the modern polar ocean during the winter, the vertical distribution of temperature promotes overturning, with colder water over warmer, while the salinity distribution typically promotes stratification, with fresher water over saltier. However, the sensitivity of seawater density to temperature is reduced as temperature approaches the freezing point, with potential consequences for global ocean circulation under cold climates. Here we present deep-sea records of biogenic opal accumulation and sedimentary nitrogen isotopic composition from the Sub-arctic North Pacific Ocean and the Southern Ocean. These records indicate that vertical stratification increased in both northern and southern high latitudes 2.7 million years ago, when Northern Hemisphere glaciation intensified in association with global cooling during the late Pliocene epoch. We propose that the cooling caused this increased stratification by weakening the role of temperature in polar ocean density structure so as to reduce its opposition to the stratifying effect of the vertical salinity distribution. The shift towards stratification in the polar ocean 2.7 million years ago may have increased the quantity of carbon dioxide trapped in the abyss, amplifying the global cooling.

在寒冷气候下极地海洋的分层作用

Sigman D. M., Jaccard S. L., Haug G. H.

(美国普林斯顿大学地质科学系)

摘要：低纬度海洋因其表面水的温暖而强烈分层。因此，大量的深海水容易通过极地海洋表面与大气接触。在现代极地海洋的冬季，温度的垂直分布促进翻转，使较冷海水在较温暖的海水之上，与此同时盐度分布也典型地促进分层作用，使淡水在咸水之上。然而，当温度接近冰点时，海水密度对温度(响应的)敏感性降低，对寒冷气候下的全球海洋环流可能产生一些后果。在这里，我们报道来自亚北极北太平洋和南大洋的生物成因蛋白石积累和沉积的氮同位素组成的深海记录。这些记录表明，270 万年前，当北半球冰川作用随着上新世晚期全球变冷而加强时，垂直分层在北部和南部的高纬度地区都增加。我们认为，冷却通过削弱温度在极地海洋密度结构中的作用，减少其盐度垂直分布的分层效应的反作用，导致了这种分层作用的变强。这种 270 万年前对分层作用在极地海洋的转变可能增加被固定在深水中的二氧化碳的量，加强了全球变冷的趋势。

28. Coxall H K, Wilson P A, Palike H, et al. Rapid stepwise onset of Antarctic glaciation and deeper calcite compensation in the Pacific Ocean[J]. Nature, 2005, 433(7021): 53-57.

Rapid stepwise onset of Antarctic glaciation and deeper calcite compensation in the Pacific Ocean

Coxall H. K., Wilson P. A., Palike H., Lear C. H., Backman J.

(Southampton Oceanography Centre, School of Ocean and Earth Science, European Way, Southampton SO14 3ZH, UK)

Abstract: The ocean depth at which the rate of calcium carbonate input from surface waters equals the rate of dissolution is termed the calcite compensation depth. At present, this depth is 4500 m, with some variation between and within ocean basins. The calcite compensation depth is linked to ocean acidity, which is in turn linked to atmospheric carbon dioxide concentrations and hence global climate. Geological records of changes in the calcite compensation depth show a prominent deepening of more than 1 km near the Eocene/Oligocene boundary (similar to 34 million years ago) when significant permanent ice sheets first appeared on Antarctica, but the relationship between these two events is poorly understood. Here we present ocean sediment records of calcium carbonate content as well as carbon and oxygen isotopic compositions from the tropical Pacific Ocean that cover the Eocene/Oligocene boundary. We find that the deepening of the calcite compensation depth was more rapid than previously documented and occurred in two jumps of about 40000 years each, synchronous with the stepwise onset of Antarctic ice-sheet growth. The glaciation was initiated, after climatic preconditioning, by an interval when the Earth's orbit of the Sun favoured cool summers. The changes in oxygen-isotope composition across the Eocene/Oligocene boundary are too large to be explained by Antarctic ice-sheet growth alone and must therefore also indicate contemporaneous global cooling and/or Northern Hemisphere glaciation.

南极冰川作用和太平洋方解石补偿深度加深的快速阶梯式开始

Coxall H. K., Wilson P. A., Palike H., Lear C. H., Backman J.

(英国南安普顿海洋与地球科学学院)

摘要：来自地表水的碳酸钙输入率等于溶出速率的海洋深度被称为方解石补偿深度。目前，这个深度是4500 m，在大洋盆地内外存在一些变动。方解石补偿深度与海洋酸度相关联，这又依次与大气中二氧化碳的浓度和全球气候相关。方解石补偿深度的变化的地质记录表明，在始新世/渐新世边界附近(约3400万年前)存在超过1 km的显著加深，同时在南极首次出现显著的永久性冰盖，但人类对这两个事件之间的关系却知之甚少。在这里，我们的报道包含了始新世/渐新世边界时期的热带太平洋的碳酸钙含量以及碳和氧同位素组成的海洋沉积物记录。我们发现，方解石补偿深度的加深比以往记录更加迅速且发生在均持续约40000年的两个跳跃阶段，并与南极冰盖增长逐步发生同步。在当地球公转轨道有利于夏季凉爽的间隔期，气候预先调节后，冰川作用开始。跨越始新世/渐新世的边界的氧同位素组成变化太大导致无法独立解释南极冰盖增长，因此也表明当时的全球变冷和/或北半球冰川作用。

29. Pahnke K，Zahn R. Southern hemisphere water mass conversion linked with North Atlantic climate variability[J]. Science，2005，307(5716)：1741-1746.

Southern hemisphere water mass conversion linked with North Atlantic climate variability

Pahnke K.，Zahn R.

(School of Earth，Ocean，and Planetary Sciences，Cardiff University，Park Place，Cardiff CF10 3YE，UK)

Abstract：Intermediate water variability at multicentennial scales is documented by 340000-year-long isotope time series from bottom-dwelling foraminifers at a mid-depth core site in the southwest Pacific. Periods of sudden increases in intermediate water production are linked with transient Southern Hemisphere warm episodes，which implies direct control of climate warming on intermediate water conversion at high southern latitudes. Coincidence with episodes of climate cooling and minimum or halted deepwater convection in the North Atlantic provides striking evidence for interdependence of water mass conversion in both hemispheres，with implications for interhemispheric forcing of ocean thermohaline circulation and climate instability.

南半球水体转换与北大西洋气候变化的关系

Pahnke K.，Zahn R.

(英国卡迪夫大学地球海洋与行星科学学院)

摘要：多世纪尺度上的中层水变化被340000年之久西南太平洋中等深度沉积柱中的底栖有孔虫同位素时间系列所记录。中层水生产量突然增加与短暂南半球变暖存在联系，这意味着在高南纬气候变暖对中层水转变中的直接控制。气候变冷和北大西洋深海对流达到最低或停止的同时发生为两个半球水体转换的相互依存提供了有力的证据，并指示了海洋温盐环流和气候不稳定的半球间驱动力。

30. Sachs J P，Anderson R F. Increased productivity in the subantarctic ocean during Heinrich events [J]. Nature，2005，434(7037)：1118-1121.

Increased productivity in the subantarctic ocean during Heinrich events

Sachs J. P.，Anderson R. F.

(Department of Earth，Atmospheric and Planetary Sciences，Massachusetts Institute of Technology，77 Massachusetts Avenue，Room E34-254，Cambridge，Massachusetts 02139，USA)

Abstract：Massive iceberg discharges from the Northern Hemisphere ice sheets，“Heinrich events”，coincided with the coldest periods of the last ice age. There is widespread evidence for Heinrich events and their profound impact on the climate and circulation of the North Atlantic Ocean，but their influence beyond that region remains uncertain. Here we use a combination of molecular fingerprints of algal productivity and radioisotope tracers of sedimentation to document eight periods of increased productivity in the subpolar Southern Ocean during the past 70000 years that occurred within 1000-2000 years of a Northern Hemisphere Heinrich event. We discuss possible causes for such a link, including increased supply of iron from upwelling and increased stratification during the growing season, which imply an alteration of the global ocean circulation during Heinrich events. The mechanisms linking North Atlantic iceberg discharges with subantarctic productivity remain unclear at this point. We suggest that understanding how the Southern Ocean was altered during these extreme climate perturbations is critical to understanding the role of the ocean in climate change.

在 Heinrich 事件时期亚南极海域生产力的增长

Sachs J. P., Anderson R. F.

(美国麻省理工学院地球大气与行星科学系)

摘要:冰山从北半球冰盖大规模的释放,即"Heinrich 事件",恰逢最后一个冰期的最冷时期。有广泛的证据说明了 Heinrich 事件及其对北大西洋气候和环流产生了深远的影响,但它们对其他地区的影响力仍然不明朗。在这里,我们将藻类生产力分子指纹和沉积物的放射性同位素示踪相结合,表征了过去 70000 年间亚极地南大洋地区八个生产力提高时期,其发生在北半球 Heinrich 事件的 1000～2000 年之间。我们讨论了这样的联系可能存在的原因,其中包括海洋上升流铁供给的增加和生长季节增加的分层作用,这意味着在 Heinrich 事件时期全球海洋环流的改变。在这一点上,联系着北大西洋冰山释放和亚南极生产力的机制仍然不清楚。我们认为,了解南大洋是如何在这些极端气候扰动时发生改变的,对于理解海洋在气候变化中的作用是至关重要的。

31. Tripati A, Elderfield H. Deep-sea temperature and circulation changes at the Paleocene-Eocene thermal maximum[J]. Science, 2005, 308(5730): 1894-1898.

Deep-sea temperature and circulation changes at the Paleocene-Eocene thermal maximum

Tripati A., Elderfield H.

(Department of Earth Sciences, University of Cambridge, Downing Street, CB2 3EQ, UK)

Abstract: A rapid increase in greenhouse gas levels is thought to have fueled global warming at the Paleocene-Eocene Thermal Maximum (PETM). Foraminiferal magnesium/calcium ratios indicate that bottom waters warmed by 4 degrees to 5 degrees C, similar to tropical and subtropical surface ocean waters, implying no amplification of warming in high-latitude regions of deep-water formation under ice-free conditions. Intermediate waters warmed before the carbon isotope excursion, in association with downwelling in the North Pacific and reduced Southern Ocean convection, supporting changing circulation as the trigger for methane hydrate release. A switch to deep convection in the North Pacific at the PETM onset could have amplified and sustained warming.

在古新世-始新世最热期的深海温度和环流变化

Tripati A. , Elderfield H.

（英国剑桥大学地球科学系）

摘要：在古新世-始新世最热期（PETM）中温室气体水平迅速增加被认为是助长了全球变暖。有孔虫的 Mg/Ca 比值表明底部水域升温 4～5 ℃，这与热带和亚热带海洋表面水域相似，意味着在无冰情况下的深层水形成并没有放大高纬度地区的气候变暖。中层水域在碳同位素偏移前变热，伴随着北太平洋沉降流及南大洋对流减少，这支持了环流变化触发了甲烷水合物的释放。PETM 时期在北太平洋深层对流切换的开始将可能放大和维持变暖。

32. Nunes F，Norris R D. Abrupt reversal in ocean overturning during the Palaeocene/Eocene warm period[J]. Nature，2006，439(7072)：60-63.

Abrupt reversal in ocean overturning during the Palaeocene/Eocene warm period

Nunes F. , Norris R. D.

(Scripps Institution of Oceanography, University of California, San Diego. LaJolla, California 92092-0208, USA)

Abstract: An exceptional analogue for the study of the causes and consequences of global warming occurs at the Palaeocene/Eocene Thermal Maximum, 55 millions ago. A rapid rise of global temperatures during this event accompanied turnovers in both marine and terrestrial biota, as well as significant changes in ocean chemistry and circulation. Here we present evidence for an abrupt shift in deep-ocean circulation using carbon isotope records from fourteen sites. These records indicate that deep-ocean circulation patterns changed from Southern Hemisphere overturning to Northern Hemisphere overturning at the start of the Palaeocene/Eocene Thermal Maximum. This shift in the location of deep-water formation persisted for at least 40000 years, but eventually recovered to original circulation patterns. These results corroborate climate model inferences that a shift in deep-ocean circulation would deliver relatively warmer waters to the deep sea, thus producing further warming. Greenhouse conditions can thus initiate abrupt deep-ocean circulation changes in less than a few thousand years, but may have lasting effects; in this case taking 100000 years to revert to background conditions.

大洋翻转流在古新世/始新世的温暖期发生突然逆转

Nunes F. , Norris R. D.

（美国加州大学圣地亚哥分校斯克里普斯海洋研究所）

摘要：5500 万年前古新世/始新世最热期为研究全球变暖的原因和后果提供一个特殊的类似物。这一时期的全球气温迅速上升，伴随着海洋和陆地生物群的翻转以及在海洋化学和循环的显著变化。在这里，我们用^{14}C记录提供了深海洋环流突然转变的证据。这些记录表明，在古新世/始新世最热期开始时，深海洋环流模式从南半球翻转流变为北半球翻转流。这种深层水形成的位置的转变持续了至少40000年，但最终恢复到原来的环流模式。这些结果证实了气候模型的推断，即深海环流的转变将提供深海相对温暖的水，从而产生进一步升温。因此，温室条件下可以在不到几千年里引起突然的深海环流的变化，但可能有持久的影响；这一事件需要10万年来恢复背景条件。

33. Palike H, Norris R D, Herrle J O, et al. The heartbeat of the oligocene climate system[J]. Science, 2006, 314(5807): 1894-1898.

The heartbeat of the oligocene climate system

Palike H., Norris R. D., Herrle J. O., Wilson P. A., Coxall H. K., Lear C. H., Shackleton N. J., Tripati A. K., Wade B. S.

(Department of Earth Sciences, University of Cambridge, Cambridge, England, United Kingdom)

Abstract: A 13-million-year continuous record of Oligocene climate from the equatorial Pacific reveals a pronounced "heartbeat" in the global carbon cycle and periodicity of glaciations. This heartbeat consists of 405000-127000 and 96000-year eccentricity cycles and 1. 2-million-year obliquity cycles in periodically recurring glacial and carbon cycle events. That climate system response to intricate orbital variations suggests a fundamental interaction of the carbon cycle, solar forcing, and glacial events. Box modeling shows that the interaction of the carbon cycle and solar forcing modulates deep ocean acidity as well as the production and burial of global biomass. The pronounced 405000-year eccentricity cycle is amplified by the long residence time of carbon in the oceans.

渐新世气候系统的“心跳”(周期性变化)

Palike H., Norris R. D., Herrle J. O., Wilson P. A., Coxall H. K., Lear C. H., Shackleton N. J., Tripati A. K., Wade B. S.

(英国剑桥大学地球科学系)

摘要：一个13万年的赤道太平洋渐新世气候的连续记录揭示出在全球碳循环和周期性冰期中有着一个显著的“心跳”。这个心跳包含405000～127000年和96000年的偏心率周期和120万年的黄赤交角周期并伴随着定期冰川和碳循环事件的重现。该气候系统对错综复杂的轨道变化的响应表明碳循环、太阳辐射和冰川活动的基本相互作用。盒模型显示，碳循环和太阳辐射的相互作用调节深海酸度以及全球生物量的生产和降解。显著的405000年的偏心率周期被海洋中的碳长时间停留所放大。

34. Wolff E W, Fischer H, Fundel F, et al. Southern Ocean sea-ice extent, productivity and iron flux over the past eight glacial cycles[J]. Nature, 2006, 440(7083): 491-496.

Southern Ocean sea-ice extent, productivity and iron flux over the past eight glacial cycles

Wolff E. W., Fischer H., Fundel F., Ruth U., Twarloh B., Littot G. C., Mulvaney R., Rothlisberger R., de Angelis M., Boutron C. F., Hansson M., Jonsell U., Hutterli M. A., Lambert F., Kaufmann P., Stauffer B., Stocker T. F., Steffensen J. P., Bigler M., Siggaard-Andersen M. L., Udisti R., Becagli S., Castellano E., Severi M., Wagenbach D., Barbante C., Gabrielli P., Gaspari V.

(British Antarctic Survey, High Cross, Madingley Road, Cambridge, CB3 0ET, UK)

Abstract: Sea ice and dust flux increased greatly in the Southern Ocean during the last glacial period. Palaeorecords provide contradictory evidence about marine productivity in this region, but beyond one glacial cycle, data were sparse. Here we present continuous chemical proxy data spanning the last eight glacial cycles (740000 years) from the Dome C Antarctic ice core. These data constrain winter sea-ice extent in the Indian Ocean, Southern Ocean biogenic productivity and Patagonian climatic conditions. We found that maximum sea-ice extent is closely tied to Antarctic temperature on multi-millennial timescales, but less so on shorter timescales. Biological dimethylsulphide emissions south of the polar front seem to have changed little with climate, suggesting that sulphur compounds were not active in climate regulation. We observe large glacial-interglacial contrasts in iron deposition, which we infer reflects strongly changing Patagonian conditions. During glacial terminations, changes in Patagonia apparently preceded sea-ice reduction, indicating that multiple mechanisms may be responsible for different phases of CO_2 increase during glacial terminations. We observe no changes in internal climatic feedbacks that could have caused the change in amplitude of Antarctic temperature variations observed 440000 years ago.

过去8个冰期旋回中南大洋的海冰范围、生产力和铁通量

Wolff E. W., Fischer H., Fundel F., Ruth U., Twarloh B., Littot G. C., Mulvaney R., Rothlisberger R., de Angelis M., Boutron C. F., Hansson M., Jonsell U., Hutterli M. A., Lambert F., Kaufmann P., Stauffer B., Stocker T. F., Steffensen J. P., Bigler M., Siggaard-Andersen M. L., Udisti R., Becagli S., Castellano E., Severi M., Wagenbach D., Barbante C., Gabrielli P., Gaspari V.

(英国南极调查局)

摘要:在末次冰期,南大洋的海冰和粉尘通量大大增加。古环境记录提供了有关在这一地区的海洋生产力的相互矛盾的证据,但一个冰期旋回以外的数据很少。在这里,我们利用来自南极 Dome C 地区的冰芯获得了连续化学替代性指标数据,横跨了最近的 8 个冰期旋回(740000 年)。这些数据驱动了在印度洋的冬季海冰扩张、南大洋的生物成因的生产力和 Patagonian 的气候条件。我们发现,最大的海冰范围与南极几千年时间尺度上的温度息息相关,但在较短的时间尺度上相关性较差。在极锋以南,生物二甲硫醚二甲硫排放似乎没有随气候变化,这表明,含硫化合物在气候调节中并不活跃。我们观察到铁沉积在冰期和间冰期的巨大反差,从而可以推断太阳反射强烈改变了 Patagonian 的(气候)条件。在冰

期结束期,在 Patagonian 的(气候)变化显然领先于海冰减少,说明冰期结束期的二氧化碳增加的不同阶段可能由多种机制控制着。我们并没有观察到可能导致 440000 年前南极温度变化幅度变化的内部气候反馈。

35. Abram N J,Gagan M K,Liu Z Y,et al. Seasonal characteristics of the Indian Ocean Dipole during the Holocene epoch[J]. Nature,2007,445(7125):299-302.

Seasonal characteristics of the Indian Ocean Dipole during the Holocene epoch

Abram N. J. , Gagan M. K. , Liu Z. Y. , Hantoro W. S. , McCulloch M. T. , Suwargadi B. W.

(Research School of Earth Sciences, The Australian National University, Canberra, Australian Capital Territory 0200, Australia)

Abstract: The Indian Ocean Dipole (IOD)—an oscillatory mode of coupled ocean-atmosphere variability—causes climatic extremes and socio-economic hardship throughout the tropical Indian Ocean region. There is much debate about how the IOD interacts with the El Nino/Southern Oscillation (ENSO) and the Asian monsoon, and recent changes in the historic ENSO-monsoon relationship raise the possibility that the properties of the IOD may also be evolving. Improving our understanding of IOD events and their climatic impacts thus requires the development of records defining IOD activity in different climatic settings, including prehistoric times when ENSO and the Asian monsoon behaved differently from the present day. Here we use coral geochemical records from the equatorial eastern Indian Ocean to reconstruct surface-ocean cooling and drought during individual IOD events over the past similar to 6500 years. We find that IOD events during the middle Holocene were characterized by a longer duration of strong surface ocean cooling, together with droughts that peaked later than those expected by El Nino forcing alone. Climate model simulations suggest that this enhanced cooling and drying was the result of strong cross-equatorial winds driven by the strengthened Asian monsoon of the middle Holocene. These IOD-monsoon connections imply that the socioeconomic impacts of projected future changes in Asian monsoon strength may extend throughout Australasia.

全新世期间印度洋偶极现象的季节性特征

Abram N. J. , Gagan M. K. , Liu Z. Y. , Hantoro W. S. , McCulloch M. T. , Suwargadi B. W.

(澳大利亚国立大学地球科学研究学院)

摘要:印度洋偶极现象(IOD),一种海洋-大气变化耦合振动模式,会导致整个热带印度洋地区极端气候和社会经济困难。争论焦点在于 IOD 与厄尔尼诺/南方涛动(ENSO)和亚洲季风的相互作用,历史时期 ENSO-季风关系在最近的变化中显示 IOD 的性质可能也在变化。要提高我们对 IOD 及其气候影响的理解,需要在不同的气候环境中 IOD 活动的发展记录,包括 ENSO 与亚洲季风和现代不同的史前时代。在这里,我们利用赤道东印度洋珊瑚地球化学记录重建在过去约 6500 年独立 IOD 事件中海洋表面降温和干旱。我们发现,在全新世中期 IOD 事件表现为长时间强烈的海洋表面降温的特点,同时伴随着干旱，其达到高峰比预计单独由厄尔尼诺现象驱动的要迟。气候模型模拟结果表明，这种增强的变冷和干燥是由中全新世亚洲季风驱动的强大的跨赤道风加强的结果。这些IOD-季风联系意味着，预计未来亚洲季风强度变化对社会经济的影响可能扩大到整个大洋洲。

36. Barrows T T, Lehman S J, Fifield L K, et al. Absence of cooling in New Zealand and the adjacent ocean during the Younger Dryas chronozone[J]. Science, 2007, 318(5847): 86-89.

Absence of cooling in New Zealand and the adjacent ocean during the Younger Dryas chronozone

Barrows T. T., Lehman S. J., Fifield L. K., De Deckker P.

(Department of Nuclear Physics, Research School of Physical Sciences and Engineering, Australian National University, Canberra, ACT 0200, Australia)

Abstract: As the climate warmed at the end of the last glacial period, a rapid reversal in temperature, the Younger Dryas (YD) event, briefly returned much of the North Atlantic region to near full-glacial conditions. The event was associated with climate reversals in many other areas of the Northern Hemisphere and also with warming over and near Antarctica. However, the expression of the YD in the mid to low latitudes of the Southern Hemisphere (and the southwest Pacific region in particular) is much more controversial. Here we show that the Waiho Loop advance of the Franz Josef Glacier in New Zealand was not a YD event, as previously thought, and that the adjacent ocean warmed throughout the YD.

新西兰及其周边海域在新仙女木时期没有变冷

Barrows T. T., Lehman S. J., Fifield L. K., De Deckker P.

(澳大利亚国立大学核物理研究所物理科学和工程学院)

摘要：在末次冰期结束时，当气候变暖时，一个温度急剧逆转即新仙女木事件(YD)，将大多数北大西洋地区短暂地带回接近完全冰川化的条件。该事件与北半球许多其他的气候逆转相关联，也与南极洲附近气候变暖相关。然而，YD事件在南半球中低纬度地区(特别是西南太平洋地区)的表现更是有争议的。在这里，我们展示了Waiho循环前进到新西兰的Franz Josef冰川不是一个YD事件，这正如之前认为的那样，而且整个YD时期临近的海洋均变暖了。

37. Donnelly J P, Woodruff J D. Intense hurricane activity over the past 5000 years controlled by El Nino and the West African monsoon[J]. Nature, 2007, 447(7143): 465-468.

Intense hurricane activity over the past 5000 years controlled by El Nino and the West African monsoon

Donnelly J. P., Woodruff J. D.

(Coastal Systems Group, Woods Hole Oceanographic Institution, 360 Woods Hole Road, Woods Hole, Massachusetts 02543, USA)

Abstract: The processes that control the formation, intensity and track of hurricanes are poorly understood. It has been proposed that an increase in sea surface temperatures caused by anthropogenic climate change has led to an increase in the frequency of intense tropical cyclones, but this proposal has been challenged on the basis that the instrumental record is too short and unreliable to reveal trends in intense tropical cyclone activity. Storm-induced deposits preserved in the sediments of coastal lagoons offer the opportunity to study the links between climatic conditions and hurricane activity on longer timescales, because they provide centennial-to millennial-scale records of past hurricane landfalls. Here we present a record of intense hurricane activity in the western North Atlantic Ocean over the past 5000 years based on sediment cores from a Caribbean lagoon that contain coarse-grained deposits associated with intense hurricane landfalls. The record indicates that the frequency of intense hurricane landfalls has varied on centennial to millennial scales over this interval. Comparison of the sediment record with palaeo-climate records indicates that this variability was probably modulated by atmospheric dynamics associated with variations in the El Nino/Southern Oscillation and the strength of the West African monsoon, and suggests that sea surface temperatures as high as at present are not necessary to support intervals of frequent intense hurricanes. To accurately predict changes in intense hurricane activity, it is therefore important to understand how the El Nino/Southern Oscillation and the West African monsoon will respond to future climate change.

过去5000年受厄尔尼诺和非洲季风控制的强飓风活动

Donnelly J. P., Woodruff J. D.

（美国伍兹霍尔海洋研究所）

摘要：人们对控制飓风的形成、强度和跟踪的过程知之甚少。有人曾提出，人为造成的气候变化引起的海洋表面温度的增加已导致强热带气旋的频率增加，但这个看法被挑战，基于在揭示在强烈的热带气旋活动的趋势中仪器记录太短且不可靠。在沿海潟湖沉积物中保存的风暴引起的沉积物提供了研究更长时间尺度上的气候条件和飓风活动之间联系的机会，因为它们提供了百年至千年尺度的飓风登陆记录。在这里，我们报道了一个来自加勒比海的潟湖、含有强飓风登陆相关的粗粒沉积、提供了在过去的5000年在北大西洋西部的强飓风活动的记录。记录表明，在这个区间，飓风登陆的频率存在上百年至千年尺度变化。沉积记录与古气候记录的对比表明，这种变化可能为厄尔尼诺/南方涛动和西非季风强度的变化相关的大气动力学所调节，同时也指出，同现代一样高海表面温度并非是支持频繁的强飓风间隔的必要条件。为了准确地预测强飓风活动的变化，重要的是了解厄尔尼诺/南方涛动和西非季风将如何对未来的气候变化做出响应。

38. Duplessy J C, Roche D M, Kageyama M. The deep ocean during the last interglacial period[J].

Science,2007,316(5821):89-91.

The deep ocean during the last interglacial period

Duplessy J. C., Roche D. M., Kageyama M.

(Laboratoire des Sciences du Climat et l'Environnement, Gif, Île-de-France, France)

Abstract: Oxygen isotope analysis of benthic foraminifera in deep sea cores from the Atlantic and Southern Oceans shows that during the last interglacial period, North Atlantic Deep Water (NADW) was 0.4 degrees ±0.2 degrees C warmer than today, whereas Antarctic Bottom Water temperatures were unchanged. Model simulations show that this distribution of deep water temperatures can be explained as a response of the ocean to forcing by high-latitude insolation. The warming of NADW was transferred to the Circumpolar Deep Water, providing additional heat around Antarctica, which may have been responsible for partial melting of the West Antarctic Ice Sheet.

末次间冰期的深海

Duplessy J. C., Roche D. M., Kageyama M.

(法国凡尔赛大学)

摘要:对大西洋和南大洋深海沉积柱底栖有孔虫氧同位素分析显示,在末次间冰期,北大西洋深层水(NADW)比今天暖和0.4±0.2℃,而南极底层水的温度没有变化。模型模拟结果表明,这种深水温度分布可以解释为海洋对由高纬度地区的日照驱动力的响应。NADW变暖被转移到绕极地的深水,为南极洲周围提供额外的热量,这可能导致西南极冰盖的部分融化。

39. Garabato A C N, Stevens D P, Watson A J, et al. Short-circuiting of the overturning circulation in the Antarctic Circumpolar Current[J]. Nature, 2007, 447(7141):194-197.

Short-circuiting of the overturning circulation in the Antarctic Circumpolar Current

Garabato A. C. N., Stevens D. P., Watson A. J., Roether W.

(University of Southampton)

Abstract: The oceanic overturning circulation has a central role in the Earth's climate system and in biogeochemical cycling, as it transports heat, carbon and nutrients around the globe and regulates their storage in the deep ocean. Mixing processes in the Antarctic Circumpolar Current are key to this circulation, because they control the rate at which water sinking at high latitudes returns to the surface in the Southern Ocean. Yet estimates of the rates of these processes and of the upwelling that they induce are poorly constrained by observations. Here we take advantage of a natural tracer-release experiment—an injection of mantle helium from hydrothermal vents into the Circumpolar Current near Drake Passage—to measure the rates of mixing and upwelling in the current's intermediate layers over a sector that spans nearly one-tenth of its circumpolar path. Dispersion of the tracer reveals rapid upwelling along density surfaces and intense mixing across density surfaces, both occurring at rates that are an order of magnitude greater than rates implicit in models of the average Southern Ocean overturning. These findings support the view that deep-water pathways along and across density surfaces intensify and intertwine as the Antarctic Circumpolar Current flows over complex ocean-floor topography, giving rise to a short circuit of the overturning circulation in these regions.

南极绕极流中翻转环流的短路(变短)

Garabato A. C. N., Stevens D. P., Watson A. J., Roether W.

(英国南安普顿大学)

摘要：大洋翻转环流在地球气候系统和生物地球化学循环中起核心作用，因为它绕地球传输热、碳和养分并调节其在深海的存储。在南极绕极流的混合过程是这种循环的关键，因为它们控制高纬度地区返回南大洋的表面水下沉的速率。然而，观测结果不能很好地对这些过程速率和引发的上涌进行估计。下面我们就利用一个天然示踪剂-释放实验——注射来自深海热泉的地幔氦到 Drake 海峡附近的极地环流——来测量跨越约十分之一极地环流的中间层的混合和上涌的速率。示踪剂的分布揭示沿密度表面的快速上涌和跨密度表面的激烈混合，两者发生率比南大洋翻转模型所隐含的速率大一个数量级。这些发现支持了沿着或穿越密度表面的深海流会加强并缠绕南极环流流过非常复杂的洋底地形，从而引发了在这些地区翻转环流的短路(变短)的观点。

40. Matthews A J, Singhruck P K, Heywood K J. Deep ocean impact of a Madden-Julian Oscillation observed by Argo floats Madden-Julian[J]. Science, 2007, 318(5857): 1765-1769.

Deep ocean impact of a Madden-Julian Oscillation observed by Argo floats Madden-Julian

Matthews A. J., Singhruck P. K., Heywood K. J.

(School of Environment Science, University of East Anglia, Norwich NR4 7TJ, UK)

Abstract: Using the new Argo array of profiling floats that gives unprecedented space-time coverage of the upper 2000 meters of the global ocean, we present definitive evidence of a deep tropical ocean component of the Madden-Julian Oscillation (MJO). The surface wind stress anomalies associated with the MJO force eastward-propagating oceanic equatorial Kelvin waves that extend downward to 1500 meters. The amplitude of the deep ocean anomalies is up to six times the amplitude of the observed annual cycle. This deep ocean sink of energy input from the wind is potentially important for understanding phenomena such as El Nino-Southern Oscillation and for interpreting deep ocean measurements made from ships.

海洋实时观测系统观测的季节内震荡的深海效应

Matthews A. J. , Singhruck P. K. , Heywood K. J.

（英国东英吉利大学环境科学学院）

摘要：新海洋实时观测系统的阵列前所未有地覆盖了全球海洋上层2000米的时空，我们通过使用这个阵列，给出了Madden-Julian震荡(MJO)的一种热带深海洋组成部分的确切证据。表面风应力异常与MJO驱动的向东传播的海洋赤道Kelvin波向下延伸到1500米相关。深海异常的幅度是所观察到的年度周期振幅的6倍。这种风输入的深海能量汇对ENSO现象的理解和解释深海船舶测量数据具有潜在重要性。

41. Partin J W, Cobb K M, Adkins J F, et al. Millennial-scale trends in west Pacific warm pool hydrology since the Last Glacial Maximum[J]. Nature, 2007, 449(7161): 452-453.

Millennial-scale trends in west Pacific warm pool hydrology since the Last Glacial Maximum

Partin J. W. , Cobb K. M. , Adkins J. F. , Clark B. , Fernandez D. P.

(School of Earth and Atmospheric Sciences, Georgia Institute of Technology, Atlanta, Georgia 30332, USA)

Abstract: Models and palaeoclimate data suggest that the tropical Pacific climate system plays a key part in the mechanisms underlying orbital-scale and abrupt climate change. Atmospheric convection over the western tropical Pacific is a major source of heat and moisture to extratropical regions, and may therefore influence the global climate response to a variety of forcing factors. The response of tropical Pacific convection to changes in global climate boundary conditions, abrupt climate changes and radiative forcing remains uncertain, however. Here we present three absolutely dated oxygen isotope records from stalagmites in northern Borneo that reflect changes in west Pacific warm pool hydrology over the past 27000 years. Our results suggest that convection over the western tropical Pacific weakened 18000-20000 years ago, as tropical Pacific and Antarctic temperatures began to rise during the early stages of deglaciation. Convective activity, as inferred from oxygen isotopes, reached a minimum during Heinrich event 1, when the Atlantic meridional overturning circulation was weak, pointing to feed-backs between the strength of the overturning circulation and tropical Pacific hydrology. There is no evidence of the Younger Dryas event in the stalagmite records, however, suggesting that different mechanisms operated during these two abrupt deglacial climate events. During the Holocene epoch, convective activity appears to track changes in spring and autumn insolation, highlighting the sensitivity of tropical Pacific convection to external radiative forcing. Together, these findings demonstrate that the tropical Pacific hydrological cycle is sensitive to high-latitude climate processes in both hemispheres, as well as to external radiative forcing, and that it may have a central role in abrupt climate change events.

末次冰盛期以来西太平洋暖池水文(水汽)千年尺度变化趋势

Partin J. W., Cobb K. M., Adkins J. F., Clark B., Fernandez D. P.

(美国佐治亚理工学院地球和大气科学学院)

摘要:模型和古气候资料表明,热带太平洋气候系统在轨道尺度和突然气候变化的机制中起着关键作用。在热带西太平洋的大气对流是热带以外(温带)地区热量和水分的主要来源,因此可能会影响全球气候对各种驱动因子的响应。但是,热带太平洋对流对全球气候变化的边界条件、气候突变和辐射驱动的响应仍然不明朗。在这里,我们报道三份在婆罗洲北部绝对年代的石笋氧同位素记录,反映了过去27000年西太平洋暖池的水文变化。我们的研究结果表明,当在末次间冰期的早期阶段热带太平洋和南极气温开始回升,热带西太平洋对流在18000～20000年前减弱。从氧同位素推断对流活动在Heinrich事件1达到最小,同时大西洋经向翻转环流减弱,表明翻转环流和热带太平洋水汽的强度反馈。然而在石笋记录中并没有新仙女木事件的证据,这表明了两个间冰期气候突变事件中由不同的机制在驱动。在全新世期间,对流活动似乎伴随着春季和秋季的日照变化,突出了热带太平洋对流对外部辐射驱动力的敏感性。总之,这些研究结果表明,热带太平洋水文循环对两个半球高纬度的气候过程和外部辐射驱动是敏感的,它可能在气候突变事件中起到核心作用。

42. Stott L, Timmermann A, Thunell R. Southern hemisphere and deep-sea warming led deglacial atmospheric CO_2 rise and tropical warming[J]. Science, 2007, 318(5849): 435-438.

Southern hemisphere and deep-sea warming led deglacial atmospheric CO_2 rise and tropical warming

Stott L., Timmermann A., Thunell R.

(Department of Earth Science, University of Southern California Los Angeles, CA90089, USA)

Abstract: Establishing what caused Earth's largest climatic changes in the past requires a precise knowledge of both the forcing and the regional responses. We determined the chronology of high- and low-latitude climate change at the last glacial termination by radiocarbon dating benthic and planktonic foraminiferal stable isotope and magnesium/calcium records from a marine core collected in the western tropical Pacific. Deep-sea temperatures warmed by similar to 2 degrees C between 19 and 17 thousands before the present (ky B. P.), leading the rise in atmospheric CO_2 and tropical-surface-ocean warming by similar to 1000s. The cause of this deglacial deep-water warming does not lie within the tropics, nor can its early onset between 19 and 17 ky B. P. be attributed to CO_2 forcing. Increasing austral-spring insolation combined with sea-ice albedo feedbacks appear to be the key factors responsible for this warming.

南半球和深海暖化导致间冰期大气CO_2上升和热带变暖

Stott L. , Timmermann A. , Thunell R.

（美国南加州大学地球科学系）

摘要：确定是什么原因造成过去地球上最大的气候变化需要驱动力和区域响应的精确知识。我们通过研究西热带太平洋的一根海洋沉积柱的底栖和浮游有孔虫的稳定同位素的放射性碳定年与 Mg/Ca 记录来确定末次冰期结束时高低纬度气候变化的年代学记录。深海温度在距今 19000～17000 年间变暖了约 2 ℃，导致约 1000 年的大气二氧化碳上升和热带海洋表面变暖。这种间冰期深水变暖的原因不在于在热带，也不能将其在距今 19000～17000 年间的早期发展归咎于二氧化碳的驱动力。而将南半球春天日照的增加与海冰反射率的反馈结合，似乎是导致这种变暖的关键因素。

43. Wolff E W, Fischer H, Fundel F, et al. Southern Ocean sea-ice extent, productivity and iron flux over the past eight glacial cycles[J]. Nature, 2007, 449(7163): 748.

Southern Ocean sea-ice extent, productivity and iron flux over the past eight glacial cycles

Wolff E. W. , Fischer H. , Fundel F. , Ruth U. , Twarloh B. , Littot G. C. , Mulvaney R. , Rothlisberger R. , de Angelis M. , Boutron C. F. , Hansson M. , Jonsell U. , Hutterli M. A. , Lambert F. , Kaufmann P. , Stauffer B. , Stocker T. F. , Steffensen J. P. , Bigler M. , Siggaard-Andersen M. L. , Udisti R. , Becagli S. , Castellano E. , Severi M. , Wagenbach D. , Barbante C. , Gabrielli P. , Gaspari V.

(British Antarctic Survey, High Cross, Madingley Road, Cambridge, CB3 0ET, UK)

Abstract: Sea ice and dust flux increased greatly in the Southern Ocean during the last glacial period. Palaeorecords provide contradictory evidence about marine productivity in this region, but beyond one glacial cycle, data were sparse. Here we present continuous chemical proxy data spanning the last eight glacial cycles (740000 years) from the Dome C Antarctic ice core. These data constrain winter sea-ice extent in the Indian Ocean, Southern Ocean biogenic productivity and Patagonian climatic conditions. We found that maximum sea-ice extent is closely tied to Antarctic temperature on multi-millennial timescales, but less so on shorter timescales. Biological dimethylsulphide emissions south of the polar front seem to have changed little with climate, suggesting that sulphur compounds were not active in climate regulation. We observe large glacial-interglacial contrasts in iron deposition, which we infer reflects strongly changing Patagonian conditions. During glacial terminations, changes in Patagonia apparently preceded sea-ice reduction, indicating that multiple mechanisms may be responsible for different phases of CO_2 increase during glacial terminations. We observe no changes in internal climatic feedbacks that could have caused the change in amplitude of Antarctic temperature variations observed 440000 years ago.

在过去八个冰期旋回中南大洋的海冰范围、生产力和铁通量

Wolff E. W., Fischer H., Fundel F., Ruth U., Twarloh B., Littot G. C., Mulvaney R., Rothlisberger R., de Angelis M., Boutron C. F., Hansson M., Jonsell U., Hutterli M. A., Lambert F., Kaufmann P., Stauffer B., Stocker T. F., Steffensen J. P., Bigler M., Siggaard-Andersen M. L., Udisti R., Becagli S., Castellano E., Severi M., Wagenbach D., Barbante C., Gabrielli P., Gaspari V.

(英国南极调查局)

摘要:末次冰期的时候,海冰和灰尘输入量在南大洋都出现了增加。古气候记录对于这一时期生产力的变化存在着争议。而且在末次冰期以前的冰期旋回中,这方面的数据还很缺失。在本文中,我们从Dome C冰芯中做了过去八个冰期旋回的化学替代性指标。这些数据包括了冬季印度洋的海冰面积、南大洋的生产力以及南美的气候条件等。我们发现海冰在大尺度上与南极温度有关,但小尺度不一定。而生物生产力跟气候的关系则不是很明确。在极锋以南,生物二甲硫醚二甲硫的排放似乎没有随气候变化,这表明,含硫化合物在气候调节中并不活跃。我们观察到铁沉积在冰期和间冰期的巨大反差,从而可以推断太阳反射强烈改变了Patagonian的气候条件。在冰期结束期,在Patagonian的气候变化显然领先于海冰减少,说明冰期结束期的二氧化碳增加的不同阶段可能由多种机制控制着。我们并没有观察到可能导致440000年前南极温度幅度变化的内部气候反馈。

3.3 摘要翻译——北极

1. Walsh Je. Climate change: the elusive Arctic warming[J]. Nature, 1993, 361(6410): 300-301.

Climate change—the elusive Arctic warming

Walsh Je

(Department of Atmospheric Sciences, University of Illinois American, Vrbara, Illinois 61801, USA)

Abstract: Contrary to the greenhouse-induced climate predictions of global climate models, temperature data obtained from drifting ice stations and aircraft over the Arctic Ocean yield no evidence of surface warming in that region over the past four decades (Kahl et al., 1993). This paper shows that, when viewed in a broader context, these findings are not necessarily at odds with climate-model projections; for instance, it is noted the drifting ice stations from which the balloon-borne radio sondes for temperature measurements were launched, do not sample the potentially sensitive marginal ice zone where the ice retreat generally results in the greatest warming in CO_2-doubling experiments of climate models.

气候变化——不易确定的北极变暖

Walsh Je

(美国伊利诺大学大气科学系)

摘要:与气候模型预报的温室气体诱发变暖相反,流动冰站和航空观测得到的温度数据显示在过去的40年里北极地区并没有出现明显的增暖(Kahl et al.,1993)。但是在更为广阔的角度上来讲,这一现象未必能反驳气候模型预测的结果。例如,采用气球运载的流动冰站观测台并没有在冰盖边缘这一对气温变化敏感的区域进行温度测试,而两倍 CO_2 气候模型实验表明该处的冰退缩正是由此导致的。

2. Stein R, Nam S I, Schubert C, et al. The last deglaciation event in the eastern central Arctic-Ocean [J]. Science, 1994, 264(5159): 692-696.

The last deglaciation event in the eastern central Arctic-Ocean

Stein R., Nam S. I., Schubert C., Vogt C., Futterer D., Heinemeier J.

(Alfred Wegener Institute Helmholtz Centre for Pola and Marine Research Bremerhaven, Germany)

Abstract: Oxygen isotope records of cores from the central Arctic Ocean yield evidence for a major influx of meltwater at the beginning of the last deglaciation 15.7 thousand years ago (16650 calendar years B. C.). The almost parallel trends of the isotope records from the Arctic Ocean, the Fram Strait, and the east Greenland continental margin suggest contemporaneous variations of the Eurasian Arctic and Greenland (Laurentide) ice sheets or increased export of low-saline waters from the Arctic within the East Greenland Current during the last deglaciation. On the basis of isotope and carbon data, the modern surface and deep-water characteristics and seasonally open-ice conditions with increased surface-water productivity were established in the central Arctic at the end of Termination Ib (about 7.2 thousand years ago or 6000 calendar years B. C.).

北冰洋东部核心区域的末次冰消期事件

Stein R., Nam S. I., Schubert C., Vogt C., Futterer D., Heinemeier J.

(德国阿尔弗雷德魏格纳极地海洋研究所)

摘要:来自北冰洋东部核心区域的沉积柱氧同位素的记录显示,在末次冰消期开始的时候,大约15.7 ka以前,这一区域曾有大量的淡水注入。来自北冰洋、弗莱曼海峡以及东格陵兰陆地边缘的记录表现出一致的变化趋势,说明这末次冰消期期间,欧亚大陆冰盖、北极冰川以及格陵兰冰盖是同步变化的。根据同位素和碳的数据,现代表层-深层水特征和季节性开放冰原条件是大约在距今7200年左右建立起来的。

3. Overpeck J, Hughen K, Hardy D, et al. Arctic environmental change of the last four centuries[J]. Science, 1997, 278(5341): 1251-1256.

Arctic environmental change of the last four centuries

Overpeck J., Hughen K., Hardy D., Bradley R., Case R., Douglas M.,
Finney B., Gajewski K., Jacoby G., Jennings A., Lamoureux S., Lasca A.,
MacDonald G., Moore J., Retelle M., Smith S., Wolfe A., Zielinski G.

(NOAA National Geophysical Date Center(NGDC) Paleoclimatology Program, 325 Broadway, Boulder, Co80303, USA)

Abstract: A compilation of paleoclimate records from lake sediments, trees, glaciers, and marine sediments provides a view of circum-Arctic environmental variability over the last 400 years. From 1840 to the mid-20th century, the Arctic warmed to the highest temperatures in four centuries. This warming ended the Little Ice Age in the Arctic and has caused retreats of glaciers, melting of permafrost and sea ice, and alteration of terrestrial and lake ecosystems. Although warming, particularly after 1920, was likely caused by increases in atmospheric trace gases, the initiation of the warming in the mid-19th century suggests that increased solar irradiance, decreased volcanic activity, and feedbacks internal to the climate system played roles.

过去四个世纪里北极气候环境变化

Overpeck J., Hughen K., Hardy D., Bradley R., Case R., Douglas M., Finney B., Gajewski K., Jacoby G., Jennings A., Lamoureux S., Lasca A., MacDonald G., Moore J., Retelle M., Smith S., Wolfe A., Zielinski G.

(美国 NOAA-NGDC 古气候研究项目组)

摘要:通过对古气候记录如湖泊沉积物、树轮、冰川和海洋沉积物的分析,获得了过去400年环北极地区的环境变化。从1840年到20世纪中期,北极变暖并达到了过去4个世纪的最高温度。这一变暖终止了环北极地区的小冰期,并导致了冰盖的消退、永久冻土和海冰的融化以及陆地和湖泊生态系统的改变。虽然这一阶段的变暖,尤其是1920年之后的变暖,很有可能是由大气温室气体所引起的,但是19世纪中期的变暖则表明太阳辐射增加、火山活动下降以及气候系统内部的反馈作用也在气候变化中起到了重要作用。

4. Bacon S. Decadal variability in the outflow from the Nordic seas to the deep Atlantic Ocean[J]. Nature,1998,394(6696):871-874.

Decadal variability in the outflow from the Nordic seas to the deep Atlantic Ocean

Bacon S.

(National Oceanography Centre, Southampton, Southampton, England, United Kingdom)

Abstract: The global thermohaline circulation is the oceanic overturning mode, which is manifested in the North Atlantic Ocean as northward-flowing surface waters which sink in the Nordic (Greenland, Iceland and Norwegian) seas and return southwards-after overflowing the Greenland-Scotland ridge-as deep water. This process has been termed the conveyor belt: and is believed to keep Europe 5-8 degrees C warmer than it would be if the conveyor were to shut down. The variability of today's conveyor belt is therefore an important component of climate regulation. The Nordic seas are the only Northern Hemisphere source of deep water and a previous study has revealed no long-term variability in the outflow of deep water from the Nordic seas to the Atlantic Ocean. Here I use flows derived from hydrographic data to show that this outflow has approximately doubled, and then returned to previous values, over the past four decades. I present evidence which suggests that this variability is forced by variability in polar air temperature, which in turn may be connected to the recently reported Arctic warming.

从诺迪克海到大西洋深水的外流的十年际变化

Bacon S.

(英国南安普顿海洋研究中心)

摘要:全球大洋温盐环流是一个翻转流,在北大西洋它以向北表层流的形式到达格陵兰-挪威等地的诺迪克海,然后再下沉为深部流向南运行。这一过程被叫作海洋传送带,这一传送带的存在使得欧洲温度比没有传送带的情况下高出 5～8 ℃。因此,当今传送带的变化被认为是气候模式的重要组成部分。诺迪克海是北半球唯一的深水源,先前的研究表明从诺迪克海到大西洋的深水外流没有长期变化。这里,我使用水文学证明这一深水流在过去的四十年里曾经增大过一倍,然后又恢复到了开始的状态。这篇文章还列出了一些证据证明这一过程主要受极地温度的影响。

5. Driscoll N W, Haug G H. A short circuit in thermohaline circulation: a cause for Northern Hemisphere glaciation? [J]. Science, 1998, 282(5388): 436-438.

A short circuit in thermohaline circulation: a cause for Northern Hemisphere glaciation?

Driscoll N. W., Haug G. H.

(Woods Hole Oceanographic Institution, Woods Hole, MA 02543, USA)

Abstract: The cause of Northern Hemisphere glaciation about 3 million years ago remains uncertain. Closing the Panamanian Isthmus increased thermohaline circulation and enhanced moisture supply to high Latitudes, but the accompanying heat would have inhibited ice growth. One possible solution is that enhanced moisture transported to Eurasia also enhanced freshwater delivery to the Arctic via Siberian rivers. Freshwater input to the Arctic would facilitate sea ice formation, increase the albedo, and isolate the high heat capacity of the ocean from the atmosphere. It would also act as a negative feedback on the efficiency of the "conveyor belt" heat pump.

温盐环流短路:北半球冰期的原因?

Driscoll N. W., Haug G. H.

(美国伍兹霍尔海洋研究所)

摘要:300 万年前北半球冰期的原因至今仍然不清楚。巴拿马海道的关闭使得温盐环流增强,增加了向高纬度地区输送的水汽,但是同时输送的热量会阻止冰盖的扩张。一种可能的方式就是,增加的水汽输送到了欧亚大陆,导致通过西伯利亚河流注入北冰洋的淡水增加,淡水的增加可以促进海冰的发展,增加反射率,隔离了大气来的热量。这一过程同时还可以导致海洋"传送带"发生负反馈过程。

6. Hu F S, Slawinski D, Wright H E, et al. Abrupt changes in North American climate during early Holocene times[J]. Nature, 1999, 400(6743): 437-440.

Abrupt changes in North American climate during early Holocene times

Hu F. S., Slawinski D., Wright H. E., Ito E., Johnson R. G., Kelts K. R., McEwan R. F., Boedigheimer A.

(Department of Geology, University of Illinois, Urbana-Champaign, Urbana, Illinois, United States)

Abstract: Recent studies of the Greenland ice cores have offered many insights into Holocene climatic dynamics at decadal to century timescales. Despite the abundance of continental records of Holocene climate, few have sufficient chronological control and sampling resolution to compare with the Greenland findings. But annually laminated sediments (varves) from lakes can provide high-resolution continental palaeoclimate data with secure chronologies. Here we present analyses of varved sediments from Deep Lake in Minnesota, USA. Trends in the stable oxygen-isotope composition of the sedimentary carbonate indicate a pronounced climate cooling from 8.9 to 8.3 kyr before present, probably characterized by increased outbreaks of polar air, decreased precipitation temperatures, and a higher fraction of the annual precipitation falling as snow. The abrupt onset of this climate reversal, over several decades, was probably caused by a reorganization of atmospheric circulation and cooling of the Arctic air mass in summer that resulted from the final collapse of the Laurentide ice near Hudson Bay and the discharge of icebergs from the Quebec and Keewatin centres into the Tyrell Sea. The timing and duration of this climate reversal suggest that it is distinct from the prominent widespread cold snap that occurred 8200 years ago in Greenland and other regions. No shifts in the oxygen-isotope composition of sediment carbonate occurred at 8.2 kyr before present at Deep Lake, but varve thickness increased dramatically, probably as a result of increased deposition of aeolian dust. Taken together, our data suggest that two separate regional-scale climate reversals occurred between 9000 and 8000 years ago, and that they were driven by different mechanisms.

早全新世期间北美快速气候变化

Hu F. S., Slawinski D., Wright H. E., Ito E., Johnson R. G., Kelts K. R., McEwan R. F., Boedigheimer A.

(美国伊利诺伊大学香槟分校地质系)

摘要：最近有关格陵兰冰芯的研究结果对全新世年代际-世纪尺度气候变化提出了很多深刻的见解。虽然陆地上也有很多全新世的气候记录，但是很少有记录能够在年代的准确度和分辨率上跟格陵兰冰芯进行对比。具有纹层的湖泊沉积物有可能提供高分辨率而且年代准确的陆地气候记录。这里我们报道一个采自美国明尼苏达的高分辨率湖泊沉积物。碳酸盐氧同位素记录显示在距今 8.9 千年～8.3 千年的时候这里曾经出现了一次明显降温事件，主要特征为极地来的冷空气加强，降雨温度下降，降雪量增加。这次突然的气候变化有可能是由大气环流改变和北极夏季降温(可能主要由于北极附近哈德森湾冰盖坍塌造成的)引起的。这次快速发生，持续几十年的冷事件与发生在格陵兰和其他地区著名的 8200 年事件是截然不同的。这个沉积柱碳酸盐的氧同位素在 8.2 kyr 时没有发生跳跃性变化，但是纹层的厚度在这个时间段出现了明显的增加，可能是这个时间段风尘沉积增加造成的。总的来说，我们的数据显示了发生在距今 9000～8000 年之间的两次区域尺度的气候反转，并且它们的机制是不一样的。

7. Khodri M, Leclainche Y, Ramstein G, et al. Simulating the amplification of orbital forcing by ocean feedbacks in the last glaciation[J]. Nature, 2001, 410(6828): 570-574.

Simulating the amplification of orbital forcing by ocean feedbacks in the last glaciation

Khodri M., Leclainche Y., Ramstein G., Braconnot P., Marti O., Cortijo E.

(LSCE, CF-Saclay, 91191, Gif sur Yvette, France)

Abstract: According to Milankovitch theory, the lower summer insolation at high latitudes about 115000 years ago allowed winter snow to persist throughout summer, leading to ice-sheet build-up and glaciation. But attempts to simulate the last glaciation using global atmospheric models have failed to produce this outcome when forced by insolation changes only. These results point towards the importance of feedback effects-for example, through changes in vegetation or the ocean circulation-for the amplification of solar forcing. Here we present a fully coupled ocean-atmosphere model of the last glaciation that produces a build-up of perennial snow cover at known locations of ice sheets during this period. We show that ocean feedbacks lead to a cooling of the high northern latitudes, along with an increase in atmospheric moisture transport from the Equator to the poles. These changes agree with available geological data and, together, they lead to an increased delivery of snow to high northern latitudes, The mechanism we present explains the onset of glaciation—which would be amplified by changes in vegetation-in response to weak orbital forcing.

模拟末次冰期辐射强迫-海洋反馈的放大作用

Khodri M., Leclainche Y., Ramstein G., Braconnot P., Marti O., Cortijo E.

(法国高等科学研究所)

摘要:根据米兰科维奇理论,距今115000年前北半球高纬度太阳辐射量较低,这一结果使得北半球冬季的冰雪可以度过整个夏季而不融化,导致陆地冰盖的发展和冰期的到来。但是,如果仅仅依靠太阳辐射的变化的话,大气环流模型很难模拟出这样的过程。这一结果说明反馈,比如植被和海洋环流变化,在放大太阳辐射的过程中起了重要的作用。在这篇文章中,我们利用耦合大气-海洋模型模拟出了冰期陆地冰雪覆盖的情况。我们的结果显示,海洋反馈导致了北半球高纬度的变冷,并伴随有较多的水汽从热带输送到极地。这一变化与已有的地质证据相吻合,表明更多的降雪被输送到了高纬度地区。我们的机制解释了冰期的爆发——陆地植被对弱太阳辐射的响应可以放大这一过程。

8. Thompson D W J, Wallace J M. Regional climate impacts of the Northern Hemisphere annular mode [J]. Science, 2001, 293(5527): 85-89.

Regional climate impacts of the Northern Hemisphere annular mode

Thompson D. W. J., Wallace J. M.

(Department of Atmospheric Science, Colorado State University, Fort Collins, CO 80523, USA)

Abstract: The Northern Hemisphere annular mode (NAM) (also known as the North Atlantic

Oscillation) is shown to exert a strong influence on wintertime climate, not only over the Euro-Atlantic half of the hemisphere as documented in previous studies, but over the Pacific half as well. It affects not only the mean conditions, but also the day-to-day variability, modulating the intensity of mid-latitude storms and the frequency of occurrence of high-latitude blocking and cold air outbreaks throughout the hemisphere. The recent trend in the NAM toward its high-index polarity with stronger subpolar westerlies has tended to reduce the severity of winter weather over most middle- and high-latitude Northern Hemisphere continental legions.

北半球环流模式对区域气候的影响

Thompson D. W. J., Wallace J. M.

(美国科罗拉多大学大气科学系)

摘要:北半球大气环流模式如北大西洋涛动被证明对冬季气候有重要影响,不仅仅是对欧亚-大西洋地区,对太平洋地区也同样重要。它不仅影响气候的平均态,对于天-天之间的变化同样如此。而且还能影响中纬度风暴、高纬度反气旋以及冷空气的爆发等。最近北半球大气环流模式趋向于高指数,对应强的北半球西风,这将减弱北半球大部分中高纬度地区的冬季严重程度。

9. Adkins J F, McIntyre K, Schrag D P. The salinity, temperature, and delta ^{18}O of the glacial deep ocean[J]. Science, 2002, 298(5599): 1769-1773.

The salinity, temperature, and delta ^{18}O of the glacial deep ocean

Adkins J. F., McIntyre K., Schrag D. P.

(MS 100-23, Department of Geological and Planetary Sciences, California Institute of Technology, Pasadena, CA 91125, USA)

Abstract: We use pore fluid measurements of the chloride concentration and the oxygen isotopic composition from Ocean Drilling Program cores to reconstruct salinity and temperature of the deep ocean during the Last Glacial Maximum (LGM). Our data show that the temperatures of the deep Pacific, Southern, and Atlantic oceans during the LGM were relatively homogeneous and within error of the freezing point of seawater at the ocean's surface. Our chloride data show that the glacial stratication was dominated by salinity variations, in contrast with the modern ocean, for which temperature plays a primary role. During the LGM the Southern Ocean contained the saltiest water in the deep ocean. This reversal of the modern salinity contrast between the North and South Atlantic implies that the freshwater budget at the poles must have been quite different. A strict conversion of mean salinity at the LGM to equivalent sea-level change yields a value in excess of 140 meters. However, the storage of fresh water in ice shelves and/or groundwater reserves implies that glacial salinity is a poor predictor of mean sea level.

冰期深海的盐度、温度和氧同位素

Adkins J. F. , McIntyre K. , Schrag D. P.

(美国加州理工学院地球与行星科学系)

摘要:我们对大洋深钻计划的深海沉积柱(ODP)的氯浓度进行孔流体测定,并对其进行氧同位素分析,重建末次冰期最盛期的深海温度和盐度。我们的结果显示在冰盛期的太平洋、南大洋和大西洋的温带相对均匀,都接近于表层水的冰点。对氯化物的分析显示冰期深海氯离子主要受盐度变化控制,与现代海洋的温度控制相反。在末次冰盛期,南大洋深海水盐度最高。这与现代的北大西洋和南大西洋盐度特征也是相反的,说明当时极地附近淡水输入跟现代也不一致。通过盐度计算,我们估计在末次冰盛期的时候海平面变化了大约 140 m。但是,由于冰期时冰架和地下水同样储存了大量的淡水,使得利用盐度估计海平面变化显得不那么可靠。

10. Clark P U,Pisias N G,Stocker T F,et al. The role of the thermohaline circulation in abrupt climate change[J]. Nature,2002,415(6874):863-869.

The role of the thermohaline circulation in abrupt climate change

Clark P. U. , Pisias N. G. , Stocker T. F. , Weaver A. J.

(Department of Geosciences, Oregon State University, Corvallis, Oregon 97331, USA)

Abstract: The possibility of a reduced Atlantic thermohaline circulation in response to increases in greenhouse-gas concentrations has been demonstrated in a number of simulations with general circulation models of the coupled ocean-atmosphere system. But it remains difficult to assess the likelihood of future changes in the thermohaline circulation, mainly owing to poorly constrained model parameterizations and uncertainties in the response of the climate system to greenhouse warming. Analyses of past abrupt climate changes help to solve these problems. Data and models both suggest that abrupt climate change during the last glaciation originated through changes in the Atlantic thermohaline circulation in response to small changes in the hydrological cycle. Atmospheric and oceanic responses to these changes were then transmitted globally through a number of feedbacks. The palaeoclimate data and the model results also indicate that the stability of the thermohaline circulation depends on the mean climate state.

温盐环流在快速气候变化中的作用

Clark P. U. , Pisias N. G. , Stocker T. F. , Weaver A. J.

(美国俄勒冈州立大学地质科学系)

摘要:海气耦合的大气环流模型证明随着大气二氧化碳浓度的增加,大西洋温盐环流可能会出现减弱。但是目前仍然很难评估未来温盐环流可能的变化,主要是由于模型参数不够确定以及气候系统对温室气体响应的不确定性。分析过去的快速气候变化将帮助我们解决这些问题。数据和模型同时表明，末次冰期时的快速气候变化主要起源于水文变化引起的大西洋温盐环流变化。大气和海洋随后会将这些变化通过一系列的反馈转移到全球。古气候数据和模型结果同时还指出温盐环流的稳定性取决于平均气候状态。

11. Moritz R E, Bitz C M, Steig E J. Dynamics of recent climate change in the Arctic[J]. Science, 2002, 297(5586): 1497-1502.

Dynamics of recent climate change in the Arctic

Moritz R. E., Bitz C. M., Steig E. J.

(Polor Science Center, University of Washington Seattle, Seattle, Washington, United States)

Abstract: The pattern of recent surface warming observed in the Arctic exhibits both polar amplification and a strong relation with trends in the Arctic Oscillation mode of atmospheric circulation. Paleoclimate analyses indicate that Arctic surface temperatures were higher during the 20th century than during the preceding few centuries and that polar amplification is a common feature of the past. Paleoclimate evidence for Holocene variations in the Arctic Oscillation is mixed. Current understanding of physical mechanisms controlling atmospheric dynamics suggests that anthropogenic influences could have forced the recent trend in the Arctic Oscillation, but simulations with global climate models do not agree. In most simulations, the trend in the Arctic Oscillation is much weaker than observed. In addition, the simulated warming tends to be largest in autumn over the Arctic Ocean, whereas observed warming appears to be largest in winter and spring over the continents.

北极最近气候变化的动力学机制

Moritz R. E., Bitz C. M., Steig E. J.

(美国华盛顿大学极地科学中心)

摘要：最近北极地表变暖模式观测显示极地有幅度放大的情况，并且这种变化与大气环流的北极涛动模式有关。古气候记录显示北极20世纪的温度比之前的几个世纪要高，并且极地幅度放大的特征也一直存在。北极涛动在全新世的变化目前还不清楚。当前大气动力学的物理机制认为人类影响的强迫效应有可能是最近北极涛动变化的原因，但是全球大气模拟的结果却与此不吻合。在大多数模拟中，北极涛动的变化要比实际观测的弱。另外，模拟的结果认为北极的变暖主要在秋季，但实际上主要发生在冬季和春季。

12. Peterson B J, Holmes R M, McClelland J W, et al. Increasing river discharge to the Arctic Ocean[J]. Science, 2002, 298(5601): 2171-2173.

Increasing river discharge to the Arctic Ocean

Peterson B. J., Holmes R. M., McClelland J. W., Vorosmarty C. J.,
Lammers R. B., Shiklomanov A. I., Shiklomanov I. A., Rahmstorf S.

(The Ecosystems Center, Marine Biological Laboratory, Woods Hole, MA 02543, USA)

Abstract: Synthesis of river-monitoring data reveals that the average annual discharge of fresh water from the six largest Eurasian rivers to the Arctic Ocean increased by 7% from 1936 to 1999. The average annual rate of increase was 2.0±0.7 cubic kilometers per year. Consequently, average annual discharge from the six rivers is now about 128 cubic kilometers per year greater than it was when routine measurements of discharge began. Discharge was correlated with changes in both the North Atlantic Oscillation and global mean surface air temperature. The observed large-scale change in freshwater flux has potentially important implications for ocean circulation and climate.

北冰洋河流输入的增加

Peterson B. J., Holmes R. M., McClelland J. W., Vorosmarty C. J.,
Lammers R. B., Shiklomanov A. I., Shiklomanov I. A., Rahmstorf S.

(美国伍兹霍尔海洋生物实验室)

摘要:综合的河流数据显示欧亚大陆的 6 条最大入北冰洋河流的平均年淡水输入量从 1936~1999 年增加了 7%。平均每年增加速率为 2±0.7 km^3。因此 6 条河流现在的平均年输入量要比最初观测的时候多 128 km^3。这一变化与北大西洋涛动和全球平均温度相关。观测到的大尺度淡水输入变化对大洋环流和气候有着潜在的重要影响。

13. Alley R B, Marotzke J, Nordhaus W D, et al. Abrupt climate change[J]. Science, 2003, 299(5615): 2005-2010.

Abrupt climate change

Alley R. B., Marotzke J., Nordhaus W. D., Overpeck J. T., Peteet D. M., Pielke R. A.,
Pierrehumbert R. T., Rhines P. B., Stocker T. F., Talley L. D., Wallace J. M.

(Department of Geosciences and EMS Environment Institute, Penn state university, University Park, PA16802, USA)

Abstract: Large, abrupt, and widespread climate changes with major impacts have occurred repeatedly in the past, when the Earth system was forced across thresholds. Although abrupt climate changes can occur for many reasons, it is conceivable that human forcing of climate change is increasing the probability of large, abrupt events. Were such an event to recur, the economic and ecological impacts could be large and potentially serious. Unpredictability exhibited near climate thresholds in simple models shows that some uncertainty will always be associated with projections. In light of these uncertainties, policy-makers should consider expanding research into abrupt climate change, improving monitoring systems, and taking actions designed to enhance the adaptability and resilience of ecosystems and economies.

快速气候变化

Alley R. B., Marotzke J., Nordhaus W. D., Overpeck J. T., Peteet D. M., Pielke R. A., Pierrehumbert R. T., Rhines P. B., Stocker T. F., Talley L. D., Wallace J. M.

(美国宾夕法尼亚州立大学地球科学和 EMS 环境研究所)

摘要:幅度大、快速、分布广、影响大的气候变化在地球历史上,每当地球系统经历大的驱动力时,就会发生。虽然快速气候变化的原因有多种可能,但是人类活动导致的强迫效应增加了大幅度快速气候变化的可能性。如果这样的快速气候变化再发生,那么对经济和生态的潜在影响将是大而严峻的。这种快速气候变化在接近气候阈值时用简单的模型不可预测。考虑到这种不确定性,政策制定者应该考虑加大快速气候变化的研究,增加观测,并调整经济和生态的适应能力。

14. Nakagawa T, Kitagawa H, Yasuda Y, et al. Asynchronous climate changes in the North Atlantic and Japan during the last termination[J]. Science, 2003, 299(5607): 688-691.

Asynchronous climate changes in the North Atlantic and Japan during the last termination

Nakagawa T., Kitagawa H., Yasuda Y., Tarasov P. E., Nishida K., Gotanda K., Sawai Y., Program Y. R. C.

(Department of Mathematics and Computer Science, Kagoshima University, Kagosima, Kagoshima, Japan)

Abstract: Pollen records from the annually laminated sediment sequence in Lake Suigetsu, Japan, suggest a sequence of climate changes during the Last Termination that resembles that of the North Atlantic region but with noticeable differences in timing. An interstadial interval commenced a few centuries earlier (similar to 15000 years before the present (yr B. P.)) than the North Atlantic GI-1 (Bolling) event. Conversely, the onset of a Younger Dryas (YD)-like cold reversal (112300 to 11250 yr B. P.) postdated the North Atlantic GS-1 (YD) event by a few centuries. Climate in the Far East during the Last Termination reflected solar insolation changes as much as Atlantic influences.

北大西洋和日本在末次冰期结束时的气候变化不同步特征

Nakagawa T., Kitagawa H., Yasuda Y., Tarasov P. E., Nishida K., Gotanda K., Sawai Y., Program Y. R. C.

(日本国立大学数学与计算机科学系)

摘要:来自日本湖泊年层沉积物的花粉记录显示了末次冰期结束时的一系列气候变化,类似于北大西洋区域的变化，但是时间上有差别。次间间隔比北大西洋的 Bolling 事件早几个世纪。与此相反，类似新仙女木的事件则落后北大西洋几个世纪。末次冰期结束时远东地区的气候变化同时反映了太阳辐射的变化和北大西洋的影响。

15. Visser K, Thunell R, Stott L. Magnitude and timing of temperature change in the Indo-Pacific warm pool during deglaciation[J]. Nature, 2003, 421(6919): 152-155.

Magnitude and timing of temperature change in the Indo-Pacific warm pool during deglaciation

Visser K., Thunell R., Stott L.

(Department of Geological Sciences, University of South Carolina, Columbia, South Carolina 29205, USA)

Abstract: Ocean-atmosphere interactions in the tropical Pacific region have a strong influence on global heat and water vapour transport and thus constitute an important component of the climate system. Changes in sea surface temperatures and convection in the tropical Indo-Pacific region are thought to be responsible for the interannual to decadal climate variability observed in extratropical regions, but the role of the tropics in climate changes on millennial and orbital timescales is less clear. Here we analyse oxygen isotopes and Mg/Ca ratios of foraminiferal shells from the Makassar strait in the heart of the Indo-Pacific warm pool, to obtain synchronous estimates of sea surface temperatures and ice volume. We find that sea surface temperatures increased by 3. 5-4. 0 degrees C during the last two glacial-interglacial transitions, synchronous with the global increase in atmospheric CO_2 and Antarctic warming, but the temperature increase occurred 2000-3000 years before the Northern Hemisphere ice sheets melted. Our observations suggest that the tropical Pacific region plays an important role in driving glacial-interglacial cycles, possibly through a system similar to how El Nino/ Southern Oscillation regulates the poleward flux of heat and water vapour.

Indo-Pacific 暖池间冰期期间温度变化的幅度和时间点

Visser K., Thunell R., Stott L.

(美国南卡莱罗纳大学地质科学系)

摘要：热带太平洋地区的海气耦合对全球水汽和热量输送有着重要的影响，因此其对地球气候系统非常重要。暖池地区海表面温度和对流变化被认为是热带以外地区年际和年代际气候变化的主要原因，但是它在千年-轨道尺度上的作用目前还不清楚。这篇文章主要报道暖池区域浮游有孔虫氧同位素和 Mg/Ca 值，以获得同步变化的海表面温度和冰量数据。我们发现海表面温度在过去两次冰期-间冰期转换过程中增加了 3.5～4 ℃，与大气二氧化碳浓度增加以及南极变暖同步，但是温度增加要超前于北半球冰盖融化 2000～3000 年。我们的结果表明热带地区在驱动冰期-间冰期循环中起到了重要的作用，可能类似于当前的 ENSO 影响向中高纬度地区的水汽和热量输送过程。

16. Jenkyns H C, Forster A, Schouten S, et al. High temperatures in the Late Cretaceous Arctic Ocean [J]. Nature, 2004, 432(7019): 888-892.

High temperatures in the Late Cretaceous Arctic Ocean

Jenkyns H. C. , Forster A. , Schouten S. , Damste J. S. S.

(Department of Earth Sciences, University of Oxford, Parks Road, Oxford OX1 3PR, UK)

Abstract: To understand the climate dynamics of the warm, equable greenhouse world of the Late Cretaceous period, it is important to determine polar palaeotemperatures. The early palaeoceanographic history of the Arctic Ocean has, however, remained largely unknown, because the sea floor and underlying deposits are usually inaccessible beneath a cover of floating ice. A shallow piston core taken from a drifting ice island in 1970 fortuitously retrieved unconsolidated Upper Cretaceous organic-rich sediment from Alpha ridge, a submarine elevated feature of probable oceanic origin. A lack of carbonate in the sediments from this core has prevented the use of traditional oxygenisotope palaeothermometry. Here we determine Arctic palaeotemperatures from these Upper Cretaceous deposits using TEX86, a new palaeothermometer that is based on the composition of membrane lipids derived from a ubiquitous component of marine plankton, Crenarchaeota. From these analyses we infer an average sea surface temperature of similar to 15 degrees C for the Arctic Ocean about 70 million years ago. This calibration point implies an Equator-to-pole gradient in sea surface temperatures of similar to 15 degrees C during this interval and, by extrapolation, we suggest that polar waters were generally warmer than 20 degrees C during the middle Cretaceous (similar to 90 million years ago).

晚白垩纪北冰洋高温

Jenkyns H. C. , Forster A. , Schouten S. , Damste J. S. S.

(英国牛津大学地球科学系)

摘要:了解极地古温度对理解晚白垩纪(温暖,温室气体浓度稳定)的气候动力学特征非常重要。但是北冰洋早期的古海洋历史目前还不知道,浮冰之下的海底沉积物通常无法获得。1970 年在阿尔法脊附近一个流动的冰岛采集的活塞钻孔很偶然地包含了早白垩纪的富含有机质的沉积物。由于碳酸盐的缺乏,这个沉积柱很难用传统的同位素温度计进行温度恢复。在这里,我们采用 TEX86,一种新的基于浮游生物脂类的温度计,来恢复北极早白垩纪的温度。通过测试分析,我们发现在距今 7000 万年前,北极的海表面温度接近 15 ℃。这一结果暗示当时从赤道到极地的温度梯度大约为 15 ℃。进一步推算,我们估计极地水温在中白垩纪(大约 9000 万年前)可能超过了 20 ℃。

17. Sigman D M,Jaccard S L,Haug G H. Polar ocean stratification in a cold climate[J]. Nature,2004,428(6978):59-63.

Polar ocean stratification in a cold climate

Sigman D. M. , Jaccard S. L. , Haug G. H.

(Department of Geosciences, Princeton University, Princeton, New Jersey 08544, USA)

Abstract: The low-latitude ocean is strongly stratified by the warmth of its surface water. As a result, the great volume of the deep ocean has easiest access to the atmosphere through the polar surface ocean. In the modern polar ocean during the winter, the vertical distribution of temperature promotes overturning, with colder water over warmer, while the salinity distribution typically promotes stratification, with fresher water over saltier. However, the sensitivity of seawater density to temperature is reduced as temperature approaches the freezing point, with potential consequences for global ocean circulation under cold climates. Here we present deep-sea records of biogenic opal accumulation and sedimentary nitrogen isotopic composition from the Sub-arctic North Pacific Ocean and the Southern Ocean. These records indicate that vertical stratification increased in both northern and southern high latitudes 2.7 million years ago, when Northern Hemisphere glaciation intensified in association with global cooling during the late Pliocene epoch. We propose that the cooling caused this increased stratification by weakening the role of temperature in polar ocean density structure so as to reduce its opposition to the stratifying effect of the vertical salinity distribution. The shift towards stratification in the polar ocean 2.7 million years ago may have increased the quantity of carbon dioxide trapped in the abyss, amplifying the global cooling.

冷期极地海洋的分层特征

Sigman D. M., Jaccard S. L., Haug G. H.

(美国普林斯顿大学地质科学系)

摘要:低纬度海洋的海水因表层水的热量而出现了非常显著的分层特征。分层的结果是大量的深海海水通过极地海表面与大气接触。在当前冬季的极地海洋,温度的垂直分布会发生翻转,使得冷海水在暖海水之上,同时盐度也会发生分层,低盐度的海水在高盐度的海水之上。但是,海水密度对温度的敏感程度会随着温度接近冰点而降低,因此冷期的时候海洋环流有发生变化的潜在可能。这里我们报道亚极地北太平洋地区和南大洋地区蛋白石和氮同位素的深海记录。这些记录显示在270万年前的晚上新世时期,伴随着北半球冰川的发展和全球变冷,海水垂直分层的现象在南半球和北半球高纬度地区同时存在。我们认为当时的变冷使得海洋温度在海洋密度分层中的作用减弱是这种分层加强的原因。这种分层作用的强化会增加海洋截留大气二氧化碳的能力,从而放大当时的全球变冷。

18. Hatun H, Sando A B, Drange H, et al. Influence of the Atlantic subpolar gyre on the thermohaline circulation[J]. Science, 2005, 309(5742): 1841-1844.

Influence of the Atlantic subpolar gyre on the thermohaline circulation

Hatun H., Sando A. B., Drange H., Hansen B., Valdimarsson H.

(Havstovan Faroe Marine Researc Institute)

Abstract: During the past decade, record-high salinities have been observed in the Atlantic Inflow to the Nordic Seas and the Arctic Ocean, which feeds the North Atlantic thermohaline circulation (THC). This may counteract the observed long-term increase in freshwater supply to the area and tend to stabilize the North Atlantic THC. Here we show that the salinity of the Atlantic Inflow is tightly linked to the dynamics of the North Atlantic subpolar gyre circulation. Therefore, when assessing the future of the North Atlantic THC, it is essential that the dynamics of the subpolar gyre and its influence on the salinity are taken into account.

副极地大西洋回旋对温盐环流的影响

Hatun H. , Sando A. B. , Drange H. , Hansen B. , Valdimarsson H.

(丹麦法罗群岛渔业实验室)

摘要:在过去十年,从大西洋流入罗迪克海和北冰洋的高盐度海水被观测所证实,这些高盐度海水流入北大西洋温盐环流。这一结果可能抵消掉观测到的淡水输入增加,从而稳定北大西洋温盐环流。这里,我们证明北大西洋流的盐度与北大西洋-副极地回旋直接存在密切的关联。因此,要评估北大西洋温盐环流的未来,必须要考虑副极地回旋以及它对盐度的影响。

19. Tarasov L,Peltier W R. Arctic freshwater forcing of the Younger Dryas cold reversal[J]. Nature, 2005,435(7042):662-665.

Arctic freshwater forcing of the Younger Dryas cold reversal

Tarasov L. , Peltier W. R.

(Department of Physics, University of Toronto, Toronto, Ontario, Canada M5S 1A7.)

Abstract: The last deglaciation was abruptly interrupted by a millennial-scale reversal to glacial conditions, the Younger Dryas cold event. This cold interval has been connected to a decrease in the rate of North Atlantic Deep Water formation and to a resulting weakening of the meridional overturning circulation owing to surface water freshening. In contrast, an earlier input of fresh water (meltwater pulse 1a), whose origin is disputed, apparently did not lead to a reduction of the meridional overturning circulation. Here we analyse an ensemble of simulations of the drainage chronology of the North American ice sheet in order to identify the geographical release points of freshwater forcing during deglaciation. According to the simulations with our calibrated glacial systems model, the North American ice sheet contributed about half the fresh water of meltwater pulse 1a. During the onset of the Younger Dryas, we find that the largest combined meltwater/iceberg discharge was directed into the Arctic Ocean. Given that the only drainage outlet from the Arctic Ocean was via the Fram Strait into the Greenland-Iceland-Norwegian seas, where North Atlantic Deep Water is formed today, we hypothesize that it was this Arctic freshwater flux that triggered the Younger Dryas cold reversal.

北冰洋淡水驱动新仙女木冷事件

Tarasov L. , Peltier W. R.

(加拿大多伦多大学物理系)

摘要:末次冰期的消退过程被一次千年尺度的快速气候反转所中断,这就是著名的新仙女木冷事件。这次冷事件与北大西洋深海流减弱和由于海表面淡化所引起的北大西洋经向环流减弱相关。相反,早期的淡水输入(冰融水),却并没有导致经向环流的减弱。这里我们模拟研究了北美冰盖淡水排放,确定在冰消间淡水驱动的地理释放点。根据我们校准的冰川系统模型的结果,北美冰盖输入了大约一半的冰融淡水。在新仙女木事件爆发期间,我们发现最大的冰融水输入到了北冰洋。由于北大西洋温盐环流附近的格陵兰-冰岛-挪威海是北冰洋唯一的水道出口,因此,我们认为北冰洋的大规模淡水输入可能是新仙女木事件的促发原因。

20. Brinkhuis H,Schouten S,Collinson M E,et al. Episodic fresh surface waters in the Eocene Arctic Ocean[J]. Nature,2006,441(7093):606-609.

Episodic fresh surface waters in the Eocene Arctic Ocean

Brinkhuis H., Schouten S., Collinson M. E., Sluijs A., Damste J. S. S., Dickens G. R., Huber M., Cronin T. M., Onodera J., Takahashi K., Bujak J. P., Stein R., van der Burgh J., Eldrett J. S., Harding I. C., Lotter A. F., Sangiorgi F., Cittert H. V. V., de Leeuw J. W., Matthiessen J., Backman J., Moran K., Scientists E.

(Palaeoecology, Institute of Environmental Biology, Utrecht University, Laboratory of Palaeobotany and Palynology, Budapestlaan 4, 3584 CD Utrecht, The Netherlands)

Abstract: It has been suggested, on the basis of modern hydrology and fully coupled palaeoclimate simulations, that the warm greenhouse conditions that characterized the early Palaeogene period (55-45 Myr ago) probably induced an intensified hydrological cycle with precipitation exceeding evaporation at high latitudes. Little field evidence, however, has been available to constrain oceanic conditions in the Arctic during this period. Here we analyse Palaeogene sediments obtained during the Arctic Coring Expedition, showing that large quantities of the free-floating fern Azolla grew and reproduced in the Arctic Ocean by the onset of the middle Eocene epoch (similar to 50 Myr ago). The Azolla and accompanying abundant freshwater organic and siliceous microfossils indicate an episodic freshening of Arctic surface waters during an similar to 800000-year interval. The abundant remains of Azolla that characterize basal middle Eocene marine deposits of all Nordic seas probably represent transported assemblages resulting from freshwater spills from the Arctic Ocean that reached as far south as the North Sea. The termination of the Azolla phase in the Arctic coincides with a local sea surface temperature rise from similar to 10 degrees C to 13 degrees C, pointing to simultaneous increases in salt and heat supply owing to the influx of waters from adjacent oceans. We suggest that onset and termination of the Azolla phase depended on the degree of oceanic exchange between Arctic Ocean and adjacent seas.

始新世北冰洋海表的间插性淡水

Brinkhuis H., Schouten S., Collinson M. E., Sluijs A., Damste J. S. S., Dickens G. R., Huber M., Cronin T. M., Onodera J., Takahashi K., Bujak J. P., Stein R., van der Burgh J., Eldrett J. S., Harding I. C., Lotter A. F., Sangiorgi F., Cittert H. V. V., de Leeuw J. W., Matthiessen J., Backman J., Moran K., Scientists E.

(荷兰乌德勒支大学古生态学、古环境生物研究所植物学和孢粉实验室)

摘要:基于现代水文学和古气候学的模拟认为,在早第三纪期间的温室条件可增强水文循环,并导致高纬度地区降雨要大于蒸发。但是还很少有野外证据可以用来证明这一时期的北极海洋条件。这里我们分析了北极钻孔的早第三纪沉积物，结果表明在中始新世期间，这里存在大量的蕨类植物。这些植物和丰富的淡水有机体以及硅酸盐微体化石，表明在大约80万年中，北极海表面水存在间插性淡化。丰富的蕨类植物在中始新世期间覆盖了整个罗迪克海，表明北冰洋海淡水溢出向南扩散，远至北海。这些蕨类植物的终止与当地海表面温度上升到10～13℃一致，表明来自临近海水的盐度和热量同时增加。我们认为Azolla蕨类植物的爆发和终止主要依赖于北冰洋与邻近海的交换程度。

21. Moran K, Backman J, Brinkhuis H, et al. The Cenozoic palaeoenvironment of the Arctic Ocean[J]. Nature, 2006, 441(7093): 601-605.

The Cenozoic palaeoenvironment of the Arctic Ocean

Moran K., Backman J., Brinkhuis H., Clemens S. C., Cronin T., Dickens G. R., Eynaud F., Gattacceca J., Jakobsson M., Jordan R. W., Kaminski M., King J., Koc N., Krylov A., Martinez N., Matthiessen J., McInroy D., Moore T. C., Onodera J., O'Regan M., Palike H., Rea B., Rio D., Sakamoto T., Smith D. C., Stein R., St John K., Suto I., Suzuki N., Takahashi K., Watanabe M., Yamamoto M., Farrell J., Frank M., Kubik P., Jokat W., Kristoffersen Y.

(Graduate School of Oceanography & Department of Ocean Engineering, University of Rhode Island, Narragansett, Rhode Island 02882, USA)

Abstract: The history of the Arctic Ocean during the Cenozoic era (0-65 million years ago) is largely unknown from direct evidence. Here we present a Cenozoic palaeoceanographic record constructed from >400 m of sediment core from a recent drilling expedition to the Lomonosov ridge in the Arctic Ocean. Our record shows a palaeoenvironmental transition from a warm "greenhouse" world, during the late Palaeocene and early Eocene epochs, to a colder "icehouse" world influenced by sea ice and icebergs from the middle Eocene epoch to the present. For the most recent similar to 14 Myr, we find sedimentation rates of 1-2 cm per thousand years, in stark contrast to the substantially lower rates proposed in earlier studies; this record of the Neogene reveals cooling of the Arctic that was synchronous with the expansion of Greenland ice (similar to 3. 2 Myr ago) and East Antarctic ice (similar to 14 Myr ago). We find evidence for the first occurrence of ice-rafted debris in the middle Eocene epoch (similar to 45 Myr ago), some 35 Myr earlier than previously thought; fresh surface waters were present at, 49 Myr ago, before the onset of ice-rafted debris. Also, the temperatures of surface waters during the Palaeocene/Eocene thermal maximum (similar to 55 Myr ago) appear to have been substantially warmer than previously estimated. The revised timing of the earliest Arctic cooling events coincides with those from Antarctica, supporting arguments for bipolar symmetry in climate change.

北冰洋新生代古环境

Moran K., Backman J., Brinkhuis H., Clemens S. C., Cronin T., Dickens G. R., Eynaud F., Gattacceca J., Jakobsson M., Jordan R. W., Kaminski M., King J., Koc N., Krylov A., Martinez N., Matthiessen J., McInroy D., Moore T. C., Onodera J., O'Regan M., Palike H., Rea B., Rio D., Sakamoto T., Smith D. C., Stein R., St John K., Suto I., Suzuki N., Takahashi K., Watanabe M., Yamamoto M., Farrell J., Frank M., Kubik P., Jokat W., Kristoffersen Y.

(美国罗德岛大学海洋研究所)

摘要：目前还很少有直接证据证明新生代北冰洋历史。这里我们报道一个基于北冰洋海洋钻孔的新生代古海洋沉积柱，其长度大于400 m。我们的结果表明北冰洋由晚古新世-早始新世时期的温暖状态，受中始新世的海冰和冰山所影响，转变为中始新世-现代的寒冷态。在距今1400万年时期，我们发现沉积速率是1～2 cm/千年，这与先前研究证明的低沉积速率很不相符。这个晚第三纪的记录表明北极变冷与格陵兰和东南极冰盖的扩张同步。我们发现中始新世的首次冰砾的出现时间比先前报道的早3500万年。表面淡水在4900万年前就存在，在冰砾出现之前。另外，在晚古新世-早始新世温暖期，海表面温度也要比先前估计的高。改进的早北极变冷时间与南极一致，支持了两极气候同步变化的争论。

22. Sluijs A, Schouten S, Pagani M, et al. Subtropical Arctic ocean temperatures during the Palaeocene/Eocene thermal maximum[J]. Nature, 2006, 441(7093): 610-613.

Subtropical Arctic ocean temperatures during the Palaeocene/Eocene thermal maximum

Sluijs A., Schouten S., Pagani M., Woltering M., Brinkhuis H., Damste J. S. S., Dickens G. R., Huber M., Reichart G. J., Stein R., Matthiessen J., Lourens L. J., Pedentchouk N., Backman J., Moran K., Scientists E.

(Palaeoecology, Institute of Environmental Biology, Utrecht University, Laboratory of Palaeobotany and Palynology, Budapestlaan 4, 3584 CD Utrecht, The Netherlands)

Abstract: The Palaeocene/Eocene thermal maximum, similar to 55 million years ago, was a brief period of widespread, extreme climatic warming, that was associated with massive atmospheric greenhouse gas input. Although aspects of the resulting environmental changes are well documented at low latitudes, no data were available to quantify simultaneous changes in the Arctic region. Here we identify the Palaeocene/Eocene thermal maximum in a marine sedimentary sequence obtained during the Arctic Coring Expedition. We show that sea surface temperatures near the North Pole increased from similar to 18 degrees C to over 23 degrees C during this event. Such warm values imply the absence of ice and thus exclude the influence of ice-albedo feedbacks on this Arctic warming. At the same time, sea level rose while anoxic and euxinic conditions developed in the ocean's bottom waters and photic zone, respectively. Increasing temperature and sea level match expectations based on palaeoclimate model simulations, but the absolute polar temperatures that we derive before, during and after the event are more than 10 degrees C warmer than those model-predicted. This suggests that higher-than-modern greenhouse gas concentrations must have operated in conjunction with other feedback mechanisms-perhaps polar stratospheric clouds or hurricane-induced ocean mixing to amplify early Palaeogene polar temperatures.

始新世温暖期副热带北极海温度

Sluijs A., Schouten S., Pagani M., Woltering M., Brinkhuis H., Damste J. S. S., Dickens G. R., Huber M., Reichart G. J., Stein R., Matthiessen J., Lourens L. J., Pedentchouk N., Backman J., Moran K., Scientists E.

(荷兰乌德勒支大学古生态学古环境生物研究所植物学和孢粉实验室)

摘要:始新世温暖期,大约 5500 万年之前,是一个短时间的广泛的极端温暖期,与这一时期大气二氧化碳的高浓度同步。虽然这一时期的环境特征在低纬度地区已经有较好的研究,但是在北极地区还没有可用的数据来定量当时的变化。这里我们分析了一个来自北极深钻的始新世温暖期海洋沉积物。我们的结果表明,北极附近的海表面温度在这一时期从大约 18 ℃到超过 23 ℃。这样的温暖条件暗示当时没有冰盖的存在,因此也不存在冰面反射反馈作用。同时,海表面上升以及海洋底部缺氧的静海条件发展。增加的温度和海平面与古气候模拟的预期相吻合,但是我们得到的绝对温度要比预测的高 10 ℃左右。这说明比现代高的温室气体浓度必须和其他反馈机制结合(比如极地中层云或者飓风导致的海洋混合层增加)一起来放大古新世的极地温度。

四、生物科学类文献

4.1 分析概述

自1990年以来,生物科学类在三个主要期刊 *Nature*, *Science*, *PNAS* 上共发表文献109篇(2008～2010年无相关文献)。经过仔细核对及检索原文,南极文献共59篇,北极文献共50篇。现在对这109篇文献进行分析小结。

1. 期刊来源分析

从发表刊物类型看,*Nature* 最多,为48篇,*PNAS* 上发表相关论文34篇,*Science* 上27篇(图4.1)。可能原因是 *Nature* 上发表的文章多为生物类相关领域,同样在极地生物科学领域也表现了这一特点。

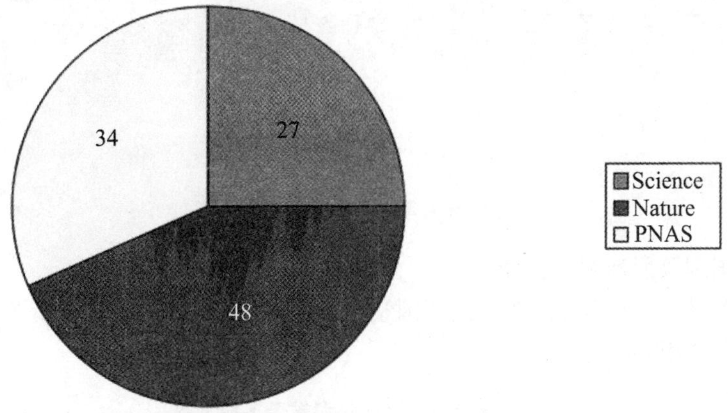

图 4.1 期刊来源分布

2. 年代分析

从图4.2中可以看出,生物科学的研究在2000～2002年形成了一次高峰,北极生物科学研究的文献在1998年以后开始增加,在2004年以前的研究比重都超过南极,1998年北极冰量减少迅速,为生物科学调查研究提供了良好的条件。南极生物科学在1990～2007年文献量较为稳定,均在2篇以上,最多达到6篇,而在2008～2010年期间未检索到南极相关文献。

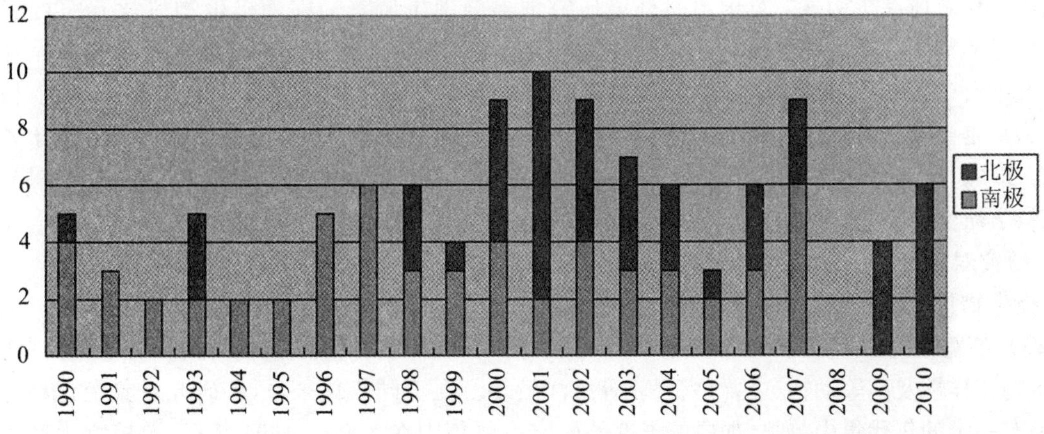

图 4.2 期刊论文年代分布

3. 第一作者国别

从图 4.3 可以看出,美国、英国两国是极地生物科学研究方面的引领者,其文献数量分别为 46 篇、16 篇。紧接着是加拿大 11 篇、德国 7 篇。美国在极地生物方面的贡献卓著,而我国到目前为止尚属空白,急需加强生物领域方面的研究。

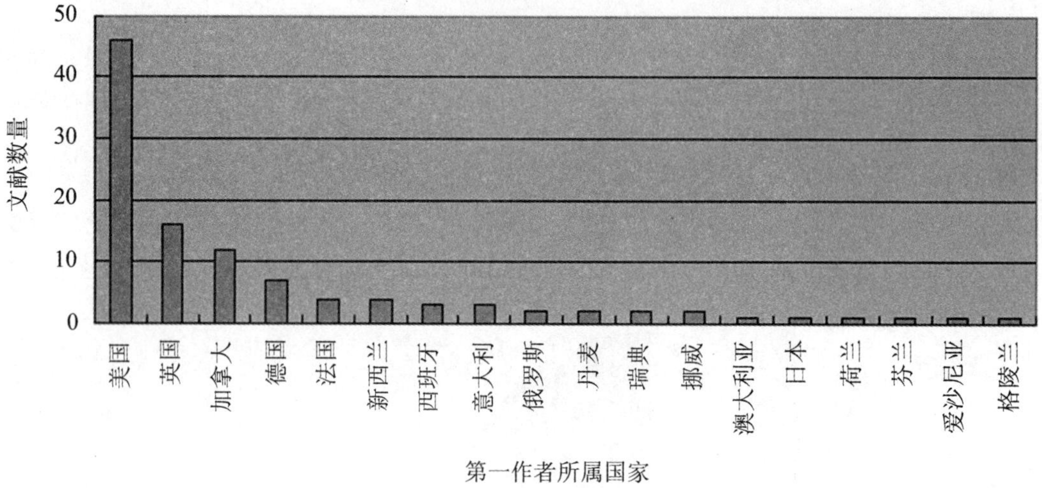

图 4.3 第一作者所属国家

4. 第一作者国别及所属机构分析

1990～2007 年,仅 2 位美国学者、1 位英国学者及 1 位德国学者发表了多篇文章(表 4.1),可见这 3 位学者在科研上具有很强的实力。

表 4.1 同一作者发表多篇文章

作 者	国别	机 构	发表文章年代
Chen L. B.	美国	美国伊利诺伊大学分子与整合生理学系	1997,1997
Meskhidze N.	美国	美国亚特南大乔治亚理工学院地球和大气科学系	2006,2007,2007
Darling K. F.	英国	英国爱丁堡大学地质学和地球物理系	2000,2004
Quirin Schiermeie	德国	*Nature* 德国慕尼黑办公室(记者)	2007,2007

从第一作者所在机构看,英国自然环境研究理事会南极调查局在南极生物科学方面贡献最大(6 篇),其次是美国阿拉斯加费尔班克斯大学(4 篇)、英国爱丁堡大学地质学和地球物理系(3 篇)、法国国家科学研究中心(3 篇)、美国亚特兰大乔治亚理工学院地球和大气科学系(3 篇)(图 4.4)。可见,极地的主要研究力量是国家科研院所和高等院校。从美国第一作者所在机构看,其极地研究力量分散于各高等院校,共有 36 家单位,其中高等院校有 32 家。英国极地生物科学研究的主力军是英国南极调查局,法国的是国家科学研究中心。这些特点与各自国家的科技体制和经济体制有一定的相关性。

5. 研究对象分析

南极生物科学按研究对象分为磷虾丰度(6 篇)、分子类(12 篇)、浮游植物生产力(13 篇)、生物多样性(11 篇)、高等动物(11 篇)和其他(6 篇)6 个方面(图 4.5)。可见,科学家们对南极分子类、浮游植物生产力、生物多样性及高等动物(主要是企鹅)等关注较多。在不同年代来看,各研究对象变化趋势不是很明显,但有一定的年代集中趋势,如磷虾丰度的研究主要集中在 1990～1991 年;浮游植物生产力研究多在 1996～2000 年,随后在 2006～2007 年又取得了一定的成果。

北极生物科学按研究对象分为分子及遗传类(13 篇)、生态和生物多样性(13 篇)、高等动物(7

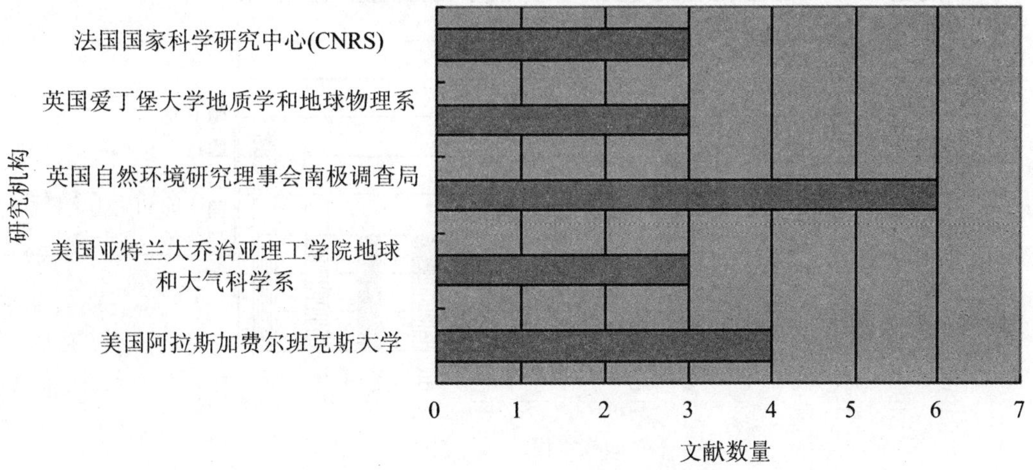

图 4.4 第一作者所属研究机构文献数量

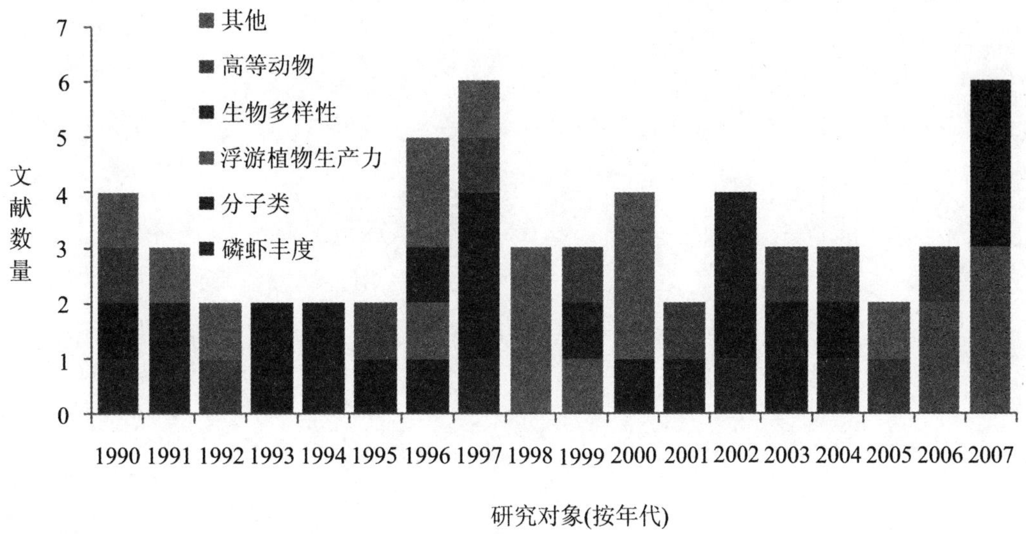

图 4.5 南极生物科学文献研究对象分布

篇)、浮游植物(3 篇)和其他(14 篇)5 个方面(图 4.6)。可见,科学家们对北极分子类、生物多样性及高等动物等关注较多。从不同年代来看,各研究对象变化趋势不是很明显,但也有一定的年代集中趋势,如分子及遗传类的研究主要集中在 2000 年、2001 年和 2010 年;其他研究主要指生物与环境气候的相互影响方面,文献集中在 1998 年、2002 年和 2007 年;而生态和生物多样性的研究则从 2000 年开始逐渐受到重视。

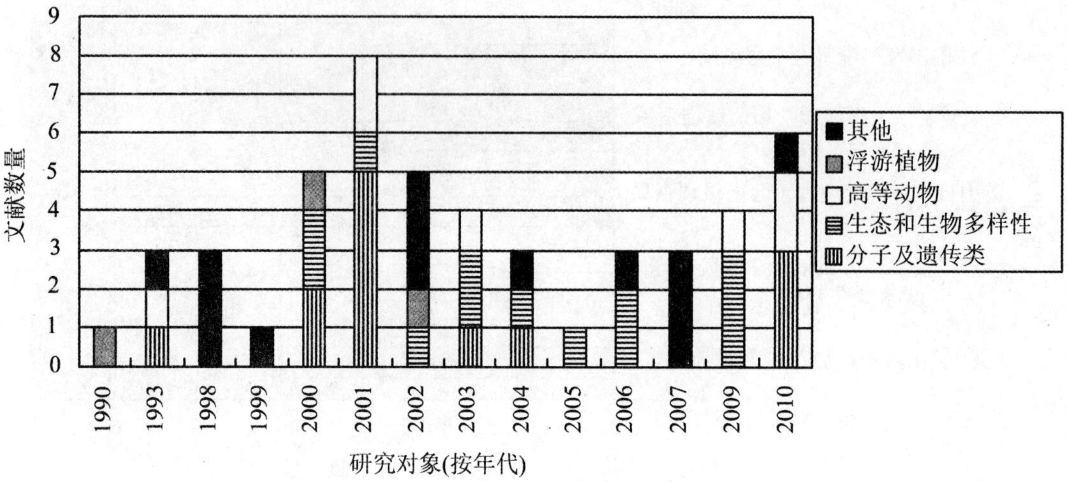

图 4.6 北极生物科学文献研究对象分布

4.2 生物科学类摘要翻译——南极

1. Barinaga M. Eco-quandary: what killed the skuas[J]. Science, 1990, 249(4966): 243.

Eco-quandary: what killed the skuas

Barinaga M.

(Journal Article)

Abstract: Reports on the controversy that has arisen over the cause of a massive die-off of South Polar skua chicks in 1989, and the possible role an oil spill played in the deaths. Periodic die-offs of young skuas: Oil as a trigger; Regional silverfish shortage; Role of storms.

生态困境:是什么杀死了贼鸥

Barinaga M.

(杂志文章)

摘要:本文报道了起源于1989年的南极贼鸥幼鸟大规模死亡的争议事件,以及石油泄漏在其死亡中可能的作用。贼鸥幼鸟周期性死亡的原因应该包括石油诱因、当地银鱼的减少、暴风雨的作用等三方面因素。

2. Inigo Everson, Jonathan L Watkins, Douglas G Bone, et al. Implications of a new acoustic target strength for abundance estimates of Antarctic krill[J]. Nature, 1990, 345(6273): 338-340.

Implications of a new acoustic target strength for abundance estimates of Antarctic krill

Inigo Everson, Jonathan L. Watkins, Douglas G. Bone, Kenneth G. Foote

(British Antarctic Survey, High Cross, Madingley Road, Cambridge CB3 GET, UK)

Abstract: Antarctic krill (*Euphausia superba*) is the dominant component of the diet of many whales, seals, birds, fish and squid, and their survival could be affected by a reduction in krill abundance due to fishing. Commercial fishing takes nearly half a million tonnes of krill annually and accurate estimates of abundance are needed for rational management of this resource. Indirect estimates of abundance based on predator consumption rates give a roughly estimated total annual production of several hundred million tonnes. The life-span of krill is at least two and maybe five years, so the standing stock would need to be of at least the same magnitude as the annual production. But direct estimates of abundance using nets and acoustics have indicated biomass figures lower by an order of magnitude. Acoustic estimates are sensitive to the scaling factor or target strength (TS) used to convert echo energy to absolute abundance. Previous published values for TS, when applied to survey data, gave estimates of krill abundance that were much too low to account for local bird and seal predation rates near South Georgia, and were also lower than expected when compared with density estimates from net hauls. We therefore sought to determine TS using direct measurements developed for fish studies, and also by applying models developed for other crustacean zooplankton. Our results show that krill TS is much lower than previously thought, and consequently that acoustic estimates of krill abundance are likely to have been gross underestimates.

应用新的声学目标强度方法估算南极磷虾的丰度

Inigo Everson, Jonathan L. Watkins, Douglas G. Bone, Kenneth G. Foote

(英国南极调查局)

摘要：南极磷虾(*Euphausia superba*)是鲸、海豹、鸟类、鱼类和鱿鱼饮食的主要组成部分，由于捕捞导致磷虾数量减少，这些动物的生存也受到影响。商业捕捞每年带走约50万吨的磷虾，准确估计其丰度对合理管理利用磷虾资源是很有必要的。间接估计磷虾丰度的方法基于食肉动物消费速率，并粗略估计其总的年生产力为数亿吨。磷虾的寿命至少2年，也可能5年，因此现存量至少和年生产量一致；但是，用网和声学方法直接估计丰度已经表明实际生物量要低一个数量级。声学估计对用来转换回波能量到绝对丰度的比例因子或目标强度(TS)是敏感的。先前发表的目标强度数值，当应用到调查数据中时，其估算的磷虾数量明显过低而不能解释南乔治亚地区鸟类和海豹捕食率，也比用拖网法估算的密度低。因此，我们尝试着用在渔业上应用成熟的直接测定方法测定目标强度，并应用其他用之于甲壳类浮游动物的模型。我们的研究结果表明，磷虾目标强度比以前想象的要低得多，因此用声学评估磷虾丰度很可能使结果总体偏低。

3. Hsiao K C, Cheng C H, Fernandes I E, et al. An Antifreeze glycopeptide gene from the Antarctic cod Notothenia-coriiceps-neglecta encodes a polyprotein of high peptide copy number[J]. PNAS, 1990, 87(23):9265-9269.

An Antifreeze glycopeptide gene from the Antarctic cod Notothenia-coriiceps-neglecta encodes a polyprotein of high peptide copy number

Hsiao K. C., Cheng C. H., Fernandes I. E., Detrich H. W., DeVries A. L.

(Department of Physiology and Biophysics, University of Illinois, Urbana, IL 61801)

Abstract: The antarctic fish *Notothenia coriceps neglecta* synthesizes eight antifreeze glycopeptides (AFGP 1-8; Mr 2600-34000) to avoid freezing in its ice-laden freezing habitat. We report here the sequence of one of its AFGP genes. The structural gene contains 46 tandemly repeated segments, each encoding one AFGP peptide plus a 3-amino acid spacer. Most of the repeats (44/46) code for peptides of AFGP 8; the remaining 2 code for peptides of AFGP 7. At least 2 of the 3 amino acids in the spacers could act as substrate for chymotrypsin-like proteases. The nucleotide sequence between the translation initiation codon (ATG) and the first AFGP-coding segment is G+T-rich and encodes a presumptive 37-

residue signal peptide of unusual sequence. Primer extension establishes the transcription start site at nucleotide 43 upstream from ATG. CAAT and TATA boxes begin at nucleotides 53 and 49, respectively, upstream from the transcription start site. The polyadenlylation signal, AATAAA, is located 240 nucleotides downstream from the termination codon. AmRNA (3 kilobases) was found that matches the size of this AFGP gene. Thus, this AFGP gene encodes a secreted, high-copy number polyprotein that is processed posttranslationally to produce active AFGPs.

南极鳕鱼 Notothenia coriiceps neglecta 抗冻糖肽基因编码高拷贝数的聚合蛋白

Hsiao K. C., Cheng C. H., Fernandes I. E., Detrich H. W., DeVries A. L.

(美国伊利诺斯大学生理学和生物物理学系)

摘要：南极鳕鱼 *Notothenia coriceps neglecta* 合成了 8 种抗冻糖肽(AFGP 1-8; Mr 2600-34000)，来避免其在冰覆盖环境下的冻结。我们在这里报告的是其中之一 AFGP 的基因序列。它的结构基因包括 46 个重复衔接的片段，每个编码一个 AFGP 肽加上 3-氨基酸的间隔。大部分重复(44/46)编码 AFGP 8 多肽，其余 2 个编码是 AFGP 7 多肽。3 个氨基酸间隔中至少有 2 个可以充当胰凝乳蛋白酶底物。翻译起始密码子(ATG)和第一 AFGP 编码片段之间的核苷酸序列是 G+T(鸟嘌呤+胸腺嘧啶)型丰富，可能编码一个 37 个残基的不寻常的信号肽。引物延伸方法证明转录起始点在 ATG 上游的第 43 个碱基。CAAT 和 TATA 盒开始点为始点在 ATG 上游的第 53 个和 49 个碱基。加尾信号 AATAAA 位于终止密码子的下游，距离为 240 个核苷酸。结果发现 AmRNA(3 个碱基)符合这个 AFGP 基因的大小。因此，这个 AFGP 基因编码一种分泌的、高拷贝数多蛋白，它在后翻译过程将产生活性 AFGPs。

4. James B Mcclintock, John Janssen. Pteropod abduction as a chemical defense in a pelagic Antarctic amphipod[J]. Nature, 1990, 346(6283): 462-464.

Pteropod abduction as a chemical defense in a pelagic Antarctic amphipod

James B. Mcclintock., John Janssen

(Department of Biology, University of Alabama at Birmingham, Birmingham, Alabama 35294, USA)

Abstract: This study documents an example of an invertebrate that cannot defend itself chemically (an amphipod) increasing its chances of survival by capturing and carrying a species that can (a pteropod). Although chemical defences are found in a wide variety of marine invertebrates, few studies have established the extent to which these chemicals will deter predators, and even fewer have investigated how one organism might exploit another's chemical defence to protect itself. These chemicals are usually ingested or sequestered in the host's tissue. Several species indiscriminately decorate themselves with potentially toxic organisms or are passively fouled with chemically defended organisms. These seem to be commensalisms or sometimes mutualisms. Our example has benefits and costs for the carrier, but only costs to the captive. The antarctic marine food web not only has a variety of chemically defended organisms, contrary to earlier predictions, but has at least one unusual symbiosis that makes use of noxiousness.

诱拐翼足目是远洋南极片脚类动物的一个化学防御措施

James B. Mcclintock, John Janssen

(美国阿拉巴马伯明翰大学生物学系)

摘要:本文阐述了一个不能用化学方法保护自己的无脊椎动物(一个片脚类动物)通过捕捉和携带一个能用化学方法保护自己的无脊椎动物(一个翼足目)基因来增加其生存的机会。尽管已发现化学防御存在于多种海洋无脊椎动物体内,但是能证明这些化学品阻止食肉/食虫动物程度的研究较少,而研究一个生物体如何利用另一个生物体的化学防御来保护自己的则更少。这些化学物质通常被摄入或隐藏在寄主组织内。少数几个物种能够无差别地利用潜在有毒有机体来修饰自己,或是被动利用化学方法保护自己。这些似乎是共栖或某种程度的互利共生。我们的例子对载体有利有弊,但对俘虏者仅有弊。南极海洋食物网不但有多样的化学防御生物(这与早期预测相反),而且至少有一个独特的利用有毒物种的共栖作用。

5. Charles H Greene, Timothy K Stanton, Peter H Wie, et al. Acoustic estimates of Antarctic krill[J]. Nature, 1991, 349(6305): 110.

Acoustic estimates of Antarctic krill

Charles H. Greene, Timothy K. Stanton, Peter H. Wie, Sam Mcciatchi

(Ocean Resources and Ecosystems Program, Cornell University, Ithaca, New York 14853, USA)

Abstract: Sir-Everson et al discussed the implications of new measurements of target strength for estimating the abundance of krill in the southern ocean. Their conclusions were first, that previously used equations relating target strength to physical size of these animals were greatly in error and, second, that use of these equations has resulted in gross underestimates of krill abundance in the southern ocean. As krill provides the basis of a larger fishery and is the main component of the diet of many marine predators, accurate estimates are essential for management of this resource. We have collected data covering a broad size range of crustacean zooplankton and micro-nekton of Everson et al. We present new target-strength-by-size relationships over the full size range of krill at the acoustical frequencies commonly used in field studies.

南极磷虾的声学估算

Charles H. Greene, Timothy K. Stanton, Peter H. Wie, Sam Mcciatchi

(美国康奈尔大学海洋资源和生态系统计划部)

摘要:Everson 等讨论了目标强度(TS)这个新的测量方法在估算南大洋磷虾丰度上的意义。他们的结论是:第一,以前使用的把目标强度和这些动物个体大小联系起来的方程有很大的错误;第二,利用这些方程会在总体上低估南大洋的磷虾丰度。因为磷虾是大型渔业的基础,是许多海洋食肉动物的主要饮食组成,准确地估算磷虾丰度对管理这一资源至关重要。我们已收集的数据涵盖了 Everson 等阐述的甲壳类浮游动物和游泳动物。这里我们报道新的目标强度-磷虾大小关系，这个关系适用于不同大小的磷虾和野外调查常用的声学频率。

6. Hewitt R P, Demer D A. Krill abundance[J]. Nature, 1991, 353(6342): 310.

Krill abundance

Hewitt R. P., Demer D. A.

(Southwest Fisheries Science Center, La Jolla, California 92038, USA)

Abstract: Antarctic krill (*Euphausia superba*) is the primary food source for many animals in the southern ocean and is also the basis of a large fishery. To manage this resource, krill abundance has been estimated directly with acoustics and in-directly with estimates of predator consumption rates. However, the abundance estimates using acoustics are often an order of magnitude less than those based on predator consumption rates. The acoustic method converts echo energy to absolute biomass by assuming that the echo return is the sum of individual scatterers, and assuming an empirical or modeled acoustic target strength (TS) for individual krill. Everson et al and Greene et al reported new TS measurements of experimentally constrained krill, but to date no corroborating field data have been published. We present new in situ TS measurements of krill obtained in March 1991 off Elephant Island, Antarctica.

磷虾丰度

Hewitt R. P., Demer D. A.

(美国加州拉霍亚西南渔业科学中心)

摘要：南极磷虾(*Euphausia superba*)是南大洋许多动物的主要食物来源，也是一个大型渔业基础。为了管理这一资源，可以直接用声学或间接用捕食者消费速率估计磷虾丰度。然而，使用声学估算丰度的方法常比基于捕食者消费速率这个方法少一个数量级。通过假定返回的回声是分散个体的总和，以及假定一个经验的或模型模拟的声学目标强度(TS)，声学方法转换回声能量为绝对生物量。Everson 等和 Greene 等报道了新的目标强度测量方法测量实验控制下的磷虾丰度，但是到目前为止，还没有可佐证的野外数据发表。我们报道 1991 年 3 月在南极象岛用现场测定目标强度的方法获得的磷虾丰度的新数据。

7. Karentz S, Cleaver J E, Mitchell D L. DNA damage in the Antarctic[J]. Nature, 1991, 350(6313): 28.

DNA damage in the Antarctic

Karentz S., Cleaver J. E., Mitchell D. L.

(Laboratory of Radiobiology and Environmental Health, University of California, San Francisco, California 94143, USA)

Abstract: The Antarctic ozone hole represents a cycle of annual springtime depletion of stratospheric ozone over the continent and surrounding ocean areas, with consequent increases in levels of ultraviolet-B radiation (UV-B, 280-320 nm). The ecological impact of this increased radiation is of particular concern for Antarctic marine communities, which depend on the productivity of phytoplankton at the base of short food chains. There has been much speculation about ecological effects based on a priori assumptions about unique photobiological properties of Antarctic organisms. A common misconception is that because these organisms inhabit an area of relatively low ambient ultraviolet intensity, Antarctic species may be more sensitive to UV-B exposure than their temperate of tropical counterparts. But Antarctic species are derived from temperate and tropical stocks and should therefore retain a capacity for biochemical protection from ultraviolet exposure and for repair of radiation-induced damage.

南极 DNA 损伤

Karentz S., Cleaver J. E., Mitchell D. L.

(美国加州大学旧金山分校放射生物学和环境健康实验室)

摘要：南极臭氧洞每年春季出现在大陆和周围海域，对应着平流层臭氧耗竭，随后 UV-B 辐射水平(280～320 nm)就增加了。因辐射增加所导致的生态影响，对南极海洋群落特别重要，因为这些海洋群落依赖于建立在短食物链基础上的浮游植物生产力。对于南极生物独特的光生物学特性的先验假设的生态效应，人们有很多的猜测。一个常见的误解是，由于这些生物栖息环境周围的紫外线强度相对较低，南极的物种可能比相应的温带、热带物种对 UV-B 辐射环境更敏感。但是南极的物种是来自温带和热带，因此保留了一种对紫外的生化防护能力和修复诱导损伤的能力。

8. Ancel A, Kooyman G L, Ponganis P J, et al. Foraging behavior of emperor penguins as a resource detector in winter and summer[J]. Nature, 1992, 360(6402): 336-339.

Foraging behavior of emperor penguins as a resource detector in winter and summer

Ancel A., Kooyman G. L., Ponganis P. J., Gendner J.-P., Lignon J., Mestre X., Huin N., Thorson P. H., Robisson P., Le Maho Y.

(Centre d'Ecologie et Physiologie Energétiqu. es, Centre National de la Recherche Scientifique, 23 rue Becquerel, 67087 Strasbourg, France)

Abstract: The emperor penguin (*Aptenodytes forsteri*), which feeds only at sea, is restricted to the higher latitudes of the antarctic sea-ice habitat. It breeds on the winter fast ice when temperatures are −30-degrees-C and high winds are frequent. Assuming entirely the task of incubating the single egg, the male fasts for about 120 days in the most severe conditions. When it is relieved by the female around hatching time, the distance between the colony and the open sea may be 100 km or more, but where emperors go to forage at that time or during the summer is unknown. The polynias are areas of open water in sea-ice and during winter, with the under-ice habitats at any time of the year, they are among the most difficult of all Antarctic areas to sample. Here we monitor by satellite the routes taken by emperor penguins for foraging and compare them with satellite images of sea-ice. Winter birds walking over fast ice travelled up to 296 km to feed in polynias, whereas those swimming in light pack-ice travelled as far as 895 km from the breeding colony. One record of diving showed that although most dives are to mid-water depths, some are near the bottom. Obtaining such detailed information on foraging in emperor penguins means that this bird now offers a unique opportunity to investigate the Antarctic sea-ice habitat.

冬季和夏季帝企鹅觅食行为可作为一个资源探测器

Ancel A., Kooyman G. L., Ponganis P. J., Gendner J.-P., Lignon J., Mestre X., Huin N., Thorson P. H., Robisson P., Le Maho Y.

(法国国家科学研究中心(CNRS)生态与生理学中心)

摘要:仅在海中取食的帝企鹅(*Aptenodytes forsteri*)生活在南极高纬度海冰生境。当温度在−30 ℃且时常有大风时,它在冬季固定冰上繁殖。假设完全是孵化单卵,雄性企鹅在最严峻的环境下大约要禁食120天。在孵化期如果有雌企鹅接替的话,聚居地和大海的距离可能有100 km,甚至更远,但是,在孵化期间或夏季,帝企鹅去哪里取食尚属未知。冰间湖是海冰中的开放水域,冬季时长期为冰下生境,是南极区域采样最困难的地区。在这里,我们通过卫星监测帝企鹅觅食路径,并与海冰卫星图像相比较。冬季,企鹅穿过固定冰,步行长达296 km到冰间湖取食。而那些在浅的固定冰中游泳前行的,会离繁殖聚居地达895 km之远来取食。一个潜水记录显示,大多数企鹅潜水深度是中水,也有一些接近底部。取得这些有关帝企鹅觅食的详细信息,意味着这些海鸟为调查南极海冰生境提供了一个独特的机会。

9. Cullen J J, Neale P J, Lesser M P. Biological weighting function for the inhibition of phytoplankton photosynthesis by ultraviolet-radiation[J]. Science, 1992, 258(5082): 646-650.

Biological weighting function for the inhibition of phytoplankton photosynthesis by ultraviolet-radiation

Cullen J. J., Neale P. J., Lesser M. P.

(Department of Oceanography, Dalhousie University, Halifax, Nova Scotia, Canada B3H 4J1, and Bigelow Laboratory for Ocean Sciences, McKown Point, West Boothbay Harbor, ME 04575)

Abstract: Severe reduction of stratospheric ozone over Antarctica has focused increasing concern on the biological effects of ultraviolet-B (UVB) radiation (280 to 320 nanometers). Measurements of photosynthesis from an experimental system, in which phytoplankton are exposed to a broad range of irradiance treatments, are fit to an analytical model to provide the spectral biological weighting function that can be used to predict the short-term effects of ozone depletion on aquatic photosynthesis. Results show that UV-A (320 to 400 nanometers) significantly inhibits the photosynthesis of a marine diatom and a dinoflagellate, and that the effects of UVB are even more severe. Application of the model suggests that the Antarctic ozone hole might reduce near-surface photosynthesis by 12 to 15 percent, but less so at depth. The experimental system makes possible routine estimation of spectral weightings for natural phytoplankton.

紫外辐射抑制浮游植物光合作用的生物加权函数

Cullen J. J., Neale P. J., Lesser M. P.

(加拿大达尔豪西大学海洋系和比奇洛海洋科学实验室)

摘要：南极上空平流层臭氧严重减少增加了人们对紫外线-B(UV-B)辐射(280～320 nm)生物效应的关注。对暴露于一系列辐射处理的浮游植物进行光合作用的测量，并用一个分析模型来拟合以获得光谱生物加权函数，这种生物加权函数可以预测臭氧消耗对水草光合作用的短期效应。结果表明，紫外线-A(UV-A)(320～400 nm)明显地抑制海洋硅藻和甲藻的光合作用，而UV-B的抑制作用更强烈。模型应用表明，南极上空的臭氧洞可能会减少近地表光合作用的12%～15%，但随着深度增加臭氧洞对光合作用的抑制程度则降低。这个实验方法使常规的光谱加权在评估自然浮游植物时成为可能。

10. Baker C S, Perry A, Bannister J L, et al. Abundant mitochondrial-DNA variation and worldwide population-structure in humpback whales[J]. PNAS, 1993, 90(17): 8239-8243.

Abundant mitochondrial-DNA variation and worldwide population-structure in humpback whales

Baker C. S., Perry A., Bannister J. L., Weinrich M. T., Abernethy R. B., Calambokidis J., Lien J., Lambertsen R. H., Urban Ramlrez J., Vasquez O., Clapham P. J., Alling A., O'brien S. J., Palumbi S. R.

(Department of Zoology and Kewalo Marine Laboratory, University of Hawaii, Honolulu, HI 96822)

Abstract: Hunting during the last 200 years reduced many populations of mysticete whales to near extinction. To evaluate potential genetic bottlenecks in these exploited populations, we examined mitochondrial DNA control region sequences from 90 individual humpback whales (*Megaptera novaeangliae*) representing six subpopulations in three ocean basins. Comparisons of relative nucleotide and nucler type diversity reveal an abundance of genetic variation in all but one of the oceanic subpopulations. Phylogenetic reconstruction of nucleotypes and analysis of maternal gene flow show that current genetic variation is not due to postexploitation migration between oceans but is a relic of past population variability. Calibration of the rate of control region evolution across three families of whales suggests that existing humpback whale lineages are of ancient origin. Preservation of preexploitation variation in humpback whales may be attributed to their long life-span and overlapping generations and to an effective, though perhaps not timely, international prohibition against hunting.

座头鲸丰富的线粒体 DNA 变异和世界范围内种群结构

Baker C. S. , Perry A. , Bannister J. L. , Weinrich M. T. , Abernethy R. B. , Calambokidis J. , Lien J. , Lambertsen R. H. , Urban Ramlrez J. , Vasquez O. , Clapham P. J. , Alling A. , O'brien S. J. , Palumbi S. R.

(美国夏威夷大学动物学系和 Kewalo 海洋实验室)

摘要:过去 200 年来狩猎使许多须鲸种群濒临灭绝。为了评估这些被狩猎种群潜在的遗传瓶颈,我们测定了 90 个座头鲸(*Megaptera novaeangliae*)线粒体 DNA 控制区序列,它们代表了 3 个海洋盆地的 6 个亚种群。相关核苷酸和核型多样性的比较结果表明,除了一个海洋亚种群之外,所有亚种群都具有丰富的基因变化。核型进化史重建以及母性基因流动分析表明,当前基因变化并不是由于狩猎引起动物在海洋间的迁移导致的,而是过去种群变化的遗迹。对 3 个鲸鱼家族控制区进化速率的计算表明,现有的座头鲸谱系学是古代起源。狩猎前座头鲸基因变化的保存,可以归因于它们的长寿命、世代重叠,以及尽管也许不及时但有效的国际禁止狩猎行为。

11. Langin D,Laurell H,Holst L S,et al. Gene organization and primary structure of human hormone-sensitive lipase: possible significance of a sequence homology with a lipase of moraxella Ta144, an Antarctic bacterium[J]. PNAS,1993,90(11):4897-4901.

Gene organization and primary structure of human hormone-sensitive lipase: possible significance of a sequence homology with a lipase of moraxella Ta144, an Antarctic bacterium

Langin D. , Laurell H. , Holst L. S. , Belfrage P. , Holm C.

(Department of Medical and Physiological Chemistry 4, Lund University, P. O. Box 94, S-221 00 Lund, Sweden)

Abstract: The human hormone-sensitive lipase (HSL) gene encodes a 786-aa polypeptide (85. 5 kDa). It is composed of nine exons spanning almost-equal-to 11 kb, with exons 2-5 clustered in a 1. 1 kb region. The putative catalytic site (Ser423) and a possible lipid-binding region in the C-terminal part are encoded by exons 6 and 9, respectively. Exon 8 encodes the phosphorylation site (Ser551) that controls cAMP-mediated activity and a second site (Ser553) that is phosphorylated by 5'-AMP-activated protein kinase. Human HSL showed 83% identity with the rat enzyme and contained a 12-aa deletion immediately upstream of the phosphorylation sites with an unknown effect on the activity control. Besides the catalytic site motif (Gly-Xaa-Ser-Xaa-Gly) found in most lipases, HSL shows no homology with other known lipases or proteins, except for a recently reported unexpected homology between the region surrounding its catalytic site and that of the lipase 2 of Moraxella TA144, an Antarctic psychrotrophic bacterium. The gene of lipase 2, which catalyses lipolysis below 4-degrees-C, was absent in the genomic DNA of five other Moraxella strains living at 37-degrees-C. The lipase 2-like sequence in HSL may reflect an evolutionarily conserved cold adaptability that might be of critical survival value when low-temperature-mobilized endogenous lipids are the primary energy source (e. g. , in poikilotherms or hibernators). The finding that HSL at 10-degrees-C retained 3- to 5-fold more of its 37-degrees-C catalytic activity than lipoprotein lipase or carboxyl ester lipase is consistent with this hypothesis.

人类激素敏感酯酶的基因组织和基本结构:与南极细菌(摩拉克氏菌属)moraxella Ta144 的序列同源性的可能意义

Langin D., Laurell H., Holst L. S., Belfrage P., Holm C.

(瑞典隆德大学医学和生理化学部)

摘要:人类激素敏感酯酶(HSL)基因编码 786-氨基酸肽(85.5 kDa),是由 9 个长几乎为 1.1 kb 的外显子组成的,外显子 2～5 群集在一个 1.1 kb 的区域。C-端部分可能的催化部位(Ser423)和一个可能的脂质结合区域由外显子 6 和 9 分别编码。外显子 8 编码用于控制环磷酸腺苷(cAMP)介导活动的磷酸化位点(Ser551),以及由 5'-AMP 活化蛋白激酶磷酸化的次等位点(Ser553)。人类 HSL 与老鼠酶 83%同一,并载有直接在磷酸化位点上游的 12 个氨基酸缺失,它们对活动控制的效果是未知的。除了在许多脂肪酶中发现的催化部位模块(甘氨酸-Xaa-丝氨酸 Xaa-甘氨酸)外,HSL 没有显示出与其他已知的脂肪酶或蛋白质的同源性,一个例外是最近报道的其催化部位和 Moraxella TA144 脂肪酶 2(一个南极嗜冷菌)的意外同源。脂肪酶 2 的基因,能催化低于 4 ℃的脂肪分解作用,存活于 37 ℃下的 5 个其他摩拉克氏菌属的染色体基因不存在。HSL 中的脂肪酶 2 样序列可以反映一个进化上保守的冷适应能力,当低温移动内生脂质是基本能量来源时,这种冷适应能力具有重要的生存价值。与脂蛋白脂肪酶或羧基酯脂肪酶催化活性相比,10 ℃时 HSL 保留了其在 37 ℃催化活性的 3～5 倍甚至更多,这一研究结果与此假说是一致的。

12. Brey T,Klages M,Dahm C,et al. Antarctic benthic diversity[J]. Nature,1994,368(6469):297.

Antarctic benthic diversity

Brey T., Klages M., Dahm C., Gorny M., Gutt J., Hain S., Stiller M., Arntz W. E.

(Alfred Wegener Institute for Polar-and Marine Research, Postfach 120161, D-27515 Bremerhaven, Germany)

Abstract: Poore and Wilson, and Rex et al, give evidence for a distince decline in species diversity in deep-sea communities from the tropics towards north polar regions; and for a less clear trend but high interregions variability in the southern Hemisphere. Poore and Wilson included some isopod data from the Scotia Basin and the Weddell Sea (Atlantic sector of the sourthern ocean), but the data of Rex et al are limited to regions north of 40°S.

南极底栖生物多样性

Brey T., Klages M., Dahm C., Gorny M., Gutt J., Hain S., Stiller M., Arntz W. E.

(德国阿尔弗雷德魏格纳极地和海洋研究所)

摘要:Poore 和 Wilson,以及 Rex 等为以下两个趋势提供了证据:(1) 深海群落物种多样性从热带地区到北部极区呈现递减的趋势;(2) 南半球一个不明显但区域间高度变化的趋势。Poore 和 Wilson 包括斯科舍海盆和威德尔海(南大洋的大西洋区域)等脚类动物资料,但 Rex 等的资料仅限于 40°S 以北区域。

13. Delong E F, Wu K Y, Prezelin B B, et al. High abundance of archaea in Antarctic marine picoplankton[J]. Nature, 1994, 371(6499): 695-697.

High abundance of archaea in Antarctic marine picoplankton

Delong E. F., Wu K. Y., Prezelin B. B., Jovine R. V. M.

(Biology Department, Division of Ecology, Evolution and Marine Biology, University of California, Santa Barbara, Santa Barbara, California 93106, USA)

Abstract: Archaea (archaebacteria) constitute one of the three major evolutionary lineages of life on Earth. Previously these prokaryotes were thought to predominate in only a few unusual and disparate niches, characterized by hypersaline, extremely hot, or strictly anoxic conditions. Recently, novel (uncultivated) phylotypes of Archaea have been detected in coastal and subsurface marine waters, but their abundance, distribution, physiology and ecology remain largely undescribed. Here we report exceptionally high archaeal abundance in frigid marine surface waters of Antarctica. Pelagic Archaea constituted up to 34% of the prokaryotic biomass in coastal Antarctic surface waters, and they were also abundant in a variety of other cold, pelagic marine environments. Because they can make up a significant fraction of picoplankton biomass in the vast habitats encompassed by cold and deep marine waters, these pelagic Archaea represent an unexpectedly abundant component of the Earth's biota.

南极海洋超微型浮游生物古细菌的高丰度

Delong E. F., Wu K. Y., Prezelin B. B., Jovine R. V. M.

(美国加州大学圣巴巴拉分校生物学系生态学、进化和海洋生物学部)

摘要:古细菌(原始细菌)构成了地球上生命的三个主要进化谱系学之一。以前认为这些原核生物只在少数几个不寻常的和截然不同的生境(如高盐、极热或严格缺氧环境)占主导地位。最近,已在沿海和次表层海水发现新的(未培养的)古细菌种系型,但是尚缺乏对它们丰度、分布、生理学和生态学的描述。在这里,我们特别地报道在南极寒冷的表层海水具有很高的古细菌丰度。远洋古细菌占了南极沿海表层海水原核生物量的 34%,并且它们在其他多样的冷的远海环境中也大量存在。因为它们在由冷的和深的海水包围的极端生境中构成了超微型浮游生物生物量的重要部分,这些远洋古细菌的存在代表了地球生物区系不寻常的丰度组成。

14. Cocca E, Ratnayake Lecamwasam M, Parker S K, et al. Genomic remnants of alpha-globin genes in the hemoglobinless Antarctic icefishes[J]. PNAS, 1995, 92(6): 1817-1821.

Genomic remnants of alpha-globin genes in the hemoglobinless Antarctic icefishes

Cocca E., Ratnayake Lecamwasam M., Parker S. K., Camardella L., Ciaramella M., Diprisco G., Detrich H. W.

(Istituto di Biochimica delle Proteine ed Enzimologia, Consiglio Nazionale delle Ricerche, 80125 Naples, Italy)

Abstract: Alone among piscine taxa, the antarctic icefishes (family Channichthyidae, suborder Notothenioidei) have evolved compensatory adaptations that maintain normal metabolic functions in the absence of erythrocytes and the respiratory oxygen transporter hemoglobin. Although the uniquely "colorless" or "white" condition of the blood of icefishes has been recognized since the early 20th century, the status of globin genes in the icefish genomes has, surprisingly, remained unexplored. Using alpha- and beta-globin cDNAs from the antarctic *rockcod Notothenia coriiceps* (family Nototheniidae, suborder Notothenioidei), we have probed the genomes of three white blooded icefishes and four red-blooded notothenioid relatives (three antarctic, one temperate) for globin-related DNA sequences. We detect specific, high-stringency hybridization of the alpha-globin probe to genomic DNAs of both white and red blooded species, whereas the beta-globin cDNA hybridizes only to the genomes of the red-blooded fishes. Our results suggest that icefishes retain inactive genomic remnants of alpha-globin genes but have lost, either through deletion or through rapid mutation, the gene that encodes beta-globin. We propose that the hemoglobinless phenotype of extant icefishes is the result of deletion of the single adult beta-globin locus prior to the diversification of the clade.

在无血红蛋白的南极银鱼中α-球蛋白基因染色体组的残留

Cocca E., Ratnayake Lecamwasam M., Parker S. K., Camardella L., Ciaramella M., Diprisco G., Detrich H. W.

(意大利国家研究理事会蛋白质和酶学生物化学研究所)

摘要:在鱼的类群中,独一的南极冰鱼(科:Channichthyidae,亚目:Notothenioidei)已经进化了补偿适应机制,即在缺乏红细胞和运输氧的传递体血红蛋白时仍能保持正常的代谢功能。虽然自20世纪初冰鱼独特的"无色"或"白色"血液状况已经被人们所认识,但是冰鱼基因组的球蛋白基因状况仍然未知。用南极 *rockcod Notothenia coriiceps*(科:Nototheniidae,亚目:Notothenioidei)α和β cDNAs,我们探讨了三个白色纯种冰鱼和四个红血 notothenioid 亲属基因组(三个南极,一个温带)相关球蛋白的 DNA 序列。我们检测到在α-球蛋白探针和白色的、红色的冰鱼基因组 DNA 之间特定的高度杂交。我们的研究结果表明,冰鱼保留了α-球蛋白基因的惰性染色体组残留,但是通过删除或快速突变失去了编码β球蛋白的基因。我们分析,现存冰鱼无血红蛋白的表型是在进化分支的多元化前删除了单个成熟β-球蛋白基因的结果。

15. Nevitt G A, Veit R R, Kareiva P. Dimethyl sulfide as a foraging cue for Antarctic procellariiformes seabirds[J]. Nature, 1995, 376(6542): 680-682.

Dimethyl sulfide as a foraging cue for Antarctic procellariiformes seabirds

Nevitt G. A., Veit R. R., Kareiva P.

(Section of Neurobiology, Physiology and Behavior, Division of Biological Sciences, University of California, Davis, California 95616, USA)

Abstract: Many Procellariiformes seabirds make their living flying over vast expanses of seemingly featureless ocean waters in search of food. The secret of their success is a mystery, but an ability to hunt by smell has long been suspected. Here we present experimental evidence that Procellariiform seabirds can use a naturally occurring scented compound, dimethyl sulphide, as an orientation cue. Dimethyl sulphide has been studied intensely for its role in regulating global climate and is produced by phytoplankton in response to zooplankton grazing. Zooplankton, including Antarctic krill (*Euphausia superba*), are in turn eaten by seabirds and other animals. Results from controlled behavioural experiments performed at sea show that many Procellariiforms can detect dimethyl sulphide, and that some species (for example, storm petrels) are highly attracted to it. To our knowledge, this constitutes the first evidence that dimethyl sulphide is part of the natural olfactory landscape overlying the southern oceans.

二甲基硫醚作为南极鹱形目海鸟觅食的信号

Nevitt G. A., Veit R. R., Kareiva P.

(美国加州大学戴维斯分校生物科学部神经生物学,生理学和行为学系)

摘要:许多鹱形目海鸟在广阔的看似平常的海域飞行寻找食物。它们成功的秘诀是一个谜,但是猜测它们具有靠嗅觉捕食猎物的能力。在这里,我们提供了实验证据,证明鹱形目海鸟可以使用天然香味化合物二甲基硫醚作为一个方向信号。二甲基硫醚是由浮游植物对被浮游动物取食时产生的,人们大量地研究了它在调节全球气候方面的作用。浮游动物,其中包括南极磷虾(*Euphausia superba*),反过来被海鸟和其他动物取食。从海上进行的控制行为实验表明,许多鹱形目能发觉二甲基硫醚,而有些物种(例如,暴海燕)能被二甲基硫醚高度吸引。据我们所知,这构成了第一个证据,即二甲基硫醚是覆盖在南大洋上的自然嗅觉景观的一部分。

16. Paul Arthur Berkman, Michael L Prentice. Pliocene extinction of Antarctic pectinid mollusks[J]. Science, 1996, 271(5255): 1606-1607.

Pliocene extinction of Antarctic pectinid mollusks

Paul Arthur Berkman, Michael L. Prentice

(Byrd Polar Research Center, Ohio State University, Columbus, OH 43210, USA Michael L. Prentice)

Abstract: The report by Edward J. Petuch about a two-stage Pliocene-Pleistocene mass extinction that decreased the diversity of stenothermal molluscan genera in Florida raises the question of where the climatic cooling events propagated. It is accepted that the Northern Hemisphere ice sheets began developing at the end of the Pliocene, but their feedback and late Neogene connection with changes in the Antarctic ice sheets have not been resolved.

上新世时期南极扇贝科软体动物的灭绝

Paul Arthur Berkman, Michael L. Prentice

(美国俄亥俄州立大学哥伦布分校伯德极地研究中心)

摘要:Edward报道了上新世-更新世两个阶段大规模生物灭绝,减少了佛罗里达窄温性软体动物属的生物多样性。这引起了一个疑问,即气候变冷事件是如何传播的。众所周知,北半球冰架开始形成于上新世末期,但是它们的反馈信息和新第三纪晚期南极冰架变化之间的联系尚未解决。

17. Cattaneo Vietti R,Bavestrello G,Cerrano C,et al. Optical fibres in an Antarctic sponge[J]. Nature, 1996,383(6599):397-398.

Optical fibres in an Antarctic sponge

Cattaneo Vietti R., Bavestrello G., Cerrano C., Sara M., Benatti U., Giovine M., Gaino E.

(Istituto di Zoologia dell' Universita di Genova, Via Balbi 5, 1-16126 Genova, Italy)

Abstract: SIR-The skeleton of demosponges and hexactinellids is generally composed of spicules consisting of amorphous hydrated silica (opal) laid out around a proteic axial filament. Spicules constitute a mechanical support and act as deterrents to predators. The presence of a filamentous green alga (Ostreobium) closely associated with the spicule bundles of a demosponge (*Tethya seychellensis*) suggested to us that light energy might reach the inside of the sponge body visa siliceousspicules. Such a natural optical-fibre system could influence sponge/symbiont interactions and wider aspects of sponge biology.

南极海绵动物体内的光纤

Cattaneo Vietti R., Bavestrello G., Cerrano C., Sara M., Benatti U., Giovine M., Gaino E.

(意大利热那亚大学动物学研究室)

摘要:寻常海绵纲和六放海绵纲动物的骨骼一般由骨针组成,骨针由含有围绕蛋白质轴丝排列的无定形水合二氧化硅(蛋白石)组成。骨针构成了机械支撑体,并起着威慑捕食者的作用。某种与寻常海绵纲动物(*Tethya seychellensis*)的骨针束密切相关的丝状绿藻(蚝壳藻属,Ostreobium)的存在暗示我们光能可能通过硅质的骨针进入了海绵动物体内。这种自然的光纤系统会影响到海绵动物/共生体间的相互作用及海绵动物生物特征的其他方面。

18. Marshall W A. Biological particles over Antarctica[J]. Nature,1996,383(6602):680.

Biological particles over Antarctica

Marshall W. A.

(British Antarctic Survey, Natural Environment Research Council, High Cross, Madingley Road, Cambridge CB3 OET, UK)

Abstract: SIR-Antarctica is viewed by almost all as a continent in biological isolation. But recent aerobiological sampling in the maritime Antarctic has revealed a dramatic influx of material from south America when, within a single 24-hour period, the density of the air-spora increased 20-fold over normal levels. This was associated with specific weather pattern which occurs with an estimated mean annual frequency of 1.5. This is the first time such an impressive transfer of biological material from another continent has been observed and it provides direct evidence of one way in which the Antarctic may have been recolonized since the last glacial maximum, about 18000 years ago. Such events also provide a mechanism of organisms to extend their range into Antarctica as mean annual isotherms move south with regional warming, as in recent years.

南极上空的生物微粒

Marshall W. A.

(英国自然环境研究理事会南极调查局)

摘要:几乎所有人都认为南极是一个与生物隔离的大陆。但是最近在南极近海进行的一次大气生物学采样却发现,有来自南美的物质大量涌入,在 24 小时的检测时间段内空气中花粉的密度较正常水平增加 20 倍。这与特殊的天气模式有关,其发生频率估计为每年 1.5 次。这是第一次观察到来自另一个大陆的生物物质发生大规模迁移,它也为末次冰盛期(约 18000 年前)以来生物重新移生到南极提供了直接证据。这类事件也为近年来因局部变暖,年均等温线南移,生物随之向南极扩展其生存范围提供了一个解释的机制。

19. Pakulski J D,Coffin R B,Kelley C A,et al. Iron stimulation of Antarctic bacteria[J]. Nature,1996, 383(6596):133-134.

Iron stimulation of Antarctic bacteria

Pakulski J. D., Coffin R. B., Kelley C. A., Downer R., Holder S. L., Aas P., Lyons M. M., Jeffrey W. H.

(US EPA Gulf Ecology Division, 1 Sabine Island Drive, Gulf Breeze, Florida 32561, USA)

Abstract: Recent investigations of the ocean's iron cycle have focused primarily on the response of phytoplankton to iron enrichment. Bacteria, however, are important in the trophodynamics and elemental cycles of marine ecosystems. With the exception of phototrophic prokaryotes, the response of bacteria to iron enrichment has largely been ignored. Here we report the results of an iron-enrichment experiment suggesting that the growth of heterotrophic bacteria in Antarctic waters is stimulated by low-concentration additions of iron.

铁元素对南极细菌的促进作用

Pakulski J. D. , Coffin R. B. , Kelley C. A. , Downer R. , Holder S. L. , Aas P. , Lyons M. M. , Jeffrey W. H.

(美国环保局海湾生态分局)

摘要:最近关于海洋中铁循环的研究主要集中在浮游植物对铁富集的反应上。然而,细菌对于海洋生态系统中的营养动力学和元素循环也很重要。除了光合营养的原核生物以外,细菌对于铁富集的反应在很大程度上被忽视了。在此,我们对所做的一项铁富集实验的结果进行报告,结果显示增加低浓度的铁元素就对南极海水中异养菌的生长有促进作用。

20. Peck L S,Brey T. Bomb signals in old Antarctic brachiopods[J]. Nature,1996,380(6571):207-208.

Bomb signals in old Antarctic brachiopods

Peck L. S. , Brey T.

(Natural Environment Research Coucil, British Antarctic Survey, High Cross, Madingley Road, Cambridge CB3 OET, UK)

Abstract: SIR-Skeletal check-marks are commonly used to assess the age and growth of organisms. They are usually assumed to be formed annually. By using radiocarbon bomb signals to calibrate growth checks in shells of Antarctic brachiopods, we show that they were laid down with a subbiennial periodicity. The data also indicate that low Southern Ocean ^{14}C signals are probably not caused by upwelling deep water, but are more probably due to reduced atmospheric supply and long-term radiocarbon deposition in ice.

南极古老腕足类动物的核弹实验信号

Peck L. S. , Brey T.

(英国自然环境研究理事会南极调查局)

摘要:骨骼校验标志常被用来估算生物的年龄和生长情况,通常认为它们每年都会形成。通过用放射性碳同位素核弹实验信号来标定南极腕足类动物壳中的生长标志,我们发现,它们以略小于两年的周期产生。实验数据也表明南大洋中较低的^{14}C信号可能不是由来自深海的上升流造成的,而更可能是由不断减少的大气供应和冰中长期的放射性碳同位素沉积引起的。

21. Chen L B, De Vries A L, Cheng C H C. Evolution of antifreeze glycoprotein gene from a trypsinogen gene in Antarctic notothenioid fish[J]. PNAS,1997,94(8):3811-3816.

Evolution of antifreeze glycoprotein gene from a trypsinogen gene in Antarctic notothenioid fish

Chen L. B. , De Vries A. L. , Cheng C. H. C.

(Department of Molecular and Integrative Physiology, University of Illinois, Urbana, IL 61801)

Abstract: Freezing avoidance conferred by different types of antifreeze proteins in various polar and subpolar fishes represents a remarkable example of cold adaptation, but how these unique proteins arose is unknown. We have found that the antifreeze glycoproteins (AFGPs) of the predominant Antarctic fish taxon, the notothenioids, evolved from a pancreatic trypsinogen. We have determined the likely evolutionary process by which this occurred through characterization and analyses of notothenioid AFGP and trypsinogen genes. The primordial AFGP gene apparently arose through recruitment of the 5' and 3' ends of an ancestral trypsinogen gene, which provided the secretory signal and the 3' untranslated region, respectively, plus ne novo amplification of a 9-nt Thr-Ala-Ala coding element from the trypsinogen progenitor to create a new protein coding region for the repetitive tripeptide backbone of the antifreeze protein. The small sequence divergence (4%～7%) between notothenioid AFGP and trypsinogen genes indicates that the transformation of the proteinase gene into the novel ice-binding protein gene occurred quite recently, about 5-14 million years ago (mya), which is highly consistent with the estimated times of the freezing of the Antarctic Ocean at 10-14 mya, and of the main phyletic divergence of the AFGP-bearing notothenioid families at 7-15 mya. The notothenioid trypsinogen to AFGP conversion is the first clear example of how an old protein gene spawned a new gene for an entirely new protein with a new function. It also represents a rare instance in which protein evolution, organismal adaptation, and environmental conditions can be linked directly.

南极鱼(notothenioid)胰蛋白酶原基因向抗冻糖蛋白基因的进化转化

Chen L. B. , De Vries A. L. , Cheng C. H. C.

(美国伊利诺伊大学香槟分校分子与综合生理学系)

摘要:极地和亚极地地区不同鱼类因体内含有不同的抗冻蛋白而具有抗冻能力,这是生物适应严寒的一个生动的例子,然而人们对这些特殊的抗冻蛋白是如何产生的却依然未知。我们发现南极鱼类中的优势种南极鱼体内的抗冻糖蛋白(AFGPs)是从一种胰蛋白酶原进化而来的。通过鉴定及分析南极鱼AFGP和胰蛋白酶原基因,我们测定了可能的进化过程。最初的AFGP基因看起来是由一种古老的胰蛋白酶原基因5'端和3'端募集而产生的,这些胰蛋白酶原基因分别产生分泌信号和3'非编码区,同时通过扩增一段来自原始胰蛋白酶原9-nt的苏氨酸—丙氨酸—丙氨酸的编码元件为抗冻蛋白重复的三肽主链产生了一段新的蛋白编码区。南极鱼AFGP和胰蛋白酶原基因间较小的序列差异(4%～7%)表明由蛋白酶基因向新的抗冻蛋白基因的转化就发生在最近,500万～1400万年前(mya),这与所估计的南冰洋冻结时间(10～14 mya)和容忍AFGP的南极鱼家族主要种群的分异时间(7～15 mya)高度一致。南极鱼胰蛋白酶原向AFGP转化是第一个清晰地向我们展示一种古老的基因如何产生一种具有新功能的蛋白质基因的实例。同时,它也是一个有关蛋白质演化、生物适应和环境状况发生直接联系的特例。

22. Chen L B, De Vries A L, Cheng C H C. Convergent evolution of antifreeze glycoproteins in Antarctic notothenioid fish and Arctic cod[J]. PNAS,1997,94(8):3817-3822.

Convergent evolution of antifreeze glycoproteins in Antarctic notothenioid fish and Arctic cod

Chen L. B. , De Vries A. L. , Cheng C. H. C.

(Department of Molecular and Integrative Physiology, University of Illinois, Urbana, IL 61801)

Abstract: Antarctic notothenioid fishes and several northern cods are phylogenetically distant (in different orders and superorders), yet produce near-identical antifreeze glycoproteins (AFGPs) to survive in their respective freezing environments. AFGPs in both fishes are made as a family of discretely sized polymers composed of a simple glycotripeptide monomeric repeat. Characterizations of the AFGP genes from notothenioids and the Arctic cod show that their AFGPs are both encoded by a family of polyprotein genes, with each gene encoding multiple AFGP molecules linked in tandem by small cleavable spacers. Despite these apparent similarities, detailed analyses of the AFGP gene sequences and substructures provide strong evidence that AFGPs in these two polar fishes in fact evolved independently. First, although Antarctic notothenioid AFGP genes have been shown to originate from a pancreatic trypsinogen, Arctic cod AFGP genes share no sequence identity with the trypsinogen gene, indicating trypsinogen is not the progenitor. Second, the AFGP genes of the two fish have different intron-exon organizations and different spacer sequences and, thus, different processing of the polyprotein precursors, consistent with separate genomic origins. Third, the repetitive AFGP tripeptide (Thr-Ala/Pro-Ala) coding sequences are drastically different in the two groups of genes, suggesting that they arose from duplications of two distinct, short ancestral sequences with a different permutation of three codons for the same tripeptide. The molecular evidence for separate ancestry is supported by morphological, paleontological, and paleoclimatic evidence, which collectively indicate that these two polar fishes evolved their respective AFGPs separately and thus arrived at the same AFGPs through convergent evolution.

南极鱼(notothenioid)和北极鳕鱼抗冻糖蛋白的协同进化

Chen L. B. , De Vries A. L. , Cheng C. H. C.

(美国伊利诺伊大学香槟分校分子与综合生理学系)

摘要：南极的南极鱼和几种北方鳕鱼在系统发育上相差甚远(属于不同的序列和亚群)，但是在各自的冷环境中产生几乎相同的抗冻糖蛋白(AFGPs)用以生存。两种鱼类的AFGPs家庭由大小不一的聚合物，由一个简单的甘氨酸三肽单体重复组成。来自南极鱼和北极鳕鱼的AFGP基因表明，它们的AFGPs都是多蛋白基因家族编码，每个基因编码多个AFGP分子，这些分子由小裂解间隙串联连接。尽管有这些明显的相似之处，AFGP基因序列和亚结构的详细分析提供的强烈的证据表明：事实上两个极地鱼类AFGPs是独立进化的。首先，虽然南极鱼AFGP基因已被证实来自胰腺胰蛋白酶原，北极鳕鱼AFGP基因并没有相似的胰蛋白酶原基因序列，这说明胰蛋白酶并不是其祖先。其次，两种鱼AFGP基因有不同的内含子-外显子构架和不同的间隔序列，因此，蛋白前体的加工不同，具有各自的基因起源。第三，两组基因的重复AFGP肽(苏氨酸-丙氨酸/脯氨酸-丙氨酸)编码序列也是明显不同的，表明它们来自两个不同的短祖先序列,有相同的三肽但不同的密码子。各自祖先的分子证据,由形态学、古生物学、古气候的证据所支持,它们共同暗示了这两个极地鱼类 AFGPs 是各自进化的,因此,是通过不同的进化路径达到同样 AFGPs 的。

23. Gardner H, Kerry K, Riddle M, et al. Poultry virus infection in Antarctic penguins[J]. Nature, 1997, 387(6630): 245.

Poultry virus infection in Antarctic penguins

Gardner H., Kerry K., Riddle M., Brouwer S., Gleeson L.

(Antarctic Division, Department of Environment, Sport and Territories, Channel Highway, Kingston 7050, Tasmania, Australia)

Abstract: Antarctic penguins appear to be relatively free of infectious disease, although there is serological evidence of infection with a number of avian diseases found elsewhere. An infectious agent is suspected in one case of mass morality in Adelie penguin chicks, *Pygoscelis adeliae*, but there have been no confirmed reports of any major outbreaks of infectious disease. Here we report evidence for the occurance of an avian pathogen, infectious bursal disease virus (IBDV), in wild Antarctic penguins. This raises concern for the conservation of avian wildlife in Anatarctic.

南极企鹅病毒感染

Gardner H., Kerry K., Riddle M., Brouwer S., Gleeson L.

(澳大利亚环境、运动与领地部南极司)

摘要:虽然在有的地方存在一些南极企鹅感染多种鸟类疾病的血清学证据,但它们看上去并没有传染病。尽管在一次阿德雷幼企鹅(*Pygoscelis adeliae*)的集体死亡事件中怀疑有传染源,但目前还没有关于传染病大规模暴发的可靠报道。在此我们报道在南极野生企鹅中出现一种叫作传染性法氏囊病病毒(IBDV)的鸟类病原体的证据,希望这能引起人们对南极野生鸟类保护的关注。

24. Loeb V, Siegel V, Holm Hansen O, et al. Effects of sea-ice extent and krill or salp dominance on the Antarctic food web[J]. Nature, 1997, 387(6636): 897-900.

Effects of sea-ice extent and krill or salp dominance on the Antarctic food web

Loeb V., Siegel V., Holm Hansen O., Hewitt R., Fraser W., Trivelpiece W., Trivelpiece S.

(Moss Landing Marine Laboratories, P. O. Box 450, Moss Landing, California 95039-450, USA)

Abstract: Krill (*Euphausia superba*) provide a direct link between primary producers and higher trophic levels in the Antarctic marine food web. The pelagic tunicate *Salpa thompsoni* can also be important during spring and summer through the formation of extensive and dense blooms. Although salps are not a major dietary item for Antarctic vertebrate predators, their blooms can affect adult krill reproduction and survival of krill larvae. Here we provide data from 1995 and 1996 that support hypothesized relationships between krill, salps and region-wide sea-ice conditions. We have assessed salp consumption as a proportion of net primary production, and found correlations between herbivore densities and integrated chlorophyll-a that indicate that there is a degree of competition between krill and salps. Our analysis of the relationship between annual sea-ice cover and a longer time series of air temperature measurements indicates a decreased frequency of winters with extensive sea-ice development over the last five decades. Our data suggest that decreased krill availability may affect the levels of their vertebrate predators. Regional warming and reduced krill abundance therefore affect the marine food web and krill resource management.

海冰覆盖范围和磷虾或樽海鞘的优势地位对南极食物网的影响

Loeb V., Siegel V., Holm Hansen O., Hewitt R., Fraser W., Trivelpiece W., Trivelpiece S.

(美国加州莫斯兰丁海洋实验室)

摘要:磷虾(*Euphausia superba*)将南极海洋食物网的初级生产者和更高营养级直接联系了起来,带有被囊的浮游生物纽鳃樽(*Salpa thompsoni*)因在春季和夏季时范围和密度迅速扩大,也扮演着重要角色。尽管樽海鞘不是南极脊椎类捕食者的主要食物,但它们的大量繁殖会影响到成年磷虾的繁殖和小磷虾的生存。现在,我们给出一些1995年和1996年的资料,它们都支持磷虾、樽海鞘和区域海冰条件之间某种猜测的关系。我们估测了樽海鞘作为净初级生产量中一部分的消耗量,发现在食草动物密度和总叶绿素-a间存在关联关系,这种关系表明磷虾和樽海鞘在一定程度上存在竞争。我们对每年海冰覆盖情况和一系列时间更长的大气温度测量间的关系进行了分析,结果显示过去50年中随着海冰大量形成,冬季频率开始下降。我们的资料表明磷虾数量下降可能会影响其脊椎类捕食者的营养级。区域性变暖和磷虾数量降低随之影响到海洋食物网和磷虾资源的管理。

25. Malloy K D, Holman M A, Mitchell D, et al. Solar UVB-induced DNA damage and photoenzymatic DNA repair in Antarctic zooplankton[J]. PNAS, 1997, 94(4): 1258-1263.

Solar UVB-induced DNA damage and photoenzymatic DNA repair in Antarctic zooplankton

Malloy K. D., Holman M. A., Mitchell D., Detrich H. W.

(Department of Biology, Northeastern University, Boston, MA 02115)

Abstract: The detrimental effects of elevated intensities of mid-UV radiation (UVB), a result of stratospheric ozone depletion during the austral spring, on the primary producers of the Antarctic marine ecosystem have been well documented. Here we report that natural populations of Antarctic zooplankton also sustain significant DNA damage [measured as cyclobutane pyrimidine dimers (CPDs)] during periods of increased UVB flux. This is the first direct evidence that increased solar UVB may result in damage to marine organisms other than primary producers in Antarctica. The extent of DNA damage in pelagic icefish eggs correlated with daily incident UVB irradiance, reflecting the difference between acquisition and repair of. Patterns of DNA damage in fish larvae did not correlate with daily

UVB flux, possibly due to different depth distributions and/or different capacities for DNA repair. Clearance of CPDs by Antarctic fish and krill was mediated primarily by the photoenzymatic repair system. Although repair rates were large for all species evaluated, they were apparently inadequate to prevent the transient accumulation of substantial CPD burdens. The capacity for DNA repair in Antarctic organisms was highest in those species whose early life history stages occupy the water column during periods of ozone depletion (austral spring) and lowest in fish species whose eggs and larvae are abundant during winter. Although the potential reduction in fitness of Antarctic zooplankton resulting from DNA damage is unknown, we suggest that increased solar UV may reduce recruitment and adversely affect trophic transfer of productivity by affecting heterotrophic species as well as primary producers.

太阳紫外线引起的南极浮游动物 DNA 损伤和光酶修复

Malloy K. D., Holman M. A., Mitchell D., Detrich H. W.

(美国东北大学生物系)

摘要:南半球春季平流层臭氧耗损的结果之一是中波紫外线辐射(UV-B)强度的增加,这种增加对南极海洋生态系统初级生产者的致命作用已经有了很好的研究。在这里,我们报道了在紫外线通量增加期间,南极浮游动物的自然种群 DNA 也受到严重损伤[测定指标为环丁烷嘧啶二聚体(CPDs)]。这是第一个直接证据,即除了南极初级生产者外,增加太阳 UV-B 也可能导致对南极海洋生物的伤害。远洋冰鱼(icefish)卵 DNA 受损伤的程度与日常 UV-B 辐射相关,反映了获得和修复 CPDs 的不同。鱼卵 DNA 损伤模式和日常 UV-B 辐射通量不相关,这可能由于不同深度的分布和/或不同 DNA 修复能力。南极鱼和磷虾 CPDs 清除主要由光酶修复系统调节。虽然修复速率对所有评估物种来说很大,它们显然不足以防止大量的 CPD 短暂积累所引起的负担。DNA 修复能力以那些在南极春季臭氧耗损期间,早期生活史阶段在水柱中的南极生物最高,冬季卵和幼体丰度大的鱼类物种中最低。虽然 DNA 损伤导致南极浮游动物潜在适应性的降低尚属未知,我们认为,太阳紫外线的增加可能会减少新幼体的产生,并通过影响异氧物种以及初级生产者,从而对生产力营养转移产生不利的影响。

26. Sidell B D, Vayda M E, Small D J, et al. Variable expression of myoglobin among the hemoglobinless Antarctic icefishes[J]. PNAS, 1997, 94(7): 3420-3424.

Variable expression of myoglobin among the hemoglobinless Antarctic icefishes

Sidell B. D., Vayda M. E., Small D. J., Moylan T. J., Londraville R. L., Yuan M. L., Rodnick K. J., Eppley Z. A., Costello L.

(University of Marine, Orne, Me 04469, USA)

Abstract: The important intracellular oxygen-binding protein, myoglobin (Mb), is thought to be absent from oxidative muscle tissues of the family of hemoglobinless Antarctic icefishes, Channichthyidac. Within this family of fishes, which is endemic to the Southern Ocean surrounding Antarctica, there exist 15 known species and Ⅱ genera. To date, we have examined eight species of icefish (representing seven genera) using immunoblot analyses. Results indicate that Mb is present in heart ventricles from five of these species of icefish, Mb is absent from heart auricle and oxidative skeletal muscle of all species, We have identified a 0. 9-kb mRNA in Mb-expressing species that hybridizes with a Mb cDNA probe from the closely related red-blooded Antarctic nototheniid fish, *Notothenia coriiceps*. In confirmation that the 0. 9-kb mRNA encodes Mb, we report the full-length Mb cDNA sequence of the ocellated icefish, *Chionodraco rastrospinosus*. Of the eight icefish species examined, three lack Mb polypeptide in heart ventricle, although one of these expresses the Mb mRNA. All species of icefish retain the Mb gene in their genomic DNA. Based on phylogeny of the icefishes, loss of Mb expression has occurred independently at least three times and by at least two distinct molecular mechanisms during speciation of the family.

在无血红蛋白的南极冰鱼中肌红蛋白的变量表达

Sidell B. D., Vayda M. E., Small D. J., Moylan T. J., Londraville R. L., Yuan M. L., Rodnick K. J., Eppley Z. A., Costello L.

(美国缅因大学海洋科学院和动物学系)

摘要:最重要的细胞内氧结合蛋白,肌红蛋白(Mb),被认为在无血红蛋白的南极冰鱼(icefishes)科中鳄冰鱼科氧化肌肉组织中是不存在的。这种鱼家族中的鱼类,是围绕南极洲的南大洋所特有的,现存在15个种和2个属。到目前为止,我们已经用免疫分析研究了8种冰鱼物种(代表7个属)。结果表明,这些冰鱼中的5个肌红蛋白在心脏心室中是存在的,但在所有物种的心耳廓和氧化骨骼肌没有发现肌红蛋白。我们已经鉴定出一个肌红蛋白表达的物种,其0.9 kb的mRNA与探测到的与南极冰鱼、*Notothenia coriiceps* 紧密联系的Mb cDNA杂交。为了确认0.9 kb的mRNA编码肌红蛋白,我们报道了单眼冰鱼 *Chionodraco rastrospinosus* 完整长度的Mb cDNA序列。在8个检测的冰鱼物种中,3个心室中缺乏Mb多肽,尽管它们其中之一Mb mRNA已表达。冰鱼所有物种在其染色体组DNA中保留了Mb的基因。基于冰鱼发展史,Mb表达的损失至少独立发生了3次,并在该家族物种形成的过程中至少有2个不同的分子机制。

27. Falkowski P G, Barber R T, Smetacek V. Biogeochemical controls and feedbacks on ocean primary production[J]. Science,1998,281(5374):200-206.

Biogeochemical controls and feedbacks on ocean primary production

Falkowski P. G., Barber R. T., Smetacek V.

(Rutgers State Univ, Inst Marine & Coastal Sci, New Brunswick, NJ 08901 USA)

Abstract: Changes in oceanic primary production, linked to changes in the network of global biogeochemical cycles, have profoundly influenced the geochemistry of Earth for over 3 billion years. In the contemporary ocean, photosynthetic carbon fixation by marine phytoplankton leads to formation of similar to 45 gigatons of organic carbon per annum, of which 16 gigatons are exported to the ocean interior. Changes in the magnitude of total and export production can strongly influence atmospheric CO_2 levels(and hence climate) on geological time scales, as well as set upper bounds for sustainable fisheries harvest. The two fluxes are critically dependent on geophysical processes that determine mixed-layer depth, nutrient fluxes to and within the ocean, and food-web structure. Because the average turnover time of phytoplankton carbon in the ocean is on the order of a week or less, total and export production are extremely sensitive to external forcing and consequently are seldom in steady state. Elucidating the biogeochemical controls and feedbacks on primary production is essential to understanding how oceanic biota responded to and affected natural climatic variability in the geological past, and will respond to anthropogenically influenced changes in coming decades. One of the most crucial feedbacks results from changes in radiative forcing on the hydrological cycle, which influences the aeolian iron flux and, in turn, affects nitrogen fixation and primary production in the oceans.

海洋初级生产量的生物地球化学控制和反馈

Falkowski P. G., Barber R. T., Smetacek V.

（美国罗格斯州立大学海洋与海岸科学研究所）

摘要：海洋初级生产力的变化是与全球地球化学循环网络变化相联系的。在过去的三十多亿年，海洋初级生产力的变化深刻地影响地球上的地球化学。在现代海洋，通过海洋浮游植物的光合碳固定，每年有近450亿吨的有机碳形成，其中有160亿吨输出到海洋内部。总量和输出生产力大小的变化，在地质时间尺度上强烈影响大气中的CO_2含量（以及气候），并影响到可持续渔业丰收的上限。这两个通量紧紧地依赖于地球物理过程，地球物理过程决定了混合层深度、流入海洋内部的营养通量，以及食物网结构。因为海洋中浮游植物碳的平均周转时间，是以一周或少于一周轮回的，总量和输出生产力对外部力量极其敏感，因此极少处于稳定态。为了深入了解海洋生物群如何在过去地质学上对自然气候变化进行响应和影响，以及对未来数十年人类影响的变化如何响应，阐明生物地球化学控制及其对初级生产力的反馈是必不可少的。最关键的反馈之一起因于辐射力量变化对水文循环的影响，它影响风成铁通量，进而影响海洋中的固氮作用和初级生产力。

28. Hutchins D A, Bruland K W. Iron-limited diatom growth and Si∶N uptake ratios in a coastal upwelling regime[J]. Nature,1998,393(6685):561-564.

Iron-limited diatom growth and Si∶N uptake ratios in a coastal upwelling regime

Hutchins D. A., Bruland K. W.

(Univ Delaware, Coll Marine Studies, Lewes, DE 19958 USA)

Abstract: There is compelling evidence that phytoplankton growth is limited by iron availability in the subarctic pacific, and equatorial Pacific and Southern oceans. A lack of iron prevents the complete biological utilization of the ambient nitrate and influences phytoplankton species composition in these open-ocean "high-nitrate, low-chlorophyll" (HNLC) regimes. But the effects of iron availability on coastal primary productivity and nutrient biogeochemistry are unknown. Here we present the results of shipboard seawater incubation experiments which demonstrate that phytoplankton are iron-limited in parts of the California coastal upwelling region. As in offshore HNLC regimes, the addition of iron to these nearshore HNLC waters promotes blooms of large chain-forming diatoms. The silicic acid∶nitrate (Si∶N) uptake ratios in control incubations are two to three times higher than those in iron incubations. Diatoms stressed by a lack of iron should therefore deplete surface waters of silicic acid before nitrate, leading to a secondary silicic acid limitation of the phytoplankton community. Higher Si : cell, Si : C and Si : pigment ratios in diatoms in the control incubations suggest that iron limitation leads to more silicified, faster-sinking diatom biomass. These results raise fundamental questions about the nature of nutrient-limitation interactions in marine ecosystems, palaeoproductivity estimates based on the sedimentary accumulation of biogenic opal, and the controls on carbon export from some of the world's most productive surface waters.

沿海岸上升流区铁限制硅藻的生长和硅：氮吸收比

Hutchins D. A. , Bruland K. W.

(美国特拉华州大学科尔海洋研究所)

摘要：有令人信服的证据表明，在亚北极太平洋、赤道太平洋、南大洋浮游植物的生长受铁的可利用性的限制。铁的缺乏，限制了生物对周围环境中硝酸盐的充分利用，影响"高硝酸盐低叶绿素"(HNLC)机制下的开放海域中浮游植物的物种组成。但铁的可用性对沿海初级生产力和营养的生物地球化学的影响是未知的。目前船上海水培养实验表明，在加州沿海上升流的部分地区浮游植物的生长受铁限制。正如在离岸"高硝酸盐低叶绿素"HNLC区域一样，近岸HNLC水域铁的增加促进大型硅藻暴发。控制培养条件下硅酸：硝酸(Si：N)的吸收比例比铁培养时高出2～3倍。因此，硅藻受到缺铁限制，会在耗尽表层海水中硝酸盐之前耗尽硅酸，导致浮游植物群落受硅酸限制。控制培养条件下硅藻中Si：细胞，Si：C，Si：色素的高比例，表明铁限制导致更多的硅化，较快的硅藻生物量下沉。这些结果对海洋生态系统中营养-限制相互作用的本质，基于生物硅沉积积累古生产力估计，以及一些世界上最富饶的表层海水对碳输出的控制的本质提出了重要的疑问。

29. Takeda S. Influence of iron availability on nutrient consumption ratio of diatoms in oceanic waters [J]. Nature, 1998, 393(6687): 774-777.

Influence of iron availability on nutrient consumption ratio of diatoms in oceanic waters

Takeda S.

(Cent Res Inst Elect Power Ind, Dept Biol, Chiba 27011, Japan)

Abstract: The major nutrients (nitrate, phosphate and silicate) needed for phytoplankton growth are abundant in the surface waters of the subarctic Pacific, equatorial Pacific and Southern oceans, but this growth is limited by the availability of iron. Under iron-deficient conditions, phytoplankton exhibit reduced uptake of nitrate and lower cellular levels of carbon, nitrogen and phosphorus. Here I describe seawater and culture experiments which show that iron limitation can also affect the ratio of consumed silicate to nitrate and phosphate. In iron-limited waters from all three of the aforementioned environments, addition of iron to phytoplankton assemblages in incubation bottles halved the silicate : nitrate and silicate : phosphate consumption ratios, in spite of the preferential growth of diatoms (silica-shelled phytoplankton). The nutrient consumption ratios of the phytoplankton assemblage from the Southern Ocean were similar to those of an iron-deficient laboratory culture of Antarctic diatoms, which exhibit increased cellular silicon or decreased cellular nitrogen and phosphorus in response to iron limitation. Iron limitation therefore increases the export of biogenic silicon, relative to nitrogen and phosphorus, from the surface to deeper waters. These findings suggest how the sedimentary records of carbon and silicon deposition in the glacial Southern Oceans can be consistent with the idea that changes in productivity, and thus in drawdown of atmospheric CO_2, during the last glaciation were stimulated by changes in iron inputs from atmospheric dust.

海水中铁的可利用性对硅藻营养物质消耗比例的影响

Takeda S.

(日本电力工业研究中心生物学系)

摘要：亚北极太平洋、赤道太平洋和南大洋的表层海水中浮游植物生长所需的主要营养物质（硝酸盐、磷酸盐和硅酸盐）丰富，但其生长受铁的可用性的限制。在缺铁条件下，浮游植物表现出硝酸盐摄取量减少，细胞水平里碳、氮和磷的含量较低的现象。海水和培养实验显示，铁的限制也可以影响硅酸盐对硝酸盐和磷酸盐的消耗比例。从上述三个环境中的铁限制水域来看，向培养瓶中浮游植物群落添加铁，尽管硅藻（硅胶壳浮游植物）的生长更好，硅酸盐∶硝酸盐、硅酸盐∶磷酸盐的消耗比率却减半。南大洋浮游植物群落的营养消耗比例与缺铁实验室里培养的南极硅藻相似，作为对铁限制的响应，均表现出细胞中硅含量增加或细胞中氮、磷含量减少的现象。因此，从表层到更深水域，相比氮和磷，铁限制会增加生源硅的输出。这些结果表明，南大洋冰期碳、硅沉积的沉积记录如何与生产力变化一致的想法，从而在末次冰期中大气中 CO_2 含量减少是由大气尘埃铁输入量变化所致的。

30. Arrigo K R, Robinson D H, Worthen D L, et al. Phytoplankton community structure and the drawdown of nutrients and CO_2 in the Southern Ocean[J]. Science, 1999, 283(5400): 365-367.

Phytoplankton community structure and the drawdown of nutrients and CO_2 in the Southern Ocean

Arrigo K. R., Robinson D. H., Worthen D. L., Dunbar R. B., Di Tullio G. R., Van Woert M., Lizotte M. P.

(NASA Goddard Space Flight Center, Code 971.0, Greenbelt, MD 20771, USA)

Abstract: Data from recent oceanographic cruises show that phytoplankton community structure in the Ross Sea is related to mixed layer depth. Diatoms dominate in highly stratified waters, whereas Phaeocystis antarctica assemblages dominate where waters an more deeply mixed. The drawdown of both carbon dioxide (CO_2) and nitrate per mole of phosphate and the rate of new production by diatoms are much lower than that measured for P. antarctica. Consequently, the capacity of the biological community to draw down atmospheric CO_2 and transport it to the deep ocean could diminish dramatically if predicted increases in upper ocean stratification due to climate warming should occur.

南大洋中浮游植物群落结构和养分及 CO_2 的减少

Arrigo K. R., Robinson D. H., Worthen D. L., Dunbar R. B.,
Di Tullio G. R., Van Woert M., Lizotte M. P.

(美国国家航空和航天局戈达德宇宙飞行中心)

摘要：最近海洋巡航的数据显示，罗斯海浮游植物群落结构与混合层深度有关。在高度分层的水域里硅藻占主导地位；而在充分混合的水域里南极棕囊藻属群落占主导地位，CO_2 和每摩尔磷酸盐中所含硝酸盐这两者的减少量和硅藻的新生产率，都比南极棕囊藻的测量值低得多。因此，如果气候变暖将导致上层海洋分层增加的预测成为事实的话，生物群落吸收大气中的二氧化碳并将其运输到深海的能力会显著降低。

31. Davis R W, Fuiman L A, Williams T M, et al. Hunting behavior of a marine mammal beneath the Antarctic fast ice[J]. Science, 1999, 283(5404): 993-996.

Hunting behavior of a marine mammal beneath the Antarctic fast ice

Davis R. W., Fuiman L. A., Williams T. M., Collier S. O., Hagey W. P.,
Kanatous S. B., Kohin S., Horning M.

(Department of Marine Biology, Texas A&M University, 5007 Avenue U, Galveston, TX 77553, USA)

Abstract: The hunting behavior of a marine mammal was studied beneath the Antarctic fast ice with an animal-borne video system and data recorder. Weddell seals stalked large Antarctic cod and the smaller subice fish *Pagothenia borchgrevinki*, often with the under-ice surface for backlighting, which implies that vision is important for hunting. They approached to within centimeters of cod without startling the fish. Seals flushed *P. borchgrevinki* by blowing air into subice crevices or pursued them into the platelet ice. These observations highlight the broad range of insights that are possible with simultaneous recordings of video, audio, three-dimensional dive paths, and locomotor effort.

南极冰下海洋哺乳动物的狩猎行为

Davis R. W., Fuiman L. A., Williams T. M., Collier S. O., Hagey W. P.,
Kanatous S. B., Kohin S., Horning M.

(美国得克萨斯 A&M 大学海洋生物系)

摘要：我们用动物背载的视觉系统和数据记录器，研究南极冰下海洋哺乳动物的狩猎行为。通常在冰面下背光的地方，威德尔海豹悄悄地追捕大型南极鳕鱼和一种小的冰下南极鱼 *Pagothenia borchgrevinki*，这意味着视觉对于狩猎是很重要的。在距离鳕鱼几厘米的地方，威德尔海豹偷偷潜近却不惊动它们。海豹通过将空气吹入冰下缝隙或将之逼入薄片冰层中的方法追击南极鱼 *P. borchgrevinki*。这些观察表明同时记录视频、音频、三维潜水途径和运动能力是可能的。

32. Vincent W F. Icy life on a hidden lake[J]. Science,1999,286(5447):2094-2095.

Icy life on a hidden lake

Vincent W. F.

(Centre d'études Nordiques, Laval University, Sainte-Foy, QC Canada, G1K 7P4)

Abstract: Lake Vostok, one of Earth's most hidden lakes, is located underneath more than 3 km of ice in the coldest, most remote part of Antarctica. An ice core that almost reaches the lake has been recovered. Jouzel et al show that the lowest parts of this ice core consist of lake ice, in contrast to the overlying ice, which is glacial in origin. Karl et al and Priscu et al have analyzed bacteria from the lake ice that give clues to life in the underlying lake.

一个隐蔽湖泊的冰冷生活

Vincent W. F.

(加拿大拉瓦尔大学 d'études Nordiques 中心)

摘要:沃斯托克湖是地球上最隐蔽的湖泊之一,位于南极洲最遥远、最冷的时候在冰下 3000 米多的地方。一个几乎接近它湖面的冰芯已经取得。Jouzel 等研究表明,这个冰芯的最低部由湖冰组成,不同于起源于冰期的覆冰。Karl 等和 Priscu 等已经对来自湖冰的一种细菌进行了分析,这为我们探索底层湖的生命提供了线索。

33. Edward R Abraham,Cliff S Law,Philip W Boyd, et al. Importance of stirring in the development of an iron-fertilized phytoplankton bloom[J]. Nature,2000,407(6805):727-730.

Importance of stirring in the development of an iron-fertilized phytoplankton bloom

Edward R. Abraham, Cliff S. Law, Philip W. Boyd, Samantha J. Lavender, Maria T. Maldonadok, Andrew R. Bowie

(National Institute for Water and Atmospheric Research (NIWA), PO Box 14-901, Kilbirnie, Wellington, New Zealand)

Abstract: The growth of populations is known to be influenced by dispersal, which has often been described as purely diffusive. In the open ocean, however, the tendrils and filaments of phytoplankton populations provide evidence for dispersal by stirring. Despite the apparent importance of horizontal stirring for plankton ecology, this process remains poorly characterized. Here we investigate the development of a discrete phytoplankton bloom, which was initiated by the iron fertilization of a patch of water (7 km in diameter) in the Southern Ocean. Satellite images show a striking, 150-km-long bloom near the experimental site, six weeks after the initial fertilization. We argue that the ribbon-like bloom was produced from the fertilized patch through stirring, growth and diffusion, and we derive an estimate of the stirring rate. In this case, stirring acts as an important control on bloom development, mixing phytoplankton and iron out of the patch, but also entraining silicate. This may have prevented the onset of silicate limitation, and so allowed the bloom to continue for as long as there was sufficient iron. Stirring in the ocean is likely to be variable, so blooms that are initially similar may develop very differently.

搅拌对于富铁水体里浮游植物暴发的重要性

Edward R. Abraham, Cliff S. Law, Philip W. Boyd, Samantha J. Lavender, Maria T. Maldonadok, Andrew R. Bowie

(新西兰国家水和大气研究所(NIWA))

摘要：我们知道，种群增长受散布的影响，散布通常也被描述为纯粹的扩散。然而在开放的海域中，浮游植物种群的卷须和细丝为搅拌引起分散提供了证据。尽管横向搅拌对于浮游生物生态学具有十分重要的作用，这一过程仍很不清楚。这里我们调查了一个独立的浮游植物暴发的形成过程，它是始于南大洋一个直径为 7 km 的富铁的块状水域。卫星图像显示的结果十分惊人，最初施肥后的第六个星期，在试验区域附近 150 km 长的水体中均有暴发发生。我们认为那个带状的暴发是该块状富铁水体通过搅拌、增长和扩散而产生的，并且我们还估算了搅拌速率。在这种情况下，搅拌作为一个控制暴发形成、浮游植物混合，以及铁游离出斑块，并夹着硅的重要手段。这可能会阻止硅酸盐限制的开始，所以，只要铁足够，暴发就会持续发展。海洋中的搅拌可能是多种多样的，所以即使一开始相似，暴发也会朝着不同的方向发展。

34. Philip W Boyd, Andrew J Watson, Cliff S Law, et al. A mesoscale phytoplankton bloom in the polar Southern Ocean stimulated by iron fertilization[J]. Nature, 2000, 407(6805): 695-702.

A mesoscale phytoplankton bloom in the polar Southern Ocean stimulated by iron fertilization

Philip W. Boyd, Andrew J. Watson, Cliff S. Law, Edward R. Abraham, Thomas Trull, Rob Murdoch, Dorothee C. E. Bakker, Andrew R. Bowie, K. O. Buesseler, Hoe Chang, Matthew Charette, Peter Croot, Ken Downing, Russell Frew, Mark Gall, Mark Hadfield, Julie Hall, Mike Harvey, Greg Jameson, Julie La Roche, Malcolm Liddicoat, Roger Ling, Maria T. Maldonado, Michael McKay R. , Scott Nodder, Stu Pickmere, Rick Pridmore, Steve Rintoul, Karl Safi, Philip Sutton, Robert Strzepek, Kim Tanneberger, Suzanne Turner, Anya Waite, John Zeldis

(National Institute of Water and Atmosphere, Centre for Chemical and Physical Oceanography, Department of Chemistry, University of Otago, Dunedin, New Zealand)

Abstract: Changes in iron supply to oceanic plankton are thought to have a significant effect on concentrations of atmospheric carbon dioxide by altering rates of carbon sequestration, a theory known as the "iron hypothesis". For this reason, it is important to understand the response of pelagic biota to increased iron supply. Here we report the results of a mesoscale iron fertilization experiment in the polar Southern Ocean, where the potential to sequester iron-elevated algal carbon is probably greatest. Increased iron supply led to elevated phytoplankton biomass and rates of photosynthesis in surface waters, causing a large drawdown of carbon dioxide and macronutrients, and elevated dimethyl sulphide levels after 13 days. This drawdown was mostly due to the proliferation of diatom stocks. But downward export of biogenic carbon was not increased. Moreover, satellite observations of this massive bloom 30 days later, suggest that a sufficient proportion of the added iron was retained in surface waters. Our findings demonstrate that iron supply controls phytoplankton growth and community composition during summer in these polar Southern Ocean waters, but the fate of algal carbon remains unknown and depends on the interplay between the processes controlling export, remineralisation and timescales of water mass subduction.

极地南大洋中由富铁引起的一个中尺度浮游植物暴发现象

Philip W. Boyd, Andrew J. Watson, Cliff S. Law, Edward R. Abraham, Thomas Trull, Rob Murdoch, Dorothee C. E. Bakker, Andrew R. Bowie, K. O. Buesseler, Hoe Chang, Matthew Charette, Peter Croot, Ken Downing, Russell Frew, Mark Gall, Mark Hadfield, Julie Hall, Mike Harvey, Greg Jameson, Julie La Roche, Malcolm Liddicoat, Roger Ling, Maria T. Maldonado, Michael McKay R. , Scott Nodder, Stu Pickmere, Rick Pridmore, Steve Rintoul, Karl Safi, Philip Sutton, Robert Strzepek, Kim Tanneberger, Suzanne Turner, Anya Waite, John Zeldis

(新西兰奥塔哥大学化学部化学和物理海洋学研究中心)

摘要:“铁假说”理论认为,海洋浮游生物铁供应量的变化通过改变碳排除速率,显著影响到大气中CO_2的浓度。出于这个原因,了解远洋生物群对增加铁供应的响应极为重要。这里我们报道了在极地南大洋一个中尺度水域里铁施肥试验的结果,南大洋铁增加导致藻类碳的排除潜力可能是最大的。铁供应增加导致表层海水中浮游植物生物量和光合速率升高,从而使得13天后CO_2和大量营养素显著减少,同时二甲基硫化物含量升高。这种减少主要是硅藻库存的扩散引起的。但是生物成因的碳输出量的下降并没有增加。此外,30天后用卫星对这种大规模暴发现象进行观测,结果表明,增加的铁中有很大一部分被保留在了表层海水里。我们的发现证实,在极地南大洋夏季期间,铁供应控制着海水中浮游植物生长和其群落组成,但是藻类碳何去何从仍然是个不解之谜,它取决于控制输出、再矿化的过程和水团下沉时间尺度之间的相互作用。

35. Darling K F, Wade C M, Stewart I A, et al. Molecular evidence for genetic mixing of Arctic and Antarctic subpolar populations of planktonic foraminifers[J]. Nature, 2000, 405(6782): 43-47.

Molecular evidence for genetic mixing of Arctic and Antarctic subpolar populations of planktonic foraminifers

Darling K. F., Wade C. M., Stewart I. A., Kroon D., Dingle R., Brown A. J. L.

(Department of Geology and Geophysics, University of Edinburgh, Edinburgh EH9 3JW, UK)

Abstract: Bipolarity, the presence of a species in the high latitudes separated by a gap in distribution across the tropics, is a well-known pattern of global species distribution. But the question of whether bipolar species have evolved independently at the poles since the establishment of the cold-water provinces 16-8 million years ago, or if genes have been transferred across the tropics since that time, has not been addressed. Here we examine genetic variation in the small subunit ribosomal RNA gene of three bipolar planktonic foraminiferal morphospecies. We identify at least one identical genotype in all three morphospecies in both the Arctic and Antarctic subpolar provinces, indicating that trans-tropical gene flow must have occurred. Our genetic analysis also reveals that foraminiferal morphospecies can consist of a complex of genetic types. Such occurrences of genetically distinct populations within one morphospecies may affect the use of planktonic foraminifers as a palaeoceanographic proxy for climate change and necessitate a reassessment of the species concept for the group.

北极和南极亚极地浮游有孔虫种群基因混合的分子证据

Darling K. F., Wade C. M., Stewart I. A., Kroon D., Dingle R., Brown A. J. L.

(英国爱丁堡大学地质与地球物理系)

摘要:两极性,存在于高纬度地区的一个物种在分布上被穿过热带地区的缺口所隔离,这是一个著名的全球物种分布模式。但是这个问题并没有解决,即自从800万至1600万年前的冷水区域建立以来,两极地区的双极物种是否已经独立进化了,或者从那个时候起它们的基因是否越过热带地区而转移了?这里,我们检测三种两极的浮游有孔虫物种小亚基核糖体RNA基因的遗传变异,发现在北极和南极亚极地区三种物种至少有一种基因型相同,这表明曾经一定发生过跨热带的基因流动。基因分析还揭示,有孔虫物种由一系列复杂的基因型组成。同一物种不同种群基因却存在差异,这种可能影响浮游有孔虫作为气候变化海洋学指标的使用,并且有必要重新考虑这一物种概念。

36. Di Tullio G R, Grebmeier J M, Arrigo K R, et al. Rapid and early export of Phaeocystis Antarctica blooms in the Ross Sea, Antarctica[J]. Nature, 2000, 404(6778): 595-598.

Rapid and early export of Phaeocystis Antarctica blooms in the Ross Sea, Antarctica

Di Tullio G. R., Grebmeier J. M., Arrigo K. R., Lizotte M. P., Robinson D. H., Leventer A., Barry J. B., Van Woert M. L., Dunbar R. B.

(University of Charleston, Grice Marine Laboratory, 205 Fort Johnson, Charleston, South Carolina 29412, USA)

Abstract: The Southern Ocean is very important for the potential sequestration of carbon dioxide in the oceans and is expected to be vulnerable to changes in carbon export forced by anthropogenic climate warming. Annual phytoplankton blooms in seasonal ice zones are highly productive and are thought to contribute significantly to pCO drawdown in the Southern Ocean. Diatoms are assumed to be the most important phytoplankton class with respect to export production in the Southern Ocean; however, the colonial prymnesiophyte *Phaeocystis antarctica* regularly forms huge blooms in seasonal ice zones and coastal Antarctic waters. There is little evidence regarding the fate of carbon produced by P. antarctica in the Southern Ocean, although remineralization in the upper water column has been proposed to be the main pathway in polar waters. Here we present evidence for early and rapid carbon export from P. antarctica blooms to deep water and sediments in the Ross Sea. Carbon sequestration from P. antarctica blooms may influence the carbon cycle in the Southern Ocean, especially if projected climatic changes lead to an alteration in the structure of the phytoplankton community.

南极罗斯海中南极棕囊藻大量繁殖和其早期的快速输出

Di Tullio G. R., Grebmeier J. M., Arrigo K. R., Lizotte M. P., Robinson D. H., Leventer A., Barry J. B., Van Woert M. L., Dunbar R. B.

(美国查尔斯顿大学 Grice 海洋实验室)

摘要:南大洋对海洋中潜在的 CO_2 去除作用有着非常重要的意义,可以预料南大洋对人为气候变暖促使的碳输出的变化来说是比较脆弱的。季节性冰区每年浮游植物暴发时,生产力极高,这对南大洋中二氧化碳分压的下降有重要贡献。关于南大洋生产力的输出,硅藻被认为是最重要的浮游植物。然而,在季节性冰区和南极沿岸水域中,拓殖的南极棕囊藻(*Phaeocystis antarctica*)常常形成大规模暴发。尽管人们已经提出在两极水域的上层水柱中,再矿化过程是主要途径,还是很难指出南大洋中南极棕囊藻产生的碳的去向。这里我们列出罗斯海南极棕囊藻早期快速碳输出到底部海水和沉积物中的证据。南极棕囊藻暴发对碳的去除作用可能影响到南大洋的碳循环,尤其是在气候变化造成浮游植物群落结构改变的情况下。

37. Gaston K J, Chown S L, Mercer R D. The animal species-body size distribution of Marion Island [J]. PNAS,2001,98(25):14493-14496.

The animal species-body size distribution of Marion Island

Gaston K. J., Chown S. L., Mercer R. D.

(Biodiversity and Macroecology Group, Department of Animal and Plant Sciences, University of Sheffield, Sheffield S10 2TN, United Kingdom)

Abstract: Body size is one of the most significant features of animals. Not only is it correlated with many life history and ecological traits, but it also may influence the abundance of species within, and their membership of, assemblages. Understanding of the latter processes is frequently based on a comparison of model outcomes with the frequency of species of different body mass within natural assemblages. Consequently, the form of these frequency distributions has been much debated. Empirical data usually concern taxonomically delineated groups, such as classes or orders, whereas the processes ultimately apply to whole assemblages. Here, we report the most complete animal species-body size distribution to date for those free-living species breeding on sub-Antarctic Marion Island and using the terrestrial environment. Extending over 15 orders of magnitude of variation in body mass, this distribution is bimodal, with separate peaks for invertebrates and vertebrates. Under logarithmic transformation, the distribution for vertebrates is not significantly skewed, whereas that for invertebrates is right-skewed. Contrary to expectation based on a fractal or pseudofractal environmental structure, the decline in the richness of species at the smallest body sizes is a real effect and not a consequence of unrecorded species or of species introductions to the island. The scarcity of small species might well be a consequence of their large geographic ranges.

马里恩岛动物物种-体型分布

Gaston K. J., Chown S. L., Mercer R. D.

(英国谢菲尔德大学动物与植物科学系)

摘要:体型是动物最重要的特征之一,它不仅与很多生活史和生态特性相关,而且可能影响群体内物种的丰度和成员关系。搞清楚后者常用的方法是应用模型对自然群体内不同个体重量的频率算出结果,再与自然结果进行比较。因此,这些频率分布的形式也就常会引起争论。因为实验数据通常只考虑分类学上描述的群体,比如纲、目,而实验程序最终却是应用于整个生物群体。在本文中,我们选择了那些在亚南极马里恩岛上繁殖并且利用那里的陆地环境自由生活的物种,报告了迄今为止关于它们的最完整的动物物种体型分布结果。包括的动物物种在个体重量上差15个量级,其分布曲线呈双峰状,两个峰分别对应无脊椎动物和脊椎动物。经对数变换,发现脊椎动物没有呈现显著的偏态分布,而无脊椎动物则呈右偏分布。与基于分形或伪分形环境结构的预期相反的是,最小体型物种的丰富程度在减少是真实存在的,并不是物种未被记载或有其他物种引入该岛的结果。小型物种稀少可能是因为它们所在的地理范围非常巨大。

38. Lopez-Garcia P, Rodriguez-Valera F, Pedros-Alio C, et al. Unexpected diversity of small eukaryotes in deep-sea Antarctic plankton[J]. Nature, 2001, 409(6820): 603-607.

Unexpected diversity of small eukaryotes in deep-sea Antarctic plankton

Lopez-Garcia P., Rodriguez-Valera F., Pedros-Alio C., Moreira D.

(Univ Miguel Hernandez, Div Microbiol, San Juan De Alicante 03550, Spain)

Abstract: Phylogenetic information from ribosomal RNA genes directly amplified from the environment changed our view of the biosphere, revealing an extraordinary diversity of previously undetected prokaryotic lineages. Using ribosomal RNA genes from marine picoplankton, several new groups of bacteria and archaea have been identified, some of which are abundant. Little is known, however, about the diversity of the smallest planktonic eukaryotes, and available information in general concerns the phytoplankton of the euphotic region. Here we recover eukaryotes in the size fraction 0.2-5 mum from the aphotic zone (250-3000 m deep) in the Antarctic polar front. The most diverse and relatively abundant were two new groups of alveolate sequences, related to dinoflagellates that are found at all studied depths. These may be important components of the microbial community in the deep ocean. Their phylogenetic position suggests a radiation early in the evolution of alveolates.

南极深海浮游生物中的小型真核生物出乎意料的多样性

Lopez-Garcia P., Rodriguez-Valera F., Pedros-Alio C., Moreira D.

(西班牙米盖尔·埃尔南德斯大学微生物学系)

摘要：来自核糖体 RNA 基因(环境中直接扩增得到)的系统进化信息改变了我们对生物圈的观念，揭示了以前未被发现的原核生物世系的非寻常多样性。应用来自海洋超微型浮游生物的核糖体 RNA 基因，几类全新的细菌和古菌被鉴别了出来，其中有的种类还很丰富。然而，人们对于最微小的浮游真核生物的多样性依然知之甚少，已知的信息基本上是关于透光层中浮游植物的。我们从南极极前锋无光层(250～3000 m 深)获得了粒度为 0.2～5 μm 的真核生物。多样性最丰富及数量较多的是两组新的囊泡虫类序列，与沟鞭藻类有亲缘关系，该藻在所有已做过研究的深度中都能找到。它们可能是深海微生物群落的重要成员，其系统进化地位表明在囊泡虫类进化过程中曾有过早期辐射。

39. Andrew S Brierley, Paul G Fernandes, Mark A Brandon, et al. Antarctic krill under sea ice: elevated abundance in a narrow band just south of ice edge[J]. Science, 2002, 295(5561): 1890-1892.

Antarctic krill under sea ice: elevated abundance in a narrow band just south of ice edge

Andrew S. Brierley, Paul G. Fernandes, Mark A. Brandon, Frederick Armstrong, Nicholas W. Millard, Steven D. McPhail, Peter Stevenson, Miles Pebody, James Perrett, Mark Squires, Douglas G. Bone, Gwyn Griffiths

(British Antarctic Survey, High Cross, Madingley Road, Cambridge CB3 0ET, UK)

Abstract: We surveyed Antarctic krill (*Euphausia superba*) under sea ice using the autonomous underwater vehicle Autosub-2. Krill were concentrated within a band under ice between 1 and 13 kilometers south of the ice edge. Within this band, krill densities were fivefold greater than that of open water. The under-ice environment has tong been considered an important habitat for krill, but sampling difficulties have previously prevented direct observations under ice over the scale necessary for robust krill density estimation. Autosub-2 enabled us to make continuous high-resolution measurements of krill density under ice reaching 27 kilometers beyond the ice edge.

南极海冰下的磷虾:在冰缘线南部的狭窄地带数量在上升

Andrew S. Brierley, Paul G. Fernandes, Mark A. Brandon, Frederick Armstrong, Nicholas W. Millard, Steven D. McPhail, Peter Stevenson, Miles Pebody, James Perrett, Mark Squires, Douglas G. Bone, Gwyn Griffiths

(英国南极调查局)

摘要:我们使用自主式水下航行器 Autosub-2 调查了南极海冰下的磷虾(*Euphausia superba*)。磷虾集中在距冰缘线南边 1~13 km 的冰下,呈带状分布。在这条分布带内,磷虾的密度是无冰水域的 5 倍。长期以来冰下环境被认为是磷虾的重要栖息地,但是采样上的困难一直阻碍着在冰下对磷虾密度进行大尺度、可靠的直接观测。Autosub-2 则能让我们在距离冰缘线 27 km 的范围内对冰下磷虾的密度进行持续的高分辨率测量。

40. Gavaghan H. Life in the deep freeze[J]. Nature, 2002, 415(6874): 828-830.

Life in the deep freeze

Gavaghan H.

(A free journalist based in Yorkshire)

Abstract: Unknown ecosystems and untapped records of the Earth's past may lie hidden in the lonely waters of Antarctica's Lake Vostok. But the lake's millions of years of isolation may be about to end, as Helen Gavaghan reveals.

严寒下的生命

Gavaghan H.

(约克郡的自由新闻记者)

摘要:在南极沃斯托克湖(Lake Vostok)荒凉的水体中可能隐藏着未知的生态系统和尚未触及到的有关地球过去的记录,不过它那掩盖了上百万年的神秘面纱将要被 Helen Gavaghan 揭去.

41. Antonio Riccio, Luigi Vitagliano, Guido di Prisco, et al. The crystal structure of a tetrameric hemoglobin in a partial hemichrome state[J]. PNAS, 2002, 99(15): 9801-9806.

The crystal structure of a tetrameric hemoglobin in a partial hemichrome state

Antonio Riccio, Luigi Vitagliano, Guido di Prisco, Adriana Zagari, Lelio Mazzarella

(Istituto di Biochimica delle Proteine ed Enzimologia, Consiglio Nazionale delle Ricerche, I-80125 Naples, Italy)

Abstract: Tetrameric hemoglobins are the most widely used systems in studying protein cooperativity. Allosteric effects in hemoglobins arise from the switch between a relaxed (R) state and a tense (T) state occurring upon oxygen release. Here we report the 2.0-Angstrom crystal structure of the main hemoglobin component of the Antarctic fish *Trematomus newnesi*, in a partial hemichrome form. The two alpha-subunit iron atoms are bound to a CO molecule, whereas in the beta subunits the distal histidine residue is the sixth ligand of the heme iron. This structure, a tetrameric hemoglobin in the hemichrome state, demonstrates that the iron coordination by the distal histidine, usually associated with denaturing states, may be tolerated in a native-like hemoglobin structure. In addition, several features of the tertiary and quaternary organization of this structure are intermediate between the R and T states and agree well with the R → T transition state properties obtained by spectroscopic and kinetic techniques. The analysis of this structure provides a detailed pathway of heme-heme communication and it indicates that the plasticity of the beta heme pocket plays a role in the R → T transition of tetrameric hemoglobins.

一个部分高铁血色原状态的血红蛋白四聚体的晶体结构

Antonio Riccio, Luigi Vitagliano, Guido di Prisco, Adriana Zagari, Lelio Mazzarella

(意大利那不勒斯蛋白质生物化学研究所)

摘要:血红蛋白四聚体在研究蛋白质协同性时是最广泛使用的系统。血红蛋白的变构效应产生于氧气释放发生的放松(R)状态与紧张(T)状态的转换。在这里,我们报道了南极鱼 *Trematomus newnesi* 主要血红蛋白组成的 2.0 埃的透明结构,处于部分高铁血色原形式。有两个 α-亚基铁原子结合到一氧化碳分子上,而 β 亚基的组氨酸残基形成亚血红素铁的第六个配体。处于高铁血原状态的血红蛋白四聚体结构,表明了由通常与变性状态相联系的末端组氨酸与铁的配位,可能为天然的血红蛋白结构所容许。此外,这种结构的三级和四级组织中有几个特点是介于 R 和 T 之间的中间状态,并与光谱和动力学技术获得的 R→T 转换态属性相一致。这个结构的分析提供了亚铁血红素-亚铁血红素沟通的详细路径,并表明了 β 亚铁血红素口袋的可塑性在血红蛋白四聚体 R→T 转换过程所起的作用。

42. Thomas D N, Dieckmann G S. Ocean science-Antarctic Sea ice: a habitat for extremophiles[J]. Science, 2002, 295(5555): 641-644.

Ocean science-Antarctic Sea ice:a habitat for extremophiles

Thomas D. N., Dieckmann G. S.

(Univ Wales, Sch Ocean Sci, Menai Bridge LL59 5EY, Anglesey, Wales)

Abstract: The pack ice of Earth's polar oceans appears to be frozen white desert, devoid of life. However, beneath the snow ties a unique habitat for a group of bacteria and microscopic plants and animals that are encased in an ice matrix at low temperatures and light levels, with the only liquid being pockets of concentrated brines. Survival in these conditions requires a complex suite of physiological and metabolic adaptations, but sea-ice organisms thrive in the ice, and their prolific growth ensures they play a fundamental role in polar ecosystems. Apart from their ecological importance, the bacterial and algae species found in sea ice have become the focus for novel biotechnology, as well as being considered proxies for possible life forms on ice-covered extraterrestrial bodies.

海洋科学-南极海冰：极端生物的一个栖息地

Thomas D. N., Dieckmann G. S.

(英国威尔士大学海洋科学院)

摘要：地球上极地海洋浮冰似乎是被冻结的白色沙漠，缺乏生命。然而，雪下是一群奇特的细菌、极小的植物和动物的栖息地，它们被包裹在低温、低光的冰盒中，其现有的唯一液体是孔隙中的高浓度盐水。在这些条件下生存，需要一套复杂的生理和代谢适应机制，但海冰微生物在冰中茁壮成长，它们繁茂地增长，确保它们在极地生态系统中起着基础性作用。除了其生态重要性外，在海冰发现的细菌和藻类物种已成为新的生物技术重点，并被认为是冰雪覆盖的地球外机体可能生命形式的替代指标。

43. Allan C Ashworth, Christian Thompson F. A fly in the biogeographic ointment[J]. Nature, 2003, 423(6936): 135-136.

A fly in the biogeographic ointment

Allan C. Ashworth, Christian Thompson F.

(Department of Geosciences, North Dakota State University, Fargo, North Dakota 58105-5517, USA)

Abstract: We have discovered a fossil of a higher fly (Diptera: Cyclorrhapha) from Antarctica, a finding that goes against the long-held belief that the continent was never inhabited by these insects. The fly must either have colonized Antarctica during a warm interval in the Neogene epoch, between 3 million and 17 million years ago, or it was an original member of the Gondwana fauna that survived in Antarctica for tens of millions of years before becoming extinct.

在生物地理软质地层中的一只苍蝇

Allan C. Ashworth, Christian Thompson F.

(美国北达科他州立大学地质学系)

摘要:我们已经在南极发现了一个更高级的苍蝇(双翅目:环裂亚目)化石,这一发现打破了长期持有的南极大陆没有这类昆虫居住的观点。这些苍蝇要么在 300 万～1700 万年前的新第三纪时代温暖间期在南极洲聚居,要么是 1000 万年前在开始灭绝前存活于南极洲的冈瓦纳大陆(古陆)动物区系的原始成员。

44. José R de la Torre, Lynne M Christianson, Oded Béjà, et al. Proteorhodopsin genes are distributed among divergent marine bacterial taxa[J]. PNAS, 2003, 100(22): 12830-12835.

Proteorhodopsin genes are distributed among divergent marine bacterial taxa

José R. de la Torre, Lynne M. Christianson, Oded Béjà, Marcelino T. Suzuki

(Monterey Bay Aquarium Research Institute, 7700 Sandholdt Road, Moss Landing, CA 95039)

Abstract: Proteorhodopsin (PR) is a retinal-binding bacterial integral membrane protein that functions as a light-driven proton pump. The gene encoding this photoprotein was originally discovered on a large genome fragment derived from an uncultured marine gamma-proteobacterium of the SAR86 group. Subsequently, many variants of the PR gene have been detected in marine plankton, via PCR-based gene surveys. It has not been clear, however, whether these different PR genes are widely distributed among different bacterial groups, or whether they have a restricted taxonomic distribution. We report here comparative analyses of PR-bearing genomic fragments recovered directly from planktonic bacteria inhabiting the California coast, the central Pacific Ocean, and waters offshore the Antarctica Peninsula. Sequence analysis of an Antarctic genome fragment harboring PR (ANT32C12) revealed moderate conservation in gene order and identity, compared with a previously reported PR-containing genome fragment from a Monterey Bay gamma-proteobacterium (EBAC31A08). Outside the limited region of synteny shared between these clones, however, no significant DNA or protein identity was evident. Analysis of a third PR-containing genome fragment (HOT2C01) from the North Pacific subtropical gyre showed even more divergence from the gamma-proteobacterial PR-flanking region. Subsequent phylogenetic and comparative genomic analyses revealed that the Central North Pacific PR-containing genome fragment (HOT2C01) originated from a planktonic alpha-proteobacterium. These data indicate that PR genes are distributed among a variety of divergent marine bacterial taxa, including both alpha- and gamma-proteobacteria. Our analyses also demonstrate the utility of cultivation-independent comparative genomic approaches for assessing gene content and distribution in naturally occurring microbes.

Proteorhodopsin 基因分布在多种多样的海洋细菌类群中

José R. de la Torre, Lynne M. Christianson, Oded Béjà, Marcelino T. Suzuki

(美国加州蒙特里湾水族馆研究所)

摘要：Proteorhodopsin(PR)是一个视网膜结合的细菌整合细胞膜蛋白，它作为光驱动质子泵起作用。这个发光蛋白的基因编码早在一个来源于 SAR86 组未培养海洋 γ-蛋白质细菌的大的基因组片段中就发现了。随后，通过基于 PCR 的基因研究，在海洋浮游生物中发现了许多 PR 基因变种。但是，这些不同的 PR 基因是否广泛分布于不同的细菌群，还是它们在分类学上的分布具有限制性，目前还不清楚。在这里，我们比较分析了直接来自栖息于太平洋中心的加州海岸浮游细菌以及南极半岛近岸海水的 PR-bearing(关联)基因组片段。与以前报道的来自蒙特里湾 γ-蛋白质菌(EBAC31A08)PR-containing 基因片段相比，包含 PR(ANT32C12)的南极基因组片段，其基因序列和特征适度保守。这些克隆具有共同的外部同线性限定地区。但是，没有发现重要的 DNA 或蛋白质的等同。对第三个包含 PR 的来自北太平洋亚热带环流的基因组片段(HOT2C01)的分析显示，其与 γ-蛋白质菌侧翼区的分歧更大。随后的动植物演化史和比较基因组分析表明，北太平洋中部含 PR 的基因组片段(HOT2C01)来自浮游的 α-变形杆菌。这些数据表明，PR 基因分布在多样的海洋细菌的品种类群，包括 α-和 γ-变形杆菌。我们的分析还显示了独立培养的可比较的基因方法，用来评估自然存在的微生物基因含量和分布。

45. Moller P R, Nielsen J G, Fossen I. Fish migration: patagonian toothfish found off Greenland: this catch is evidence of transequatorial migration by a cold-water Antarctic fish[J]. Nature, 2003, 421 (6923): 599.

Fish migration: Patagonian toothfish found off Greenland —this catch is evidence of transequatorial migration by a cold-water Antarctic fish

Moller P. R., Nielsen J. G., Fossen I.

(Zoological Museum, University of Copenhagen, 2100 Copenhagen, Denmark)

Abstract: A large (1. 80 metres in length and weighing 70 kg) Patagonian toothfish (*Dissostichtus eleginoides* Smitt, 1898) has been caught in the northwest Atlantic, representing the first Northern Hemisphere record of the diverse, abundant and mainly Antarctic suborder Notothenioidei. This extraordinary catch indicates that large, cold-temperate fishes may occasionally migrate from sub-Antarctic to sub-Arctic waters by using deep, cold water, supporting a widely accepted but unproven proposal that the anti-tropical distribution patterns of many marine biota could be explained by transequatorial migration.

鱼的迁徙:格陵兰岛不远处发现巴塔哥尼亚齿鱼——这个发现是南极冷水鱼类跨赤道迁徙的证据

Moller P. R., Nielsen J. G., Fossen I.

(丹麦哥本哈根大学动物学博物馆)

摘要:在大西洋西北部抓到了大的(长 1.80 m,体重 70 kg)巴塔哥尼亚齿鱼(*Dissostichtus eleginoides* Smitt,1898 年),首次记录在北半球存在多样的、众多的南极亚目 Notothenioidei。这次难得的捕捞说明,大的、冷温带鱼类可能会从亚南极游到亚北极水域中,支持了已广泛接受但未经证实的观点,即许多海洋生物群反热带分布格局可能解释跨赤道的迁移。

46. Angus Atkinson, Volker Siegel, Evgeny Pakhomov, et al. Long-term decline in krill stock and increase in salps within the Southern Ocean[J]. Nature, 2004, 432(7013): 100-103.

Long-term decline in krill stock and increase in salps within the Southern Ocean

Angus Atkinson, Volker Siegel, Evgeny Pakhomov, Peter Rothery

(British Antarctic Survey, Natural Environment Research Council, High Cross, Madingley Road, Cambridge CB3 OET, UK)

Abstract: Antarctic krill (*Euphausia superba*) and salps (mainly *Salpa thompsoni*) are major grazers in the Southern Ocean, and krill support commercial fisheries. Their density distributions have been described in the period 1926-51, while recent localized studies suggest short-term changes. To examine spatial and temporal changes over larger scales, we have combined all available scientific net sampling data from 1926 to 2003. This database shows that the productive southwest Atlantic sector contains > 50% of Southern Ocean krill stocks, but here their density has declined since the 1970s. Spatially, within their habitat, summer krill density correlates positively with chlorophyll concentrations. Temporally, within the southwest Atlantic, summer krill densities correlate positively with sea-ice extent the previous winter. Summer food and the extent of winter sea ice are thus key factors in the high krill densities observed in the southwest Atlantic Ocean. Krill need the summer phytoplankton blooms of this sector, where winters of extensive sea ice mean plentiful winter food from ice algae, promoting larval recruitment and replenishing the stock. Salps, by contrast, occupy the extensive lower-productivity regions of the Southern Ocean and tolerate warmer water than krill. As krill densities decreased last century, salps appear to have increased in the southern part of their range. These changes have had profound effects within the Southern Ocean food web.

南大洋磷虾数量长期下降和樽海鞘数量增加

Angus Atkinson, Volker Siegel, Evgeny Pakhomov, Peter Rothery

（英国自然环境研究理事会南极调查局）

摘要：南极磷虾（*Euphausia superba*）和樽海鞘（主要是纽鳃樽 *Salpa thompsoni*）是南大洋主要的滤食者，磷虾支持了商业捕鱼业。1926 年至 1951 年期间，它们的密度分布已有研究，而近期局部研究表明了短期的变化。为了调查更大尺度的空间和时间的变化，我们结合了 1926 年至 2003 年所有可利用的科学的网法采样数据。这个数据库显示，大西洋西南地区分布着大于 50％的南大洋磷虾库存，但其密度自 1970 年代以来有所下降。空间上，在它们的栖息地，夏季磷虾密度与叶绿素浓度呈正相关。时间上，大西洋西南地区，夏季磷虾密度与上季冬天的海冰范围呈正相关。因此，夏季食物和冬季海冰范围是我们观察到的大西洋西南地区海域磷虾高密度分布的关键因素。磷虾需要该区域夏季浮游植物的暴发，而在这里冬季海冰的扩大，意味着有大量的冬季食物冰藻，从而促进了幼体的繁殖，补充了库存。相反，樽海鞘主要分布在南大洋低生产力地区，比磷虾耐温暖的海水。因为 20 世纪以来磷虾密度下降，樽海鞘在其南部的范围显得有些增加。这些变化已在南大洋的食物网中产生深远的影响。

47. Francesco Bonadonna, Gabrielle A Nevitt. Partner-specific odor recognition in an Antarctic seabird [J]. Science, 2004, 306(5697): 835.

Partner-specific odor recognition in an Antarctic seabird

Francesco Bonadonna, Gabrielle A. Nevitt

(Behavioural Ecology Group, CNRS-Centre d'Ecologie Fonctionnelle et Evolutive, 1919 route de Mende, F-34293 Montpellier, Cedex 5, France)

Abstract: Among birds, the Procellariiform seabirds (petrels, albatrosses, and shearwaters) are prime candidates for using chemical cues for individual recognition. These birds have an excellent olfactory sense, and a variety of species nest in burrows that they can recognize by smell. However, the nature of the olfactory signature—the scent that makes one burrow smell more like home than another—has not been established for any species. Here, we explore the use of intraspecific chemical cues in burrow recognition and present evidence for partner-specific odor recognition in a bird.

南极海鸟对特定伙伴气味的识别

Francesco Bonadonna, Gabrielle A. Nevitt

（法国国家科学研究中心功能与进化生态学中心）

摘要：在鸟类中，南极鹱形目（Procellariiform）海鸟（海燕、信天翁和海鸥）是利用化学线索进行个体识别的首要候选者。这些鸟类有极好的嗅觉判断力，它们可以靠嗅觉辨别挖掘筑巢的多个物种。然而，嗅觉特征的本质——气味使得一个洞穴比另一个更像家的味道，还尚未为其他任何物种所建立。在这里，我们用一种内化学信号的方法进行洞穴识别，并报道一个鸟对特定伙伴气味识别的证据。

48. Kate F Darling, Michal Kucera, Carol J Pudsey, et al. Molecular evidence links cryptic diversification in polar planktonic protists to quaternary climate dynamics[J]. PNAS, 2004, 101(20): 7657-7662.

Molecular evidence links cryptic diversification in polar planktonic protists to quaternary climate dynamics

Kate F. Darling, Michal Kucera, Carol J. Pudsey, Christopher M. Wade

(School of GeoSciences and Institute of Cell, Animal and Population Biology, King's Buildings, University of Edinburgh, Edinburgh EH9 3JW, United Kingdom)

Abstract: It is unknown how pelagic marine protists undergo diversification and speciation. Superficially, the open ocean appears homogeneous, with few clear barriers to gene flow, allowing extensive, even global, dispersal. Yet, despite the apparent lack of opportunity for genetic isolation, diversity is prevalent within marine taxa. A lack of candidate isolating mechanisms would seem to favor sympatric over allopatric speciation models to explain the diversity and biogeographic patterns observed in the oceans today. However, the ocean is a dynamic system, and both current and past circulation patterns must be considered in concert to gain a true perspective of gene flow through time. We have derived a comprehensive picture of the mechanisms potentially at play in the high latitudes by combining molecular, biogeographic, fossil, and paleoceanographic data to reconstruct the evolutionary history of the polar planktonic foraminifer *Neogloboquadrina pachyderma sinistral*. We have discovered extensive genetic diversity within this morphospecies and that its current "extreme" polar affinity did not appear until late in its evolutionary history. The molecular data demonstrate a stepwise progression of diversification starting with the allopatric isolation of Atlantic Arctic and Antarctic populations after the onset of the Northern Hemisphere glaciation. Further diversification occurred only in the Southern Hemisphere and seems to have been linked to glacial-interglacial climate dynamics. Our findings demonstrate the role of Quaternary climate instability in shaping the modern high-latitude plankton. The divergent evolutionary history of *N. pachyderma sinistral* genotypes implies that paleoceanographic proxies based on this taxon should be calibrated independently.

连接神秘的极地浮游原生生物多样化与第四纪气候动力学的分子证据

Kate F. Darling, Michal Kucera, Carol J. Pudsey, Christopher M. Wade

(英国爱丁堡大学 King's Buildings 校区地质科学学院和细胞、动物、种群生物学研究所)

摘要：目前还不清楚远洋海洋原生生物如何进行多样化和物种形成。表面上看，开放的海洋似乎是同质的，对基因流没有多少明确的界限，它允许广泛的甚至全球的基因分布。然而，尽管没有明显的遗传隔离的机制，海洋生物多样性仍是很普遍的。隔离机制的缺乏，似乎有利于用同域(生态区重叠)的模型而不是异域(生态区不重叠)的模型来解释今天海洋的多样性和生物地理格局。然而，海洋是一个动态的系统，必须同步考虑当前和过去的循环模式，以获得基因流随着时间变化的正确观点。通过在高纬度地区结合分子、生物地理、化石和古海洋学数据，以重建在极地的浮游有孔虫 *Neogloboquadrina pachyderma* sinistral 进化的历史，我们得到一个综合的潜在的机制。我们发现这个形态种有丰富的遗传多样性，而且直到进化历史的晚期，其目前的"极端"亲极地的性质才表现出来。分子资料表明，多元化的逐步发展始于北半球冰河作用开始时北极和南极大西洋种群分布区不重叠。进一步多样化只发生在南半球，似乎与冰期-间冰期气候动力学相联系。我们的研究结果证明了第四纪气候不稳定性在形成现代高纬度浮游生物中的作用。有孔虫 *N. pachyderma* sinistral 多样的进化历史暗示基于这个分类的古海洋学指标应该独立进行校准。

49. Shepherd L D, Millar C D, Ballard G, et al. Microevolution and mega-icebergs in the Antarctic[J]. PNAS, 2005, 102(46): 16717-16722.

Microevolution and mega-icebergs in the Antarctic

Shepherd L. D., Millar C. D., Ballard G., Ainley D. G., Wilson P. R.,
Haynes G. D., Baroni C., Lambert D. M.

(Allan Wilson Centre for Molecular Ecology and Evolution, Institute of Molecular BioSciences, Massey University, Private Bag 102904, NSMC, Albany, Auckland, New Zealand)

Abstract: Microevolution is regarded as changes in the frequencies of genes in populations over time. Ancient DNA technology now provides an opportunity to demonstrate evolution over a geological time frame and to possibly identify the causal factors in any such evolutionary event. Using nine nuclear microsatellite DNA loci, we genotyped an ancient population of Adelie penguins (*Pygoscelis adeliae*) aged approximate to 6000 years B. P. Subfossil bones from this population were excavated by using an accurate stratigraphic method that allowed the identification of individuals even within the same layer. We compared the allele frequencies in the ancient population with those recorded from the modern population at the same site in Antarctica. We report significant changes in the frequencies of alleles between these two time points, hence demonstrating microevolutionary change. This study demonstrates a nuclear gene-frequency change over such a geological time frame. We discuss the possible causes of such a change, including the role of mutation, genetic drift, and the effects of gene mixing among different penguin populations. The latter is likely to be precipitated by mega-icebergs that act to promote migration among penguin colonies that typically show strong natal return.

微观进化和南极巨型冰山

Shepherd L. D., Millar C. D., Ballard G., Ainley D. G., Wilson P. R.,
Haynes G. D., Baroni C., Lambert D. M.

(新西兰梅西大学分子生物科学研究所 Allan Wilson 分子生态与进化研究中心)

摘要：微观进化是种群随着时间推移基因频率发生的变化。古老的 DNA 技术提供了一个说明地质学时间尺度上进化和可能识别任何进化事件的起因的机会。利用 9 个核心序列微卫星标记 DNA 位点，我们测定古代阿德利企鹅群(*Pygoscelis adeliae*)基因型，其年龄距今约为 6000 年。用准确的地层方法挖出这个种群亚化石骨骼，这些亚化石骨骼可以用来识别甚至是同一层的个体。我们比较了南极洲古代和现代同一地点种群的等位基因频率。这两个时间点之间的等位基因频率有显著的变化，从而表现出微进化的变化。这项研究显示核心基因频率在地质学时间尺度上的变化。我们讨论了这种变化的可能原因，包括突变作用、遗传漂变，以及不同企鹅种群基因混合的影响。后者很可能是由于巨型冰山促成的。

50. Victor Smetacek, Stephen Nicol. Polar ocean ecosystems in a changing world[J]. Nature, 2005, 437 (7057): 362-368.

Polar ocean ecosystems in a changing world

Victor Smetacek, Stephen Nicol

(Alfred Wegener Institute for Polar and Marine Research, Am Handelshafen 12, 27570 Bremerhaven, Germany)

Abstract: Polar organisms have adapted their seasonal cycles to the dynamic interface between ice and water. This interface ranges from the micrometre-sized brine channels within sea ice to the planetary-scale advance and retreat of sea ice. Polar marine ecosystems are particularly sensitive to climate change because small temperature differences can have large effects on the extent and thickness of sea ice. Little is known about the interactions between large, long-lived organisms and their planktonic food supply. Disentangling the effects of human exploitation of upper trophic levels from basin-wide, decade-scale climate cycles to identify long-term, global trends is a daunting challenge facing polar bio-oceanography.

变化中的极地海洋生态系统

Victor Smetacek, Stephen Nicol

(德国阿尔弗雷德·魏格纳极地和海洋研究中心)

摘要:极地生物已经适应了动态的冰-水界面,形成了季节性循环。这个界面从海冰内微米大小的盐水通道到行星尺度上的海冰进退。极地海洋生态系统对气候变化特别敏感,因为小的温度变化也对海冰范围和厚度产生巨大的影响。目前对大型长期生存的生物和它们的浮游生物食物供给的交互作用了解得较少。从全流域,十年尺度的气候循环,解开人类开发利用对上层营养级的影响,以识别长期的全球趋势是极地生物海洋学面临的一项艰巨的挑战。

51. Barbraud C, Weimerskirch H. Antarctic birds breed later in response to climate change[J]. PNAS, 2006, 103(16): 6248-6251.

Antarctic birds breed later in response to climate change

Barbraud C., Weimerskirch H.

(Centre d'Etudes Biologiques de Chizé, Unité Propre de Recherche 1934 du)

Abstract: In the northern hemisphere, there is compelling evidence for climate-related advances of spring events, but no such long-term biological time series exist for the southern hemisphere. We have studied a unique data set of dates of first arrival and laying of first eggs over a 55-year period for the entire community of Antarctic seabirds in East Antarctica. The records over this long period show a general unexpected tendency toward later arrival and laying, an inverse trend to those observed in the northern hemisphere. Overall, species now arrive at their colonies 9.1 days later, on average, and lay eggs an average of 2. 1 days later than in the early 1950s. Furthermore, these delays are linked to a decrease in sea ice extent that has occurred in eastern Antarctica, which underlies the contrasted effects of global climate change on species in Antarctica.

应对气候变化——南极海鸟繁殖推迟

Barbraud C. , Weimerskirch H.

(法国德拉科学研究所 Chizé 生物中心)

摘要：在北半球，有与春季事件气候进展相关的令人信服的证据，但在南半球没有这样长期的生物学上的时间序列存在。历时 55 年，我们研究了整个东南极海鸟种群在首次到达无冰区的时间以及开始产卵的时间，这是独一无二的数据资料。这漫长时期的记录表明，与北半球观测趋势相反，南半球一般体现出晚到达无冰区，且产卵时间也晚。总体而言，与 20 世纪 50 年代初相比，现在海鸟到达聚居地的时间平均晚 9. 1 天，产卵平均晚 2. 1 天。而且，这些延迟是与东南极海冰减少程度联系在一起的，这是全球气候变化对南极物种有差别的响应的基础。

52. Marinov I, Gnanadesikan A, Toggweiler J R, et al. The Southern Ocean biogeochemical divide[J]. Nature, 2006, 441(7096): 964-967.

The Southern Ocean biogeochemical divide

Marinov I. , Gnanadesikan A. , Toggweiler J. R. , Sarmiento J. L.

(Atmospheric and Oceanic Sciences Program, Princeton University, Princeton, New Jersey 08540, USA)

Abstract: Modelling studies have demonstrated that the nutrient and carbon cycles in the Southern Ocean play a central role in setting the air-sea balance of CO_2 and global biological production. Box model studies first pointed out that an increase in nutrient utilization in the high latitudes results in a strong decrease in the atmospheric carbon dioxide partial pressure ($p(CO_2)$). This early research led to two important ideas: high latitude regions are more important in determining atmospheric $p(CO_2)$ than low latitudes, despite their much smaller area, and nutrient utilization and atmospheric $p(CO_2)$ are tightly linked. Subsequent general circulation model simulations show that the Southern Ocean is the most important high latitude region in controlling preindustrial atmospheric CO_2 because it serves as a lid to a larger volume of the deep ocean. Other studies point out the crucial role of the Southern Ocean in the uptake and storage of anthropogenic carbon dioxide and in controlling global biological production. Here we probe the system to determine whether certain regions of the Southern Ocean are more critical than others for air-sea CO_2 balance and the biological export production, by increasing surface nutrient drawdown in an ocean general circulation model. We demonstrate that atmospheric CO_2 and global biological export production are controlled by different regions of the Southern Ocean. The air-sea balance of carbon dioxide is controlled mainly by the biological pump and circulation in the Antarctic deep-water formation region, whereas global export production is controlled mainly by the biological pump and circulation in the Subantarctic intermediate and mode water formation region. The existence of this biogeochemical divide separating the Antarctic from the Subantarctic suggests that it may be possible for climate change or human intervention to modify one of these without greatly altering the other.

南大洋生物地球化学分界线

Marinov I.，Gnanadesikan A.，Toggweiler J. R.，Sarmiento J. L.

（美国普林斯顿大学大气和海洋科学项目）

摘要：模拟研究表明，南大洋营养和碳循环在调节大气-海洋界面 CO_2 平衡和全球生物生产力方面发挥着核心作用。箱式模型的研究首次指出，高纬度地区营养可利用性的增加导致大气 CO_2 分压（$p(CO_2)$）有明显的降低。这种早期的研究提出了两个重要思想：(1) 在决定大气 CO_2 分压（$p(CO_2)$）方面，尽管高纬度地区面积小，也比低纬度地区更重要；(2) 营养可利用性与大气 CO_2 分压（$p(CO_2)$）是紧密联系的。随后环流模式模拟表明，南大洋是控制工业化前大气二氧化碳浓度的最重要的高纬度地区，因为它是体积巨大的深海海洋的一个盖子。其他研究指出南大洋在吸收和存储人类源的二氧化碳以及控制全球的生物产量方面起着关键作用。

在这里，用通用的海洋环流模型，通过增加表层营养消耗的方式，我们决定调查南大洋的某些区域在大气-海洋界面 CO_2 平衡和生物输出生产方面是否有比其他地区更重要的机制。我们的研究结果证明，大气中 CO_2 和全球生物输出生产是由南大洋不同区域控制的。大气-海洋 CO_2 平衡主要由生物泵和南极深层水形成的区域循环所控制，而全球输出生产由生物泵和亚南极中部和模式水形成的区域所控制。这种生物地球化学鸿沟的存在隔开了南极和亚南极，表明气候变化或人类干扰在更改其中之一而不极大地改变其他方面是可能的。

53. Nicholas Meskhidze, Athanasios Nenes. Phytoplankton and cloudiness in the Southern Ocean[J]. Science, 2006, 314(5804): 1419-1423.

Phytoplankton and cloudiness in the Southern Ocean

Nicholas Meskhidze, Athanasios Nenes

(School of Earth and Atmospheric Sciences, Georgia Institute of Technology, Atlanta, GA 30332, USA)

Abstract: The effect of ocean biological productivity on marine clouds is explored over a large phytoplankton bloom in the Southern Ocean with the use of remotely sensed data. Cloud droplet number concentration over the bloom was twice what it was away from the bloom, and cloud effective radius was reduced by 30%. The resulting change in the short-wave radiative flux at the top of the atmosphere was-15 watts per square meter, comparable to the aerosol indirect effect over highly polluted regions. This observed impact of phytoplankton on clouds is attributed to changes in the size distribution and chemical composition of cloud condensation nuclei. We propose that secondary organic aerosol, formed from the oxidation of phytoplankton-produced isoprene, can affect chemical composition of marine cloud condensation nuclei and influence cloud droplet number. Model simulations support this hypothesis, indicating that 100% of the observed changes in cloud properties can be attributed to the isoprene secondary organic aerosol.

南大洋浮游植物和云量

Nicholas Meskhidze, Athanasios Nenes

(美国亚特兰大乔治亚理工学院地球和大气科学系)

摘要:使用遥感数据研究了大型浮游植物暴发,探讨了海洋植物生产力对海洋云的影响。暴发区域中云滴数量浓度是没有暴发区域的两倍,且云的有效半径减少了30%。与气溶胶对高污染区域的间接作用类似,大气顶端短波辐射通量为−15 W/m^2。观测的浮游植物对云的影响归因于云凝结核的尺度分布和化学组成。我们认为,浮游植物产生的异戊二烯氧化所形成的二次有机气溶胶,能影响海洋云凝结核的化学组成以及云滴数目。模型模拟支持这一假设,表明观测到的100%云特性的变化可以归因于异戊二烯产生的二次有机气溶胶。

54. Stéphane Blain, Bernard Quéguiner, Leanne Armand, et al. Effect of natural iron fertilization on carbon sequestration in the Southern Ocean[J]. Nature, 2007, 446(7139): 1070-1074.

Effect of natural iron fertilization on carbon sequestration in the Southern Ocean

Stéphane Blain, Bernard Quéguiner, Leanne Armand, Sauveur Belviso, Bruno Bombled, Laurent Bopp, Andrew Bowie, Christian Brunet, Corina Brussaard, François Carlotti, Urania Christaki, Antoine Corbière, Isabelle Durand, Frederike Ebersbach, Jean-Luc Fuda, Nicole Garcia, Loes Gerringa, Brian Griffiths, Catherine Guigue, Christophe Guillerm, Stéphanie Jacquet, Catherine Jeandel, Patrick Laan, Dominique Lefèvre, Claire Lo Monaco, Andrea Malits, Julie Mosseri, Ingrid Obernosterer, Young-Hyang Park, Marc Picheral, Philippe Pondaven, Thomas Remenyi, Valérie Sandroni, Géraldine Sarthou, Nicolas Savoye, Lionel Scouarnec, Marc Souhaut, Doris Thuiller, Klaas Timmermans, Thomas Trull, Julia Uitz, Pieter van Beek, Marcel Veldhuis, Dorothée Vincent, Eric Viollier, Lilita Vong, Thibaut Wagener

(Laboratoire d'Océanographie et de Biogéochimie, Centre Océanologique de Marseille, CNRS, Université de la Méditerranée, campus de Luminy, case 901, 13288 Marseille Cedex 09, France)

Abstract: The availability of iron limits primary productivity and the associated uptake of carbon over large areas of the ocean. Iron thus plays an important role in the carbon cycle, and changes in its supply to the surface ocean may have had a significant effect on atmospheric carbon dioxide concentrations over glacial-interglacial cycles. To date, the role of iron in carbon cycling has largely been assessed using short-term iron-addition experiments. It is difficult, however, to reliably assess the magnitude of carbon export to the ocean interior using such methods, and the short observational periods preclude extrapolation of the results to longer timescales. Here we report observations of a phytoplankton bloom induced by natural iron fertilization-an approach that offers the opportunity to overcome some of the limitations of short-term experiments. We found that a large phytoplankton bloom over the Kerguelen plateau in the Southern Ocean was sustained by the supply of iron and major nutrients to surface waters from iron-rich deep water below. The efficiency of fertilization, defined as the ratio of the carbon export to the amount of iron supplied, was at least ten times higher than previous estimates from short-term blooms induced by iron-addition experiments. This result sheds new light on the effect of long-term fertilization by iron and macronutrients on carbon sequestration, suggesting that changes in iron supply from below—as invoked in some palaeoclimatic and future climate change scenarios—may have a more significant effect on atmospheric carbon dioxide concentrations than previously thought.

自然“铁施肥”对南大洋碳螯合作用的影响

Stéphane Blain, Bernard Quéguiner, Leanne Armand, Sauveur Belviso, Bruno Bombled, Laurent Bopp, Andrew Bowie, Christian Brunet, Corina Brussaard, François Carlotti, Urania Christaki, Antoine Corbière, Isabelle Durand, Frederike Ebersbach, Jean-Luc Fuda, Nicole Garcia, Loes Gerringa, Brian Griffiths, Catherine Guigue, Christophe Guillerm, Stéphanie Jacquet, Catherine Jeandel, Patrick Laan, Dominique Lefèvre, Claire Lo Monaco, Andrea Malits, Julie Mosseri, Ingrid Obernosterer, Young-Hyang Park, Marc Picheral, Philippe Pondaven, Thomas Remenyi, Valérie Sandroni, Géraldine Sarthou, Nicolas Savoye, Lionel Scouarnec, Marc Souhaut, Doris Thuiller, Klaas Timmermans, Thomas Trull, Julia Uitz, Pieter van Beek, Marcel Veldhuis, Dorothée Vincent, Eric Viollier, Lilita Vong, Thibaut Wagener

(法国国家科学研究中心(CNRS)海洋与生物地球化学中心)

摘要:在辽阔的海洋中,铁的供应限制初级生产力及与之相关的碳的吸收。因此,铁在碳循环中起着重要的作用,它对海洋表层供给的变化可能对大气 CO_2 浓度在冰期-间冰期的循环有着重要的影响。迄今为止,铁在碳循环中的作用在很大程度上用短期增铁的实验来进行评估。但是,使用这种方法来可靠地评估碳输出到海洋内部是很困难的,短的观测周期妨碍了将结果外推到较长的时间尺度。在这里,我们报道了浮游生物暴发诱导自然铁施肥—— 一个提供了克服短期实验一些局限性的方法。我们发现,南大洋凯尔盖朗高原(Kerguelen plateau)大型浮游生物暴发得以持久是由于富含铁的深层海水供给到表层,使表层海水中富集了铁和大量营养元素的原因。施肥效率,定义为碳输出与铁供给的比值,比以前用增铁试验诱导短期浮游生物暴发的方法来评估至少高 10 倍。这一结果揭示了长期施肥,如铁和大量营养元素对碳的去除作用,表明从海洋深层供应铁的变化(由一些古气候和未来气候变化所引起)可能比以前认为的对大气 CO_2 浓度作用效果更为显著。

55. Angelika Brandt, Andrew J Gooday, Simone N Brandao, et al. First insights into the biodiversity and biogeography of the Southern Ocean deep sea[J]. Nature, 2007, 447(7142): 307-311.

First insights into the biodiversity and biogeography of the Southern Ocean deep sea

Angelika Brandt, Andrew J. Gooday, Simone N. Brandao, Saskia Brix, Wiebke Brokeland, Tomas Cedhagen, Madhumita Choudhury, Nils Cornelius, Bruno Danis, Ilse De Mesel, Robert J. Diaz, David C. Gillan, Brigitte Ebbe, John A. Howe, Dorte Janussen, Stefanie Kaiser, Katrin Linse, Marina Malyutina, Jan Pawlowski, Michael Raupach, Ann Vanreuse

(Zoological Museum Hamburg, Martin-Luther-King-Platz 3, 20146 Hamburg, Germany)

Abstract: Shallow marine benthic communities around Antarctica show high levels of endemism, gigantism, slow growth, longevity and late maturity, as well as adaptive radiations that have generated considerable biodiversity in some taxa. The deeper parts of the Southern Ocean exhibit some unique environmental features, including a very deep continental shelf and a weakly stratified water column, and are the source for much of the deep water in the world ocean. These features suggest that deep-sea faunas around the Antarctic may be related both to adjacent shelf communities and to those in other oceans. Unlike shallow-water Antarctic benthic communities, however, little is known about life in this vast deep-sea region. Here, we report new data from recent sampling expeditions in the deep Weddell Sea and adjacent areas (748-6348 m water depth) that reveal high levels of new biodiversity; for example, 674 isopods species, of which 585 were new to science. Bathymetric and biogeographic trends varied between taxa. In groups such as the isopods and polychaetes, slope assemblages included species that have invaded from the shelf. In other taxa, the shelf and slope assemblages were more distinct. Abyssal faunas tended to have stronger links to other oceans, particularly the Atlantic, but mainly in taxa with good dispersal capabilities, such as the Foraminifera. The isopods, ostracods and nematodes, which are poor dispersers, include many species currently known only from the Southern Ocean. Our findings challenge suggestions that deep-sea diversity is depressed in the Southern Ocean and provide a basis for exploring the evolutionary significance of the varied biogeographic patterns observed in this remote environment.

对南大洋深海生物多样性和生物地理学的初步研究

Angelika Brandt, Andrew J. Gooday, Simone N. Brandao, Saskia Brix, Wiebke Brokeland, Tomas Cedhagen, Madhumita Choudhury, Nils Cornelius, Bruno Danis, Ilse De Mesel, Robert J. Diaz, David C. Gillan, Brigitte Ebbe, John A. Howe, Dorte Janussen, Stefanie Kaiser, Katrin Linse, Marina Malyutina, Jan Pawlowski, Michael Raupach, Ann Vanreuse

(德国汉堡动物学博物馆)

摘要:南极洲周围浅海底栖生物群落表现出较高水平的地方性生长、巨大发育、生长缓慢、长寿和晚熟,以及适应辐射的能力,因而在某些分类上有可观的生物多样性。南大洋深层部位表现出一些独特的环境特征(包括非常深的大陆架和弱分层水柱),这些深层部分也是世界海洋深层水的来源。这些特征表明,南极洲周围的深海动物群可能既与相邻的大陆架群落相联系,也与其他海洋群落相联系。然而,不同于浅水南极底栖生物群落，很少有人知道这片辽阔的深海区域的生命。我们报道了最近在威德尔海和邻近地区海洋深处(748～6348 m)抽样考察报告的最新数据，这些数据表明，新的较高水平的生物多样性，例如，等脚类动物674种，其中585种在科学上是新的。分类单元之间在海洋测探学和生态地理类群上有变化趋势。在一些群体，如等脚类和多毛类动物，斜坡群落包括已从陆架入侵的物种。在其他分类上，陆架和斜坡群落更是截然不同。深海动物群往往与其他海洋，特别是大西洋，有较强的联系，但主要在分类上具有良好的分散能力，如有孔虫类群。等脚类动物、介形虫和线虫类，包括目前认为是仅来自南大洋的许多物种分散性均较低。我们的研究结果挑战了认为南大洋深海多样性是少量的观点，并为在这个偏僻环境中探索变化的生物地理格局进化重要性提供了基础。

56. Nicolas Cassar, Michael L Bender, Bruce A Barnett, et al. The Southern Ocean biological response to Aeolian iron deposition[J]. Science, 2007, 317(5841): 1067-1070.

The Southern Ocean biological response to Aeolian iron deposition

Nicolas Cassar, Michael L. Bender, Bruce A. Barnett, Songmiao Fan, Walter J. Moxim, Hiram Levy II, Bronte Tilbrook

(Department of Geosciences, Princeton University, Princeton, NJ 08544, USA)

Abstract: Biogeochemical rate processes in the Southern Ocean have an important impact on the global environment. Here, we summarize an extensive set of published and new data that establishes the pattern of gross primary production and net community production over large areas of the Southern Ocean. We compare these rates with model estimates of dissolved iron that is added to surface waters by aerosols. This comparison shows that net community production, which is comparable to export production, is proportional to modeled input of soluble iron in aerosols. Our results strengthen the evidence that the addition of aerosol iron fertilizes export production in the Southern Ocean. The data also show that aerosol iron input particularly enhances gross primary production over the large area of the Southern Ocean downwind of dry continental areas.

南大洋对风成铁沉降的生态响应

Nicolas Cassar, Michael L. Bender, Bruce A. Barnett, Songmiao Fan, Walter J. Moxim, Hiram Levy II, Bronte Tilbrook

(美国普林斯顿大学地学系)

摘要：南大洋生物地球化学速度进程对全球环境具有重要影响。这里，我们总结了建立南大洋大面积总初级生产力和净群落生产力格局的已发和新的数据。我们用模型比较评估通过气溶胶方式增加溶解铁到表层海水的速率。比较结果表明，与外源生产力可比的净群落生产力，与模拟气溶胶中的溶解铁是成比例的。我们的结果加强了这一证据，也就是，气溶胶中的铁使南大洋生产力的输出多产化。此外，气溶胶中铁的输入尤其增加了干旱大陆下风向的南大洋大区域尺度上的总初级生产力。

57. Convey P, Stevens M I. Antarctic biodiversity[J]. Science, 2007, 317(5846): 1877-1878.

Antarctic biodiversity

Convey P. , Stevens M. I.

(The British Antarctic Survey, Natural Environment Research Council, High Cross, Madingley Road, Cambridge CB3 0ET, UK)

Abstract: Only about 0.3% of Antarctica is free of ice. The terrestrial and freshwater ecosystems in this tiny fraction are generally small and isolated, and are populated by small invertebrates, lower plants, and microbes. Recent studies have shown that these biota are of ancient origin and have persisted in isolation for tens of millions of years. However, ice sheet modeling of the Last Glacial Maximum (～20000 years ago) and previous ice maxima in the Miocene (23 to 5 million years ago), along with reconstructions of previous glacial extent, suggest that most or all currently ice-free low-altitude surfaces would have been covered with ice during previous glacial maxima. These models leave no ice-free refuges for most terrestrial biota, and they require recolonization after each glacial maximum.

南极生物多样性

Convey P. , Stevens M. I.

(英国自然环境研究理事会南极调查局)

摘要:南极洲仅有约0.3%为无冰区,这小部分区域上陆地和淡水生态系统通常小而孤立,由小型无脊椎动物、低等植物和微生物占据。最近的研究表明,这些生物是古老的起源,孤立了数千万年依然存在。不过,末次盛冰期(20000年前)的冰盖模拟,以往中新世的冰期峰值(2300万～500万年前),还有以前冰期覆盖度重建表明,大部分或全部无冰区低海拔表面可能在此前的冰期最大化时覆盖着冰。这些模型没有考虑大多数陆地生物群的庇护所,它们需要在每个冰盛期后重新拓殖。

58. Meskhidze N. Isoprene, cloud droplets, and phytoplankton[J]. Science,2007,317(5834):42-43.

Isoprene, cloud droplets, and phytoplankton

Meskhidze N.

(School of Earth and Atmospheric Sciences Georgia Institute of Technology Atlanta, GA 30332, USA)

Abstract: There is an error that may invalidate the main conclusion of the Research Article "Phytoplankton and cloudiness in the Southern Ocean" by N. Meskhidze and A. Nenes (1 Dec. 2006, p. 1419). The authors report an increase in cloud reflectivity resulting from a 30% decrease in cloud droplet effective radius and a doubling of cloud droplet number concentration over a large phytoplankton bloom in the Southern Ocean, resulting in an extra 15 W $\cdot$ m^{-2} of energy reflected back to space. They attribute these changes to enhanced isoprene produced in the bloom. Our measurements made during the Southern Ocean Iron Experiments (SOFeX) were used by Meskhidze and Nenes to scale seawater isoprene values based on measured chlorophyll-a concentrations. Unfortunately, they converted our isoprene concentrations incorrectly, resulting in a three-order-of-magnitude overestimation and hence a much greater calculated isoprene flux.

异戊二烯、云滴和浮游生物

Meskhidze N.

(美国亚特兰大乔治亚理工学院地球和大气科学系)

摘要:Meskhidze N. 和 Nenes A. 的研究文章《南大洋浮游植物和云量》(2006 年 12 月 1 日,第 1419 页)犯了一个错误,可能会使其主要结论无效。该作者报道在大量浮游生物暴发的南大洋云滴有效性半径减少了 30%,南大洋云滴数量浓度增加一倍,从而导致云反射率的增加和额外 15 W · m^{-2} 的能量反射到太空。他们把这些变化归因于浮游生物暴发导致的异戊二烯的增加。Meskhidze 和 Nenes 用我们在南大洋铁实验(SOFeX)期间的测定数据,通过测定叶绿素 a 浓度,计算海水中异戊二烯含量。不幸的是,他们错误地转换了我们测定的异戊二烯浓度,导致高估了 3 个数量级,从而高估了异戊二烯通量。

59. Odling-Smee L. Letting the light in on Antarctic ecosystems[J]. Nature,2007,446(7137):713.

Letting the light in on Antarctic ecosystems

Odling-Smee L.

(Institute of Cell, Animal and Population Biology, The University of Edinburgh, UK)

Abstract: In just over a decade, two major ice shelves have collapsed on the eastern side of the Antarctic peninsula, uncovering a part of the sea floor that had not seen sunlight for several thousand years. A ten-week expedition that ended in late January has shed light on the biology of these waters, and has recovered samples of some 1000 species from the region, several of which may be new to science.

让阳光洒向南极生态系统

Odling-Smee L.

(英国爱丁堡大学细胞、动物和种群生态学研究所)

摘要:仅在过去的十年中,南极半岛东侧的两个主要冰架倒塌了,暴露了一部分几千年未见到阳光的海底。为期十周的探险在 1 月底结束,研究了这些海水的生物学,恢复了该地区大约 1000 个物种,其中有若干从科学来说可能是新的物种。

4.3 摘要翻译——北极

1. Butterfield N J,Knoll A H,Swett K. A Bangiophyte red alga from the proterozoic of Arctic Canada [J]. Science,1990,250(4977):104-107.

A Bangiophyte red alga from the proterozoic of Arctic Canada

Butterfield N. J. , Knoll A. H. , Swett K.

(Department of Earth and Planetary Sciences, Harvard University, Cambridge, MA 02138, USA)

Abstract: Silicified peritidal carbonate rocks of the 1250- to 750-million-year-old Hunting Formation, Somerset Island, arctic Canada, contain fossils of well-preserved bangiophyte red algae. Morphological details, especially the presence of multiseriate filaments composed of radially arranged wedge-shaped cells derived by longitudinal divisions from disc-shaped cells in uniseriate filaments, indicate that the fossils are related to extant species in the genus Bangia. Such taxonomic resolution distinguishes these fossils from other pre-Ediacaran eukaryotes and contributes to growing evidence that multicellular algae diversified well before the Ediacaran radiation of large animals.

加拿大北极地区一种原生代的 Bangiophyte 红藻

Butterfield N. J. , Knoll A. H. , Swett K.

(美国哈佛大学地球与行星科学系)

摘要:硅酸化的潮缘硅酸盐岩石属于 1.25 亿～7.5 亿年群组,位于萨默塞特岛,加拿大北极地区,包括保存完好的红藻化石 bangiophyte。形态学的细节,特别是放射状排列的由单列细丝在纵向分裂细胞中的细丝组成的多列的存在,表明该化石属和现存种属 Bangia 有关。这种分类结果将这些化石与来自其他的前埃迪卡拉真核生物区分开,并有助于进一步证明多细胞藻类在大型动物的埃迪卡拉辐射前也是很多元化的。

2. F Stuart Chapin Ⅲ,Lori Moilanen,Knut Kielland. Preferential use of organic nitrogen for growth by a nonmycorrhizal Arctic sedge[J]. Nature,1993,361(6408):150-153.

Preferential use of organic nitrogen for growth by a nonmycorrhizal Arctic sedge

F. Stuart Chapin Ⅲ, Lori Moilanen, Knut Kielland

(University of Alaska Fairbanks, Fairbanks, Alaska, United States)

Abstract: Plant growth in arctic tundra is strongly nitrogen-limited despite large pools of soil organic nitrogen. Here we report that field-collected roots of Eriophorum vaginatum, an arctic sedge, rapidly absorb free amino acids, accounting for at least 60% of the nitrogen absorbed by this species in the field. In solution culture, Eriophorum accumulates more nitrogen and biomass when supplied with amino acids than when grown on inorganic nitrogen, whereas Hordeum vulgare (a cereal adapted to mineral soils) grows least when nitrogen is supplied as amino acids. To our knowledge, this is the first documentation of preferential absorption and use of organic nitrogen by a non-mycorrhizal vascular plant. The direct absorption of amino acids by Eriophorum short-circuits the bottle-neck in arctic nitrogen cycles imposd by temperature-limited mineralization.

非菌根北极莎草生长中对有机氮的优先使用

F. Stuart Chapin Ⅲ, Lori Moilanen, Knut Kielland

(美国阿拉斯加州费尔班克斯大学)

摘要:在北极苔原植物的生长强烈受到氮的影响,尽管土壤中有机氮很多。本文我们研究,实地收集了有鞘的羊胡子草,一种北极莎草的根,它们能迅速吸收游离氨基酸,土壤中至少60%的氮被这种属植物吸收。在氨基酸作为氮来源的溶液培养中,羊胡子草生物量积累的氮和生物质比在无机氮作为氮来源的溶液培养中积累得多,而大麦属(一种适应矿物的土壤的谷物)生长至少需要无机氮供应。据我们所知,这是首次通过非维管束植物菌根优先吸收和利用有机氮的研究。氨基酸通过羊胡子草亚属捷径进行直接吸收,而不需经过北极氮循环中受到温度限制的矿化的瓶颈。

3. Hanski I, Turchin P, Korpimaeki E, et al. Population oscillations of boreal rodents: regulation by mustelid predators leads to chaos[J]. Nature, 1993, 364(6434): 232-235.

Population oscillations of boreal rodents: regulation by mustelid predators leads to chaos

Hanski I., Turchin P., Korpimaeki E., Henttonen H.

(Department of Zoology, University of Helsinki, Finland)

Abstract: The four-year cycle of microtine rodents in boreal and arctic regions was first described in 1924. Competing hypotheses on the mechanisms underlying the small mammal cycle have been extensively tested, but so far the sustained rodent oscillations are unexplained. Here we use two mutually supportive approaches to investigate this question. First, building on studies of the interaction between rodents and their mustelid predators, we construct a predator-prey model with seasonality. Second, we use a new technique of nonlinear analysis to examine empirical time-series data, and compare them with the model dynamics. The model parameterized with field data predicts dynamics that closely resemble the observed dynamics of boreal rodent populations. Both the predicted and observed dynamics are chaotic, albeit with a statistically significant periodic component. Our results suggest that the multiannual oscillations of rodent populations in Fennoscandia are due to delayed density dependence imposed by mustelid predators, and are chaotic.

北方鼠害周期受鼬鼠天敌调节而导致混乱

Hanski I. , Turchin P. , Korpimaeki E. , Henttonen H.

(芬兰赫尔辛基大学动物学系)

摘要：在北部和北极地区四年周期的田鼠鼠害最初记录是在1924年。底层的小哺乳动物周期竞争机制假说已进行大量实验，但至今持续鼠害周期的原因不明。本文中，我们使用两个相互支持的方法来研究这个问题。首先，基于啮齿类动物和它们的天敌相互作用的研究，构建具有季节性捕食者-猎物模型。其次，我们使用了非线性分析的新技术研究实验的时间序列数据，并将它们与模型相对比。现场数据参数化的模型预测的动态，十分类似于北方鼠类种群动态观察。预测和观察到的动态虽然有显著的周期性，但是也很混乱。我们的研究结果表明，鼠类在芬诺斯堪的亚种群的多年度周期混乱变动是由于其密度依赖于鼬鼠天敌的影响。

4. Wenink P W, Baker A J, Tilanus M G. Hypervariable-control-region sequences reveal global population structuring in a long-distance migrant shorebird, the Dunlin (Calidris-Alpina)[J]. PNAS, 1993,90(1):94-98.

Hypervariable-control-region sequences reveal global population structuring in a long-distance migrant shorebird, the Dunlin (*calidris-alpina*)

Wenink P. W. , Baker A. J. , Tilanus M. G.

(Canada's royal Ontario museum)

Abstract: Hypervariable segments of the control region of mtDNA as well as part of the cytochrome b gene of Dunlins were amplified with PCR and sequenced directly. The 910 base pairs (bp) obtained for each of 73 individuals complete another of the few sequencing studies that examine the global range of a vertebrate species. A total of 35 types of mtDNA were detected, 33 of which were defined by the hypervariable-control-region segments. Thirty of the latter were specific to populations of different geographic origin in the circumpolar breeding range of the species. The remaining three types indicate dispersal between populations in southern Norway and Siberia, but female-mediated flow of mtDNA apparently is too low to overcome the effects of high mutation rates of the control-region sequences, as well as population subdivision associated with historical range disjunctions. A genealogical tree relating the types grouped them into five populations: Alaska, West Coast of North America, Gulf of Mexico, western Europe, and the Taymyr Peninsula. The Dunlin is thus highly structured geographically, with measures of mutational divergence approaching 1.0 for fixation of alternative types in different populations. High diversity of types within populations as well as moderate long-term effective population sizes argue against severe population bottlenecks in promoting this differentiation. Instead, population fragmentation in Pleistocene refuges is the most plausible mechanism of mtDNA differentiation but at a much earlier time scale than suggested previously with morphometric data.

在高变控制区域的远距离迁徙的黑腹滨鹬海岸鸟类(*calidris-alpina*)揭示全球数量结构

Wenink P. W., Baker A. J., Tilanus M. G.

(加拿大安大略皇家博物馆)

摘要:用 PCR 将滨鹬的 mtDNA 控制区以及细胞色素 b 部分的高变片段进行扩增和测序。从 73 个个体获得的 910 个碱基对(bp)完成了几个审查全球范围内脊椎动物物种的另一项测序研究。检测了全部 35 种 mtDNA,33 种被定义为高变控制区域片段。其中 30 种是环极地不同地理区域来源繁殖的特殊种群,其余 3 种类型表明在挪威南部和西伯利亚种群间的分散,但雌性介导的 mtDNA 流动太低,不足以克服在控制区序列的高突变率的影响,以及历史种群细分。家谱分析将这些种类分为 5 类:阿拉斯加、北美西海岸墨西哥湾、西欧海岸和泰梅尔半岛。因此,黑腹滨鹬在地理区域上高度结构化,在不同种群,其不同类型固定化的突变分歧指数接近 1.0。种群高多样性以及温和长期有效的种群数量规模不支持种群数量瓶颈促进分化观点。更新世种群分散庇护所是 mtDNA 差异最可能的机制,但在时间尺度上要比以前的那些数据所建议的更早。

5. Alison Butler. Acquisition and utilization of transition metal ions by marine organisms[J]. Science, 1998,281(5374):207-210.

Acquisition and utilization of transition metal ions by marine organisms

Alison Butler

(Department of Chemistry, University of Calfornia, Santa Barbara, CA 93106-9510, USA)

Abstract: Recent research has revealed that trace metals, particularly transition metals, play important roles in marine productivity. Most of the work has been on iron, which shows a nutrient-depleted profile in the upper ocean. Marine organisms have a variety of means for acquiring iron and other transition metal ions that differ from those of terrestrial organisms.

海洋生物体获得和利用过渡金属离子

Alison Butler

(美国加利福尼亚大学圣巴巴拉分校化学系)

摘要:近期研究显示痕量金属,尤其是过渡金属元素,在海洋生产力中起着重要的作用。大部分的研究都关注于铁离子,它显示了海洋上部的营养消耗变化。海洋生物体有多种与陆地生物体不同的获得铁离子和其他过渡金属离子的方法。

6. Goulden M L, Wofsy S C, Harden J W, et al. Sensitivity of boreal forest carbon balance to soil thaw [J]. Science, 1998, 279(5348):214-217.

Sensitivity of boreal forest carbon balance to soil thaw

Goulden M. L., Wofsy S. C., Harden J. W., Trumbore S. E., Crill P. M., Gower S. T., Fries T., Daube B. C., Fan S.-M., Sutton D. J., Bazzaz A., Munger J. W.

(Department of Earth and Planetary Sciences, Harvard University, Cambridge, MA 02138, USA)

Abstract: We used eddy covariance; gas-exchange chambers; radiocarbon analysis; wood, moss, and soil inventories; and laboratory incubations to measure the carbon balance of a 120-year-old black spruce forest in Manitoba, Canada. The site lost 0.3 +/−0.5 metric ton of carbon per hectare per year (ton C ha(−1) year(−1)) from 1994 to 1997, with a gain of 0.6 +/−0.2 ton C ha(−1) year(−1) in moss and wood offset by a loss of 0.8 +/−0.5 ton C ha(−1) year(−1) from the soil. The soil remained frozen most of the year, and the decomposition of organic matter in the soil increased 10-fold upon thawing. The stability of the soil carbon pool (similar to 150 tons C ha(−1)) appears sensitive to the depth and duration of thaw, and climatic changes that promote thaw are likely to cause a net efflux of carbon dioxide from the site.

敏感性寒带森林的碳平衡对土壤解冻的敏感性

Goulden M. L., Wofsy S. C., Harden J. W., Trumbore S. E., Crill P. M., Gower S. T., Fries T., Daube B. C., Fan S.-M., Sutton D. J., Bazzaz A., Munger J. W.

(美国哈佛大学地球与行星科学系)

摘要:我们用涡度协方差,气体交换室,放射性碳分析,树林、苔藓和土壤的详细存量及实验室培养来衡量加拿大马尼托省一片120年黑云杉森林的碳平衡。1994年至1997年该区每公顷每年损失0.3±0.5吨碳,苔藓和树木的增益每公顷每年为0.6±0.2吨碳,但来自土壤的亏损每公顷每年为0.8±0.5吨碳。土壤在一年的大部分时间是冻结的,解冻后土壤中的有机质分解增加了10倍以上。土壤的稳定碳库(大约每公顷150吨碳)对解冻的深度和时间敏感,促进解冻的气候变化可能会导致此地二氧化碳净流出。

7. Tarduno J A, Brinkman D B, Renne P R, et al. Evidence for extreme climatic warmth from Late Cretaceous Arctic vertebrates[J]. Science, 1998, 282(5397): 2241-2244.

Evidence for extreme climatic warmth from Late Cretaceous Arctic vertebrates

Tarduno J. A., Brinkman D. B., Renne P. R., Cottrell R. D., Scher H., Castillo P.

(Department of Earth and Environmental Sciences, University of Rochester, Rochester, NY, 14627, USA)

Abstract: A Late Cretaceous (92 to 86 million years ago) vertebrate assemblage from the high Canadian Arctic (Axel Heiberg Island) implies that polar climates were warm (mean annual temperature exceeding 14 degrees C) rather than near freezing. The assemblage includes large (2.4 meters long) champsosaurs, which are extinct crocodilelike reptiles. Magmatism at six large igneous provinces at this time suggests that volcanic carbon dioxide emissions helped cause the global warmth.

北极晚白垩世脊椎动物作为极端温暖气候的证据

Tarduno J. A., Brinkman D. B., Renne P. R., Cottrell R. D., Scher H., Castillo P.

(美国罗彻斯特大学地球与环境科学系)

摘要：一晚白垩世(9200万~8600万年前)加拿大北极区(阿克塞尔海贝格岛)脊椎动物组合意味着极地气候是温暖的(年平均温度超过14℃)，而不是接近冰点。该组合包括大鳄(2.4 m)，这是灭绝的鳄鱼类爬行动物。此时期六个大火成岩省岩浆活动表明，火山岩浆活动的二氧化碳排放导致全球温暖。

8. Ian D Campbell, Karen McDonald, Michael D Flannigan, et al. Long-distance transport of pollen into the Arctic[J]. Nature, 1999, 399(6731): 29-30.

Long-distance transport of pollen into the Arctic

Ian D. Campbell, Karen McDonald, Michael D. Flannigan, Joanni Kringayark

(Canadian Forest Service, 5320-122 Street, Edmonton, Alberta, T6H 3S5, Canada)

Abstract: Airborne particulates can be carried over long distances, but for significant quantities of particulates larger than a few micrometres in diameter to be transported more than a few kilometres usually requires a means of injecting the material high into the atmosphere, such as a volcanic eruption, forest fire or desert windstorm. But an unusual event occurred in the Canadian Arctic last year, in which significant amounts of pine and spruce pollen (30~55 μm long) were transported roughly 3000 km.

花粉长距离输入北极

Ian D. Campbell, Karen McDonald, Michael D. Flannigan, Joanni Kringayark

(加拿大自然资源部林务局)

摘要：大气飘尘可以远距离进行，但对颗粒直径大于几个微米较大的微粒的运输超过数千公里，通常需要物质被注射到较高的大气中，如火山喷发、森林火灾或沙漠风暴。但是，一个不寻常的事件发生在去年的加拿大北极地区，其中大量的松树和云杉花粉(30~55 μm)被运送大约3000公里。

9. Richard J Abbott, Lisa C Smith, Richard I Milne, et al. Molecular analysis of plant migration and refugia in the Arctic[J]. Science, 2000, 289(5483): 1343-1346.

Molecular analysis of plant migration and refugia in the Arctic

Richard J. Abbott, Lisa C. Smith, Richard I. Milne, Robert M. M. Crawford, Kirsten Wolff, Jean Balfour

(Division of Environmental and Evolutionary Biology, Harold Mitchell Building, School of Biology, University of St. Andrews, St. Andrews, Fife KY16 9TH, UK)

Abstract: The arctic flora is thought to have originated during the Late Tertiary, approximately 3

million years ago. Plant migration routes during colonization of the Arctic are currently unknown, and uncertainty remains over where arctic plants survived Pleistocene glaciations. A phylogenetic analysis of chloroplast DNA variation in the purple saxifrage (*Saxifraga oppositifolia*) indicates that this plant first occurred in the Arctic in western Beringia before it migrated east and west to achieve a circumpolar distribution. The geographical distribution of chloroplast DNA variation in the species supports the hypothesis that, during Pleistocene glaciations, some plant refugia were Located in the Arctic as well as at more southern latitudes.

北极植物迁移和植物残遗种的分子研究

Richard J. Abbott, Lisa C. Smith, Richard I. Milne, Robert M. M. Crawford, Kirsten Wolff, Jean Balfour

(英国圣安德鲁斯大学生物学院环境与进化生物学部)

摘要:北极植物群落被认为始于晚第三纪,大约 300 万年前。北极植物化期间的植物迁移路线普遍未知,而且在更新世冰期中存活下来的北极植物现存于北极何处也是不确定的。通过系统地对紫色虎耳草属植物(*Saxifraga oppositifolia*)叶绿体中的 DNA 变化进行研究,显示这种植物最先出现在北极的西白令陆桥,随后它迁移到东部和西部形成了环极地分布。叶绿体 DNA 变化的地理分布支持了一个假设:更新世冰期时期,一些植物残遗种在北极和较为偏南的纬度均存在。

10. Philip W Boyd, Andrew J Watson, Cliff S Law, et al. A mesoscale phytoplankton bloom in the polar Southern Ocean stimulated by iron fertilization[J]. Nature, 2000, 407(6805): 695-702.

A mesoscale phytoplankton bloom in the polar Southern Ocean stimulated by iron fertilization

Philip W. Boyd, Andrew J. Watson, Cliff S. Law, Edward R. Abraham, Thomas Trull, Rob Murdoch, Dorothee C. E. Bakker, Andrew R. Bowie, K. O. Buesseler, Hoe Chang, Matthew Charette, Peter Croot, Ken Downing, Russell Frew, Mark Gall, Mark Hadfield, Julie Hall, Mike Harvey, Greg Jameson, Julie LaRoche, Malcolm Liddicoat, Roger Ling, Maria T. Maldonado, Michael McKay R. , Scott Nodder, Stu Pickmere, Rick Pridmore, Steve Rintoul, Karl Safi, Philip Sutton, Robert Strzepek, Kim Tanneberger, Suzanne Turner, Anya Waite, John Zeldis

(National Institute of Water and Atmosphere, Centre for Chemical and Physical Oceanography, Department of Chemistry, University of Otago, Dunedin, New Zealand)

Abstract: Changes in iron supply to oceanic plankton are thought to have a significant effect on concentrations of atmospheric carbon dioxide by altering rates of carbon sequestration, a theory known as the "iron hypothesis". For this reason, it is important to understand the response of pelagic biota to increased iron supply. Here we report the results of a mesoscale iron fertilization experiment in the polar Southern Ocean, where the potential to sequester iron-elevated algal carbon is probably greatest. Increased iron supply led to elevated phytoplankton biomass and rates of photosynthesis in surface waters, causing a large drawdown of carbon dioxide and macronutrients, and elevated dimethyl sulphide levels after 13 days. This drawdown was mostly due to the proliferation of diatom stocks. But downward export of biogenic carbon was not increased. Moreover, satellite observations of this massive bloom 30 days later, suggest that a sufficient proportion of the added iron was retained in surface waters. Our findings demonstrate that iron supply controls phytoplankton growth and community composition during summer in these polar Southern Ocean waters, but the fate of algal carbon remains unknown and depends on the interplay between the processes controlling export, remineralisation and timescales of water mass subduction.

极地南部海域由于铁营养带来的中尺度浮游植物繁盛

Philip W. Boyd, Andrew J. Watson, Cliff S. Law, Edward R. Abraham, Thomas Trull, Rob Murdoch, Dorothee C. E. Bakker, Andrew R. Bowie, K. O. Buesseler, Hoe Chang, Matthew Charette, Peter Croot, Ken Downing, Russell Frew, Mark Gall, Mark Hadfield, Julie Hall, Mike Harvey, Greg Jameson, Julie LaRoche, Malcolm Liddicoat, Roger Ling, Maria T. Maldonado, Michael McKay R., Scott Nodder, Stu Pickmere, Rick Pridmore, Steve Rintoul, Karl Safi, Philip Sutton, Robert Strzepek, Kim Tanneberger, Suzanne Turner, Anya Waite, John Zeldis

(新西兰奥塔哥大学化学系水和大气研究所化学和物理海洋学中心)

摘要：铁供应海洋浮游生物的变化通过改变碳汇，在我们所知的"铁假说"理论中被认为是对二氧化碳浓度的大气效果影响显著。基于这个原因，了解远洋生物对铁增加的响应更加重要了。本文中，我们研究了在极地南大洋中尺度铁营养的实验结果，极地南大洋研究铁增加带来的藻类碳增加的潜力可能是最大的。增加铁的供给，导致浮游植物生物量增加和表层水光合作用速率增加，造成了大量二氧化碳和营养素的下降，并在13天后导致了二甲基硫化物的水平升高。这种下降主要是由于硅藻类的增殖。但是，生物碳输出并未增加。此外，人造卫星对此区域30天后的观测表明，在表层水中的多余铁比例很充足。我们的研究证明极地南大洋水域铁的供应量在夏季控制了浮游植物的生长和群落组成，但藻类碳存在过程仍然未知，它依赖于控制输出的过程、矿质的补充和水团俯冲时间的相互作用。

11. Kate F Darling, Christopher M Wade, Iain A Stewart, et al. Molecular evidence for genetic mixing of Arctic and Antarctic subpolar populations of planktonic foraminifers[J]. Nature, 2000, 405(6782): 43-47.

Molecular evidence for genetic mixing of Arctic and Antarctic subpolar populations of planktonic foraminifers

Kate F. Darling, Christopher M. Wade, Iain A. Stewart, Dick Kroon, Richard Dingle, Andrew J. Leigh Brown

(Department of Geology and Geophysics, University of Edinburgh, Edinburgh EH9 3JW, UK)

Abstract: Bipolarity, the presence of a species in the high latitudes separated by a gap in distribution across the tropics, is a well-known pattern of global species distribution. But the question of whether bipolar species have evolved independently at the poles since the establishment of the cold-water provinces 16-8 million years ago, or if genes have been transferred across the tropics since that time, has not been addressed. Here we examine genetic variation in the small subunit ribosomal RNA gene of three bipolar planktonic foraminiferal morphospecies. We identify at least one identical genotype in all three morphospecies in both the Arctic and Antarctic subpolar provinces, indicating that trans-tropical gene flow must have occurred. Our genetic analysis also reveals that foraminiferal morphospecies can consist of a complex of genetic types. Such occurrences of genetically distinct populations within one morphospecies may affect the use of planktonic foraminifers as a palaeoceanographic proxy for climate change and necessitate a reassessment of the species concept for the group.

两极的亚极地地区浮游有孔虫混合种群遗传的分子证据

Kate F. Darling, Christopher M. Wade, Iain A. Stewart, Dick Kroon, Richard Dingle, Andrew J. Leigh Brown

(英国爱丁堡大学地质与地球物理系)

摘要：两极高纬度地区物种的存在由于整个热带地区分配差距而分离，是全球物种分布的知名格局。但是问题在于是否两极地区物种自从1600万～800万年前冷水地区的确定开始独立进化，还是在同时期越过热带地区进行了基因转变。本文中我们测试了三个两极地区浮游有孔虫形态种基因小亚基核糖体RNA的遗传变异。我们确定在亚极地地区都有的三种形态种中至少有一个相同的基因型，这表明转热带基因流动肯定有发生。我们的基因分析还表明，有孔虫目形态种可以由复杂遗传类型组成。形态种内基因的不同类型的产生可能影响浮游有孔虫作为古海洋气候变化指标剂的使用，所以有必要对生物团的物种概念重新评估。

12. Tim J Karels, Rudy Boonstra. Concurrent density dependence and independence in populations of Arctic ground squirrels[J]. Nature, 2000, 408(6811): 460-463.

Concurrent density dependence and independence in populations of Arctic ground squirrels

Tim J. Karels, Rudy Boonstra

(Division of Life Sciences, University of Toronto at Scarborough, 1265 Military Trail, Scarborough, Ontario, M1C 1A4, Canada)

Abstract: No population increases without limit. The processes that prevent this can operate in either a density-dependent way (acting with increasing severity to increase mortality rates or decrease reproductive rates as density increases), a density-independent way, or in both ways simultaneously. However, ecologists disagree for two main reasons about the relative roles and influences that density-dependent and density-independent processes have in determining population size. First, empirical studies showing both processes operating simultaneously are rare. Second, time-series analyses of long-term census data sometimes overestimate dependence. By using a density-perturbation experiment on arctic ground squirrels, we show concurrent density-dependent and density-independent declines in weaning rates, followed by density-dependent declines in overwinter survival during hibernation. These two processes result in strong, density-dependent convergence of experimentally increased populations to those of control populations that had been at low, stable levels.

北极松鼠种群并发密度依赖和独立性

Tim J. Karels, Rudy Boonstra

(加拿大多伦多大学生命科学系)

摘要:没有数量限制的增长。防止这种情况的进程可以以密度依赖的方式(密度增加时导致繁殖率减少和日益严重的死亡率增加),也可以以密度无关的方式,或两种方式同时进行。然而,生态学家对密度依赖和密度无关过程的相对作用和影响意见不一,主要有两个原因。首先,经验研究表明这两个进程同时进行并不多见。其次,长期数量普查时间序列数据分析有时高估依赖性。通过北极地松鼠的种群密度扰动实验,我们展示了在(幼崽)断奶率下降时密度依赖和密度无关的一致性,随之在冬眠期存活率下降时表现为密度依赖。这两个过程的结果表明,通过实验增长的种群比那些被控制的低且稳定水平上的种群表现为强而种群依赖的趋同现象。

13. Moore P D. Ecology:bats about the Arctic[J]. Nature,2000,404:446-447.

Ecology: bats about the Arctic

Moore P. D.

(The department of life sciences at king's college London)

Abstract: Why do bats fly at night rather than during the day? There are exceptions, of course, but most bats confine their activities to dusk and the depths of night. The insect-eating bats of the higher latitudes, in particular, are night-time hunters and are equipped with sophisticated techniques for the non-visual location of prey. There are several possible advantages that could be associated with nocturnal feeding, such as the avoidance of predators or mobbing by birds, the tapping of a food resource unexploited by other insectivores, or sensitivity to the problems of overheating when active in daylight. In a study of bats, insectivorous birds and predators in northern Norway, Speakman and colleagues, reporting in Oikos, conclude that the avoidance of competition for food is the most likely answer.

生态:北极蝙蝠

Moore P. D.

(英国伦敦国王学院生命科学学部)

摘要:为什么蝙蝠在夜间飞行而不是在白天?当然有些例外,但大部分蝙蝠在黄昏和深夜活动。特别是高纬地区的食虫蝙蝠是夜晚捕食者且有着精良的技术完成非视觉定位捕食。一些可能的优势或许与夜间捕食相关,如避开捕食者或被鸟类围攻,有着未被其他食昆虫物种开发的食物资源,或是在白天应用会过热的灵敏性问题。Speakman和同事们在奥伊科斯报道,在北挪威一项针对蝙蝠、食虫鸟类和捕食者的研究中,得出的结论是躲避食物竞争是最可能的答案。

14. Thomas Alerstam,Gudmundur A Gudmundsson,Martin Green,et al. Migration along orthodromic sun compass routes by Arctic birds[J]. Science,2001,291(5502):300-303.

Migration along orthodromic sun compass routes by Arctic birds

Thomas Alerstam, Gudmundur A. Gudmundsson, Martin Green, Anders Hedenström

(Department of Animal Ecology, Lund University, Ecology Building, SE-22362 Lund, Sweden)

Abstract: Flight directions of birds migrating at high geographic and magnetic latitudes can be used to test bird orientation by celestial or geomagnetic compass systems under polar conditions. Migration patterns of arctic shorebirds, revealed by tracking radar studies during an icebreaker expedition along the Northwest Passage in 1999, support predicted sun compass trajectories but cannot be reconciled with orientation along either geographic or magnetic loxodromes (rhumb Lines). Sun compass routes are similar to orthodromes (great circle routes) at high latitudes, showing changing geographic courses as the birds traverse longitudes and their internal clock gets out of phase with Local time. These routes bring the shorebirds from high arctic Canada to the east coast of North America, from which they make transoceanic flights to South America. The observations are also consistent with a migration Link between Siberia and the Beaufort Sea region by way of sun compass routes across the Arctic Ocean.

北极鸟类沿着日光罗盘方向迁移

Thomas Alerstam, Gudmundur A. Gudmundsson, Martin Green, Anders Hedenström

(瑞典隆德大学动物生态学系)

摘要:在高地理和磁纬度地区,鸟类迁徙的飞行方向可以用来测试极地条件下天文或地磁的罗盘系统对鸟类的定向。北极海岸鸟类的迁徙模式,于 1999 年沿西北通道的破冰航行期间通过跟踪雷达显示来进行研究,支持预测的日光罗盘轨道,但与地理和磁航线(罗盘航线)定向不一致。日光罗盘轨道在高纬度地区类似大圆弧形(大圆弧轨迹),当鸟类横断经度,它们的生物时钟和当地时钟不同,显示了地理路线的变化。这些路线使海岸鸟类从高北极圈加拿大地区飞向北美洲的东海岸,使它们横越海洋飞行至南美洲。此项观察资料也和鸟类沿日光罗盘轨道线横越北冰洋从西伯利亚迁徙至波弗特海地区的研究结果相符。

15. Marianne Barrier, Robert H Robichaux, Michael D Purugganan. Accelerated regulatory gene evolution in an adaptive radiation[J]. PNAS,2001,98(18):10208-10213.

Accelerated regulatory gene evolution in an adaptive radiation

Marianne Barrier, Robert H. Robichaux, Michael D. Purugganan

(Department of Genetics, Box 7614, North Carolina State University, Raleigh, NC 27695, USA)

Abstract: The disparity between rates of morphological and molecular evolution remains a key paradox in evolutionary genetics. A proposed resolution to this paradox has been the conjecture that morphological evolution proceeds via diversification in regulatory loci, and that phenotypic evolution may correlate better with regulatory gene divergence. This conjecture can be tested by examining rates of regulatory gene evolution in species that display rapid morphological diversification within adaptive radiations. We have isolated homologues to the Arabidopsis APETALA3 (ASAP3/TM6) and APETALA1 (ASAP1) floral regulatory genes and the CHLOROPHYLL A/B BINDING PROTEIN9 (ASCAB9) photosynthetic structural gene from species in the Hawaiian silversword alliance, a premier example of plant adaptive radiation. We have compared rates of regulatory and structural gene evolution in the Hawaiian species to those in related species of North American tarweeds. Molecular evolutionary analyses indicate significant increases in nonsynonymous relative to synonymous nucleotide substitution rates in the ASAP3/TM6 and ASAP1 regulatory genes in the rapidly evolving Hawaiian species. By contrast, no general increase is evident in neutral mutation rates for these loci in the Hawaiian species. An increase in nonsynonymous relative to synonymous nucleotide substitution rate is also evident in the ASCAB9 structural gene in the Hawaiian species, but not to the extent displayed in the regulatory loci. The significantly accelerated rates of regulatory gene evolution in the Hawaiian species may reflect the influence of allopolyploidy or of selection and adaptive divergence. The analyses suggest that accelerated rates of regulatory gene evolution may accompany rapid morphological diversification in adaptive radiations.

在适应辐射中调控基因的加速进化

Marianne Barrier, Robert H. Robichaux, Michael D. Purugganan

(美国北卡罗来纳州立大学遗传学系)

摘要:形态学和分子进化速率的差距仍然是遗传进化学的一个无定论关键论题。拟议的解决这一问题的猜测是形态学进化通过调控基因位点的变异,表型的进化和调控基因的变异关系更大。这个猜想可以通过检查在辐射适应中显示快速适应形态多样化物种的调控基因进化速率来检测。我们已经分离出的拟南芥 APETALA3(ASAP3/TM6)和 APETALA1(ASAP1)花的调控基因和叶绿素 a / b 结合蛋白 9 (ASCAB9)光合结构基因的类似基因,这些物种来自夏威夷锡尔弗斯沃联盟,是植物适应辐射重要的样例。我们比较了夏威夷物种和北美树脂腺植物相关物种的调控基因和结构基因的进化速率。分子进化分析表明在迅速进化的夏威夷物种中,ASAP3/TM6 和 ASAP1 调控基因,非同义核苷酸相较同义核苷酸替代速率明显地增加。相比之下,在夏威夷物种中在这些位点上中性突变明显没有增加。在 ASCAB9 在夏威夷种结构基因中,非同义核苷酸相对于同义的替代率增加也是显而易见的,但在调控基因位点上并不广泛显示。夏威夷物种基因进化调整速度可能反映了适应和选择趋异的异源多倍性影响。这些分析表明,调控基因进化调整加速可能会伴随辐射适应快速形态多样化。

16. Klaassen M, Lindström A A, Meltofte H, et al. Ornithology: Arctic waders are not capital breeders [J]. Nature, 2001, 413(6858): 794.

Ornithology—Arctic waders are not capital breeders

Klaassen M., Lindström A. A., Meltofte H., Piersma T.

(Centre for Ecological and Evolutionary Studies (CEES), University of Groningen, Groningen, Netherlands)

Abstract: Birds prepare their eggs from recently ingested nutrients ("income" breeders) or from body stores ("capital" breeders). As summers are short at Arctic latitudes, Arctic migrants have been presumed to bring nutrients for egg production from their previous habitats, so that they can start breeding immediately upon arrival. But we show here that eggs laid by 10 different wader species from 12 localities in northeast Greenland and Arctic Canada are produced from nutrients originating from tundra habitats, as inferred from carbon stable-isotope ratios in eggs, natal down, and juvenile and adult feathers.

鸟类学——北极涉禽类不是资本种畜

Klaassen M., Lindström A. A., Meltofte H., Piersma T.

(荷兰格罗宁根大学生态与进化研究所(CEES))

摘要:现代鸟类产卵通过摄取营养物质("收入"种畜)或来自身体存粮("资本"种畜)。由于北极纬度原因夏天时间短,人们认为北极候鸟从以前的栖息地为它们产卵带来生产营养,让它们可以在到达繁殖地后迅速繁殖。但是,我们通过鸟蛋、出生绒毛和幼年成年的羽毛中稳定碳同位素比值来推断,来自12个格陵兰岛东北部和加拿大北极地点的10种不同候鸟种类的鸟蛋营养来自北极苔原。

17. Mestres F, Balanyà J, Arenas C, et al. Colonization of America by Drosophila subobscura: heterotic effect of chromosomal arrangements revealed by the persistence of lethal genes[J]. PNAS, 2001, 98(16): 9167-9170.

Colonization of America by Drosophila subobscura: heterotic effect of chromosomal arrangements revealed by the persistence of lethal genes

Mestres F., Balanyà J., Arenas C., Solé E., Serra L.

(Departaments de Genètica and d'Estadistica, Facultat de Biologia, Universitat de Barcelona, Diagonal 645, 08071 Barcelona, Spain)

Abstract: About 20 years ago *Drosophila subobscura*, a native Palearctic species, colonized both North and South America. In Palearctic populations lethal genes are not associated in general with particular chromosomal arrangements. In colonizing populations they are not randomly distributed and usually are associated to a different degree with chromosomal arrangements caused by the founder event. The persistence of two lethal genes in the colonizing populations, one completely associated with the O-5 inversion and the other partially associated with the O3+4+7 arrangement, has been analyzed. In all populations studied (five North American and six South American) the observed frequency of the lethal gene completely associated with the O-5 inversion is higher than expected, the difference being statistically significant in all South American and one North American populations. The observed frequency of the lethal gene partially associated with the O3+4+7 arrangement is also significantly higher than expected. Taking into account that the O-5 inversion exhibits significant latitudinal clines both in North and South America, an overdominant model favoring the heterokaryotypes seems to be in operation. From this model, a polynomial expression has been developed that allows us to estimate the relative fitness and the coefficient of selection against all karyotypes not carrying the O-5 inversion. The relative fitness of the O-5 heterokaryotypes is higher in South American than in North American populations. Furthermore, the observed frequencies of the lethal genes studied are in general very close to those of the equilibrium. This case is an outstanding demonstration in nature of an heterotic effect of chromosomal segments associated with lethal genes on a large geographic scale.

殖民美国的果蝇致命基因的持久性所揭示的染色体排列的杂种优势效应

Mestres F., Balanyà J., Arenas C., Solé E., Serra L.

(西班牙巴塞罗那大学生物学系)

摘要:大约 20 年前,果蝇(*Drosophila subobscura*),原生古北区种,殖民北美和南美。在古北区的种群,致死基因一般不与特定染色体的排序相关。殖民的种群致死基因不是随机分布的,通常在不同程度上和一个首创事件造成的染色体排序相关。本文分析殖民种群的两个致死基因的持续存在,一个完全与 O-5 倒置相关,另一个与 O3+4+7 排序部分相关。所有研究观测种群(五个北美和六个南美)中,致死基因完全与 O-5 倒置相关的发生次数高于预期,这种统计学差异存在于所有南美和一个北美种群中。与 O3+4+7 序列部分相关的致死基因的发生次数也明显高于预期。考虑在北美和南美 O-5 倒置显示出明显的纬度渐变群,有利于直核型的绝对优势模式可能起作用。从这个模型中,我们建立多项式表达来评估相对适应性,以及排除携带 O-5 倒置的所有染色体组型的选择系数,和 O-5 杂种优势效应的适应性在南美要高于北美种群。此外,所观测的致死基因发生次数一般非常接近平衡状态。这种例子是与大范围致死基因和染色体排列杂种优势效应相关的显著实例。

18. David E Reich, Michele Cargill, Stacey Bolk, et al. Linkage disequilibrium in the human genome[J]. Nature, 2001, 411(6834): 199-204.

Linkage disequilibrium in the human genome

David E. Reich, Michele Cargill, Stacey Bolk, James Ireland, Pardis C. Sabeti, Daniel J. Richter, Thomas Lavery, Rose Kouyoumjian, Shelli F. Farhadian, Ryk Ward, Eric S. Lander

(Whitehead Institute/MIT Center for Genome Research, Nine Cambridge Center, Cambridge, Massachusetts 02142, USA)

Abstract: With the availability of a dense genome-wide map of single nucleotide polymorphisms

(SNPs), a central issue in human genetics is whether it is now possible to use linkage disequilibrium (LD) to map genes that cause disease. LD refers to correlations among neighbouring alleles, reflecting "haplotypes" descended from single, ancestral chromosomes. The size of LD blocks has been the subject of considerable debate. Computer simulations and empirical data have suggested that LD extends only a few kilobases (kb) around common SNPs, whereas other data have suggested that it can extend much further, in some cases greater than 100 kb. It has been difficult to obtain a systematic picture of LD because past studies have been based on only a few loci and different populations. Here, we report a large-scale experiment using a uniform protocol to examine 19 randomly selected genomic regions. LD in a United States population of north-European descent typically extends 60 kb from common alleles, implying that LD mapping is likely to be practical in this population. By contrast, LD in a Nigerian population extends markedly less far. The results illuminate human history, suggesting that LD in northern Europeans is shaped by a marked demographic event about 27000-53000 years ago.

人类基因组连锁不平衡

David E. Reich, Michele Cargill, Stacey Bolk, James Ireland,
Pardis C. Sabeti, Daniel J. Richter, Thomas Lavery, Rose Kouyoumjian,
Shelli F. Farhadian, Ryk Ward, Eric S. Lander

(美国麻省怀特基因组研究所)

摘要:随着密集的全基因组图谱的单核苷酸多态性(SNPs)的可用性,人类遗传学的一个核心问题是现在是否可以使用连锁不平衡(LD)来找到导致疾病的基因。连锁不平衡指的是相邻的等位基因的相关性,反映从单个祖先染色体的单倍型遗传。连锁不平衡模块的大小,一直有很多争论。计算机模拟及经验数据表明,连锁不平衡模块只有几个千碱基对(kb)围绕共同的单核苷酸多态性,而其他数据显示,它可以在某些情况下更进一步扩展至大到100个千碱基对。取得连锁不平衡的系统性理解一直很难,因为过去的研究只涉及少数几个位点和种群。本文中,我们做了大量实验,使用统一的协议审查19个随机选择的基因组区域。在北欧洲裔美国人的连锁不平衡基因通常从等位基因延伸60 kb,这意味着连锁不平衡图可能在这个人类种群中很实用。相反,尼日利亚人口的连锁不平衡延伸明显要少得多。我们的结果揭示人类历史中北欧连锁不平衡是在27000～53000年前因显著的人口事件而形成的。

19. Mark K Sears, Richard L Hellmich, Diane E Stanley-Horn, et al. Impact of Bt corn pollen on monarch butterfly populations: a risk assessment[J]. PNAS, 2001, 98(21): 11937-11942.

Impact of Bt corn pollen on monarch butterfly populations: a risk assessment

Mark K. Sears, Richard L. Hellmich, Diane E. Stanley-Horn, Karen S. Oberhauser,
John M. Pleasants, Heather R. Mattila, Blair D. Siegfried, Galen P. Dively

(Department of Environmental Biology, University of Guelph, Guelph, ON, Canada N1G 2W1)

Abstract: A collaborative research effort by scientists in several states and in Canada has produced information to develop a formal risk assessment of the impact of Bt corn on monarch butterfly (*Danaus* *plexippus*) populations. Information was sought on the acute toxic effects of Bt corn pollen and the degree to which monarch larvae would be exposed to toxic amounts of Bt pollen on its host plant, the common milkweed, *Asclepias syriaca*, found in and around cornfields. Expression of Cry proteins, the active toxicant found in Bt corn tissues, differed among hybrids, and especially so in the concentrations found in pollen of different events. In most commercial hybrids, Bt expression in pollen is low, and laboratory and field studies show no acute toxic effects at any pollen density that would be encountered in the field. Other factors mitigating exposure of larvae include the variable and limited overlap between pollen shed and larval activity periods, the fact that only a portion of the monarch population utilizes milkweed stands in and near cornfields, and the current adoption rate of Bt corn at 19% of North American corn-growing areas. This 2-year study suggests that the impact of Bt corn pollen from current commercial hybrids on monarch butterfly populations is negligible.

转 Bt 基因玉米授粉对于帝王蝶数量的影响:风险评估

Mark K. Sears, Richard L. Hellmich, Diane E. Stanley-Horn, Karen S. Oberhauser, John M. Pleasants, Heather R. Mattila, Blair D. Siegfried, Galen P. Dively

(加拿大圭尔夫大学环境生物学系)

摘要:一个由多个州和加拿大科学家合作研究的工作已做出了转 Bt 基因玉米对帝王蝶(*Danaus plexippus*)数量影响的正式的风险评估。信息被用来寻找转 Bt 基因玉米花粉的急性毒性作用和帝王蝶幼虫在它主要寄主的玉米田地附近的植物——普通乳草属植物 *Asclepias syriaca* 上,暴露在有毒转基因花粉中的中毒程度。Cry 蛋白质是在转 Bt 基因玉米组织中发现的活跃的毒物,在和不同混合种类,尤其是和不同的事件中发现的花粉中的浓度不同。在大多数商业杂交种中,转 Bt 基因在花粉中表达很低,实验室和现场研究表明在该地区任何花粉密度没有急性毒性作用。减轻幼虫暴露的其他因素包括花粉分散时期和幼虫活动时期只有若干有限重叠,事实上只有一部分帝王蝶利用乳草停留在玉米地里或附近,目前采用转 Bt 基因玉米的在北美国玉米种植区只占 19%。这两年的研究表明,目前从商业化程度来看,转 Bt 基因玉米花粉对帝王蝶种群的影响是微不足道的。

20. Karin Valmsen, Ivar Järving, William E Boeglin, et al. The origin of 15R-prostaglandins in the Caribbean coral *Plexaura homomalla*: molecular cloning and expression of a novel cyclooxygenase[J]. PNAS, 2001, 98(14): 7700-7705.

The origin of 15R-prostaglandins in the Caribbean coral *Plexaura homomalla*: molecular cloning and expression of a novel cyclooxygenase

Karin Valmsen, Ivar Järving, William E. Boeglin, Külliki Varvas, Reet Koljak, Tõnis Pehk, Alan R. Brash, Nigulas Samel

(Department of Bioorganic Chemistry, Institute of Chemistry at Tallinn Technical University, Akadeemia tee 15, Tallinn 12618, Estonia)

Abstract: The highest concentrations of prostaglandins in nature are found in the Caribbean gorgonian *Plexaura homomalla*. Depending on its geographical location, this coral contains prostaglandins with typical mammalian stereochemistry (15S-hydroxy) or the unusual 15R-prostaglandins. Their metabolic origin has remained the subject of mechanistic speculations for three decades. Here, we report the structure of a type of cyclooxygenase (COX) that catalyzes transformation of arachidonic acid into 15R-prostaglandins. Using a homology-based reverse transcriptase-PCR strategy, we cloned a cDNA corresponding to a COX protein from the R variety of *P. homomalla*. The deduced peptide sequence shows 80% identity with the 15S-specific coral COX from the Arctic soft coral *Gersemia fruticosa* and approximate to 50% identity to mammalian COX-1 and COX-2. The predicted tertiary structure shows high homology with mammalian COX isozymes having all of the characteristic structural units and the amino acid residues important in catalysis. Some structural differences are apparent around the peroxidase active site, in the membrane-binding domain, and in the pattern of glycosylation. When expressed in Sf9 cells, the *P. homomalla* enzyme forms a 15R-prostaglandin endoperoxide together with 11R-hydroxyeicosatetraenoic acid and 15R-hydroxyeicosatetraenoic acid as byproducts. The endoperoxide gives rise to 15R-prostaglandins and 12R-hydroxyheptadecatrienoic acid, identified by comparison to authentic standards. Evaluation of the structural differences of this 15R-COX isozyme should provide new insights into the substrate binding and stereospecificity of the dioxygenation reaction of arachidonic acid in the cyclooxygenase active site.

加勒比海珊瑚虫(*Plexaura homomalla*)的15R前列腺素的来源:分子克隆和一种新型环氧合酶的表达

Karin Valmsen, Ivar Järving, William E. Boeglin, Külliki Varvas, Reet Koljak, Tõnis Pehk, Alan R. Brash, Nigulas Samel

(爱沙尼亚塔林理工大学生物有机化学研究所)

摘要:在自然界中前列腺素的最高含量在加勒比海柳珊瑚 *Plexaura homomalla* 中发现。取决于地理位置的不同,这种珊瑚虫有与典型的哺乳动物相同的立体化学体(15S-羟基)或不同的R15-前列腺素。30年来人们仍然在猜测其代谢起源的机制。在本文中,我们研究一种催化花生四烯酸转化为R15-前列腺素的环氧合酶(COX)结构。利用同源性为基础的逆转录聚合酶链反应策略,我们通过将R转化为P,克隆了 *P. homomalla* 的R变体COX蛋白cDNA。推导的氨基酸序列显示了和来自北极软珊瑚 *Gersemia fruticosa* 15S种的珊瑚环氧合酶有80%相似,和哺乳动物COX-1和COX-2有50%相似。预测的三级结构显示其具有与哺乳动物COX同工酶的特征结构单元以及在催化时有重要作用的氨基酸残基。在过氧化物酶的活性部位附近,在膜结合结构域和在糖基化模式中有些结构上的差异。在昆虫细胞中表达的*P. homomalla* 酶形成了一个R15-前列腺素内过氧化物,并以11R-羟酸、15R-羟酸为副产品。通过和标准比对,内过氧化物产生了15R-前列腺素和12R-羟酸。对15R-COX同工酶结构性差异的评价为底物结合和中环氧合酶活性部位的花生四烯酸定向性分子氧反应提供了新的见解。

21. Wirth T, Bernatchez L. Genetic evidence against panmixia in the European eel[J]. Nature, 2001, 409 (6823): 1037-1040.

Genetic evidence against panmixia in the European eel

Wirth T. , Bernatchez L.

(GIROQ, Département de Biologie, Université Laval, Ste-Foy, Québec, Canada)

Abstract: The panmixia hypothesis-that all European eel (*Anguilla anguilla*) migrate to the Sargasso Sea for reproduction and comprise a single, randomly mating population-is widely accepted. If true, then this peculiar life history strategy would directly impact the population genetics of this species, and eels from European and north African rivers should belong to the same breeding population through the random dispersal of larvae. To date, the panmixia hypothesis has remained unchallenged: genetic studies realized on eel's mitochondrial DNA failed to detect any genetic structure; and a similar lack of structure was found using allozymes, with the exception of clinal variation imposed by selection. Here we have used highly polymorphic genetic markers that provide better resolution to investigate genetic structure in European eel. Analysis of seven microsatellite loci among 13 samples from the north Atlantic, the Baltic Sea and the Mediterranean Sea basins reveals that there is global genetic differentiation. Moreover, pairwise Cavalli-Sforza and Edwards' 13 chord distances correlate significantly with coastal geographical distance. This pattern of genetic structure implies non-random mating and restricted gene flow among eels from different sampled locations, which therefore refute the hypothesis of panmixia. Consequently, the reproductive biology of European eel must be reconsidered.

遗传证据否定欧洲鳗随机交配

Wirth T. , Bernatchez L.

(加拿大拉瓦尔大学生物系)

摘要:被广泛接受的随机交配群体假设指所有的欧洲鳗(*Anguilla anguilla*)迁徙至马尾藻海来繁殖,并且它们大部分都是单一的随机交配。如果情况属实,那么这个奇特的生活史策略将直接影响这一物种的群体遗传学,欧洲和北非河流的鳗鱼应该通过幼虫随机分散属于相同的繁殖种群。到今天为止,随机交配假说一直未受到质疑:对鳗鱼的线粒体 DNA 的遗传研究没有发现任何基因结构;利用异型酶结构也没有发现任何结构,除非通过利用选择渐变群变异。本文我们用能提供更好分辨率的高度多态标志来研究欧洲鳗遗传结构的遗传标记。13 个来自北大西洋、波罗的海和地中海盆地样品的微卫星标记显示了全球性的遗传分化。此外,成对卡瓦利-斯福尔扎和爱德华兹 13 腱的距离与沿海的地理距离联系显著。这种遗传结构模式意味着非随机交配和来自不同采样地区的鳗鱼基因受限制流动,反驳了随机交配的假说。因此,欧洲鳗生殖生物学必须重新考虑。

22. James K B Bishop, Russ E Davis, Jeffrey T Sherman. Robotic observations of dust storm enhancement of carbon biomass in the North Pacific[J]. Science, 2002, 298(5594): 817-821.

Robotic observations of dust storm enhancement of carbon biomass in the North Pacific

James K. B. Bishop, Russ E. Davis, Jeffrey T. Sherman

(Earth Sciences Division, Lawrence Berkeley National Laboratory, 1 Cyclotron Road, MS 90-1116, Berkeley, CA 94708, USA)

Abstract: Two autonomous robotic pro ling oats deployed in the subarctic North Pacific on 10 April 2001 provided direct records of carbon biomass variability from surface to 1000 meters below surface at daily and diurnal time scales. Eight months of real-time data documented the marine biological response to natural events, including hydrographic changes, multiple storms, and the April 2001 dust event. High-frequency observations of upper ocean particulate organic carbon variability show a near doubling of biomass in the mixed layer over a 2-week period after the passage of a cloud of Gobi desert dust. The temporal evolution of particulate organic carbon enhancement and an increase in chlorophyll use efficiency after the dust storm suggest a biotic response to a natural iron fertilization by the dust.

对北太平洋沙尘暴中生物碳量增加的遥控观测

James K. B. Bishop, Russ E. Davis, Jeffrey T. Sherman

（美国劳伦斯伯克利国家实验室地球科学部）

摘要：2001 年 4 月 10 日，两个自动遥控设备压型浮标被配置在北太平洋地区，记录了碳量从表层到表层以下 1000 m 的日变化。8 个月的实时数据记录了海洋对于自然事件的生物响应，包括水文变化、多风暴，以及 2001 年 4 月的沙尘天气。海洋表面微粒有机碳变化的高频观测表明在戈壁沙漠尘埃通过的 2 周中生物量接近翻倍。沙尘暴后颗粒有机碳的时空演化和叶绿素使用效率的提高，表明生物对于自然尘埃中铁营养的响应。

23. Martin Holmstrup, Mark Bayley, Hans Ramløv. Supercool or dehydrate? An experimental analysis of overwintering strategies in small permeable Arctic invertebrates[J]. PNAS, 2002, 99(8): 5716-5720.

Supercool or dehydrate? An experimental analysis of overwintering strategies in small permeable Arctic invertebrates

Martin Holmstrup, Mark Bayley, Hans Ramløv

(National Environmental Research Institute, Denmark)

Abstract: Soil invertebrate survival in freezing temperatures has generally been considered in the light of the physiological adaptations seen in surface living insects. These adaptations, notably the ability to supercool, have evolved in concert with surface invertebrates' ability to retain body water in a dry environment. However, most soil invertebrates are orders of magnitude less resistant to desiccation than these truly terrestrial insects, opening the possibility that the mechanisms involved in their cold-hardiness are also of a radically different nature. Permeable soil invertebrates dehydrate when exposed in frozen soil. This dehydration occurs because the water vapor pressure of supercooled water is higher than that of ice at the same temperature. The force of this vapor pressure difference is so large that even a few degrees of supercooling will result in substantial water loss, continuing until the vapor pressure of body fluids equals that of the surrounding ice. At this stage, the risk of tissue ice formation has been eliminated, and subzero survival is ensured. Here we show that these soil invertebrates do not base their winter survival on supercooling, as do many other ectotherms, but instead dehydrate and equilibrate their body-fluid melting point to the ambient temperature. They can achieve this equilibration even at the extreme cooling rates seen in polar soils.

过度冷却还是脱水？实验分析小型有渗透作用的北极无脊椎动物过冬策略

Martin Holmstrup, Mark Bayley, Hans Ramløv

（丹麦国家环境研究所）

摘要：土壤无脊椎动物在冻结温度的生存机制一般被认为与生活在表面的昆虫的生理适应相似。这些适应，特别是对过度冷却的适应能力，都与表面无脊椎动物进化的在干燥环境留住体内水分的能力相关。然而，大多数土壤无脊椎动物比这些真正的陆地昆虫的干燥抗性差几个数量级，它们的抗寒性所涉及的机制也可能完全不同。渗透土壤无脊椎动物在冻土暴露出来时脱水。这种脱水的原因是过冷的水蒸气压力大于冰在相同温度的压力。这种蒸汽压力差是如此之大，即使是过冷几摄氏度都将导致大量失水，一直持续到体液的蒸汽压和周围的冰相同。在这个阶段，组织结冰的危险已经消除，保障了零摄氏度以下的生存。本文中，我们证明，这些土壤无脊椎动物的冬天生存没有像许多其他冷血动物一样依赖过冷，而是改为脱水和平衡自己的身体流动性熔点和周围环境温度。它们甚至可以在极端的冷却速率下在极地土壤中达到这个平衡。

24. David Johnson, Colin D Campbell, John A Lee, et al. Arctic microorganisms respond more to elevated UV-B radiation than CO_2[J]. Nature, 2002, 416(6876): 82-83.

Arctic microorganisms respond more to elevated UV-B radiation than CO_2

David Johnson, Colin D. Campbell, John A. Lee, Terry V. Callaghan, Dylan Gwynn-Jones

(Department of Animal and Plant Sciences, University of Sheffield, Sheffield S10 2TN, UK)

Abstract: Surface ultraviolet-B radiation and atmospheric CO_2 concentrations have increased as a result of ozone depletion and burning of fossil fuels. The effects are likely to be most apparent in polar regions where ozone holes have developed and ecosystems are particularly sensitive to disturbance. Polar plant communities are dependent on nutrient cycling by soil microorganisms, which represent a significant and highly labile portion of soil carbon (C) and nitrogen (N). It was thought that the soil microbial biomass was unlikely to be affected by exposure of their associated plant communities to increased UV-B. In contrast, increasing atmospheric CO_2 concentrations were thought to have a strong effect as a result of greater below-ground C allocation. In addition, there is a growing belief that ozone depletion is of only minor environmental concern because the impacts of UV-B radiation on plant communities are often very subtle. Here we show that 5 years of exposure of a subarctic heath to enhanced UV-B radiation both alone and in combination with elevated CO_2 resulted in significant changes in the C∶N ratio and in the bacterial community structure of the soil microbial biomass.

北极微生物对 UV-B 增加的反应强于 CO_2

David Johnson, Colin D. Campbell, John A. Lee, Terry V. Callaghan, Dylan Gwynn-Jones

(英国谢菲尔德大学动物和植物科学系)

摘要：地表 UV-B 辐射和大气中的二氧化碳浓度增加是臭氧损耗和燃烧化石燃料的结果。它们的影响可能会在已出现臭氧洞和生态系统对干扰特别敏感的极地地区最为明显。极地植物群落依赖于土壤中微生物形成的养分循环生存，土壤微生物代表了土壤碳(C)和氮(N)的高度不稳定部分。一般认为，土壤微生物生物量不太可能受其关联的植物群落暴露于增加的 UV-B 影响。相比之下，大气中二氧化碳的浓度增加被认为对地下碳分配有强烈作用。此外，人们越来越相信，臭氧消耗的环境问题轻微，因为 UV-B 辐射对植物群落的影响往往很小。本文中，我们研究了 5 年的亚北极石楠接触增强 UV-B 辐射分别在单独条件和与 CO_2 浓度升高结合的条件下，导致 C∶N 和土壤的微生物生物量的细菌群落结构的巨大变化。

25. Robert B McKane, Loretta C Johnson, Gaius R Shaver, et al. Resource-based niches provide a basis for plant species diversity and dominance in Arctic tundra[J]. Nature, 2002, 415(6867): 68-71.

Resource-based niches provide a basis for plant species diversity and dominance in Arctic tundra

Robert B. McKane, Loretta C. Johnson, Gaius R. Shaver, Knute J. Nadelhoffer, Edward B. Rastetter, Brian Fry, Anne E. Giblin, Knut Kiellandk, Bonnie L. Kwiatkowski, James A. Laundre, Georgia Murray

(US Environmental Protection Agency, Corvallis, Oregon 97333, USA)

Abstract: Ecologists have long been intrigued by the ways co-occurring species divide limiting resources. Such resource partitioning, or niche differentiation, may promote species diversity by reducing competition. Although resource partitioning is an important determinant of species diversity and composition in animal communities, its importance in structuring plant communities has been difficult to resolve. This is due mainly to difficulties in studying how plants compete for belowground resources. Here we provide evidence from a N-15-tracer field experiment showing that plant species in a nitrogen-limited, Arctic tundra community were differentiated in timing, depth and chemical form of nitrogen uptake, and that species dominance was strongly correlated with uptake of the most available soil nitrogen forms. That is, the most productive species used the most abundant nitrogen forms, and less productive species used less abundant forms. To our knowledge, this is the first documentation that the composition of a plant community is related to partitioning of differentially available forms of a single limiting resource.

北极苔原资源型小环境提供了植物物种多样性和优势的基础

Robert B. McKane, Loretta C. Johnson, Gaius R. Shaver, Knute J. Nadelhoffer,
Edward B. Rastetter, Brian Fry, Anne E. Giblin, Knut Kiellandk,
Bonnie L. Kwiatkowski, James A. Laundre, Georgia Murray

(美国环境保护局)

摘要:生态学家一直好奇共生物种分享有限资源的方式。这样的资源分割,或小环境分化,可能通过减少竞争促发物种的多样性。虽然资源分区在物种多样性和动物群落组成中起重要决定作用,但是它在构筑植物群落的重要性方面还难以了解。这主要是由于很难研究植物如何竞争地下资源。本文中,我们提供了利用^{15}N示踪剂野外试验显示结果,在氮限制的北极苔原区,植物种类在时间、深度和化学氮的吸收形式分化,物种的优势和土壤氮形态的摄取方式密切相关。也就是说,最有生产力的物种利用了最丰富的氮形态,生产力小的物种能利用的氮形态较少。据我们所知,这是第一个关于植物群落组成和单一限制资源形式差异关系的研究。

26. Michael L Moody, Donald H Les. Evidence of hybridity in invasive watermilfoil (Myriophyllum) populations[J]. PNAS,2002,99(23):14867-14871.

Evidence of hybridity in invasive watermilfoil (Myriophyllum) populations

Michael L. Moody, Donald H. Les

(Department of Ecology and Evolutionary Biology, University of Connecticut, Storrs 06269-3043, USA)

Abstract: Invasions of nonindigenous species have caused ecological devastation to natural communities worldwide, yet the biological bases for invasiveness remain poorly understood. Our studies of invasive watermilfoil (Myriophyllum) populations revealed widespread polymorphisms in biparentally inherited nuclear ribosomal DNA sequences, which were not detected in populations of native North American species. Subclones of the polymorphic regions revealed the occurrence of distinct sequences matching those acquired from both nonindigenous and native North American species. Molecular data demonstrate clearly that invasive watermilfoil populations in North America have resulted from hybridization between nonindigenous and native species. These observations suggest that invasiveness in these aggressive aquatic weeds may be linked to heterosis maintained by vegetative propagation.

入侵型水生藻类(狐尾藻属)杂种性证据

Michael L. Moody, Donald H. Les

(美国康涅狄格大学生态与进化生物学系)

摘要:对非本地物种入侵对世界范围的自然群落造成的生态破坏,其侵袭性的生物学基础仍然知之甚少。我们研究了入侵水藻(狐尾藻属)种群,发现广泛分布的双亲遗传核糖体DNA序列的多形性,这些都没有在北美本土物种的种群中检测到。亚克隆多态地区显示了北美非本地和本地物种不同序列匹配的发生。分子的数据清楚地表明,北美入侵性水藻是由非本地和本地物种杂交结果产生的。这些观察资料表明,这些水藻的侵袭可能与无性繁殖杂种优势有关。

27. James S Clark, Jason S McLachlan. Stability of forest biodiversity[J]. Nature, 2003, 423(6940): 635-638.

Stability of forest biodiversity

James S. Clark, Jason S. McLachlan

(Center on Global Change, Biology, and Nicholas School of the Environment, Duke University, Durham, North Carolina 27708, USA)

Abstract: Two hypotheses to explain potentially high forest biodiversity have different implications for the number and kinds of species that can coexist and the potential loss of biodiversity in the absence of speciation. The first hypothesis involves stabilizing mechanisms, which include tradeoffs between species in terms of their capacities to disperse to sites where competition is weak, to exploit abundant resources effectively and to compete for scarce resources. Stabilization results because competitors thrive at different times and places. An alternative, "neutral model" suggests that stabilizing mechanisms may be superfluous. This explanation emphasizes "equalizing" mechanisms, because competitive exclusion of similar species is slow. Lack of ecologically relevant differences means that abundances experience random "neutral drift", with slow extinction. The relative importance of these two mechanisms is unknown, because assumptions and predictions involve broad temporal and spatial scales. Here we demonstrate that predictions of neutral drift are testable using palaeodata. The results demonstrate strong stabilizing forces. By contrast with the neutral prediction of increasing variance among sites over time, we show that variances in post-Glacial tree abundances among sites stabilize rapidly, and abundances remain coherent over broad geographical scales.

森林生物多样性的稳定性

James S. Clark, Jason S. McLachlan

(美国杜克大学“尼古拉斯”环境与地球科学学院生物学和全球变化中心)

摘要:两个来解释潜在用材林生物多样性的假说,在缺少物种形成的情况下,对于生物数量、可以共存种类和潜在损失多样性有不同的见解。第一个假说涉及稳定机制,包括在物种之间权衡的能力,将它们分散在竞争较弱的地域,有效地利用丰富的资源,并争夺稀缺资源。稳定来自于竞争者在不同时间和地点的繁荣。另一种假说,“中性模型”表明,稳定的机制可能是多余的。这一解释强调“平衡”机制,因为类似的物种竞争排斥缓慢。生态相关差异的缺乏意味着物种随机“中性漂移”,并伴随着缓慢灭绝。这两个机制的相对重要性是未知的,因为假设和预测涉及广泛的时间和空间。本文,我们使用古生物数据证明中性漂移预测的可测试性。结果显示强稳定性。通过对比随着时间推移渐渐增加变化的中立预测,我们发现,在冰期后不同地点树的繁茂差异会很快稳定,它们的繁茂在广泛的地域环境得以维持。

28. Olivier Gilg, Ilkka Hanski, Benoît Sittler. Cyclic dynamics in a simple vertebrate predator-prey

community[J]. Science, 2003, 302(5646): 866-868.

Cyclic dynamics in a simple vertebrate predator-prey community

Olivier Gilg, Ilkka Hanski, Benoît Sittler

(Department of Ecology and Systematics, Division of Population Biology, Post Office Box 65, 00014 University of Helsinki, Finland)

Abstract: The collared lemming in the high-Arctic tundra in Greenland is preyed upon by four species of predators that show marked differences in the numbers of lemmings each consumes and in the dependence of their dynamics on lemming density. A predator-prey model based on the field-estimated predator responses robustly predicts 4-year periodicity in lemming dynamics, in agreement with long-term empirical data. There is no indication in the field that food or space limits lemming population growth, nor is there need in the model to consider those factors. The cyclic dynamics are driven by a 1-year delay in the numerical response of the stoat and stabilized by strongly density-dependent predation by the arctic fox, the snowy owl, and the long-tailed skua.

一个简单脊椎动物捕食群落的循环动态

Olivier Gilg, Ilkka Hanski, Benoît Sittler

(芬兰赫尔辛基大学生态系统学和人口生物学系)

摘要:在格陵兰岛的高纬度北极苔原地区领旅鼠是 4 个物种的捕食对象,每个物种对它的捕食数量表现出明显的不同,依赖它们对旅鼠的动态需求。一个基于区域中捕食者响应的捕食模型稳定地预测了 4 年周期内旅鼠的动态,并和长期经验数据一致。没有任何迹象表明在该领域食物或空间限制了旅鼠数量增长,因此也没有必要在模型中考虑这些因素。循环动态是由白鼬的 1 年迟延数值响应驱动并由有强烈密度依赖来捕食的北极狐、雪鸮、长尾鸥稳定。

29. Dan Mishmar, Eduardo Ruiz-Pesini, Pawel Golik, et al. Natural selection shaped regional mtDNA variation in humans[J]. PNAS, 2003, 100(1): 171-176.

Natural selection shaped regional mtDNA variation in humans

Dan Mishmar, Eduardo Ruiz-Pesini, Pawel Golik, Vincent Macaulay, Andrew G. Clark, Seyed Hosseini, Martin Brandon, Kirk Easley, Estella Chen, Michael D. Brown, Rem I. Sukernik, Antonel Olckers, Douglas C. Wallace

(Center for Molecular and Mitochondrial Medicine and Genetics, University of California, Irvine, CA 92697-3940)

Abstract: Human mtDNA shows striking regional variation, traditionally attributed to genetic drift. However, it is not easy to account for the fact that only two mtDNA lineages (M and N) left Africa to colonize Eurasia and that lineages A, C, D, and G show a 5-fold enrichment from central Asia to Siberia. As an alternative to drift, natural selection might have enriched for certain mtDNA lineages as people migrated north into colder climates. To test this hypothesis we analyzed 104 complete mtDNA sequences from all global regions and lineages. African mtDNA variation did not significantly deviate from the standard neutral model, but European, Asian, and Siberian plus Native American variations did. Analysis of amino acid substitution mutations (nonsynonymous, Ka) versus neutral mutations (synonymous, Ks) (ka/ks) for all 13 mtDNA protein-coding genes revealed that the ATP6 gene had the highest amino acid sequence variation of any human mtDNA gene, even though ATP6 is one of the more conserved mtDNA proteins. Comparison of the ka/ks ratios for each mtDNA gene from the tropical, temperate, and arctic zones revealed that ATP6 was highly variable in the mtDNAs from the arctic zone, cytochrome b was particularly variable in the temperate zone, and cytochrome oxidase I was notably more variable in the tropics. Moreover, multiple amino acid changes found in ATP6, cytochrome b, and cytochrome oxidase I appeared to be functionally significant. From these analyses we conclude that selection may have played a role in shaping human regional mtDNA variation and that one of the selective influences was climate.

自然选择造成人类区域性 mtDNA 变异

Dan Mishmar, Eduardo Ruiz-Pesini, Pawel Golik, Vincent Macaulay, Andrew G. Clark, Seyed Hosseini, Martin Brandon, Kirk Easley, Estella Chen, Michael D. Brown, Rem I. Sukernik, Antonel Olckers, Douglas C. Wallace

(美国加利福尼亚大学尔湾分校分子和线粒体医学与遗传学中心)

摘要:人类 mtDNA 显示了惊人的地区差异,传统上归因于遗传漂变。但是,它不能简单说明只有两条 mtDNA(M 和 N)谱系离开非洲殖民欧亚大陆的事实,且宗族 A,C,D 和 G 从中亚到西伯利亚丰度增加 5 倍。作为另一假设,当人们迁移到北方寒冷的气候,自然选择可能使得某些 mtDNA 谱系丰富。为了验证这一假说,我们分析了来自全球所有地区和谱系的 104 条完整的 mtDNA 序列。非洲 mtDNA 变化并没有显著偏离标准的中性模型,但欧洲、亚洲和西伯利亚加上美国本地人的 mtDNA 变化则偏离了。分析全部 13 个 mtDNA 蛋白质编码基因氨基酸替代突变(非同义,*Ka*)与中性突变(同义,*Ks*)的比值(*Ka*/*Ks*)发现,所有人类 mtDNA 基因上的 ATP6 基因都有最高的氨基酸序列变异,即使 ATP6 是比较稳定的 mtDNA 蛋白之一。比较热带、温带和极地每个 mtDNA 基因 *Ka*/*Ks* 比值发现,极区 mtDNAs ATP6 突变高,细胞色素 b 在温带突变高,细胞色素氧化酶在热带地区更易突变。此外,发现 ATP6、细胞色素 b 和细胞色素氧化酶中多种氨基酸变化,似乎在功能上显著。从这些分析,我们得出结论,选择可能都起到形成区域人类 mtDNA 变异的作用,而且影响选择的因素之一就是气候。

30. Springer M, Estes J A, van Vliet G B, et al. Sequential megafaunal collapse in the North Pacific Ocean: an ongoing legacy of industrial whaling? [J]. PNAS, 2003, 100(21): 12223-12228.

Sequential megafauna collapse in the North Pacific Ocean: an ongoing legacy of industrial whaling?

Springer M., Estes J. A., van Vliet G. B., Williams T. M., Doak D. F., Danner E. M., Forney K. A., Pfister B.

(Institute of Marine Science, University of Alaska, Fairbanks, AK 99775, USA)

Abstract: Populations of seals, sea lions, and sea otters have sequentially collapsed over large areas of the northern North Pacific Ocean and southern Bering Sea during the last several decades. A bottom-up nutritional limitation mechanism induced by physical oceanographic change or competition with fisheries was long thought to be largely responsible for these declines. The current weight of evidence is more consistent with top-down forcing. Increased predation by killer whales probably drove the sea otter collapse and may have been responsible for the earlier pinniped declines as well. We propose that decimation of the great whales by post-World War Ⅱ industrial whaling caused the great whales' foremost natural predators, killer whales, to begin feeding more intensively on the smaller marine mammals, thus "fishing-down" this element of the marine food web. The timing of these events, information on the abundance, diet, and foraging behavior of both predators and prey, and feasibility analyses based on demographic and energetic modeling are all consistent with this hypothesis.

北太平洋巨型动物群落的逐渐崩溃：是否由于持续的工业捕鲸？

Springer M., Estes J. A., van Vliet G. B., Williams T. M., Doak D. F., Danner E. M., Forney K. A., Pfister B.

(美国阿拉斯加大学费尔班克斯分校海洋科学研究所)

摘要：过去几十年，海豹、海狮、海獭数量已经在北太平洋和白令海南部的广大区域连续下降。一个由物理海洋变化或渔业竞争产生的自下而上的营养限制机制长期被认为是这些下降的主要原因。目前重要的证据更符合自上而下的强迫机制。逆戟鲸捕食增加可能推动了海獭数量崩溃，并可能导致了较早的鳍足类数量下降。我们认为，二战时期的工业捕鲸对大鲸鱼造成大批杀害，导致大鲸鱼的最先自然捕杀者逆戟鲸开始更多地以小型海洋哺乳动物为食，因此海洋食物网的方式为"捕杀降级"。这些事件发生的时间，数量、饮食以及捕食者和猎物的觅食信息，和以统计数量和能量的模式为基础的可行性分析，都和这一假说结果相一致。

31. Kate F Darling, Michal Kucera, Carol J Pudsey, et al. Molecular evidence links cryptic diversification in polar planktonic protists to quaternary climate dynamics[J]. PNAS, 2004, 101(20): 7657-7662.

Molecular evidence links cryptic diversification in polar planktonic protists to quaternary climate dynamics

Kate F. Darling, Michal Kucera, Carol J. Pudsey, Christopher M. Wade

(School of GeoSciences and Institute of Cell, Animal, and Population Biology, King's Buildings, University of Edinburgh, Edinburgh EH9 3JW, United Kingdom)

Abstract: It is unknown how pelagic marine protists undergo diversification and speciation. Superficially, the open ocean appears homogeneous, with few clear barriers to gene flow, allowing extensive, even global, dispersal. Yet, despite the apparent lack of opportunity for genetic isolation, diversity is prevalent within marine taxa. A lack of candidate isolating mechanisms would seem to favor sympatric over allopatric speciation models to explain the diversity and biogeographic patterns observed in the oceans today. However, the ocean is a dynamic system, and both current and past circulation patterns must be considered in concert to gain a true perspective of gene flow through time. We have derived a comprehensive picture of the mechanisms potentially at play in the high latitudes by combining molecular, biogeographic, fossil, and paleoceanographic data to reconstruct the evolutionary history of the polar planktonic foraminifer *Neogloboquadrina pachyderma* sinistral. We have discovered extensive genetic diversity within this morphospecies and that its current "extreme" polar affinity did not appear until late in its evolutionary history. The molecular data demonstrate a stepwise progression of diversification starting with the allopatric isolation of Atlantic Arctic and Antarctic populations after the onset of the Northern Hemisphere glaciation. Further diversification occurred only in the Southern Hemisphere and seems to have been linked to glacial-interglacial climate dynamics. Our findings demonstrate the role of Quaternary climate instability in shaping the modern high-latitude plankton. The divergent evolutionary history of *N. pachyderma* sinistral genotypes implies that paleoceanographic proxies based on this taxon should be calibrated independently.

分子证据表明不可理解的极地浮游原生生物多样化和第四纪气候动态相关

Kate F. Darling, Michal Kucera, Carol J. Pudsey, Christopher M. Wade

(英国爱丁堡大学地球科学学院与动物和人口生物学研究所)

摘要：目前还不清楚浮游的海洋原生生物是如何进行多样化和物种形成的。从表面上看，开放的海洋似乎是同质的，几乎没有阻碍基因流动的障碍物，允许其广泛地，甚至全球地传播。然而，尽管没有明显的遗传隔离的机会，在海洋生物分类中多样性是普遍的。缺乏隔离机制偏爱同域，而不是异域模型，来解释现在海洋观测到的多样性和生物地理格局。然而，海洋是一个动态的系统，我们必须考虑目前和过去的流通模式来审视随时间变化的基因流动。结合分子、生物地理、化石和古海洋学数据，我们重建了极地浮游有孔虫 *Neogloboquadrina pachyderma* 左旋的进化历史，得到了一个在高纬度起作用的潜在机制。我们发现这个形态种有广泛的基因多样性，且其目前的"极端"的极地亲和力直到晚期才在其进化历史上出现。分子数据表明了逐步发展的多元化，始于北半球冰期开始后与北极和南极的大西洋种群异域隔离后。进一步多样化只发生在南半球，似乎已与冰期-间冰期的气候驱动相关。我们的研究结果证明了第四纪气候不稳定在塑造现代高纬度的浮游生物中的作用。*N. pachyderma* 左旋基因型的不同进化历史表明，以此种群为古海洋学指标时需要特别校准。

32. Marianne S V Douglas，John P Smol，James M Savelle，et al. Prehistoric Inuit whalers affected Arctic freshwater ecosystems[J]. PNAS，2004，101(6)：1613-1617.

Prehistoric Inuit whalers affected Arctic freshwater ecosystems

Marianne S. V. Douglas，John P. Smol，James M. Savelle，Jules M. Blais

(Paleoenvironmental Assessment Laboratory，Department of Geology，University of Toronto，22 Russell Street，Toronto，ON，Canada M5S 3B1)

Abstract：It is commonly assumed that High Arctic lakes and ponds were not affected by direct local human activities before the arrival of Europeans，because most native peoples were primarily nomadic，maintained relatively low population densities，and practiced unintrusive hunting and gathering technologies. Our archeological and paleolimnological data show that this was not always the case. Thule Inuit whalers，whose winter settlements consisted of houses constructed from the bones of bowhead whales on Somerset Island between about anno Domini 1200 and 1600，markedly changed pond water quality and ecology. The arrival of whalers 8 centuries ago caused marked changes in water chemistry and the expansion of moss substrates. Although whalers abandoned the area ＞4 centuries ago，the legacy of these human disturbances is still evident in the pond's present-day limnology and is characterized by elevated nutrient concentrations and atypical biota. This is the earliest reported paleolimnological record of changes in aquatic ecology associated with local human activities in Canada or the United States，or for any circumpolar ecosystem.

史前因纽特人捕鲸对北极淡水生态系统的影响

Marianne S. V. Douglas，John P. Smol，James M. Savelle，Jules M. Blais

(加拿大多伦多大学地质系古环境评估实验室)

摘要：一般认为，高纬度北极湖泊和池塘，在欧洲人到来之前没有受到当地人类活动的直接影响，因为原住民主要是游牧民族，保持相对较低的人口密度，实行非侵入性的狩猎和采集技术。我们的考古和古湖沼学的数据显示，实际并非总是如此。极北部的因纽特捕鲸人，他们约公元 1200 年和 1600 年在萨默塞特岛的冬季定居点房屋由北极露脊鲸骨骼造成，明显改变了池塘水质和生态。8 世纪前的捕鲸活动引起了显著水化学和苔藓基板扩展变化。虽然放弃了 4 个世纪前的捕鲸区，人类的干扰遗迹仍然在当今的池塘湖藻中存在：明显的营养浓度升高和非典型生物群特征。本文是最早的古湖泊学记录，记录了水生生态系统的变化和当地人在加拿大或美国，或任何的环极地地区活动相关的生态系统变化。

33. Pitulko V V，Nikolsky P A，Girya E Yu，et al. The Yana RHS site：humans in the Arctic before the Last Glacial Maximum[J]. Science，2004，303(5654)：52-56.

The Yana RHS site: humans in the Arctic before the Last Glacial Maximum

Pitulko V. V., Nikolsky P. A., Girya E. Yu., Basilyan A. E., Tumskoy V. E., Koulakov S. A., Astakhov S. N., Pavlova E. Yu., Anisimov M. A.

(Institute for the History of Material Culture, Russian Academy of Sciences, 18 Dvortsovaya nab., St. Petersburg 191186, Russia)

Abstract: A newly discovered Paleolithic site on the Yana River, Siberia, at 71 degrees N, lies well above the Arctic circle and dates to 27000 radiocarbon years before present, during glacial times. This age is twice that of other known human occupations in any Arctic region. Artifacts at the site include a rare rhinoceros foreshaft, other mammoth foreshafts, and a wide variety of tools and flakes. This site shows that people adapted to this harsh, high-latitude, Late Pleistocene environment much earlier than previously thought.

亚纳 RHS 点:末次冰期前在北极地区的人类

Pitulko V. V., Nikolsky P. A., Girya E. Yu., Basilyan A. E., Tumskoy V. E., Koulakov S. A., Astakhov S. N., Pavlova E. Yu., Anisimov M. A.

(俄罗斯科学院物质文化学院)

摘要:一个新发现的旧石器时代遗址,位于西伯利亚亚纳河,北纬 71°,远高于北极圈,距今 27000 放射性碳年的冰川时代。这个年龄是在任何其他已知的北极地区人类存在时间的两倍。现场文物包括一个罕见的犀牛前角、猛犸象牙,以及各种工具和薄片。这个地区显示,人们适应这种恶劣、高纬度、晚更新世环境比以前认为的要早得多。(注:RHS 点为犀牛角发现点)

34. Croll D A, Maron J L, Estes J A, et al. Introduced predators transform subarctic islands from grassland to tundra[J]. Science, 2005, 307(5717): 1959-1961.

Introduced predators transform subarctic islands from grassland to tundra

Croll D. A., Maron J. L., Estes J. A., Danner E. M., Byrd G. V.

(Department of Ecology and Evolutionary Biology, Island Conservation, University of California-Santa Cruz, Santa Cruz, CA 95060, USA)

Abstract: Top predators often have powerful direct effects on prey populations, but whether these direct effects propagate to the base of terrestrial food webs is debated. There are few examples of trophic cascades strong enough to alter the abundance and composition of entire plant communities. We show that the introduction of arctic foxes (Alopex lagopus) to the Aleutian archipelago induced strong shifts in plant productivity and community structure via a previously unknown pathway. By preying on seabirds, foxes reduced nutrient transport from ocean to land, affecting soil fertility and transforming grasslands to dwarf shrub/forb-dominated ecosystems.

引入的捕食者使得亚北极岛屿由草原向苔原区转变

Croll D. A., Maron J. L., Estes J. A., Danner E. M., Byrd G. V.

(美国加州大学圣克鲁兹分校生态学和进化生物学系)

摘要:顶级捕食者往往对猎物种群有强大的直接影响,但究竟这些直接影响是否会传播到陆地食物网还不得知。营养级链强大到足以改变整个植物群落的丰富性和组成的例子很少。我们发现,引进北极狐(蓝狐)到阿留申群岛,以前所未知的途径导致植物生产力和群落结构的强烈变化。通过对海鸟捕食,狐狸减少养分从海洋到陆地运输,影响土壤肥力和使草原转化到矮灌木草原/杂类草为主的生态系统。

35. F Stuart Chapin Ⅲ, Amy L Lovecraft, Erika S Zavaleta, et al. Policy strategies to address sustainability of Alaskan boreal forests in response to a directionally changing climate[J]. PNAS, 2006, 103(45):16637-16643.

Policy strategies to address sustainability of Alaskan boreal forests in response to a directionally changing climate

F. Stuart Chapin Ⅲ, Amy L. Lovecraft, Erika S. Zavaleta, Joanna Nelson, Martin D. Robards, Gary P. Kofinas, Sarah F. Trainor, Garry D. Peterson, Henry P. Huntington, Rosamond L. Naylor

(Institute of Arctic Biology and Department of Political Science, University of Alaska, Fairbanks, AK 99775, USA)

Abstract: Human activities are altering many factors that determine the fundamental properties of ecological and social systems. Is sustainability a realistic goal in a world in which many key process controls are directionally changing? To address this issue, we integrate several disparate sources of theory to address sustainability in directionally changing social-ecological systems, apply this framework to climate-warming impacts in Interior Alaska, and describe a suite of policy strategies that emerge from these analyses. Climate warming in Interior Alaska has profoundly affected factors that influence landscape processes (climate regulation and disturbance spread) and natural hazards, but has only indirectly influenced ecosystem goods such as food, water, and wood that receive most management attention. Warming has reduced cultural services provided by ecosystems, leading to some of the few institutional responses that directly address the causes of climate warming, e. g., indigenous initiatives to the Arctic Council. Four broad policy strategies emerge: (i) enhancing human adaptability through learning and innovation in the context of changes occurring at multiple scales; (ii) increasing resilience by strengthening negative (stabilizing) feedbacks that buffer the system from change and increasing options for adaptation through biological, cultural, and economic diversity; (iii) reducing vulnerability by strengthening institutions that link the high-latitude impacts of climate warming to their low-latitude causes; and (iv) facilitating transformation to new, potentially more beneficial states by taking advantage of opportunities created by crisis. Each strategy provides societal benefits, and we suggest that all of them be pursued simultaneously.

关于解决阿拉斯加北部森林对于定向气候变化响应的可持续性政策战略

F. Stuart Chapin Ⅲ, Amy L. Lovecraft, Erika S. Zavaleta, Joanna Nelson, Martin D. Robards, Gary P. Kofinas, Sarah F. Trainor, Garry D. Peterson, Henry P. Huntington, Rosamond L. Naylor

(美国阿拉斯加大学保罗北极生物研究所和政治科学系)

摘要:人类活动正在改变很多决定生态和社会系统根本性质的因素。在这个许多关键过程的控制是方向性改变的世界里,持续性是否是一个现实的目标? 为了解决这个问题,我们结合了几个不同的理论来解决定向不断变化的社会生态系统的可持续性,应用此框架研究阿拉斯加气候变暖影响,并描述了这些分析中出现的政策、战略。气候变暖在阿拉斯加内部有深刻的影响,影响景观过程(气候调节和干扰的传播)和自然灾害,但只是间接地影响生态系统,如食物、水和木材等受到多数管理者关注的物品。气候变暖减少了生态系统所提供的文化服务,导致少数机构产生响应来直接解决气候变暖的原因,例如,对北极理事会发挥的本土主观能动性。四大政策战略出现:① 通过学习,提高人的适应能力和在不同规模的变化背景下发生的创新;② 通过增加生物、文化和经济的多样性,提高抗灾能力,加强负(稳定)反馈系统从该缓冲区的变化和适应办法;③ 通过加强高纬度对气候变暖的影响和气候变暖的低纬度原因之间的联系来降低系统脆弱性;④ 利用危机创造机会,促进转变到更有利的状态。每个战略都提供了社会福利,我们建议,这些策略可以同时进行。

36. Hanne Hegre Grundt, Siri Kjølner, Liv Borgen, et al. High biological species diversity in the Arctic flora[J]. PNAS, 2006, 103(4): 972-975.

High biological species diversity in the Arctic flora

Hanne Hegre Grundt, Siri Kjølner, Liv Borgen, Loren H. Rieseberg, Christian Brochmann

(National Centre for Biosystematics, Natural History Museum, University of Oslo, P. O. Box 1172, Blindern NO-0318 Oslo, Norway)

Abstract: The arctic flora is considered to be impoverished, but estimates of species diversity are based on morphological assessments, which may not provide accurate counts of biological species. Here we report on crossing relationships within three diploid circumpolar plant species in the genus Draba (Brassicaceae). Although 99% of parental individuals were fully fertile, the fertility of intraspecific crosses was surprisingly low. Hybrids from crosses within populations were mostly fertile (63%), but only 8% of the hybrids from crosses within and among geographic regions (Alaska, Greenland, Svalbard, and Norway) were fertile. The frequent occurrence of intraspecific crossing barriers is not accompanied by significant morphological or ecological differentiation, indicating that numerous cryptic biological species have arisen within each taxonomic species despite their recent (Pleistocene) origin.

北极植物中生物物种的高多样性

Hanne Hegre Grundt, Siri Kjølner, Liv Borgen, Loren H. Rieseberg, Christian Brochmann

(挪威奥斯陆大学国家生物系统学中心、自然历史博物馆)

摘要:北极植物被认为是种类贫乏的,但当地物种多样性是根据形态评估的,这可能无法提供准确的生物物种数量。本文我们研究了环北极三种二倍体属植物物种葶(十字花科)的交叉关系。虽然99%母体是有繁殖力的,但种内杂交体有繁殖力的比例出奇的低。种群内部交叉杂交体大多数有繁殖力(63%),但在一些地理区域(阿拉斯加、格陵兰、斯瓦尔巴群岛和挪威)内和之间,只有8%有繁殖力。种内交叉繁殖的障碍不会伴随着明显的形态学或生态学分化,这表明有许多还不清楚的生物物种出现在每一个分类种中,而不管它们最近时代(更新世)的起源是什么。

37. Marilyn D Walker, C Henrik Wahren, Robert D Hollister, et al. Plant community responses to experimental warming across the tundra biome[J]. PNAS, 2006, 103(5): 1342-1346.

Plant community responses to experimental warming across the tundra biome

Marilyn D. Walker, C. Henrik Wahren, Robert D. Hollister, Greg H. R. Henry, Lorraine E. Ahlquist, Juha M. Alatalo, M. Syndonia Bret-Harte, Monika P. Calef, Terry V. Callaghan, Amy B. Carroll, Howard E. Epstein, Ingibjörg S. Jónsdóttir, Julia A. Klein, Borgtór Magnússon, Ulf Molau, Steven F. Oberbauer, Steven P. Rewa, Clare H. Robinson, Gaius R. Shaver, Katharine N. Suding, Catharine C. Thompson, Anne Tolvanen, Ørjan Totland, P. Lee Turner, Craig E. Tweedie, Patrick J. Webber, Philip A. Wookey

(Boreal Ecology Cooperative Research Unit, U. S. Department of Agriculture Forest Service Pacific Northwest Research Station, University of Alaska, P. O. Box 756780, Fairbanks, AK 99775-6780, USA)

Abstract: Recent observations of changes in some tundra ecosystems appear to be responses to a warming climate. Several experimental studies have shown that tundra plants and ecosystems can respond strongly to environmental change, including warming; however, most studies were limited to a single location and were of short duration and based on a variety of experimental designs. In addition, comparisons among studies are difficult because a variety of techniques have been used to achieve experimental warming and different measurements have been used to assess responses. We used metaanalysis on plant community measurements from standardized warming experiments at 11 locations across the tundra biome involved in the International Tundra Experiment. The passive warming treatment increased plant-level air temperature by 1-3 degrees C, which is in the range of predicted and observed warming for tundra regions. Responses were rapid and detected in whole plant communities after only two growing seasons. Overall, warming increased height and cover of deciduous shrubs and graminoids, decreased cover of mosses and lichens, and decreased species diversity and evenness. These results predict that warming will cause a decline in biodiversity across a wide variety of tundra, at least in the short term. They also provide rigorous experimental evidence that recently observed increases in shrub cover in many tundra regions are in response to climate warming. These changes have important implications for processes and interactions within tundra ecosystems and between tundra and the atmosphere.

苔原生物群落中植物群落对实验性气候变暖的响应

Marilyn D. Walker, C. Henrik Wahren, Robert D. Hollister, Greg H. R. Henry, Lorraine E. Ahlquist, Juha M. Alatalo, M. Syndonia Bret-Harte, Monika P. Calef, Terry V. Callaghan, Amy B. Carroll, Howard E. Epstein, Ingibjörg S. Jónsdóttir, Julia A. Klein, Borgtór Magnússon, Ulf Molau, Steven F. Oberbauer, Steven P. Rewa, Clare H. Robinson, Gaius R. Shaver, Katharine N. Suding, Catharine C. Thompson, Anne Tolvanen, Ørjan Totland, P. Lee Turner, Craig E. Tweedie, Patrick J. Webber, Philip A. Wookey

(美国阿拉斯加大学)

摘要:最近观测的一些苔原生态系统变化似乎是对气候变暖的反应。一些实验研究表明,冻原植物和生态系统能对环境变化包括气候变暖有强烈反应,但大多数研究受限于一个位置、持续时间短和有限的实验设计。此外,对比研究结果比较困难,因为用各种不同的技术来实现测量变暖,用不同的测量尺度来评估环境响应。我们对在国际苔原实验中 11 个横跨苔原冻土带的区域,用标准化变暖方法进行的植物群落测定结果做了分析。气候被动变暖实验使植物空气温度上升 1～3 ℃,这在预测和观察苔原地区变暖的温度范围内。在两个生长季节后,所有的植物群落被观测到迅速响应。总体而言,气候变暖增加了落叶灌木和禾草的高度和密集度,减少了苔藓和地衣的覆盖度,并降低了物种的多样性和均匀度。这些结果预测气候变暖将导致各种各样的苔原生物多样性减少,至少在短期内如此。它们还提供了严格的实验证据,表明最近发现在许多苔原地区的灌木盖增加是对气候变暖的响应。这些变化对苔原生态系统之间以及苔原和大气之间的相互作用和进程有重要的作用。

38. Inger Greve Alsos, Pernille Bronken Eidesen, Dorothee Ehrich, et al. Frequent long-distance plant colonization in the changing Arctic[J]. Science, 2007, 316(5831): 1606-1609.

Frequent long-distance plant colonization in the changing Arctic

Inger Greve Alsos, Pernille Bronken Eidesen, Dorothee Ehrich, Inger Skrede, Kristine Westergaard, Gro Hilde Jacobsen, Jon Y. Landvik, Pierre Taberlet, Christian Brochmann

(National Centre for Biosystematics, Natural History Museum, University of Oslo, Post Office Box 1172 Blindern, NO-0318 Oslo, Norway)

Abstract: The ability of species to track their ecological niche after climate change is a major source of uncertainty in predicting their future distribution. By analyzing DNA fingerprinting (amplified fragment-length polymorphism) of nine plant species, we show that long-distance colonization of a remote arctic archipelago, Svalbard, has occurred repeatedly and from several source regions. Propagules are likely carried by wind and drifting sea ice. The genetic effect of restricted colonization was strongly correlated with the temperature requirements of the species, indicating that establishment limits distribution more than dispersal. Thus, it may be appropriate to assume unlimited dispersal when predicting long-term range shifts in the Arctic.

在不断变化的北极植物频繁的长距离迁移

Inger Greve Alsos, Pernille Bronken Eidesen, Dorothee Ehrich, Inger Skrede, Kristine Westergaard, Gro Hilde Jacobsen, Jon Y. Landvik, Pierre Taberlet, Christian Brochmann

(挪威奥斯陆大学国家生物系统学中心、自然历史博物馆)

摘要:在气候变化后物种循迹它们的生态龛位的能力,是预测它们未来分布不确定性的主要来源。通过分析9种植物的DNA指纹识别(扩增片段长度多态性),我们发现,来自几个源区的植物,重复地长距离地迁至斯瓦尔巴遥远的北极群岛繁殖。这种繁殖可能是由风和漂流海冰的携带造成的。受限制迁移物种的遗传作用和物种的温度需求有强烈的相关性,表明稳定的受限分布要多于分散分布。因此,在预测北极长期迁移变化范围时,可以合理假定为无约束的散布。

39. Quirin Schiermeie. The new face of the Arctic[J]. Nature, 2007, 446(7132): 133-135.

The new face of the Arctic

Quirin Schiermeie

(Nature's Munich office Germany)

Abstract: Due to global warming and climate change the Arctic Ocean loses more ice every summer, affecting all those living in the north. Possible effects include: the listing as polar bears as endangered species; exploitation of hydrocarbon deposits; overfishing of Arctic cod (*Arctogadus glacialis*) and its eventual displacement by temperate fishes such ast the Atlantic cod (*Gadus morhua*); and, an increase in vector-borne diseases amongst the Inuit as mosquitos migrate to the Arctic. Other species which may be effected by the "Atlantification" of the Arctic include black guillemot (*Cepphus grylle*), capelin fish (*Mallotus villosus*), seals, whales and the Pacific walrus (*Odobenus rosmanus divergens*).

极地研究:北极新面貌

Quirin Schiermeie

(德国 Nature 慕尼黑办事处)

摘要:由于全球变暖和气候变化,北冰洋每年夏天失去更多的冰,影响到所有那些在北部生活的生物。可能的影响包括:作为濒危物种清单上的北极熊;油气开采;过度捕捞北极鳕鱼(*Arctogadus* *glacialis*)和其最终被温带的大西洋鳕鱼(*Gadus morhua*)所取代;以及在因纽特人之间传播疾病的蚊子迁移到北极。可能被北极"大西洋化"影响到的其他物种,包括黑色鸠(*Cepphus grylle*)、毛鳞鱼(*Mallotus villosus*)、海豹、鲸和太平洋海象(*Odobenus rosmanus divergens*)。

40. Quirin Schiermeier. Life in a warming world[J]. Nature,2007,446(7132):135.

Life in a warming world

Quirin Schiermeier

(Correspondent in Nature's Munich office Germany)

Abstract: While climate change and global warming is leading to unpredictable ice conditions in the Arctic Circle, some Inuit believe the story is not all bad, and that it is just another situation that they will have to adjust to. A warming Arctic has some advantages including extending the hunting season for beluga whales and easier travel by boat. On the down side, the Inuit depend on roads across the ice for transport of food and other supplies, and less ice may increase their dependence on helicopters and cost of living. Many Inuit are working with scientists to investigate the changes in the Arctic, including the migration of Arctic cod (*Arctogadus glacialis*) and are being taught to handle meteorological instruments.

变暖世界中的生命

Quirin Schiermeier

(德国 Nature 慕尼黑办事处记者)

摘要:虽然气候变化和全球变暖正在导致不可预测的北极圈冰情,一些因纽特人相信这个故事并不全是坏事,而只是另一种他们将不得不适应的情况。北极变暖有一些优势,包括延长白鲸狩猎的季节,更容易乘船旅行。它的缺点是因纽特人需要依靠穿越冰道来运送食品和其他用品,少冰可能会增加其对直升机的消费和生活费。许多因纽特人正与科学家一道研究北极的变化,包括北极鳕鱼(*Arctogadus glacialis*)迁移,并正在学习使用气象仪器。

41. Byron C Crump, Bruce J Peterson, Peter A Raymond, et al. Circumpolar synchrony in big river bacterioplankton[J]. PNAS,2009,106(50):21208-21212.

Circumpolar synchrony in big river bacterioplankton

Byron C. Crump, Bruce J. Peterson, Peter A. Raymond, Rainer M. W. Amon, Amanda Rinehart, James W. McClelland, Robert M. Holmes

(Horn Point Laboratory, University of Maryland Center for Environmental Science, Cambridge, MD 21613, USA)

Abstract: Natural bacterial communities are extremely diverse and highly dynamic, but evidence is mounting that the compositions of these communities follow predictable temporal patterns. We investigated these patterns with a 3-year, circumpolar study of bacterio-plankton communities in the six largest rivers of the pan-arctic watershed (Ob', Yenisey, Lena, Kolyma, Yukon, and Mackenzie), five of which are among Earth's 25 largest rivers. Communities in the six rivers shifted synchronously over time, correlating with seasonal shifts in hydrology and biogeochemistry and clustering into three groups: winter/spring, spring freshet, and summer/fall. This synchrony indicates that hemisphere-scale variation in seasonal climate sets the pace of variation in microbial diversity. Moreover, these seasonal communities reassembled each year in all six rivers, suggesting a long-term, predictable succession in the composition of big river bacterioplankton communities.

极地附近大河流中的浮游细菌的同步性

Byron C. Crump, Bruce J. Peterson, Peter A. Raymond, Rainer M. W. Amon, Amanda Rinehart, James W. McClelland, Robert M. Holmes

(美国马里兰大学环境科学中心 Horn Point 实验室)

摘要:天然细菌群落极其多样且极具活力,但越来越多的证据显示,这些群落的组成具有可预测的时间模式。我们研究了这些模式 3 年,对极地附近的 6 条最大的河流(Ob', Yenisey, Lena, Kolyma, Yukon, Mackenzie)浮游细菌群落进行研究,其中 5 条属于全球 25 条最大的河流。在 6 条河流中的群落随时间同步转变,随着水文学和生物地球化学的季节变化,可以聚类关联分成 3 组:冬季/春季,春天河流上涨,夏季/秋季。同步变化表明,半球的季节性气候变化影响了微生物物种多样性变化的速度。此外,这 6 条河流每年季节性群落重组,意味着大河流中浮游细菌群落组成长期的可预见性。

42. Pierre E Galand, Emilio O Casamayor, David L Kirchman, et al. Ecology of the rare microbial biosphere of the Arctic Ocean[J]. PNAS, 2009, 106(52): 22427-22432.

Ecology of the rare microbial biosphere of the Arctic Ocean

Pierre E. Galand, Emilio O. Casamayor, David L. Kirchman, Connie Lovejoy

(Department of Continental Ecology-Limnology, Centre d'Estudis Avançats de Blanes-CSIC, Blanes, Spain)

Abstract: Understanding the role of microbes in the oceans has focused on taxa that occur in high abundance; yet most of the marine microbial diversity is largely determined by a long tail of low-abundance taxa. This rare biosphere may have a cosmopolitan distribution because of high dispersal and low loss rates, and possibly represents a source of phylotypes that become abundant when environmental conditions change. However, the true ecological role of rare marine microorganisms is still not known. Here, we use pyrosequencing to describe the structure and composition of the rare biosphere and to test whether it represents cosmopolitan taxa or whether, similar to abundant phylotypes, the rare community has a biogeography. Our examination of 740353 16S rRNA gene sequences from 32 bacterial and archaeal communities from various locations of the Arctic Ocean showed that rare phylotypes did not have a cosmopolitan distribution but, rather, followed patterns similar to those of the most abundant members of the community and of the entire community. The abundance distributions of rare and abundant phylotypes were different, following a log-series and log-normal model, respectively, and the taxonomic composition of the rare biosphere was similar to the composition of the abundant phylotypes. We conclude that the rare biosphere has a biogeography and that its tremendous diversity is most likely subjected to ecological processes.

北冰洋罕见微生物圈的生态学

Pierre E. Galand, Emilio O. Casamayor, David L. Kirchman, Connie Lovejoy

(西班牙布拉内斯高级研究中心)

摘要:了解海洋中的微生物作用到目前为止主要集中在高富集种类;然而,大多数的海洋微生物多样性主要由低富集、狭长分布的类群决定。这种罕见的生物圈,因其高分散和低损失率可能有着世界性的分布,并可能代表着那些环境条件变化后变得更为丰富的种系类型的一种来源。然而,真正的稀有海洋微生物生态的作用仍然不得而知。本文中,我们用焦磷酸测序来描述罕见的生物圈的结构和组成,并测试它是否代表了世界的种属分类或是否和丰富的种系型相似,具有地理分布特点。我们对来自北冰洋各地32种细菌和古细菌的740353个16S rRNA基因序列研究表明,罕见种系型没有一个世界性分布,而是遵循和那些具有最丰富成员的生物群落类似的模式。罕见和丰富的种系型分布不同,分别遵循日常系列和日常正常模式,罕见生物圈的分类组成类似于丰富种系型组成。我们的结论是罕见的生物圈有自己的地理区域,它大量的多样性最有可能受到生态学进程,如选择、物种形成和灭绝的影响。

43. Casey Hubert, Alexander Loy, Maren Nickel, et al. A constant flux of diverse thermophilic bacteria into the cold Arctic seabed[J]. Science, 2009, 325(5947): 1541-1544.

A constant flux of diverse thermophilic bacteria into the cold Arctic seabed

Casey Hubert, Alexander Loy, Maren Nickel, Carol Arnosti, Christian Baranyi, Volker Brüchert, Timothy Ferdelman, Kai Finster, Flemming Mønsted Christensen, Júlia Rosa de Rezende, Verona Vandieken, Bo Barker Jørgensen

(Biogeochemistry Group, Max Planck Institute for Marine Microbiology, Celsiusstrasse 1, D-28359 Bremen, Germany)

Abstract: Microorganisms have been repeatedly discovered in environments that do not support their metabolic activity. Identifying and quantifying these misplaced organisms can reveal dispersal mechanisms that shape natural microbial diversity. Using endospore germination experiments, we estimated a stable supply of thermophilic bacteria into permanently cold Arctic marine sediment at a rate exceeding 10^8 spores per square meter per year. These metabolically and phylogenetically diverse Firmicutes show no detectable activity at cold in situ temperatures but rapidly mineralize organic matter by hydrolysis, fermentation, and sulfate reduction upon induction at 50 degrees C. The closest relatives to these bacteria come from warm subsurface petroleum reservoir and ocean crust ecosystems, suggesting that seabed fluid flow from these environments is delivering thermophiles to the cold ocean. These transport pathways may broadly influence microbial community composition in the marine environment.

多样化的嗜热细菌不断流入寒冷的北极海底

Casey Hubert, Alexander Loy, Maren Nickel, Carol Arnosti, Christian Baranyi, Volker Brüchert, Timothy Ferdelman, Kai Finster, Flemming Mønsted Christensen, Júlia Rosa de Rezende, Verona Vandieken, Bo Barker Jørgensen

(德国马普海洋微生物研究所生物地球化学组)

摘要：已多次发现微生物在不能维持它们新陈代谢活动的环境中存在。识别和量化这些错位的生物可以揭示形成天然的微生物多样性的分散机制。利用内生孢子萌发实验，我们评估嗜热菌对寒冷的北极海洋沉积物的稳定供应超过每年每平方米 10^8 个孢子。这些在新陈代谢和系统发育上分化的厚壁菌在寒冷的原位温度下无活动，但经由水解、发酵和在 50 ℃以上诱导的硫酸盐还原，可迅速矿化有机物质。这些细菌与来源自温暖的地下石油库和海洋地壳生态系统的细菌很接近，这表明这些环境海底流体流动将嗜热生物送到冷的海洋。这些运输途径可能广泛影响海洋环境的微生物群落组成。

44. Natalia Rybczynski, Mary R Dawson, Richard H Tedford. A semi-aquatic Arctic mammalian carnivore from the Miocene epoch and origin of Pinnipedia[J]. Nature, 2009, 458(7241): 1021-1024.

A semi-aquatic Arctic mammalian carnivore from the Miocene epoch and origin of Pinnipedia

Natalia Rybczynski, Mary R. Dawson, Richard H. Tedford

(Canadian Museum of Nature, PO Box 3443 STN D, Ottawa, Ontario K1P 6P4, Canada)

Abstract: Modern pinnipeds (seals, sea lions and the walrus) are semi-aquatic, generally marine carnivores the limbs of which have been modified into flippers. Recent phylogenetic studies using morphological and molecular evidence support pinniped monophyly, and suggest a sister relationship with ursoids (for example bears) or musteloids (the clade that includes skunks, badgers, weasels and otters). Although the position of pinnipeds within modern carnivores appears moderately well resolved, fossil evidence of the morphological steps leading from a terrestrial ancestor to the modern marine forms has been weak or contentious. The earliest well-represented fossil pinniped is Enaliarctos, a marine form with flippers, which had appeared on the northwestern shores of North America by the early Miocene epoch. Here we report the discovery of a nearly complete skeleton of a new semi-aquatic carnivore from an early Miocene lake deposit in Nunavut, Canada, that represents a morphological link in early pinniped evolution. The new taxon retains a long tail and the proportions of its fore-and hindlimbs are more similar to those of modern terrestrial carnivores than to modern pinnipeds. Morphological traits indicative of semi-aquatic adaptation include a forelimb with a prominent deltopectoral ridge on the humerus, a posterodorsally expanded scapula, a pelvis with relatively short ilium, a shortened femur and flattened phalanges, suggestive of webbing. The new fossil shows evidence of pinniped affinities and similarities to the early Oligocene Amphicticeps from Asia and the late Oligocene and Miocene Potamotherium from Europe. The discovery suggests that the evolution of pinnipeds included a freshwater transitional phase, and may support the hypothesis that the Arctic was an early centre of pinniped evolution.

一种起源自鳍脚亚目的中新世半水生北极食肉类哺乳动物

Natalia Rybczynski, Mary R. Dawson, Richard H. Tedford

(加拿大自然博物馆)

摘要:现代鳍足类(海豹、海狮、海象)是半水生的海洋食肉动物,其翼已进化成鳍状肢。最近动植物种类史的研究利用形态学和分子证据支持鳍足类单系,并提出了与 ursoids(例如熊)或 musteloids(该分支,其中包括臭鼬、獾、黄鼠狼、水獭)的姊妹关系。虽然鳍足在现代类食肉动物的位置表面看比较清楚,但从一个陆地祖先通往现代海洋形式形态步骤的化石证据并不清楚并受到争议。最早的良好保存的鳍足类化石是海熊兽,以一个鳍状肢的海洋形态出现在北美早期的中新世的西北海岸。本文中,我们研究了一个新的半水生食肉动物接近完整的骨架,来自加拿大努纳武特地区中新世早期湖泊沉积,这代表了早期鳍足类演化的形态链接。新类群保留一条长长的尾巴,它前肢和后肢的比例比起现代鳍足类,更接近现代陆地食肉动物。半水生的形态特征适应包括一个突出的肩胸脊的肱骨的前肢,一个后上鬃扩大的肩胛骨,有着相对较短髂骨的骨盆,缩短的大腿骨和扁平缩短指骨,带有不明显的厚边。新化石显示了鳍足类与渐新世早期来自亚洲的 Amphicticeps 和晚渐新世至中新世时期欧洲 Potamotherium 相似之处的证据。这一发现表明,鳍足类的进化包括淡水的过渡阶段,并可能支持北极是一个鳍足类早期演化的中心这一假设。

45. Steven C Amstrup, Eric T De Weaver, David C Douglas, et al. Greenhouse gas mitigation can reduce sea-ice loss and increase polar bear persistence[J]. Nature, 2010, 468(7326): 955-958.

Greenhouse gas mitigation can reduce sea-ice loss and increase polar bear persistence

Steven C. Amstrup, Eric T. De Weaver, David C. Douglas, Bruce G. Marcot, George M. Durner, Cecilia M. Bitz, David A. Bailey

(US Geological Survey, Alaska Science Center, 4210 University Drive, Anchorage, Alaska 99508, USA)

Abstract: On the basis of projected losses of their essential sea-ice habitats, a United States Geological Survey research team concluded in 2007 that two-thirds of the world's polar bears (*Ursus maritimus*) could disappear by mid-century if business-as-usual greenhouse gas emissions continue. That projection, however, did not consider the possible benefits of greenhouse gas mitigation. A key question is whether temperature increases lead to proportional losses of sea-ice habitat, or whether sea-ice cover crosses a tipping point and irreversibly collapses when temperature reaches a critical threshold. Such a tipping point would mean future greenhouse gas mitigation would confer no conservation benefits to polar bears. Here we show, using a general circulation model, that substantially more sea-ice habitat would be retained if greenhouse gas rise is mitigated. We also show, with Bayesian network model outcomes, that increased habitat retention under greenhouse gas mitigation means that polar bears could persist throughout the century in greater numbers and more areas than in the business-as-usual case. Our general circulation model outcomes did not reveal thresholds leading to irreversible loss of ice; instead, a linear relationship between global mean surface air temperature and sea-ice habitat substantiated the hypothesis that sea-ice thermodynamics can overcome albedo feedbacks proposed to cause sea-ice tipping points. Our outcomes indicate that rapid summer ice losses in models and observations represent increased volatility of a thinning sea-ice cover, rather than tipping-point behaviour. Mitigation-driven Bayesian network outcomes show that previously predicted declines in polar bear distribution and numbers are not unavoidable. Because polar bears are sentinels of the Arctic marine ecosystem and trends in their sea-ice habitats foreshadow future global changes, mitigating greenhouse gas emissions to improve polar bear status would have conservation benefits throughout and beyond the Arctic.

温室气体减排可以减少海冰的损失并改善北极熊的生存

Steven C. Amstrup, Eric T. De Weaver, David C. Douglas, Bruce G. Marcot, George M. Durner, Cecilia M. Bitz, David A. Bailey

(美国地质调查局阿拉斯加科学中心)

摘要:在其基本海冰栖息地的预计损失的基础上,一个美国地质调查研究小组2007年的结论认为在21世纪中叶,如果工业照常进行、温室气体排放量继续,世界三分之二的北极熊(*Ursus maritimus*)可能会消失。但是这一预测并没有考虑到温室气体减排带来的好处。关键的问题是,是否温度的上升会导致海冰栖息地的比例下降,还有海冰覆盖是否超过那个转折点,当温度到达临界值,会发生不可逆转的崩溃。这样一个转折点,将意味着未来温室气体减排不会对北极熊有任何的好处。在这里,我们使用大气环流模型显示,如果温室气体的增加被减轻,会大幅度增加海冰栖息地。我们使用贝叶斯网络模型的结果还表明,在温室气体减排下会增加栖息地的保留,这意味着大量的北极熊在百年中继续生存,且比在工业照常排放温室气体的时候分布更广。我们的大气环流模式的结果没有发现会导致不可逆转冰损失的转折点,而是全球平均地面气温和海冰栖息地的线性关系证实了这一假设:海冰热力学可以克服反照率反馈导致的海冰转折点。我们的结果表明:在模型和观察中,夏季冰的迅速损失表明了薄海冰覆盖的易变性,而非转折点的存在。受缓解驱动的贝叶斯网络结果表明,先前预测的北极熊分布减少和数量下降不是不可避免的。由于北极熊是北极海洋生态系统的哨兵,其海冰栖息地的变化趋势预示着未来全球变化。减少温室气体排放能改善北极熊的现状,将对整个北极以及其余地域带来好处。

46. Barry N Duplantis, Milan Osusky, Crystal L Schmerk, et al. Essential genes from Arctic bacteria used to construct stable, temperature-sensitive bacterial vaccines [J]. PNAS, 2010, 107 (30): 13456-13460.

Essential genes from Arctic bacteria used to construct stable, temperature-sensitive bacterial vaccines

Barry N. Duplantis, Milan Osusky, Crystal L. Schmerk, Darrell R. Ross, Catharine M. Bosio, Francis E. Nano

(Department of Biochemistry and Microbiology, University of Victoria, Victoria, BC, V8W 3P6, Canada)

Abstract: All bacteria share a set of evolutionarily conserved essential genes that encode products that are required for viability. The great diversity of environments that bacteria inhabit, including environments at extreme temperatures, place adaptive pressure on essential genes. We sought to use this evolutionary diversity of essential genes to engineer bacterial pathogens to be stably temperature-sensitive, and thus useful as live vaccines. We isolated essential genes from bacteria found in the Arctic and substituted them for their counterparts into pathogens of mammals. We found that substitution of nine different essential genes from psychrophilic (cold-loving) bacteria into mammalian pathogenic bacteria resulted in strains that died below their normal-temperature growth limits. Substitution of three different psychrophilic gene orthologs of ligA, which encode NAD—dependent DNA ligase, resulted in bacterial strains that died at 33, 35, and 37 degrees C. One ligA gene was shown to render Francisella tularensis, Salmonella enterica, and Mycobacterium smegmatis temperature-sensitive, demonstrating that this gene functions in both Gram-negative and Gram-positive lineage bacteria. Three temperature-sensitive *F. tularensis* strains were shown to induce protective immunity after vaccination at a cool body site. About half of the genes that could be tested were unable to mutate to temperature-resistant forms at detectable levels. These results show that psychrophilic essential genes can be used to create a unique class of bacterial temperature-sensitive vaccines for important human pathogens, such as S. enterica and Mycobacterium tuberculosis.

用北极细菌的基础基因构建稳定的温度敏感的细菌疫苗

Barry N. Duplantis, Milan Osusky, Crystal L. Schmerk, Darrell R. Ross, Catharine M. Bosio, Francis E. Nano

(加拿大维多利亚大学生物化学和微生物学系)

摘要：所有的细菌都有一个进化中被保存的生存必需的基础基因。细菌的栖息环境差异很大，包括极端温度的环境，对基础基因产生自适应压力。我们试图利用这一重要的基因工程细菌病原体的进化多样性来生产稳定温度敏感且有用的活疫苗。我们分离在北极发现的细菌基础基因，并用它们来置换哺乳类动物病原体中的相似物。我们发现利用9种来自喜寒细菌的不同的基础基因替代物替换哺乳动物病原体的基因，会导致病原体在低于它们正常生长温度受限下死亡。置换3种不同的喜寒基因ligA的直系同源基因，其编码的NAD——依赖的DNA连接酶，导致菌株在33℃、35℃和37℃下死亡。一种ligA基因使土拉热杆菌、肠道沙门氏菌和阴垢分枝杆菌对温度敏感，表明这个基因在革兰氏阴性和革兰氏阳性菌中均有作用。3个温度敏感的*F. tularensis*在身体的冷部位接种疫苗后会显示出免疫保护。约一半的可以检测到这种基因在可检测水平不会变异成耐温形式。这些结果表明，嗜冷基础基因，可用于创建独特的细菌对温度敏感疫苗来医治重要的人类病原体，如肠道沙门氏菌和结核分枝杆菌。

47. Carsten Egevang, Iain J Stenhouse, Richard A Phillips, et al. Tracking of Arctic terns Sterna paradisaea reveals longest animal migration[J]. PNAS, 2010, 107(5): 2078-2081.

Tracking of Arctic terns Sterna paradisaea reveals longest animal migration

Carsten Egevang, Iain J. Stenhouse, Richard A. Phillips, Aevar Petersen, James W. Fox, Janet R. D. Silk

(Greenland Institute of Natural Resources, DK-3900 Nuuk, Greenland)

Abstract: The study of long-distance migration provides insights into the habits and performance of organisms at the limit of their physical abilities. The Arctic tern *Sterna paradisaea* is the epitome of such behavior; despite its small size (<125 g), banding recoveries and at-sea surveys suggest that its annual migration from boreal and high Arctic breeding grounds to the Southern Ocean may be the longest seasonal movement of any animal. Our tracking of 11 Arctic terns fitted with miniature (1.4 g) geolocators revealed that these birds do indeed travel huge distances (more than 80000 km annually for some individuals). As well as confirming the location of the main wintering region, we also identified a previously unknown oceanic stopover area in the North Atlantic used by birds from at least two breeding populations (from Greenland and Iceland). Although birds from the same colony took one of two alternative southbound migration routes following the African or South American coast, all returned on a broadly similar, sigmoidal trajectory, crossing from east to west in the Atlantic in the region of the equatorial Intertropical Convergence Zone. Arctic terns clearly target regions of high marine productivity both as stopover and wintering areas, and exploit prevailing global wind systems to reduce flight costs on long-distance commutes.

跟踪北极燕鸥揭示最长时间的动物迁徙

Carsten Egevang, Iain J. Stenhouse, Richard A. Phillips, Aevar Petersen, James W. Fox, Janet R. D. Silk

(格陵兰自然资源研究所)

摘要:长途迁徙的研究提供了对生物在其极限状态下习性和体能的了解。北极燕鸥(*Sterna paradisaea*)是这种行为的一个缩影,尽管它的尺寸小(<125 g),环标回收和海上调查显示,它们每年从北极高纬度地区迁移到南大洋繁殖地,这可能是动物最长的季节迁徙。我们对11个北极燕鸥安装微型(1.4 g)地理位置跟踪显示工具,显示出这些鸟类确实旅行极远的距离(对某些个体来说每年超过8万公里)。我们不仅确认了它们主要的越冬区域的位置,还确定了以前未知的位于北大西洋的海洋停留点为至少两种种群(来自格陵兰和冰岛)提供繁殖地。尽管来自同一栖息地的鸟类会在沿着非洲或南美洲海岸向南的两条迁移路线中选择一条,所有返回的路线大致相同,S形曲线轨迹,从东到西穿越大西洋赤道辐合中心带。北极燕鸥明确以海洋生产力高的地区为目标,既作为中途停留又为越冬地,并利用随时的全球风力系统来减少长途飞行成本。

48. Charlotte Lindqvist, Stephan C Schuster, Yazhou Sun, et al. Complete mitochondrial genome of a Pleistocene jawbone unveils the origin of polar bear[J]. PNAS, 2010, 107(11): 5053-5057.

Complete mitochondrial genome of a Pleistocene jawbone unveils the origin of polar bear

Charlotte Lindqvist, Stephan C. Schuster, Yazhou Sun, Sandra L. Talbot, Ji Qi, Aakrosh Ratan, Lynn P. Tomsho, Lindsay Kasson, Eve Zeyl, Jon Aars, Webb Miller, Ólafur Ingólfsson, Lutz Bachmann, Øystein Wiig

(Department of Biological Sciences, University at Buffalo, Buffalo, NY 14260, USA)

Abstract: The polar bear has become the flagship species in the climate-change discussion. However, little is known about how past climate impacted its evolution and persistence, given an extremely poor fossil record. Although it is undisputed from analyses of mitochondrial (mt) DNA that polar bears constitute a lineage within the genetic diversity of brown bears, timing estimates of their divergence have differed considerably. Using next-generation sequencing technology, we have generated a complete, high-quality mt genome from a stratigraphically validated 130000 to 110000 year-old polar bear jawbone. In addition, six mt genomes were generated of extant polar bears from Alaska and brown bears from the Admiralty and Baranof islands of the Alexander Archipelago of southeastern Alaska and Kodiak Island. We show that the phylogenetic position of the ancient polar bear lies almost directly at the branching point between polar bears and brown bears, elucidating a unique morphologically and molecularly documented fossil link between living mammal species. Molecular dating and stable isotope analyses also show that by very early in their evolutionary history, polar bears were already inhabitants of the Artic sea ice and had adapted very rapidly to their current and unique ecology at the top of the Arctic marine food chain. As such, polar bears provide an excellent example of evolutionary opportunism within a widespread mammalian lineage.

完整的更新世颌骨线粒体基因组揭示了北极熊的起源

Charlotte Lindqvist, Stephan C. Schuster, Yazhou Sun, Sandra L. Talbot, Ji Qi, Aakrosh Ratan, Lynn P. Tomsho, Lindsay Kasson, Eve Zeyl, Jon Aars, Webb Miller, Ólafur Ingólfsson, Lutz Bachmann, Øystein Wiig

(美国纽约州立大学布法罗分校生物科学系)

摘要:北极熊已经成为气候变化讨论的指示物种。然而,因化石记录极端贫乏,很少有人知道过去的气候如何影响其进化和持续性。虽然线粒体(mt)DNA 分析无可争议地表明北极熊构成棕熊遗传多样性的一个分支,但它们的分化时间评估差异很大。采用新一代测序技术,我们已经从一个地层学验证年龄为 130000～110000 年的北极熊颌骨中测定了一个完整的、高质量的 mt 基因组。此外,还测定了现存的阿拉斯加北极熊与来自金钟的科迪亚克岛东南部阿拉斯加和亚历山大群岛巴拉诺夫岛的棕熊等 6 个 mt 基因组。我们证明了古代北极熊的进化点差不多就是北极熊和棕熊的分支点,阐明一个独特的形态和分子化石哺乳类动物联系的记录。分子数据和稳定同位素分析还显示,在北极熊进化历史中非常早的时期,北极熊已经是北极海冰的居民,并很快成为它们现在独特的在北极海洋生态食物链的顶端地位。因此,北极熊为广泛的哺乳动物进化的机会主义谱系提供了很好的例子。

49. McKinnon L,Smith P A,Nol E,et al. Lower predation risk for migratory birds at high latitudes[J]. Science,2010,327(5963):326-327.

Lower predation risk for migratory birds at high latitudes

McKinnon L., Smith P. A., Nol E., Martin J. L., Doyle F. I., Abraham K. F., Gilchrist H. G., Morrison R. I. G., Bêty J.

(Département de Biologie, Université du Québec à Rimouski and Centre d'Etudes Nordiques, Rimouski, Québec, G5L3A1, Canada)

Abstract: Quantifying the costs and benefits of migration distance is critical to understanding the evolution of long-distance migration. In migratory birds, life history theory predicts that the potential survival costs of migrating longer distances should be balanced by benefits to lifetime reproductive success, yet quantification of these reproductive benefits in a controlled manner along a large geographical gradient is challenging. We measured a controlled effect of predation risk along a 3350 kilometer south-north gradient in the Arctic and found that nest predation risk declined more than twofold along the latitudinal gradient. These results provide evidence that birds migrating farther north may acquire reproductive benefits in the form of lower nest predation risk.

高纬度候鸟的低捕食风险

McKinnon L., Smith P. A., Nol E., Martin J. L., Doyle F. I., Abraham K. F., Gilchrist H. G., Morrison R. I. G., Bêty J.

(加拿大魁北克大学里穆斯基分校生物学系)

摘要:迁移距离的量化成本和效益是理解长途迁移演变至关重要的点。生活史理论预测,候鸟潜在较长距离迁移的生存费用应和终生繁殖成功率相平衡,但在控制方式下沿着一个大的地理梯度量化繁殖收益具有挑战性。我们测量在北极沿着一条 3350 km 长的南北梯度的捕食风险控制的效果,发现巢捕食风险沿纬度梯度下降超过 50%。这些结果提供的证据表明,鸟类迁徙至更远的北部可能获得低巢捕食风险带来的繁殖受益。

50. Abigail L Swann, Inez Y Fung, Samuel Levis, et al. Changes in Arctic vegetation amplify high-latitude warming through the greenhouse effect[J]. PNAS, 2010, 107(4): 1295-1300.

Changes in Arctic vegetation amplify high-latitude warming through the greenhouse effect

Abigail L. Swann, Inez Y. Fung, Samuel Levis, Gordon B. Bonan, Scott C. Doney

(Department of Earth & Planetary Science, University of California, Berkeley, CA 94720, USA)

Abstract: Arctic climate is projected to change dramatically in the next 100 years and increases in temperature will likely lead to changes in the distribution and makeup of the Arctic biosphere. A largely deciduous ecosystem has been suggested as a possible landscape for future Arctic vegetation and is seen in paleo-records of warm times in the past. Here we use a global climate model with an interactive terrestrial biosphere to investigate the effects of adding deciduous trees on bare ground at high northern latitudes. We find that the top-of-atmosphere radiative imbalance from enhanced transpiration (associated with the expanded forest cover) is up to 1.5 times larger than the forcing due to albedo change from the forest. Furthermore, the greenhouse warming by additional water vapor melts sea-ice and triggers a positive feedback through changes in ocean albedo and evaporation. Land surface albedo change is considered to be the dominant mechanism by which trees directly modify climate at high-latitudes, but our findings suggest an additional mechanism through transpiration of water vapor and feedbacks from the ocean and sea-ice.

北极植被变化放大了温室效应造成的高纬度升温

Abigail L. Swann, Inez Y. Fung, Samuel Levis, Gordon B. Bonan, Scott C. Doney

(美国加利福尼亚大学伯克利分校地球与行星科学系)

摘要：北极气候预计将在未来100年急剧变化，温度急剧上升可能会导致北极生物圈分布和结构的变化。一个在很大程度上的落叶生态系统被认为是未来北极的植被景观，此景观在古记录的温暖时期出现过。本文中，我们使用一个交互式陆地生物圈的全球气候模型来研究裸露地面上添加落叶树在北部高纬度地区的影响。我们发现，因增加了蒸发量(与扩大森林覆盖面积相关)导致的大气顶级的辐射不平衡比森林反照率变化所导致的大1.5倍。此外，额外的水汽导致的温室效应融化海冰并触发通过海洋反照率和蒸发量变化带来的正响应。地表反照率的变化被认为是通过高纬度地区树木直接改变气候的主要机制，但我们发现表明存在通过水蒸气和海洋、海冰反馈的额外机制。

五、现代海洋科学类文献

5.1 分析概述

根据统计结果，从1990年到2010年期间 *Nature*，*Science*，*PNAS* 三家刊物上与现代海洋科学相关的文献有89篇，其中以南极为主或与这一区域相关的文献有58篇，以北极海域为主或与北极相关的文献有31篇。海洋科学是一门综合科学，表现在研究区域的空间范围广大和涉及学科门类复杂两方面。这些文献中有很大一部分并不局限在单一海域，特别是与南大洋相关的文献中就有相当一部分涉及其他海域，甚至与全球变化紧密联系。

上述三家刊物平均每年发表4.2篇有关海洋科学方面的文章，*Nature* 上发表相关论文48篇，*Science* 上发表相关论文41篇，*PNAS* 上无现代海洋科学方面的文章。

将这89篇文章按照研究区域和发表年代进行归类（图5.1），两刊物在1991～1994年和2002～2004年间比较关注南北极海洋科学，发表该类文章较多。而且其关注重点都集中在远洋海域的铁肥实验，以及与之相关的生态环境效应，乃至对气候变化的影响。而在两刊物上在1994年、1997年、2005年和2009年均无此类文章发表。2004年之后发表的文章数整体减少，说明上述区域的海洋科学热点研究热潮已过。2000年之前，发表的文章明显比之后要多，而且集中在南极地区。而2010年有3篇关于北极大洋的文章发表，似乎预示着北极地区海洋科学研究的升温。

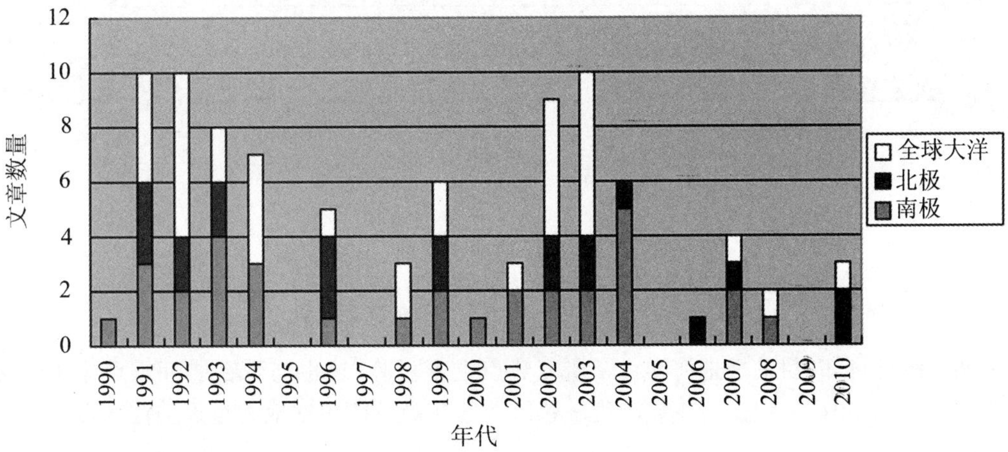

图5.1　文章数量随年代的变化

现代海洋科学大体可以分为海洋物理学、海洋化学、气候变化和生态地质环境四个大的方向（图5.2）。其中海洋化学共32篇，主要集中在1991～1993年和2004年，内容多集中在铁肥实验的观测和环境气候效应研究。海洋物理学共有28篇，1992年、2002年和2003年较多地发表了这一方面的文章，主要关注洋流及水团物理性质的示踪研究。气候变化方面主要集中关注全球尺度的变化状况，其关键区域主要是北大西洋海域海洋状态对全球气候的作用，其他大多将南北极海域作为对全球尺度气候变化的敏感响应区域研究，共20篇。其余为生态地质方面文章，热点较为分散。

将刊物中简讯和新闻报道排除，将文章数量按所署第一单位所在国家进行排列（图5.3），美国发表的文章数最多，高达53篇，超过总数量的50%，其次为英国，论文量为10篇。德国、加拿大发表6篇。值得注意的是除去这几个科研强国，澳大利亚、瑞士、丹麦、新西兰等近极圈国家由于地理优势，也对这些区域的研究有所贡献。近20年来中国以第一署名国在该领域还没有一篇。说明我国对南北极海洋学的研究还相对落后，或者说是影响力不够。

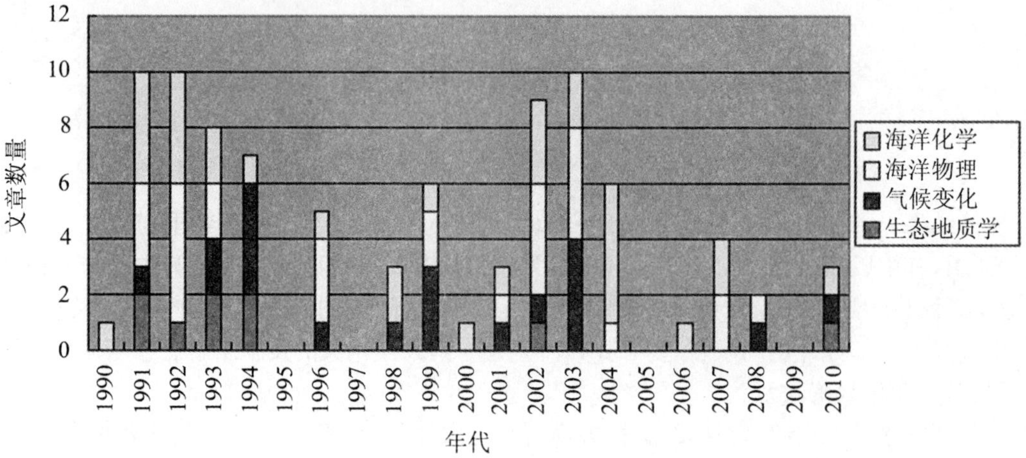

图 5.2　各研究方向文章数量年代分布

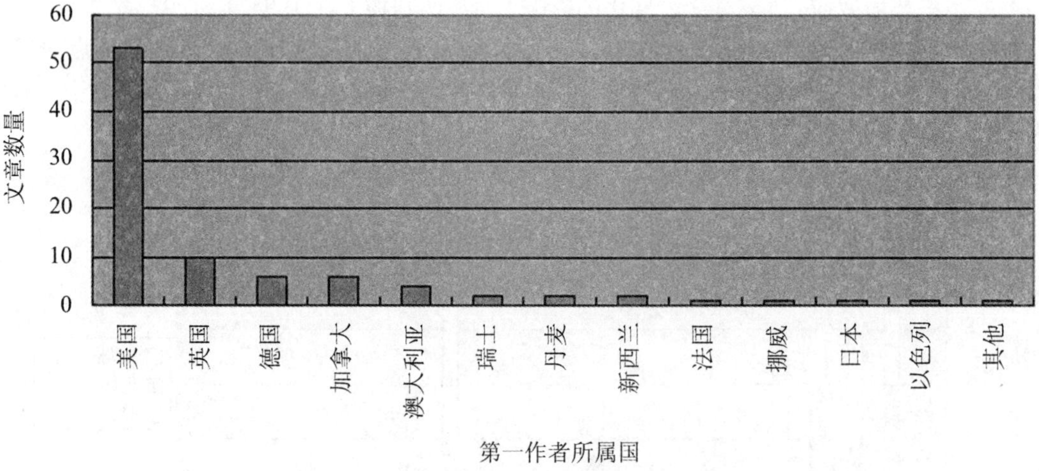

图 5.3　第一作者所在国家发表的文章数量

由以上初步统计可知,目前极地海洋科学研究热点乏善可陈,这也与现在研究重点集中于热带海域对全球气候变化的作用有关,但是对气候响应敏感区域的北极可能再次成为重点关注区域。

5.2 摘要翻译——南极

1. Martin J H, Gordon R M, Fitzwater S E. Iron in Antarctic waters[J]. Nature, 1990, 345: 156-158.

Iron in Antarctic waters

Martin J. H., Gordon R. M., Fitzwater S. E.

(Moss Landing Marine Laboratories, California USA)

Abstract: We have performed an in situ test of the iron limitation hypothesis in the subarctic North Pacific Ocean. A single enrichment of dissolved iron caused a large increase in phytoplankton standing stock and decreases in macronutrients and dissolved carbon dioxide. The dominant phytoplankton species shifted after the iron addition from pennate diatoms to a centric diatom, *Chaetoceros debilis*, that showed a very high growth rate, 2.6 doublings per day. We conclude that the bioavailability of iron regulates the magnitude of the phytoplankton biomass and the key phytoplankton species that determine the biogeochemical sensitivity to iron supply of high-nitrate, low-chlorophyll waters.

南极海水中的铁

Martin J. H., Gordon R. M., Fitzwater S. E.

(美国莫斯兰丁海洋实验室)

摘要:我们现场验证了一个假说:铁缺乏会限制亚北极北太平洋的浮游植物生长。溶解铁的补充会导致浮游植物大量增长,以及宏量营养素和溶解 CO_2 的减少。加铁后,主导的浮游植物种从羽状转至中心状的 *Chaetoceros debilis*,其生长速度快,每天翻 2.6 倍。我们的结论是铁的生物利用度控制浮游植物的生物总量以及关键的浮游植物种类,这些种类在高氮、低叶绿素的水域,对铁具有生物地球化学敏感性。

2. Huntley M E, Lopez M D G, Karl D M. Top predators in the Southern-Ocean—a major leak in the biological carbon pump[J]. Science, 1991, 253: 64-66.

Top predators in the Southern-Ocean—a major leak in the biological carbon pump

Huntley M. E., Lopez M. D. G., Karl D. M.

(Scripps Institution of Oceanography, University of California, San Diego, La Jolla 92093)

Abstract: Primary productivity in the Southern Ocean is approximately 3.5 gigatons of carbon per year, which accounts for nearly 15 percent of the global total. The presence of high concentrations of nitrate in Antarctic waters suggests that it might be possible to increase primary production significantly and thereby alleviate the net accumulation of atmospheric carbon dioxide. An analysis of the food web for these waters implies that the Southern Ocean may be remarkably inefficient as a carbon sink. This inefficiency is caused by the large flux of carbon respired to the atmosphere by air-breathing birds and mammals, dominant predators in the unusually simple food web of Antarctic waters. These top predators may transfer into the atmosphere as much as 20 to 25 percent of photosynthetically fixed carbon.

南大洋的顶级捕食者——生物碳泵的主要漏洞

Huntley M. E., Lopez M. D. G., Karl D. M.

(美国加州大学圣迭戈分校斯克里普斯海洋研究所)

摘要:在南大洋初级生产力约为每年35亿吨碳,大约占到全球总量的百分之十五。在南极海域存在高浓度的硝酸盐表明,有可能利用显著增加初级生产力来减轻对大气中二氧化碳的增加。对这些水域的食物网分析显示南大洋作为碳汇效率并不是很高的。这种低效率是由于鸟类和哺乳动物的呼吸作用将碳大量带到了空气中,这些顶级食肉动物可能将光合作用固定的碳的20%～25%重新释放。

3. Ittekkot V, Nair R R, Honjo S, et al. Enhanced particle fluxes in Bay of Bengal induced by injection of fresh-water[J]. Nature, 1991, 351: 385-387.

Enhanced particle fluxes in Bay of Bengal induced by injection of fresh-water

Ittekkot V., Nair R. R., Honjo S., Ramaswamy V., Bartsch M., Manganini S., Desai B. N.

(Institute of Biogeochemistry and Marine Chemistry, University of Hamburg, Bundesstrasse 55, 2000 Hamburg 13, Germany)

Abstract: The melting of ice sheets during deglaciation results in the injection of large amounts of fresh water into the oceans. To investigate how such injections might influence particle fluxes in the ocean, and hence the uptake of atmospheric CO_2, we deployed three sediment-trap moorings (two traps in each mooring) in the northern, central and southern parts of the Bay of Bengal, respectively. The Bay of Bengal is suitable for such a study, because some of the world's largest rivers supply pulses of fresh water and sediment to the bay, resulting in large seasonal changes in surface salinity. We find that the maximum river discharge, which occurs during the southwest monsoon, coincides with the maximum observed flux of particulate matter. From north to south, the carbonate flux increases, whereas fluxes of opal, organic carbon and particulate matter decrease. The overall flux pattern seems to be controlled by the seasonally varying input from the rivers and the accompanying shift in marine biogenic production. We conclude that freshwater pulses during deglaciation may therefore have caused similar shifts in marine biogenic production, resulting in short-term episodes of increased oceanic uptake of atmospheric CO_2.

淡水输入引发的孟加拉湾海域颗粒物通量增加

Ittekkot V. , Nair R. R. , Honjo S. , Ramaswamy V. , Bartsch M. , Manganini S. , Desai B. N.

(德国海德堡大学生物地球化学和海洋化学所)

摘要:间冰期冰盖融化,大量淡水注入海洋。为了探讨淡水注入影响海洋颗粒物通量,从而影响大气二氧化碳的问题,我们在孟加拉湾的北部、中部和南部地区部署了三个沉积物捕集系泊(每个停泊位 2 个捕获器)。孟加拉湾是适合这种研究的地方,因为世界上最大的一些河流为海湾带来了大量的脉冲式的淡水和沉积物,导致表层盐度大的季节变化。我们发现,发生在西南季风爆发期最大的河流流量,与最大的颗粒物通量观测时间相一致。从北到南,碳酸盐通量增加而蛋白石、有机碳通量以及颗粒物减少。总的流量模式似乎是为河流的季节性输入变化和海洋生物生产变化所控制。我们得出结论,在末次冰消期淡水注入,可能造成海洋生物生产力发生类似的变化,海洋吸收大气二氧化碳在短期内会增加。

4. Joos F, Sarmiento J L, Siegenthaler U. Estimates of the effect of Southern-Ocean iron fertilization on atmospheric CO_2 concentrations[J]. Nature, 1991, 349: 772-775.

Estimates of the effect of Southern-Ocean iron fertilization on atmospheric CO_2 concentrations

Joos F. , Sarmiento J. L. , Siegenthaler U.

(Physics Institute, University of Bern, CH-3012 Bern, Switzerland)

Abstract: It has been suggested that fertilizing the ocean with iron might offset the continuing increase in atmospheric CO_2 by enhancing the biological uptake of carbon, thereby decreasing the surface-ocean partial pressure of CO_2 and drawing down CO_2 from the atmosphere. Using a box model, we present estimates of the maximum possible effect of iron fertilization, assuming that iron is continuously added to the phosphate-rich waters of the Southern Ocean, which corresponds to 16% of the world ocean surface. We find that after 100 years of fertilization, the atmospheric CO_2 concentration would be 59 p. p. m. below what it would have been with no fertilization, assuming no anthropogenic CO_2 emissions, and 90-107 p. p. m. less when anthropogenic emissions are included in the calculation. Such a large uptake of CO_2 is unlikely to be achieved in practice, owing to a variety of constraints that require further study; the effect of iron fertilization on the ecology of the Southern Ocean also remains to be evaluated. Thus, the most effective and reliable strategy for reducing future increases in atmospheric CO_2 continues to be control of anthropogenic emissions.

关于南大洋铁肥实验对大气中二氧化碳含量影响的估计

Joos F. , Sarmiento J. L. , Siegenthaler U.

(瑞士波恩大学物理研究院)

摘要:有人认为对海洋施肥可以提高生物对碳的固定,降低表面海洋中二氧化碳分压,吸收大气中的二氧化碳，从而减缓大气中的二氧化碳的不断增加。使用盒子模型，我们假设将铁不断加入有足量磷的相当于16%的世界海洋面积的南大洋，我们估计铁施肥可能产生的最大影响。假设没有人为的二氧化碳排放，我们发现经过100多年的施肥，大气二氧化碳浓度将比不进行施肥实验的值低约59 ppm。而考虑人为排放的话，会低90～107 ppm。这样大的二氧化碳的吸收在实践中，由于各种制约因素是不可能达到的；铁施肥对南大洋生态效应也有待评估。因此，为减少大气中二氧化碳的增加，未来最有效和可靠的战略仍然是对人为排放的控制。

5. Peng T H, Broecker W S. Dynamical limitations on the Antarctic iron fertilization strategy[J]. Nature, 1991, 349: 227-229.

Dynamical limitations on the Antarctic iron fertilization strategy

Peng T. H., Broecker W. S.

(Environmental Sciences Division, Oak Ridge National Laboratory, Oak Ridge, Tennessee 37831, USA)

Abstract: Martin et al have proposed an ingenious means by which the rise in atmospheric CO_2 content generated by the burning of fossil fuels and deforesation might be partially compensated. The idea is that plant production in the nutrient-rich surface waters of the Antarctic could be stimulated by the addition of dissolved iron, thereby reducing the CO_2 partial pressure in these waters and allowing CO_2 to flow from the atmosphere into the Antarctic Ocean. We have used a box model calibrated with transient tracer data to examine the dynamical aspects of this proposal, and conclude that after 100 years of totally successful fertilization the CO_2 content of the atmosphere would be lowered by only 10 +/−5% below what it would have been in the absence of fertilization. So if after 100 years the CO_2 content of the atmosphere were 500 μatm without fertilization, it would be between 425 and 475 μatm with full fertilization. In other words, if our model calibration is correct, even if iron fertilization worked perfectly it would not significantly reduce the atmospheric CO_2 content.

南极铁肥实验的动力学限制

Peng T. H., Broecker W. S.

(美国橡树岭国家实验室环境科学部)

摘要：马丁等人提出了一种弥补化石燃料的燃烧和森林退化带来的大气二氧化碳上升的手段。这个想法是，通过对环南极营养丰富海域中加入可溶性Fe，达到刺激浮游植物的生长，从而减少二氧化碳在表层水中的分压，让二氧化碳从大气中流入南极海洋。我们用一个经瞬态示踪数据校正的盒子模型，审查这一方法的动力学约束，并得出结论：在完全成功的条件下，100年后大气中的二氧化碳含量相对于不施肥的情况将会降低(10±5)%。因此，如果100年的大气中的二氧化碳含量在不施肥下达到500 μatm，而施肥的话，会在425～475 μatm之间。换句话说，如果我们的模型校正是正确的，即使铁施肥工作很成功，也不会大幅减少大气中二氧化碳的含量。

6. Prospero J M, Savoie D L, Saltzman E S, et al. Impact of oceanic sources of biogenic sulphur on sulfate aerosol concentrations at Mawson, Antarctica[J]. Nature, 1991, 350: 221-223.

Impact of oceanic sources of biogenic sulphur on sulphate aerosol concentrations at Mawson, Antarctica

Prospero J. M. , Savoie D. L. , Saltzman E. S. , Larsen R.

(University of Miami, RSMAS/MAC, 4600 Ricken backer Cause way, Miami, Florida 33149, USA)

Abstract: Sulphate is the dominant aerosol species in the Antarctic atmosphere and an important constituent in Antarctic snow and ice. Various sources have been suggested for Antarctic non-sea-salt sulphate (n. s. s. SO_4^{2-}): volcanic emissions, stratospheric injection, pollutants transported from the low latitudes and biogenic dimethylsulphide (DMS) from the ocean. Although the oceanic source is now believed to be especially important, there has been no strong chemical evidence directly linking oceanic DMS with the Antarctic n. s. s. SO_4^{2-} concentrations. Here we present extended measurements from the Antarctic for both n. s. s. SO_4^{2-} and methanesulphonate (MSA), an oxidation product of DMS. Both species have a very strong seasonal cycle with a maximum in the austral summer; this cycle parallels that of the oceanic biogenic sulphur producers, thereby suggesting a strong link between the Antarctic atmospheric sulphur cycle and biological processes in the Southern Ocean.

南极莫森地区海洋源生物硫对硫酸盐气溶胶浓度的影响

Prospero J. M. , Savoie D. L. , Saltzman E. S. , Larsen R.

(美国迈阿密大学)

摘要:硫酸盐是南极大气气溶胶和南极冰雪的重要组成部分。各种南极非海盐(n. s. s. SO_4^{2-})硫酸盐的来源广泛:火山排放,平流层注入,从低纬度运输来的污染物,以及海洋生物产生的二甲基硫醚(DMS)。虽然洋源现在被认为特别重要,但是没有很强的化学证据直接将南极海洋DMS与非海盐硫酸盐的浓度关联。这里,我们报道从南极测量的非海盐硫酸盐和甲烷磺酸盐(MSA),DMS的氧化产物。这两种物质均有很强的季节性周期,在南半球夏季最高;这种周期与海洋生物硫生产者的周期相似,从而表明了大气硫循环与生物过程在南大洋的密切联系。

7. Richards K J,Pollard R T. Structure of the upper ocean in the western equatorial Pacific[J]. Nature, 1991,350:48-50.

Structure of the upper ocean in the western equatorial Pacific

Richards K. J. , Pollard R. T.

(Department of Oceanography, The University, Southampton S09 5NH, UK)

Abstract: The western equatorial Pacific has an important role in the El Nino/Southern Oscillation climate cycle, and sea-surface temperature anomalies in this area are known to affect the global atmospheric circulation. Little is known, however, about the factors that control the response of the upper ocean, and hence sea surface temperature, to changes in atmospheric conditions. Here we present observations of the density and current structure of the upper layers of the western equatorial Pacific Ocean. We observed frontal features with temperature differences of 1-degrees-C extending from the surface down to a depth of 150 m, a strong salinity front associated with a surface convergence, and interleaving of high- and low-salinity waters in 10-m-thick layers that extended for several hundred kilometres in the north-south direction. The large spatial and temporal variability in temperature and salinity observed must influence heat transport, and hence the ocean response to atmospheric forcing (in particular the response to westerly wind bursts, which are thought to be precursors to an El Nino). We estimate the horizontal diffusion coefficient of the interleaving, which indicates that the observed layering may contribute to horizontal mixing within the thermocline.

赤道西太平洋上层海水结构

Richards K. J., Pollard R. T.

(英国南安普顿大学海洋科学系)

摘要:赤道西太平洋地区是厄尔尼诺/南方涛动气候循环中一个重要的环节,这一海域的海水表面温度异常能对全球大气环流造成影响。鲜为人知的是,到底是什么因素控制海洋上层的反应,而改变海洋表面温度和大气条件的变化,在这里,我们报道西部赤道太平洋上层海水的密度和流场结构的观察结果。我们观察到 1 ℃的温度差的前沿,从表面向下一直延伸到 150 m 水深,具有很强的高盐度锋面结合表层的收敛面,表层 10 m 的高盐与低盐水团在南北方向上交错延伸达到数百公里。温度和盐度的巨大的空间和时间差异肯定会影响热量传递,而显示为海洋对大气强迫的响应(尤其是对西风暴发,这被认为是厄尔尼诺现象前兆响应)。我们估计这种交互模式的水平扩散系数,结果表明,所观察到的分层可能有助于在温跃层混合水平。

8. Ancel A, Kooyman G L, Ponganis P J, et al. Foraging behaviour of emperor penguins as a resource detector in winter and summer[J]. Nature, 1992, 360: 336-339.

Foraging behaviour of emperor penguins as a resource detector in winter and summer

Ancel A., Kooyman G. L., Ponganis P. J., Gendner J. P., Lignon J., Mestre X., Huin N., Thorson P. H., Robisson P., Lemaho Y.

(Centre d'Ecologie et Physiologie Energétiqu. es, Centre National de la Recherche Scientifique, 23 rue Becquerel, 67087 Stras-bourg, France)

Abstract: The emperor penguin (*Aptenodytes forsteri*), which feeds only at sea, is restricted to the higher latitudes of the antarctic sea-ice habitat. It breeds on the winter fast ice when temperatures are −30-degrees-C and high winds are frequent. Assuming entirely the task of incubating the single egg, the male fasts for about 120 days in the most severe conditions. When it is relieved by the female around hatching time, the distance between the colony and the open sea may be 100 km or more, but where emperors go to forage at that time or during the summer is unknown. The polynias are areas of open water in sea-ice and during winter, with the under-ice habitats at any time of the year, they are among the most difficult of all Antarctic areas to sample. Here we monitor by satellite the routes taken by emperor penguins for foraging and compare them with satellite images of sea-ice. Winter birds walking over fast ice travelled up to 296 km to feed in polynias, whereas those swimming in light pack-ice travelled as far as 895 km from the breeding colony. One record of diving showed that although most dives are to mid-water depths, some are near the bottom. Obtaining such detailed information on foraging in emperor penguins means that this bird now offers a unique opportunity to investigate the Antarctic sea-ice habitat.

帝企鹅的觅食行为——冬夏季节的资源探测器

Ancel A., Kooyman G. L., Ponganis P. J., Gendner J. P., Lignon J., Mestre X., Huin N., Thorson P. H., Robisson P., Lemaho Y.

(法国国家科学研究中心)

摘要：帝企鹅(*Aptenodytes forsteri*)，在海中捕食，仅生活在高纬度南极海冰区。在温度－30 ℃和强风盛行的冬季，它们都能在冰层上哺育后代。假设完全只进行单卵孵化，雄性企鹅在大约120天的最严峻的气候条件下独自完成孵化任务。当接近孵化时，雄性企鹅被雌性企鹅接替，其孵卵地点与海洋的距离可能超过100公里。但是这时，帝企鹅在哪觅食和夏季的觅食地都是未知的。冰穴指海冰中的开放水域；冰穴在冬季，以及冰下的栖息地在一年的任何时间，都是南极地区采样最困难的地方。在这里，我们通过卫星监测帝企鹅的觅食路线，并与海冰的卫星图像比较。冬季鸟在紧密冰上需步行296公里到冰穴去觅食，而那些较松散的冰上，需游895公里。一个潜水记录显示，大多数帝企鹅潜水至中水深处，一些可达底部。这些关于帝企鹅在觅食的详细信息，提供了一个独特的角度，来探讨南极海冰栖息地。

9. Bindoff N L, Church J A. Warming of the water column in the Southwest Pacific-Ocean[J]. Nature, 1992, 357: 59-62.

Warming of the water column in the Southwest Pacific-Ocean

Bindoff N. L., Church J. A.

(CSIRO Division of Oceanography, and Co-operative Research Center for Antarctic and Southern Ocean Studies, GPO Box 1538 Hobart, Tasmania 7001, Australia)

Abstract: The response of the deep ocean to long-period temperature variations at the ocean surface is a crucial issue in understanding climate change. There are, however, very few observations available for studying changes in the thermal structure of ocean interiors. On the basis of measurements made 22 years apart of full-depth temperature sections in the Pacific Ocean between Australia and New Zealand, we show here that there has been a depth-averaged warming of 0. 04-degrees-C and 0. 03-degrees-C at 43-degrees-S and 28-degrees-S, respectively, throughout most of the water column below the mixed layer. The sea-level rise caused by expansion between a depth of 300 m and the ocean floor is 2-3 cm, consistent with the observed rate of global sea-level rise. In the main thermocline there is a coherent cooling and freshening on density surfaces, consistent with surface warming in the Southern Ocean where these waters originate. Similar observations in the North Atlantic show comparable changes in the thermal structure and water-mass volumes, but further measurements in other regions are required before firm conclusions can be drawn about the global significance of these changes.

西南太平洋水体变暖

Bindoff N. L. , Church J. A.

(澳大利亚联邦科学与工业研究组织)

摘要：深海对长阶段海洋表面温度变化的响应是理解气候变化的一个关键问题。然而用于研究海洋内部热结构变化的观测数据很少。我们对位于澳大利亚和新西兰之间太平洋全深度温度系列观测了22年。我们在这里报道，在43°S和28°S混合层以下的水柱分别有平均0.04 ℃和0.03 ℃的升温。在深度300 m和洋底之间的升温导致的扩张造成海平面上升2～3 cm，这与可观测到的全球海平面上升的速率是一致的。在主温跃层有一个连贯的冷却和新鲜的密度表面，与这些水的来源——南大洋表面温度上升一致。在北大西洋热结构和水质量有类似的变化，但在得出关于这些变化的全球意义结论之前，需要进一步观测其他地区。

10. Broecker W S, Peng T H. Interhemispheric transport of carbon-dioxide by ocean circulation[J]. Nature, 1992, 356: 587-589.

Interhemispheric transport of carbon-dioxide by ocean circulation

Broecker W. S. , Peng T. H.

(Lamont-Doherty Geological Observatory of Columbia University, Palisades, New York 10964, USA)

Abstract: Although anthropogenic emissions of carbon dioxide have today created a greater atmospheric CO_2 concentration in the Northern than in the Southern Hemisphere, a comparison of interhemispheric CO_2 profiles from 1980 and 1962 led Keeling and Heimann to conclude that, before the Industrial Revolution, natural CO_2 sources and sinks acted to set up a reverse (south to north) gradient which drove about one gigatonne of carbon each year through the atmosphere from the Southern to the Northern Hemisphere. At steady state, this flux must have been balanced by a counter flow of carbon from north to south through the ocean. Here we present a means to estimate this natural flux by a separation of oceanic carbon anomalies into those created by biogenic processes and those created by CO_2 exchange between the ocean and atmosphere. We find that before the Industrial Revolution, deep water formed in the northern Atlantic Ocean carried about 0. 6 gigatonnes of carbon annually to the Southern Hemisphere, providing support for Keeling and Heimann's proposal. The existence of this oceanic carbon pump also raises questions about the need for a large terrestrial carbon sink in the Northern Hemisphere, as postulated by Tans et al to balance the present global carbon budget.

二氧化碳通过洋流在南北半球之间传送

Broecker W. S. , Peng T. H.

(美国哥伦比亚大学拉蒙特-多尔蒂地质观测站)

摘要：虽然现今人为排放的二氧化碳在北半球的浓度远远高于南半球。在1980年和1962年对南北半球二氧化碳浓度进行了比较后，基林和海曼得出这样的结论：在工业革命以前，每年自然二氧化碳的源和汇构成了反向(南向北)梯度，使得约10亿吨的碳从南半球大气输送到北半球大气中。在稳定状态下，这种通量必须由一个反向的由北到南的过程经过海洋得到平衡。在这里，我们提出一个方法来估计由海洋碳异常到由生物过程创建的，并通过海洋和大气之间的二氧化碳交换所建立的这种自然流量。我们发现，在工业革命以前，深层水在北大西洋将约6亿吨的碳向南半球输送，为基林和海曼的理论提供了证据。这个海洋碳泵的存在也使人们意识到北半球陆地必然存在一个巨大的碳汇去平衡现在的全球碳循环，就如Tans等的假设。

11. Frew R D, Hunter K A. Influence of Southern-Ocean waters on the cadmium phosphate properties of the global ocean[J]. Nature, 1992, 360: 144-146.

Influence of Southern-Ocean waters on the cadmium phosphate properties of the global ocean

Frew R. D., Hunter K. A.

(Department of Chemistry, University of Otago, Dunedin, New Zealand)

Abstract: Measurements of the cadmium content of marine $CaCO_3$ deposits, such as aragonitic corals and foraminiferal shells, have been used to trace past ocean circulation and as an indicator of labile nutrient concentrations. In particular, studies of foraminiferal cadmium have provided insight into the oceanic processes controlling atmospheric CO_2 concentration in glacial times. The use of Cd/Ca ratios in $CaCO_3$ as a labile nutrient indicator is made possible by a relationship between seawater cadmium and phosphate content that is remarkably uniform throughout the present ocean. In deep waters of the open ocean, [Cd] at a given value of [PO_4^{3-}] is consistent within about +/−7%; However, there is a unusual kink in the relationship at [PO_4^{3-}] almost-equal-to 1.3 μmol • kg^{-1} whose cause is not known but which has been suggested to result from a slightly deeper regeneration cycle for Cd relative to PO_4^{3-}. Points to the right of the kink correspond mainly to intermediate and deep waters of the North Pacific, whereas low PO_4^{3-} concentrations to the left of the kink represent mainly Atlantic and upper-ocean waters. Waters of the Southern Ocean have a strong influence on the composition of the global ocean but few Cd-PO_4^{3-} measurements are available for this region. Here we present recent data for Cd and PO_4^{3-} in Southern Ocean waters, which suggest that the kink is created by the input of Cd-depleted Subantarctic waters into the intermediate waters of the global ocean.

南部大洋海水对全球海洋的镉磷酸盐属性的影响

Frew R. D., Hunter K. A.

(新西兰奥塔哥大学化学系)

摘要：海洋沉积中的碳酸钙(如珊瑚文石和有孔虫壳体)中的Cd被用来跟踪过去洋流和作为一个不稳定的营养盐浓度的指标。特别是，有孔虫镉的研究有助于人们对在冰期控制大气中二氧化碳浓度其海洋过程的理解。碳酸钙中的Cd/Ca比率作为不稳定的营养指标成为可能的前提是，当前海水中镉、磷之间的关系在整个海洋中是一致的。在开放的海洋深层水，[Cd]在一个给定的[PO_4^{3-}]下，其波动在±7%之间。但是，当[PO_4^{3-}]浓度在1.3 μmol • kg^{-1}时，这个关系有一个例外点，其原因尚不清楚，一种观点认

为相对于 PO_4^{3-}，Cd 再生周期要更深些。例外点的右边主要对应于北太平洋中深水域，例外点左边的低 PO_4^{3-} 浓度主要代表大西洋和上层海洋。南大洋海域对全球海洋的组成具有很大的影响力，但很少有本地区 Cd-PO_4^{3-} 测量。在这里，我们发表南大洋海域 Cd 和 PO_4^{3-} 数据，表明这个例外点是由于亚南极水域镉缺损的海水输入到全球中层海水所导致的。

12. Martin J H, Fitzwater S E. Dissolved organic-carbon in the Atlantic, Southern and Pacific Oceans [J]. Nature, 1992, 356: 699-700.

Dissolved organic-carbon in the Atlantic, Southern and Pacific Oceans

Martin J. H., Fitzwater S. E.

(Moss Landing Marine Labs, California City, California, United States)

Abstract: The amount of dissolved organic carbon (DOC) in sea water is controversial. Using a high-temperature catalytic oxidation (HTCO) technique, Sugimura and Suzuki reported that surface waters contained 2-4 times as much DOC as that measured previously using wet chemistry and ultraviolet oxidation techniques. They also observed a relationship between DOC content and apparent oxygen utilization suggesting that the consumption of DOC is responsible for oxygen depletion in the deep sea. How to reconcile the apparent differences between these techniques has not been clear. Here we provide independent confirmation of the findings of Sugimura and Suzuki. We collected surface and deep waters from the equatorial Pacific Ocean, the Drake passage and the Atlantic Ocean south of Iceland, and analysed their DOC content using the HTCO methodology. We found DOC concentrations 2-3 times higher than those measured previously. These results imply that the carbon content of the oceans has previously been underestimated by 10^{12} (1000 billion) tonnes, and that the new estimated total of 1800 billion tonnes represents one of the largest carbon reservoirs on Earth. We found no evidence of a cause-and-effect relationship between DOC and apparent oxygen utilization.

大西洋、南大洋和太平洋中的溶解有机碳

Martin J. H., Fitzwater S. E.

(美国莫斯兰丁海洋实验室)

摘要：海水中溶解的有机碳(DOC)的量是有争议的。使用一个高温催化氧化技术(HTCO)，杉村和铃木证明表层水包含的 DOC 是采用湿化学和紫外线氧化技术测量的 2～4 倍。他们还观察到 DOC 含量和明显的氧利用率之间的关系，暗示了 DOC 的消耗是深海氧匮乏的原因。如何协调这些技术之间的明显差异，一直没有明确。在这里，我们提供对 Sugimura 和 Suzuki 结果的独立确认。我们在赤道太平洋、德雷克海峡和大西洋的冰岛南部表层和深层水域，用 HTCO 方法分析他们的 DOC 含量，发现 DOC 浓度比以前测量的高出 2～3 倍。这些结果意味着海洋的碳含量先前被低估了 10^{12} 吨，新估计的 1.8×10^{12} 吨代表地球上最大的碳库之一。我们没有发现一个 DOC 和明显的氧利用率之间的原因和因果关系的证据。

13. Morgan J P, Smith W H F. Flattening of the sea-floor depth age curve as a response to asthenospheric flow[J]. Nature, 1992, 359: 524-527.

Flattening of the sea-floor depth age curve as a response to asthenospheric flow

Morgan J. P., Smith W. H. F.

(Institute of Geophysics and Planetary Physics, Scripps Institution of Oceanography, 9500 Oilman Drive, La Jolla, California 92093-0225, USA)

Abstract: The flattening of sea-floor depths from the square-root age dependence predicted by considering the cooling plate as a growing thermal boundary layer is a fundamental constraint on the evolution of oceanic lithosphere. Previous explanations for the flattening have included reheating from convective instabilities that form beneath lithosphere older than approximately 80 Myr, thermal rejuvenation of lithosphere that passes over stationary hotspots, or a whole-mantle flow forced by two converging plates. We suggest here that the flattening of old ocean floors can perhaps best be explained as a dynamic phenomenon reflecting flow in asthenosphere underlying the oceanic lithosphere. Applying the model to the Pacific plate, we show that a solution for flow in an asthenosphere low-viscosity channel which is "consumed" by plate accretion and the subduction of lithosphere at trenches, and replenished by near-ridge upwelling and near-ridge hotspots, generates the observed approximately 1 km of sea-floor flattening for asthenospheric viscosities of 2×10^{18} Pa • s and an absolute Pacific plate motion of 100 mm • yr^{-1}. Our model can also explain the asymmetric subsidence of the South American and African plates away from the Mid-Atlantic Ridge.

海底深度的年龄曲线的扁平化作为对一个软流圈流动的响应

Morgan J. P., Smith W. H. F.

(美国斯克里普斯海洋研究所地球和行星物理研究所)

摘要:将一个冷却板块考虑为一个成长的热边界层会预测到海底深度对年龄平方根依赖性的扁平化,它是对海洋地壳演变的根本制约。之前,对扁平化的解释包括老于8000万年岩石圈下部的对流不稳定性产生的热,穿过固定热点的岩石圈的热复兴,或由于两个板块汇合造成的整个地幔流的不稳定。我们在这里提出,老的洋底的扁平化或许可以解释为一个动态的现象,反映大洋岩石圈之下软流圈的流动。软流圈低黏性的流通途径的流动为吸积盘和在沟壕里的岩石圈俯冲中所"消耗",由附近的洋脊上涌和附近的洋脊热点所补充。将这个模型应用到太平洋板块,我们对软流圈的流动,找到一个解;这个解产生了所观察到的具 2×10^{18} Pa • s 黏度的软流圈和每年 100 mm 移动速度的绝对太平洋板块大约 1 公里的扁平化。我们的模型也可以解释离大西洋中脊的南美和非洲板块的不对称沉降。

14. Morrow R, Church J, Coleman R, et al. Eddy momentum flux and its contribution to the Southern-Ocean momentum balance[J]. Nature, 1992, 357: 482-484.

Eddy momentum flux and its contribution to the Southern-Ocean momentum balance

Morrow R., Church J., Coleman R., Chelton D., White N.

(Marine Studies Centre, University of Sydney, New South Wales 2006, Australia)

Abstract: A large amount of momentum is transferred to the Southern Ocean by strong westerly winds. Analytical and numerical models have suggested that transient eddies may be important in transporting this momentum away from the region of wind forcing, either horizontally or vertically downwards where it is balanced by bottom topographic drag. There are, however, few long-term in situ observations of horizontal eddy momentum flux, and no large-scale measurements of vertical eddy fluxes, to test these models. As a result, the momentum balance of the Antarctic circumpolar current (ACC) remains uncertain, and the role of eddies controversial. Here we use Geosat satellite altimeter data to resolve directional eddy kinetic energy and horizontal eddy momentum flux in the ACC on fine spatial and temporal scales. The complex spatial distribution of surface eddy momentum flux is strongly influenced by bottom topography. The horizontal eddy momentum flux tends generally to concentrate the mean flow, although some regions of divergence are observed. Our results show that the zonally averaged horizontal eddy momentum flux from transient eddies is an order of magnitude too small, and in the wrong direction to directly balance the eastward momentum input from wind.

涡流动量通量及其对南大洋的动量平衡的贡献

Morrow R., Church J., Coleman R., Chelton D., White N.

(澳大利亚悉尼大学海洋研究中心)

摘要：大量的动量被强西风转移到南大洋。分析和数字模型提出瞬态涡旋在将这一动量从风动力区域水平转移出去，或垂直向下运输至大洋底部被地形拖力所平衡的过程中可能起重要作用。然而，对水平涡流动量通量并没有几个长期的实地观测；对垂直涡流通量则没有大规模的测量来检验这些模型。因此，南极绕极流(ACC)的动量平衡仍然不确定，漩涡的角色有争议。在这里，我们使用 Geosat 卫星测高数据，在精细的空间和时间尺度上测定 ACC 的定向涡流动能和水平涡流动量通量。表面涡流动量通量的复杂的空间分布受到海底地形的强烈影响。尽管观测到一些地区的分歧，一般水平涡流动量通量趋于集中的平均流量。我们的研究结果表明，纬向平均水平涡流瞬时涡动量通量幅度太小，而且在错误的方向，不足以直接平衡来自于风的东向动量输入。

15. Oerter H, Kipfstuhl J, Determann J, et al. Evidence for basal marine ice in the Filchner-Ronne ice shelf[J]. Nature, 1992, 358: 399-401.

Evidence for basal marine ice in the Filchner-Ronne ice shelf

Oerter H., Kipfstuhl J., Determann J., Miller H., Wagenbach D., Minikin A., Graf W.

(Alfred-Wegener-Institut für Polar-und Meeresforschung, Postfach 120161, D-W2850 Bremerhaven, Germany)

Abstract: The Filchner-Ronne ice shelf, which drains most of the marine-based portions of the West Antarctic ice sheet, is the largest ice shelf on Earth by volume. The origin and properties of the ice that constitutes this shelf are poorly understood, because a strong reflecting interface within the ice and the diffuse nature of the ice-ocean interface make seismic and radio echo sounding data difficult to interpret. Ice in the upper part of the shelf is of meteoric origin, but it has been proposed that a basal layer of saline ice accumulates from below. Here we present the results of an analysis of the physical and chemical characteristics of an ice core drilled almost to the bottom of the Ronne ice shelf. We observe a change in ice properties at about 150 m depth, which we ascribe to a change from meteoric ice to basal marine ice. The basal ice is very different from sea ice formed at the ocean surface, and we propose a formation mechanism in which ice platelets in the water column accrete to the bottom of the ice shelf.

在 Filchner-Ronne 冰架上基底海洋冰存在的证据

Oerter H., Kipfstuhl J., Determann J., Miller H., Wagenbach D., Minikin A., Graf W.

(德国阿尔弗雷德魏格纳极地海洋研究所)

摘要:Filchner-Ronne 冰架,排除南极西部冰盖海洋部分的大多数,是地球上以体积计算的最大的冰盖。对这个冰盖的起源和构成的属性,人们知之甚少,这是因为冰内界面的强烈反射以及冰-海洋界面的弥漫性使地震和无线电回声测深数据很难解释。冰盖上部的冰是大气的起源,但有证据指出它下面的基底层是由盐水冰积累形成的。在这里,我们报道了几乎达到 Ronne 冰架底部钻冰芯的物理和化学特性的分析结果。我们观察到在深约 150 m 处的冰属性的变化,我们认为是大气冰和海洋冰的交界。基底冰与在海洋表面形成的冰有很大不同,我们提出了基底冰是由水柱中的冰盘积聚在大气冰底部而形成的。

16. Cole J E, Fairbanks R G, Shen G T. Recent variability in the Southern Oscillation—isotopic results from a Tarawa atoll coral[J]. Science, 1993, 260: 1790-1793.

Recent variability in the Southern Oscillation—isotopic results from a Tarawa atoll coral

Cole J. E., Fairbanks R. G., Shen G. T.

(Lamont-Doherty Earth Observatory, Palisades, NY 10964 and Department of Geological Science, Columbia University, NewYork, NY10027)

Abstract: In the western tropical Pacific, the interannual migration of the Indonesian Low convective system causes changes in rainfall that dominate the regional signature of the El Nino-

Southern Oscillation (ENSO) system. A 96-year oxygen isotope record from a Tarawa Atoll coral (1-degrees-N, 172-degrees-E) reflects regional convective activity through rainfall-induced salinity changes. This monthly resolution record spans twice the length of the local climatological record and provides a history of ENSO variability comparable in quality with those derived from instrumental climate data. Comparison of this coral record with a historical chronology of El Nino events indicates that climate anomalies in coastal South America are occasionally decoupled from Pacific-wide ENSO extremes. Spectral analysis suggests that the distribution of variance in this record has shifted among annual to interannual periods during the present century, concurrent with observed changes in the strength of the Southern Oscillation.

南方涛动的近期变化——来自塔拉瓦环礁珊瑚的同位素记录

Cole J. E., Fairbanks R. G., Shen G. T.

(美国拉蒙特-多尔蒂地质观测站和哥伦比亚大学地质科学系)

摘要：在热带西太平洋，印度尼西亚的低对流系统的年际迁移导致了主导厄尔尼诺-南方涛动(ENSO)系统区域变化特性的降水变化。一个96年的来自塔拉瓦环礁珊瑚(1°N，172°E)氧同位素记录反映通过降雨诱发的盐度变化的区域对流活动。这种在月分辨率的记录时间是当地气候记录时间的两倍，并提供与来自仪器气候数据相媲美的ENSO变化的历史。这个珊瑚记录与厄尔尼诺事件的历史年表的比较表明，在南美沿海的气候异常偶尔与泛太平洋ENSO的极端条件分离。谱分析表明，在21世纪期间，记录的方差分布在年际之间转换，与观察到的南方涛动的强度一致。

17. Gordon A L, Huber B A, Hellmer H H, et al. Deep and bottom water of the Weddell Seas western rim[J]. Science, 1993, 262: 95-97.

Deep and bottom water of the Weddell Seas western rim

Gordon A. L., Huber B. A., Hellmer H. H., Ffield A.

(Lamont-Doherty Earth Observatory, Palisades, NY 10964)

Abstract: Oceanographic observations from the Ice Station Weddell show that the western rim of the Weddell Gyre contributes to Weddell Sea Bottom Water. A thin (<300 meters), highly oxygenated benthic layer is composed of a low-salinity type of bottom water overlying a high-salinity component. This complex layering disappears near 66°S because of vertical mixing and further inflow from the continental margin. The bottom water flowing out of the western rim is a blend of the two types. Additionally, the data show that a narrow band of warmer Weddell Deep Water hugged the continental margin as it flowed into the western rim, providing the continental margin with the salt required for bottom-water production.

威德尔海西部边缘的底层水

Gordon A. L., Huber B. A., Hellmer H. H., Ffield A.

（美国拉蒙特-多尔蒂地质观测站）

摘要：威德尔海冰站海洋研究观测表明，威德尔海西部边缘的环流对威德尔海底水有补偿作用。薄（小于 300 m）而高含氧的底栖层是由低盐的底水和高盐的上覆水组成的。这种复杂的层次在接近南纬 66 度消失了，这与从大陆边缘垂直混合以及直接输入的水有关。底层水流淌的西部边缘是两种类型的混合。此外，数据显示，一个温暖的威德尔海深水窄带在流入西部边缘时在陆缘汇和，提供大陆边底水生产所需的盐度。

18. Kumar N, Gwiazda R, Anderson R F, et al. $^{231}Pa/^{230}Th$ ratios in sediments as a proxy for past changes in Southern-Ocean productivity[J]. Nature, 1993, 362: 45-48.

$^{231}Pa/^{230}Th$ ratios in sediments as a proxy for past changes in Southern-Ocean productivity

Kumar N., Gwiazda R., Anderson R. F., Froelich P. N.

(Lamont-Doherty Earth Observatory of Columbia University, Palisades, New York 10964, USA)

Abstract: The biological productivity of the oceans is sensitive to changes in climate, which can affect essential factors such as nutrient and light availability. In turn, ocean productivity may influence climate by regulating the partitioning of carbon dioxide, a greenhouse gas, between the ocean and the atmosphere. Investigators have attempted to link variations in atmospheric CO_2 content, recorded in ice cores, to the productivity of the Southern Ocean, but an unambiguous means of assessing past changes in ocean productivity has been lacking. Here we exploit established relationships between $^{231}Pa/^{230}Th$ ratios and particle flux to infer, from the analysis of dated sediment cores, variability through time of fluxes of particulate biogenic material exported from surface waters. Records from two cores in the Atlantic sector of the Southern Ocean indicate that ocean productivity during glacial periods was lower than at present south of the Antarctic polar front, and support earlier conclusions that the zone of maximum productivity migrated northwards during glacial conditions. Although further work at other sites is needed for an assessment of changes in total Antarctic productivity, our technique has the potential to provide this information while avoiding some of the limitations of other productivity proxies.

沉积物中 $^{231}Pa/^{230}Th$ 的比率作为过去南大洋生产力变化的指标

Kumar N., Gwiazda R., Anderson R. F., Froelich P. N.

（美国哥伦比亚大学拉蒙特-多尔蒂地质观测站）

摘要：气候变化可以影响一些基本因素，例如营养和见光度，所以海洋生产力对气候变化是敏感的。

反过来，海洋生产力可以通过调节大气和海洋中二氧化碳的比例来影响气候。研究人员曾试图把在冰芯中记录的大气二氧化碳含量的变化和南部海洋中生产力的变化联系起来，但缺乏一个明确的手段来评估过去海洋生产力的改变。在这里，我们利用$^{231}Pa/^{230}Th$比率和颗粒通量之间建立关系，从沉积柱的定年分析，来推断表层水输出的生物来源的微小物质随着时间的变化。南部海洋在大西洋区域的沉积记录显示冰川期间海洋生产力低于现在南极南部边缘海洋生产力，并支持早期的结论：海洋生产力的极大区域在冰川期间向北迁移。尽管需要在其他地点对北极总生产力做进一步评估，我们的技术有能力提供这些信息，同时避免其他生产力指标的一些局限性。

19. Sager W W, Han H C. Rapid formation of the Shatsky Rise Oceanic plateau inferred from its magnetic anomaly[J]. Nature, 1993, 364: 610-613.

Rapid formation of the Shatsky Rise Oceanic plateau inferred from its magnetic anomaly

Sager W. W., Han H. C.

(Departments of Oceanography and Geophysics, Texas A&M University, College Station, Texas 77843, USA)

Abstract: Shatsky Rise, in the northwest Pacific Ocean, is probably the oldest extant oceanic plateau, and as with most such features, its origin is uncertain. Both oceanic plateaus and continental flood basalts are thought to be formed by rapid, voluminous eruptions that occur when the "head" of a newly born mantle plume ascends to the base of the lithosphere. High eruption rates have been estimated for flood basalts (for example, 1.5 $km^3 \cdot yr^{-1}$ for the Deccan Traps) from dating of lava flows, but the inaccessibility of oceanic plateaus makes it necessary to extrapolate dating information from a small number of samples and sites. Here we estimate the eruption rate of Shatsky Rise by a method that is indirect, but has the virtue of "sampling" the entire volume of the plateau above the surrounding sea floor. The main, southern part of the plateau has a positive magnetic anomaly, corresponding to a reversed geomagnetic polarity at the time of eruption. Using age constraints to identify the longest period of reversed polarity during which the plateau could have formed, we estimate that 2×10^6 km^3 of material erupted at a minimum rate of 1.7 $km^3 \cdot yr^{-1}$. This is somewhat less than the rate of 8-22 $km^3 \cdot yr^{-1}$ estimated for the Ontong-Java Plateau, but still represents a massive eruption, consistent with the plume-head hypothesis.

从磁异常推断 Shatsky Rise 海洋高原的迅速形成

Sager W. W., Han H. C.

(美国得克萨斯大学海洋和地球物理学系)

摘要：Shatsky Rise，在西北太平洋，可能是现存最古老的大洋高原，和大多数具有这样特征的高原类似，它的起源是不确定的。大洋高原和大陆溢流玄武岩被认为是因迅速的大量喷发形成的，大多数的喷发发生在一个刚出生的地幔柱的“头部”上升到地壳的底部。玄武岩流量的高喷发率已经用来通过岩浆定年来估计(例如，1.5立方公里每年的德干高原)，但是由于海洋高原的不可接近，所以通过少量的样品和地点来外推是有必要的。在这里，我们是用间接的方法来估计 Shatsky Rise 的喷发率，对高于周围海底的整个高原区域取样。重要的是，高原南部的部分有一个显著的磁异常，与喷发期间地磁的扭转一致。

使用年龄限制来确定反极性的最长期限，在此期间，可能已经形成了高原，我们估计，2×10^6 立方公里的材料喷发，具有 1.7 立方公里每年的最低喷发率。这个比 Ontong-Java 高原 8～22 立方公里每年的喷发速率低，但仍然代表着一个大规模的火山喷发，与地幔柱顶假说相一致。

20. Siegenthaler U, Sarmiento J L. Atmospheric carbon-dioxide and the ocean[J]. Nature, 1993, 365: 119-125.

Atmospheric carbon-dioxide and the ocean

Siegenthaler U., Sarmiento J. L.

(Physics Institute, University of Bern, 3012 Bern, Switzerland)

Abstract: The ocean is a significant sink for anthropogenic carbon dioxide, taking up about a third of the emissions arising from fossil-fuel use and tropical deforestation. Increases in the atmospheric carbon dioxide concentration account for most of the remaining emissions, but there still appears to be a "missing sink" which may be located in the terrestrial biosphere.

大气中的二氧化碳和海洋

Siegenthaler U., Sarmiento J. L.

（瑞士伯尔尼大学物理研究所）

摘要：海洋是一个显著的人为来源的二氧化碳的汇，约占使用化石燃料和砍伐热带森林所产生的排放量的三分之一。大部分剩余的排放导致了在大气中的二氧化碳浓度的增加，但似乎仍然有一个"消失的沉降"，可能位于陆地生物圈。

21. Sullivan C W, Arrigo K R, Mcclain C R, et al. Distributions of phytoplankton blooms in the Southern-Ocean[J]. Science, 1993, 262: 1832-1837.

Distributions of phytoplankton blooms in the Southern-Ocean

Sullivan C. W., Arrigo K. R., Mcclain C. R., Comiso J. C., Firestone J.

(Graduate Program in Ocean Science, Hancock Institute for Marine Studies, University of South California, Los Angeles, CA 900890373)

Abstract: A regional pigment retrieval algorithm for the Nimbus Coastal Zone Color Scanner (CZCS) has been tested for the Southern Ocean. The pigment concentrations estimated with this algorithm agree to within 5 percent with in situ values and are more than twice as high as those previously reported. The CZCS data also revealed an asymmetric distribution of enhanced pigments in the waters surrounding Antarctica; in contrast, most surface geophysical properties are symmetrically distributed. The asymmetry is coherent with circumpolar current patterns and the availability of silicic acid in surface waters. Intense blooms (> 1 milligram of pigment per cubic meter) that occur downcurrent from continental masses result from dissolved trace elements such as iron derived from shelf sediments and glacial melt.

南大洋大量浮游植物的分布

Sullivan C. W., Arrigo K. R., Mcclain C. R., Comiso J. C., Firestone J.

(美国南加州大学 Hancock 海洋研究所)

摘要：用南大海(洋)测试海岸带彩色扫描仪(CZCS)的区域性色素恢复算法。这种色素浓度估计算法的结果与原位值相比误差在5%以内，比以前报道的结果高两倍以上。CZCS的数据还显示增强色素在南极周围水域的不对称分布；相比之下，最表面的地球物理属性是对称分布的。不对称与环流模式和在地表水中的硅酸供应协调一致。来源于陆地物质的剧烈增加(大于 1 mg · m^{-3} 色素)是由于从地壳和冰川融化物中产生了示踪元素，例如铁。

22. Fritsen C H, Lytle V I, Ackley S F, et al. Autumn bloom of Antarctic Pack-Ice algae[J]. Science, 1994, 266: 782-784.

Autumn bloom of Antarctic Pack-Ice algae

Fritsen C. H., Lytle V. I., Ackley S. F., Sullivan C. W.

(Department of Biological Sciences, University of Southern California, Los Angeles, CA 90089-0371, USA)

Abstract: An autumn bloom of sea-ice algae was observed from February to June of 1992 within the upper 0.4 meter of multiyear ice in the Western Weddell Sea, Antarctica. The bloom was reliant on the freezing of porous areas within the ice that initiated a vertical exchange of nutrient-depleted brine with nutrient-rich seawater. This replenishment of nutrients to the algal community allowed the net production of 1760 milligrams of carbon and 200 milligrams of nitrogen per square meter of ice. The location of this autumn bloom is unlike that of spring blooms previously observed in both polar regions.

南极浮冰藻在秋季爆发性增长

Fritsen C. H., Lytle V. I., Ackley S. F., Sullivan C. W.

(美国南加州大学生物科学系)

摘要：西南极洲威德尔海多年冰表层 0.4 m 中的海冰藻类在 1992 年从 2 月至 6 月的爆发被观测。爆发是由于多孔结构的冰冻结，而这个冻结过程引发了贫营养和富营养海水的垂直交换作用。这种对藻类群落的营养物质的补充，使得每平方米冰可以容纳 1760 mg 碳和 200 mg 氮的净生产力。今年秋天爆发位置不同于先前在两极地区观测到的春天位置。

23. Jacobs G A, Hurlburt H E, Kindle J C, et al. Decade-scale trans-Pacific propagation and warming effects of an El-Nino anomaly[J]. Nature, 1994, 370: 360-363.

Decade-scale trans-Pacific propagation and warming effects of an El-Nino anomaly

Jacobs G. A., Hurlburt H. E., Kindle J. C., Metzger E. J., Mitchell J. L., Teague W. J., Wallcraft A. J.

(Naval Research Laboratory, Stennis Space Center, Mississippi 39529, USA)

Abstract: El Nino events in the Pacific Ocean can have significant local effects lasting up to two years. For example the 1982-83 El Nino caused increases in the sea-surface height and temperature at the coasts of Ecuador and Peru, with important consequences for fish populations and local rainfall. But it has been believed that the long-range effects of El Nino events are restricted to changes transmitted through the atmosphere, for example causing precipitation anomalies over the Sahel. Here we present evidence from modelling and observations that planetary-scale oceanic waves, generated by reflection of equatorial shallow-water waves from the American coasts during the 1982-83 El Nino, have crossed the North Pacific and a decade later caused northward re-routing of the Kuroshio Extension-a strong current that normally advects large amounts of heat from the southern coast of Japan eastwards into the mid-latitude Pacific. This has led to significant increases in sea surface temperature at high latitudes in the northwestern Pacific, of the same amplitude and with the same spatial extent as those seen in the tropics during important El Nino events. These changes may have influenced weather patterns over the North American continent during the past decade, and demonstrate that the oceanic effects of El Nino events can be extremely long-lived.

一次厄尔尼诺异常对十年尺度的跨太平洋传播和气候变暖的影响

Jacobs G. A., Hurlburt H. E., Kindle J. C., Metzger E. J., Mitchell J. L., Teague W. J., Wallcraft A. J.

(美国密西西比州斯坦尼斯空间中心海军研究实验室)

摘要:在太平洋的厄尔尼诺事件可以有显著的长达两年的局部效应。例如1982～1983年的厄尔尼诺现象造成了在厄瓜多尔和秘鲁海岸海面高度和温度的增加,对鱼类种群和当地的降雨产生重要影响。但人们一直认为,由于大气传输过程中的变化,厄尔尼诺事件的远距离影响受到限制,例如在萨赫勒地区造成的降水异常。在这里,我们报道模拟和观测结果。1982～1983年厄尔尼诺期间的美国海岸浅水波的反射形成的海洋波浪,越过北太平洋并在十年后造成黑潮延伸流的重新北上。黑潮延伸流是强大的洋流,通常从日本南部海岸向东进入太平洋中纬度传输大量的热量。这导致了在西北太平洋高纬度地区海表面温度的显著上升,与在热带地区重要的厄尔尼诺事件具有相同的上升幅度和空间范围。这些变化可能会影响过去十年中北美大陆的天气模式,并表明,厄尔尼诺事件对海洋影响可以非常长。

24. Jin F F, Neelin J D, Ghil M. El-Nino on the devils staircase: annual subharmonic steps to chaos[J]. Science, 1994, 264: 70-72.

El-Nino on the devils staircase—annual subharmonic steps to chaos

Jin F. F. , Neelin J. D. , Ghil M.

(University of Southern California Department of Atmospheric Sciences, University of California at Los Angeles, Los Angeles, Ca90024-1565, USA)

Abstract: The source of irregularity in El Nino, the large interannual climate variation of the Pacific ocean-atmosphere system, has remained elusive. Results from an El Nino model exhibit transition to chaos through a series of frequency-locked steps created by nonlinear resonance with the Earth's annual cycle. The overlapping of these resonances leads to the chaotic behavior. This transition scenario explains a number of climate model results and produces spectral characteristics consistent with currently available data.

厄尔尼诺困扰阶梯——年度次谐波从有序到混乱

Jin F. F. , Neelin J. D. , Ghil M.

(美国加州大学洛杉矶分校大气科学系)

摘要:厄尔尼诺现象,太平洋海洋-大气系统的年际气候变化,其不规范的来源仍然难以琢磨。通过对地球年度周期的非线性响应产生的一系列频率锁定步骤,厄尔尼诺现象的模型展示了从有序到混乱的转变。这些共振重叠导致混乱行为。这种转变的过程解释一些气候模型的结果并产生与现有的数据相一致的谱分析特征。

25. Mcminn A, Heijnis H, Hodgson D. Minimal effects of Uvb-radiation on Antarctic diatoms over the past 20 years[J]. Nature, 1994, 370:547-549.

Minimal effects of Uvb-radiation on Antarctic diatoms over the past 20 years

Mcminn A. , Heijnis H. , Hodgson D.

(Antarctic CRC and Institute of Antarctic and Southern Ocean Studies, University of Tasmania, Box 252C, Hobart 7001, Tasmania, Australia)

Abstract: It has been suggested that increased springtime WB radiation caused by stratospheric ozone depletion is likely to reduce primary production and induce changes in the species composition of Antarctic marine phytoplankton. Experiments conducted at Arthur Harbour in the Antarctic Peninsula revealed a reduction in primary productivity at both ambient and increased levels of Uvb. Laboratory studies have shown that most species in culture are sensitive to high Uvb levels, although the level at which either growth or photosynthesis is inhibited is variable. Stratospheric ozone depletion, with resultant increased springtime Uvb irradiance, has been occurring with increasing severity since the late 1970s. Thus the phytoplankton community has already experienced about 20 years' exposure to increasing levels of Uvb radiation. Here we present analyses of diatom assemblages from high-resolution stratigraphic sequences from anoxic basins in fjords of the Vestfold Hills, Antarctica. We find that compositional changes in the diatom component of the phytoplankton community over the past 20 years cannot be distinguished from long-term natural variability, although there is some indication of a decline in the production of some sea-ice diatoms. We anticipate that our results are applicable to other Antarctic coastal regions, where thick ice cover and the timing of the phytoplankton bloom protect the phytoplankton from the effects of increased Uvb radiation.

过去 20 年紫外线辐射对南极硅藻的微小影响

Mcminn A., Heijnis H., Hodgson D.

(澳大利亚塔斯曼尼亚大学南极 CRC 和南极南大洋研究所)

摘要:有人曾提出,臭氧层缺陷导致的春天 WB 辐射增加可能会减少初级产品生产和诱导南极海洋浮游植物的物种组成的变化。在南极半岛的阿瑟港进行的实验发现在自然和提高的 Uvb 水平周围初级生产力会减少。实验室研究表明,大多数物种对高水平 Uvb 敏感,但导致成长或光合作用被抑制的 Uvb 水平是可变的。平流层臭氧耗损,由此导致春天紫外线辐射增加,自 20 世纪 70 年代发生以来日益严重。因此,浮游植物群落已经经历了约 20 年暴露在水平不断提高的紫外线辐射之下。在这里,我们报道在西福尔山,南极洲的峡湾缺氧盆地高分辨率地层序列的硅藻组合的分析。我们发现,在过去 20 年中浮游植物群落的硅藻组成部分成分变化与长期的自然变化无显著差别,虽然有一些迹象显示一些海冰硅藻生产下降。我们预计我们的结果适用于其他南极沿海地区,其中厚的冰层覆盖和浮游植物爆发时间使浮游植物免受 Uvb 辐射增强的影响。

26. Rahmstorf S. Rapid climate transitions in a coupled ocean-atmosphere model[J]. Nature,1994,372: 82-85.

Rapid climate transitions in a coupled ocean-atmosphere model

Rahmstorf S.

(Institut für Meereskunde, Düsternbrooker Weg 20, 24105 Kiel, Germany)

Abstract: Recent geochemical data have challenged the view that rapid climate fluctuations in the North Atlantic at the end of the last glacial mere caused by the thermohaline circulation of the ocean being switched "on" or "off". Instead, these data suggest that the circulation pattern must have switched between a warm, deep mode and a cold, shallow mode, probably associated with different sites of deep convection. Here I present simulations with a three-dimensional ocean model, coupled to an idealized atmosphere, which show this kind of transition. The mechanism for the transition is a rearrangement of convection in the North Atlantic, triggered by a brief freshwater pulse. This results in a drop in sea surface temperature in the North Atlantic by up to 5 degrees C within less than 10 years. The rate of North Atlantic Deep Water (NADW) formation is the same in the cold climate as in the warm climate, but NADW sinks to intermediate depths only, while Antarctic Bottom Water pushes north to fill the entire abyssal Atlantic.

海气耦合模型中的快速气候变化

Rahmstorf S.

(德国 Meereskunde 研究所)

摘要：最近的地球化学数据挑战了在末次冰期北大西洋的快速气候波动是由于海洋温盐循环“开启”或“关闭”引起的这个观点。相反，这些数据表明，流通模式是在一个温暖、深的模式和一个寒冷、浅的模式之间转变，可能与不同地点的深对流有关。在这里，我报道一个立体的与一个理想化的大气偶联海洋模式的模拟，可呈现这种过渡。过渡的机制是通过一个简短的淡水脉冲触发，使得北大西洋的对流发生转变。结果显示在不到 10 年内在北大西洋海面温度下降了 5 ℃。寒冷气候的北大西洋深层水(NADW)形成的速度在与温暖和冷的气候下是一样的，但 NADW 只下沉到中间深度，同时南极底层水向北推动，填补了整个大西洋深海。

27. Tziperman E, Stone L, Cane M A, et al. El-Nino chaos: overlapping of resonances between the seasonal cycle and the Pacific Ocean-atmosphere oscillator[J]. Science, 1994, 264: 72-74.

El-Nino chaos—overlapping of resonances between the seasonal cycle and the Pacific Ocean-atmosphere oscillator

Tziperman E., Stone L., Cane M. A., Jarosh H.

(Environmental Sciences and Energy Research, The Weizmann Institute of Science, Rehovot 76100, Israel)

Abstract: The El Nino-Southern Oscillation (ENSO) cycle is modeled as a low-order chaotic process driven by the seasonal cycle. A simple model suggests that the equatorial Pacific ocean-atmosphere oscillator can go into nonlinear resonance with the seasonal cycle and that with strong enough coupling between the ocean and the atmosphere, the system may become chaotic as a result of irregular jumping of the ocean-atmosphere system among different nonlinear resonances. An analysis of a time series from an ENSO prediction model is consistent with the low-order chaos mechanism.

厄尔尼诺混乱——在季节性周期和太平洋-大气振荡系统之间的响应重叠

Tziperman E., Stone L., Cane M. A., Jarosh H.

(以色列威兹曼科学研究所环境科学和能源研究部)

摘要：厄尔尼诺-南方涛动(ENSO)循环被认为是一个低级混乱的季节性周期驱动的过程。一个简单的模型表明，赤道太平洋海洋-大气振荡系统可以与季节性周期形成非线性共振，并且，海洋和大气之间存在很强的耦合，系统可能会由于海洋-大气在不同的非线性共振下产生的不规则跳跃而变得混乱。从 ENSO 预测模型对厄尔尼诺的一个时间序列分析与低级混乱的机制是一致的。

28. White W B, Peterson R G. An Antarctic circumpolar wave in surface pressure, wind, temperature

and sea-ice extent[J]. Nature, 1996, 380: 699-702.

An Antarctic circumpolar wave in surface pressure, wind, temperature and sea-ice extent

White W. B., Peterson R. G.

(Scripes Institution of Oceanography, University of California-San Diego, La Jolla, California 92093-0230, USA)

Abstract: The Southern Ocean is the only oceanic domain encircling the globe. It contains the strong eastward flow of the Antarctic Circumpolar Current, acid is the unifying link for exchanges of water masses at all depths between the world's major ocean basins. As these exchanges are an important control on mean global climate, the Southern Ocean is expected to play an important role in transmitting climate anomalies around the globe. Interannual variability has been often observed at high southern latitudes, and observations of sea-ice extent suggest that such features propagate eastwards around the Southern Ocean. Here we use data from a variety of observational techniques to identify significant interannual variations in the atmospheric pressure at sea level, wind stress, sea surface temperature and sea-ice extent over the Southern Ocean. These anomalies propagate eastward with the circumpolar flow, with a period of 4-5 years and taking 8-10 years to encircle the pole. This system of coupled anomalies, which we call the Antarctic Circumpolar Wave, is likely to play an important role in climate regulation and dynamics both within and beyond the Southern Ocean.

南极环极地表面压力、风、温度和海冰张力的绕极波

White W. B., Peterson R. G.

(美国加州大学圣地亚哥分校斯克里普斯海洋研究所)

摘要:南大洋是唯一环绕地球的洋域。它包含强烈向东流动的南极绕极流,酸性水是世界主要海洋盆地所有深处的水交换的统一链接。由于这些交换是对全球气候的重要控制,南大洋被认为在传输全球气候异常中发挥重要的作用。年际变化已经在南部高纬度地区经常观测到,海冰扩张观测显示着这些特征环绕南大洋向东扩张。在这里,我们使用各种观测数据来确定海平面定在大气压力、风应力,海面温度在南大洋海冰范围的年际变化。这些异常随着环流向东传播,用4~5年的时间,并用8~10年包围极地。该系统异常,我们称之为南极绕极波,可能在南大洋内外气候调节和动态变化中发挥重要作用。

29. Ganopolski A, Kubatzki C, Claussen M, et al. The influence of vegetation-atmosphere-ocean interaction on climate during the mid-Holocene[J]. Science, 1998, 280: 1916-1919.

The influence of vegetation-atmosphere-ocean interaction on climate during the mid-Holocene

Ganopolski A., Kubatzki C., Claussen M., Brovkin V., Petoukhov V.

(Potsdam-Institut für Klimafolgenforschung, Postfach 601203, D-14412 Potsdam, Germany)

Abstract: Simulations with a synchronously coupled atmosphere-ocean-vegetation model show that changes in vegetation cover during the mid-Holocene, some 6000 years ago, modify and amplify the climate system response to an enhanced seasonal cycle of solar insolation in the Northern Hemisphere both directly (primarily through the changes in surface albedo) and indirectly (through changes in oceanic temperature, sea-ice cover, and oceanic circulation). The model results indicate strong synergistic effects of changes in vegetation cover, ocean temperature, and sea ice at boreal latitudes, but in the subtropics, the atmosphere-vegetation feedback is most important. Moreover, a reduction of the thermohaline circulation in the Atlantic Ocean leads to a warming of the Southern Hemisphere.

在全新世中期植被-大气-海洋相互作用对气候的影响

Ganopolski A., Kubatzki C., Claussen M., Brovkin V., Petoukhov V.

(德国波茨坦研究所)

摘要：同步耦合的大气-海洋-植被模型显示，植被覆盖变化，在全新世中期，大约6000年前，通过直接(主要通过在地表反照率的变化)和间接(通过在海洋温度的变化，海冰覆盖，海洋环流)的方式影响和强化气候系统对增强的太阳辐射的季节性周期的响应。模型的结果表明，植被覆盖、海洋温度和在北半球纬度海冰变化具有强大的协同效应，但在亚热带地区，大气植被反馈是最重要的。此外，在大西洋温盐环流的减弱导致了南半球变暖。

30. Hansell D A, Carlson C A. Deep-ocean gradients in the concentration of dissolved organic carbon [J]. Nature, 1998, 395: 263-266.

Deep-ocean gradients in the concentration of dissolved organic carbon

Hansell D. A., Carlson C. A.

(Bermuda Biological Station for Research, Inc., St Georges, GE01, Bermuda)

Abstract: There is as much carbon in dissolved organic material in the oceans as there is CO_2 in the atmosphere, but the role of dissolved organic carbon (DOC) in the global carbon cycle is poorly understood. DOC in the deep ocean has long been considered to be uniformly distributed and hence largely refractory to biological decay, But the turnover of DOC, and therefore its contribution to the carbon cycle, has been evident from radiocarbon dating studies. Here we report the results of a global survey of deep-ocean DOC concentrations, including the region of deep-water formation in the North Atlantic Ocean, the Circumpolar Current of the Southern Ocean, and the Indian and Pacific oceans. DOC concentrations decreased by 14 micromolar from the northern North Atlantic Ocean to the northern North Pacific Ocean, representing a 29% reduction in concentration. We evaluate the spatial patterns in terms of source/sink processes. Inputs of DOC to the deep ocean are identifiable in the mid-latitudes of the Southern Hemisphere, but the mechanisms have not been identified with certainty.

深海中的溶解有机碳浓度梯度

Hansell D. A., Carlson C. A.

（百慕大生物科技有限公司）

摘要：就像大气中的二氧化碳，有很多的碳在海洋中溶解在有机物质中，但对溶解有机碳(DOC)在全球碳循环中的作用知之甚少。在深海中的 DOC 长期被认为是均匀分布的，然而对生物降解很难理解，但 DOC 的周转在其碳循环中的贡献，已经被放射性碳测年研究证实。在这里，我们报告深海 DOC 浓度，包括在北大西洋、南大洋环流，以及印度洋和太平洋的深水区域形成的全球性调查结果。从北部大西洋到北太平洋北部 DOC 浓度下降 14 mmol，浓度减少 29%，我们评估其在源/汇过程中的空间格局。在南半球中纬度地区的 DOC 输入到深海是可识别的，但机制尚未确定。

31. Sarmiento J L, Hughes T M C, Stouffer R J, et al. Simulated response of the ocean carbon cycle to anthropogenic climate warming[J]. Nature, 1998, 393: 245-249.

Simulated response of the ocean carbon cycle to anthropogenic climate warming

Sarmiento J. L., Hughes T. M. C., Stouffer R. J., Manabe S.

(Program in Atmospheric and Oceanic Sciences, Princeton University, PO Box CN710, Princeton, New Jersey 08544, USA)

Abstract: A 1995 report of the Intergovernmental Panel on Climate Change provides a set of illustrative anthropogenic CO_2 emission models leading to stabilization of atmospheric CO_2 concentrations ranging from 350 to 1000 p. p. m. Ocean carbon-cycle models used in calculating these scenarios assume that oceanic circulation and biology remain unchanged through time. Here we examine the importance of this assumption by using a coupled atmosphere-ocean model of global warming for the period 1765 to 2065. We find a large potential modification to the ocean carbon sink in a vast region of the Southern Ocean where increased rainfall leads to surface freshening and increased stratification. The increased stratification reduces the downward flux of carbon and the loss of heat to the atmosphere, both of which decrease the oceanic uptake of anthropogenic CO_2 relative to a constant-climate control scenario. Changes in the formation, transport and cycling of biological material may counteract the reduced uptake, but the response of the biological community to the climate change is difficult to predict on present understanding. Our simulation suggests that such physical and biological changes might already be occurring, and that they could substantially affect the ocean carbon sink over the next few decades.

海洋碳循环对人为造成的气候变暖的模拟响应

Sarmiento J. L., Hughes T. M. C., Stouffer R. J., Manabe S.

(美国普林斯顿大学大气与海洋科学计划)

摘要:1995 年的政府间气候变化专门委员会的报告提供了一个说明人为二氧化碳排放量模型,导致大气中二氧化碳浓度稳定在 350～1000 ppm。用于计算这些场景的海洋碳循环模型假设大洋环流和生物学是保持不变的。在这里,我们用大气-海洋耦合模型来检测这个假设在 1765～2065 年全球变暖中的重要性。我们发现一个巨大的南大洋,海洋碳汇有一个潜在的大问题,降雨增加导致表面水的淡化和增加分层。增加的分层减少向下的碳通量和热量流失到大气中,这两者都导致了在相对恒定的气候控制的情况下海洋减少吸收人为二氧化碳。形成、运输和生物材料的循环变动可能会抵消减少海洋吸收,但生物群落对气候变化的响应目前很难理解和预测。我们的模拟结果表明,物理和生物变化可能已经发生,并且可能在未来几十年大大影响海洋碳汇。

32. Gnanadesikan A. A simple predictive model for the structure of the oceanic pycnocline[J]. Science, 1999,283:2077-2079.

A simple predictive model for the structure of the oceanic pycnocline

Gnanadesikan A.

((NOAA) Geophysical Fluid Dynamics Laboratory and Atmospheric and Oceanic Sciences Program, Princeton University, Post Office Box CN710, Princeton, NJ 08544, USA)

Abstract: A simple theory for the large-scale oceanic circulation is developed, relating pycnocline depth, Northern Hemisphere sinking, and low-latitude upwelling to pycnocline diffusivity and Southern Ocean winds and eddies. The results show that Southern Ocean processes help maintain the global ocean structure and that pycnocline diffusion controls low-latitude upwelling.

对海洋密度跃层结构的一个简单的预测模型

Gnanadesikan A.

(美国 NOAA 国家海洋和大气局普林斯顿大学地球物理流体动力学实验室和大气与海洋科学计划)

摘要:开发关于大型海洋环流的一个简单的理论,将密度跃层的深度、北半球下沉、低纬度向密度跃层扩散、南部的海风和漩涡联系起来。结果表明,南大洋的过程,有助于保持全球海洋结构,密度跃层的扩散控制低纬度上涌。

33. Stramski D, Reynolds R A, Kahru M, et al. Estimation of particulate organic carbon in the ocean from satellite remote sensing[J]. Science, 1999, 285: 239-242.

Estimation of particulate organic carbon in the ocean from satellite remote sensing

Stramski D., Reynolds R. A., Kahru M., Mitchell B. G.

(Marine Physical Laboratory and Marine Research Division, Scripps Institution of Oceanography, University of California-San Diego, La Jolla, CA 92093-0238, USA)

Abstract: Measurements from the Southern Ocean show that particulate organic carbon (POC) concentration is welt correlated with the optical backscattering by particles suspended in seawater. This relation, in conjunction with retrieval of the backscattering coefficient from remote-sensing reflectance, provides an algorithm for estimating surface POC from Satellite data of ocean color. Satellite imagery from SeaWiFS reveals the seasonal progression of POC, with a zonal band of elevated POC concentrations in December coinciding with the Antarctic Polar Front Zone. At that time, the POC pool within the top 100 meters of the entire Southern Ocean south of 40 degrees S exceeded 0.8 gigatons.

从卫星遥感来估计海洋中的颗粒有机碳

Stramski D., Reynolds R. A., Kahru M., Mitchell B. G.

(美国加州大学圣地亚哥分校斯克里普斯海洋研究所)

摘要:对南大洋的测量显示,颗粒有机碳(POC)浓度与海水中的悬浮颗粒的光学后散射相关。这种关系,与从遥感反射后的散射系数,提供了一个用海洋颜色卫星数据估算表面 POC 的算法。SeaWiFS 的卫星图像显示,POC 浓度在 12 月的升高呈带状分布,与南极极地锋区的纬向带一致,揭示了 POC 的季节性进展。当时,整个南大洋南部 40°S 前 100 米内的 POC 超过 0.8 亿吨。

34. Caldeira K, Duffy P B. The role of the Southern Ocean in uptake and storage of anthropogenic carbon dioxide[J]. Science, 2000, 287: 620-622.

The role of the Southern Ocean in uptake and storage of anthropogenic carbon dioxide

Caldeira K., Duffy P. B.

(Climate System Modeling Group, Lawrence Livermore National Laboratory, Livermore, CA 94550, USA)

Abstract: An ocean-climate model that shows high fluxes of anthropogenic carbon dioxide into the Southern Ocean, but very low storage of anthropogenic carbon there, agrees with observation-based estimates of ocean storage of anthropogenic carbon dioxide. This low simulated storage indicates a subordinate role for deep convection in the present-day Southern Ocean. The primary mechanism transporting anthropogenic carbon out of the Southern Ocean is isopycnal transport. These results imply that if global climate change reduces the density of surface waters in the Southern Ocean, isopycnal surfaces that now outcrop may become isolated from the atmosphere, tending to diminish

Southern Ocean carbon uptake.

南大洋在人为源二氧化碳的吸收和储存中的作用

Caldeira K., Duffy P. B.

(美国 Lawrence Livermore 国家实验室气候系统建模组)

摘要:海洋气候模型显示了高通量的人为二氧化碳进入南大洋,但人为碳有非常低的存储,与观察为基础估计的海洋的人为二氧化碳存储一致。这样低的模拟存储,表示在现今的南大洋深对流的从属地位。在南大洋人为碳的主要运输机制是等密面运输。这些结果意味着,如果全球气候变化,减小地表水的密度,在南大洋等密面的表面,现在露出的等密度面可能从大气中分离出来,趋于减弱南大洋碳吸收。

35. Barbraud C, Weimerskirch H. Emperor penguins and climate change[J]. Nature, 2001, 411: 183-186.

Emperor penguins and climate change

Barbraud C., Weimerskirch H.

(Centre d'Études Biologiques de Chizé, Centre National de la Recherche Scientifique, 79360 Villiers en Bois, France)

Abstract: Variations in ocean-atmosphere coupling over time in the Southern Ocean have dominant effects on sea-ice extent and ecosystem structure, but the ultimate consequences of such environmental changes for large marine predators cannot be accurately predicted because of the absence of long-term data series on key demographic parameters. Here, we use the longest time series available on demographic parameters of an Antarctic large predator breeding on fast ice and relying on food resources from the Southern Ocean. We show that over the past 50 years, the population of emperor penguins (*Aptenodytes forsteri*) in Terre Adelie has declined by 50% because of a decrease in adult survival during the late 1970s. At this time there was a prolonged abnormally warm period with reduced sea-ice extent. Mortality rates increased when warm sea-surface temperatures occurred in the foraging area and when annual sea-ice extent was reduced, and were higher for males than for females. In contrast with survival, emperor penguins hatched fewer eggs when winter sea-ice was extended. These results indicate strong and contrasting effects of large-scale oceanographic processes and sea-ice extent on the demography of emperor penguins, and their potential high susceptibility to climate change.

帝企鹅与气候变化

Barbraud C., Weimerskirch H.

(法国国家研究中心(CNRS))

摘要:南大洋随时间的海洋-大气耦合的变化,对海冰和生态系统的结构起主导作用,但由于缺乏一系列关键的种群统计参数的长期数据,我们不能准确地预测大型海洋食肉动物等环境变化的最终后果。在这里,我们使用可得到的最长时间系列的在快速冰上繁殖并依赖南大洋食物资源的南极大型捕食者的种群参数。结果表明,在过去的50年中,由于在20世纪70年代末成年帝企鹅存活数目的减少,阿德利地区帝企鹅(*Aptenodytes forsteri*)数目下降了50%。同时,有异常温暖的时期和海冰扩展的减少。发生在觅食区海洋表面温度变暖时,和当年度海冰减少时,死亡率增加,尤其雄性比雌性高。与存活数目相比,在冬季海冰扩张时期,帝企鹅孵化较少的企鹅蛋。这些结果表明,大型海洋过程和海冰的扩张对帝企鹅的数目以及它们对气候变化的潜在的高敏感性有着强烈和相反的影响。

36. Hall A, Stouffer R J. An abrupt climate event in a coupled ocean-atmosphere simulation without external forcing[J]. Nature, 2001, 409: 171-174.

An abrupt climate event in a coupled ocean-atmosphere simulation without external forcing

Hall A., Stouffer R. J.

(Lamont-Doherty Earth Observatory, Palisades, New York 10964, USA)

Abstract: Temperature reconstructions from the North Atlantic region indicate frequent abrupt and severe climate fluctuations during the last glacial and Holocene periods. The driving forces for these events are unclear and coupled atmosphere-ocean models of global circulation have only simulated such events by inserting large amounts of fresh water into the northern North Atlantic Ocean. Here we report a drastic cooling event in a 15000 yr simulation of global circulation with present-day climate conditions without the use of such external forcing. In our simulation, the annual average surface temperature near southern Greenland spontaneously fell 6-10 standard deviations below its mean value for a period of 30-40 yr. The event was triggered by a persistent northwesterly wind that transported large amounts of buoyant cold and fresh water into the northern North Atlantic Ocean. Oceanic convection shut down in response to this flow, concentrating the entire cooling of the northern North Atlantic by the colder atmosphere in the uppermost ocean layer. Given the similarity between our simulation and observed records of rapid cooling events, our results indicate that internal atmospheric variability alone could have generated the extreme climate disruptions in this region.

在没有外部强迫情况下海洋-大气耦合模拟中一个突发的气候事件

Hall A., Stouffer R. J.

(美国 Lamont-Doherty 地球观测中心)

摘要:北大西洋地区的温度重建表明在末次冰期和全新世时期有频繁的突然爆发并且剧烈的气候波动。这些事件的驱动力是不清楚的,全球大气环流的大气-海洋耦合模型仅仅通过向北大西洋中注入大量的淡水来模拟这些事件。在这里,我们报告一个不使用外界力量来模拟一个距今15000年、与现在气候类似条件下、急剧变冷的事件。在我们的模拟中,南部格陵兰岛年平均表面温度30～40年期间自发降低比平均值低6～10个标准偏差。触发事件的是一个持久的西北风,传输大量浮动的寒冷和新鲜水到北部大西洋。为了应对这种流动,大洋传送带关闭,通过冷却海洋最上层水上方的大气来集中北大西洋北部的整体寒冷。由于我们的模拟和快速冷却事件的观察记录之间的相似性,我们的研究结果表明,单靠内部的大气变异可能产生在这一地区的极端气候扰动。

37. Bidle K D, Manganelli M, Azam F. Regulation of oceanic silicon and carbon preservation by

temperature control on bacteria[J]. Science,2002,298:1980-1984.

Regulation of oceanic silicon and carbon preservation by temperature control on bacteria

Bidle K. D., Manganelli M., Azam F.

(Marine Biology Research Division, Scripps Institution of Oceanography, University of California San Diego, La Jolla, CA 92093-0202, USA)

Abstract: We demonstrated in laboratory experiments that temperature control of marine bacteria action on diatoms strongly influences the coupling of biogenic silica and organic carbon preservation. Low temperature intensified the selective regeneration of organic matter by marine bacteria as the silicon: carbon preservation ratio gradually increased from similar to 1 at 33 degrees C to similar to 6 at −1.8 degrees C. Temperature control of bacteria-mediated selective preservation of silicon versus carbon should help to interpret and model the variable coupling of silicon and carbon sinking fluxes and the spatial patterns of opal accumulation in oceanic systems with different temperature regimes.

温度控制细菌情况下大洋硅和碳保存的调控

Bidle K. D., Manganelli M., Azam F.

(美国加州大学圣地亚哥分校斯克里普斯海洋研究所海洋生物研究室)

摘要:我们通过在实验室做实验证明了温度控制海洋细菌在硅藻属上的活动强烈地影响着生物硅和有机碳保存的耦合。低温加剧海洋细菌选择性重新产生有机物质,当温度从 33 ℃降低到−1.8 ℃时,硅和碳保存的比率从 1 增加到 6。温度控制的细菌调节的硅和碳的选择性保存有助于理解和模拟硅和碳的不同耦合,以及不同空间格局下海洋系统的蛋白石堆积。

38. del Giorgio P A,Duarte C M. Respiration in the open ocean[J]. Nature,2002,420:379-384.

Respiration in the open ocean

del Giorgio P. A., Duarte C. M.

(Départment des sciences biologiques, Université du Québec à Montréal, CP 8888, succ Centre Ville, Montréal, Québec H3C 3P8, Canada)

Abstract: A key question when trying to understand the global carbon cycle is whether the oceans are net sources or sinks of carbon. This will depend on the production of organic matter relative to the decomposition due to biological respiration. Estimates of respiration are available for the top layers, the mesopelagic layer, and the abyssal waters and sediments of various ocean regions. Although the total open ocean respiration is uncertain, it is probably substantially greater than most current estimates of particulate organic matter production. Nevertheless, whether the biota act as a net source or sink of carbon remains an open question.

开放大洋的呼吸

del Giorgio P. A. , Duarte C. M.

（加拿大魁北克大学生物学系）

摘要：试图理解全球碳循环时，一个关键的问题是海洋是碳源还是碳汇。这取决于有机物质的产生和由于生物呼吸导致的降解。在上层、中层、深海海水和不同海域的沉积物中都已有呼吸的估计。虽然总的开放的海洋呼吸是不确定的，它可能大大高于颗粒有机物质生产的最新估计。然而，生物群活动作为一个排放源还是碳汇仍然是一个悬而未决的问题。

39. Heywood K J, Garabato A C N, Stevens D P. High mixing rates in the abyssal Southern Ocean[J]. Nature, 2002, 415: 1011-1014.

High mixing rates in the abyssal Southern Ocean

Heywood K. J. , Garabato A. C. N. , Stevens D. P.

(School of Environmental Sciences, University of East Anglia, Norwich NR4 7TJ, UK)

Abstract: Mixing of water masses from the deep ocean to the layers above can be estimated from considerations of continuity in the global ocean overturning circulation. But averaged over ocean basins, diffusivity has been observed to be too small to account for the global upward flux of water, and high mixing intensities have only been found in the restricted areas close to sills and narrow gaps. Here we present observations from the Scotia Sea, a deep ocean basin between the Antarctic peninsula and the tip of South America, showing a high intensity of mixing that is unprecedented over such a large area. Using a budget calculation over the whole basin, we find a diffusivity of $(39\pm10)\times10^4\ m^2\cdot s^{-1}$, averaged over an area of $7\times10^5\ km^2$. The Scotia Sea is a basin with a rough topography, situated just east of the Drake passage where the strong flow of the Antarctic Circumpolar Current is constricted in width. The high basin-wide mixing intensity in this area of the Southern Ocean may help resolve the question of where the abyssal water masses are mixed towards the surface.

南大洋深海的高混合率

Heywood K. J. , Garabato A. C. N. , Stevens D. P.

（英国东英吉利大学环境科学学院）

摘要：深海海洋物质和海水表面物质的混合可以从考虑全球海洋翻转循环的连续性来估计。但在海洋盆地平均后，已观察到的扩散过小而不能用来计算占全球向上的水通量，高强度混合只有在邻近基石和狭窄的沟壑的限制区域发现。在这里，我们报道斯科舍海，它在南极半岛之间以及南美洲尖端的深海盆地深海盆之间，呈现出高强度的混合，在这样一个大面积中是史无前例的。使用全流域的预算计算，我们发现在面积 $7\times10^5\ km^2$ 上扩散率为$(39\pm10)\times10^4\ m^2\cdot s^{-1}$。斯科舍海是一个粗略地形的盆地，恰好位于东部的德雷克海峡，南极绕极流在宽度强烈收缩。在这南大洋区域高全流域的混合强度可能解释深海海水物质在哪里混合到表层的问题。

40. Jacobs S S, Giulivi C F, Mele P A. Freshening of the Ross Sea during the late 20th century[J]. Science, 2002, 297: 386-389.

Freshening of the Ross Sea during the late 20th century

Jacobs S. S., Giulivi C. F., Mele P. A.

(Lamont-Doherty Earth Observatory of Columbia University, Palisades, NY 10964, USA)

Abstract: Ocean measurements in the Ross Sea over the past four decades, one of the longest records near Antarctica, reveal marked decreases in shelf water salinity and the surface salinity within the Ross Gyre. These changes have been accompanied by atmospheric warming on Ross Island, ocean warming at depths of similar to 300 meters north of the continental shelf, a more negative Southern Oscillation Index, and thinning of southeast Pacific ice shelves. The freshening appears to have resulted from a combination of factors, including increased precipitation, reduced sea ice production, and increased melting of the West Antarctic Ice Sheet.

在20世纪后期罗斯海的海水淡化

Jacobs S. S., Giulivi C. F., Mele P. A.

(美国哥伦比亚大学拉蒙特-多尔蒂地质观测站)

摘要：在罗斯海的过去超过40年的海洋测量，是南极洲附近的最长记录，揭示了表层海水盐度和罗斯环流表水盐度的显著下降。这些变化伴随着罗斯岛大气变暖，海洋在北部大陆架深度接近300 m的变暖，南方涛动指数更偏负，以及东南太平洋冰架的变薄。淡化可能是多因素综合作用的结果，包括降水增加、海冰产量减少和南极西部冰盖的融化增加。

41. Keeling R F, Garcia H E. The change in oceanic O_2 inventory associated with recent global warming [J]. PNAS, 2002, 99: 7848-7853.

The change in oceanic O_2 inventory associated with recent global warming

Keeling R. F., Garcia H. E.

(Scripps Institution of Oceanography, University of California at San Diego, La Jolla, CA 92093-0244)

Abstract: Oceans general circulation models predict that global warming may cause a decrease in the oceanic O_2 inventory and an associated O_2 outgassing. An independent argument is presented here in support of this prediction based on observational evidence of the ocean's biogeochemical response to natural warming. On time scales from seasonal to centennial, natural O_2 flux/heat flux ratios are shown to occur in a range of 2 to 10 nmol of O_2 per joule of warming, with larger ratios typically occurring at higher latitudes and overlongertime scales. The ratios are several times larger than would be expected solely from the effect of heating on the O_2 solubility, indicating that most of the O_2 exchange is biologically mediated through links between heating and stratification. The change in oceanic O_2 inventory through the 1990s is estimated to be $(0.3\pm0.4)\times10^{14}$ mol of O_2 per year based on scaling the observed anomalous long-term ocean warming by natural O_2 flux/heating ratios and allowing for uncertainty due to decadal variability. Implications are discussed for carbon budgets based on observed changes in atmospheric O_2/N_2 ratio and based on observed changes in ocean dissolved inorganic carbon.

海洋的氧气储量变化和最近全球变暖之间的联系

Keeling R. F. , Garcia H. E.

（美国加州大学圣地亚哥分校斯克里普斯海洋研究所）

摘要：海洋环流模型预测，全球变暖可能会导致海洋的 O_2 的库存减少和一个相关的 O_2 排放。基于海洋生物化学对自然变暖的响应的观测，在这里我们报道支持这一预测的论据。从季节性至百年时间尺度上，自然的 O_2 流量/热流量比值在 2～10 nmol O_2 每焦耳范围内，较大的比例通常发生在高纬度地区和过长的时间尺度上。这个比率比仅仅考虑溶解氧的热效应所得到的预期值要大几倍。这表明，大部分的 O_2 交换是生物调节的，通过加热和分层之间的联系，用自然的 O_2 流量/热流量比率来调整观测到的反常的长时期的海洋变暖并考虑到因年纪变化产生的不确定性，20 世纪 90 年代海洋的 O_2 库存的变化估计为每年$(0.3\pm0.4)\times10^{14}$ mol。基于观察到的在大气中 O_2/N_2 的比例变化和海洋溶解无机碳的变化，我们讨论了碳预算的意义。

42. Munk W. Twentieth century sea level:an enigma[J]. PNAS,2002,99:6550-6555.

Twentieth century sea level:an enigma

Munk W.

（Scripps Institution of Oceanography, University of California at San Diego, La Jolla, CA 92093-0244）

Abstract: Changes in sea level (relative to the moving crust) are associated with changes in ocean volume (mostly thermal expansion) and in ocean mass (melting and continental storage): zeta(t) = zeta(steric)(t) + zeta(eustatic)(t). Recent compilations of global ocean temperatures by Levitus and coworkers are in accord with coupled ocean/atmosphere modeling of greenhouse warming; they yield an increase in 20th century ocean heat content by 2×10^{23} J (compared to 0.1×10^{23} J of atmospheric storage), which corresponds to zeta(greenhouse)(2000) = 3 cm. The greenhouse-related rate is accelerating, with a present value zeta(greenhouse)(2000) approximate to 6 cm/century. Tide records going back to the 19th century show no measurable acceleration throughout the late 19th and first half of the 20th century; we take,zeta(historic) = 18 cm/century. The Intergovernmental Panel on Climate Change attributes about 6 cm/century to melting and other eustatic processes, leaving a residual of 12 cm of 20th century rise to be accounted for, The Levitus compilation has virtually foreclosed the attribution of the residual rise to ocean warming (notwithstanding our ignorance of the abyssal and Southern Oceans): the historic rise started too early, has too linear a trend, and is too large. Melting of polar ice sheets at the upper limit of the intergovernmental panel on climate change estimates could close the gap, but severe limits are imposed by the observed perturbations in Earth rotation. Among possible resolutions of the enigma are: a substantial reduction from traditional estimates (including ours) of 1.5-2 mm/y global sea level rise; a substantial increase in the estimates of 20th century ocean heat storage; and a substantial change in the interpretation of the astronomic record.

20 世纪的海平面:一个谜

Munk W.

(美国加州大学圣地亚哥分校斯克里普斯海洋研究所)

摘要:海平面(相对于地壳的运动)的变化与海洋容量(主要是热膨胀)和海洋质量变化(熔化和大陆存储)有关:zeta(t)= zeta(立体)(t)+ zeta(海平面)(t)。Levitus 和他的同事进行的最近全球海洋温度的分析与全球变暖的海洋/大气耦合模型一致;在 20 世纪海洋热含量增加 2×10^{23} J(0.1×10^{23} J 大气中的存储),对应的 Zeta(温室)(2000)= 3 cm。温室的速度正在加快,现值 zeta(温室)(2000)近似为 6 cm/世纪。追溯到 19 世纪的潮汐记录显示 19 世纪末和 20 世纪前半部分没有可以测量到的加速,其 zeta(历史)= 18 cm/世纪。政府间气候变化对融化和海平面其他过程的贡献约 6 cm/世纪,留下一个 20 世纪残余的 12 cm 的上升尚需解释,Levitus 的分析实际上已经排除了残留的 12 cm 是海洋变暖导致的(不包括我们的未知的深海和南部海洋):历史上的海平面上升开始太早,有太多的线性的趋势,而且太大。极地冰盖的融化在政府间气候变化估计的上限,可以解释这一差距,但所观察到的地球自转扰动大大限制了这一可能。其中可能解决这个谜的解释有:从传统估计(包括我们)的 1.5～2 mm/年全球海平面上升的大幅减少,估计在 20 世纪的海洋热储存的一个大幅增加天文记录的实质性的变化。

43. Gillett N P,Zwiers F W,Weaver A J,et al. Detection of human influence on sea-level pressure[J]. Nature,2003,422:292-294.

Detection of human influence on sea-level pressure

Gillett N. P., Zwiers F. W., Weaver A. J., Stott P. A.

(School of Earth and Ocean Sciences, University of Victoria, PO Box 3055, Victoria, British Columbia, V8W 3P6, Canada)

Abstract: Greenhouse gases and tropospheric sulphate aerosols-the main human influences on climate-have been shown to have had a detectable effect on surface air temperature, the temperature of the free troposphere and stratosphere and ocean temperature. Nevertheless, the question remains as to whether human influence is detectable in any variable other than temperature. Here we detect an influence of anthropogenic greenhouse gases and sulphate aerosols in observations of winter sea-level pressure (December to February), using combined simulations from four climate models. We find increases in sea-level pressure over the subtropical North Atlantic Ocean, southern Europe and North Africa, and decreases in the polar regions and the North Pacific Ocean, in response to human influence. Our analysis also indicates that the climate models substantially underestimate the magnitude of the sea-level pressure response. This discrepancy suggests that the upward trend in the North Atlantic Oscillation index (corresponding to strengthened westerlies in the North Atlantic region), as simulated in a number of global warming scenarios, may be too small, leading to an underestimation of the impacts of anthropogenic climate change on European climate.

人类活动对海平面气压影响的探测

Gillett N. P., Zwiers F. W., Weaver A. J., Stott P. A.

（加拿大维多利亚大学地球和海洋科学学院）

摘要：温室气体和对流层硫酸盐气溶胶——主要人类活动影响气候已被证明对表面空气温度，对自由对流层与平流层的温度和海洋温度有可以探测到的影响。然而，问题仍然是人类的活动是否影响非温度的变量。在这里，结合使用4个气候模型的模拟，我们发现在冬季海平面气压(12月～2月)受人为温室气体和硫酸盐气溶胶的影响。我们发现受到人类的影响，亚热带北大西洋、欧洲南部和北非海平面的压力增加，在极地地区和北太平洋地区减少。我们的分析还表明，气候模型大大低估了海平面气压响应的幅度。这种差异表明，在北大西洋涛动指数(对应在北大西洋地区的西风加强)在全球变暖的情况下模拟中，上升的趋势可能会太小，导致对人为因素对欧洲气候变化的影响估计不足。

44. Gordon A L, Susanto R D, Vranes K. Cool Indonesian throughflow as a consequence of restricted surface layer flow[J]. Nature, 2003, 425: 824-828.

Cool Indonesian throughflow as a consequence of restricted surface layer flow

Gordon A. L., Susanto R. D., Vranes K.

(Lamont-Doherty Earth Observatory, Columbia University, Palisades, New York 10964, USA)

Abstract: Approximately 10 million $m^3 \cdot s^{-1}$ of water flow from the Pacific Ocean into the Indian Ocean through the Indonesian seas. Within the Makassar Strait, the primary pathway of the flow, the Indonesian throughflow is far cooler than estimated earlier, as pointed out recently on the basis of ocean current and temperature measurements. Here we analyse ocean current and stratification data along with satellite-derived wind measurements, and find that during the boreal winter monsoon, the wind drives buoyant, low-salinity Java Sea surface water into the southern Makassar Strait, creating a northward pressure gradient in the surface layer of the strait. This surface layer "freshwater plug" inhibits the warm surface water from the Pacific Ocean from flowing southward into the Indian Ocean, leading to a cooler Indian Ocean sea surface, which in turnmay weaken the Asian monsoon. The summer wind reversal eliminates the obstructing pressure gradient, by transferring more-saline Banda Sea surface water into the southern Makassar Strait. The coupling of the southeast Asian freshwater budget to the Pacific and Indian Ocean surface temperatures by the proposed mechanism may represent an important negative feedback within the climate system.

冷的印度尼西亚贯穿流作为限制表面层流动的结果

Gordon A. L., Susanto R. D., Vranes K.

(美国哥伦比亚大学拉蒙特-多尔蒂地质观测站)

摘要:约每秒1000万立方米的水流量通过印度尼西亚海洋从太平洋流到印度洋。望加锡海峡是印度尼西亚贯穿流的主要途径,在最近的以海洋洋流和温度测量为基础的工作中指出,印度尼西亚贯穿流远远比早期估计的更冷。在这里,我们分析了洋流和卫星对风的监测数据,发现在北半球冬季季风,风驱动丰富、低盐度的爪哇海表层海水进入南加锡海峡,在海峡表层形成了一个北向的气压梯度。这表面层的"淡水插入"抑制温暖的表层水从太平洋向南流入印度洋,造成了印度洋表层水的变冷,这个反过来可能会减弱亚洲季风。夏季风逆转消除压力梯度的阻碍,通过转移更多的班达海表层盐水进入南加锡海峡。用提出的东南亚到太平洋的淡水预算和印度洋表水温度耦合机制可能代表气候系统的一个重要负反馈。

45. Huber M, Caballero R. Eocene El Nino: evidence for robust tropical dynamics in the "hothouse"[J]. Science, 2003, 299: 877-881.

Eocene El Nino: evidence for robust tropical dynamics in the "hothouse"

Huber M., Caballero R.

(Danish Center for Earth System Science, Niels Bohr Institute for Astronomy, Physics, and Geophysics, University of Copenhagen, Juliane Maries Vej 30, DK-2100, Copenhagen, Denmark)

Abstract: Much uncertainty surrounds the interactions between the El Nino-Southern Oscillation (ENSO) and long-term global change. Past periods of extreme global warmth, exemplified by the Eocene (55 to 35 million years ago), provide a good testing ground for theories for this interaction. Here, we compare Eocene coupled climate model simulations with annually resolved variability records preserved in lake sediments. The simulations show Pacific deep-ocean and high-latitude surface warming of similar to 10 degrees C but little change in the tropical thermocline structure, atmosphere-ocean dynamics, and ENSO, in agreement with proxies. This result contrasts with theories linking past and future "hothouse" climates with a shift toward a permanent El Nino-like state.

始新世的厄尔尼诺现象:在"热室"的有力热带动力学证据

Huber M., Caballero R.

(丹麦哥本哈根大学地球系统科学中心;尼尔斯玻尔天文物理地球物理学研究所)

摘要:厄尔尼诺-南方涛动(ENSO)和长期的全球气候变化的相互作用有很多不确定的地方。过去极端的全球温暖时期,以始新世(5500万~3500万年前)为例,为关于这种作用的研究提供了一个很好的测试背景。在这里,我们把始新世的气候耦合模拟模型与湖泊沉积物中每年变化的记录做对比。模拟结果表明太平洋深海和高纬度的表面温度近似 10 ℃,但是在热带温跃层结构,大气-海洋的动力作用,厄尔尼诺-南方涛动变化不大,与替代物的结果一致。这一结果与联系过去和未来"热室"气候与永久的类似厄尔尼诺状态的理论相反。

46. Knorr G, Lohmann G. Southern Ocean origin for the resumption of Atlantic thermohaline circulation during deglaciation[J]. Nature,2003,424:532-536.

Southern Ocean origin for the resumption of Atlantic thermohaline circulation during deglaciation

Knorr G. , Lohmann G.

(Institut für Meteorologie, Universität Hamburg, Bundesstrasse 55, 20146 Hamburg, Germany)

Abstract: During the two most recent deglaciations, the Southern Hemisphere warmed before Greenland. At the same time, the northern Atlantic Ocean was exposed to meltwater discharge, which is generally assumed to reduce the formation of North Atlantic Deep Water. Yet during deglaciation, the Atlantic thermohaline circulation became more vigorous, in the transition from a weak glacial to a strong interglacial mode. Here we use a three-dimensional ocean circulation model to investigate the impact of Southern Ocean warming and the associated sea-ice retreat on the Atlantic thermohaline circulation. We find that a gradual warming in the Southern Ocean during deglaciation induces an abrupt resumption of the interglacial mode of the thermohaline circulation, triggered by increased mass transport into the Atlantic Ocean via the warm (Indian Ocean) and cold (Pacific Ocean) water route. This effect prevails over the influence of meltwater discharge, which would oppose a strengthening of the thermohaline circulation. A Southern Ocean trigger for the transition into an interglacial mode of circulation provides a consistent picture of Southern and Northern hemispheric climate change at times of deglaciation, in agreement with the available proxy records.

在末次冰消期大西洋温盐环流恢复起源于南大洋

Knorr G. , Lohmann G.

(德国汉堡大学气象研究所)

摘要:在最近两次冰消期,南半球的回暖在格陵兰岛之前。与此同时,大西洋北部暴露于融水输入,这被普遍认为降低了北大西洋深层水的形成。然而,在末次冰消期,大西洋温盐环流变得更加强盛,从弱冰期过渡到强烈的间冰期模式。这里我们用一个三维海洋环流模型来研究南大洋变暖和相关的海冰退却对大西洋温盐环流的影响。我们发现,在末次冰消期南大洋的逐渐升温,通过温水或冷水途径增加进入大西洋的物质,诱导间冰期温盐环流模式的突然恢复。这种影响超过了融水排放的影响,可能抑制温盐环流的加强。南大洋触发一个转变到间冰期的循环模式提供了南半球和北半球地区在末次冰消期的气候变化的连续过程,与可得到的替代气候指标的结果一致。

47. Shcherbina A Y, Talley L D, Rudnick D L. Direct observations of North Pacific ventilation: brine rejection in the Okhotsk Sea[J]. Science,2003,302:1952-1955.

Direct observations of North Pacific ventilation: brine rejection in the Okhotsk Sea

Shcherbina A. Y., Talley L. D., Rudnick D. L.

(Scripps Institution of Oceanography, University of California, San Diego, USA)

Abstract: Brine rejection that accompanies ice formation in coastal polynyas is responsible for ventilating several globally important water masses in the Arctic and Antarctic. However, most previous studies of this process have been indirect, based on heat budget analyses or on warm-season water column inventories. Here, we present direct measurements of brine rejection and formation of North Pacific Intermediate Water in the Okhotsk Sea from moored winter observations. A steady, nearly linear salinity increase unambiguously caused by local ice formation was observed for more than a month.

北太平洋流场的直接观察:盐水在鄂霍次克海回涌

Shcherbina A. Y., Talley L. D., Rudnick D. L.

(美国加州大学圣地亚哥分校斯克里普斯海洋研究所)

摘要:盐水排斥,伴随着开放水域海冰的形成,造成了几次北极和南极全球水流动。然而,许多以前对这个过程的研究都是间接的,以热预算分析或暖季水储量为基础。在这里,我们根据固定的冬季观察提供对盐水排斥和北太平洋在鄂霍次克海中间水形成的直接测量。一个稳定的、接近线性的盐度增加,明确是被当地海冰的形成所引起的,观察已经超过一个月的时间。

48. Tsuda A, Takeda S, Saito H, et al. A mesoscale iron enrichment in the western Subarctic Pacific induces a large centric diatom bloom[J]. Science, 2003, 300: 958-961.

A mesoscale iron enrichment in the western Subarctic Pacific induces a large centric diatom bloom

Tsuda A., Takeda S., Saito H., Nishioka J.

(Hokkaido National Research Institute, Kushiro, Hokkaido 085-0802, Japan)

Abstract: We have performed an in situ test of the iron limitation hypothesis in the subarctic North Pacific Ocean. A single enrichment of dissolved iron caused a large increase in phytoplankton standing stock and decreases in macronutrients and dissolved carbon dioxide. The dominant phytoplankton species shifted after the iron addition from pennate diatoms to a centric diatom, Chaetoceros debilis, that showed a very high growth rate, 2.6 doublings per day. We conclude that the bioavailability of iron regulates the magnitude of the phytoplankton biomass and the key phytoplankton species that determine the biogeochemical sensitivity to iron supply of high-nitrate, low-chlorophyll waters.

在亚北极太平洋西部的一个中尺度铁富集引起大范围硅藻爆发

Tsuda A. , Takeda S. , Saito H. , Nishioka J.

(日本北海道国立研究所)

摘要:我们已经在亚北极北太平洋通过原位测试检验了铁限制假说。一个单一溶解铁富集,造成在浮游植物储量的大量增加和营养元素与溶解的二氧化碳的减少。当铁增加时,占主导地位的浮游植物物种转移从羽毛藻转变为中心藻、角毛藻,表现出非常高的增长速度,平均每天以2.6倍速度增长。我们的结论是生物可利用铁调节浮游植物生物量及关键浮游植物种类,并决定了高氮低叶绿素海域铁供应的生物地球化学敏感性。

49. Bishop J K B, Wood T J, Davis R E, et al. Robotic observations of enhanced carbon biomass and export at 55 degrees S during SOFeX[J]. Science, 2004, 304: 417-420.

Robotic observations of enhanced carbon biomass and export at 55 degrees S during SOFeX

Bishop J. K. B. , Wood T. J. , Davis R. E. , Sherman J. T.

(Earth Sciences Division, Lawrence Berkeley National Laboratory, 1 Cyclotron Road, MS 90-1116, Berkeley, CA 94720, USA)

Abstract: Autonomous. oats pro. ling in high-nitrate low-silicate waters of the Southern Ocean observed carbon biomass variability and carbon exported to depths of 100 m during the 2002 Southern Ocean Iron Experiment (SOFeX) to detect the effects of iron fertilization of surface water there. Control and "in-patch" measurements documented a greater than fourfold enhancement of carbon biomass in the iron-amended waters. Carbon export through 100 m increased two-to sixfold as the patch subducted below a front. The molar ratio of iron added to carbon exported ranged between 10^4 and 10^5. The biomass buildup and export were much higher than expected for iron-amended low-silicate waters.

SOFeX 期间在南纬 55°海域机器人观察到的碳生物量的增加和输出

Bishop J. K. B. , Wood T. J. , Davis R. E. , Sherman J. T.

(美国加州伯克利大学劳伦斯伯克利国家实验室地球科学部)

摘要:在2002年南部海洋铁实验中,在南大洋高氮低硅水域,用机器人观察碳生物量的变化和对100 m深度水层碳的输出,以探测表层水铁肥实验(SOFeX)的效果。控制和"补丁"测量记录了在铁添加水域的生物量碳大约增长4倍。向100 m以下碳输出增加了2～6倍。铁输入和碳输出的摩尔比值在10^4～10^5之间。低硅水域的铁补偿使得生物质的积累和输出均远高于预期。

50. Buesseler K O, Andrews J E, Pike S M, et al. The effects of iron fertilization on carbon sequestration in the Southern Ocean[J]. Science, 2004, 304: 414-417.

The effects of iron fertilization on carbon sequestration in the Southern Ocean

Buesseler K. O., Andrews J. E., Pike S. M., Charette M. A.

(Department of Marine Chemistry and Geochemistry, Woods Hole Oceanographic Institution, Woods Hole, MA 02543, USA)

Abstract: An unresolved issue in ocean and climate sciences is whether changes to the surface ocean input of the micronutrient iron can alter the flux of carbon to the deep ocean. During the Southern Ocean Iron Experiment, we measured an increase in the flux of particulate carbon from the surface mixed layer, as well as changes in particle cycling below the iron-fertilized patch. The flux of carbon was similar in magnitude to that of natural blooms in the Southern Ocean and thus small relative to global carbon budgets and proposed geoengineering plans to sequester atmospheric carbon dioxide in the deep sea.

在南大洋铁肥对碳汇的影响

Buesseler K. O., Andrews J. E., Pike S. M., Charette M. A.

(美国伍兹霍尔海洋研究所海洋化学和地球化学系)

摘要：海洋和气候科学一个悬而未决的问题是，改变海洋表面的微量营养素铁输入是否可以改变深海碳通量。在南大洋铁实验中，我们测量从表面混合层中颗粒碳通量的增加，以及铁肥施加区域微粒循环的变化。碳的变化类似于南大洋的自然爆发，因而与全球碳预算和地球工程计划中把大气中二氧化碳封存到深海的关系不大。

51. Coale K H, Ohnson K S, Chavez F P, et al. Southern ocean iron enrichment experiment: carbon cycling in high and low-Si waters[J]. Science, 2004, 304: 408-414.

Southern ocean iron enrichment experiment: carbon cycling in high and low-Si waters

Coale K. H., Ohnson K. S., Chavez F. P., Buesseler K. O., Barber R. T.

(Moss Landing Marine Laboratories, 8272 Moss Landing Road, Moss Landing, CA 95039-9647, USA)

Abstract: The availability of iron is known to exert a controlling influence on biological productivity in surface waters over large areas of the ocean and may have been an important factor in the variation of the concentration of atmospheric carbon dioxide over glacial cycles. The effect of iron in the Southern Ocean is particularly important because of its large area and abundant nitrate, yet iron-enhanced growth of phytoplankton may be differentially expressed between waters with high silicic acid in the south and low silicic acid in the north, where diatom growth may be limited by both silicic acid and iron. Two mesoscale experiments, designed to investigate the effects of iron enrichment in regions with high and low concentrations of silicic acid, were performed in the Southern Ocean. These experiments demonstrate iron's pivotal role in controlling carbon uptake and regulating atmospheric partial pressure of carbon dioxide.

南大洋铁富集实验:高、低硅水域中的碳循环

Coale K. H., Ohnson K. S., Chavez F. P., Buesseler K. O., Barber R. T.

(美国莫斯兰丁海洋实验室)

摘要:铁对控制大面积的海洋表面水域上生物生产力的作用是众所周知的,并可能是在冰期-间冰期旋回中影响大气中二氧化碳浓度变化的一个重要因素。铁在南大洋的作用尤为重要,因为它广阔和丰富的营养盐,但铁增加浮游植物的生长的效率可能会在南部高硅酸盐和北部低硅酸盐水域中表现不同。在南大洋进行的两个中间尺度的实验,旨在探讨在高、低浓度的硅酸的地区铁富集的影响。这些实验表明铁在控制碳的吸收和调节大气中二氧化碳的分压中起到举足轻重的作用。

52. Garabato A C N, Polzin K L, King B A, et al. Widespread intense turbulent mixing in the Southern Ocean[J]. Science, 2004, 303: 210-213.

Widespread intense turbulent mixing in the Southern Ocean

Garabato A. C. N., Polzin K. L., King B. A., Heywood K. J., Visbeck M.

(School of Environmental Sciences, University of East Anglia, Norwich NR4 7TJ, UK)

Abstract: Observations of internal wave velocity fluctuations show that enhanced turbulent mixing over rough topography in the Southern Ocean is remarkably intense and widespread. Mixing rates exceeding background values by a factor of 10 to 1000 are common above complex bathymetry over a distance of 2000 to 3000 kilometers at depths greater than 500 to 1000 meters. This suggests that turbulent mixing in the Southern Ocean may contribute crucially to driving the upward transport of water closing the ocean's meridional overturning circulation, and thus needs to be represented in numerical simulations of the global ocean circulation and the spreading of biogeochemical tracers.

在南大洋广泛强烈的湍流混合

Garabato A. C. N., Polzin K. L., King B. A., Heywood K. J., Visbeck M.

(英国东英吉利大学环境科学学院)

摘要:内部波速波动的观测表明,在南大洋的粗糙海床混合是非常强烈而广泛存在的。超过背景值10～1000倍的混合率,范围普遍超过了2000～3000 km,深度达到500～1000 m。这表明,在南大洋的湍流混合,可能会导致至关重要的水向上传输的驱动,关闭海洋的经向翻转环流,因此在全球海洋环流的数值模拟和生物地球化学示踪中需要加以考虑。

53. Wingenter O W, Haase K B, Strutton P, et al. Changing concentrations of CO, CH_4, C_5H_8, CH_3Br, CH_3I,

and dimethyl sulfide during the southern ocean iron enrichment experiments[J]. PNAS,2004,101:8537-8541.

Changing concentrations of CO, CH_4, C_5H_8, CH_3Br, CH_3I, and dimethyl sulfide during the southern ocean iron enrichment experiments

Wingenter O. W., Haase K. B., Strutton P., Friederich G., Meinardi S., Blake D. R., Rowland F. S.

(Department of Chemistry, New Mexico Institute of Mining and Technology, Socorro, NM 87801)

Abstract: Oceanic iron (Fe) fertilization experiments have advanced the understanding of how Fe regulates biological productivity and air-sea carbon dioxide (CO_2) exchange. However, little is known about the production and consumption of halocarbons and other gases as a result of Fe addition. Besides metabolizing inorganic carbon, marine microorganisms produce and consume many other trace gases. Several of these gases, which individually impact global climate, stratospheric ozone concentration, or local photochemistry, have not been previously quantified during an Fe-enrichment experiment. We describe results for selected dissolved trace gases including methane (CH_4), isoprene (C_5H_8), methyl bromide (CH_3Br), dimethyl sulfide, and oxygen (O_2), which increased subsequent to Fe fertilization, and the associated decreases in concentrations of carbon monoxide (CO), methyl iodide (CH_3I), and CO_2 observed during the Southern Ocean Iron Enrichment Experiments.

在南大洋铁富集实验中，CO、CH_4、C_5H_8、CH_3Br、CH_3I 和二甲基硫的浓度的改变

Wingenter O. W., Haase K. B., Strutton P., Friederich G., Meinardi S., Blake D. R., Rowland F. S.

(美国新墨西哥海洋和技术研究所化学系)

摘要:海洋铁(Fe)施肥实验,有利于进一步理解铁是如何调节生物生产力和海-气二氧化碳(CO_2)交流的。然而,很少有人知道铁增加如何影响卤烃和其他气体的产生和消耗。除了代谢无机碳,海洋微生物还生产和消耗许多其他微量气体。这些气体中的几种,独自影响全球气候、平流层臭氧浓度,或局部光化学,但未曾在一个铁富集实验里得到量化。我们选定微量溶解的气体包括甲烷(CH_4)、异戊二烯(C_5H_8)、甲基溴(CH_3Br)、二甲基硫醚和氧气(O_2),这些气体在铁施肥后增加,而且在南大洋铁富集实验中,可观察到一氧化碳(CO)、碘甲烷(CH_3I)和二氧化碳的相应减少。

54. Bohannon J. Sailing the southern sea[J]. Science,2007,315:1520-1521.

Sailing the southern sea

Bohannon J.

(News)

Abstract: When John Martin was a researcher at Moss Landing Marine Laboratory in California, he proposed that massive blooms of photosynthetic plankton in the frigid Southern Ocean around Antarctica and other nutrient-rich but iron-starved waters could be an antidote to global warming. By pulling carbon dioxide out of the atmosphere to build their tiny bodies and then sequestering that carbon as they die and drift to the bottom of the ocean, these microscopic algae could reduce greenhouse gases and cool Earth.

南大洋航行

Bohannon J.

(新闻)

摘要:当约翰马丁在加利福尼亚州的莫斯兰丁海洋实验室当研究员的时候,他提出了大量光合浮游生物在南大洋靠近南极洲区域大量繁殖,其他营养丰富但是铁缺乏水域可以用来减缓全球变暖。通过吸收大气中的二氧化碳来生长,然后当它们死亡之后隐藏起那些碳,使其沉积到海底,这种显微藻可以减少温室气体并使地球降温。

55. Boyd P W, Jickells T, Law C S, et al. Mesoscale iron enrichment experiments 1993-2005: synthesis and future directions[J]. Science, 2007, 315: 612-617.

Mesoscale iron enrichment experiments 1993-2005: synthesis and future directions

Boyd P. W., Jickells T., Law C. S., Blain S., Boyle E. A., Buesseler K. O.

(National Institute for Water and Atmospheric Research (NIWA) Centre for Chemical and Physical Oceanography, Department of Chemistry, University of Otago, Dunedin, New Zealand)

Abstract: Since the mid-1980s, our understanding of nutrient limitation of oceanic primary production has radically changed. Mesoscale iron addition experiments (FeAXs) have unequivocally shown that iron supply limits production in one-third of the world ocean, where surface macronutrient concentrations are perennially high. The findings of these 12 FeAXs also reveal that iron supply exerts controls on the dynamics of plankton blooms, which in turn affect the biogeochemical cycles of carbon, nitrogen, silicon, and sulfur and ultimately influence the Earth climate system. However, extrapolation of the key results of FeAXs to regional and seasonal scales in some cases is limited because of differing modes of iron supply in FeAXs and in the modern and paleo-oceans. New research directions include quantification of the coupling of oceanic iron and carbon biogeochemistry.

1993～2005 年中尺度铁富集实验:综合和未来发展方向

Boyd P. W., Jickells T., Law C. S., Blain S., Boyle E. A., Buesseler K. O.

(新西兰奥塔哥大学国立水与大气研究中心化学和物理海洋学系)

摘要:20 世纪 80 年代中期以来,我们对海洋初级生产的养分限制的认识已经发生了根本改变。中尺度铁增加实验(FeAXs)已明确表明,在世界三分之一海洋铁的供给限制生产,那些海域表面营养素浓度常年都很高。这些铁增加实验调查结果还显示,铁的供给施加控制浮游生物大量繁殖,这反过来又影响了碳、氮、硅、硫,最终影响地球气候系统的生物地球化学的动态循环。然而,在某些情况下,区域和季节尺度 FeAXs 的主要结果的推断是有限的,因为铁的供给在 FeAXs 以及现代和古海洋模式不同。新的研究方向包括洋铁和碳的生物地球化学耦合的量化。

56. Roemmich D. Physical oceanography—super spin in the southern seas[J]. Nature,2007,449:34-35.

Physical oceanography—super spin in the southern seas

Roemmich D.

(Dean Roemmich is at the Scripps Institution of Oceanography, University of California, San Diego, La Jolla, California 92093-0230, USA)

Abstract: The southern oceans are generally considered as isolated systems, much like their northern counterparts. But a combination of historical data and new density profiles suggests that they may be connected on a global scale.

物理海洋学——在南部海域极好的拓展

Roemmich D.

(美国加州大学圣地亚哥分校斯克里普斯海洋研究所)

摘要:南大洋一般被认为是一个孤立的体系,就像它们的北部区域一样。但历史数据和新的密度外形的组合揭示了它们也许在全球尺度范围内是一个体系。

57. Smale D A,Brown K M,Barnes D K A,et al. Ice scour disturbance in Antarctic waters[J]. Science, 2008,321:371.

Ice scour disturbance in Antarctic waters

Smale D. A. , Brown K. M. , Barnes D. K. A. , Fraser K. P. P. , Clarke A.

(British Antarctic Survey (BAS), Natural Environment Research Council, High Cross, Madingley Road, Cambridge CB3 0ET, UK)

Abstract: The West Antarctic Peninsula is one of the fastest warming regions on Earth, and, as a consequence, most maritime glaciers and ice shelves in the region have significantly retreated over the past few decades. We collected a multiyear data set on ice scouring frequency from Antarctica by using unique experimental markers and scuba diving surveys. We show that the annual intensity of ice scouring is negatively correlated with the duration of the winter fast ice season. Because fast ice extent and duration is currently in decline in the region after recent rapid warming, it is likely that marine benthic communities are set for even more scouring in the near future.

冰冲入对南极水域的扰动

Smale D. A. , Brown K. M. , Barnes D. K. A. , Fraser K. P. P. , Clarke A.

（英国自然环境研究理事会南极调查局）

摘要：南极半岛西部是在地球上气候变暖速度最快的地区之一，因此，该地区大多数的海洋冰川和冰架在过去的几十年间已经显著撤退。我们收集了在南极洲采用独特的实验指标和潜水调查获得的多年的冰冲刷频率数据。我们展示的年度冰冲刷强度与冬季冰期持续时间呈负相关。因为海冰范围和持续时间在该地区近期快速变暖后呈现快速下跌，很可能海洋底栖群落在不久的将来会产生更多的冲刷。

58. Toggweiler J R，Russell J. Ocean circulation in a warming climate[J]. Nature，2008，451：286-288.

Ocean circulation in a warming climate

Toggweiler J. R. , Russell J.

（Geophysical Fluid Dynamics Laboratory，National Oceanic and Atmospheric Administration，Princeton，New Jersey 08542，USA）

Abstract：Climate models predict that the ocean's circulation will weaken in response to global warming，but the warming at the end of the last ice age suggests a different outcome.

温暖气候中的海洋循环

Toggweiler J. R. , Russell J.

（美国国家海洋和大气管理局地球物理流体动力学实验室）

摘要：气候模型预测，为了应对全球变暖，海洋环流将减弱，但在最后一个冰期的变暖显示了一个不同的结果。

5.3 摘要翻译——北极

1. Czipott P V, Levine M. D., Paulson. C. A. et al. Ice flexure forced by internal wave-packets in the Arctic Ocean[J]. Science, 1991, 254: 832-835.

Ice flexure forced by internal wave-packets in the Arctic Ocean

Czipott P. V., Levine M. D., Paulson. C. A., Menemenlis D., Farmer D. M., Williams R. G.

(SQM Technology, Inc., Post Office Box 2225, La Jolla, CA 92038)

Abstract: Tiltmeters on the Arctic Ocean were used to measure flexure of the ice forced by an energetic packet of internal waves riding the crest of diurnal internal bores emanating from the Yermak Plateau, north of the Svalbard Archipelago. The waves forced an oscillatory excursion of 36 microradians in tilt of the ice, corresponding to an excursion of 16 micrometers per second in vertical velocity at the surface and of 3.5 millimeters in surface displacement. Strainmeters embedded in the ice measured an excursion of 3×10^{-7} in strain, consistent with ice flexure rather than compression. The measured tilt is consistent with direct measurements of excursions in horizontal current near the surface (12 centimeters per second) and in vertical displacement (36 meters) of the pycnocline 100 meters below the surface.

北冰洋内波包控制的冰弯曲

Czipott P. V., Levine M. D., Paulson. C. A., Menemenlis D., Farmer D. M., Williams R. G.

(加拿大 SQM 技术公司)

摘要：北冰洋海域倾斜常被用来测定受到每日发源于北斯瓦尔巴群岛 Yermak 高原的大潮能量控制的冰扭曲状态。冰层顶部在波的强迫下，发生 36 微弧度的弯曲，相当于在水面发生每秒 16 μm 的漂移和水面 3.5 mm 的迁移。在冰中安放的应变计测得了 3×10^{-7} 的变化，这与冰扭曲，而不是压缩一致。测量的倾斜与水面的流动速度(12 cm/s)和密度跃变层的在水面 100 m 下的垂直迁移(36 m)一致。

2. Hebbeln D, Wefer G. Effects of ice coverage and ice-rafted material on sedimentation in the Fram strait[J]. Nature, 1991, 350: 409-411.

Effects of ice coverage and ice-rafted material on sedimentation in the Fram strait

Hebbeln D., Wefer G.

(Universität Bremen, Geowissenschaften, Postfach 330440, 2800 Bremen 33, FRG)

Abstract: As little is known about pelagic sedimentation processes in Arctic environments, the interpretation of biological and chemical processes, as well as the reconstruction of ancient conditions, including those in the glacial North Atlantic, is difficult. Here we provide sediment-trap results, which show that the position of the sea-ice boundary significantly influences the particle flux. The seasonal variability of the particle flux differed markedly in the various sediment-trap sites in Fram Strait, depending on the behaviour of the sea ice. Under complete ice cover, sedimentation is very low, whereas maximum sedimentation is found at the ice margin. The highest particle flux observed, showing a large lithogenic component, was observed at the ice edge where the water was warmer (>2-degrees-C). We find that high biogenic opal fluxes are characteristic of the summer ice margin, indicating that the sedimentary record of opal fluxes may allow the position of ice margins in the past to be reconstructed.

冰覆盖和冰筏物对弗拉姆海峡沉积物的影响

Hebbeln D. , Wefer G.

(德国不莱梅大学)

摘要:北冰洋远洋沉积过程很少为人熟知,试图解释其生物和化学过程,以及重建古代环境条件,包括那些在北大西洋的冰川很困难。在这里,我们报道沉积物捕集的结果,结果表明,海冰边界的位置显著影响颗粒通量。弗拉姆海峡的站位在不同季节,沉积物捕集的颗粒通量的变化明显不同,可能是由于海冰的行为变化所致。完整的冰层覆盖下,沉积率是非常低的,而最大的沉积率发生在冰的边缘。最高的颗粒通量发生在温暖的冰面边缘(>2 ℃)并出现大颗粒成岩碎屑。我们发现,高的生源蛋白石通量与夏季冰盖边缘特征性相关,表明蛋白石通量的沉积记录,可能重建过去冰边缘的位置。

3. Macdonald R W, Carmack E C. Age of Canada basin deep waters—a way to estimate primary production for the Arctic Ocean[J]. Science, 1991, 254:1348-1350.

Age of Canada basin deep waters—a way to estimate primary production for the Arctic Ocean

Macdonald R. W. , Carmack E. C.

(Institute of Ocean Sciences, Post Office Box 6000, Sidney, British Columbia, Canada V8L 4B2)

Abstract: An empirical model of carbon flux and ^{14}C derived ages of the water in the Canada Basin of the Arctic Ocean as a function of depth was used to estimate the long-term rate of primary production within this region. An estimate can be made because the deep waters of the Canadian Basin are isolated from the world oceans by the Lomonosov Ridge (sill depth about 1500 meters). Below the sill, the age of the water correlates with increased nutrients and oxygen utilization and thus provides a way to model the average flux of organic material into the deep basin over a long time period. The ^{14}C ages of the deep water in the Canada Basin were about 1000 years, the carbon flux across the 1500 meter isobath was 0. 3 gram of carbon per square meter per year, and the total production was 9 to 14 grams of carbon per square meter per year. Such estimates provide a baseline for understanding the role of the Arctic Ocean in global carbon cycling.

加拿大海盆深海水的年龄——一种估计北极海洋初级生产的方式

Macdonald R. W. , Carmack E. C.

(加拿大悉尼海洋研究所)

摘要:一个从深度估算北冰洋加拿大海盆的碳通量和^{14}C年龄的经验模型被用来估算在本地区的初级生产力。这个因为加拿大海盆的深海水域被罗蒙诺索夫山脊(1500 m 的深度)从其他大洋隔离是可能的。在此高度下,水的年龄与增加营养物质和氧的利用率相关,这提供了一种在较长时间内模拟深海洋盆有机物质的平均通量。加拿大海盆水^{14}C年龄大约为1000年,跨越1500 m等深线的碳通量大约为0.3 g/(m^3·年)碳,总产量为9～14 g/(m^3·年)碳。这种估计为了解北冰洋在全球碳循环中的作用提供了一个基准。

4. Schlosser P, Bonisch G, Rhein M, et al. Reduction of deep-water formation in the Greenland Sea during the 1980s—evidence from tracer data[J]. Science, 1991, 251: 1054-1056.

Reduction of deep-water formation in the Greenland Sea during the 1980s—evidence from tracer data

Schlosser P. , Bonisch G. , Rhein M. , Bayer R.

(Lamont-Doherty Geological Observatory of Columbia University and Department of Geological Sciences, Palisades, NY 10964)

Abstract: Hydrographic observations and measurements of the concentrations of chlorofluorocarbons (CFCs) have suggested that the formation of Greenland Sea deep water (GSDW) slowed down considerably during the 1980s. Such a decrease is related to weakened convection in the Greenland Sea and thus could have significant impact on the properties of the waters flowing over the Scotland-Iceland-Greenland ridge system into the deep Atlantic. Study of the variability of GSDW formation is relevant for understanding the impact of the circulation in the European Polar seas on regional and global deep water characteristics. New long-term multitracer observations from the Greenland Sea show that GSDW formation indeed was greatly reduced during the 1980s. A box model of deepwater formation and exchange in the European Polar seas tuned by the tracer data indicates that the reduction rate of GSDW formation was about 80 percent and that the start date of the reduction was between 1978 and 1982.

20世纪80年代格陵兰海深水形成的减少——来自跟踪物的证据

Schlosser P. , Bonisch G. , Rhein M. , Bayer R.

(美国哥伦比亚大学拉蒙特-多尔蒂地质观测站)

摘要:氯氟烃(CFCs)的浓度水文观察和测量显示20世纪80年代形成的格陵兰海深水(GSDW)大大减缓。这种下跌与格陵兰海对流减弱相关,从而可以对苏格兰-冰岛-格陵兰脊系统流入大西洋的深水域的属性产生重大影响。GSDW 形成的变异的研究有助于理解欧洲极地海洋环流对区域和全球的深层水特性的影响。新的长期多重示踪研究显示,20 世纪 80 年代格陵兰海 GSDW 形成确实是大大减少了。基于示踪数据的欧洲极地海洋深水形成和交换的盒模型表明，GSDW 形成减少了约 80%,减少的开始日期在 1978～1982 年之间。

5. Mclaren A S,Walsh J E,Bourke R H,et al. Variability in sea-ice thickness over the North-Pole from 1977 to 1990[J]. Nature,1992,358:224-226.

Variability in sea-ice thickness over the North-Pole from 1977 to 1990

Mclaren A. S. , Walsh J. E. , Bourke R. H. , Weaver R. L. , Wittmann W.

(Lamont-Doherty Observatory, Columbia University, Palisades, New York 10964 and Science Service, Inc. , Washington DC, 20036, USA)

Abstract: Changes in the thickness of polar sea-ice have the potential to provide a signal of climate change, but attempts to identify trends must take into account the range of natural variability. Here we present an analysis of measurements of the subsurface ice thickness of sea-ice around the North Pole made from 1977 to 1990. These data were collected during six submarine cruises in late April/early May of 1977, 1979, 1986, 1987, 1988 and 1990, and represent the most extensive dataset so far for ice draft in the central Arctic at the same season and location. The results reveal considerable interannual variability both in mean ice draft (±1. 0 m) and in open-water extent (±2. 5%). This variability limits the confidence that can be placed in any apparent trends observed for sea-ice thickness or type since the late 1970s, and illustrates the need for a reliable baseline against which to assess future trends.

1977～1990 年北极海冰厚度的变化

Mclaren A. S. , Walsh J. E. , Bourke R. H. , Weaver R. L. , Wittmann W.

(美国哥伦比亚大学拉蒙特-多尔蒂地质观测站)

摘要:极地海冰厚度变化能够提供气候变化的信号,但这种趋势必须考虑到自然变化本身的范围。在这里,我们公布从 1977～1990 年的北极海冰(图案)的水面下冰层厚度的测量分析结果。这些数据来自 6 个潜艇在 4 月底、5 月初巡航的调查结果,年代包括 1977 年、1979 年、1986 年、1987 年、1988 年和 1990 年,代表同一季节和地点在中央北极冰迄今为止最广泛的数据集。结果表明无论是在平均浮冰(±1. 0 m),或在开阔水域范围(±2. 5%)都有相当大的年际变化。这种变化限制了对 20 世纪 70 年代后期以来海冰厚度或类型中观察到任何明显趋势的可信度,说明需要一个可靠的基线,以评估未来的趋势。

6. Sturges W T,Cota G F,Buckley P T. Bromoform emission from Arctic ice algae[J]. Nature,1992,358:660-662.

Bromoform emission from Arctic ice algae

Sturges W. T., Cota G. F., Buckley P. T.

(Cooperative Institute for Research in Environmental Sciences, University of Colorado, Boulder, Colorado, 80309-0449, USA)

Abstract: Destruction of surface ozone in the Arctic environment during the spring is thought to be caused by photochemical reactions involving bromine compounds. Berg et al. reported a pulse of bromine particles and gases in the Arctic lower atmosphere in spring, which may be responsible for this surface ozone destruction and for which biogenic sources have been hypothesized. Here we report laboratory and in situ measurements which indicate that Arctic ice microalgae emit significant quantities of bromoform ($CHBr_3$), which may be converted photochemically into active forms of bromine. Our estimates of total annual bromoform release indicate that polar ice algae might contribute globally significant amounts of organic bromine compounds, comparable with anthropogenic and macrophyte sources.

北极冰藻释放的三溴甲烷

Sturges W. T., Cota G. F., Buckley P. T.

(美国科罗拉多大学环境科学合作研究所)

摘要：在春季，北极表层臭氧的破坏被认为与涉及溴代分子的光化学反应有关。Berg 等报道了春季北极地区低层大气中溴颗粒和气体的波动，这可能就是这个地区表面臭氧的破坏被假定是生物来源的原因。在这里，我们报告的实验室模拟和现场测量显示，北极海冰微藻放出大量的三溴甲烷($CHBr_3$)，通过光化学转化为主动形式的溴。我们的估计表明，对年度总溴甲烷释放，极地冰藻相比于人类和大型植物可能更大地影响全球范围内有机溴化合物的释放。

7. Kahl J D, Charlevoix D J, Zaitseva N A, et al. Absence of evidence for greenhouse warming over the Arctic-Ocean in the past 40 years[J]. Nature, 1993, 361: 335-337.

Absence of evidence for greenhouse warming over the Arctic-Ocean in the past 40 years

Kahl J. D., Charlevoix D. J., Zaitseva N. A., Schnell R. C., Serreze M. C.

(Department of Geosciences, university of Wisconsin-Miluaukee PO Box 413, Milwaukee, Wisconsin 53201, USA)

Abstract: Atmospheric general circulation models predict enhanced greenhouse warming at high latitudes owing to positive feedbacks between air temperature, ice extent and surface albedo. Previous analyses of Arctic temperature trends have been restricted to land-based measurements on the periphery of the Arctic Ocean. Here we present temperatures measured in the lower troposphere over the Arctic Ocean during the period 1950-90. We have analysed more than 27000 temperature profiles, measured by radiosonde at Russian drifting ice stations and by dropsonde from US "Ptarmigan" weather reconnaissance aircraft, for trends as a function of season and altitude. Most of the trends are not statistically significant. In particular, we do not observe the large surface warming trends predicted by models; indeed, we detect significant surface cooling trends over the western Arctic Ocean during winter and autumn. This discrepancy suggests that present climate models do not adequately incorporate the physical processes that affect the polar regions.

过去 40 年在北极海洋的温室效应缺少证据

Kahl J. D., Charlevoix D. J., Zaitseva N. A., Schnell R. C., Serreze M. C.

（美国威斯康星大学地质科学系）

摘要：大气环流模型预测高纬度地区由于空气温度、冰面积和表面返照率会有增强的温室效应。以前对北极地区的温度变化趋势的分析被限制在北冰洋周围陆地。在这里，我们公布北冰洋对流层低层 1950～1990 年期间测量的温度数据。我们已经使用俄罗斯浮冰站和美国雷鸟飞机无线电测空仪测定的超过 27000 条的温度曲线，分析了季节和海拔高度对温度变化趋势的影响。大多数变化趋势并不显著。特别是，我们没有观察到趋势模型预测的大的表面变暖，事实上，我们发现在冬季和秋季北冰洋西部重要的地面有冷却的趋势。这种差异表明，目前的气候模型没有充分纳入影响极地地区的物理过程。

8. Macdonald R W, Carmack E C, Wallace D W R. Tritium and radiocarbon dating of Canada basin deep waters[J]. Science, 1993, 259: 103-104.

Tritium and radiocarbon dating of Canada basin deep waters

Macdonald R. W., Carmack E. C., Wallace D. W. R.

(Marine research institute, Canada)

Abstract: Radiocarbon data were applied to determine the rates of nitrate (NO_3) regeneration and oxygen (O_2) depletion in the Canada Basin of the Arctic Ocean. A one-dimensinal, time-dependent diffusion model was used to caliabrate the age of the deep water. The combination of this model with a model of an organic carbon (C) flux, suggested how C fluxes into the basin might reflect primary productivity within the central Arctic Ocean. Errors in the age estimates of basin waters are corrected and suggestions made concerning where the basin-derived estimate of C flux might fit into the larger scheme of primary production in the Arctic Ocean.

加拿大洋盆深水的氚和放射性碳定年

Macdonald R. W., Carmack E. C., Wallace D. W. R.

（加拿大海洋研究所）

摘要：放射性碳测定年代数据应用于确定北冰洋加拿大盆地硝酸盐(NO_3)的再生率和氧气(O_2)消耗。时变扩散模型被用来计算深水的年龄。这个模型组合模拟出有机碳(C)通量，提出 C 通量到盆地如何反映在北冰洋中央初级生产力上。流域水域中的年龄估计进行了纠错，同时提出在北冰洋关于C通量的估计可能融入更大的主生产计划。

9. Martin J H, Coale K H, Johnson K S, et al. Testing the iron hypothesis in ecosystems of the equatorial Pacific-Ocean[J]. Nature, 1994, 371: 123-129.

Testing the iron hypothesis in ecosystems of the equatorial Pacific-Ocean

Martin J. H., Coale K. H., Johnson K. S., Fitzwater S. E., Gordon R. M., Tanner S. J., Hunter C. N., Elrod V. A., Nowick J. L., Colex T. L., Barber R. T., Lindley S., Watson A. J., Van Scoy K., Law C. S., Liddicoat M. I., Ling R., Stanton T., Stockel J., Collins C., Anderson A., Bidigare R., Ondrusek M., Latasa M., Millero F. J., Lee K., Yao W., Zhang J. Z., Friederich G., Sakamoto C., Charez F., Buck K., Kolber Z., Greene R., Falkowski P., Chisholm S. W., Hoge F., Swift R., Yungle J., Turner S., Nightingale P., Hatton A., Liss P., Tindale N. W.

(Moss Landing Marine Laboratories, Moss Landing Marine Labs, California City, California, United States)

Abstract: The idea that iron might limit phytoplankton growth in large regions of the ocean has been tested by enriching an area of 64 km^2 in the open equatorial Pacific Ocean with iron. This resulted in a doubling of plant biomass, a threefold increase in chlorophyll and a fourfold increase in plant production. Similar increases were found in a chlorophyll-rich plume downstream of the Galapagos Islands, which was naturally enriched in iron. These findings indicate that iron limitation can control rates of phytoplankton productivity and biomass in the ocean.

赤道太平洋生态系统的铁盐假说

Martin J. H., Coale K. H., Johnson K. S., Fitzwater S. E., Gordon R. M., Tanner S. J., Hunter C. N., Elrod V. A., Nowick J. L., Colex T. L., Barber R. T., Lindley S., Watson A. J., Van Scoy K., Law C. S., Liddicoat M. I., Ling R., Stanton T., Stockel J., Collins C., Anderson A., Bidigare R., Ondrusek M., Latasa M., Millero F. J., Lee K., Yao W., Zhang J. Z., Friederich G., Sakamoto C., Charez F., Buck K., Kolber Z., Greene R., Falkowski P., Chisholm S. W., Hoge F., Swift R., Yungle J., Turner S., Nightingale P., Hatton A., Liss P., Tindale N. W.

(美国莫斯兰丁海洋实验室)

摘要：在64 km^2 赤道的太平洋开放海域，铁可能会限制中大区域的海洋浮游植物生长的想法已经过测试。这一结果导致植物生物量增加一倍、叶绿素增加三倍和植物生产率增加四倍。类似的升幅出现在加拉巴哥群岛富含叶绿素的羽流下游，其中含有自然丰富的铁。这些研究结果表明铁可以控制浮游植物的生产力和生物量在海洋中的生产效率。

10. Johannessen O M, Bjorgo E, Miles M W. Global warming and the Arctic[J]. Science, 1996, 271: 129.

Global warming and the Arctic

Johannessen O. M. , Bjorgo E. , Miles M. W.

(Nansen Environmental and Remote Sensing Center, Bergen)

Abstract: In the News article"Polar regions give cold shoulder to theories"by Dennis Normile (8 Dec. p. 1566), John Walsh of the University of Minois is said to note the absence of retreating sea ice, and H. Jay Zwally of Goddard Space Flight Center says his search for long-term trends in ice cover, based on a review ofsatellite-based remote sensing of polar ice, "has given ambiguous results". These comments are contrary to the enhanced warming in the Arctic region predicted by the global climate model developed by the Hadley Centre for Climate Prediction and Research in Bracknell, United Kongdom(J). These statements are also contrary to what we find.

北极与全球变暖

Johannessen O. M. , Bjorgo E. , Miles M. W.

(挪威卑尔根曼森环境和遥感中心)

摘要:在 12 月 8 日第 1566 页的新闻文章《极地冷落了理论》中,伊利诺伊大学的 Dennis Normile 和 John Walsh 提出要注意海冰消融的停滞。同时戈达德宇宙飞行中心的 H. Jay Zwally 提出,基于卫星得到的极地冰盖遥感数据,对冰盖的长期变化趋势"给了不明确的结果"。这些评论与英国布拉克内尔的哈德利气候预测和研究中心通过全球气候模型预测的北极地区变暖增强不一致。这些声明也与我们的发现不同。

11. Macdonald R. Awakenings in the Arctic[J]. Nature, 1996, 380: 286-287.

Awakenings in the Arctic

Macdonald R.

(Institute of Ocean Sciences, PO Box 6000, Sidney, British Columbia, Canada V8L 4BZ)

Abstract: Until only a few years ago, the Arctic Ocean was considered to be a quiet, unproductive backwater, hardly changing or causing change in the oceans to the sparseness of observations in the interior ocean, inaccessible because of the ice, but there seems also to be a consensus emerging that a change has recently occurred in the Arctic Ocean's circulation. A layer of water that derives from the Atlantic is warming up, and the exchange between the Arctic Ocean's two major basins-Canadian and Eurasian-has somehow altered.

北极苏醒

Macdonald R.

(加拿大海洋科学研究所)

摘要:直到数年前,北冰洋一直被认为是一个安静且没有生产力的不流动水体,几乎不能改变或引起海洋的变化,这是根据因海冰覆盖导致稀少的内部海洋观测得出的。但近期似乎一致显露北冰洋环流出现的一个变化。一个来自大西洋的水层在升温,而北冰洋两个主要的流域——加拿大和欧亚流域的交换以某种方式发生了改变。

12. Polzin K L,Speer K G,Toole J M,et al. Intense mixing of Antarctic bottom water in the equatorial Atlantic Ocean[J]. Nature,1996,380:54-57.

Intense mixing of Antarctic bottom water in the equatorial Atlantic Ocean

Polzin K. L., Speer K. G., Toole J. M., Schmitt R. W.

(University of Washington, School of Oceanography, Box 357940, Seattle, Washington 98195, USA)

Abstract: The spreading of Antarctic Bottom Water—the densest global-scale water mass—is highly constrained by ocean-floor topography. In the Atlantic Ocean, the Mid-Atlantic Ridge confines this water mass mainly to the western basins, the bottom waters in the eastern basins being renewed by flows through gaps in the ridge. One such gap is the Romanche fracture zone, a large offset of the ridge which straddles the Equator. It has been observed that sills within this fracture zone block the passage of waters colder than similar to 0. 9 degrees C; warmer, less dense waters passing over the sills appear to cascade downslope where they are modified by mixing. Here we present direct measurements which quantify these processes. The flow is vertically sheared and exhibits remarkably intense turbulence, comparable to that seen at the ocean surface in the presence of winds of similar to 10 $m \cdot s^{-1}$. This turbulence mixes the densest waters passing through the fracture zone with the warmer, overlying waters, so that the coldest waters exiting this region have been warmed by similar to 0. 6 degrees C during transit, Topographic obstructions and turbulent mixing together thus determine the properties of the flows renewing the deepest waters of the Atlantic Ocean's eastern basins.

南极洋底水在大西洋近赤道海域的强烈混合

Polzin K. L., Speer K. G., Toole J. M., Schmitt R. W.

(美国华盛顿大学海洋学院)

摘要:南极底层水(全球密度最大的水团)的扩散受到洋底地形的限制。在大西洋,中部洋脊把水团限制在西部洋盆,而东部洋盆的底水通过洋脊的沟得到更新。一个这样的沟是罗曼断裂带,据观察该脊跨越赤道。这个断裂带台面阻止低于 0. 9 ℃的水通过;较温暖的高密度的水,通过像瀑布一样流过豁口后,被混合。在这里,我们发表对这些过程的直接测量结果。洋流是垂直剪切流,并表现出显著的激烈扰动,这与海洋表面存在 10 m/s 风时差不多。这个不稳定将通过断裂带高密度水与暖的上覆水域混合,所以这水在传输过程中冷水升温了 0. 6 ℃。地形障碍物和湍流混合在一起,确定大西洋东部盆地深水流动更新。

13. Wheeler P A, Gosselin M, Sherr E, et al. Active cycling of organic carbon in the central Arctic Ocean[J]. Nature,1996,380:697-699.

Active cycling of organic carbon in the central Arctic Ocean

Wheeler P. A., Gosselin M., Sherr E., Thibault D., Kirchman D. L., Benner R., Whitledge T. E.

(College of Oceanic and Atmospheric Sciences, Oregon State University, Corvallis, Oregon 97331, USA)

Abstract: The notion of a barren central Arctic Ocean has been accepted since English's pioneering work on drifting ice-islands. The year-round presence of ice, a short photosynthetic season and low temperatures were thought to severely limit biological production, although the paucity of data was often noted. Because primary production appeared to be low, subsequent studies assumed that most organic carbon was either derived from river inputs or imported from adjacent continental-shelf regions. Here we present shipboard measurements of biological production, biomass and organic carbon standing-stocks made during a cruise through the ice covering the central Arctic Ocean. Our results indicate that the central Arctic region is not a biological desert. Although it is less productive than oligotrophic ocean regions not covered by ice, it supports an active biological community which contributes to the cycling of organic carbon through dissolved and particulate pools.

北冰洋中心活跃的有机碳循环

Wheeler P. A., Gosselin M., Sherr E., Thibault D., Kirchman D. L., Benner R., Whitledge T. E.

(美国俄勒冈州立大学海洋和大气科学学院)

摘要:因为英国人在漂浮的冰岛上的开创性工作,一个荒芜的中央北冰洋的概念已被广泛接受。虽然经常指出的数据不足,但全年覆冰,简短的光合季节,气温低,还是被认为严重限制了生物的生产力。由于初级生产力低,随后的研究认为,大多数有机碳要么来自河流输入,要么从邻近的大陆架地区输入。在这里,我们发布船载仪器测量的北冰洋中心区域的生物生产力、生物量和有机碳库。我们的研究结果表明,北极中部地区不是一个生命沙漠。虽然这是一个比贫营养海域更贫瘠的海区,但通过溶解颗粒物,它拥有一个活跃的有机碳循环的生物群落。

14. Dickson B, Meincke J, Vassie I, et al. Possible predictability in overflow from the Denmark Strait [J]. Nature, 1999, 397: 243-246.

Possible predictability in overflow from the Denmark Strait

Dickson B., Meincke J., Vassie I., Jungclaus J., Osterhus S.

(Centre for Environment, Fisheries and Aquaculture Science, Lowestoft, Suffolk NR33 0HT, UK)

Abstract: The overflow and descent of cold dense water from the Denmark Strait sill-a submarine passage between Greenland and Iceland-is a principal means by which the deep ocean is ventilated, and is an important element in the global thermohaline circulation. Previous investigations of its variability-in particular, direct current measurements in the overflow core since 1986-have shown surprisingly little evidence of long-term changes in now speed. Here we report significant changes in the overflow characteristics during the winter of 1996-97, measured using two current-meter moorings and an inverted echo sounder located at different depths in the fastest part of the now. The overflow warmed to the highest monthly value yet recorded (2.4 degrees C), and showed a pronounced slowing and thinning at its lower margin. We believe that the extreme warmth of the overflow caused it to run higher on the continental slope off east Greenland, so that the lower current meters and the echo sounder were temporarily outside and deeper than the fast-flowing core; model simulations appear to confirm this interpretation, we suggest that the extreme warmth of the overflow is a lagged response to a warming upstream in the Fram Strait three years earlier (caused by an exceptional amplification of the winter North Atlantic Oscillation). If this is so, over-now characteristics may be predictable.

丹麦海峡溢流可能的可预测性

Dickson B., Meincke J., Vassie I., Jungclaus J., Osterhus S.

(英国渔业和水产养殖科学中心)

摘要:丹麦海峡冷重水溢出和下沉通过格陵兰岛和冰岛之间是深海水交换的主要方式,并且是全球温盐环流的重要组成部分。自 1986 年以前的调查,尤其直接测量的流量变化数据,表明长期变化出奇地小。在这里,我们报道来源于现在流速最快的位置停泊处不同深度的两个倒置的回声测深仪测试结果,显示在 1996～1997 年冬季其溢出特征有显著变化。流出水暖到从没有记录的最高月温度(2.4 ℃),并在边际区域呈现出一个显著放缓变薄的趋势。我们认为,极端温暖使得洋流可以在格陵兰岛的大陆坡更快流过,使电流测试器和回声测深仪暂时离开或者深于水流湍急的核心;模型模拟似乎证实了这一点解释,我们建议,洋流的极端温暖是 3 年前弗拉姆海峡的气候变暖的延后反应(造成冬季北大西洋涛动的放大)。如果是这样,现在一些环境因子是可预测的。

15. Shindell D T, Miller R L, Schmidt G A, et al. Simulation of recent northern winter climate trends by greenhouse-gas forcing[J]. Nature, 1999, 399: 452-455.

Simulation of recent northern winter climate trends by greenhouse-gas forcing

Shindell D. T., Miller R. L., Schmidt G. A., Pandolfo L.

(NASA Goddard Institute for Space Studies, 2880 Broadway, New York, New York 10025, USA)

Abstract: The temperature of air at the Earth's surface has risen during the past century, but the fraction of the warming that can be attributed to anthropogenic greenhouse gases remains controversial. The strongest warming bends have been over Northern Hemisphere land masses during winter, and are closely related to changes in atmospheric circulation. These circulation changes are manifested by a gradual reduction in high-latitude sea-level pressure, and an increase in mid-latitude sea-level pressure associated with one phase of the Arctic Oscillation (a hemisphere-scale version of the North Atlantic Oscillation). Here we use several different climate-model versions to demonstrate that the observed sea-level-pressure trends, including their magnitude, can be simulated by realistic increases in greenhouse-gas concentrations, Thus, although the warming appears through a naturally occurring mode of atmospheric variability, it may be anthropogenically induced and may continue to rise. The Arctic

Oscillation trend is captured only in climate models that include a realistic representation of the stratosphere, while changes in ozone concentrations are not necessary to simulate the observed climate trends. The proper representation of stratospheric dynamics appears to be important to the attribution of climate change, at least on a broad regional scale.

温室气体对近期北方地区气候变化趋势驱动的模拟

Shindell D. T., Miller R. L., Schmidt G. A., Pandolfo L.

(美国航空航天局戈达德宇宙飞行中心)

摘要：过去一个世纪以来，人类活动所导致地球气温的上升的部分始终有争议。而北半球陆地冬季气温的上升是最为强烈的，这与大气环流模式的变化有关。主要的转变是在一个北极振荡周期内，高纬度海平面压力的梯度减少和中纬度气压的增加(半球尺度的北大西洋振荡)。我们使用不同气候模型去演示观测海平面压力变化趋势，包括其由于温室气体增加造成的变化。在模型中，真实的温室气体增加可导致这些变化。因此，气温上升似乎来自自然发生的大气变化，但是它可能是人类活动导致的，并可能继续变暖。北极振荡趋势仅在包含有平流层状态描述的气候模式时得到体现，不包括臭氧层的变化。平流层动力学的合适表述在一个很广泛范围内对气候变化起到重要作用。

16. Watson A J, Messias M J, Fogelqvist E, et al. Mixing and convection in the Greenland Sea from a tracer-release experiment[J]. Nature, 1999, 401: 902-904.

Mixing and convection in the Greenland Sea from a tracer-release experiment

Watson A. J., Messias M. J., Fogelqvist E., Van Scoy K. A., Johannessen T., Oliver K. I. C., Stevens D. P., Rey F., Tanhua T., Olsson K. A., Carse F., Simonsen K., Ledwell J. R., Jansen E., Cooper D. J., Kruepke J. A., Guilyardi E.

(School of Environmental Sciences, University of East Anglia, Norwich NR4 7TJ, UK)

Abstract: Convective vertical mixing in restricted areas of the subpolar oceans, such as the Greenland Sea, is thought to be the process responsible for forming much of the dense water of the ocean interior. Deep-water formation varies substantially on annual and decadal timescales, and responds to regional climate signals such as the North Atlantic Oscillation; its variations may therefore give early warning of changes in the thermohaline circulation that may accompany climate changes. Here we report direct measurements of vertical mixing, by convection and by turbulence, from a sulphur hexafluoride tracer-release experiment in the central Greenland Sea gyre. In summer, we found rapid turbulent vertical mixing of about 1.1 $cm^2 \cdot s^{-1}$. In the following late winter, part of the water column was mixed more vigorously by convection, indicated by the rising and vertical redistribution of the tracer patch in the centre of the gyre. At the same time, mixing outside the gyre centre was only slightly greater than in summer. The results suggest that about 10% of the water in the gyre centre was vertically transported in convective plumes, which reached from the surface to, at their deepest, 1200-1400 m. Convection was limited to a very restricted area, however, and smaller volumes of water were transported to depth than previously estimated. Our results imply that it may be the rapid year-round turbulent mixing, rather than convection, that dominates vertical mixing in the region as a whole.

通过示踪剂释放试验研究格陵兰海域混合传送过程

Watson A. J., Messias M. J., Fogelqvist E., Van Scoy K. A., Johannessen T., Oliver K. I. C., Stevens D. P., Rey F., Tanhua T., Olsson K. A., Carse F., Simonsen K., Ledwell J. R., Jansen E., Cooper D. J., Kruepke J. A., Guilyardi E.

(英国东安格利亚大学环境科学学院)

摘要：在副极地海域，如格陵兰海，垂直方向的混合传送被认为是海区内部重海水形成的重要推手。深部海水的形成有明显的年际和年代际区，这与区域气候信号，如北大西洋振荡有关。所以这种变化会改变温盐环流状态，从而为气候变化提供早期预警。我们通过格陵兰中心海区的六氟化硫的示踪试验测量传送和湍流进行的垂直混合。在夏季，快速的湍流垂直混合速度达到 1.1 cm/s。在接下来的冬末，水柱的传送混合更剧烈，说明环流中心上升和示踪剂垂直再分配。而混合环流中心以外区域，仅略高于夏季。结果表明，环流中心约 10%的水从表面被对流垂直传送到 1200～1400 m 深度。对流被限制在一个很严格的区域，但是小水柱可以被传送到比前面更深的位置。我们认为快速的年际的挛流混合比对流更强地控制了这一区域的垂直混合。

17. Wuethrich B. Climate change—new center gives Japan an Arctic toehold[J]. Science, 1999, 285:1827.

Climate change—new center gives Japan an Arctic toehold

Wuethrich B.

(An exhibit writer at the Smithsonian's National Museum of Natural History)

Abstract: Fairbanks, Alaska—Japan and the United States have launched a $32 million research center to plumb the consequences of climate change in the Far North. Late last month, researchers from the two countries gathered here on the campus of the University of Alaska, Fairbanks, to dedicate the International Arctic Research Center. Guided by representatives from several Japanese and U. S. organizations, the center plans to do "big picture, big science".

By compiling satellite data on everything from sea ice to vegetation patterns in the Arctic, where the effects of global warming are being felt first, IARC hopes to do some informed crystal-ball gazing. Most of the center's funding is expected to come from Japanese and U. S. agencies and universities. Besides supporting outside scientists, the center hopes to expand its 50 person research staff to 150 by 2005. The upshot, says Syun-Ichi Akasofu, IARC's U. S. director, will be clearer forecasts of global climate and of the ways perturbations in the Arctic might influence northern countries. "The ultimate purpose of IARC is to make it possible to predict global change," adds Taro Matsuno, director-general of Japan's Frontier Research Program, a key player at IARC.

气候变化——日本立足北极的新中心

Wuethrich B.

(史密斯国家自然历史博物馆)

摘要：费尔班克斯，阿拉斯加——日本和美国斥资3200万美元启动了一个研究中心来探测远北极地区气候变化的影响。上月下旬，两国科学家聚集在费尔班克斯的阿拉斯加大学校园内致力于成立国际北极研究中心（IARC）。在一些来自日本和美国组织机构代表的领导下，中心计划进行“大图像，大科学”。

通过编辑北冰洋各种各样从海冰到植物模式的卫星数据，全球变暖的效果首先被发现，而IARC希望完成一些详尽的预测。中心大部分研究经费期望来自日本和美国的机构和大学。除支持外部科学家外，中心希望到2005年能将研究人员规模从50人扩大至150人。IARC美国主任Syun-Ichi Akasofu说，研究结果将会明晰全球变化的预测和北冰洋扰动可能对北部国家的影响方式。“IARC的最终目的是使得预测全球变化成为可能”，IARC主要成员、日本前沿研究计划总干事Taro Matsuno补充道。

18. Shen Y A，Buick R，Canfield D E. Isotopic evidence for microbial sulphate reduction in the early Archaean era[J]. Nature，2001，410：77-81.

Isotopic evidence for microbial sulphate reduction in the early Archaean era

Shen Y. A.，Buick R.，Canfield D. E.

(Danish Center for Earth System Science (DCESS) and Institute of Biology，Odense University，SDU，Campusvej 55，5230 Odense M，Denmark)

Abstract：Sulphate-reducing microbes affect the modern sulphur cycle，and may be quite ancient，though when they evolved is uncertain. These organisms produce sulphide while oxidizing organic matter or hydrogen with sulphate. At sulphate concentrations greater than 1 mM，the sulphides are isotopically fractionated (depleted in S-34) by 10-40 parts per thousand compared to the sulphate，with fractionations decreasing to near 0 parts per thousand at lower concentrations. The isotope record of sedimentary sulphides shows large fractionations relative to seawater sulphate by 2.7 Gyr ago，indicating microbial sulphate reduction. In older rocks，however，much smaller fractionations are of equivocal origin，possibly biogenic but also possibly volcanogenic. Here we report microscopic sulphides in similar to 3.47 Gyr-old barites from North Pole，Australia，with maximum fractionations of 21.1 parts per thousand，about a mean of 11.6 parts per thousand，clearly indicating microbial sulphate reduction. Our results extend the geological record of microbial sulphate reduction back more than 750 million years，and represent direct evidence of an early specific metabolic pathway—allowing time calibration of a deep node on the tree of life.

早太古代微生物硫酸还原的同位素证据

Shen Y. A. , Buick R. , Canfield D. E.

（丹麦欧登塞大学丹麦地球系统科学中心和生物学研究所）

摘要：硫酸还原菌会影响到现代硫循环，而且可以追溯到古代，虽然其进化过程还不清楚。这些生物体在氧化有机物或者给还原硫酸盐过程中会产生硫化物，在硫酸盐浓度高于 1 mmol 时，硫化物发生分馏（S-34 亏损），其速度一般为 10‰～40‰，但是在较低浓度下分馏速度接近于 0。沉积物中的硫化物同位素结果显示大约 27 亿年前相对于海水发生了大的分馏，表明了硫酸还原菌的存在。然而，在老的岩石中，分馏要小得多，这可能是生物原因，也可能是火山过程。澳大利亚重晶石中小的硫化物具有 34.7 亿年前的北极重晶石同样的分馏值，最大为 21.1‰，平均为 11.6‰，清楚表明微生物硫酸还原的存在。我们的结果使得微生物硫酸还原的地质记录拓展到 7.5 亿年前，表明这是一个早期特定的代谢过程——这就为我们对“生命树”节点的时间校准提供了依据。

19. Bishop J K B, Davis R E, Sherman J T. Robotic observations of dust storm enhancement of carbon biomass in the North Pacific[J]. Science, 2002, 298: 817-821.

Robotic observations of dust storm enhancement of carbon biomass in the North Pacific

Bishop J. K. B. , Davis R. E. , Sherman J. T.

(Earth Sciences Division, Lawrence Berkeley National Laboratory, 1 Cyclotron Road, MS 90-1116, Berkeley, CA 94708, USA)

Abstract: Two autonomous robotic prolingoats deployed in the subarctic North Pacific on 10 April 2001 provided direct records of carbon biomass variability from surface to 1000 meters below surface at daily and diurnal time scales. Eight months of real-time data documented the marine biological response to natural events, including hydrographic changes, multiple storms, and the April 2001 dust event. High-frequency observations of upper ocean particulate organic carbon variability show a near doubling of biomass in the mixed layer over a 2 week period after the passage of a cloud of Gobi desert dust. The temporal evolution of particulate organic carbon enhancement and an increase in chlorophyll use efficiency after the dust storm suggest a biotic response to a natural iron fertilization by the dust.

尘暴增加北太平洋碳生物量的机器人观测

Bishop J. K. B. , Davis R. E. , Sherman J. T.

（美国劳伦斯伯克利国家实验室地球科学部）

摘要：2001 年 4 月 10 日部署在亚北极北太平洋的两个自动观测仪器提供了直接从海表面到 1000 m 以下在白天和昼夜时间尺度上生物碳变化的直接记录。8 个月的实时数据记录了海洋生物对各种自然事件，包括水文变化、风暴和 2001 年 4 月沙尘天气的反应。高频观测表明在当戈壁沙漠的尘埃通过的两周内,上层海洋颗粒有机碳的变化接近混合层的生物量的2倍。沙尘暴后,颗粒有机碳随时间的提高和叶绿素增加暗示对一种灰尘带来铁施肥的一种生物反应。

20. Goldman E. Even in the high Arctic, nothing is permanent[J]. Science,2002,297:1493-1494.

Even in the high Arctic, nothing is permanent

Goldman E.

(Cambridge, U. K.)

Abstract: The Rising temperatures are thawing vast swaths of northern land that had been frozen for millennia, creating headaches-and hazards-for communities perched above unstable ground: roads are caving in, airport runways are fracturing, and buildings are cracking, tilting, and sometimes falling down. And the situation, it appears, will only become direr.

在北极,啥都不永久

Goldman E.

(英国剑桥)

摘要:气温升高解冻大片北部已被冻结了数千年的土地,对在不稳定地面以上的社区,不断产生头痛的事和危险:道路崩落压裂,机场跑道破碎,建筑物开裂倾斜,甚至倒塌。这种情况将越来越可怕。

21. Kerr R A. A warmer Arctic means change for all[J]. Science,2002,297:1490-1493.

A warmer Arctic means change for all

Kerr R. A.

(Cambridge, U. K.)

Abstract: The seeming inevitability of shrinking ice on the Arctic Ocean means hard times for polar bears, a threat to an indigenous way of life, and an age-old dream come true for sailors, who may be able to navigate an ice-free Northwest Passage in the summer months. This new polar frontier might well set the stage for conflict, however, as naval powers stake out competing claims in the newly open waters of a new "global commons". A hundred years from now, life around the Arctic Ocean will go on—but it will not be the same.

一个温暖的北极意味着一切都会变

Kerr R. A.

(英国剑桥)

摘要:北冰洋的冰表面的萎缩意味着北极熊艰难的时刻,这威胁到它们原本的生活方式。但是一个古老的水手梦想或许成真,人们可以在夏季享受这里的一个无冰的西北航道。这个新的极地前沿可能为未来冲突提供好舞台,从现在起,一百多年后,北冰洋周围的生活还会继续——但它不会是相同的。

22. Alford M H. Redistribution of energy available for ocean mixing by long-range propagation of internal waves[J]. Nature,2003,423:159-162.

Redistribution of energy available for ocean mixing by long-range propagation of internal waves

Alford M. H.

(Applied Physics Laboratory and School of Oceanography, University of Washington, 1013 NE 40th Street, Seattle, Washington 98105, USA)

Abastract: Ocean mixing, which affects pollutant dispersal, marine productivity and global climate, largely results from the breaking of internal gravity waves-disturbances propagating along the ocean's internal stratification. A global map of internal-wave dissipation would be useful in improving climate models, but would require knowledge of the sources of internal gravity waves and their propagation. Towards this goal, I present here computations of horizontal internal-wave propagation from 60 historical moorings and relate them to the source terms of internal waves as computed previously. Analysis of the two most energetic frequency ranges-near-inertial frequencies and semidiurnal tidal frequencies-reveals that the fluxes in both frequency bands are of the order of 1 kW · m^{-1} (that is, 15-50%of the energy input) and are directed away from their respective source regions. However, the energy flux due to near-inertial waves is stronger in winter, whereas the tidal fluxes are uniform throughout the year. Both varieties of internal waves can thus significantly affect the space-time distribution of energy available for global mixing.

在海洋混合过程中的长期范围内波传播的能量再分配能源

Alford M. H.

(美国华盛顿大学应用物理实验室和海洋学院)

摘要:海洋混合,从而影响污染物的扩散,海洋生产力和全球气候,在很大程度上是从内部重力波沿海洋的内部分层传播过程中干扰的结果。内部波耗散的全球地图将有助于改善气候模型,但需要了解重力波的来源及其传播知识。为了达到这个目标,我把 60 个系泊和与它们以前相关的计算作为内波源项,对水平内部波的传播做计算。两个最有活力的频率范围:近惯性频率和半日潮汐的分析频率显示,在这两个频段的通量是 1 kW · m^{-1}(15%～50%的能量输入),并远离各自的源区。然而,由于近惯性波的能量通量是在冬季强,而潮汐通量全年统一。内波的这两个品种,从而显著地影响全球混合能源的时空分布。

23. Broecker W S. Does the trigger for abrupt climate change reside in the ocean or in the atmosphere? [J]. Science,2003,300:1519-1522.

Does the trigger for abrupt climate change reside in the ocean or in the atmosphere?

Broecker W. S.

(Lamont-Doherty Earth Observatory of Columbia University, 61 Route 9W, Post Office Box 1000, Palisades, NY 10964-8000, USA)

Abstract: Two hypotheses have been put forward to explain the large and abrupt climate changes that punctuated glacial time. One attributes such changes to reorganizations of the ocean's thermohaline circulation and the other to changes in tropical atmosphere-ocean dynamics. In an attempt to distinguish between these hypotheses, two lines of evidence are examined. The first involves the timing of the freshwater injections to the northern Atlantic that have been suggested as triggers for the global impacts associated with the Younger Dryas and Heinrich events. The second has to do with evidence for precursory events associated with the Heinrich ice-rafted debris layers in the northern Atlantic and with the abrupt Dansgaard-Oeschger warmings recorded in the Santa Barbara Basin.

气候突变的触发在海洋还是在大气?

Broecker W. S.

(美国哥伦比亚大学拉蒙特-多尔蒂地质观测站)

摘要:对于冰期爆发性的气候突变有两种假说。一种归因于海洋温盐环流,另一种则认为是热带海洋-大气动力学的变化。为尝试区分这两种假说,两种证据被拿来检验。首先涉及向北大西洋输入淡水的时间,这被认为触发了新仙女木事件和 Heinrich 事件。第二个就是北大西洋冰筏碎屑层和发生在圣巴巴拉洋盆的 D-O 变暖事件之前的先导事件。

24. Roman J, Palumbi S R. Whales before whaling in the North Atlantic[J]. Science, 2003, 301: 508-510.

Whales before whaling in the North Atlantic

Roman J., Palumbi S. R.

(Department of Organismic and Evolutionary Biology, Harvard University, 16 Divinity Avenue, Cambridge, MA 02138, USA)

Abstract: It is well known that hunting dramatically reduced all baleen whale populations, yet reliable estimates of former whale abundances are elusive. Based on coalescent models for mitochondrial DNA sequence variation, the genetic diversity of North Atlantic whales suggests population sizes of approximately 240000 humpback, 360000 fin, and 265000 minke whales. Estimates for fin and humpback whales are far greater than those previously calculated for prewhaling populations and 6 to 20 times higher than present-day population estimates. Such discrepancies suggest the need for a quantitative reevaluation of historical whale populations and a fundamental revision in our conception of the natural state of the oceans.

捕鲸之前北大西洋的鲸

Roman, J. , Palumbi S. R.

(美国哈佛大学生命进化生物学系)

摘要:捕鲸会严重减少所有须鲸种群数量,然而捕鲸之前头鲸数量的估算是不可靠的。基于对线粒体 DNA 序列的变化的模拟,北大西洋鲸的遗传多样性表明大约有 240000 头座头鲸、360000 头鳍鲸和 265000 头小须鲸。鳍鲸和座头鲸的估算远比早先对捕鲸之前的估算要高,而且比现在的数量高 6~20 倍。这种差异说明需要对历史上鲸的数量进行重新定量估算和我们对海洋数据定量演算有一个根本的革新。

25. Shindell D. Whither Arctic climate? [J]. Science, 2003, 299:215-216.

Whither Arctic climate?

Shindell D.

(The author is at NASA-Goddard Institute for Space Studies, New York, NY 10025, USA)

Abstract: Large changes in Arctic climate have been observed. In his Perspective, Shindell shows that models have difficulty reproducing these changes, raising doubt over their ability to elucidate the causes of the observed changes or predict future trends. He argues that these difficulties can largely be attributed to gravity waves, which are not yet well represented in the models. These waves are generated by atmospheric disturbances such as storm fronts, strong wind shears, and flow over mountains and play a key role in many of the observed changes.

北极气候何去何从?

Shindell D.

(美国国家航空和航天局戈达德宇宙飞行中心)

摘要:北极气候的变化较大且显著。Shindell 认为,模型中再现这些变化有困难,因此他怀疑这些模型研究变化的原因,或预测未来趋势的能力。他认为,这些困难在很大程度上归因于引力波,它尚未在模型中得到表现。这些波产生的大气扰动,如风暴前端、强风和山地流场,在许多观测到的变化中起到关键作用。

26. Hansell D A, Kadko D, Bates N R. Degradation of terrigenous dissolved organic carbon in the western Arctic Ocean[J]. Science, 2004, 304:858-861.

Degradation of terrigenous dissolved organic carbon in the western Arctic Ocean

Hansell D. A., Kadko D., Bates N. R.

(Rosenstiel School of Marine and Atmospheric Science, University of Miami, Miami, FL 33149, USA)

Abstract: The largest flux of terrigenous organic carbon into the ocean occurs in dissolved form by way of rivers. The fate of this material is enigmatic; there are numerous reports of conservative behavior over continental shelves, but the only knowledge we have about removal is that it occurs on long unknown time scales in the deep ocean. To investigate the removal process, we evaluated terrigenous dissolved organic carbon concentration gradients in the Beaufort Gyre of the western Arctic Ocean, which allowed us to observe the carbon's slow degradation. Using isotopic tracers of water-mass age, we determined that terrigenous dissolved organic carbon is mineralized with a half-life of 7.1 ±3.0 years, thus allowing only 21-32% of it to be exported to the North Atlantic Ocean.

北冰洋西部陆源溶解态有机碳的降解

Hansell D. A., Kadko D., Bates N. R.

(美国迈阿密大学"罗森蒂尔"海洋和大气科学学院)

摘要:进入海洋的陆源有机碳大多来自河流的溶解态有机碳。这种物质的命运是神秘莫测的;有许多关于大陆架的固碳行为的报告,而我们只知道这些碳的去除在未知的时间尺度上发生在深海。为探讨去除过程,我们估算在北冰洋西部博福特环流陆相溶解态有机碳的浓度梯度,这使我们能够观察碳的缓慢降解。使用同位素示踪剂跟踪水团年龄,我们认为陆源溶解态有机碳的埋藏半衰期为 7.1±3.0 年,这就使得可能有 21%～32%的碳被输送到北大西洋。

27. Yamamoto-Kawai M, Carmack E, McLaughlin F. Nitrogen balance and Arctic throughflow[J]. Nature,2006,443:43.

Nitrogen balance and Arctic throughflow

Yamamoto-Kawai M., Carmack E., McLaughlin F.

(Department of Fisheries and Oceans, Institute of Ocean Sciences, Sidney, British Columbia V8L 4B2, Canada)

Abstract: Waters moving east through the Arctic Ocean significantly contribute tonitrogen fixation in the Atlantic.

北极流和氮平衡

Yamamoto-Kawai M., Carmack E., McLaughlin F.

(加拿大海洋科学研究所渔业海洋科学部)

摘要:洋流通过北冰洋海域向东至大西洋固氮有显著贡献。

28. Serreze M C, Holland M M, Stroeve J. Perspectives on the Arctic's shrinking sea-ice cover[J]. Science, 2007, 315: 1533-1536.

Perspectives on the Arctic's shrinking sea-ice cover

Serreze M. C., Holland M. M., Stroeve J.

(Cooperative Institute for Research in Environmental Sciences, National Snow and Ice Data Center, Campus Box 449, University of Colorado, Boulder, CO 80309-0449, USA)

Abstract: Linear trends in arctic sea-ice extent over the period 1979 to 2006 are negative in every month. This ice loss is best viewed as a combination of strong natural variability in the coupled ice-ocean-atmosphere system and a growing radiative forcing associated with rising concentrations of atmospheric greenhouse gases, the latter supported by evidence of qualitative consistency between observed trends and those simulated by climate models over the same period. Although the large scatter between individual model simulations leads to much uncertainty as to when a seasonally ice-free Arctic Ocean might be realized, this transition to a new arctic state may be rapid once the ice thins to a more vulnerable state. Loss of the ice cover is expected to affect the Arctic's freshwater system and surface energy budget and could be manifested in middle latitudes as altered patterns of atmospheric circulation and precipitation.

北极退缩海冰覆盖的观点

Serreze M. C., Holland M. M., Stroeve J.

（美国科罗拉多大学环境科学合作研究所、国家冰雪数据中心）

摘要：1979～2006年每个月北极海冰面积都呈线性下降趋势。这个冰损失最可能是海洋-大气耦合系统的自然变异和因大气中的温室气体的浓度上升而导致不断增长的辐射强迫的结果。辐射强迫的增加与同时期观察到的趋势和气候模型模拟的结果一致。当涉及什么时候北冰洋会有完全无冰的季节时，不同模型的结果出现很大的分歧和不确定性。一旦冰层出现变薄或一个比较脆弱的状态时，北极地区可能会快速过渡到一个新的状态。冰雪覆盖的丢失预计将影响北极地区的淡水系统和表面能量估算，并影响中纬度的大气环流和降水模式。

29. Murton J B, Bateman M D, Dallimore S R, et al. Identification of Younger Dryas outburst flood path from Lake Agassiz to the Arctic Ocean[J]. Nature, 2010, 464: 740-743.

Identification of Younger Dryas outburst flood path from Lake Agassiz to the Arctic Ocean

Murton J. B., Bateman M. D., Dallimore S. R., Teller J. T., Yang Z. R.

(Permafrost Laboratory, Department of Geography, University of Sussex, Brighton BN1 9QJ, UK)

Abstract: The melting Laurentide Ice Sheet discharged thousands of cubic kilometres of freshwater each year into surrounding oceans, at times suppressing the Atlantic meridional overturning circulation and triggering abrupt climate change. Understanding the physical mechanisms leading to events such as the Younger Dryas cold interval requires identification of the paths and timing of the freshwater discharges. Although Broecker et al hypothesized in 1989 that an outburst from glacial Lake Agassiz triggered the Younger Dryas, specific evidence has so far proved elusive, leading Broecker to conclude in 2006 that "our inability to identify the path taken by the flood is disconcerting". Here we identify the missing flood path-evident from gravels and a regional erosion surface-running through the Mackenzie River system in the Canadian Arctic Coastal Plain. Our modelling of the isostatically adjusted surface in the upstream Fort McMurray region, and a slight revision of the ice margin at this time, allows Lake Agassiz to spill into the Mackenzie drainage basin. From optically stimulated luminescence dating we have determined the approximate age of this Mackenzie River flood into the Arctic Ocean to be shortly after 13000 years ago, near the start of the Younger Dryas. We attribute to this flood a boulder terrace near Fort McMurray with calibrated radiocarbon dates of over 11500 years ago. A large flood into the Arctic Ocean at the start of the Younger Dryas leads us to reject the widespread view that Agassiz overflow at this time was solely eastward into the North Atlantic Ocean.

从阿加西斯湖和北冰洋之间"新仙女木"洪水暴发路径的识别

Murton J. B., Bateman M. D., Dallimore S. R., Teller J. T., Yang Z. R.

(英国苏塞克斯大学地理系多年冻土实验室)

摘要:劳伦泰德冰盖每年融出数千立方千米的淡水,输入到周围的海洋,有时抑制大西洋经向翻转环流和触发突然的气候变化。了解如新仙女木冷事件的物理机制,需要识别的淡水输出路径和时间。虽然布勒克等在1989年假设从冰川湖阿加西溃决引发新仙女木事件,具体的证据至今证明是难以被人确定的,使得布勒克在2006年得出结论:"我们无法确定洪水路径是令人不安的。"在这里,我们确定在加拿大北极地区的海岸平原的麦肯齐河系统中有洪水砾石和一个表层侵蚀区域,明显是其通过的路径。在模型中,我们调整上游麦克默里堡地区的表面,轻微修改冰边缘,从而允许湖阿加西蔓延到麦肯齐流域。从光释定年结果来看,麦肯齐河道行洪进入北冰洋大致年龄在13000年之前,接近"新仙女木"事件的开始时间。我们认为产生于这次水灾的一个麦克默里堡附近的巨石露台的放射性碳年龄大约为11500年。"新仙女木"开始时,一次流入北冰洋的大洪水使我们否定阿加西向东溢出到北大西洋的普遍看法。

30. Shakhova N, Semiletov I, Salyuk A et al. Extensive methane venting to the atmosphere from sediments of the East Siberian Arctic Shelf[J]. Science, 2010, 327: 1246-1250.

Extensive methane venting to the atmosphere from sediments of the East Siberian Arctic Shelf

Shakhova N., Semiletov I., Salyuk A., Yusupov V., Kosmach D., Gustafsson Ö.

(International Arctic Research Centre, University of Alaska, Fairbanks, AK 99709, USA)

Abstract: Remobilization to the atmosphere of only a small fraction of the methane held in East Siberian Arctic Shelf (ESAS) sediments could trigger abrupt climate warming, yet it is believed that sub-sea permafrost acts as a lid to keep this shallow methane reservoir in place. Here, we show that more than 5000 at-sea observations of dissolved methane demonstrates that greater than 80% of ESAS bottom waters and greater than 50% of surface waters are supersaturated with methane regarding to the atmosphere. The current atmospheric venting flux, which is composed of a diffusive component and a gradual ebullition component, is on par with previous estimates of methane venting from the entire World Ocean. Leakage of methane through shallow ESAS waters needs to be considered in interactions between the biogeosphere and a warming Arctic climate.

东西伯利亚北极大陆架沉积物有大量的甲烷排放到大气

Shakhova N., Semiletov I., Salyuk A., Yusupov V., Kosmach D., Gustafsson Ö.

(美国阿拉斯加大学国际北极研究中心)

摘要:只占东西伯利亚的北极大陆架沉积物所固定一小部分的甲烷如再迁移到大气,也可能引发突然的气候变暖。当前子海冻土层只作为一个盖子,来保证浅层甲烷库安全。这里我们的研究显示,超过5000个海上溶解甲烷的观察表明,超过80%的东西伯利亚的北极大陆架的底部水域和50%的表层水中甲烷相对于大气处于过饱和状态。目前大气通量,由扩散的组成部分和渐进的暴发部分组成,与原先估计的整个世界海洋甲烷排放量相当。东西伯利亚的北极大陆架浅水域的甲烷泄漏需要被考虑到生物岩石圈和北极变暖之间的关系中去。

31. Zahn M, von Storch H. Decreased frequency of North Atlantic polar lows associated with future climate warming[J]. Nature, 2010, 467: 309-312.

Decreased frequency of North Atlantic polar lows associated with future climate warming

Zahn M., von Storch H.

(Environmental Systems Science Centre, University of Reading, 3 Earley Gate, Reading, Berkshire RG6 6AL, UK)

Abstract: Every winter, the high-latitude oceans are struck by severe storms that are considerably smaller than the weather-dominating synoptic depressions. Accompanied by strong winds and heavy precipitation, these often explosively developing mesoscale cyclones-termed polar lows—constitute a threat to offshore activities such as shipping or oil and gas exploitation. Yet owing to their small scale, polar lows are poorly represented in the observational and global reanalysis data often used for climatological investigations of atmospheric features and cannot be assessed in coarse-resolution global simulations of possible future climates. Here we show that in a future anthropogenically warmed climate, the frequency of polar lows is projected to decline. We used a series of regional climate model simulations to downscale a set of global climate change scenarios from the Intergovernmental Panel of Climate Change. In this process, we first simulated the formation of polar low systems in the North Atlantic and then counted the individual cases. A previous study using NCEP/NCAR re-analysis data revealed that polar low frequency from 1948 to 2005 did not systematically change. Now, in projections for the end of the twenty-first century, we found a significantly lower number of polar lows and a northward shift of their mean genesis region in response to elevated atmospheric greenhouse gas concentration. This change can be related to changes in the North Atlantic sea surface temperature and mid-troposphere temperature; the latter is found to rise faster than the former so that the resulting stability is increased, hindering the formation or intensification of polar lows. Our results provide a rare example of a climate change effect in which a type of extreme weather is likely to decrease, rather than increase.

北大西洋极地低压震荡频率降低与未来气候变暖相关

Zahn M. , von Storch H.

(英国雷丁大学环境系统科学中心)

摘要：每到冬天，高纬度海洋会遭到严重的暴风雨袭击；暴风雨的规模比气象为主的低压要小得多。伴随着大风和强降雨，往往是爆炸式发展成中尺度气旋性极地低压——构成对航运、石油和天然气开采等近海活动的威胁。然而，由于其规模小，很少用于气候调查的观测全球再分析数据和极地低压的观测中，并且也不会用在粗分辨率模拟全球未来气候变化的预测上。我们认为，未来，人类活动造成的气候变暖，极地低压震荡的频率将下降。我们使用了一系列区域气候模式，将政府间气候变化专门委员会的全球气候变化情景小规模化。在这个过程中，我们首先模拟了在北大西洋极地低压系统的形成，然后计算个案。先前的研究使用 NCEP/NCAR 再分析显示，从 1948～2005 年的极地低压没有系统的变化。现在，预测在 21 世纪结束时，我们发现，作为大气中温室气体浓度升高的响应，极地低压数量上将减少，其平均形成地区也向北转移。这种变化与北大西洋海面温度和对流层中的温度变化相关。而中对流层温度增加速度明显高于北大西洋海表温度，所以稳定性将增加，并阻碍了极地低压的形成或加强。我们的研究结果提供了一个气候变化的罕特例——极端天气在减少而不是增加。

六、地质、测绘类文献

6.1 分析概述

1990年以来 *Nature* 和 *Science* 上共有"地质、测绘类"文献38篇(*PNAS* 上没有此类文献),其中南极方面文献有22篇,北极方面文献有15篇,南北极都涉及的有1篇。

将文章数量分区域并按照年代进行分类(图6.1)可以发现,整体上两刊物上每年发表地质、测绘类文章在3篇左右,仅在1997年和2008年发表该类文章较多,分别为5篇和8篇。两刊物上在1994年、1998～2000年和2002年均无此类文章发表。2000年之后发表的文章数整体高于之前,且2000年之前的文章以南极为主,2000年之后北极相关文章发表数量增长较多,南极相关文章数量则相对减少。由此可见,南极地区地质、测绘等工作已较好开展,北极地区此类工作正在出现新的科学增生点。

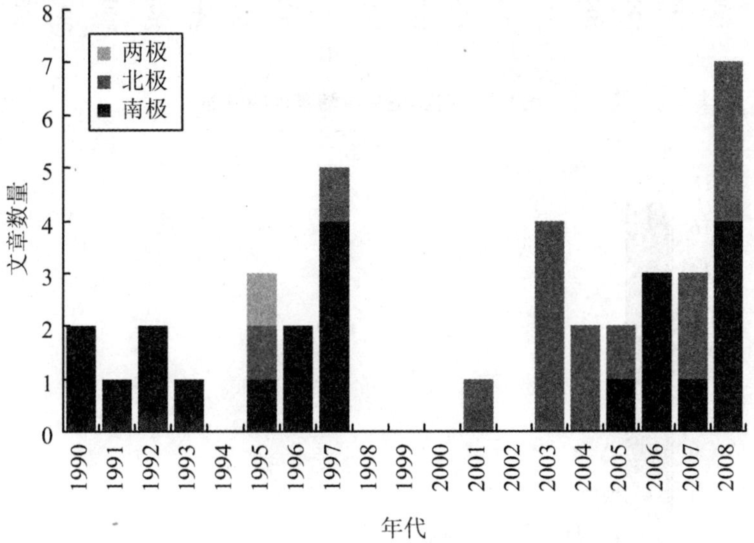

图6.1 文章数量随年代的变化

将文章研究的区域及过程按照空间可分为深部、表层、大气及太空、测绘四类,其中深部类主要包括深海洋中脊以及地幔等深部过程;表层类主要为近地表过程,包括冰川、地貌和湖泊等;大气及太空类则主要包括与陨石和极光相关的研究。由各类文章随时间的分布(图6.2)可以看出,2000年之前各类研究较为平均,表层相关的研究稍多;而2000年之后主要为表层和深部相关的研究,其中以洋中脊和冰川相关的研究为主。这可能体现了:(1) 随着科技的进步,人类对深海的探索开始加强;(2) 气候变暖促进了冰川等相关研究的发展。

除去两篇 *Nature* 的报道,将文章数量按所署第一作者所在国家进行排列(图6.3),美国发表的文章数最多,高达17篇,接近总数量的50%,其次为英国。近20年来中国以第一署名国在该领域于2008年在 *Nature* 发表文章1篇,为洋中脊相关研究。

由以上初步统计,目前极地地质、测绘类研究发展的重心已由南极逐渐向北极转移,深部过程(尤其是深海构造)发展迅速,与全球气候变化相关的表层研究也已成为热点。未来,与气候相关的表层地质过程和深海地质可能成为极地地质、测绘研究的增长点。中国在极地地质的高水平研究正逐渐开展,目前已占有一席之地,但与美国等仍有相当大的差距。

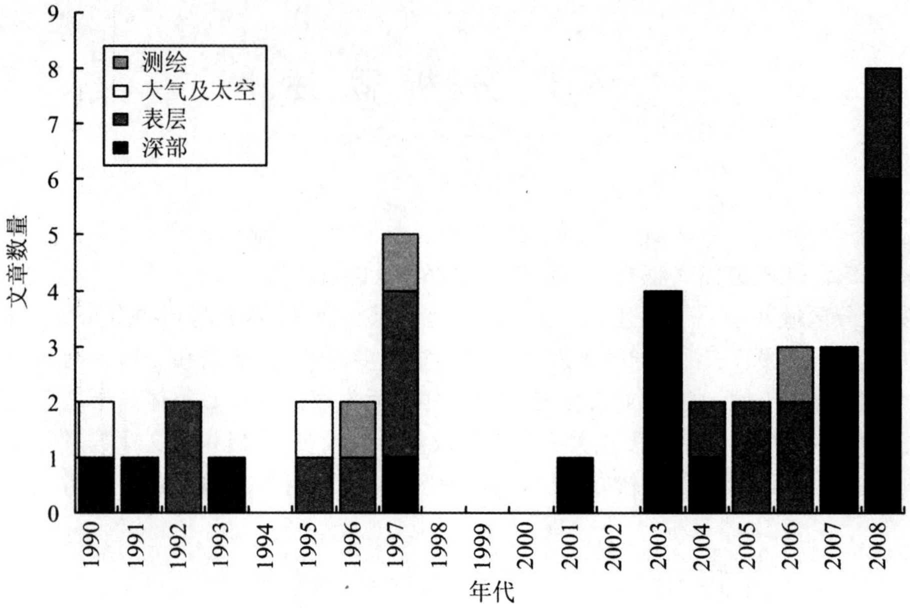

图 6.2　各类型文章随年代的分布

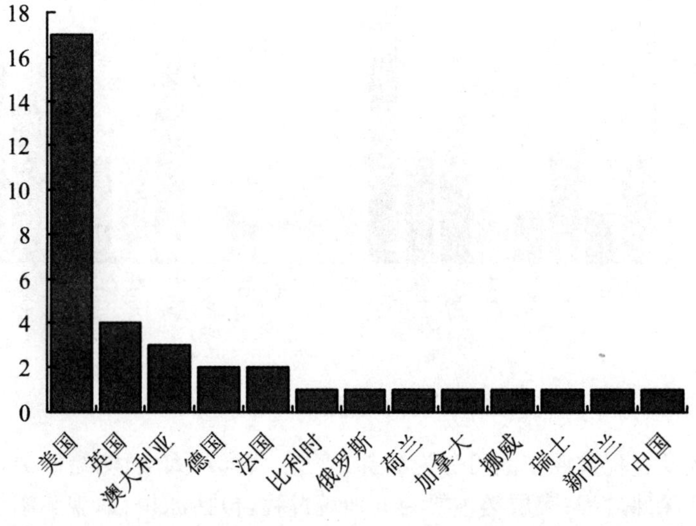

图 6.3　第一作者所在国家发表的文章数量

6.2 摘要翻译——南极

1. Barling J, Goldstein S L. Extreme isotopic variations in Heard-Island lavas and the nature of mantle reservoirs[J]. Nature, 1990, 348(6296): 59-62.

Extreme isotopic variations in Heard-Island lavas and the nature of mantle reservoirs

Barling J., Goldstein S. L.

(Max-Planck-Institut für Chemie, Postfach 3060, D-6500 Mainz, Germany Department of Earth Sciences, Monash University, Clayton, Victoria 3168, Australia)

Abstract: Oceanic basalts are produced by melting of the Earth's mantle and are widely used to probe its composition. The observation of systematic, coupled variations in the neodymium, strontium and lead isotopic compositions of these basalts has been widely interpreted as reflecting the mixing of identifiable mantle components. White and Zindler and Hart have suggested that the isotopic compositions of mid-ocean-ridge and ocean island basalts (MORB and OIB) can be considered as mixtures of a depleted MORB-type mantle component with a few other components reflecting long-term enrichment of Rb/Sr, Nd/Sm and/or U/Pb ratios. It remains unclear whether these components should be considered as hypothetical mixing endmembers, or whether they exist physically as mantle reservoirs. This question may be addressed if the mixing endmembers of individual mantle plumes can be identified. To this end, we report linear Pb-Pb and curvilinear Pb-Nd and Pb-Sr isotopic covariations in recent lavas of Heard Island, southernIndian Ocean, indicating binary mixing. The hyperbolic relationships are unique among oceanic basalts and tightly constrain the isotopic compositions of the plume source components, which do not coincide with the mantle endmember components of ref. 3. If our results are generally applicable to other plumes, they call into question the existence of large mantle reservoirs corresponding to these components, and indicate that actual oceanic basalt source reservoirs have intermediate isotopic compositions.

赫德岛熔岩同位素的极端变化与地幔层的性质

Barling J., Goldstein S. L.

(德国马普化学研究所;澳大利亚莫纳什大学地球科学系)

摘要:大洋玄武岩由地幔熔融形成,被广泛用于研究地幔的组成特征。其 Nd、Sr 和 Pb 同位素的系统变化及其耦合变化可反映地幔物质的混合作用。White、Zindler 和 Hart 认为,大洋中脊玄武岩(MORB)和洋岛玄武岩(OIB)的同位素组成可认为是亏损 MORB 型地貌与少量其他一些组分的混合,反映了Rb/Sr、Nd/Sm 和/或 U/Pb 值的长期富集。但这些组成是否能够作为混合的端元,或者说它们是否是地幔的组成部分,至今仍不清楚。解决这一问题的方法之一是鉴定出单个地幔柱的混合端元。为此,本文报道赫德岛和南印度洋现代熔岩 Pb-Pb 同位素的线性共变和 Pb-Nd 以及 Pb-Sr 同位素的曲线共变特征,具有了二元混合的特征。大洋玄武岩普遍具有独特的双曲关系,其同位素组成与地幔柱源紧密相关,但与参考文献 3 中的地幔端元组成不符。如果本文结果能够扩展应用到其他地幔柱,则提出一个问题,即是否存在与这些组分对应的大地幔层,且指示了大洋玄武岩的源具有中等的同位素组成。

2. Wasson J T. Ungrouped iron-meteorites in Antarctica-Origin of anomalously high abundance[J]. Science,1990,249(4971):900-902.

Ungrouped iron-meteorites in Antarctica-Origin of anomalously high abundance

Wasson J. T.

(The University of California, USA)

Abstract: Eighty-five percent of the iron meteorites collected outside Antarctica are assigned to 13 compositionaily and structurally defined groups; the remaining 15 percent are ungrouped. Of the 31 iron meteorites recovered from Antarctica, 39 percent are ungrouped. This major difference in the two sets is almost certainly not a stochastic variation, a latitudinal effect, or an effect associated with differences in terrestrial ages. It seems to be related to the median mass of Antarctic irons, which is about 1/100 that of non-Antarctic irons. During impacts on asteroids, smaller fragments tend to be ejected into space at higher velocities than larger fragments, and, on average, small meteoroids have undergone more changes in orbital velocity than large ones. As a result, the set of asteroids that contributes small meteoroids to Earth-crossing orbits is larger than the set that contributes large meteoroids. Most small iron meteorites may escape from the asteroid belt as a result of impact-induced changes in velocity that reduce their perihelia to values less than the aphelion of Mars.

南极洲未分类铁陨石数量异常多的原因

Wasson J. T.

(美国加利福尼亚大学)

摘要:南极以外发现的 85%的铁陨石已根据组成和结构分为 13 组,其余 15%尚未分类。在南极发现的 31 个铁陨石中,39%尚未分类。两组数据如此大的差别基本上不可归因于随机误差、纬度效应或大陆年龄差异效应。这似乎和南极铁陨石的平均质量有关,它是非南极铁陨石的百分之一。受小行星的影响,小的碎片进入太空的速度比大的碎片高得多,同时,小陨石在轨道速度上的变化也较大陨石大。因此,能够在地球交叉轨道上释放小陨石的流星比释放大陨石的多。大部分从行星带脱离的小铁陨石的速度在火星的近日点受到的影响远小于在远日点受到的影响。

3. Chadwick W W, Embley R W, Fox C G. Evidence for volcanic-eruption on the southern Juan-De-Fuca ridge between 1981 and 1987[J]. Nature,1991,350(6317):416-418.

Evidence for volcanic-eruption on the southern Juan-De-Fuca ridge between 1981 and 1987

Chadwick W. W. , Embley R. W. , Fox C. G.

(OSU/CIMRS, Hatfield Marine Science Center, Newport, Oregon 97365, USA)

Abstract: The formation of new ocean crust at mid-ocean ridges is known to be a discontinuous process in both space and time, but little is known about the frequency and duration of eruptions along an active ridge segment. Here we present evidence, from Sea Beam surveys and underwater photography, for the eruption of lavas along a segment of the Juan de Fuca ridge between 1981 and 1987. Although previous studies have inferred volcanic activity on ridges in areas where recent seismicity or young lava flows have been observed, none has yet had direct evidence to date such a recent submarine eruption. The temporal coincidence between this eruptive episode and the megaplumes (huge, sudden emissions of hot mineral-laden water) observed over this part of the ridge in 1986 and 1987 supports previous suggestions that megaplumes are caused by sea-floor spreading events.

1981～1987 年间南胡安·德·富卡洋中脊火山喷发的证据

Chadwick W. W. , Embley R. W. , Fox C. G.

(美国俄勒冈州立大学海洋资源合作研究所、俄勒冈州立大学哈特菲尔德海洋科学中心)

摘要:新的洋壳在洋中脊的形成在空间和时间上都不连续,但目前对活动海岭块段火山喷发的频率和持续时间仍所知甚少。本文根据波段探测和水下拍摄,给出了 1981～1987 年间胡安·德·富卡洋中脊海岭块段的岩浆喷发证据。虽然过去的研究从现代地震及岩浆流的区域推论出洋中脊存在火山活动,但没有海底火山喷发的直接证据。这一喷发事件在时间上与 1986 年和 1987 年观察到的该区域海底热液一致,支持热液由洋底扩张导致的说法。

4. Barrett P J, Adams C J. Geochronological evidence supporting Antarctic deglaciation 3 million years ago[J]. Nature, 1992, 359(6398): 816-818.

Geochronological evidence supporting Antarctic deglaciation 3 million years ago

Barrett P. J. , Adams C. J.

(Research School of Earth Sciences, Victoria University of Wellington, PO Box 600, Wellington, New Zealand)

Abstract: The response of the Antarctic ice sheets to increased global temperatures is an important unresolved issue in the assessment of future climate change. In particular, considerable controversy exists as to whether the East Antarctic ice sheet suffered extensive deglaciation during the mid-Pliocene epoch (approximately 3 Myr ago), when temperatures were only slightly warmer than today. Although the ice sheet is widely assumed to have existed in something like its present form for the past 14 Myr, marine diatoms eroded from the Antarctic interior have been found in glacial till deposits high in the Transantarctic Mountains, and have been biostratigraphically dated at approximately 3 Myr before present. This age has been disputed because it implies marine deposition in the Antarctic interior, and hence substantial deglaciation, at a time when other evidence has been marshalled for the persistence of cold, polar conditions. Here we report K-Ar and $^{40}Ar/^{39}Ar$ ages for a volcanic ash bed in diatom-bearing glaciomarine strata cored in Ferrar Fiord (East Antarctica) by the CIROS-2 drill-holes, which confirm the age of the diatoms at approximately 3 Myr, and hence also confirm the mid-Pliocene deglaciation.

地质年代证据支持300万年前南极存在冰消期

Barrett P. J., Adams C. J.

(新西兰惠灵顿维多利亚大学地球科学学院)

摘要：南极冰盖对全球升温的响应是估计未来气候变化研究中尚未解决的问题。然而，在温度略高于现代的中上新世(约300万年前)时期，目前对东南极冰盖是否存在冰消过程仍存在争议。虽然普遍认为南极冰盖在过去的1400万年中基本与现在类似，在Transantarctic山上的冰碛物中却发现了来自南极内陆的海洋硅藻类，生物地层定年为约300万年之前。由于该研究指示了南极内陆的海相沉积，进而反映了冰消期的存在，而其他证据却显示该时期仍处于冷的极地环境，因此这一年龄备受争议。本文利用K-Ar和$^{40}Ar/^{39}Ar$年代学方法测定了东南极Ferrar Fiord处CIROS-2所钻岩芯中含硅藻冰川-海相地层中火山灰层的年代，证实了硅藻存在的年代约为300万年，进而证实了中上新世冰消期的存在。

5. Dehairs F, Baeyens W, Goeyens L. Accumulation of suspended barite at mesopelagic depths and export production in the Southern-Ocean[J]. Science, 1992, 258(5086): 1332-1335.

Accumulation of suspended barite at mesopelagic depths and export production in the Southern-Ocean

Dehairs F., Baeyens W., Goeyens L.

(Free University of Brussels)

Abstract: The relation between the accumulation of barite ($BaSO_4$) microcrystals in suspended matter from the mesopelagic depth region (100 to 600 meters) and the type of production in the euphotic layer (new versus recycled) was studied for different Southern Ocean environments. Considerable subsurface barite accumulated in waters characterized by maintained new production and limited grazing pressure during the growth season. On the other hand, little if any barite accumulated in areas where relatively large amounts of photosynthetically fixed carbon were transferred to the microheterotrophic community and where recycled production became predominant.

南大洋中层悬浮重晶石的聚集与输出生产力的关系

Dehairs F. , Baeyens W. , Goeyens L.

(比利时布鲁塞尔自由大学分析化学系)

摘要:本文研究了南大洋不同环境下海洋中层(100～600 m)悬浮物中重晶石微晶的聚集与透光层生产力类型之间的关系。在生长季,中层大量重晶石的聚集能够反映新生产力的持续发展及对食物压力的限制。反过来说,如果某一区域的重晶石含量较少,则大量光合作用汇集的碳会被异养群落转化,在此区域再生生产力占主要部分。

6. Ferguson E M, Klein E M. Fresh basalts from the Pacific Antarctic ridge extend the Pacific geochemical province[J]. Nature, 1993, 366(6453): 330-333.

Fresh basalts from the Pacific Antarctic ridge extend the Pacific geochemical province

Ferguson E. M. , Klein E. M.

(Department of Geology, Duke University, Durham, North Carolina 27708, USA)

Abstract: Over the past several decades, the examination of volcanic rocks recovered from mid-ocean ridges and ocean islands worldwide has led to the identification of large-scale geochemical provinces reflecting compositionally distinct domains in the earth's mantle. The spatial distribution of these domains may reveal global patterns of upper-mantle convection and mixing. Previous studies have shown, for example, that the distinct Indian Ocean isotope province has a relatively sharp eastern boundary within the Australian-Antarctic Discordance (AAD) south of Australia. The juxtaposition of Indian Ocean basalt compositions west of this boundary and Pacific compositions to the east suggests that the AAD may overlie a zone of convergence between ocean-basin-scale upper-mantle convection regimes: if this is so, Pacific isotope compositions should occur continuously along the length of the Pacific-Antarctic Ridge (PAR). Fresh basaltic glasses have now been recovered from the southernmost portion of this previously unsampled ridge axis, and we report their major element, trace element and isotopic compositions. The chemical systematics of these rocks suggest that the Pacific Ocean geochemical province includes the PAR, extends to the AAD south of Australia, and thus is one of the largest chemically coherent mantle domains on the Earth.

太平洋-南极洲洋脊的新鲜玄武岩扩大了太平洋地球化学的范围

Ferguson E. M, Klein E. M.

(美国杜克大学地质系)

摘要:在过去的几十年间,对全球洋中脊和海岛火山岩的研究鉴定出了大范围的地球化学,反映了地幔组成的区域差异。这些区域的空间分布可能反映了上地幔的对流和混合的全球模式特征。例如,之前的研究显示,独特的印度洋同位素省与澳大利亚南部的澳大利亚-南极洲不整合区域(AAD)内存在相对明显的东部边界。边界西部印度洋玄武岩组成和东部太平洋组成的拼合表明,AAD可能位于一个处于海洋-盆地规模的上地幔对流区的交汇区:若果真如此,太平洋同位素组成应沿着太平洋-南极洲脊(PAR)连续出现。目前已在这个之前未被取样的脊轴的最南部找到了新鲜的玄武岩质玻璃,这里我们已报道了其主量元素、痕量元素和同位素组成。这些岩石的化学组成表明,太平洋的地球化学省(包括PAR)正向澳大利亚南部的AAD地区扩展,因而是地球上最大的具有化学一致性的地幔区域之一。

7. Sugden D E, Marchant D R, JR N P, et al. Preservation of miocene glacier ice in East Antarctica[J]. Nature, 1995, 376(6539): 412-414.

Preservation of miocene glacier ice in East Antarctica

Sugden D. E., Marchant D. R., JR N. P., Souchez R. A., Denton G. H., Swisher C. C., Tison J. L. et al.

(Department of Geography, University of Edinburgh, Edinburgh EH8 9XP, UK)

Abstract: Antarctic climate during the Pliocene has been the subject of considerable debate. One view holds that, during part of the Pliocene, East Antarctica was largely free of glacier ice and that vegetation survived on the coastal mountains. An alternative viewpoint argues for the development of a stable polar ice sheet by the middle Miocene, which has persisted since then. Here we report the discovery of buried glacier ice in Beacon valley, East Antarctica, which appears to have survived for at least 8.1 million years. We have dated the ice by $^{40}Ar/^{39}Ar$ analysis of volcanic ash in the thin, overlying glacial till which, we argue, has undergone little (if any) reworking. Isotope and crystal fabric analyses of the ice show that it was derived from an ice sheet. We suggest that stable polar conditions must have persisted in this region for at least 8.1 million years for this ice to have avoided sublimation.

东南极中新世时期冰川的保存

Sugden D. E., Marchant D. R., JR N. P., Souchez R. A., Denton G. H., Swisher C. C., Tison J. L. et al.

(英国爱丁堡大学地理系)

摘要:上新世时期的南极气候一直以来都存在很大的争议。一种观点认为,在该时期内,东南极大部分地区是没有冰川存在的,在沿岸的山川上存活着植被。另一种观点认为一直到中新世时期极地冰盖都在稳定增长,在此之后不再变化。在此,我们报道在东南极Beacon谷发现的埋藏的冰川,这个冰川似乎已经存在了至少810万年。我们通过对较薄而且覆盖在上层的冰碛(这些冰碛在我们看来几乎没有或者很少经历过再造)中的火山灰进行$^{40}Ar/^{39}Ar$方法来对这些冰进行定年。该冰的同位素和晶体构造分析表明,它来自于一个冰盖。我们认为此地区稳定的极地环境一定在持续了至少810万年才使得该处的冰川免于消失。

8. Newell P T, Meng C I. Creation of theta-auroras: the isolation of plasma sheet fragments in the polar-cap[J]. Science, 1995, 270(5240): 1338-1341.

Creation of theta-auroras: the isolation of plasma sheet fragments in the polar-cap

Newell P. T., Meng C. I.

(Johns Hopkins University Applied Physics Laboratory, Laurel, MD 20723, USA)

Abstract: The auroral oval is a ring of luminosity enclosing geomagnetic field lines connected to the solar wind. Occasionally the ring has a bar across it, which seems to imply a bifurcation of the open region. Here results confirm that this theta-auroras actually does represent this odd bifurcated configuration, and they demonstrate how it happens. It has hitherto been assumed that theta-auroras occur when the interplanetary magnetic field is directed northward, because that is true of most very-high-latitude arcs. In fact, theta-auroras occur exclusively during the dynamic reconfiguration that follows when the interplanetary magnetic field turns southward after a prolonged northward interval.

ξ-极光的产生:极地冰冠处等离子层碎片的隔离

Newell P. T., Meng C. I.

(美国约翰霍普金斯大学应用物理实验室)

摘要:极光是一个闭合地磁线与极地风连接在一起形成的发光环。偶尔这个环有一个条带横跨着,这看起来在暗示着开放区域的分叉。在此,有结果证实ξ-极光确实代表了这个奇怪的分叉型构型,而且,它们证明了这是如何发生的。过去曾有过假设:ξ-极光在行星际磁场指向北时会发生,因为这符合大部分极高纬度的弧光。事实上,ξ-极光仅仅在动态重组期间发生,这个动态重组会于行星际磁场在一个持续很久的指北间隔后再指向南之后发生。

9. Kapitsa A P, Ridley J K, Robin G De Q, et al. A large deep freshwater lake beneath the ice of central East Antarctica[J]. Nature, 1996, 381(6584): 684-686.

A large deep freshwater lake beneath the ice of central East Antarctica

Kapitsa A. P., Ridley J. K., Robin G. De Q., Siegert M. J., Zotikov I. A.

(Lomonosov Moscow State University, Moskva, Moscow, Russia)

Abstract: In 1974-75, an airborne radio-echo survey of ice depths over central East Antarctica led to the discovery of a sub-ice lake of unknown depth and composition, with an area of about 10000 km^2 and lying beneathsimilar to 4 km of ice. In 1993, altimetric data from satellite measurements provided independent evidence of the lake's areal extent, thus confirming it to be the largest known sub-ice lake by an order of magnitude. Here we analyse new altimetric and radio-echo data, along with existing seismic data, to show that the lake is deep (mean depth of 125 m or more) and fresh, and that it has an area that exceeds previous estimates by about 50% dimensions comparable with those of Lake Ontario. We estimate that the residence time of the crater in the lake is of the order of tens of thousands of years, and that the mean age of water in the lake, since deposition as surface ice, is about one million years, Regional ice-dynamics can be explained in terms of steady-state ice flow along and over the lake.

东南极中部冰层下有一个大而深的淡水湖

Kapitsa A. P., Ridley J. K., Robin G. De Q., Siegert M. J., Zotikov I. A.

(俄罗斯莫斯科州立大学地理部)

摘要:1974～1975 年间,一项覆盖东南极中心区域、利用空气传播的无线电回波对冰层深部的探测发现了一个未知深度和组成的冰下湖,这个湖面积约为 10000 km^2,位于约 4 km 冰层的下方。1993 年,来自卫星的高度测量法所得的数据为这个湖泊的面积范围提供了独立的证据,证明了它比已知的最大冰下湖还要大一个数量级。在此,我们分析新的测高和无线电回波数据以及现有的地震学上的数据,结果表明湖很深(平均深度为 125 m 或更深)且是淡水,其面积超过了之前估计的 50%,超出的部分可以与安大略湖的面积相比。我们估计,这个湖中的火山口的滞留时间是几万年,湖水的平均年龄有一百万年。地区性的冰川动态学可以用沿湖和湖水上层的稳定态冰流动来解释。

10. Ellis Evans J C, Wynn Williams D. A great lake under the ice[J]. Nature, 1996, 381(6584): 644-646.

A great lake under the ice

Ellis Evans J. C., Wynn Williams D.

(Head, NERC Arctic Office British Antarctic Surve)

Abstract: The Lake Vostok present below a moving ice sheet in central Antarctica is 200 km long and up to 500 m deep. The difference in the ice melting point at the upstream and downstream ends probably produces circulation and the release of compressed atmospheric gases into the lake. The lake lacks salts and the water temperature is just below zero degrees celsius. The ice core of Vostok contains many different microbes that stay viable in the ice for over 3000 years. The lake probably contains microbes that developed more than 500000 years ago and have been isolated from local and global changes.

冰下的巨大湖泊

Ellis Evans J. C., Wynn Williams D.

(英国南极调查局)

摘要:Vostok 湖位于南极洲中部的移动冰川之下,长 200 km,深达 500 m。湖泊上下游冰川融化速率的差异可能是导致湖水流动以及释放压缩空气至湖中的主要原因。湖泊为淡水湖,温度低于 0 ℃。Vostok 冰芯中含有多种微生物,有的在冰中存在时间长达 3000 年。而湖泊中可能含有超过 50 万年的与世隔绝的微生物。

11. Wilson K, Sprent J I, Hopkins D W. Nitrification in Antarctic soils[J]. Nature, 1997, 385(6615): 404.

Nitrification in Antarctic soils

Wilson K. , Sprent J. I. ,Hopkins D. W.

(Department of Biological Sciences, University of Dundee, Dundee DD1 4HN, UK)

Abstract: Marshall found pollen from trees (Nothofagus spp. and Podocarpus spp.) and the spores of several fungi not normally native to Antarctica in air samples collected at Signy Island in the maritime Antarctic. Their presence was associated with a specific weather pattern, occurring with an estimated mean annual frequency of exotic biological particles to Antarctica from South American, as Marshall pointed out, such events provide potential mechanisms for organisms to extend their ranges into Antarctica is not a continent in biological isolation. We suggest here that bacteria may have been dispersed throughout Antarctica in a similar way.

南极土壤中的硝化作用

Wilson K. , Sprent J. I. ,Hopkins D. W.

(英国邓迪大学生物科学系)

摘要:马歇尔发现在南极近海的 Signy 岛采集的空气样本中含有的一些树(假山毛榉和罗汉松)的花粉和一些真菌的孢子不是来自于南极本地。它们的存在与一个特定的天气模式有关,这些来自南美的生物粒子具有一致的年平均频率。马歇尔指出,此类事件提出了一个可能的机制,即生物扩展到南极并不是由于大陆的生物隔离效应。我们认为细菌也可能是以类似的方式分散在南极洲的。

12. Rex Dalton. Rifts found as Antarctic ice breaks apart[J]. Nature,1997,385(6617):566.

Rifts found as Antarctic ice breaks apart

Rex Dalton

(News)

Abstract: Researchers from Greenpeace, the environmentalist group, last week came across significant rifts in the southern part of the Larsen ice shelf in the Antarctic peninsula—an area known as Larsen B (see above). Scientists from the British Antarctic Survey predicted last year that Larsen B was beginning to rift and that eventually the neighboring sector, Larsen C, "may well behave in a similar way" (see Nature 378, 328; 1996).

The researchers found that five Antarctic ice shelves out of nine they had studied have disintegrated in the past 50 years. Meterorological records over the same period show that the disintegration coincided with a 2.5 ℃ rice in atmospheric temperatures.

Accelerated distintegration is a common feature of retreating ice shelves in the Antarctic. The final stage of disintegration of Larsen A, the 4200 square kilometre northern part of the Larsen ice shelf, was marked by sudden collapse in January 1995. Some 1300 square kilometers of ice were lost in 50 days when the ice shelf broke into thousands of small icebergs, producing a plume 200 km into the Weddell Sea (see Nature 374, 108; 1995).

南极冰盖破裂形成的裂隙

Rex Dalton

(新闻)

摘要:上周,环境学家团体"绿色和平组织"的研究者们横穿了南极半岛 Larsen 冰盖南部地区——Larsen B 的一条重要裂隙。去年英国南极局的科学家预测,Larsen B 的裂隙正在扩大,且相邻的 Larsen C 也"可能会以相同的方式"分裂(见 *Nature*,1996,378:328)。研究者发现,在过去的 50 年中,他们研究的 9 个南极冰盖中的 5 个已经开始瓦解。同期的气象数据显示,冰盖瓦解过程中,气温升高了 2.5 ℃。冰盖加速瓦解是南极冰盖退缩时的一个普遍现象。Larsen 冰盖北部 4200 km^2 的 Larsen A 瓦解最终阶段的标志是 1995 年 1 月的突然崩塌,在 50 天内,1300 km^2 的冰盖分解成成千上万个冰山,在 Weddell 海中形成了 200 km 长的羽状浮冰区(见 *Nature*,1995,374:108)。

13. McAdoo D, Laxon S. Antarctic tectonics: constraints from an ERS-1 satellite marine gravity field [J]. Science, 1997, 276(5312):556-560.

Antarctic tectonics: constraints from an ERS-1 satellite marine gravity field

McAdoo D., Laxon S.

(Geosciences Laboratory, National Ocean Service, National Oceanic and Atmospheric Administration, Silver Spring, MD 20910, USA)

Abstract: A high-resolution gravity field of poorly charted and ice-covered ocean near West Antarctica, from the Ross Sea east to the Weddell Sea, has been derived with the use of satellite altimetry, including ERS-1 geodetic phase, wave-form data. This gravity field reveals regional tectonic fabric, such as gravity lineations, which are the expression of fracture zones left by early (65 to 83 million years ago) Pacific-Antarctic sea-floor spreading that separated the Campbell Plateau and New Zealand continent from West Antarctica. These lineations constrain plate motion history and confirm the hypothesis that Antarctica behaved as two distinct plates, separated from each other by an extensional Bellingshausen plate boundary active in the Amundsen Sea before about 61 million years ago.

南极构造地质学:来自 ERS-1 人造卫星海洋重力场的估计

McAdoo D., Laxon S.

(美国国家海洋与大气委员会地球科学实验室)

摘要:使用包括 ERS-1 大地测量相态和波形数据在内的卫星测量术,获得了西南极地区从罗斯海向东到威德尔海的从未被清晰绘制的被冰川覆盖海域的高分辨率重力场。这个重力场显示出区域性板块构造的特征，如反映早期(6500万～8300万年前)将坎贝尔高原和新西兰大陆从西南极处分离的太平洋-南极洲海底扩张所遗留破裂区的重力线。这些重力线驱使板块运动的历史，而且证实了南极洲一直以来都像两块独立板块的活动特征，这两个板块被6100万年以前活跃在阿蒙森海外延的白令豪山板块边缘所分开。

14. Geli L, Bougault H, Aslanian D, et al. Evolution of the Pacific-Antarctic Ridge south of the Udintsev fracture zone[J]. Science, 1997, 278(5341): 1281-1284.

Evolution of the Pacific-Antarctic Ridge south of the Udintsev fracture zone

Geli L., Bougault H., Aslanian D., Briais A., Dossol., Etoubleau J., Le Formal J. P., Maia M., Ondréas H., Olivet J. L., Richardson C., Sayanagi K., Seama N., Shah A., Vlastelic I., Yamamoto M.

(Institut Francais de Recherche pour l'Exploitation de la Mer (IFREMER), Boite Postale 70, 29280 Plouzané, France)

Abstract: Because of the proximity of the Euler poles of rotation of the Pacific and Antarctic plates, small variations in plate kinematics are fully recorded in the axial morphology and in the geometry of the Pacific-Antarctic Ridge south of the Udintsev fracture zone. Swath bathymetry and magnetic data show that clockwise rotations of the relative motion between the Pacific and Antarctic plates over the last 6 million years resulted in rift propagation or in the linkage of ridge segments, with transitions from transform faults to giant overlapping spreading centers. This bimodal axial rearrangement has propagated southward for the last 30 to 35 million years, leaving trails on the sea floor along a 1000-kilometer-long V-shaped structure south of the Udintsev fracture zone.

Udintsev断裂带南部太平洋-南极洋中脊的演化

Geli L., Bougault H., Aslanian D., Briais A., Dossol., Etoubleau J., Le Formal J. P., Maia M., Ondréas H., Olivet J. L., Richardson C., Sayanagi K., Seama N., Shah A., Vlastelic I., Yamamoto M.

(法国海洋开发研究所)

摘要：由于太平洋和南极洲板块旋转欧拉点的靠近，板块动力学的微小变化在Udintsev断裂带南部的太平洋-南极洋中脊轴形态和几何学上有了充分记载。条带测深和磁性数据显示，在过去600万年间，太平洋和南极洲板块之间的顺时针旋转导致了裂缝的增多或者脊碎片之间的链接，这其中也伴随着从转换断层到巨大的交叉重叠的扩张中心的转换。这个双峰轴线重排在过去的3000万～3500万年间已向南扩增，在沿着Udintsev断裂带南部的一个1000 km长的V形结构的海底处留下了扩张的痕迹。

15. Inman M. The plan to unlock Lake Vostok[J]. Science, 2005, 310(5748): 611-612.

The plan to unlock Lake Vostok

Inman M.

(News)

Abstract: After a 6 year pause to consider the risks of environmental contamination, a Russian research team will resume drilling through the Antarctic ice next month.

解开 Vostok 湖之谜的计划

Inman M.

（新闻）

摘要：在暂停6年以对环境污染风险进行评估后，一个俄罗斯研究组将在下个月重启钻穿南极冰盖的行动。

16. Giles J. Lakes linked beneath Antarctic ice[J]. Nature，2006，440(7087)：977.

Lakes linked beneath Antarctic ice

Giles J.

（News）

Abstract：The discovery that lakes hidden deep beneath the Antarctic ice could burst under pressure "like champagne corks" is prompting a rethink of plans to explore the watery world under the continent's four-kilometre-thick ice sheet. Any flooding might also make its way to the sea, a prospect that has sparked a debate about whether it could affect global climate.

南极冰盖下的湖泊

Giles J.

（新闻）

摘要：南极冰盖深部的湖泊在压力下会像"打开香槟软木塞"一样爆发，这一发现促进我们重新考虑对大陆4000 m厚冰盖下世界的探索计划。所有的溢流都会以其方式进入海洋，这一现象激起了关于它是否能够影响全球气候的讨论。

17. Wingham D J，Siegert M J，Shepherd A，et al. Rapid discharge connects Antarctic subglacial lakes [J]. Nature，2006，440(7087)：1033-1036.

Rapid discharge connects Antarctic subglacial lakes

Wingham D. J.，Siegert M. J.，Shepherd A.，Muir A. S.

（Centre for Polar Observation and Modelling，Department of Space and Climate Physics，Pearson Building，University College London，Gower Street，London WC1E 6BT，UK）

Abstract：The existence of many subglacial lakes provides clear evidence for the widespread presence of water beneath the East Antarctic ice sheet，but the hydrology beneath this ice mass is poorly understood. Such knowledge is critical to understanding ice flow, basal water transfer to the ice margin, glacial landform development and subglacial lake habitats. Here we present ice-sheet surface elevation changes in central East Antarctica that we interpret to represent rapid discharge from a subglacial lake. Our observations indicate that during a period of 16 months, 1.8 km^3 of water was transferred over 290 km to at least two other subglacial lakes. While viscous deformation of the ice roof above may moderate discharge, the intrinsic instability of such a system suggests that discharge events are a common mode of basal drainage. If large lakes, such as Lake Vostok or Lake Concordia, are pressurizing, it is possible that substantial discharges could reach the coast. Our observations conflict with expectations that subglacial lakes have long residence times and slow circulations, and we suggest that entire subglacial drainage basins may be flushed periodically. The rapid transfer of water between lakes would result in large-scale solute and microbe relocation, and drainage system contamination from in situ exploration is, therefore, a distinct risk.

水流快速排放的南极冰下湖

Wingham D. J., Siegert M. J., Shepherd A., Muir A. S.

(英国伦敦大学学院空间与气候物理学院极地观察与建模中心)

摘要：大量冰下湖的存在为东南极的冰盖下普遍存在水提供了显著证据，但是对于这其中的水文学原理却了解匮乏。这些原理知识对于理解冰流、底层水转移至冰缘、冰川地形地貌发育以及冰下湖栖息环境是关键的。在此我们给出东南极中心地带的冰盖表面高程变化，这个变化在我们看来可以代表来自冰下湖的快速排放。我们的观察显示在16个月的时期内，1.8 km^3 的水被转移290 km的距离而到达至少两个别的冰下湖。虽然冰顶部的黏性变形会缓和这种排放，但是该排放系统本质上的不稳定性表明排放是底层排水的普遍通用模式。如果大的湖泊，比如Vostok湖和Concordia湖，都在增压，那么大量的排放就有可能到达海岸。我们的观察与冰下湖具有长久的停留时间和缓慢的循环这一预期相矛盾，而且我们认为总的冰下流域或许是受到周期性的冲刷，湖泊之间水分的快速迁移会导致大范围的溶质和微生物的迁移，而且来自于原地探测的排水系统污染就会因此而成为一个明显的风险。

18. Velicogna I, Wahr J. Measurements of time-variable gravity show mass loss in Antarctica[J]. Science, 2006, 311(5768): 1754-1756.

Measurements of time-variable gravity show mass loss in Antarctica

Velicogna I., Wahr J.

(University of Colorado, Cooperative Institute for Research in Environmental Sciences and Department of Physics, University Campus Box 390, Boulder, Co 80309-0390, USA)

Abstract: Using measurements of time-variable gravity from the Gravity Recovery and Climate Experiment satellites, we determined mass variations of the Antarctic ice sheet during 2002-2005. We found that the mass of the ice sheet decreased significantly, at a rate of 152±80 cubic kilometers of ice per year, which is equivalent to 0.4±0.2 millimeters of global sea-level rise per year. Most of this mass loss came from the West Antarctic Ice Sheet.

时变重力测量显示南极地区质量损失

Velicogna I., Wahr J.

(美国科罗拉多大学环境科学与物理系合作研究中心)

摘要:通过分析利用来自重力恢复和气候实验卫星的时变重力测量数据,我们测定在 2002～2005 年间南极冰盖的质量变化。我们发现冰盖的质量以每年 152±80 km^3 体积冰的速率显著下降,这些体积的冰化成水后,体积相当于每年全球海平面上升 0.4±0.2 mm 所对应的水量。大部分损失的质量来自于西南极冰盖。

19. Whittaker J M, Muller R D, Leitchenkov G, et al. Major Australian-Antarctic plate reorganization at Hawaiian-Emperor bend time[J]. Science, 2007, 318(5847): 83-86.

Major Australian-Antarctic plate reorganization at Hawaiian-Emperor bend time

Whittaker J. M., Muller R. D., Leitchenkov G., Stagg H., Sdrolias M., Gaina C., Goncharov A.

(Earth Byte Group, School of Geosciences, University of Sydney, Sydney 2006, Australia)

Abstract: A marked bend in the Hawaiian-Emperor seamount chain supposedly resulted from a recent major reorganization of the plate-mantle system there 50 million years ago. Although alternative mantle-driven and plate-shifting hypotheses have been proposed, no contemporaneous circum-Pacific plate events have been identified. We report reconstructions for Australia and Antarctica that reveal a major plate reorganization between 50 and 53 million years ago. Revised Pacific Ocean sea-floor reconstructions suggest that subduction of the Pacific-Izanagi spreading ridge and subsequent Marianas/Tonga-Kermadec subduction initiation may have been the ultimate causes of these events. Thus, these plate reconstructions solve long-standing continental fit problems and improve constraints on the motion between East and West Antarctica and global plate circuit closure.

夏威夷帝王岛链转折时期澳大利亚-南极大板块的重组

Whittaker J. M., Muller R. D., Leitchenkov G., Stagg H., Sdrolias M., Gaina C., Goncharov A.

(澳大利亚悉尼大学地球科学学院)

摘要:夏威夷-帝王岛链处的一个明显的转折可能来源于约 5000 万年前一个最近的板块-地幔系统的重大重组。虽然目前存在不同的因地幔驱动和板块移动导致这一转折的猜想,但是尚无同时期的环太平洋板块活动事件来验证。在此我们报道 5000 万～5300 万年前澳大利亚-南极洲板块的重大重组事件。太平洋海底的修正重建表明,太平洋-伊邪那歧扩张山脊的俯冲和后续的马里亚纳群岛/汤加-科美迪克俯冲可能最终导致了重组事件的发生。因而,这些板块重组解决了长期未解的大陆之间的匹配问题,改进了东、西南极和全球板块回路闭合之间运动的约束条件。

20. Helsen M M, van den Broeke M R, van de wal R S W, et al. Elevation changes in Antarctica mainly determined by accumulation variability[J]. Science, 2008, 320(5883): 1626-1629.

Elevation changes in Antarctica mainly determined by accumulation variability

Helsen M. M. , van den Broeke M. R. ,van de Wal R. S. W. ,van de Berg W. J. , Meijgaard E. ,Davis C. H. ,Li Y. ,Goodwin I.

(Institute for Marine and Atmospheric Research, Utrecht University, 3584 CC Utrecht, Netherlands)

Abstract: Antarctic Ice Sheet elevation changes, which are used to estimate changes in the mass of the interior regions, are caused by variations in the depth of the firn layer. We quantified the effects of temperature and accumulation variability on firn layer thickness by simulating the 1980-2004 Antarctic firn depth variability. For most of Antarctica, the magnitudes of firn depth changes were comparable to those of observed ice sheet elevation changes. The current satellite observational period (similar to 15 years) is too short to neglect these fluctuations in firn depth when computing recent ice sheet mass changes. The amount of surface lowering in the Amundsen Sea Embayment revealed by satellite radar altimetry (1995-2003) was increased by including firn depth fluctuations, while a large area of the East Antarctic Ice Sheet slowly grew as a result of increased accumulation.

南极地区的海拔变化主要由冰雪累积导致

Helsen M. M. , van den Broeke M. R. ,van de Wal R. S. W. ,van de Berg W. J. , Meijgaard E. ,Davis C. H. ,Li Y. ,Goodwin I.

(荷兰乌特勒支大学海洋和大气研究协会)

摘要:南极冰盖的海拔变化(常用于估计内部区域的质量变化)是由积雪层的深度变化引起的。我们通过模拟1980～2004年间南极积雪层深度定量计算了温度和累积变量对积雪层厚度的影响。对南极的大部分地区来说,积雪层深度变化的数量级可与观测到的冰盖海拔变化的数量级相比拟。由于目前人造卫星观测期太短(约15年),因而在计算最近冰盖质量变化时不可以忽略了这些积雪层厚度波动的影响。人造卫星雷达高度探测(1995～2003)测定的阿蒙森海湾的表面下降,在考虑了积雪层深度变化后有所增加,同时东南极冰盖的一大片区域因不断增加的积雪累积而缓慢增长。

21. Goodge J W, Vervoort J D, Fanning C M, et al. A positive test of east Antarctica-Laurentia juxtaposition within the Rodinia supercontinent[J]. Science, 2008, 321(5886): 235-240.

A positive test of east Antarctica-Laurentia juxtaposition within the Rodinia supercontinent

Goodge J. W. , Vervoort J. D. , Fanning C. M. , Brecke D. M. , Farmer G. L. , Williams I. S. , Myrow P. M. , Depaolo D. J.

(Department of Geological Sciences, University of Minnesota-Duluth, Duluth, MN 55812, USA)

Abstract: The positions of Laurentia and other landmasses in the Precambrian supercontinent of Rodinia are controversial. Although geological and isotopic data support an East Antarctic fit with western Laurentia, alternative reconstructions favor the juxtaposition of Australia, Siberia, or South China. New geologic, age, and isotopic data provide a positive test of the juxtaposition with East Antarctica: Neodymium isotopes of Neoproterozoic rift- margin strata are similar; hafnium isotopes of similar to 1.4 billion year old Antarctic-margin detrital zircons match those in Laurentian granites of similar age; and a glacial clast of A-type granite has a uraniun- lead zircon age of similar to 1440 million years, an epsilon- hafnium initial value of +7, and an epsilon- neodymium initial value of +4. These tracers indicate the presence of granites in East Antarctica having the same age, geochemical properties, and isotopic signatures as the distinctive granites in Laurentia.

罗迪尼亚超级大陆内部东南极-劳伦古大陆拼合的实证检验

Goodge J. W., Vervoort J. D., Fanning C. M., Brecke D. M., Farmer G. L., Williams I. S., Myrow P. M., Depaolo D. J.

(美国明尼苏达-德卢斯大学地质科学系)

摘要：前寒武纪罗迪尼亚超级大陆上的劳伦古大陆和其他大陆地块的位置一直存在争议。虽然地质学和同位素数据都支持西劳伦迪亚超级大陆与东南极拼合的说法，但是也有重建结果认为其与澳大利亚、西伯利亚或南中国拼合。新的地质学、定年和同位素数据为其与东南极相毗邻提供了一个实证检验：与新元古代的裂谷-边缘地层的钕同位素数据接近；约14亿年前的南极边缘碎屑锆石的铪同位素数据符合劳伦系岩石层的类似年龄的花岗岩的数据；一个A类花岗岩的冰层碎屑岩的铀-铅锆石年龄接近于14.40亿年，ε-铪的初始值为+7，ε-钕的初始值为+4。这些示踪物质表明东南极花岗岩与劳伦迪亚超级大陆的花岗岩有着相同年龄、地球化学特性和同位素特征。

22. Whittaker J M, Muller R D, Leitchenkov G, et al. Response to comment on "major Australian-Antarctic plate reorganization at Hawaiian-Emperor bend time"[J]. Science, 2008, 321(5888): 490.

Response to comment on "major Australian-Antarctic plate reorganization at Hawaiian-Emperor bend time"

Whittaker J. M., Muller R. D., Leitchenkov G., Stagg H., Sdrolias M., Gaina C., Goncharov A.

(EarthByte Group, School of Geosciences, University of Sydney, Sydney 2006, Australia)

Abstract: Accurately locating boundaries between continental and oceanic crust is topical in view of locating offshore boundaries relevant to margin formation models, plate kinematics, and frontier resource exploration. Although we disagree with Tikku and Direen's interpretations, the associated controversies reflect an absence of agreed-upon geophysical criteria for distinguishing stretched continental from oceanic crust, and a lack of samples from nonvolcanic margins.

回复对《夏威夷帝王岛链转折时期澳大利亚-南极大板块的重组》一文的质疑

Whittaker J. M., Muller R. D., Leitchenkow G., Stagg H., Sdrolias M., Gaina C., Goncharov A.

(澳大利亚悉尼大学地球科学系)

摘要：对近岸便捷相对边缘带模型、板块动力学以及前缘资源勘查来说，大陆和洋壳的精确边界相当重要。虽然我们不同意 Tikku 和 Direen 的解释，但这些矛盾却反映了对大陆和洋壳判定的公认地球物理标准尚未建立，且缺少非火山边界的样品。

23. Tikku A A, Direen N G. Comment on "major Australian-Antarctic plate reorganization at Hawaiian-Emperor bend time"[J]. Science, 2008, 321(5888): 490.

Comment on "major Australian-Antarctic plate reorganization at Hawaiian-Emperor bend time"

Tikku A. A., Direen N. G.

(Exxon mobil upstream research company, United States)

Abstract: Whittaker et al presented reconstructions for Australia and Antarctica showing a change in relative plate motion ~53 million years ago, coincident with an inferred major global plate reorganization. This comment addresses problematic areas in their assumptions and the geological consequences of their reconstructions.

对《夏威夷帝王岛链转折时期澳大利亚-南极大板块的重组》一文的评论

Tikku A. A., Direen N. G.

(美国埃克森美孚上游研究公司)

摘要：Whittaker 等的重建结果显示澳大利亚和南极洲板块在约 5300 万年前的相对运动，与全球板块的重大重组时间一致。本评论对他们的假设以及重建的地质结果提出了一些疑问。

6.3 摘要翻译——北极

1. Bischof J F, Darby D A. Mid-to Late Pleistocene ice drift in the western Arctic Ocean: evidence for a different circulation in the past[J]. Science, 1997, 277(5322): 74-78.

Mid-to Late Pleistocene ice drift in the western Arctic Ocean: evidence for a different circulation in the past

Bischof J. F., Darby D. A.

(Applied Marine Research Laboratory, Department of Oceanography, Old Dominion University, 1034 West 45th Street, Norfolk, VA 23529, USA)

Abstract: The provenance of ice-rafted debris (IRD) in four Arctic sediment cores implies that icebergs from the northwestern Laurentide ice sheets drifted across the western Arctic Ocean along the 180 degrees-0 degrees meridian toward Fram Strait during mid to late Pleistocene deglaciations within the last 700000 years. This iceberg drift was different from the present-day Beaufort Gyre circulation and resembled a dislocated transpolar drift (TPD). Sea ice mainly followed the iceberg trajectories but also frequently drifted from the Russian shelves eastward into the Amerasian Basin.

中晚更新世北冰洋西部的冰漂移:过去环流与现在不同的证据

Bischof J. F., Darby D. A.

(美国老道明大学海洋学系应用海洋研究实验室)

摘要:北冰洋 4 个沉积岩芯中的冰筏沉积记录显示,在过去 70 万年中的中晚更新世冰消期,来自劳伦太德冰盖的浮冰顺 180°～0°经线经过弗拉姆海峡横越北冰洋西部。浮冰的漂移与现代波弗特流涡不同,而更类似于离位穿极漂流。海冰与浮冰的路径基本相似,但是也经常从俄罗斯陆架向东漂至美亚海盆。

2. Edwards M H, Kurras G J, Tolstoy M, et al. Evidence of recent volcanic activity on the ultraslow-spreading Gakkel ridge[J]. Nature, 2001, 409(6822): 808-812.

Evidence of recent volcanic activity on the ultraslow-spreading Gakkel ridge

Edwards M. H., Kurras G. J., Tolstoy M., Bohnenstiehl D. R., Coakley B. J., Cochran J. R.

(Hawaii Institute of Geophysics and Planetology, POST 815)

Abstract: Seafloor spreading is accommodated by volcanic and tectonic processes along the global mid-ocean ridge system. As spreading rate decreases the influence of volcanism also decreases, and it is unknown whether significant volcanism occurs at all at ultraslow spreading rates (<1.5 cm · yr^{-1}). Here we present three-dimensional sonar maps of the Gakkel ridge, Earth's slowest-spreading mid-ocean ridge, located in the Arctic basin under the Arctic Ocean ice canopy. We acquired this data using hull-mounted sonars attached to a nuclear-powered submarine, the USS Hawkbill. Sidescan data for the ultraslow-spreading eastern Gakkel ridge depict two young volcanoes covering approximately 720 km^2 of an otherwise heavily sedimented axial valley. The western volcano coincides with the average location of epicentres for more than 250 teleseismic events detected in 1999, suggesting that an axial eruption was imaged shortly after its occurrence. These findings demonstrate that eruptions along the ultraslow-spreading Gakkel ridge are focused at discrete locations and appear to be more voluminous and occur more frequently than was previously thought.

Gakkle 超慢速扩张中脊上现代火山活动的证据

Edwards M. H., Kurras G. J., Tolstoy M., Bohnenstiehl D. R., Coakley B. J., Cochran J. R.

(美国夏威夷地球物理与行星科学研究所)

摘要:洋盆扩张是由分布在全球洋中脊系统的火山和构造过程推动的。当扩张速率减慢时,火山活动也会减少,但目前仍不清楚在超慢速扩张速率条件(<1.5 cm · yr^{-1})下是否有大型火山活动。本文绘制了北冰洋流冰群下洋盆中地球上扩张最慢的洋中脊——Gakkel 中脊的三维声呐地图,数据由附在 USS Hawkbill 核潜艇上的舰壳声呐获得。在 Gakkel 中脊东侧的侧扫数据显示了轴向谷中有两个覆盖了约 720 km^2的年轻火山。西边的火山与 1999 年探测到的 250 次远震时间发生的平均地点相符,说明震后发生了轴向的短暂火山喷发。这些证据指示超慢速扩张 Gakkel 中脊的火山喷发发生在不同的地方,较以往认为的量更多并更为频繁。

3. Dick H J B, Lin J, Schouten H. An ultraslow-spreading class of ocean ridge[J]. Nature, 2003, 426 (6965): 405-412.

An ultraslow-spreading class of ocean ridge

Dick H. J. B., Lin J., Schouten H.

(Woods Hole Oceanographic Institution, Woods Hole, Massachusetts 02543, USA)

Abstract: New investigations of the Southwest Indian and Arctic ridges reveal an ultraslow-spreading class of ocean ridge that is characterized by intermittent volcanism and a lack of transform faults. We find that the mantle beneath such ridges is emplaced continuously to the seafloor over large regions. The differences between ultraslow- and slow-spreading ridges are as great as those between slow-and fast-spreading ridges. The ultraslow-spreading ridges usually form at full spreading rates less than about 12 mm · yr^{-1}, though their characteristics are commonly found at rates up to approximately 20 mm · yr^{-1}. The ultraslow-spreading ridges consist of linked magmatic and amagmatic accretionary ridge segments. The amagmatic segments are a previously unrecognized class of accretionary plate boundary structure and can assume any orientation, with angles relative to the spreading direction ranging from orthogonal to acute. These amagmatic segments sometimes coexist with magmatic ridge segments for millions of years to form stable plate boundaries, or may displace or be displaced by transforms and magmatic ridge segments as spreading rate, mantle thermal structure and ridge geometry change.

一种超慢速扩张类型的洋中脊

Dick H. J. B., Lin J., Schouten H.

(美国伍兹霍尔海洋研究所)

摘要:最新调查结果显示,西南印度洋和北冰洋存在一种超慢速扩张类型的洋中脊,该处火山爆发断断续续,且无转换断层。我们发现,这类洋中脊下的地幔大范围连续地分布在海底。超慢速与慢速扩张洋中脊的区别之大等同于慢速与快速扩张类型洋中脊的区别。超慢速洋中脊一般形成于扩张速率低于 12 mm • yr^{-1}的区域,虽然其扩张速率一般超过 20 mm • yr^{-1}。这种洋中脊由相互关联的岩浆型和非岩浆型生长中脊段组成。非岩浆型中脊段在以前的研究中尚无认识,它是生长板块边界的结构,生长方向不定,与板块生长方向呈锐角至直角都有可能。这些非岩浆型中脊段有时与岩浆型中脊段共存几百万年,结合形成稳定的板块边界;也可能在扩散速率、地幔热结构以及中脊形状发生改变时取代转换断层和岩浆型中脊,或被两者取代。

4. Edmonds H N, Michael P J, Baker E T, et al. Discovery of abundant hydrothermal venting on the ultraslow-spreading Gakkel ridge in the Arctic[J]. Nature, 2003, 421(6920): 252-256.

Discovery of abundant hydrothermal venting on the ultraslow-spreading Gakkel ridge in the Arctic

Edmonds H. N., Michael P. J., Baker E. T., Connelly D. P., Snow J. E., Langmuir H., Dick H. J. B., Mühe R., German C. R., Graham D. W.

(The University of Texas at Austin, Marine Science Institute, 750 Channel View Drive, Port Aransas, Texas 78373-5015, USA)

Abstract: Submarine hydrothermal venting along mid-ocean ridges is an important contributor to ridge thermal structure, and the global distribution of such vents has implications for heat and mass fluxes from the Earth's crust and mantle and for the biogeography of vent-endemic organisms. Previous studies have predicted that the incidence of hydrothermal venting would be extremely low on ultraslow-spreading ridges (ridges with full spreading rates <2 cm • yr^{-1} which make up 25 percent of the global ridge length), and that such vent systems would be hosted in ultramafic in addition to volcanic rocks. Here we present evidence for active hydrothermal venting on the Gakkel ridge, which is the slowest spreading (0. 6-1. 3 cm • yr^{-1}) and least explored mid-ocean ridge. On the basis of water column profiles of light scattering, temperature and manganese concentration along 1100 km of the rift valley, we identify hydrothermal plumes dispersing from at least nine to twelve discrete vent sites. Our discovery of such abundant venting, and its apparent localization near volcanic centres, requires a reassessment of the geologic conditions that control hydrothermal circulation on ultraslow-spreading ridges.

北冰洋 Gakkel 超慢速扩张中脊大量热液排放的发现

Edmonds H. N., Michael P. J., Baker E. T., Connelly D. P., Snow J. E., Langmuir H., Dick H. J. B., Mühe R., German C. R., Graham D. W.

(美国得克萨斯大学奥斯汀分校海洋科学研究所)

摘要:洋中脊附近的热液排放是影响中脊热结构的重要因素,这一事件的全球分布对来自地壳和地幔的热流和物质量以及排放口周边有机体的生物地理学都有重要的指示意义。过去的研究预测,超慢速扩张中脊(扩张速率小于 2 cm • yr^{-1},组成了全球 25%长度的洋中脊)热液的排放非常慢,且这一排放系统主要存在于火山岩及基性岩中。本文报道世界上扩张最慢(0. 6~1. 3 cm • yr^{-1})以及研究最少的 Gakkel 中脊活动热液排放的证据。沿着 1100 km 长的裂谷,我们测定了水文剖面中光散射、温度和锰含量,鉴别出了分布在至少 9~12 个独立排放点的热液柱。如此大量热液排放系统的发现,还有它们位于火山中心,表明需要对控制超慢速扩张中脊热液循环的地质状况进行重新分析。

5. Jokat W,Ritzmann O,Schmidt-Aursch M C,et al. Geophysical evidence for reduced melt production on the Arctic ultraslow Gakkel mid-ocean ridge[J]. Nature,2003,423(6943):962-965.

Geophysical evidence for reduced melt production on the Arctic ultraslow Gakkel mid-ocean ridge

Jokat W., Ritzmann O., Schmidt-Aursch M. C., Drachev S., Gauger S., Snow J.

(Alfred Wegener Institute Helmholtz Centre for Polar and Marine Research, Bremerhaven, Bremen, Germany)

Abstract: Most models of melt generation beneath mid-ocean ridges predict significant reduction of melt production at ultraslow spreading rates (full spreading rates < 20 mm • yr^{-1}) and consequently they predict thinned oceanic crust. The 1800-km-long Arctic Gakkel mid-ocean ridge is an ideal location to test such models, as it is by far the slowest portion of the global mid-ocean-ridge spreading system, with a full spreading rate ranging from 6 to 13 mm • yr^{-1}. Furthermore, in contrast to some other ridge systems, the spreading direction on the Gakkel ridge is not oblique and the rift valley is not offset by major transform faults. Here we present seismic evidence for the presence of exceptionally thin crust along the Gakkel ridge rift valley with crustal thicknesses varying between 1. 9 and 3. 3 km (compared to the more usual value of 7 km found on medium-to fast-spreading mid-ocean ridges). Almost 8300 km of closely spaced aeromagnetic profiles across the rift valley show the presence of discrete volcanic centres along the ridge, which we interpret as evidence for strongly focused, three-dimensional magma supply. The traces of these eruptive centres can be followed to crustal ages of ~25 Myr off-axis, implying that these magma production and transport systems have been stable over this timescale.

北极地区超慢速扩张的Gakkel洋中脊处熔融物产量减少的地球物理证据

Jokat W. , Ritzmann O. , Schmidt-Aursch M. C. , Drachev S. , Gauger S. , Snow J.

(德国阿尔弗雷德·魏格纳极地与海洋研究所)

摘要：大多数洋中脊下方的熔融物产生模型都预测到在超慢速扩张条件(小于每年 20 mm)下其产量显著减少，且洋壳随之变薄。1800 km 长的北极 Gakkel 洋中脊是一个检测这些模型的理想地区，因为这里是全球洋中脊扩张系统中迄今为止最慢的部分，其扩张速率为每年 6～13 mm。而且，与一些其他的洋中脊系统相反的是，Gakkel 洋中脊的扩张方向不是斜的，而且裂谷不能被主要的转换断层所抵消。在此，我们报道出沿着 Gakkel 洋中脊裂谷的超薄地壳存在的地震学证据，这些地壳厚度从 1.9～3.3 km 不定，这与在中速至快速扩张的洋中脊地区发现的更普遍的 7 km 形成对比。几乎长 8300 km 紧密排列的横跨这个裂谷的空中探测地磁剖面图显示沿着这个山脊分布着离散的火山中心，我们将这解释为强烈聚集的三维的岩浆供给证据。这些喷发中心的年代在约 2500 万年前，与地壳年龄一致，指示了这些岩浆的产生和输送系统在这一时间尺度上的稳定性。

6. Michael P J, Langmuir C H, Dick H J B, et al. Magmatic and amagmatic seafloor generation at the ultraslow-spreading Gakkel ridge, Arctic Ocean[J]. Nature, 2003, 423(6943): 956-961.

Magmatic and amagmatic seafloor generation at the ultraslow-spreading Gakkel ridge, Arctic Ocean

Michael P. J. , Langmuir C. H. , Dick H. J. B. , Snow J. E. , Goldstoin S. L. , Gradan D. W. , Lehnert K. , Kurras G. , Jokat W. , Mühe R. , Edmonds H. N.

(Department of Geosciences, The University of Tulsa, 600 College Avenue, Tulsa, Oklahoma 74104, USA)

Abstract: A high-resolution mapping and sampling study of the Gakkel ridge was accomplished during an international ice-breaker expedition to the high Arctic and North Pole in summer 2001. For this slowest-spreading endmember of the global mid-ocean-ridge system, predictions were that magmatism should progressively diminish as the spreading rate decreases along the ridge, and that hydrothermal activity should be rare. Instead, it was found that magmatic variations are irregular, and that hydrothermal activity is abundant. A 300-kilometre-long central amagmatic zone, where mantle peridotites are emplaced directly in the ridge axis, lies between abundant, continuous volcanism in the west, and large, widely spaced volcanic centres in the east. These observations demonstrate that the extent of mantle melting is not a simple function of spreading rate: mantle temperatures at depth or mantle chemistry (or both) must vary significantly along-axis. Highly punctuated volcanism in the absence of ridge offsets suggests that first-order ridge segmentation is controlled by mantle processes of melting and melt segregation. The strong focusing of magmatic activity coupled with faulting may account for the unexpectedly high levels of hydrothermal activity observed.

北极扩张速度超慢的Gakkel山脊处岩浆和非岩浆型海底的形成

Michael P. J., Langmuir C. H., Dick H. J. B., Snow J. E., Goldstoin S. L., Gradan D. W., Lehnert K., Kurras G., Jokat W., Mühe R., Edmonds H. N.

(美国塔尔萨大学地球科学学院)

摘要：在2001年进行的一次北极高纬和北极点的国际破冰探险考察中完成了一项对于Gakkel山脊进行的高分辨率绘图和取样。对于这一全球洋中脊系统中扩张速度最慢的终端单元，曾有预测说，岩浆作用应该随着山脊扩张速率的减慢逐渐地降低，而且热液活动应该更稀少。然而，有发现表明，岩浆活动的变化是无规律的，热液活动也十分活跃。一个300 km长的非岩浆活动区域坐落在有丰富持续火山活动的西部和广泛布满火山活动中心东部之间的地带，这个非岩浆活动区域中地幔橄榄岩直接位于洋脊轴线位置。这些观察都表明，地幔熔融程度不是简单的扩张速率的函数：地幔深处的温度或者地幔化学组成(或二者皆备)肯定沿轴线变化程度很大。在无山脊短错处严重不连续的火山活动表明一级山脊分裂被地幔熔融和熔融分离所控制。对岩浆活动与断层现象的强烈关注或许可以解释所观察到的出人意料的高强度热液活动。

7. Mack M C, Schuur E A G, Bret-Harte M S, et al. Ecosystem carbon storage in Arctic tundra reduced by long-term nutrient fertilization[J]. Nature, 2004, 431(7007): 440-443.

Ecosystem carbon storage in Arctic tundra reduced by long-term nutrient fertilization

Mack M. C., Schuur E. A. G., Bret-Harte M. S., Shaver G. R., Chapin F. S.

(Department of Botany, University of Florida, Gainesville, Florida 32611, USA)

Abstract: Global warming is predicted to be most pronounced at high latitudes, and observational evidence over the past 25 years suggests that this warming is already under way. One-third of the global soil carbon pool is stored in northern latitudes, so there is considerable interest in understanding how the carbon balance of northern ecosystems will respond to climate warming. Observations of controls over plant productivity in tundra and boreal ecosystems have been used to build a conceptual model of response to warming, where warmer soils and increased decomposition of plant litter increase nutrient availability, which, in turn, stimulates plant production and increases ecosystem carbon storage. Here we present the results of a long-term fertilization experiment in Alaskan tundra, in which increased nutrient availability caused a net ecosystem loss of almost 2000 grams of carbon per square meter over 20 years. We found that annual aboveground plant production doubled during the experiment. Losses of carbon and nitrogen from deep soil layers, however, were substantial and more than offset the increased carbon and nitrogen storage in plant biomass and litter. Our study suggests that projected release of soil nutrients associated with high-latitude warming may further amplify carbon release from soils, causing a net loss of ecosystem carbon and a positive feedback to climate warming.

北极苔原地区生态系统碳储量因长期营养性施肥而减少

Mack M. C. , Schuur E. A. G. , Bret-Harte M. S. , Shaver G. R. , Chapin F. S.

(美国佛罗里达大学植物系)

摘要：据预测，全球变暖在高纬度地区最明显，过去 25 年内的观测证据表明高纬变暖已经开始。全球土壤的碳储量有三分之一在北纬地区，因而人们对北半球生态系统中的碳平衡如何响应气候变暖有着强烈的兴趣。对苔原和极北区生态系统植物生产力对照组的实验观察已被用作建立对于气候变暖的抽象概念模式，这里较暖的土壤和植物残体的分解提高了营养物质的可利用性，反过来又促进植物的生产力和提高生态系统的碳储量。在此，我们展示出在阿拉斯加苔原地区的一项长期施肥实验的结果，在这个实验中，被提升的营养物质可利用性在 20 多年间中使得生态系统几乎每平方米净损失 2 kg 的碳。我们发现在此实验中地上植物的年生产力提高了一倍。然而，深层土壤的碳和氮的损失显著，远远抵消了植物生物量中增加的碳和氮储量。我们的研究表明，与高纬地区气候变暖相联系的土壤营养物质的释放或许会进一步增加土壤的碳流失，这会引起生态系统碳量的净流失和对气候变暖的正响应。

8. Jean-Baptiste P, Fourre E. Artic Ocean (communications arising): hydrothermal activity on Gakkel Ridge[J]. Nature, 2004, 428(6978): 36.

Artic Ocean (communications arising): hydrothermal activity on Gakkel Ridge

Jean-Baptiste P. , Fourre E.

(Laboratoire des Sciences du Climat et de l'Environnement, IPSL, CEA-CNRS, Centre d'Etudes de Saclay, 91191 Gif-sur-Yvette, France)

Abstract: In the hydrothermal circulation at mid-ocean ridges, sea water penetrates the fractured crust, becomes heated by its proximity to the hot magma, and returns to the sea floor as hot fluids enriched in various chemical elements. In contradiction to earlier results that predict diminishing hydrothermal activity with decreasing spreading rate, a survey of the ultra-slowly spreading Gakkel Ridge (Arctic Ocean) by Edmonds et al and Michael et al suggests that, instead of being rare, the hydrothermal activity is abundant—exceeding by at least a factor of two to three what would be expected by extrapolation from observation on faster spreading ridges. Here we use helium-3 (^{3}He), a hydrothermal tracer, to show that this abundance of venting sites does not translate, as would be expected, into an anomalous hydrothermal ^{3}He output from the ridge. Because of the wide implications of the submarine hydrothermal processes for mantle heat and mass fluxes to the ocean, these conflicting results call for clarification of the link between hydrothermal activity and crustal production at mid-ocean ridges.

北冰洋:Gakkel 洋中脊热液活动

Jean-Baptiste P. , Fourre E.

(法国气象及环境科学实验室)

摘要:在洋中脊热液循环过程中,海水进入裂开的地壳,在接近岩浆的过程中被加热,然后作为富含化学元素的热液回到海底。与之前研究预测的热液活动随扩张速率减少而减弱的结果相反,Edmonds 等和 Michael 等对北冰洋的超慢速扩张 Gakkel 洋脊的调查认为,该区热液活动非常丰富,而非很少,较快速扩张洋脊上发现的还多 2～3 倍。本文使用热液的指示计^{3}He 来证明这些排放点并没有像预计的那样从洋脊释放出大量热液^{3}He。由于海底热液过程对于地幔向海洋释放热量和物质通量的研究具有重要指示意义,这一矛盾的结果指示洋中脊热液活动和地壳产生量的关系需要进一步研究。

9. Jaccard S L, Haug G H, Nutman A P. Glacial/interglacial changes in subarctic North Pacific stratification[J]. Science,2005,308(5724):1003-1006.

Glacial/interglacial changes in subarctic North Pacific stratification

Jaccard S. L. , Haug G. H. , Nutman A. P.

(Department of Earth Science, Sonneggstrasse S, ETHZ, 8092 Zurich, Switzorland)

Abstract: Since the first evidence of low algal productivity during ice ages in the Antarctic Zone of the Southern Ocean was discovered, there has been debate as to whether it was associated with increased polar ocean stratification or with sea-ice cover, shortening the productive season. The sediment concentration of biogenic barium at Ocean Drilling Program site 882 indicates low algal productivity during ice ages in the Subarctic North Pacific as well. Site 882 is located southeast of the summer sea-ice extent even during glacial maxima, ruling out sea-ice-driven light limitation and supporting stratification as the explanation, with implications for the glacial cycles of atmospheric carbon dioxide concentration.

亚北极北太平洋海洋分层的冰期-间冰期变化

Jaccard S. L. , Haug G. H. , Nutman A. P.

(瑞士苏黎世联邦理工学院地球科学系)

摘要:自冰期低藻类生产力的最早证据在南半球海洋的南极地区发现以来,对其与增强的极地海洋层化有关还是与减少了可生产季节时间的海冰覆盖有关就一直存在争论。大洋钻探计划 882 站位的生物成因钡元素浓度表明,亚北极的北太平洋地区在冰期也存在着低藻类生产力现象。即便是在冰期最大范围时,882 位点位于夏季海冰范围的东南方,排除了海冰引起的光限制作用,支持了层化理论,且对解释大气二氧化碳浓度的冰期旋回具有指示意义。

10. Bennett V C,Brandon A D,Nutman A P. Coupled ^{142}Nd-^{143}Nd isotopic evidence for Hadean mantle

dynamics[J]. Science, 2007, 318(5858): 1907-1910.

Coupled ^{142}Nd-^{143}Nd isotopic evidence for Hadean mantle dynamics

Bennett V. C., Brandon A. D., Nutman A. P.

(Research School of Earth Sciences, Australian National University, Canberra ACT, 0200 Australia)

Abstract: The oldest rocks-3. 85 billion years old-from southwest Greenland have coupled neodymium-142 excesses (from decay of now-extinct samarium-146; half-life, 103 million years) and neodymium-143 excesses (from decay of samarium-147; half-life, 106 million years), relative to chondritic meteorites, that directly date the formation of chemically distinct silicate reservoirs in the first 30 million to 75 million years of Earth history. The differences in Nd-142 signatures of coeval rocks from the two most extensive crustal relicts more than 3. 6 billion years old, in Western Australia and southwest Greenland, reveal early-formed large-scale chemical heterogeneities in Earth's mantle that persisted for at least the first billion years of Earth history. Temporal variations in Nd-142 signatures track the subsequent incomplete remixing of very-early-formed mantle chemical domains.

太古宙地幔动力机制的$^{142}Nd/^{143}Nd$同位素证据

Bennett V. C., Brandon A. D., Nutman A. P.

(澳大利亚国立大学地球科学学院)

摘要：在格陵兰西南地区发现了世界上最老的岩石，其年龄为38.5亿年，与球粒陨石相比，^{142}Nd(来自目前已不存在的^{146}Sm衰变，半衰期1.03亿年)和^{143}Nd(来自^{147}Sm衰变，半衰期1.06亿年)皆过剩，表明化学性质截然不同的硅储层地层在地球形成的30～70 Ma形成。澳大利亚西部和格陵兰西南36亿年前同一时代大面积地壳残余的^{142}Nd的区别显示了早期地幔的大范围化学异质性在地球历史的最初10亿年中一直持续。^{142}Nd信号随时间的变化则记录了最早地幔主要化学成分的突然不完全融合。

11. Furnes H, de Wit M, Staudigel H, et al. A vestige of Earth's oldest ophiolite[J]. Science, 2007, 315 (5819): 1704-1707.

A vestige of Earth's oldest ophiolite

Furnes H., de Wit M., Staudigel H., Rosing M., Muehlenbachs K.

(Centre for Geobiology and Department of Earth Science, University of Bergen, Bergen, Norway)

Abstract: A sheeted-dike complex within the similar to 3. 8-billion-year-old Isua supracrustal belt (ISB) in southwest Greenland provides the oldest evidence of oceanic crustal accretion by spreading. The geochemistry of the dikes and associated pillow lavas demonstrates an intraoceanic island arc and mid-ocean ridge-like setting, and their oxygen isotopes suggest a hydrothermal ocean-floor-type metamorphism. The pillows and dikes are associated with gabbroic and ultramafic rocks that together make up an ophiolitic association: the Paleoarchean Isua ophiolite complex. These sheeted dikes offer evidence for remnants of oceanic crust formed by sea-floor spreading of the earliest intact rocks on Earth.

地球上最古老的蛇纹岩的痕迹

Furnes H. , de Wit M. , Staudigel H. , Rosing M. , Muehlenbachs K.

(挪威卑尔根大学地球生物中心与地球科学系)

摘要：西南格陵兰岛约38亿年的Isua上地壳带(ISB)的一个片状岩脉型复合体为扩张型洋壳增生提供了最古老的证据。这个岩脉及与其关联的枕状熔岩的地球化学特征证明了一个海洋内部岛弧和洋中脊岩脉结构的存在，且它们的氧同位素指示了一个热液海底型变质系统。这些枕状熔岩和岩脉与辉长岩和超基性岩相关联，且后二者一起组成了一个蛇纹岩组合——古太古代Isua蛇纹岩复合体。这些片状岩脉为洋壳残骸是由地球上最早期完整岩石的海底扩张形成的提供了证据。

12. Sohn R A, Willis C, Humphris S, et al. Explosive volcanism on the ultraslow-spreading Gakkel ridge, Arctic Ocean[J]. Nature, 2008, 453(7199): 1236-1238.

Explosive volcanism on the ultraslow-spreading Gakkel ridge, Arctic Ocean

Sohn R. A. , Willis C. , Humphris S. , Shank T. M. , Singh H. , Edmonds H. N. , Kunz C. , Hedman U. , Helmke E. , Jakuba M. , Liljebladh B. , Linder J. , Murphy C. , Nakamura K. , Sato T. , Schlindwein V. , Stranne C. , Tau senfreund M. , Upchurch L. , Winsor P. , Jakobsson M. , Soule A.

(Woods Hole Oceanographic Institution, Woods Hole, Massachusetts 02543, USA)

Abstract: Roughly 60% of the Earth's outer surface is composed of oceanic crust formed by volcanic processes at mid-ocean ridges. Although only a small fraction of this vast volcanic terrain has been visually surveyed or sampled, the available evidence suggests that explosive eruptions are rare on mid-ocean ridges, particularly at depths below the critical point for seawater (3000 m). A pyroclastic deposit has never been observed on the sea floor below 3000 m, presumably because the volatile content of mid-ocean-ridge basalts is generally too low to produce the gas fractions required for fragmenting a magma at such high hydrostatic pressure. We employed new deep submergence technologies during an International Polar Year expedition to the Gakkel ridge in theArctic Basin at 856 E, to acquire photographic and video images of "zero-age" volcanic terrain on this remote, ice-covered ridge. Here we present images revealing that the axial valley at 4000 m water depth is blanketed with unconsolidated pyroclastic deposits, including bubble wall fragments (limu o Pele), covering a large (>10 km^2) area. At least 13. 5 wt% CO_2 is necessary to fragment magma at these depths, which is about tenfold the highest values previously measured in a mid-ocean-ridge basalt. These observations raise important questions about the accumulation and discharge of magmatic volatiles at ultraslow spreading rates on the Gakkel ridge and demonstrate that large-scale pyroclastic activity is possible along even the deepest portions of the global mid-ocean ridge volcanic system.

北冰洋上扩张速率超慢的Gakkel洋脊上的爆炸式火山活动

Sohn R. A., Willis C., Humphris S., Shank T. M., Singh H., Edmonds H. N., Kunz C., Hedman U., Helmke E., Jakuba M., Liljebladh B., Linder J., Murphy C., Nakamura K., Sato T., Schlindwein V., Stranne C., Tau senfreund M., Upchurch L., Winsor P., Jakobsson M., Soule A.

(美国伍兹霍尔海洋研究所)

摘要：几乎60%的地球外表面都是由洋壳组成的，这些洋壳由洋中脊处的火山作用形成。虽然只有很少一部分这种巨大的火山地形已经做了外观调查且取样，但已有证据表明洋中脊处强烈的火山爆发极少，特别是在3000 m以下深度。3000 m以下的海底从未发现火成碎屑沉积物，大概是因为洋中脊玄武岩的易挥发性成分通常太低而无法产生在如此高的流体静力学压力下可分解岩浆的气体成分。我们在一次去北极盆地Gakkel山脊的国际极地年远征考察中，利用新的深潜技术来获得在这个遥远且覆满冰层山脊处的“0岁”火山地带的影像记录。在此我们展示一些水下4000 m处轴向山谷被松散的火成碎屑沉积物所覆盖的照片，包括覆盖了一处很大区域(>10 km^2)的一些气泡壁片。在这个深度，二氧化碳的质量分数至少为13.5%才能形成一些散碎岩浆，这些量是之前在洋中脊玄武岩处测得的最高值的10倍。这些发现对于扩张速率超慢的Gakkel山脊上的岩浆挥发物质的积累和排放提出了重要问题，而且表明，大尺度的火成碎屑活动即便是沿着全球洋中脊火山系统的最深部分也可能发生。

13. Liu C Z, Snow J E, Hellebrand E, et al. Ancient, highly heterogeneous mantle beneath Gakkel ridge, Arctic Ocean[J]. Nature, 2008, 452(7185): 311-316.

Ancient, highly heterogeneous mantle beneath Gakkel ridge, Arctic Ocean

Liu C. Z., Snow J. E., Hellebrand E., Brügmann G., Vonder Handt A., Büchl A., Hofmann A. W.

(State Key Laboratory of Lithospheric Evolution, Institute of Geology and Geophysics, Chinese Academy of Sciences, Beijing, 100029, China)

Abstract: The Earth's mantle beneath ocean ridges is widely thought to be depleted by previous melt extraction, but well homogenized by convective stirring. This inference of homogeneity has been complicated by the occurrence of portions enriched in incompatible elements. Here we show that some refractory abyssal peridotites from the ultraslow-spreading Gakkel ridge (Arctic Ocean) have very depleted $^{187}Os/^{188}Os$ ratios with model ages up to 2 billion years, implying the long-term preservation of refractory domains in the asthenospheric mantle rather than their erasure by mantle convection. The refractory domains would not be sampled by mid-ocean-ridge basalts because they contribute little to the genesis of magmas. We thus suggest that the upwelling mantle beneath mid-ocean ridges is highly heterogeneous, which makes it difficult to constrain its composition by mid-ocean-ridge basalts alone. Furthermore, the existence of ancient domains in oceanic mantle suggests that using osmium model ages to constrain the evolution of continental lithosphere should be approached with caution.

北冰洋 Gakkel 洋脊下古老且高度异相的地幔

Liu C. Z. , Snow J. E. ,Hellebrand E. ,Brügmann G. ,Vonder Handt A. ,Büchl A. ,Hofmann A. W.

(中国科学院地质与地球物理研究所岩石圈演化重点实验室)

摘要:在海洋山脉下的地幔长久以来被广泛认为是被过去的熔体萃取所消减的,但因对流混合而形成了很好的同相结构。这一同相的推断因不相容元素部分的出现而复杂化。在此,我们展示出一些来自于超慢速扩张 Gakkel 洋脊的难熔深海橄榄岩有着极度负偏的$^{187}Os/^{188}Os$,它们的模式年龄可达 20 亿年,这暗示着一些难熔区域在软流圈地幔长期保存而没有因地幔对流而消亡。这些难熔区域无法通过对洋中脊玄武岩取样进行研究,因为它们对岩浆的形成几乎没有贡献。因而我们建议,洋中脊下的上升地幔是高度异相的,这使得难以单单通过洋中脊玄武岩来研究其组成。而且,在海洋地幔中这些古老区域的存在表明,用锇的模式年龄去研究大陆岩石圈的演化应谨慎进行。

14. Froese D G,Westgate J A,Reyes A V,et al. Ancient permafrost and a future, warmer Arctic[J]. Science,2008,321(5896):1648.

Ancient permafrost and a future, warmer Arctic

Froese D. G. , Westgate J. A. ,Reyes A. V. ,Enkin R. J. ,Preece S. J.

(Department of Earth and Atmospheric Sciences, University of Alberta, Edmonton, AB T5M 0M3, Canada)

Abstract: Climate models predict extensive and severe degradation of permafrost in response to global warming, with a potential for release of large volumes of stored carbon. However, the accuracy of these models is difficult to evaluate because little is known of the history of permafrost and its response to past warm intervals of climate. We report the presence of relict ground ice in subarctic Canada that is greater than 700000 years old, with the implication that ground ice in this area has survived past interglaciations that were warmer and of longer duration than the present interglaciation.

古冻土与北极未来增暖

Froese D. G. , Westgate J. A. ,Reyes A. V. ,Enkin R. J. ,Preece S. J.

(加拿大艾伯塔大学地球和大气科学系)

摘要:气候模型预测了气候变暖情况下冻土将会出现大范围的严重退化,因此具有释放大量已储存碳的潜力。但是,由于对冻土的变化历史及其对过去气候暖期的响应了解不足,这些模型的精度仍难以估计。本文报道了在加拿大亚北极地区 70 万年前的地下冰残余,这一发现指示该区域地下冰在比目前这一间冰期更暖的过去的间冰期仍能够存在。

15. Goldstein S L, Soffer G, Langmuir C H, et al. Origin of a "Southern Hemisphere" geochemical signature in the Arctic upper mantle[J]. Nature,2008,453(7191):89-93.

Origin of a "Southern Hemisphere" geochemical signature in the Arctic upper mantle

Goldstein S. L., Soffer G., Langmuir C. H., Lehnert K. A., Graham D. W., Michael P. J.

(Lamont-Doherty Earth Observatory of Columbia University, Columbia University, 61 Route 9W, Palisades, New York 10964, USA)

Abstract: The Gakkel ridge, which extends under the Arctic ice cap for similar to 1800 km, is the slowest spreading ocean ridge on Earth. Its spreading created the Eurasian basin, which is isolated from the rest of the oceanic mantle by North America, Eurasia and the Lomonosov ridge. The Gakkel ridge thus provides unique opportunities to investigate the composition of the sub-Arctic mantle and mantle heterogeneity and melting at the lower limits of seafloor spreading. The first results of the 2001 Arctic Mid-Ocean Ridge Expedition divided the Gakkel ridge into three tectonic segments, composed of robust western and eastern volcanic zones separated by a "sparsely magmatic zone". On the basis of Sr-Nd-Pb isotope ratios and trace elements in basalts from the spreading axis, we show that the sparsely magmatic zone contains an abrupt mantle compositional boundary. Basalts to the west of the boundary display affinities to the Southern Hemisphere "Dupal" isotopic province, whereas those to the east-closest to the Eurasian continent and where the spreading rate is slowest display affinities to "Northern Hemisphere" ridges. The western zone is the only known spreading ridge outside the Southern Hemisphere that samples a significant upper-mantle region with Dupal-like characteristics. Although the cause of Dupal mantle has been long debated, we show that the source of this signature beneath the western Gakkel ridge was subcontinental lithospheric mantle that delaminated and became integrated into the convecting Arctic asthenosphere. This occurred as North Atlantic mantle propagated north into the Arctic during the separation of Svalbard and Greenland.

北极上地幔中"南半球"地球化学信号的来源

Goldstein S. L., Soffer G., Langmuir C. H., Lehnert K. A., Graham D. W., Michael P. J.

(美国哥伦比亚大学拉蒙特-多尔蒂地质观测站)

摘要:Gakkel洋脊在北极冰盖下延伸了约1800 km,是地球上扩张速率最慢的洋中脊。它的扩张创造了从北美大陆、欧亚大陆和罗蒙诺索夫脊残余大洋地幔中脱离出来的欧亚盆地。因而,Gakkel山脊为研究环北极地幔的组成、地幔异相以及在海底扩张速率较低条件下的熔融现象提供了独一无二的材料。2001年北极洋中脊考察的首批成果将Gakkel山脊划分为三个板块结构,由被"一个岩浆稀疏带"分割开的巨大的西部和东部火山带组成。基于扩张轴线玄武岩中Sr-Nd-Pb同位素比例和痕量元素数据,我们发现上述的岩浆稀疏带包含一个突变的地幔组成边界。边界西部的玄武岩与南半球的"Ddupal同位素省"较为相似,而边界东部靠近欧亚大陆且扩张速率最慢的玄武岩显现出与北半球洋脊相似的特征。西部区域是唯一已知的在南半球以外的有着Dupal特征的重要上地幔区域。虽然Dupal地幔的成因一直存在争议,但本文所示的这种位于西Gakkel洋脊下的特征来源于次大陆岩石圈地幔,这种地幔剥落后与对流的北极软流圈融为一体。这一过程发生于北大西洋地幔在斯瓦尔巴特群岛与格陵兰岛分离期间向北扩张融入北极地区时。

七、陆地古环境类文献

7.1 分 析 概 述

1990～2010 年间发表于 *Nature*，*Science* 及 *PNAS* 杂志上有关极地陆地环境领域的论文和报道共 37 篇，其中与南极相关的有 19 篇，与北极相关的有 18 篇；*Nature* 上发表 17 篇，*Science* 上发表 14 篇，*PNAS* 上发表 6 篇。

从国家论文数量分布看(图 7.1)，共有 8 个国家在陆地环境领域有贡献。美国发表 19 篇，遥遥领先，优势显而易见；英国占据第二的位置，发表 8 篇；其后是加拿大 3 篇，新西兰、挪威各 2 篇，德国、俄罗斯、中国各 1 篇。从南北极文献分布看，除英国、德国和中国外，其他 5 国均是与极地有着密切地缘关系的国家，从图 7.2 中这一点可以看得更为明显，加拿大发表的 3 篇论文均与北极地区相关，新西兰所发表的 2 篇论文则与南极地区相关，俄罗斯发表的 1 篇论文是关于北极地区的，在美国发表的 20 篇文献中也有 13 篇是关于北极地区的。英国对南极的投入很大，所以产出也最多，发表关于南极的文献有 8 篇，占据南极陆地环境研究的领先地位。值得注意的是中国在这个领域也有所贡献，在 *Nature* 上发表了 1 篇关于南极企鹅古生态的研究论文，在这个领域达到国际先进水平。

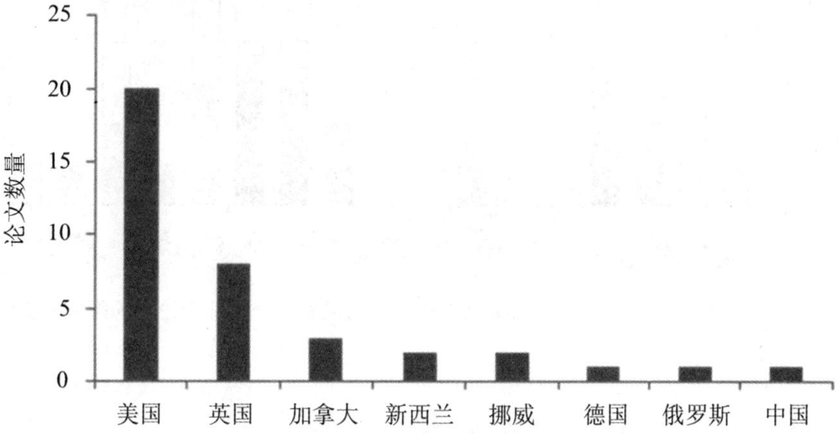

图 7.1 1990～2010 年各国在 *Nature*，*Science* 及 *PNAS* 上发表的有关极地陆地环境领域的论文

从文献具体的领域和年代分布看(图 7.3)，除 1990 年、1995 年、1998～1999 年三大杂志上未发表有关两极陆地环境方面的文献外，其余年份均有文献产出。1992～1994 年及 2001～2003 年间这一领域发表的文献居多，2002 年最多，有 7 篇文献发表，而且均来自与南极相关的研究。从更为具体的领域划分看，除有 1 篇大气化学方面的文献外，其余均是生态学、古生态学和气候变化方面的文献：古生态学文献为 17 篇，生态学为 10 篇，气候变化领域为 10 篇，反映了在极地陆地环境领域人们最关注的热点还是气候变化及其所引起的生态环境变化，随着全球或区域变化的加剧，气候生态环境变化依旧是国际极地研究的重要领域之一。

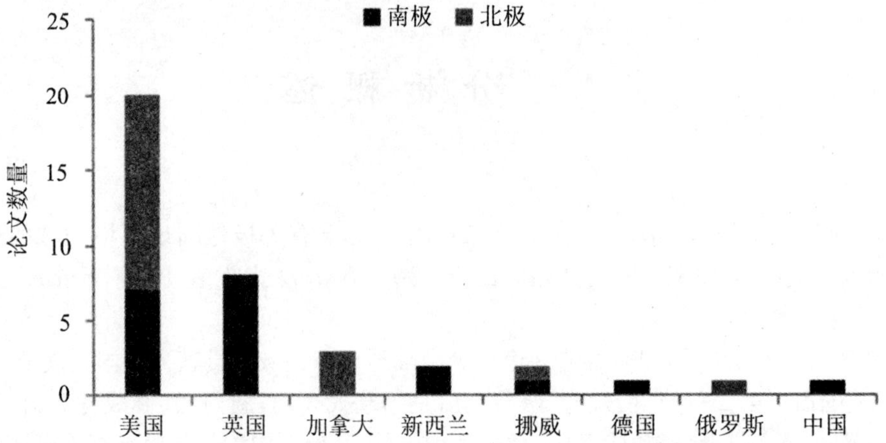

图 7.2 1990～2010 年 *Nature*,*Science* 及 *PNAS* 上有关陆地环境领域论文的两极分布情况

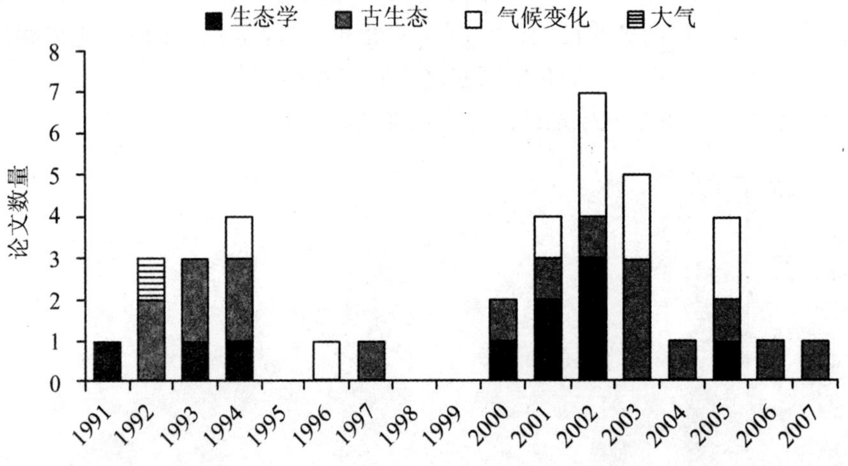

图 7.3 1990～2010 年 *Nature*,*Science* 及 *PNAS* 上有关极地陆地环境论文的领域分类统计

7.2 摘要翻译——南极

1. Craig H, Wharton R A Jr, McKay C P. Oxygen supersaturation in ice-covered Antarctic lakes: biological versus physical contributions[J]. Science, 1992, 255(5042): 318-321.

Oxygen supersaturation in ice-covered Antarctic lakes: biological versus physical contributions

Craig H., Wharton R. A. Jr., McKay C. P.

(Scripps Institution of Oceanography, University of California at San Diego, La Jolla 92093, USA)

Abstract: Freezing in ice-covered lakes causes dissolved gases to become supersaturated while at the same time removing gases trapped in the ablating ice cover. Analysis of N_2, O_2, and Ar in bubbles from Lake Hoare ice shows that, while O_2 is ~2.4 times supersaturated in the water below the ice, only 11% of the O_2 input to this lake is due to biological activity, 89% of the O_2 is derived from meltwater inflow. Trapped bubbles in a subliming ice cover provide a natural "fluxmeter" for gas exchange: in Lake Hoare as much as 70% of the total gas loss may occur by advection through the ice cover, including ~75% of the N_2, ~59% of the O_2, and ~57% of the Ar losses. The remaining gas fractions are removed by respiration at the lower boundary (O_2) and by molecular exchange with the atmosphere in the peripheral summer moat around the ice.

南极冰封湖泊过饱和氧：生物过程和物理过程的贡献

Craig H., Wharton R. A. Jr., McKay C. P.

(美国加利福尼亚大学圣地亚哥分校斯克里普斯海洋研究所)

摘要：冻结作用使得冰封湖泊溶解气体过饱和，同时去除消融冰盖中的气体。对Hoare湖冰气泡中的N_2，O_2和Ar的分析显示：冰下湖水O_2的过饱和达2.4倍，其中只有11%来自于生物活动，其余89%来自于冰融水流。升华湖面冰盖中的气泡提供了一个气体交换的"流量计"：湖面冰盖的水平运动可以导致Hoare湖中多达70%的气体流失，其中包括约75%的N_2、约59%的O_2和约57%的Ar。剩余气体的流失一部分通过水体下边界层以下的呼吸作用，另一部分由大气分子交换通过夏季湖面冰边缘处流失。

2. Roger Buick. The Antiquity of oxygenic photosynthesis: evidence from stromatolites in sulfate-deficient archean lakes[J]. Science, 1992, 255(5040): 74-77.

The Antiquity of oxygenic photosynthesis—evidence from stromatolites in sulfate-deficient archean lakes

Roger Buick

(Botanical Museum, Harvard University, Cambridge, MA 02138, USA)

Abstract: The Tumbiana Formation, about 2700 million years old, was largely deposited in ephemeral saline lakes, as judged by the unusual evaporite paragenesis of carbonate and halite with no sulfate. Stromatolites of diverse morphology occur in the lacustrine sediments, some with palimpsest fabrics after erect filaments. These stromatolites were probably accreted by phototropic microbes that, from their habitat in shallow isolated basins with negligible sulfate concentrations, almost certainly metabolized by oxygenic photosynthesis.

古老的光合作用——来自太古代缺硫酸盐湖泊中叠藻层的证据

Roger Buick

(美国哈佛大学植物博物馆)

摘要:根据独特的碳酸盐、岩盐和缺硫酸盐的共生序列来判断,Tumbiana 建造形成于 27 亿年前,主要沉积于短暂的盐湖中。叠藻层有多样的形态存在于淡水湖泊中,有些叠藻层在竖状长纤维形成变余构造。这些叠藻层与含极少硫酸盐的隔离浅水湖盆里的光和细菌合生,其中的大部分细菌是由光合作用来完成新陈代谢的。

3. Taylor E L, Taylor T N. Reproductive-biology of the permian glossopteridales and their suggested relationship to flowering plants[J]. PNAS, 1992, 89(23): 11495-11497.

Reproductive-biology of the permian glossopteridales and their suggested relationship to flowering plants

Taylor E. L., Taylor T. N.

(Byed Polar Reasearch Center and Department of Plant Biology, Ohio State University, Columbus, OH 43210)

Abstract: The discovery of permineralized glossopterid reproductive organs from Late Permian deposits in the Beardmore Glacier region (central Transantarctic Mountains) of Antarctica provides anatomical evidence for the adaxial attachment of the seeds to the megasporophyll in this important group of Late Paleozoic seed plants. The position of the seeds is in direct contradiction to many earlier descriptions, based predominantly on impression/compression remains. The attachment of the ovules on the adaxial surface of a leaf-like megasporophyll, combined with other features, such as megagametophyte development, suggests a simpler gymnospermous reproductive biology in this group than has previously been hypothesized. These findings confirm the classification of the Glossopteridales as seed ferns and are important considerations in discussions of the phylogeny of the group, including their suggested role as close relatives or possible ancestors of the angiosperms.

二叠纪舌羊齿目的生殖生物学及其与开花植物的指示关系

Taylor E. L. , Taylor T. N.

(美国俄亥俄州立大学伯德极地研究中心植物生物学部)

摘要：南极比德摩尔冰川地区(横断山脉中央)二叠纪沉积岩中矿化的舌羊齿目生殖器官的发现为晚古生代种子植物种子近轴附着大孢子叶提供了结构上的证据。种子的位置直接反驳了此前主要基于被压缩残体的描述。附着在大孢子叶近轴表面的胚珠连同其他诸如雌配子体等形态指示了相对于此前假设更为简单的裸子植物的生殖生物学特征。这些发现证实了舌羊齿目为蕨类植物，对于研究这一类的发展史具有重要的指示意义，包括其所暗示的作为被子植物的原型或近亲的角色。

4. Sellwood B W, Price G D, Valdes P J. Cooler estimates of Cretaceous temperatures[J]. Nature, 1994, 370(6489): 453-455.

Cooler estimates of Cretaceous temperatures

Sellwood B. W. , Price G. D. , Valdes P. J.

(Postgraduate Research Institute for Sedimentology, The University of Reading, Whiteknights, RG6 2AB, Berks, England)

Abstract: The Creataceous period is thought to have been warmer than the present, with higher concentrations of atmospheric greenhouse gases such as carbon dioxide. It has therefore been suggested that this time period could be used by modellers as an analogue for future climate change. But the Cretaceous Equator-to-Pole temperature gradient was flatter than today's, leading some to suggest that Cretaceous climate arose from a combination of factors, with higher atmospheric carbon dioxide concentrations leading to general warming, and other factors, such as increased ocean heat transport, leading to flattening of the latitudinal temperature gradient. Here we report new records of ocean palaeotemperature for Cenomanian sites in the Atlantic and Pacific oceans which, together with a re-evaluation of published data, cast doubt on the idea that the Cretaceous period was generally warmer. These data confirm that the latitudinal temperature gradient was flatter, but suggest that the global mean temperature was much cooler than previously believed, with minimum mean equatorial temperatures close to present values and polar temperatures close to 0 degrees C. In the light of these findings, the climatic role of atmospheric carbon dioxide in determining Cretaceous climate is unclear, suggesting that the Cretaceous cannot be used as an analogue for future climate change.

更冷的白垩纪气温的估计

Sellwood B. W. , Price G. D. , Valdes P. J.

(英国雷丁大学沉积学研究院)

摘要：白垩纪时期被认为比现在温暖，大气中含有更高量的温室气体如二氧化碳。因此这个时期可以被模型者参照来模拟未来气候变化。但是白垩纪时期赤道-极地间的温度梯度变化比现代更小，指示白垩纪温度的上升是多种联合因素造成的，更高含量的大气二氧化碳导致普遍的温暖，其他因素如海洋热量传输的加快导致沿纬线温度梯度变化更小。这里我们报道大西洋和太平洋森诺曼阶区的海洋古温度新的纪录，重新评估了此前发表的数据，并质疑白垩纪普遍更温暖的观点。我们的数据证实纬线温度梯度变化确实更小，但是指示了全球平均温度比此前认为的要冷得多，赤道平均温度和现代相当而极地温度接近 0 ℃。基于这些发现，大气二氧化碳含量在白垩纪气候中的贡献是不清楚的，暗示白垩纪不能作为模拟未来气候变化的参照。

5. Yuan Xiaojun, Mark A Cane, Douglas G Martinson. Climate variation: cycling around the South Pole [J]. Nature, 1996, 380(6576): 673-674.

Climate variation—cycling around the South Pole

Yuan Xiaojun, Mark A. Cane, Douglas G. Martinson

(Lamont Doherty Earth Observatory, Columbia University, Palisades, New York 10964, USA)

Abstract: Many of the variations in climate we all observe (and enjoy, or suffer) seem to be randomly distributed in time and space. While this view has an irreducible element of truth, there is increasing evidence that much of the Earth's climate variability arises from a structured, global, interconnected system. On page 699 of this issue, White and Peterson provide a striking example, with evidence for an interannual "Antarctic Circumpolar Wave" (ACW). This is not a water wave but a disturbance in sea-ice extent, in sea surface temperature, in surface wind speed, and in atmospheric pressure at sea level. Although the short history of measurements in the Antarctic region rules out a definitive description, the data available from the past 15 years show a two-wavelength ACW propagating around Antarctica with a frequency of about 4-5 years. This wave in the ocean, atmosphere and cryosphere(ice) is a strong and significant part of the interannual variability in the southern polar region.

气候变化——南极点附近的循环

Yuan Xiaojun, Mark A. Cane, Douglas G. Martinson

(美国哥伦比亚大学拉蒙特-多尔蒂地质观测站)

摘要：我们观察到许多气候变化的时间和空间似乎是随机分布的。虽然这种观点有一个不可缺少的因素来证实，有越来越多的证据表明，地球的气候变化起因于一个结构化的、全球性的、相互联系的系统。这个问题在本期 699 页，怀特和彼得森提供了一个特殊的例子，证实了年际南极绕极波。这不是一个水波，但扰动了海冰范围，在海洋表面温度、表面风速和海平面大气压力方面均有变动。尽管对南极地区很短的时间尺度上的测量不能得出一个明确的结论，过去 15 年的数据显示双谱绕极波环南极传播频率为 4～5 年。这个波在海洋、大气和冰冻圈的影响是南极地区年际变化一个强大和重要的部分。

6. Dominic A Hodgson, Nadine M Johnston. Inferring seal populations from lake sediments[J]. Nature, 1997, 387(6576): 30-31.

Inferring seal populations from lake sediments

Dominic A. Hodgson, Nadine M. Johnston

(British Antarctic Survey, High Cross, Madingley Road, Cambridge CB3 OET, UK)

Abstract: An explosion in the population of Antarctic fur seals (Arctocephalus gazella) has caused widespread changes to many coastal terrestrial and freshwater ecosystems in the northern maritime Antarctic islands and on the west coast of the Antarctic Peninsula. We have used seal hairs found in lake sediment cores from one maritime Antarctic island as a historical record of seal populations. This has enabled us to examine possible causes of the increasing numbers of visiting Antarctic fur seals, and has provided a historical framework from which to evaluate conservation plans to minimize the adverse effects of seals at sites of particular ecological significance.

从湖泊沉积物恢复海豹数量

Dominic A. Hodgson, Nadine M. Johnston

(英国南极调查局)

摘要:南极海狗(毛皮海狮)数量的爆炸式增长已经造成南极北部海域和南极半岛西海岸许多沿海陆地和淡水生态系统的广泛改变。我们用在湖泊沉积物岩芯中保存的海豹毛南极岛上的海豹种群的历史记录。这是我们研究南极毛皮海豹数量增加的可能原因,并提供评估保护计划,以尽量减少对有特定生态意义的这些地区海豹的不利影响。

7. Liguang Sun, Zhouqing Xie, Junlin Zhao. A 3000-year record of penguin populations[J]. Nature, 2000,407:858.

A 3000-year record of penguin populations

Liguang Sun, Zhouqing Xie, Junlin Zhao

(Institute of Polar Environment, University of Science and Technology of China, Hefei, Anhui 230026, PR China)

Abstract: There are no history records of changing penguin populations in the maritime Antarctic. Here we analyse the concentration of "bio-eiements" in a lakesediment core dating back approximately 3000 radiocarbon years. We found that the deposition of penguin droppings had a significant effect on the geochemical composition of the sediment core. Changes in sediment geochemistry reflect fluctuations in penguin numbers and suggest that variations in climate had an impact on penguin populations, which peaked somewhere between 1400 and 1800 years ago.

3000 年企鹅数量记录

Liguang Sun，Zhouqing Xie，Junlin Zhao

(中国科学技术大学极地环境研究室)

摘要：南极洲沿海区域并没有企鹅数量变化的历史记录。这里我们分析了一根约 3000 年的湖泊沉积柱的生物元素含量。我们发现企鹅粪沉积物对沉积柱的地球化学组成有显著影响。沉积物的地球化学变化反映了企鹅数量变化的波动并提示了气候变化对企鹅种群的影响使之在距今 1400～1800 的某个时间点达到峰值。

8. Vaughan D G，Marshall G J，Connolley W M，et al. Climate change：devil in the detail[J]. Science，2001，293：1777-1779.

Climate change—devil in the detail

Vaughan D. G.，Marshall G. J.，Connolley W. M.，King J. C.，Mulvaney R.

(British Antarctic Survey，Natural Environment Research Council)

Abstract：The Intergovernmental Panel on Climate Change (IPCC) this year confirmed a global mean warming of 0.6±0.2 ℃ during the 20th century and cited anthropogenic increases in greenhouse gases as the likely cause. However，this mean value conceals the complexity of observed climate change. If the recent past is a guide to the future，regional climate changes will have more profound effects than the mean global warming suggests.

气候变化——难在细节

Vaughan D. G.，Marshall G. J.，Connolley W. M.，King J. C.，Mulvaney R.

(英国自然环境研究理事会南极调查局)

摘要：政府间气候变化专门委员会(IPCC)今年确认了全球 20 世纪平均气温变暖 0.6±0.2 ℃并指出人类导致的温室气体增加是可能的因素。但是这一平均值隐藏了气候变化观测的复杂性。如果说现代是未来的指引，那么区域气候变化将会比平均全球变暖的提示有更深刻的影响。

9. Croxall J P，Trathan P N，Murphy E J. Environmental change and Antarctic seabird populations[J]. Science，2002，297：1510-1514.

Environmental change and Antarctic seabird populations

Croxall J. P.，Trathan P. N.，Murphy E. J.

(British Antarctic Survey，Natural Environment Research Council，High Cross，Madingley Road，Cambridge CB3 OET，UK)

Abstract：Recent changes in Antarctic seabird populations may reflect direct and indirect responses to regional climate change. The best long-term data for high-latitude Antarctic seabirds (Ade'lie and Emperor penguins and snow petrels) indicate that winter sea-ice has a profound influence. However, some effects are inconsistent between species and areas, some in opposite directions at different stages of breeding and life cycles, and others remain paradoxical. The combination of recent harvest driven changes and those caused by global warming may produce rapid shifts rather than gradual changes.

环境变化与南极海鸟数量

Croxall J. P., Trathan P. N., Murphy E. J.

(英国自然环境研究理事会南极调查局)

摘要:近来南极海鸟数量的变化也许反映了它们对区域气候变化直接或间接的响应。来自高纬地区南极海鸟(阿德雷企鹅、帝企鹅和雪燕)长期的数据显示冬季海冰对其有显著的影响作用。然而,一些作用在种群间和区域间是不同的,在生命周期及生殖的不同阶段甚至是相反的,另一些则是矛盾的。近来人类南大洋的捕获和气候变暖可能引起快速的突变而不是缓慢的渐变。

10. Doran P T, Priscu J C, Lyons W B, et al. Antarctic climate cooling and terrestrial ecosystem response[J]. Nature, 2002, 415: 517-520.

Antarctic climate cooling and terrestrial ecosystem response

Doran P. T., Priscu J. C., Lyons W. B., Walsh J. E., Fountain A. G., McKnight D. M., Moorhead D. L., Virginia R. A., Wall D. H., Clow G. D., Fritsen C. H., McKay C. P., Parsons A. N.

(Department of Earth and Environmental Sciences, University of Illinois at Chicago, 845 West Taylor Street, Chicago, Illinois 60607, USA)

Abstract: The average air temperature at the Earth's surface has increased by 0.06 ℃ per decade during the 20th century, and by 0.19 ℃ per decade from 1979 to 1998. Climate models generally predict amplified warming in polar regions, as observed in Antarctica's peninsula region over the second half of the 20th century. Although previous reports suggest slight recent continental warming, our spatial analysis of Antarctic meteorological data demonstrates a net cooling on the Antarctic continent between 1966 and 2000, particularly during summer and autumn. The McMurdo Dry Valleys have cooled by 0.7 ℃ per decade between 1986 and 2000, with similar pronounced seasonal trends. Summer cooling is particularly important to Antarctic terrestrial ecosystems that are poised at the interface of ice and water. Here we present data from the dry valleys representing evidence of rapid terrestrial ecosystem response to climate cooling in Antarctica, including decreased primary productivity of lakes (6-9% per year) and declining numbers of soil invertebrates (more than 10% per year). Continental Antarctic cooling, especially the seasonality of cooling, poses challenges to models of climate and ecosystem change.

南极气候变冷与陆地生态系统的响应

Doran P. T., Priscu J. C., Lyons W. B., Walsh J. E., Fountain A. G., McKnight D. M., Moorhead D. L., Virginia R. A., Wall D. H., Clow G. D., Fritsen C. H., McKay C. P., Parsons A. N.

(美国伊利诺伊大学地球与环境科学系)

摘要:20 世纪地表平均气温以 0.06 ℃/10 年的速度上升,从 1979 年到 1988 年升温的速率高达 0.19 ℃/10 年。气候模型普遍预测极区存在放大的暖化,就像南极半岛 20 世纪近 50 年来的观测结果一样。虽然此前报道暗示大陆地区近来显示出微弱的变暖,我们对南极气象数据的空间分析表明南极大陆地区 1966～2000 年间是变冷的,特别是在夏秋季节。麦克默多干谷在 1986～2000 年间以 0.7 ℃/10 年的速率变冷,且具有类似的显著季节趋势。夏季变冷对处于冰水界面的南极陆地生态系统特别重要。这里我们提供指示陆地生态系统对南极变冷的快速响应数据,包括湖泊生产力的下降(6%～9%/10 年)和无脊椎动物数量的减少(大于 10%/10 年)。南极大陆变冷,特别是季节性的变冷对气候和生态系统变化的模型化是一个挑战。

11. Lambert D M, Ritchie P A, Millar C D, et al. Rates of evolution in ancient DNA from Adelie penguins[J]. Science, 2002, 295: 2270-2273.

Rates of evolution in ancient DNA from Adelie penguins

Lambert D. M., Ritchie P. A., Millar C. D., Holland B., Drummond A. J., Baroni C.

(Institute of Molecular BioSciences, Sciences, Massey University, New Zealand)

Abstract: Well-preserved subfossil bones of Adelie penguins, Pygoscelis adeliae, underlie existing and abandoned nesting colonies in Antarctica. These bones, dating back to more than 7000 years before the present, harbor some of the best-preserved ancient DNA yet discovered. From 96 radiocarbon-aged bones, we report large numbers of mitochondrial haplotypes, some of which appear to be extinct, given the 380 living birds sampled. We demonstrate DNA sequence evolution through time and estimate the rate of evolution of the hypervariable region I using a Markov chain Monte Carlo integration and a least-squares regression analysis. Our calculated rates of evolution are approximately two to seven times higher than previous indirect phylogenetic estimates.

阿德雷企鹅古 DNA 揭示的物种演化速率

Lambert D. M., Ritchie P. A., Millar C. D., Holland B., Drummond A. J., Baroni C.

(新西兰梅西大学分子生物科学研究所)

摘要:南极阿德雷企鹅巢穴有保存完好的骨骼化石。这些测定年龄超过 7000 年的骨骼蕴含着一些保存完好的古 DNA 信息。利用 96 份放射 C 定年的骨骼样品我们报道了大量的线粒体单模标本,参照采样的 380 种活体鸟类,其中一些似乎已经灭绝。我们揭示了 DNA 序列随时间的演化并利用马尔科夫蒙特卡罗法和最小二乘回归分析估计了超变 I 区的演化速率。我们计算的演化速率大概是此前通过系统发育间接估算结果的 2～7 倍。

12. Quayle W C, Peck L S, Peat H, et al. Extreme responses to climate change in Antarctic lakes[J]. Science, 2002, 295: 645.

Extreme responses to climate change in Antarctic lakes

Quayle W. C., Peck L. S., Peat H., Ellis-Evans J. C., Harrigan P. R.

(British Antarctic Survey, Natural Environment Research Council, High Cross, Madingley Road, Cambridge CB3 OET, UK)

Abstract: We report data for maritime Antarctic lakes showing extremely fast physical ecosystem change, combined with the ecological responses to that change. Nutrient levels at some sites exhibit order of magnitude increases per decade.

南极湖泊对气候变化的极端响应

Quayle W. C., Peck L. S., Peat H., Ellis-Evans J. C., Harrigan P. R.

(英国自然环境研究理事会南极调查局)

摘要:我们的报告数据显示南极海洋湖泊对气候变化的响应表现出超快的物理生态系统变化以及生态变化。一些研究结果显示营养水平每十年增加一个量级。

13. Stenseth N C, Mysterud A, Ottersen G, et al. Ecological effects of climate fluctuations[J]. Science, 2002, 297: 1292-1296.

Ecological effects of climate fluctuations

Stenseth N. C., Mysterud A., Ottersen G., Hurrell J. W., Chan K. S., Lima M.

(Department of Biology, Division of Zoology, University of Oslo, Post office Boxloso Blindern, N-0316 OSLO, Norway)

Abstract: Climate influences a variety of ecological processes. These effects operate through local weather parameters such as temperature, wind, rain, snow, and ocean currents, as well as interactions among these. In the temperate zone, local variations in weather are often coupled over large geographic areas through the transient behavior of atmospheric planetary-scale waves. These variations drive temporally and spatially averaged exchanges of heat, momentum, and water vapor that ultimately determine growth, recruitment, and migration patterns. Recently, there have been several studies of the impact of large-scale climatic forcing on ecological systems. We review how two of the best-known climate phenomena—the North Atlantic Oscillation and the El Nino-Southern Oscillation—affect ecological patterns and processes in both marine and terrestrial systems.

气候变动的生态效应

Stenseth N. C. , Mysterud A. , Ottersen G. , Hurrell J. W. , Chan K. S. , Lima M.

(挪威奥斯陆大学生物学系)

摘要:气候影响着各种生态过程。这些影响是通过当地的气象要素如温度、风、雨、雪和洋流以及它们之间的相互作用来实现的。在温带,当地天气变化常常通过大气行星尺度波的短暂行为而影响到大范围地理区域。这些变动驱动着热、动量和水汽的时空交换并最终决定了气流生长、补募和迁移模式。近来有一些关于研究大尺度气候驱动力对生态系统的影响的研究。我们评论了两种常见的气候现象——北大西洋涛动和厄尔尼诺-南方涛动对海洋和陆地系统生态模式与过程的影响。

14. Turner J, King J C, Lachlan-Cope T A, et al. Climate change: recent temperature trends in the Antarctic[J]. Nature, 2002, 418: 291-292.

Climate change—recent temperature trends in the Antarctic

Turner J. , King J. C. , Lachlan-Cope T. A. , Jones P. D.

(British Antarctic Survey, Natural Environment Research Council)

Abstract: It is important to understand how temperatures across the Antarctic have changed in recent decades because of the huge amount of fresh water locked into the ice sheet and the impact that temperature changes may have on the ice volume. Doran et al claim that there has been a net cooling of the entire continent between 1966 and 2000, particularly during summer and autumn. We argue that this result has arisen because of an inappropriate extrapolation of station data across large, data-sparse areas of the Antarctic.

气候变化——近来南极温度趋势

Turner J. , King J. C. , Lachlan-Cope T. A. , Jones P. D.

(英国自然环境研究理事会南极调查局)

摘要:了解整个南极地区温度近几十年来如何变化非常重要,由于大量的淡水被锁定到冰原,温度变化可能影响冰的体积。Doran 称在 1966~2000 年之间有一次整个大陆的冷却,尤其是在夏季和秋季。我们质疑这个结果,因为采用的大数据外推法在数据稀疏的南极区域不适用。

15. Walther G R, Post E, Convey P, et al. Ecological responses to recent climate change[J]. Nature, 2002, 416: 389-395.

Ecological responses to recent climate change

Walther G. R., Post E., Convey P., Menzel A., Parmesan C., Beebee T. J. C., Fromentin J. M., Hoegh-Guldberg O., Bairlein F.

(Institute of Geobotany, University of Hannover, Nienburger Str. 17, 30167 Hannever, Germany)

Abstract: There is now ample evidence of the ecological impacts of recent climate change, from polar terrestrial to tropical marine environments. The responses of both flora and fauna span an array of ecosystems and organizational hierarchies, from the species to the community levels. Despite continued uncertainty as to community and ecosystem trajectories under global change, our review exposes a coherent pattern of ecological change across systems. Although we are only at an early stage in the projected trends of global warming, ecological responses to recent climate change are already clearly visible.

近来气候变化的生态响应

Walther G. R., Post E., Convey P., Menzel A., Parmesan C., Beebee T. J. C., Fromentin J. M., Hoegh-Guldberg O., Bairlein F.

(德国汉诺威大学植物地理研究所)

摘要:现在有大量关于近来气候变化对生态的影响的证据,无论是在极区陆地系统还是热带海洋系统。植物和动物的响应都贯穿着一系列的生态系统和组织的等级,从物种到群落。尽管对于全球变化下群落到生态系统的轨迹变化还有持续的不确定性,我们的评论揭示了系统中生态变化的连贯性。虽然我们只是站在全球变暖的早期阶段,生态对近来气候变化的响应仍然是清晰可见的。

16. Royer D L, Osborne C P, Beerling D J. Carbon loss by deciduous trees in a CO_2-rich ancient polar environment[J]. Nature, 2003, 424: 60-62.

Carbon loss by deciduous trees in a CO_2-rich ancient polar environment

Royer D. L., Osborne C. P., Beerling D. J.

(Department of Animal and Plant Sciences, University of Sheffield, UK)

Abstract: Fossils demonstrate that deciduous forests covered the Polar Regions for much of the past 250 million years when the climate was warm and atmospheric CO_2 high. But the evolutionary significance of their deciduous character has remained a matter of conjecture for almost a century. The leading hypothesis, argues that it was an adaptation to photoperiod, allowing the avoidance of carbon losses by respiration from a canopy of leaves unable to photosynthesize in the darkness of warm polar winters. Here we test this proposal with experiments using "living fossil" tree species grown in a simulated polar climate with and without CO_2 enrichment. We show that the quantity of carbon lost annually by shedding a deciduous canopy is significantly greater than that lost by evergreen trees through wintertime respiration and leaf litter production, irrespective of growth CO_2 concentration. Scaling up our experimental observations indicates that the greater expense of being deciduous persists in mature forests, even up to latitudes of 83 °N, where the duration of the polar winter exceeds five months. We therefore reject the carbon-loss hypothesis as an explanation for the deciduous nature of polar forests.

富 CO_2 的古极地环境落叶树的碳流失

Royer D. L., Osborne C. P., Beerling D. J.

(英国谢菲尔德大学动物和植物科学系)

摘要：化石标本显示在过去的2.5亿年气候温暖、大气 CO_2 浓度很高的大部分时间里极地地区被落叶性森林所覆盖。但是在近一个世纪以来关于这类树木的落叶特征的进化重要性仍停留在一些猜想层面上。主要的猜想主张这是对光周期的一种适应，避免由于呼吸作用导致的C流失，因为树冠的叶子在温暖的极地冬季的黑夜里无法进行光合作用。我们利用活化石树种在一个模拟的富集和亏损 CO_2 气候条件下的生长实验对上述假说进行了检验。结果显示在不考虑 CO_2 增加的条件下，每年落叶性树木叶脱落造成的C损失量显著高于常绿树木冬季呼吸作用及叶子掉落产生的C损失。实验观测结果显示成熟森林里落叶树木的消耗更大，甚至是在极区冬季超过5个月的北纬83度地区。因此我们否定了这个关于极地森林自然性落叶解释的C流失假说。

17. Shepherd L D, Millar C D, Ballard G, et al. Microevolution and mega-icebergs in the Antarctic[J]. PNAS, 2005, 102: 16717-16722.

Microevolution and mega-icebergs in the Antarctic

Shepherd L. D., Millar C. D., Ballard G., Ainley D. G., Wilson P. R., Haynes G. D., Baroni C., Lambert D. M.

(Allan Wilson Centre for Molecular Ecology and Evolution, Institute of Molecular BioSciences, Massey University)

Abstract: Microevolution is regarded as changes in the frequencies of genes in populations over time. Ancient DNA technology now provides an opportunity to demonstrate evolution over a geological time frame and to possibly identify the causal factors in any such evolutionary event. Using nine nuclear microsatellite DNA loci, we genotyped an ancient population of Adelie penguins (Pygoscelis adeliae) aged 6000 years B. P. Subfossil bones from this population were excavated by using an accurate stratigraphic method that allowed the identification of individuals even within the same layer. We compared the allele frequencies in the ancient population with those recorded from the modern population at the same site in Antarctica. We report significant changes in the frequencies of alleles between these two time points, hence demonstrating microevolutionary change. This study demonstrates a nuclear gene-frequency change over such a geological time frame. We discuss the possible causes of such a change, including the role of mutation, genetic drift, and the effects of gene mixing among different penguin populations. The latter is likely to be precipitated by mega-icebergs that act to promote migration among penguin colonies that typically show strong natal return.

南极微进化与巨型冰山

Shepherd L. D., Millar C. D., Ballard G., Ainley D. G., Wilson P. R., Haynes G. D., Baroni C., Lambert D. M.

(新西兰梅西大学分子生物科学研究所 Allan Wilson 分子生态与进化研究中心)

摘要:微进化被认为是种群基因频率随时间的改变。现在的古 DNA 技术可以揭示地质历史时期的进化并有可能确定这些进化事件的缘由。我们利用 9 个微卫星 DNA 遗传标记确定了 6000 年前阿德雷企鹅的基因型。通过可以确定个体的精确地层学方法挖掘了这类种群的化石骨骼。我们对南极同一地点古代和现代阿德雷企鹅种群的等位基因频率进行了比较,并发现了显著的变化,指示了这段时间内的微进化。本研究揭示了地质时间范围内基因频率的改变。我们讨论了这种改变可能的原因,包括突变、遗传漂变以及不同企鹅种群基因混合的作用。后者很可能是由于巨型冰山引起企鹅巢穴迁徙造成的。

18. Hall B L, Hoelzel A R, Baroni C, et al. Holocene elephant seal distribution implies warmer-than-present climate in the Ross Sea[J]. PNAS, 2006, 103: 10213-10217.

Holocene elephant seal distribution implies warmer-than-present climate in the Ross Sea

Hall B. L., Hoelzel A. R., Baroni C., Denton G. H., Le Boeuf B. J., Overturf B., Topf A. L.

(Climate Change Institute and Department of Earth Sciences, University of Maine, Orono, ME 04469)

Abstract: We show that southern elephant seal (Mirounga leonina) colonies existed proximate to the Ross Ice Shelf during the Holocene, well south of their core sub-Antarctic breeding and molting grounds. We propose that this was due to warming (including a previously unrecognized period from 1100 to 2300 ^{14}C yr B. P.) that decreased coastal sea ice and allowed penetration of warmer-than present climate conditions into the Ross Embayment. If, as proposed in the literature, the ice shelf survived this period, it would have been exposed to environments substantially warmer than present.

罗斯海全新世象海豹分布指示的比现代温暖的气候条件

Hall B. L., Hoelzel A. R., Baroni C., Denton G. H., Le Boeuf B. J., Overturf B., Topf A. L.

(美国缅因大学气候变化研究中心地球科学系)

摘要:我们发现全新世时期罗斯冰架附近有象海豹巢穴,远在它们主要生殖脱毛的亚南极的南边。我们认为这是由于气候变暖(包括此前未被承认的距今 2300～1100 年间,^{14}C 年龄)使得海岸区海冰减少,罗斯海湾存在比现代还要温暖的气候条件。如果像文献提到的那样,冰架在这个时期还存在,那也是暴露在比现在还要温暖的气候条件下。

19. Emslie S D, Patterson W P. Abrupt recent shift in $\delta^{13}C$ and $\delta^{15}N$ values in Adelie penguin eggshell in Antarctica[J]. PNAS, 2007, 104: 11666-11669.

Abrupt recent shift in $\delta^{13}C$ and $\delta^{15}N$ values in Adelie penguin eggshell in Antarctica

Emslie S. D. , Patterson W. P.

(Department of Biology and Marine Biology, University of North Carolina, 601 South College Road, Wilmington, NC 28403, USA)

Abstract: Stable isotope values of carbon (^{13}C) and nitrogen (^{15}N) in blood, feathers, eggshell, and bone have been used in seabird studies since the 1980s, providing a valuable source of information on diet, foraging patterns, and migratory behavior in these birds. These techniques can also be applied to fossil material when preservation of bone and other tissues is sufficient. Excavations of abandoned Adelie penguin (Pygoscelis adeliae) colonies in Antarctica often provide well preserved remains of bone, feathers, and eggshell dating from hundreds to thousands of years B. P. Herein we present an 38000 year time series of ^{13}C and ^{15}N values of Ade lie penguin eggshell from abandoned colonies located in three major regions of Antarctica. Results indicate an abrupt shift to lower trophic prey in penguin diets within the past 200 years. We posit that penguins only recently began to rely on krill as a major portion of their diet, in conjunction with the removal of baleen whales and krill-eating seals during the historic whaling era. Our results support the "krill surplus" hypothesis that predicts excess krill availability in the Southern Ocean after this period of exploitation.

近来南极阿德雷企鹅蛋壳^{13}C、^{15}N同位素的突变

Emslie S. D. , Patterson W. P.

(美国北卡罗来纳大学生物学与海洋生物学系)

摘要：自1980年代以来，血液、羽毛、蛋壳和骨骼中的稳定^{13}C、^{15}N同位素比值开始应用于海鸟的食谱、取食方式及迁徙行为等研究。这些技术同样可以应用到那些保存足够完整的骨骼和其他组织等化石载体中。对南极阿德雷企鹅巢穴的挖掘常可以得到保存完好且距今百年到千年的骨骼、羽毛和蛋壳。这里我们分析了南极三个主要地区阿德雷企鹅废弃巢穴中距今38000年来蛋壳片的^{13}C、^{15}N同位素序列。结果显示企鹅的食谱在近200年来突然转变为低营养级的组成。我们假定企鹅是在近来才把磷虾作为它们的主要食物的，与人类捕杀鲸和海豹的时代重合。我们的结果支持"磷虾过剩"假说。

7.3 文献摘要——北极

1. Kling G W, Kipphut G W, Miller M C. Arctic lakes and streams as gas conduits to the atmosphere: implications for tundra carbon budgets[J]. Science, 1991, 251(1991): 298-301.

Arctic lakes and streams as gas conduits to the atmosphere —implications for tundra carbon budgets

Kling G. W., Kipphut G. W., Miller M. C.

(Marine Biological Laboratory, The Ecosystems Center, Woods Hole, MA02543, USA)

Abstract: Arctic tundra has large amounts of stored carbon and is thought to be a sink for atmospheric carbon dioxide (CO_2) (0.1 to 0.3 petagram of carbon per year) (1 petagram $= 10^{15}$ grams). But this estimate of carbon balance is only for terrestrial ecosystems. Measurements of the partial pressure of CO_2 in 29 aquatic ecosystems across arctic Alaska showed that in most cases (27 of 29) CO_2 was released to the atmosphere. This CO_2 probably originates in terrestrial environments; erosion of particulate carbon plus ground-water transport of dissolved carbon from tundra contribute to the CO_2 flux from surface waters to the atmosphere. If this mechanism is typical of that of other tundra areas, then current estimates of the arctic terrestrial sink for atmospheric CO_2 may be 20 percent too high.

北极湖泊河流作为气体通向大气的通道——来自苔原 C 预算的启示

Kling G. W., Kipphut G. W., Miller M. C.

(美国生态中心海洋生物实验室)

摘要:北极苔原由于储存了大量的 C 而被认为是大气 CO_2 的汇(0.1～0.3 Pg・C/年,1 Pg$=10^{15}$ g)。但是这种对 C 平衡的估计只适用于陆地生态系统。对北极阿拉斯加 29 个水域生态系统 CO_2 分压测定的结果显示有 27 个是向大气释放 CO_2 的。这些 CO_2 可能来源于陆地环境;颗粒 C 的侵蚀加上来自于苔原的地下水溶解 C 的传输使得由水面向大气输出 CO_2。如果这种机制在其他苔原地带也是典型的,那么目前对北极陆地作为 CO_2 汇的估计可能要高出 20%。

2. Macdonald G M, Edwards T W D, Moser K A, et al. Rapid response of treeline vegetation and lakes to past climate warming[J]. Nature, 1993, 361(6409): 243-246.

Rapid response of treeline vegetation and lakes to past climate warming

Macdonald G. M., Edwards T. W. D., Moser K. A., Pienitz R., Smol J. P.

(Mcmaster Univ, Dept Geog, Hamilton L8s 4k1, Ontario, Canada)

Abstract: Future greenhouse warming is expected to be particularly pronounced in boreal regions, and consequent changes in vegetation in these regions may in turn affect global climate. It is therefore important to establish how boreal ecosystems might respond to rapid changes in climate. Here we present palaeoecological evidence for changes in terrestrial vegetation and lake characteristics during an episode of climate warming that occurred between 5000 and 4000 years ago at the boreal treeline in central Canada. The initial transformation-from tundra to forest-tundra on land, which coincided with increases in lake productivity, pH and ratio of inflow to evaporation-took only 150 years, which is roughly equivalent to the time period often used in modelling the response of boreal forests to climate warming. The timing of the treeline advance did not coincide with the maximum in high-latitude summer insolation predicted by Milankovitch theory, suggesting that northern Canada experienced regionally asynchronous middle-to-late Holocene shifts in the summer position of the Arctic front. Such Holocene climate events may provide a better analogue for the impact of future global change on northern ecosystems than the transition from glacial to nonglacial conditions.

林木线和湖泊对过去气候变暖的快速响应

Macdonald G. M., Edwards T. W. D., Moser K. A., Pienitz R., Smol J. P.

(加拿大马克马斯特大学地理系)

摘要：未来温室效应的增温被认为在北高纬地区特别显著，对这个地区植被造成的植被变化可能反过来影响全球气候。因此认识北高纬生态系统对快速气候变化可能的响应机制是十分重要的。我们提出了加拿大中心林木线距今5000～4000年间暖期时段陆地植被变化和湖泊特征的古生态证据。最初从苔原带到森林-苔原带的转变与湖泊生产力、pH及流入输出比的增加是相符合的，这个过程仅用了150年，与模拟北方森林对气候变暖响应的时间是大体一致的。林木线前进时期与由米兰科维奇理论预测北高纬太阳辐射最强期是不一致的，暗示了加拿大北部中晚全新世北极锋夏季位置的变化经历了区域性的不一致。此类全新世的气候事件也许可以为我们提供一个相对于从冰期到间冰期转变条件下更好的背景来理解未来全球变化对北部生态系统的影响。

3. Oechel W C, Hastings S J, Vourlitis G, et al. Recent change of Arctic tundra ecosystems from a net carbon: dioxide sink to a source[J]. Nature, 1993, 361(6412): 520-523.

Recent change of Arctic tundra ecosystems from a net carbon:dioxide sink to a source

Oechel W. C., Hastings S. J., Vourlitis G., Jenkins M., Riechers G., Grulke N.

(System Ecology Research Group and Department of Bidogy, San Diego State University, San Diego, California 92182, USA)

Abstract: Arctic tundra has been a net sink for carbon dioxide during historic and recent geological times, and large amounts of carbon are stored in the soils of northern ecosystems. Many regions of the Arctic are warmer now than they have been in the past, and this warming may cause the soil to change from a carbon dioxide sink to a source by lowering the water table, thereby accelerating the rate of soil decomposition (CO_2 source) so that this dominates over photosynthesis (CO_2 sink). Here we present data indicating that the tundra on the North Slope of Alaska has indeed become a source of carbon dioxide to the atmosphere. This change coincides with recent warming in the Arctic, whether this is due to increases in greenhouse gas concentrations in the atmosphere or to some other cause. Our results suggest that tundra ecosystems may exert a positive feedback on atmospheric carbon dioxide and greenhouse warming.

北极苔原生态系统近来的变化:从净 CO_2汇到源

Oechel W. C., Hastings S. J., Vourlitis G., Jenkins M., Riechers G., Grulke N.

(美国圣地亚哥州立大学系统生态学研究组)

摘要:北极苔原在历史和近来的地质时间段内已成为净的 CO_2 汇,大量 C 储存在北部生态系统的土壤里。北极很多地区较过去变得更暖,这种变暖也许可以降低土壤水位从而导致其由净的 CO_2 汇变为 CO_2 源,因此加速了土壤的分解(CO_2 源)以至于超过了光合作用(CO_2 汇)。我们的数据指示阿拉斯加北坡苔原地带已经成为向大气释放 CO_2 的源。这种变化与北极近来由温室气体浓度增加或其他的原因引起的变暖是相符合的。我们的结果暗示苔原生态系统可能对大气 CO_2 和温室效应起着正反馈效应。

4. Vartanyan S L, Garutt V E, Sher A V. Holocene dwarf mammoths from Wrangel Island in the siberian Arctic[J]. Nature, 1993, 362(6418): 337-340.

Holocene dwarf mammoths from Wrangel Island in the siberian Arctic

Vartanyan S. L., Garutt V. E., Sher A. V.

(Severtsov Institute of Evolutionary Animal Morphology and Ecology, Russian Academic Sciences, Russia)

Abstract: The cause of extinction of the woolly mammoth, Mammuthus primigenius (Blumenbach), is still debated. A major environmental change at the Pleistocene-Holocene boundary, hunting by early man, or both together are among the main explanations that have been suggested. But hardly anyone has doubted that mammoths had become extinct everywhere by around 9500 years before present (BP). We report here new discoveries on Wrangel Island in the Arctic Ocean that force this view to be revised. Along with normal-sized mammoth fossils dating to the end of the Pleistocene, numerous teeth of dwarf mammoth dated 7000-4000 yr BP have been found there. The island is thought to have become separated from the mainland by 12000 yr BP. Survival of a mammoth population may be explained by local topography and climatic features, which permitted relictual preservation of communities of steppe plants. We interpret the dwarfing of the Wrangel mammoths as a result of the insularity effect, combined with a response to the general trend towards unfavourable environment in the Holocene.

全新世西伯利亚北高纬弗兰格尔岛矮猛犸象

Vartanyan S. L., Garutt V. E., Sher A. V.

(俄罗斯科学院动物形态与生态学进化研究所)

摘要：关于猛犸象灭绝的原因依然还有争论。可能的解释包括更新世-全新世过渡期主要的环境变化、早期人类的捕猎，或者二者兼而有之，但是没有人质疑在距今 9500 年前猛犸象都已经灭绝。我们在北极海洋弗兰格尔岛新的发现将改变当前的观点。在众多更新世晚期的猛犸象化石中，大量距今 7000～4000 年间的猛犸象牙齿被发现。该岛屿被认为在距今 12000 年前与大陆分开，猛犸象种群的幸存也许是由于当地的地形和气候特征可以保存干草原植物群落，我们将弗兰格尔岛猛犸象解释为岛屿孤立效应以及对全新世不适宜环境的响应的结果。

5. Douglas M S V, Smol J P, Blake W. Marked post-18th century environmental-change in high-Arctic ecosystems[J]. Science, 1994, 266(5184): 416-419.

Marked post-18th century environmental-change in high-Arctic ecosystems

Douglas M. S. V., Smol J. P., Blake W.

(Paleoecological Environment Assessment and Research Laboratory, Department of Biology, Queen's University, Kingston Ontario K7l3n6, Canada)

Abstract: Paleolimnological data from three high-arctic ponds on Cape Herschel, Ellesmere Island, Canada, show that diatom assemblages were relatively stable over the last few millennia but then experienced unparalleled changes beginning in the 19th century. The environmental factors causing these assemblage shifts may be related to recent climatic warming. Regardless of the cause, the biota of these isolated and seemingly pristine ponds have changed dramatically in the recent past and any hopes of cataloging natural assemblages may already be fruitless.

18 世纪后期北极高纬生态系统显著的环境变化

Douglas M. S. V., Smol J. P., Blake W.

(加拿大女王大学生物学系古生态学环境评估与研究实验室)

摘要：高北极赫谢尔角、埃尔斯米尔岛和加拿大的三个湖泊古湖沼学数据显示硅藻组合在过去几千年来是相对稳定的，而从 19 世纪以来发生了空前的变化。引起硅藻组合转变的环境因素可能与近来气候变暖有关。不管原因如何，这些孤立的呈原始态的湖泊的生物系统已发生了显著的变化，而想要归类该地区自然的硅藻组合的希望可能要落空了。

6. Kunz M L, Reanier R E. Paleoindians in Beringia: evidence from Arctic Alaska[J]. Science, 1994, 263(5147): 660-662.

Paleoindians in Beringia—evidence from Arctic Alaska

Kunz M. L., Reanier R. E.

(Bureau of Land Management, 1150 University Avenue, Fairbanks, AK 99709)

Abstract: Excavations at the Mesa site in arctic Alaska provide evidence for a Paleoindian occupation of Beringia, the region adjacent to the Bering Strait. Eleven carbon-14 dates on hearths associated with Paleoindian projectile points place humans at the site between 9730 and 11660 radiocarbon years before present (years B. P.). The presence of Paleoindians in Beringia at these times challenges the notion that Paleoindian cultures arose exclusively in mid-continental North America. The age span of Paleoindians at the Mesa site overlaps with dates from two other cultural complexes in interior Alaska. A hiatus in the record of human occupation occurs between 10300 and 11000 years B. P. Late Glacial climatic fluctuations may have made northern Alaska temporarily unfavorable for humans and spurred their southward dispersal.

白令陆桥古印第安人——来自北极阿拉斯加的证据

Kunz M. L., Reanier R. E.

(美国国土管理局)

摘要：对北极阿拉斯加平顶山遗址的挖掘工作提供了古印第安人登陆与白令海峡毗邻陆桥的证据。11 个古印第安人灶台材料的 ^{14}C 年龄显示人类于距今 11660～9730 年间在这个地区活动。古印第安人这个时间段内在白令海峡的出现对古印第安文化来自北美中大陆的观点是个挑战。平顶山遗址古印第安人年龄跨度与阿拉斯加内地另外两个文化部族是重叠的。而在距今 11000～10300 年间人类登陆的记录则是空白的。晚冰河期气候的波动可能暂时使得阿拉斯加北部不适合人类居住并驱使人类向南方迁移。

7. Oechel W C, Cowles S, Grulke N, et al. Transient nature of CO_2 fertilization in Arctic tundra[J]. Nature, 1994, 371(6497): 500-503.

Transient nature of CO_2 fertilization in Arctic tundra

Oechel W. C., Cowles S., Grulke N., Hastings S. J., Lawrence B., Prudhomme T., Riechers G., Strain B., Tissue D., Vourlitis G.

(Global Change Research Group and System Ecology Research Group, San Diego State University)

Abstract: There has been much debate about the effect of increased atmospheric CO_2 concentrations on plant net primary production and on net ecosystem CO_2 flux. Apparently conflicting experimental findings could be the result of differences in genetic potential and resource availability, different experimental conditions and the fact that many studies have focused on individual components of the system rather than the whole ecosystem. Here we present results of an in situ experiment on the response of an intact native ecosystem to elevated CO_2. An undisturbed patch of tussock tundra at Toolik Lake, Alaska, was enclosed in greenhouses in which the CO_2 level, moisture and temperature could be controlled, and was subjected to ambient (340 p. p. m.) and elevated (680 p. p. m.) levels of CO_2 and temperature (+4 degrees C). Air humidity, precipitation and soil water table were maintained at ambient control levels. For a doubled CO_2 level alone, complete homeostasis of the CO_2 flux was re-established within three Sears, whereas the regions exposed to a combination of higher temperatures and doubled CO_2 showed persistent fertilization effect on net ecosystem carbon sequestration over this time. This difference may be due to enhanced sink activity from the direct effects of higher temperatures on growth and to indirect effects from enhanced nutrient supply caused by increased mineralization. These results indicate that the responses of native ecosystems to elevated CO_2 may not always be positive, and are unlikely to be straightforward. Clearly, CO_2 fertilization effects must always be considered in the context of genetic limitation, resource availability and other such factors.

北极苔原天然CO_2施肥的短暂性

Oechel W. C., Cowles S., Grulke N., Hastings S. J., Lawrence B., Prudhomme T., Riechers G., Strain B., Tissue D., Vourlitis G.

(美国圣地亚哥大学全球变化和系统生态学研究组)

摘要:关于大气 CO_2 浓度增加对植物净初级生产力和净生态系统 CO_2 通量的影响有很多争论。实验发现的明显矛盾可能是遗传潜力、资源可利用性和不同的实验条件差别的结果以及很多研究只关注于系统内的个体组分而不是整个生态系统。这里我们报道一个完整天然的生态系统对 CO_2 浓度增加的响应的原位实验结果。阿拉斯加 Toolik 湖边一块自然的苔原被封闭成为一个 CO_2 浓度、湿度和温度可以控制的温室,并将温度控制在 4 ℃而 CO_2 浓度为 340 ppm 和 680 ppm。空气湿度、降雨量和土壤水位保持与周围环境相当。当仅仅 CO_2 浓度增倍时,在三个区域内 CO_2 通量平衡重新建立,而暴露于高温和双倍 CO_2 浓度条件下的地区则显示了对这段时间内净的生态系统碳封存的持续培肥效应,这种差别也许是由于高温对生长率的直接影响和增强的矿化作用提高了营养物质供给的间接作用造成了碳汇能力的提高。这些结果指示原生态系统对 CO_2 浓度增加的响应不一定都是正效应,也不太可能是简单的。很明显,研究 CO_2 的培肥效应时,要考虑遗传限制、资源可利用性和其他因素条件。

8. Post D M, Pace M L, Hairston N G. Ecosystem size determines food-chain length in lakes[J].

Nature,2000,405(6790):1047-1049.

Ecosystem size determines food-chain length in lakes

Post D. M., Pace M. L., Hairston N. G.

(Department of Ecology and Evolutionary Biology, Cornell University)

Abstract: Food-chain length is an important characteristic of ecological communities: it influences community structure, ecosystem functions and contaminant concentrations in top predators. Since Elton first noted that food-chain length was variable among natural systems, ecologists have considered many explanatory hypotheses, but few are supported by empirical evidence. Here we test three hypotheses that predict food-chain length to be determined by productivity alone (productivity hypothesis), ecosystem size alone (ecosystem-size hypothesis) or a combination of productivity and ecosystem size (productive-space hypothesis). The productivity and productive-space hypotheses propose that food-chain length should increase with increasing resource availability; however, the productivity hypothesis does not include ecosystem size as a determinant of resource availability. The ecosystem-size hypothesis is based on the relationship between ecosystem size and species diversity, habitat availability and habitat heterogeneity. We find that food-chain length increases with ecosystem size, but that the length of the food chain is not related to productivity. Our results support the hypothesis that ecosystem size, and not resource availability, determines food-chain length in these natural ecosystems.

湖泊生态系统大小决定了食物链的长度

Post D. M., Pace M. L., Hairston N. G.

(美国康奈尔大学生态学与进化生物学系)

摘要:食物链长度是生态群落的一个重要特征:它影响到群落结构、生态系统功能和顶端捕食生物体内的污染物含量。自从 Elton 第一次注意到食物链长度在自然系统中是可变的,生态学家们发展了众多的解释假说但是很少能够得到观察或实验的支持。我们检验了食物链长度是由生产力决定(生产力假说)、生态系统空间大小决定(生态系统空间大小假说)和二者共同决定(生产力-空间假说)的三个假说。生产力假说和生产力-空间假说提出食物链长度应该随资源可利用性的增多而增大;但是生产力假说并没有包括决定资源可利用性的生态系统空间大小。生态系统空间大小假说是基于生态系统空间大小和物种多样性、栖息地可利用性及栖息地多样性之间的关系而确立的。我们发现食物链长度随生态系统空间大小增大而增长,但与生产力没有关系。我们的结果支持生态系统空间大小而不是资源可利用性的假说来决定这些自然生态系统内的食物链长度。

9. Pavlov P, Svendsen J I, Indrelid S. Human presence in the European Arctic nearly 40000 years ago [J]. Nature, 2001, 413(6851): 64-67.

Human presence in the European Arctic nearly 40000 years ago

Pavlov P., Svendsen J. I., Indrelid S.

(Centre for Studies of the Environment and Resources, University of Bergen)

Abstract: The transition from the Middle to the Upper Palaeolithic, approximately 40000-35000 radiocarbon years ago, marks a turning point in the history of human evolution in Europe. Many changes in the archaeological and fossil record at this time have been associated with the appearance of anatomically modern humans. Before this transition, the Neanderthals roamed the continent, but their remains have not been found in the northernmost part of Eurasia. It is generally believed that this vast region was not colonized by humans until the final stage of the last Ice Age some 13000-14000 years ago. Here we report the discovery of traces of human occupation nearly 40000 years old at Mamontovaya Kurya, a Palaeolithic site situated in the European part of the Russian Arctic. At this site we have uncovered stone artefacts, animal bones and a mammoth tusk with human-made marks from strata covered by thick Quaternary deposits. This is the oldest documented evidence for human presence at this high latitude; it implies that either the Neanderthals expanded much further north than previously thought or that modern humans were present in the Arctic only a few thousand years after their first appearance in Europe.

欧洲北极地区 40000 年前人类的出现

Pavlov P., Svendsen J. I., Indrelid S.

(瑞士卑尔根大学环境与资源研究中心)

摘要:从 40000～35000 年前的中旧石器时代向晚旧石器时代的转变是欧洲人类进化历史的转折点,这个时期考古和化石记录的很多变化是与现代人的出现联系起来的。在此转变之前,尼安德特人在大陆上游走,但是他们的遗物在欧亚大陆的最北边都没被发现,普遍认为这个巨大的区域直到 14000～13000 年前末次冰期的最后阶段才有人类居住。我们报道了对欧洲俄罗斯北极部分一个 Mamontovaya Kurya 旧石器时代遗址 40000 年前人类居住的发现,在遗址中我们在很厚的第四纪沉积物下发现了无盖的石头制品、动物骨骼和人类用牙齿制作的工具。这是人类在高纬地区出现的最古老的证据,它暗示了不是尼安德特人向北扩张比此前认为的更靠北就是现代人在欧洲第一次出现后的几千年后即在北极出现。

10. Sturm M, Racine C, Tape K. Increasing shrub abundance in the Arctic[J]. Nature, 2001, 411(6837): 546-547.

Increasing shrub abundance in the Arctic

Sturm M., Racine C., Tape K.

(US Army Cold Region & Engneer Laboratory)

The warming of the Alaskan Arctic during the past 150 years has accelerated over the last three decades and is expected to increase vegetation productivity in tundra if shrubs become more abundant; indeed, this transition may already be under way according to local plot studies and remote sensing. Here we present evidence for a widespread increase in shrub abundance over more than 320 km of Arctic landscape during the past 50 years, based on a comparison of historic and modern aerial photographs. This expansion will alter the partitioning of energy in summer and the trapping and distribution of snow in winter, as well as increasing the amount of carbon stored in a region that is believed to be a net source of carbon dioxide.

北极地区灌木丰度的不断增加

Sturm M. , Racine C. , Tape K.

(美国陆军冷区研究和工程实验室)

摘要:过去150年北极阿拉斯加地区的变暖在最近三十年间加速,且如果灌木丰度增加,预计将增加苔原的植物生产力。甚至,根据当地的定点研究和遥感,这种转变可能已经发生。这里我们展示了依据历史和现代航空照片对比获得的过去50年超过320 km的北极景观带灌木丰度增加的证据。这种扩张将会改变夏季能量的分区和冬季诱捕及积雪的分布,以及该地区碳储存的增加,这被认为是二氧化碳的净源。

11. Guthrie R D. Rapid body size decline in Alaskan Pleistocene horses before extinction[J]. Nature, 2003,426(6963):169-171.

Rapid body size decline in Alaskan Pleistocene horses before extinction

Guthrie R. D.

(Institute of Arctic Biology, University of Alaska)

Abstract: About 70% of North American large mammal species were lost at the end of the Pleistocene epoch. The causes of this extinction the role of humans versus that of climate-have been the focus of much controversy. Horses have figured centrally in that debate, because equid species dominated North American late Pleistocene faunas in terms of abundance, geographical distribution, and species variety, yet none survived into the Holocene epoch. The timing of these equid regional extinctions and accompanying evolutionary changes are poorly known. In an attempt to document better the decline and demise of two Alaskan Pleistocene equids, I selected a large number of fossils from the latest Pleistocene for radiocarbon dating. Here I show that horses underwent a rapid decline in body size before extinction, and I propose that the size decline and regional extinction at 12500 radiocarbon years before present are best attributed to a coincident climatic/vegetational shift. The present data do not support human overkill and several other proposed extinction causes, and also show that large mammal species responded somewhat individualistically to climate changes at the end of the Pleistocene.

阿拉斯加更新世马灭绝前躯体尺寸的快速减小

Guthrie R. D.

（美国阿拉斯加大学北极生物研究所）

摘要：更新世末期北美大型哺乳物种有70％都消失了，导致这些物种灭绝的人类论和气候论成为争论的焦点。马是争论的中心，因为马科动物是晚更新世时期北美的优势物种，不论是在丰度、地理分布和物种多样性方面，但是到了全新世时期都没能幸存下来，关于马科动物区域性灭绝的时间和伴随的进化演变了解还很少。为了试图更好地获取两种阿拉斯加更新世马科动物减少和死亡的信息，我选择了大量最近晚更新世化石来进行放射性C定年。结果显示马在灭绝前体型经历了快速的减小，我认为距今12500年前马体形的减小和随后的区域性灭绝可以很好地归因于相应的气候/植被的突变。数据不支持人类过度猎杀和其他因素引起马的灭绝，并显示大型哺乳物种对晚更新世气候变化响应的个体性特征。

12. Hu F S, Kaufman D, Yoneji S, et al. Cyclic variation and solar forcing of Holocene climate in the Alaskan subarctic[J]. Science, 2003, 301(5641): 1890-1893.

Cyclic variation and solar forcing of Holocene climate in the Alaskan subarctic

Hu F. S., Kaufman D., Yoneji S., Nelson D., Shemesh A., Huang Y., Tian J., Bond G., Clegg B., Brown T.

(Department of Plant Biology, University of Illinois)

Abstract: High-resolution analyses of lake sediment from southwestern Alaska reveal cyclic variations in climate and ecosystems during the Holocene. These variations occurred with periodicities similar to those of solar activity and appear to be coherent with time series of the cosmogenic nuclides ^{14}C and ^{10}Be as well as North Atlantic drift ice. Our results imply that small variations in solar irradiance induced pronounced cyclic changes in northern high-latitude environments. They also provide evidence that centennial-scale shifts in the Holocene climate were similar between the subpolar regions of the North Atlantic and North Pacific, possibly because of Sun-ocean-climate linkages.

阿拉斯加亚北极全新世气候的周期变化和太阳辐射强迫

Hu F. S., Kaufman D., Yoneji S., Nelson D., Shemesh A., Huang Y., Tian J., Bond G., Clegg B., Brown T.

（美国伊利诺伊大学植物生物学系）

摘要：对阿拉斯加西南部湖泊沉积物的高分辨分析揭示了全新世气候和生态系统的周期性变化，这些变化与太阳活动周期相似并与宇宙射线核素^{14}C、^{10}Be和北大西洋冰漂砾时间序列变化是一致的。我们的结果暗示太阳辐射微小的波动可以引起北高纬地区环境显著的周期性变化，结果同时显示全新世气候百年尺度的突变与北大西洋和北太平洋亚极地地区是相似的，可能源于太阳-海洋-气候之间的联系。

13. Tedford R H, Harington C R. An Arctic mammal fauna from the Early Pliocene of North America [J]. Nature, 2003, 425(6956): 388-390.

An Arctic mammal fauna from the Early Pliocene of North America

Tedford R. H., Harington C. R.

(Division of Paleontology, American Museum of Natural History)

Abstract: A peat deposit on Ellesmere Island, Nunavut, Canada, allows a unique glimpse of the Early Pliocene terrestrial biota north of the Arctic Circle. The peat accumulated in a beaver pond surrounded by boreal larch forest near regional tree line in coastal hills close to the Arctic Ocean. The ecological affinities of the plant and beetle remains contained in the peat indicate that winter temperatures on Ellesmere Island were nearly 15 degrees C higher and summer temperatures 10 degrees C higher than they are today. Here we show that the mammalian remains buried in the peat represent mainly taxa of Eurasiatic zoogeographic and phyletic affinities, including the first North American occurrence of a meline badger (Arctomeles). This deposit contains direct evidence of the composition of an Early Pliocene (4-5 million years ago) arctic mammalian fauna during an active period of interchange between Asia and North America.

一种来自于北美早上新世的北极哺乳动物

Tedford R. H., Harington C. R.

(美国自然历史博物馆古生物学部)

摘要:取自埃尔斯米尔岛、努纳武特地区和加拿大的泥炭沉积为了解早上新世北极圈以北陆地生物系统提供了独特的视角。这个泥炭沉积取自环绕着北方落叶林的 beaver 池塘,而这些落叶林在靠近北极海岸线的区域树线边。包含在泥炭里的植物和甲虫遗存的生态习性指示了埃尔斯米尔岛冬季温度比现在高接近 15 ℃,而夏季温度比现在高 10 ℃。泥炭堆积中的哺乳类遗迹代表了欧亚动物地理和种族亲缘的主要分类,包括獾在北美的第一次出现。这个沉积提供了早上新世(400 万～500 万年前)亚洲北美交替变化活跃期期间的北极哺乳动物群组成的直接证据。

14. Wilf P, Johnson K R, Huber B T. Correlated terrestrial and marine evidence for global climate changes before mass extinction at the Cretaceous-Paleogene boundary [J]. PNAS, 2003, 100 (2): 599-604.

Correlated terrestrial and marine evidence for global climate changes before mass extinction at the Cretaceous-Paleogene boundary

Wilf P., Johnson K. R., Huber B. T.

(Department of Geosciences, Pennsylvania State University)

Abstract: Terrestrial climates near the time of the end-Cretaceous mass extinction are poorly known, limiting understanding of environmentally driven changes in biodiversity that occurred before bolide impact. We estimate paleotemperatures for the last approximate to 1. 1 million years of the Cretaceous (approximate to 66. 6-65. 5 million years ago, Ma) by using fossil plants from North Dakota and employ paleomagnetic stratigraphy to correlate the results to foraminiferal paleoclimatic data from four middle- and high-latitude sites. Both plants and foraminifera indicate warming near 66. 0 Ma, a warming peak from approximate to 65. 8 to 65. 6 Mal, and cooling near 65. 6 Ma, suggesting that these were global climate shifts. The warming peak coincides with the immigration of a thermophilic flora, maximum plant diversity, and the poleward range expansion of thermophilic foraminifera. Plant data indicate the continuation of relatively cool temperatures across the Cretaceous-Paleogene boundary; there is no indication of a major warming immediately after the boundary as previously reported. Our temperature proxies correspond well with recent P_{CO_2} data from paleosol carbonate, suggesting a coupling of P_{CO_2} and temperature. To the extent that biodiversity is correlated with temperature, estimates of the severity of end-Cretaceous extinctions that are based on occurrence data from the warming peak are probably inflated, as we illustrate for North Dakota plants. However, our analysis of climate and facies considerations shows that the effects of bolide impact should be regarded as the most significant contributor to these plant extinctions.

白垩纪-早第三纪转变时期大灭绝前全球气候变化陆地和海洋的证据间的相互关系

Wilf P. , Johnson K. R. , Huber B. T.

(美国宾夕法尼亚州立大学地球科学系)

摘要：对白垩纪晚期大灭绝时的陆地气候了解很浅，限制了对火流星影响前环境驱动的生物多样性的理解。我们利用北达科特州植物化石估计了白垩纪持续近110万年的古温度(距今6660万～6650万年前)并利用古地磁地层学将结果与4个中高纬地区有孔虫古气候记录进行了相关分析。植物和有孔虫化石都指示了6600万年前后的温暖期，最暖期在6580万～6560万年间，而在接近6560万年时开始变冷，暗示了全球性的气候变迁。最暖期与喜暖植物的迁徙、最丰富的植物多样性和喜温有孔虫目向极地的扩张是一致的。植物数据指示了白垩纪-早第三纪转变时期持续相对的冷气候条件；而没有像此前报道的转变后的温暖条件。我们的温度指标与近来古土壤碳酸盐中提取的CO_2分压对应得很好，暗示了CO_2分压与温度的关联性。生物多样性程度与温度是相关的，基于最温暖期出现的数据来估计晚白垩纪生物灭绝的程度很可能是被夸大的，如我们对北达科特植物的揭示一样。然而，我们对气候的分析和演替系列的考虑显示火流星的影响应该是这些植物灭绝最重要的因素。

15. Smith L C, MacDonald G M, Velichko A A, et al. Siberian peatlands a net carbon sink and global methane source since the early Holocene[J]. Science, 2004, 303(5656): 353-356.

Siberian peatlands a net carbon sink and global methane source since the early Holocene

Smith L. C., MacDonald G. M., Velichko A. A., Beilman D. W., Borisova O. K., Frey K. E., Kremenetski K. V., Sheng Y.

(Department of Geography, University of California, Los Angeles)

Abstract: Interpolar methane gradient (IPG) data from ice cores suggest the "switching on" of a major Northern Hemisphere methane source in the early Holocene. Extensive data from Russia's West Siberian Lowland show explosive, widespread peatland establishment between 11.5 and 9 thousand years ago, predating comparable development in North America and synchronous with increased atmospheric methane concentrations and IPGs, larger carbon stocks than previously thought (70.2 Petagrams, up to similar to 26% of all terrestrial carbon accumulated since the Last Glacial Maximum), and little evidence for catastrophic oxidation, suggesting the region represents a long-term carbon dioxide sink and global methane source since the early Holocene.

全新世早期以来西伯利亚湿地净碳汇和全球甲烷来源

Smith L. C., MacDonald G. M., Velichko A. A., Beilman D. W., Borisova O. K., Frey K. E., Kremenetski K. V., Sheng Y.

（美国加州大学洛杉矶分校地理系）

摘要：来自冰芯中的极间甲烷梯度数据暗示了早全新世北半球甲烷来源的开关被打开。来自俄罗斯西西伯利亚低地大量的数据显示：距今11500～9000年前大量广泛的湿地存在，早于北美地区相对应的湿地发展和同一时期大气甲烷浓度和极间甲烷梯度的增加；远高于此前所认为的碳存量(70.2 Pg，相当于末次冰盛期以来所有陆地碳累积量的26%)以及极少关于高温剧烈氧化的证据，暗示了这个地区自早全新世以来是二氧化碳长期的汇和全球甲烷的源。

16. Chapin F S, Sturm M, Serreze M C, et al. Role of land-surface changes in Arctic summer warming [J]. Science, 2005, 310(5748): 657-660.

Role of land-surface changes in Arctic summer warming

Chapin F. S., Sturm M., Serreze M. C., McFadden J. P., Key J. R., Lloyd A. H., McGuire A. D., Rupp T. S., Lynch A. H., Schimel J. P., Beringer J., Chapman W. L., Epstein H. E., Euskirchen E. S., Hinzman L. D., Jia G., Ping C. L., Tape K. D., Thompson C. D. C., Walker D. A., Welker J. M.

(Institute of Arctic Biology, University of Alaska)

Abstract: A major challenge in predicting Earth's future climate state is to understand feedbacks that alter greenhouse-gas forcing. Here we synthesize field data from arctic Alaska, showing that terrestrial changes in summer albedo contribute substantially to recent high-latitude warming trends. Pronounced terrestrial summer warming in arctic Alaska correlates with a lengthening of the snow-free season that has increased atmospheric heating locally by about 3 watts per square meter per decade (similar in magnitude to the regional heating expected over multiple decades from a doubling of atmospheric CO_2). The continuation of current trends in shrub and tree expansion could further amplify this atmospheric heating by two to seven times.

地表变化在北极夏季变暖中的作用

Chapin F. S., Sturm M., Serreze M. C., McFadden J. P., Key J. R., Lloyd A. H., McGuire A. D., Rupp T. S., Lynch A. H., Schimel J. P., Beringer J., Chapman W. L., Epstein H. E., Euskirchen E. S., Hinzman L. D., Jia G., Ping C. L., Tape K. D., Thompson C. D. C., Walker D. A., Welker J. M.

(美国阿拉斯加大学北极生物研究所)

摘要:预测未来地球气候状态的一个主要挑战就是理解改变温室气体效应的反馈作用。综合北极阿拉斯加的野外数据显示陆地夏季反照率的变化显著地促进了近来高纬地区的变暖趋势。北极阿拉斯加夏季陆地的显著变暖与无雪季节的增长相关,而后者增加了每十年每平方米 3 瓦特的大气热量(在数量级上与大气 CO_2 增倍几十年所预期的区域热量相当),目前灌木和树木的持续扩张将进一步使大气热量增强 2～7 倍。

17. Smith L C, Sheng Y, MacDonald G M, et al. Disappearing Arctic lakes[J]. Science, 2005, 308(5727): 1429.

Disappearing Arctic lakes

Smith L. C., Sheng Y., MacDonald G. M., Hinzman L. D.

(Department of Geography, 1255 Bunche Hall, University of California-Los Angeles, Los Angeles, CA 90095, USA)

Abstract: Historical archived satellite images were compared with contemporary satellite data to track ongoing changes in more than 10000 large lakes in rapidly warming Siberia. A widespread decline in lake abundance and area has occurred since 1973, despite slight precipitation increases to the region. The spatial pattern of lake disappearance suggests (ⅰ) that thaw and "breaching" of permafrost is driving the observed losses, by enabling rapid lake draining into the subsurface; and (ⅱ) a conceptual model in which high-latitude warming of permafrost triggers an initial but transitory phase of lake and wetland expansion, followed by their widespread disappearance.

消失的北极湖泊

Smith L. C., Sheng Y., MacDonald G. M., Hinzman L. D.

(美国加州大学洛杉矶分校地理系)

摘要:历史存档的卫星图像与现代卫星数据跟踪进行比较,在迅速变暖的西伯利亚,超过 10000 个大湖正在消失。自1973年以来大部分湖泊的水量和水域发生普遍下降,尽管这一地区有轻微的降水增加。湖泊消失的空间格局表明:(1) 解冻的冻土使得湖泊快速向地表排水的损失;(2) 概念模型中高纬度变暖、冻土消融最初导致暂时的湖泊和湿地扩张,然后是湖泊普遍消失。

18. Smol J P, Wolfe A P, Birks H J B, et al. Climate-driven regime shifts in the biological communities of Arctic lakes[J]. PNAS, 2005, 102(12): 4397-4402.

Climate-driven regime shifts in the biological communities of Arctic lakes

Smol J. P., Wolfe A. P., Birks H. J. B., Douglas M. S. V., Jones V. J., Korhola A., Pienitz R., Ruhland K., Sorvari S., Antoniades D., Brooks S. J., Fallu M. A., Hughes M., Keatley B. E., Laing T. E., Michelutti N., Nazarova L., Nyman M., Paterson A. M., Perren B., Quinlan R., Rautio M., Saulnier-Talbot E., Siitoneni S., Solovieva N., Weckstrom J.

(Paleoecological Environmental Assessment and Research Laboratory, Department of Biology, Queen's University)

Abstract: Fifty-five paleolimnological records from lakes in the circumpolar Arctic reveal widespread species changes and ecological reorganizations in algae and invertebrate communities since approximately anno Domini 1850. The remoteness of these sites, coupled with the ecological characteristics of taxa involved, indicate that changes are primarily driven by climate warming through lengthening of the summer growing season and related limnological changes. The widespread distribution and similar character of these changes indicate that the opportunity to study arctic ecosystems unaffected by human influences may have disappeared.

气候驱动北极湖泊生物群落的转变

Smol J. P., Wolfe A. P., Birks H. J. B., Douglas M. S. V., Jones V. J., Korhola A., Pienitz R., Ruhland K., Sorvari S., Antoniades D., Brooks S. J., Fallu M. A., Hughes M., Keatley B. E., Laing T. E., Michelutti N., Nazarova L., Nyman M., Paterson A. M., Perren B., Quinlan R., Rautio M., Saulnier-Talbot E., Siitoneni S., Solovieva N., Weckstrom J.

(加拿大女王大学生物学系古生态环境评估与研究实验室)

摘要:55个采自环北极湖泊的古湖沼学记录揭示了自公元1850年以来藻类和无脊椎生物群落广泛的物种变化和生态重组。这些地区位置偏远加上复杂分类的生态特征指示这些变化主要是由气候变暖通过延长夏季生长季节和相关的湖沼学变化驱动的,这些变化广泛的分布和相似的特征指示了研究未受人类影响下的北极生态系统的机会可能要失去了。

八、环境污染类文献

8.1 分析概述

1990～2010 年，*Nature*，*Science* 和 *PNAS* 上共有"环境污染类"文献 22 篇（含报道），其中南极方面文献有 7 篇，北极方面文献有 15 篇。

将文章数量分区域并按照年代进行分类（图 8.1）可以发现，以上刊物上每年发表环境污染类文章 1～2 篇，仅在 1998 年、2007 年发表超过 2 篇。1990 年、1994 年、1996 年、1997 年及 2009～2010 年均无此类文章发表。2000 年之前南北极数量均等，2000 年之后以北极污染为主，南极方面仅见 2 篇（且 1 篇为评论性质文章）。由此可见，极地地区特别是南极地区由于所处地理位置的特异性，环境污染问题并不严重，而已经体现出来的现象没有在此三种期刊中表现出来；另一方面，由于北极离人类活动区较近，环境污染相对严重，也受到了人们更多的重视。

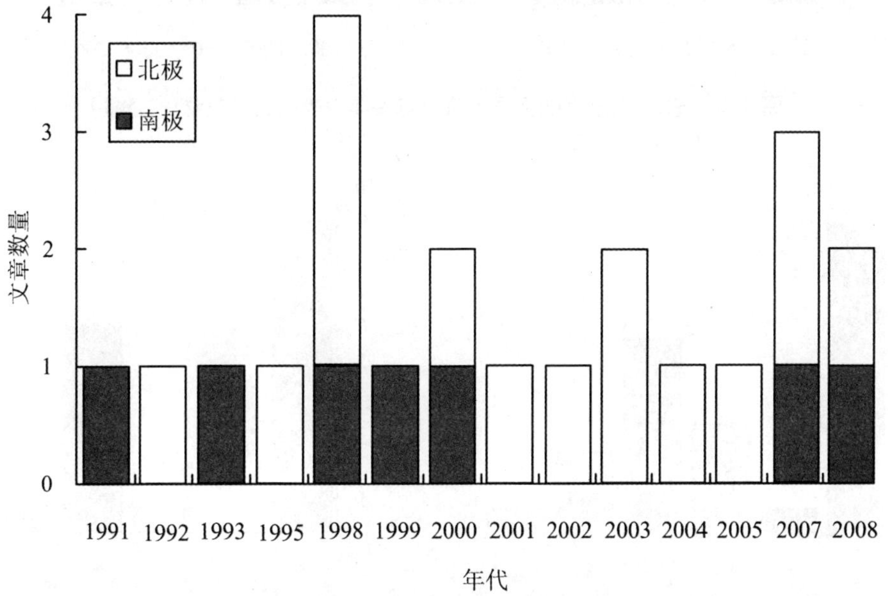

图 8.1　文章数量随年代的变化

从研究内容上来看，极地地区没有出现严重的污染现象，研究者兴趣重点大多集中于传播来源和循环的问题，特别是可以分为大气、海洋、生物三种主要传输途径。由各类文章随时间的分布（图 8.2）可以看出，2000 年之前研究主要集中在大气、海洋传播方面；而 2000 年之后研究内容细化，生物传输相关的研究成为研究重点之一。这体现了物质循环传输模型的进一步完善和环境污染方面研究的多元化发展。

将文章数量按所署单位（第一作者单位分开列出）所在国家进行排列（图 8.3），美国、加拿大发表的文章数最多，法国、挪威也有第一作者文章发表，刚果、丹麦也有提及。这可能部分因为环北极国家有更好的北极污染研究条件，在环境污染方面也有更迫切的研究需要。

由以上初步统计，目前极地环境污染类研究仍以传输为主，重点集中在北极地区，研究的发展与物质传输模型的完善相辅相成，环北极圈国家在此方面开展了许多工作。但从统计学意义而言，仅考虑 *Nature*，*Science*，*PNAS* 三种期刊不足以完全概括极地地区环境污染方面相关研究的现状。

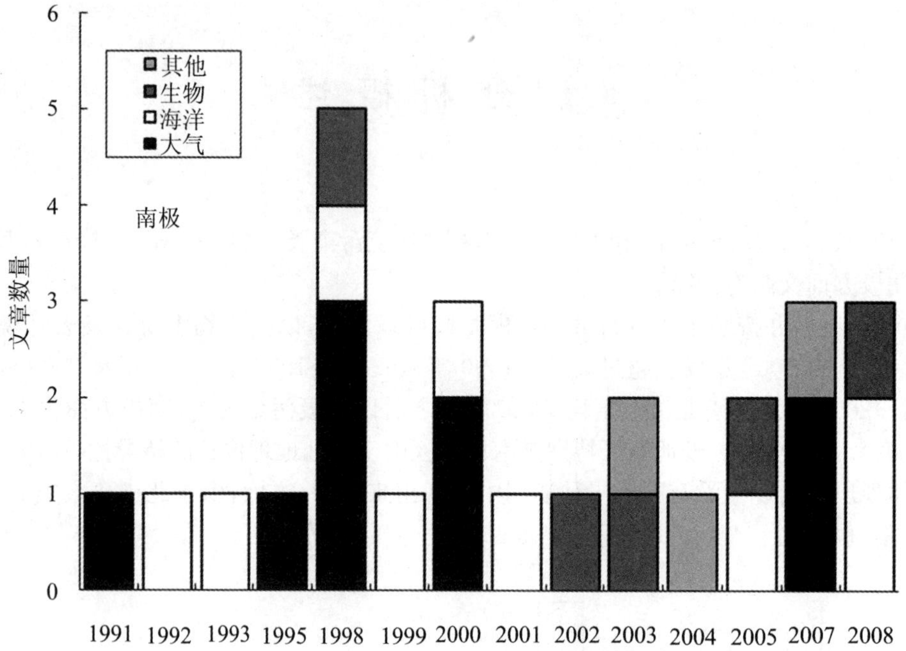

图 8.2 按传播途径划分不同研究领域每年发表的文章数量(有重叠)

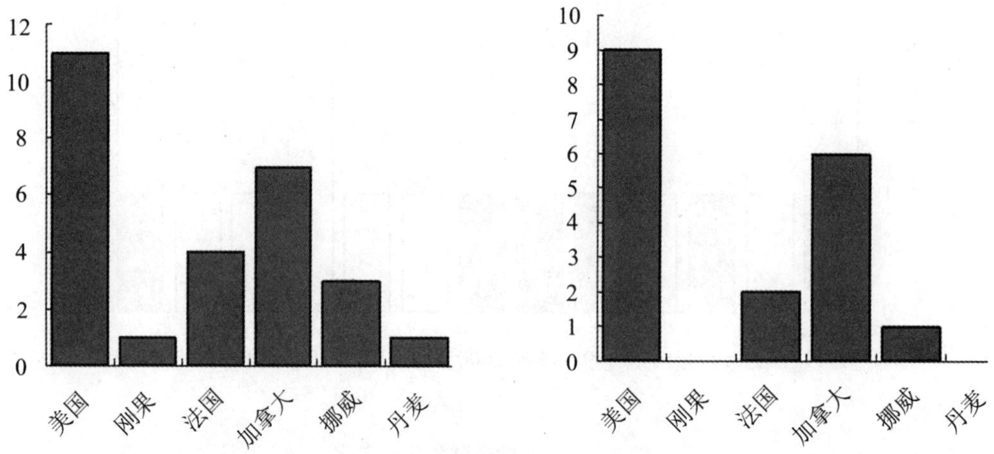

图 8.3 第一作者所在国家发表的文章数量(左,总体数量;右,第一作者数量)

8.2 摘要翻译——南极

1. Fishman J, Fakhruzzaman K, Cros B, et al. Identification of widespread pollution in the Southern-Hemisphere deduced from satellite analyses[J]. Science, 1991, 252(5013): 1693-1696.

Identification of widespread pollution in the Southern-Hemisphere deduced from satellite analyses

Fishman J., Fakhruzzaman K., Cros B., Nganga D.

(Atmospheric Sciences Division (Mail Stop 401A), National Aeronautics and Space Administration, Langley Research Center, Hampton, VA 23665)

Abstract: Vertical profiles of ozone obtained from ozonesondes in Brazzaville, Congo (4-degrees-S, 15-degrees-E), and Ascension Island (8-degrees-S, 15-degrees-W) show that large quantities of tropospheric ozone are present over southern Africa and the adjacent eastern tropical South Atlantic Ocean. The origin of this pollution is widespread biomass burning in Africa. These measurements support satellite-derived tropospheric ozone data that demonstrate that ozone originating from this region is transported throughout most of the Southern-Hemisphere. Seasonally high levels of carbon monoxide and methane observed at middle and high-latitude stations in Africa, Australia, and Antarctica likely reflect the effects of this distant biomass burning. These data suggest that even the most remote regions on this planet may be significantly more polluted than previously believed.

由卫星分析推导识别南半球广泛分布的污染物

Fishman J., Fakhruzzaman K., Cros B., Nganga D.

(美国国家航空和航天局大气科学部)

摘要：由设立在刚果的布拉柴维尔(4°S，15°E)和阿森松岛(8°S，15°W)的臭氧无线电探测仪得到了臭氧垂直廓线。结果显示，在非洲南部和毗邻的热带南大西洋的东部上空，对流层中存在大量臭氧。污染源来自非洲广泛的生物质燃烧。本研究的数据也支持来自卫星的对流层中的臭氧数据，起源于这个地区的臭氧向大部分南半球运输。在非洲、澳大利亚和南极洲中高纬度站观测到的一氧化碳、甲烷含量的季节性高值很可能反映了这种生物质燃烧的长距离传输结果。这些数据显示：即使是地球上最遥远的地方也可能比先前估计的污染水平更加严重。

2. Flegal A R, Maring H, Niemeyer S. Anthropogenic lead in Antarctic sea-water[J]. Nature, 1993, 365(6443): 242-244.

Anthropogenic lead in Antarctic sea-water

Flegal A. R., Maring H., Niemeyer S.

(Earth Sciences Board, University of California, Santa Cruz, California 95064, USA)

Abstract: Antarctica is believed to be a relatively pristine continent, mainly because of its remote location and the atmospheric circulation patterns that limit the transport of industrial aerosols into the Antarctic polar cell. This perception is apparently supported by the extremely low concentrations of lead in Antarctic surface waters-an observation that has been interpreted as showing insignificant contamination by anthropogenic lead. The isotopic composition of lead in other natural waters has been used as a tracer of the sources of lead, and in particular to identify anthropogenic inputs. Here we apply this approach to Antarctic surface waters, and show that despite the low concentrations of lead in these waters (which are confirmed by our measurements), their isotopic composition reveals a significant contribution of lead from industrial sources. The extremely low concentrations of lead in these waters appear to be due to biological scavenging of the lead during periods of intense primary production.

南极海水中的人为源铅

Flegal A. R., Maring H., Niemeyer S.

(美国加州大学圣克鲁兹分校地球科学部)

摘要:南极洲由于位置偏远,大气环流模式能够阻止工业气溶胶向其中的传输,因此一般认为南极洲是一个相对原始的大陆。南极大陆表层水中的铅含量极低,表明其中人为铅的比重更是可以忽略的。在其他天然水体中,通过分析铅同位素的组成可以用来示踪铅的来源,特别是辨别出人为源的输入。本文中,我们将这种方法应用于研究南极表层水体,测量值表明,尽管这些水体中的污染铅含量均很低,但其同位素组成揭示了工业来源的铅在其中占有很大比重。在强烈的初级生产阶段的生物清除机制是导致水体中铅含量极低的原因。

3. Shanaka L de Silva, Gregory A Zielinski. Global influence of the AD1600 eruption of Huaynaputina, Peru[J]. Nature, 1998, 393(6684): 455-458.

Global influence of the AD1600 eruption of Huaynaputina, Peru

Shanaka L. de Silva, Gregory A. Zielinski

(Department of Geography, Geology, and Anthropology, Indiana State University, Terre Haute, Indiana 47809, USA)

Abstract: It has long been estabished that gas and fine ash from large equatorial explosive eruptions can spread globally, and that the sulphuric acid that is consequently produced in the stratosphere can cause a small, but statistically significant, cooling of global temperatures. Central to revealing the ancient volcano-climate connection have been studies linking single eruptions to features of climate-proxy records such as found in ice-core and tree-ring chronologies. Such records also suggest that the known inventory of eruptions is incomplete, and that the climatic significance of unreported or poorly understood eruptions remains to be revealed. The AD1600 eruption of Huaynaputina, in southern Peru, has been speculated to be one of the largest eruptions of the past 500 years; acidity spikes from Greenland and Antarctica ice, tree-ring chronologies, along with records of atmospheric perturbations in early seventeenth-century Europe and China, implicate an eruption of similar or greater magnitude than that of Krakatau in 1883. Here we use tephra deposits to estimate the volume of the AD1600 Huaynaputina eruption, revealing that it was indeed one of the largest eruptions in historic times. The chemical characteristics of the glass from juvenile tephra allow a firm cause-effect link to be established with glass from the Antarctic ice, and thus improve on estimates of the stratospheric loading of the eruption.

公元 1600 年秘鲁 Huaynaputina 火山喷发的全球性影响

Shanaka L. de Silva, Gregory A. Zielinski

(美国印第安纳州立大学地质地理和人类学系)

摘要:赤道附近的大规模爆炸性火山爆发的火山灰细粒以及烟尘都传播至全球,随之在平流层中产生的硫酸则可以引起小幅度的但统计学意义显著的全球气温冷却。在揭示古火山-气候关系的主要研究中是将单个火山喷发和气候变化记录联系在一起的,就如同冰芯树轮年代学一样。这些记录同时也说明,已知的火山喷发记录并不完整,未经研究的或了解不深入的火山喷发事件的气候学意义有待揭示。秘鲁南部的 Huaynaputina 火山在公元 1600 年的喷发事件,被认为是过去 500 年内最强烈的火山活动之一。来自格陵兰和南极的冰芯中的酸度峰值、树轮年代学,以及 17 世纪早期欧洲和中国的大气扰动情况的记录都记录了这一次喷发事件,强度相似或者强于喀拉喀托火山(位于印度尼西亚西南部)1883 年的爆发。在这篇文章里我们利用火山灰来估测公元 1600 年 Huaynaputina 火山喷发的强度,发现它确实是历史时期最剧烈的活动之一。根据早期火山灰中的玻璃体的化学性质可以和南极冰芯中的玻璃体建立起因果关系,进而修正了对火山喷发向平流层的输入量估测。

4. Kevin Capaldo, James J Corbett, Prasad Kasibhatla, et al. Effects of ship emissions on sulphur cycling and radiative climate forcing over the ocean[J]. Nature, 1999, 400(6746): 743-746.

Effects of ship emissions on sulphur cycling and radiative climate forcing over the ocean

Kevin Capaldo, James J. Corbett, Prasad Kasibhatla, Paul Fischbeck, Spyros N. Pandis

(Departments of Chemical Engineering, Carnegie Mellon University, Pittsburgh, Pennsylvania 15213, USA)

Abstract: The atmosphere overlying the ocean is very sensitive-physically chemically and climatically-to air pollution. Given that clouds over the ocean are of great climatic significance, and that sulphate aerosols seem to be an important control on marine cloud formation, anthropogenic inputs of sulphate to the marine atmosphere could exert an important influence on climate. Recently, sulphur emissions from fossil fuel burning by international shipping have been geographically characterized, indicating that ship sulphur emissions nearly equal the natural sulphur nux from ocean to atmosphere in many areas. Here we use a global chemical transport model to show that these ship emissions can be a dominant contributor to atmospheric sulphur dioxide concentrations over much of the world's oceans and in several coastal regions. The ship emissions also contribute significantly to atmospheric non-seasalt sulphate concentrations over Northern Hemisphere ocean regions and parts of the Southern Pacific Ocean, and indirect radiative forcing due to ship-emitted particulate matter (sulphate plus organic material) is estimated to contribute a substantial fraction to the anthropogenic perturbation of the Earth's radiation budget. The quantification of emissions from international shipping forces a re-evaluation of our present understanding of sulphur cycling and radiative forcing over the ocean.

船只排放物对海洋硫循环和辐射气候强迫的影响

Kevin Capaldo, James J. Corbett, Prasad Kasibhatla, Paul Fischbeck, Spyros N. Pandis

(美国卡耐基梅隆大学化学工程系)

摘要：海洋上层大气在物理、化学性质以及气候上均对大气污染十分敏感。考虑到海洋上层的云层具有强烈的气候意义，且硫酸盐气溶胶在海洋云体形成中扮演了重要角色，人为硫酸盐向海洋系统的输入可能对气候发挥重要作用。近来，国际船运化石燃料燃烧释放的硫往往以地域性为特征，表明在很多地区船只硫的排放量和由海洋排向大气的天然硫通量相当。在本文中，我们采用全球化学输送模型证明：在大部分大洋和个别沿海地区的大气中，船只排放在大气二氧化硫浓度中占主导。在北半球海洋地区和部分南太平洋地区，船只排放同时也是大气中非海盐源的硫酸盐的主要来源，同时船只排放的特殊物质(硫酸盐和有机物)会导致间接辐射强迫，其对地球辐射平衡会产生扰动，是人为扰动的主要成分。国际航运排放物质逐渐定量化，促使了对当今对硫循环对海洋上层辐射强迫理解的重新评估。

5. Huiming Bao, Douglas A Campbell, James G Bockheim, et al. Origins of sulphate in Antarctic dry-valley soils as deduced from anomalous ^{17}O compositions[J]. Nature, 2000, 407(6803): 499-502.

Origins of sulphate in Antarctic dry-valley soils as deduced from anomalous ^{17}O compositions

Huiming Bao, Douglas A. Campbell, James G. Bockheim, Mark H. Thiemens

(Department of Chemistry & Biochemistry, Mail Code 0356, University of California San Diego, 9500 Gilman Drive, La Jolla, California 92093-0356, USA)

Abstract: The dry valleys of Antarctica are some of the oldest terrestrial surfaces on the Earth. Despite much study of soil weathering and development, ecosystem dynamics and the occurrence of life in these extreme environments, the reasons behind the exceptionally high salt content of the dry-valley soils have remained uncertain. In particular, the origins of sulphate are still controversial; proposed sources include wind-blown sea salt, chemical weathering, marine incursion, hydrothermal processes and oxidation of biogenic sulphur in the atmosphere. Here we report measurements of delta ^{18}O and delta ^{17}O values of sulphates from a range of dry-valley soils. These sulphates all have a large positive anomaly of ^{17}O, of up to 3.4 parts per thousand. This suggests that Antarctic sulphate comes not just from sea salt (which has no anomaly of ^{17}O) but also from the atmospheric oxidation of reduced gaseous sulphur compounds, the only known process that can generate the observed ^{17}O anomaly. This source is more prominent in high inland soils, suggesting that the distributions of sulphate are largely explained by differences in particle size and transport mode which exist between sea-salt aerosols and aerosols formed from biogenic sulphur emission.

由^{17}O同位素异常组成来判断南极涸谷地区土壤硫酸盐来源

Huiming Bao, Douglas A. Campbell, James G. Bockheim, Mark H. Thiemens

(美国加州大学圣地亚哥分校化学和生物化学系)

摘要:南极大陆的涸谷地区是世界上最原始的陆地表面之一。尽管关于土壤风化发育过程、生态系统动力学以及极端环境中生命的出现已经研究得十分深入了,但就涸谷地区土壤异常高盐度背后的原因仍然悬而未决。尽管针对硫酸盐的起源已经提出的来源包括风对海盐的搬运、化学风化作用、船舶侵入、热液作用以及大气中生物硫酸盐的氧化,但仍存在有很大争议。在这篇文章里,我们测量了涸谷一带土壤硫酸盐中同位素^{18}O、^{17}O的含量,发现这些硫酸盐中^{17}O含量均存在高达3.4‰的正异常。这意味着南极地区硫酸盐不仅来自海盐(海盐中没有^{17}O正异常),同时也来自大气中还原性含硫化合物的氧化(目前唯一已知的能形成已观测到的^{17}O异常的途径)。这一来源在高内陆土壤中占更大的主导地位,进而得出结论:海盐气溶胶和生物硫酸盐排放所形成的气溶胶二者之间的传输模式以及粒径大小的差异可以来解释硫酸盐的分布。

6. Melanie Baroni, Mark H Thiemens, Robert J Delmas, et al. Mass-independent sulfur isotopic compositions in stratospheric volcanic eruptions[J]. Science, 2007, 315(5808): 84-87.

Mass-independent sulfur isotopic compositions in stratospheric volcanic eruptions

Melanie Baroni, Mark H. Thiemens, Robert J. Delmas, Joel Savarino

(Laboratoire de Glaciologie et Géophysique de l'Environnement, CNRS/Université Joseph Fourier, 38400 St. Martin d'Hères, France)

Abstract: The observed mass-independent sulfur isotopic composition (Delta ^{33}S) of volcanic sulfate from the Agung (March 1963) and Pinatubo (June 1991) eruptions recorded in the Antarctic snow provides a mechanism for documenting stratospheric events. The sign of Delta ^{33}S changes over time from an initial positive component to a negative value. Delta ^{33}S is created during photochemical oxidation of sulfur dioxide to sulfuric acid on a monthly time scale, which indicates a fast process. The reproducibility of the results reveals that Delta ^{33}S is a reliable tracer to chemically identify atmospheric processes involved during stratospheric volcanism.

火山喷发导致的平流层中硫同位素非质量分馏的组成

Melanie Baroni, Mark H. Thiemens, Robert J. Delmas, Joel Savarino

(法国格勒诺布尔第一大学)

摘要:在南极冰雪中记录的来自阿贡火山(1963年3月)和皮纳图博火山(1991年6月)喷发的火山硫酸盐硫同位素非质量分馏($\delta^{33}S$)的组成,提供了这些在平流层发生的事件的产生机制。^{33}S随着时间从一开始的偏正转变成了偏负，^{33}S是在月份时间尺度上由二氧化硫经光化学催化氧化成硫酸的过程中产生的，且速度快。这一结果的可再现性保证了^{33}S是一个可靠的化学指标来确定平流层中的火山活动等大气过程。

7. Louise K Blight, David G Ainley. Southern Ocean not so pristine[J]. Science, 2008, 321(5895): 1443.

Southern Ocean not so pristine

Louise K. Blight, David G. Ainley

(Canadian environmental protection and research center)

Abstract: The Report "A global map of human impact on marine ecosystems" (Halpern B. S., et al) provides a timely overview of anthropogenic effects on even the farthest reaches of Earth's oceans. However, we contend that, for at least one region, using data from only the past decade leads to misleading results.

A widespread perception exists that waters south of the Antarctic Polar Frontal Zone—i. e., the Southern Ocean (SO)—are still nearly pristine. In fact, the northern portion of the SO saw virtually all cetacean populations removed long ago, and in subsequent years (1960s to 1980s) the largest stocks of demersal fish in the Indian Ocean and Scotia Sea/Atlantic Ocean sectors were also fished to commercial extinction. Historically exploited fish species and cetaceans show little signs of recovery in the SO, and recent legal commercial fishing activity has been correspondingly low. It is thus no surprise that the modeling used by Halpern et al shows little anthropogenic impact in these sectors apart from that of climate change. The authors acknowledge that accounting for current illegal, unregulated, and unreported fishing in these waters might show increased human impacts. The additional consideration of historical data should cause Halpern et al to temper their conclusion that for the world's oceans "large areas of relatively little human impact remain, particularly near the poles."

南大洋并非如此原始

Louise K. Blight, David G. Ainley

(加拿大环境保护研究中心)

摘要:《人类对海洋系统的影响的全球地图》一文，对地球上最遥远的海洋受到的人类影响进行了适时的综述，然而我们依然认为，至少对于一个地区，仅仅利用过去几十年的数据进行分析将会导致错误的结论。和历史时期相比，当今遭受过商业开发的鱼类种类以及鲸鱼种群在南大洋中显示出轻微的恢复迹象，与此同时最近合法商业捕捞行为相应地减弱。

一直以来，人们普遍认为南极极地锋区南部水域——也就是南大洋至今为止依然十分原始，事实上，南大洋的北部地区见证了鲸类数量长久以来的变化情况，在随后的1960～1980年期间，贮存最多的印度洋、斯科舍海、大西洋中的底栖鱼类因为遭商业捕捞而逐渐消亡。利用Halpern等人的模型表征出这些受人类活动影响小的地区在气候变化方面也表现得并不明显，这也是顺理成章的。本文作者认为，可以用对这些水域当前非法的未经报告过的捕鱼行为来解释人类影响的逐渐加重。同时，历史上的数据应该让Halpern等重新考虑他们得出的——世界上的大洋大多受到人类影响较小特别是极地地区的更加原始的结论。

8.3 摘要翻译——北极

1. Marshall E. A Scramble for data on Arctic radioactive dumping[J]. Science, 1992, 257(5070): 608-609.

A Scramble for data on Arctic radioactive dumping

Marshall E.

(Report)

Abstract: On 15 August, a Russian research vessel, the Viktor Buynitskiy, will leave the Norwegianport of Kirkenes packed with surveying equipment and scientists. Its mission will be to check out one of the more alarming environmental stories that have drifted out of the former Soviet Union since its collapse last fall: a claim that the Arctic is being polluted by tons of radioactive waste spilled or dumped by the Soviet military. Since May, US officials have been searching for information to confirm or disprove the reports about Russian radiation. The basic concern, is as follows: Arctic waters and, potentially, fisheries near Norway and Alaska are in danger of being contaminated by radioactive isotopes leaking from two major sources. One is the area around Novaya Zemlya, an archipelago where the Soviets conducted bomb tests, scuttled submarines, and disposed of waste canisters. The other is freshwater runoff into the Arctic Ocean including the Ob and Yenisey Rivers carrying isotopes from weapons plants, waste ponds, and accident sites in Siberia.

对北极放射性物质排放调查的争夺

Marshall E.

(报道)

摘要:在8月15日,一艘名为Viktor Buynitskiy的俄罗斯科研船舰将会带着随行专家和考察装备离开挪威港。此行的目的旨在调查一个苏联政府最后一次解体之后泄露的、令人吃惊的环境传言:苏联军队在北极排放或倾倒放射性废物。自从5月开始,美国官员一直在寻找更多的线索来证实或推翻苏联辐射报告。若追究本次调查的缘由,是因为北极水域,特别是挪威和阿拉斯加附近的渔业正处于来自两个主要源头的放射性同位素泄露的潜在威胁中。一个主要源是在新地岛附近,苏联政府曾在这里进行炸弹实验、凿沉潜水艇和丢弃废物罐。另外一个携带着来自武器工厂废水池以及西伯利亚事故地点放射性物质流入北冰洋的淡水径流,包括鄂毕河和叶尼塞河。

2. Karen A Kidd, David W Schindler, Derek C G Muir, et al. High-Concentrations of toxaphene in fishes from a Sub-Arctic lake[J]. Science, 1995, 269(5221): 240-242.

High-Concentrations of toxaphene in fishes from a Sub-Arctic lake

Karen A. Kidd, David W. Schindler, Derek C. G. Muir, W. Lyle Lockhart, Raymond H. Hesslein

(Department of Biological Sciences, University of Alberta, Edmonton, Alberta T6G 2E9, Canada)

Abstract: Concentrations of toxaphene and other organochlorine compounds are high in fishes from subarctic Lake Laberge, Yukon Territory, Canada. Nitrogen isotope analyses of food chains and contaminant analyses of biota, water, and dated lake sediments show that the high concentrations of toxaphene in fishes from Laberge resulted entirely from the biomagnification of atmospheric inputs. A combination of low inputs of toxaphene from the atmosphere and transfer through an exceptionally long food chain has resulted in concentrations of toxaphene in fishes that are considered hazardous to human health.

亚北极地带湖泊中鱼类体内的高浓度毒杀芬

Karen A. Kidd, David W. Schindler, Derek C. G. Muir, W. Lyle Lockhart, Raymond H. Hesslein

(加拿大阿尔伯塔大学生物科学系)

摘要:加拿大育空地区亚北极湖泊 Laberge 中鱼类体内的毒杀芬以及其他含氯杀虫剂化合物的浓度都非常高,经过对食物网进行氯同位分析,水体、种群污染物分析,定年后的湖泊沉积物分析结果都显示,毒杀芬的高浓度是由于大气输入后生物放大效应所致。大气传输输入低浓度的毒杀芬加上异常冗长的食物链二者共同导致了鱼类体内毒杀芬的高浓度,对人体健康构成危险。

3. Jules M Blais,David W Schindler,Derek C G Muir,et al. Accumulation of persistent organochlorine compounds in mountains of western Canada[J]. Nature,1998,395(6702):585-588.

Accumulation of persistent organochlorine compounds in mountains of western Canada

Jules M. Blais, David W. Schindler, Derek C. G. Muir, Lynda E. Kimpe, David B. Donald, Bruno Rosenberg

(Department of Biological Sciences, University of Alberta, Edmonton, Alberta, Canada T6G 2E9)

Abstract: Persistent, semi-volatile organochlorine compounds, including toxic industrial pollutants and agricultural pesticides, are found everywhere on Earth, including in pristine polar and near-polar locations. Higher than expected occurrences of these compounds in remote regions are the result of long-range transport in the atmosphere, precipitation and "cold condensation"—the progressive volatilization in relatively warm locations and subsequent condensation in cooler environments which leads to enhanced concentrations at high latitudes. The upper reaches of high mountains are similar to high-latitude regions in that they too are characterized by relatively low average temperatures, but the accumulation of organochlorine compounds as a function of altitude has not yet been documented. Here we report organochlorine deposition in snow from mountain ranges in western Canada that show a 10 to 100 fold increase between 770 and 3100 m altitude. In the case of less-volatile compounds, the observed increase by a factor of 10 is simply due to a 10 fold increase in snowfall over the altitude range of the sampling sites. Tn the case of the more-volatile organochlorines, cold-condensation effects further enhance the concentration of these compounds with increasing altitude. These findings demonstrate that temperate-zone mountain regions, which tend to receive high levels of precipitation while being close to pollutant sources, are particularly susceptible to the accumulation of semivolatile organochlorine compounds.

持久性有机氯化物在加拿大西部山区的聚积

Jules M. Blais, David W. Schindler, Derek C. G. Muir, Lynda E. Kimpe,
David B. Donald, Bruno Rosenberg

（加拿大阿尔伯塔大学生物科学系）

摘要：持久性或半挥发性的有机氯化物，包括有毒工业污染物和农业杀虫剂，在地球上已经无处不在，甚至已经蔓延到了原始的极地和附近的地区。这些偏远地区比估测值还要高的污染，形成原因复杂。大气长距离运输、降雨、冷凝结——在较为温暖地区强烈挥发，随后在寒冷地区又凝结——这就导致了高纬地区的污染物浓度升高。高山的上层地区以较低的平均温度为特征，这一点上和高纬地区有几分相似，但是有机氯化物聚积作为海拔的一种特征还没有被报道。在这篇文章中，我们列出了有机氯化物在加拿大西部山脉冰雪中的沉积状况，结果显示出在 770～3100 m 海拔内 10 到 100 倍的增长。至于低挥发性的化合物，已经观测到的 10 倍增长是由于山脉上取样地点降雪量 10 倍增长导致的。至于高挥发性的有机氯化物，冷凝聚效应随着海拔的升高而大大增加了这些污染物的浓度。这些发现说明，在那些温和的山地区域，倾向于接受高水平的降水而离污染源相对较近，对半挥发性的有机含氯物质特别敏感。

4. Henrietta N Edmonds, S Bradley Moran, John A Hoff, et al. Protactinium-231 and thorium-230 abundances and high scavenging rates in the western Arctic Ocean[J]. Science, 1998, 280(5362): 405-407.

Protactinium-231 and thorium-230 abundances and high scavenging rates in the western Arctic Ocean

Henrietta N. Edmonds, S. Bradley Moran, John A. Hoff, John N. Smith, R. Lawrence Edwards

(Graduate School of Oceanography, University of Rhode Island, Narragansett, RI 02882, USA)

Abstract: The Canadian Basin of the Arctic Ocean, largely ice covered and isolated from deep contact with the more dynamic Eurasian Basin by the Lomonosov Ridge, has historically been considered an area of low productivity and particle flux and sluggish circulation. High-sensitivity mass-spectrometric measurements of the naturally occurring radionuclides protactinium-231 and thorium-230 in the deep Canada Basin and on the adjacent shelf indicate high particle fluxes and scavenging rates in this region. The thorium-232 data suggest that offshore advection of particulate material from the shelves contributes to scavenging of reactive materials in areas of permanent ice cover.

西部北冰洋^{231}Pa 和^{230}Th 的丰度和高清除速率

Henrietta N. Edmonds, S. Bradley Moran, John A. Hoff, John N. Smith, R. Lawrence Edwards

(美国罗德岛大学海洋学院)

摘要:北冰洋的加拿大流域常年被大量冰雪覆盖,被罗蒙诺索夫海岭孤立,与充满活力的欧亚流域缺乏接触,曾经一度被认为是生产力低、粒子通量低、循环迟缓的荒蛮之地。对于加拿大流域深部和大陆架毗邻地区的天然放射性核素^{231}Pa 和^{230}Th 的高分辨率质谱测试结果,表征出了该地区高粒子通量和清除速率。^{232}Th 数据显示,来自大陆架的近海地区颗粒物的水平对流可以清除永久性冰盖地区的反应物质。

5. Schroeder W H, Anlauf K G, Barrie L A, et al. Arctic springtime depletion of mercury[J]. Nature, 1998, 394(6691): 331-332.

Arctic springtime depletion of mercury

Schroeder W. H., Anlauf K. G., Barrie L. A., Lu J. Y., Steffen A.

(Atmospheric Environment Service, 4905 Dufferin Street, Toronto, Ontario M3H 5T4, Canada)

Abstract: The Arctic ecosystem is showing increasing evidence of contamination by persistent, toxic substances, including metals such as mercury, that accumulate in organisms. In January 1995, we began continuous surface-level measurements of total gaseous mercury in the air at Alert, Northwest Territories, Canada (82.5°N, 62.5°W). Here we show that, during the spring (April to early June) of 1995, there were frequent episodic depletions in mercury vapour concentrations, strongly resembling depletions of ozone in Arctic surface air, during the three-month period following polar sunrise (which occurs in March).

春季北极地区汞的消减

Schroeder W. H., Anlauf K. G., Barrie L. A., Lu J. Y., Steffen A.

(加拿大大气环境局)

摘要:北极生态系统正在提供越来越多的证据,以证明持久性有毒物质包括诸如汞的重金属正在生物体内逐渐富集。1995 年 1 月在加拿大的西北特 Alert(82.5°N, 62.5°W)我们开始了持续性地对大气中总气态汞表层水平的测量。我们发现,在 1995 年 4 月份到 6 月份春季期间气态汞浓度有着频繁的不定期耗散,与北极地表大气中臭氧的耗散非常相似——这 3 个月份是极昼的时间。

6. Pierre E Biscaye, Francis E Grousset, Anders M Svensson, et al. Eurasian air pollution reaches eastern North America[J]. Science, 2000, 290(5500): 2259.

Eurasian air pollution reaches eastern North America

Pierre E. Biscaye, Francis E. Grousset, Anders M. Svensson, Aloys Bory

(Lamont-Doherty Earth Observatory of Columbia University, Palisades, NY 10964, USA)

Abstract: Pollutants of Eurasian origin cross the Pacific Ocean via the westerly winds and affect the west coast of North America, as Kenneth E. Wilkening, Leonard A. Barrie, and Marilyn Engle discuss in their Perspective (Science's Compass, 6 Oct., p. 65). However, neither anthropogenic pollutants nor natural dusts of trans-Pacific origin stop their transit at or near the western North American coast.

We have found that the continental dust in the Greenland Summit ice cores from 1500 to 44000 years before the present, and probably beyond, is of Eastern Asian provenance. We are finding the same sources in recent (past decade) dust extracted from snow pits at the North Greenland Ice core Project site (75° N, 43° W), including dust from the April 1998 outbreak shown in the figure in Wilkening et al's Perspective (see the figure). The fluxes of dust to Greenland have been generally about a few millligrams per square centimeter per thousand years, compared with fluxes about an order of magnitude greater upwind in the eastern North Pacific. Consequently, we expect that the decrease in the flux of dust and associated pollutants between western and eastern North America might be less than that factor of ~10 and therefore of significance to levels of pollution. As Asian economic expansion increases the quantities of pollutants injected into the westerly winds, their levels in western North American cities, like those cited by Wilkening et al, will also rise, as will those levels eastward and all across the continent.

欧亚大气污染到达了北美东部

Pierre E. Biscaye, Francis E. Grousset, Anders M. Svensson, Aloys Bory

(美国哥伦比亚大学拉蒙特-多尔蒂地质观测站)

摘要：来自欧亚大陆源的污染物正借助西风跨越太平洋来影响北美的西海岸，然而，人为源污染物或天然粉尘组成的这些污染物并没有在西海岸停止继续运移。

来自格陵兰距今 1500～44000 年的冰芯表明，现代大陆性粉尘可能起源自东亚。我们在最近的从北格陵兰雪坑中提取出来的冰芯也显示出一致的粉尘来源。运移到格陵兰地区的粉尘通量一般认为是每千年每平方厘米几毫克，比北太平洋东部的逆风区大一个数量级。因此，我们预测北美西部和东部之间的粉尘及相关污染物的降低量应该不到百分之十，所以仍具有显著的污染水平。随着亚洲经济的不断扩张，注入到西部季风中的污染物量不断增加，其污染水平在北美西部也会上升，而穿越整个北美大陆之后在其东部的污染水平也会随之不断增长。

7. Charles Gobeil, Robie W Macdonald, John N Smith, et al. Atlantic water flow pathways revealed by lead contamination in Arctic basin sediments[J]. Science, 2001, 293(5533): 1301-1304.

Atlantic water flow pathways revealed by lead contamination in Arctic basin sediments

Charles Gobeil, Robie W. Macdonald, John N. Smith, Luc Beaudin

(Institut Maurice-Lamontagne, Mont-Joli, QC, G5H 3Z4, Canada)

Abstract: Contaminant lead in sediments underlying boundary currents in the Arctic Ocean provides an image of current organization and stability during the past 50 years. The sediment distributions of lead, stable lead isotope ratios, and lead-210 in the major Arctic Ocean basins reveal close coupling of the Eurasian Basin with the North Atlantic during the 20th century. They indicate that the Atlantic water boundary current in the Eurasian Basin has been a prominent pathway, that contaminant lead from the Laptev Sea supplies surface water in the transpolar drift, and that the Canadian and Eurasian basins have been historically decoupled.

北极盆地沉积中铅污染情况揭示了大西洋水流路径

Charles Gobeil, Robie W. Macdonald, John N. Smith, Luc Beaudin

(加拿大莫里-斯拉蒙塔根研究所)

摘要:北冰洋边界流底部沉积物中的铅污染,提供了过去50年里现代洋流构架和稳定性的图景。沉积物中铅的分布、稳定同位素比率,以及主要北冰洋流域^{210}Pb的定年结果都揭示了20世纪欧亚流域和北大西洋的强烈耦合。这就表明现在位于欧亚流域的大西洋边界流是一条主要的水流路径,来自拉普捷夫海的污染物通过极地漂流来运送至表层海水,以及加拿大和亚欧流域历史上并不挂钩。

8. Ralph S Hames, Kenneth V Rosenberg, James D Lowe, et al. Adverse effects of acid rain on the distribution of the Wood Thrush Hylocichla mustelina in North America[J]. PNAS, 2002, 99(17): 11235-11240.

Adverse effects of acid rain on the distribution of the Wood Thrush Hylocichla mustelina in North America

Ralph S. Hames, Kenneth V. Rosenberg, James D. Lowe, Sara E. Barker, Andre A. Dhondt

(Cornell Laboratory of Ornithology, Cornell University, Ithaca, NY 14850)

Abstract: Research into population declines of North American bird species has mainly focused on the fragmentation of habitat on the breeding or wintering grounds [Robinson S. K., Thompson F. R., Donovan T. M., Whitehead D. R. & Faaborg J. (1995) Science 267, 1987-1990]. In contrast, research into declines of European species has mainly focused on intensification of agriculture [Donald P. F., Green R. E. & Heath M. F. (2001) Proc. R. Soc. London Ser. B 268, 25-29] and the role played by the atmospheric deposition of pollutants, in particular, acid rain [Graveland J. (1998) Environ. Rev. 6. 41-54]. However, despite widespread unexplained declines of bird populations in regions of heavy wet acid ion deposition [Sauer J. R., Hines J. E. & Fallon J. (2001) The North American Breeding Bird Survey Results and Analysis 1966-2000 (Patuxent Wildlife Research Center, Laurel, MD)], no North American studies have presented evidence linking such widespread terrestrial bird declines to acid rain. To address the question of the role played by acid rain in population declines of eastern North American songbird species, we combine data from several sources. We use a multiple logistic regression model to test for adverse effects of acid rain on the Wood Thrush, while controlling for regional abundance, landscape-level habitat fragmentation, elevation, soil pH, and vegetation. We show a strong, highly significant, negative effect of acid rain on the predicted probability of breeding by this species, and interactions with elevation, low pH soils, and habitat fragmentation that worsen these negative effects. Our results suggest an important role for acid rain in recent declines of some birds breeding in the eastern United States, particularly in high elevation zones with low pH soils, and show the need to consider other large-scale influences, in addition to habitat fragmentation, when addressing bird population declines.

酸雨对北美一种稀有的黄褐森鸫分布的负作用

Ralph S. Hames, Kenneth V. Rosenberg, James D. Lowe, Sara E. Barker, Andre A. Dhondt

(美国康奈尔大学鸟类学实验室)

摘要：有关北美地区鸟类种群数量下降的研究，目前主要聚焦于鸟类繁育地、越冬地或栖息地的破碎上。与此相反的是，欧洲种群数目锐减的研究则着重于农耕化不断加剧以及大气对污染物沉降上所起的作用——特别是酸雨。然而，酸性离子湿沉降十分严重的地区往往鸟类数量都有所降低，这一现象虽然在欧洲被广泛承认(尽管解释不清)，但是根据北美鸟类繁育调查结果分析来看，没有证据能把这些普遍陆生鸟类数量锐减和酸雨联系在一起。为了解释酸雨对北美东部鸣禽种群减少起的作用，我们从多个来源获取数据，同时控制地区物种丰富度、景观级别、栖息地破碎程度、海拔、土壤 pH 值以及植被覆盖情况，利用回归模型分析酸雨对黄褐森鸫的负面影响。我们发现了显著的酸雨对预测的该物种繁育的负面效应，与海拔、低土壤 pH 值和分割的栖息地的交互作用，使这种负面作用加剧。总之我们的研究结果表明，酸雨很大程度上导致美国东部一些鸟类繁殖的衰退，对于高纬度低 pH 值土壤的地区尤为严重。另外在考虑鸟类数量锐减的时候，除了考虑被分割的栖息地，我们有必要考虑其他大规模的影响。

9. Webster P. For precarious populations, pollutants present new perils[J]. Science, 2003, 299(5613): 1642-1643.

For precarious populations, pollutants present new perils

Webster P.

(Paul Webster is a writer in Moscow)

Abstract: The biggest ever study of Arctic pollutants paints a picture of an ecosystem under siege with potentially grave consequences for denizens of Earth's northernmost reaches. For the first time, long-standing villains such as pesticides, polychlorinated biphenyls, and mercury have been linked to weakened immune systems and developmental deficits in Inuit children.

处于危险中的民众,污染物带来新危险

Webster P.

(莫斯科作家)

摘要:针对北极地区污染物质的调查勾勒出了一幅生态系统遭受围攻的图画,对地球上最北部的居民可能有严重的后果。这是首次发现,农药杀虫剂、多氯联苯以及汞这些持久性污染物与因纽特孩童免疫系统和生长发育的抑制相关。

10. Marika Willerroider,Munich. Roaming polar bears reveal Arctic role of pollutants[J]. Nature,2003,426(6962):5.

Roaming polar bears reveal Arctic role of pollutants

Marika Willerroider, Munich

(University of Salzburg, Salzburg, Austria)

Abstract: Female polar bears have provided researchers with an insight into the long-term health impact of pollutants.

In a study published on 1 November, a Norwegian team reveals that bears that roam long distances in their search for food tend to accumulate relatively high levels of industrial pollutants such as polychlorinated biphenyls (PCBs) in their bodies. This could be linked to the exceptional levels of hermaphroditism seen in some bear populations, although this connection has yet to be confirmed.

PCBs, which were widely used in electrical equipment until they were banned in most countries on health grounds, are extremely resistant to biodegradation. They accumulate in the food chain, even in remote locations, and then accrue at the chain's end — in polar bears, for example, which cannot easily metabolize or excrete the compounds.

The researchers tagged 54 female bears from the Svalbard archipelago and the nearby Barents Sea in the far north of Europe with satellite transmitters, and monitored their movements for several years. They found that some bears in off-coast habitats roam huge territories each year — up to 270000 square kilometres — and accumulate significantly higher levels of PCBs in their fat, blood and milk than bears in smaller coastal or near-coastal habitats (G. H. Olsen, et al. Environ. Sci. Technol. 37, 4919-4924, 2003).

"Long-distance migration costs immense energy," says Øystein Wiig, a mammalogist at the University of Oslo's Zoological Museum and a co-author on the study. "The bears need to consume much more prey, and thus build up more PCBs."

Particularly high pollutant concentrations were found in bears migrating to the Kara Sea, north of Russia. This indicates that there is massive pollution of Siberian rivers from contaminated industrial areas upstream, says Wiig.

PCBs are part of a group of pollutants known as endocrine disruptors, which may interfere with hormone function in humans and animals at relatively low doses. Earlier research suggests that the PCBs could be the cause of sexual anomalies found in polar bears — hermaphroditism has been seen in about 2%

of the roughly 5,000-strong bear population inSvalbard (Ø. Wiig, et al. J. Wildlife Dis. 34, 792-796; 1998).

But the link is tentative. "I'd rather not jump to conclusions. Hermaphrodites have also been found among brown bears that are free of PCBs," says Aaron Fisk, a toxicologist at the University of Georgia in Athens, who investigates pollution in arctic wildlife.

The release of PCBs into the arctic environment is slowly decreasing, says Samantha Smith, who directs the arctic programme of the WWF, an environmental group. But she warns that a second wave of pollutants, now stored in arctic ice reservoirs, may be released if glaciers melt.

迁徙的北极熊揭示北极圈污染物的作用

Marika Willerroider, Munich

(奥地利萨尔茨堡大学)

摘要:研究者们通过雌性北极熊已经可以深入了解到污染物对健康的长期影响。

在 11 月 1 日发表的一项研究中,一个挪威研究组发现,那些需要迁徙较远距离以获取食物的北极熊会在体内积累相对更高水平的污染物,比如多氯联苯。这可能与一些北极熊种群中异常高的雌雄同体水平相关,尽管这种联系还有待证实。多氯联苯在出于健康目的考虑而被禁止之前,被广泛地应用于电子电气设备的生产中,它极其耐受生物降解,可以在食物链中传递(甚至到偏远区域),北极熊作为食物链的末端,难以新陈代谢或排泄这种复合物,因此在体内会大量富集。

研究者利用卫星传输器标记了 54 只来自斯瓦尔巴特群岛和附近的巴伦支海的雌性北极熊,观测它们几年内的活动情况。他们发现,和那些在海岸和近海的小型栖息地之间迁徙的北极熊相比,那些每年迁徙到远离海岸栖息地、穿越高达 270000 平方公里的大陆的北极熊,会在它们的脂肪血液哺乳乳汁中积累更高水平的多氯联苯。

来自奥斯陆大学动物博物馆的哺乳动物学家 Øystein Wiig 是该项研究的合作者,他谈道:"长途的迁徙行为会耗费巨大的能量,因此北极熊会捕食更多的猎物,因此也会积累更多的多氯联苯。研究发现,那些迁徙到俄罗斯北部喀拉海的北极熊体内的污染物浓度非常高,这表明,西伯利亚的河流被上游工业地带给污染了。"

多氯联苯是干扰内分泌系统的一组污染物质,相对较低的剂量就会干扰到人类和动物体内生理激素的正常功效,早期研究表征,在北极熊体内发现的多氯联苯可能导致性行为异常,斯瓦尔巴特群岛上 5000 只北极熊中大概有 2%的雌雄同体现象。

但是这种联系只是初步的,"我并不倾向于立刻得出结论,雌雄同体现象同时还在体内没有多氯联苯的棕熊种群中发现。"来自雅典的佐治亚大学的一名研究北极野生动物的毒理学家 Aaron Fisk 这样评价。

向北极圈中排放多氯联苯的通量在逐渐降低,但是专家仍警告,本储存在北极巨大的天然冰储库的二次污染物就要来了,待到冰川融化时,就会被释放出来重见天日。

11. Webster P. Persistent toxic substances: study finds heavy contamination across vast Russian Arctic [J]. Science, 2004, 306(5703): 1875.

Persistent toxic substances—study finds heavy contamination across vast Russian Arctic

Webster P.

(Paul Webster is a science writer in Toronto, Canada)

Abstract: The first comprehensive look at persistent toxic substances across the Russian Arctic indicates that indigenous peoples there are inordinately exposed to pesticides, industrial compounds, and heavy metals, with uncertain health effects.

研究发现广大的俄罗斯北极地区被持久性有毒物质重度污染

Webster P.

(加拿大多伦多科学作家)

摘要:在俄罗斯北极极地地区第一次全面深入了解持久性有毒物质的研究表明,当地居民都暴露在农业杀虫剂、工业化合物以及重金属污染物之下,并且带有未知的健康风险。

12. Jules M Blais, Lynda E Kimpe, Dominique McMahon, et al. Arctic seabirds transport marine-derived contaminants[J]. Science, 2005, 309(5733): 445.

Arctic seabirds transport marine-derived contaminants

Jules M. Blais, Lynda E. Kimpe, Dominique McMahon, Bronwyn E. Keatley, Mark L. Mallory, Marianne S. V. Douglas, John P. Smol

(Centre for Advanced Research in Environmental Genomics, Department of Biology, University of Ottawa, Ottawa, Ontario, K1N 6N5 Canada)

Abstract: Long-range atmospheric transport of pollutants is generally assumed to be the main vector for arctic contamination, because local pollution sources are rare. We show that arctic seabirds, which occupy high trophic levels in marine food webs, are the dominant vectors for the transport of marine-derived contaminants to coastal ponds. The sediments of ponds most affected by seabirds had 60 times higher DDT, 25 times higher mercury, and 10 times higher hexachlorobenzene concentrations than nearby control sites. Bird guano greatly stimulates biological productivity in these extreme environments but also serves as a major source of industrial and agricultural pollutants in these remote ecosystems.

北极海鸟对海洋源污染物的传输

Jules M. Blais, Lynda E. Kimpe, Dominique McMahon, Bronwyn E. Keatley,
Mark L. Mallory, Marianne S. V. Douglas, John P. Smol

(加拿大渥太华大学生物系)

摘要:由于北极地区的污染源稀少,对污染物的长尺度大气传输一直被认为是北极污染的主要来源途径。我们发现,由于北极海鸟占据海洋食物网中较高的营养级,是将海洋源污染物传输至近海湖泊的主要生物矢量。受海鸟影响最大的湖泊沉积物比附近的对照组 DDT 含量高 60 倍,比汞含量高 25 倍,比六氯代苯含量高 10 倍。在这些极端环境里海鸟粪大大刺激了生物的生产力,同时也成为了工农业污染物输入到偏远生态系统的主要来源。

13. Kathy S Law, Andreas Stohl. Arctic air pollution: origins and impacts[J]. Science, 2007, 315(5818): 1537-1540.

Arctic air pollution: origins and impacts

Kathy S. Law, Andreas Stohl

(Service d' Aéronomie, CNRS, IPSL/Université Pierre et Marie Curie, Boitê 102, 4 Place Jussieu, Paris Cedex 05, 75252 France)

Abstract: Notable warming trends have been observed in the Arctic. Although increased human-induced emissions of long-lived greenhouse gases are certainly the main driving factor, air pollutants, such as aerosols and ozone, are also important. Air pollutants are transported to the Arctic, primarily from Eurasia, leading to high concentrations in winter and spring (Arctic haze). Local ship emissions and summertime boreal forest fires may also be important pollution sources. Aerosols and ozone could be perturbing the radiative budget of the Arctic through processes specific to the region: absorption of solar radiation by aerosols is enhanced by highly reflective snow and ice surfaces; deposition of light-absorbing aerosols on snow or ice can decrease surface albedo; and tropospheric ozone forcing may also be contributing to warming in this region. Future increases in pollutant emissions locally or in mid-latitudes could further accelerate global warming in the Arctic.

北极大气污染:来源与影响

Kathy S. Law, Andreas Stohl

(法国巴黎第六大学国家科学研究中心)

摘要:北极地区已经观测到显著的变暖趋势,虽然日益增长的人类排放的长效温室气体是北极变暖的主要驱动力,但同时诸如气溶胶、臭氧等大气污染的作用也同样十分重要。主要从亚欧大陆运送而来的大气污染物会导致春季冬季污染物浓度达到峰值——我们称之为北极霾。另外,当地船只排放物和夏季的北方森林大火也是重要的污染源。气溶胶和臭氧可以通过针对区域性的特殊方式来扰乱北极地区的辐射通量——高反射率的冰雪覆盖面积的增加会提高气溶胶对太阳辐射的吸收,而能吸收光的气溶胶在冰雪上面的沉降会降低冰雪反射率,同时对流层臭氧强迫也会促进这一地区的变暖。在未来,局地污染物排放的增加以及中纬度的排放都会加强北极地区的全球变暖速度。

14. Joseph R McConnell, Ross Edwards, Gregory L Kok, et al. 20th-century industrial black carbon emissions altered Arctic climate forcing[J]. Science, 2007, 317(5843): 1381-1384.

20th-century industrial black carbon emissions altered Arctic climate forcing

Joseph R. McConnell, Ross Edwards, Gregory L. Kok, Mark G. Flanner, Charles S. Zender, Eric S. Saltzman, J. Ryan Banta, Daniel R. Pasteris, Megan M. Carter, Jonathan D. W. Kahl

(Desert Research Institute, Nevada System of Higher Education, Reno, NV 89512, USA)

Abstract: Black carbon (BC) from biomass and fossil fuel combustion alters chemical and physical properties of the atmosphere and snow albedo, yet little is known about its emission or deposition histories. Measurements of BC, vanillic acid, and non-sea-salt sulfur in ice cores indicate that sources and concentrations of BC in Greenland precipitation varied greatly since 1788 as a result of boreal forest fires and industrial activities. Beginning about 1850, industrial emissions resulted in a sevenfold increase in ice-core BC concentrations, with most change occurring in winter. BC concentrations after about 1951 were lower but increasing. At its maximum from 1906 to 1910, estimated surface climate forcing in early summer from BC in Arctic snow was about 3 watts per square meter, which is eight times the typical preindustrial forcing value.

20世纪工业黑炭的排放改变了北极地区的气候效应

Joseph R. McConnell, Ross Edwards, Gregory L. Kok, Mark G. Flanner, Charles S. Zender, Eric S. Saltzman, J. Ryan Banta, Daniel R. Pasteris, Megan M. Carter, Jonathan D. W. Kahl

(美国内华达州高等教育体系沙漠研究所)

摘要:来自生物体和化石燃料燃烧产生的黑炭能够改变大气的物理化学性质和冰雪反射率,然而关于它的排放和沉降我们知之甚少。冰芯中记载的黑炭的测量数据、香草酸以及非海盐硫等都表明了格陵兰沉积中的黑炭的主要来源和浓度自1788年发生过大的变化,原因可能是由于北方森林的火灾和工业活动的日益频繁。从大约1850年开始,工业排放导致了冰芯中黑炭浓度7倍的增加,而且大部分变化发生在冬季。1951年之后黑炭浓度较低,但仍增加,峰值在1906~1910年之间达到,估计那时的北极圈地表气候营力在夏季初期达到过3 W/m^2,为工业化以前典型营力的8倍之多。

15. Thingstad T F, Bellerby R G J, Bratbak G, et al. Counterintuitive carbon-to-nutrient coupling in an Arctic pelagic ecosystem[J]. Nature, 2008, 455(7211): 387-390.

Counterintuitive carbon-to-nutrient coupling in an Arctic pelagic ecosystem

Thingstad T. F., Bellerby R. G. J., Bratbak G., Børsheim K. Y., Egge J. K., Heldal M., Larsen A., Neill C., Nejstgaard J., Norland S., Sandaa R.-A., Skjoldal E. F., Tanaka T., Thyrhaug R., Töpper B.

(Department of Biology, University of Bergen, Jahnebakken 5PO Box 7800, 5020 Bergen, Norway)

Abstract: Predicting the ocean's role in the global carbon cycle requires an understanding of the stoichiometric coupling between carbon and growth-limiting elements in biogeochemical processes. A recent addition to such knowledge is that the carbon/nitrogen ratio of inorganic consumption and release of dissolved organic matter may increase in a high CO_2 world. This will, however, yield a negative feedback on atmospheric CO_2 only if the extra organic material escapes mineralization within the photic zone. Here we show, in the context of an Arctic pelagic ecosystem, how the fate and effects of added degradable organic carbon depend critically on the state of the microbial food web. When bacterial growth rate was limited by mineral nutrients, extra organic carbon accumulated in the system. When bacteria were limited by organic carbon, however, addition of labile dissolved organic carbon reduced phytoplankton biomass and activity and also the rate at which total organic carbon accumulated, explained as the result of stimulated bacterial competition for mineral nutrients. This counterintuitive "more organic carbon gives less organic carbon" effect was particularly pronounced in diatom-dominated systems where the carbon/ mineral nutrient ratio in phytoplankton production was high. Our results highlight how descriptions of present and future states of the oceanic carbon cycle require detailed understanding of the stoichiometric coupling between carbon and growth-limiting mineral nutrients in both autotrophic and heterotrophic processes.

在北极浮游生态系统营养物质和碳的反常耦合

Thingstad T. F., Bellerby R. G. J., Bratbak G., Børsheim K. Y., Egge J. K., Heldal M., Larsen A., Neill C., Nejstgaard J., Norland S., Sandaa R.-A., Skjoldal E. F., Tanaka T., Thyrhaug R., Töpper B.

(挪威卑尔根大学生物系)

摘要:要想评价海洋在全球碳循环过程中起到的作用,就需要我们对生物地球化学循环中碳和生长限制因子之间的化学计量耦合有深入了解。一个新的认识是在高二氧化碳浓度下,无机代谢中的C/N比值以及可溶解有机物的释放会升高。这样,如果富余的有机物质能在透光区中逃过矿化作用,就会产生一个对大气二氧化碳的负反馈。在北极浮游生态系统的背景下,富余的可降解有机碳的归宿和效应依赖于微生物食物网状态。当细菌的生长速率被矿物营养物质限制,多余的有机碳就会在系统中富集。但在另一方面,如果细菌的生长被有机碳限制,不稳定的可溶解有机碳会减少浮游生物生物量和活动,以及总有机碳的积累速率,这被解释为受刺激的微生物竞争矿物营养物质的结果。这种反常的有机碳越多导致有机碳越少的现象,在以硅藻为主的生态系统中呈现得非常显著,因为这种生态系统中代表浮游生物生产力的C/N比一般较高。总之,我们的结论说明了,为描述现在和未来海洋碳循环机制的状态,就需要对碳和矿物质营养物质生长限制因素间的化学计量耦合深入了解,无论是自养过程还是异养过程。

九、天文、空间物理类文献

9.1 分析概述

1990～2010年，*Nature* 和 *Science* 上共有“天文、空间物理类”文献21篇，其中南极方面文献有6篇，北极方面文献有3篇，南北极均包括的文献有12篇。

将文章数量分区域并按照年代进行分类(图9.1)可以发现，两刊物上每年一般发表天文、空间方面的文章1篇，2002年涉及两极的文章较多，为4篇。1990～1991年、1996年、1999年以及2006～2010年未见两刊物有关于天文、空间的文章。

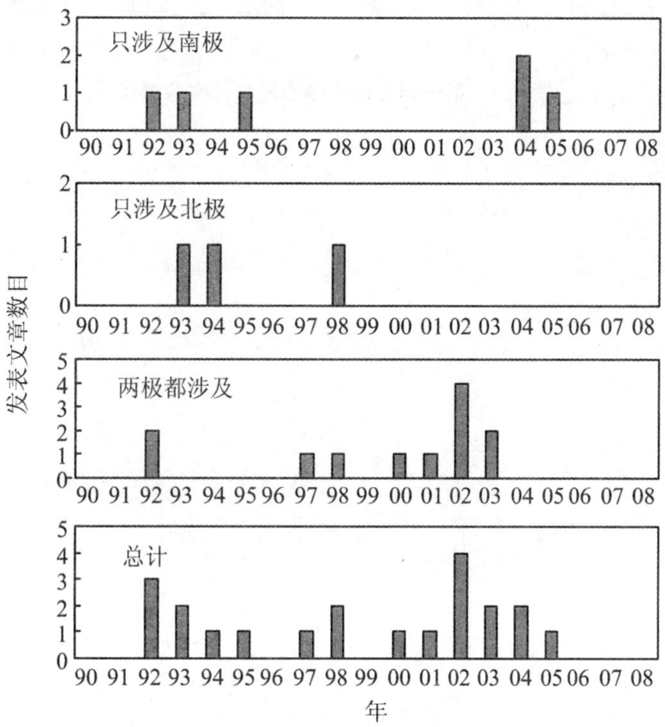

图9.1 文章数量随年代的变化

在21篇文献中，关于陨石的有2篇，关于地球两极地磁场的有12篇，关于太阳对地球两极作用的有3篇，关于地球两极中高层大气的有2篇。从这21篇文献的类别可以看出，极区空间天文的研究重心是两极的地磁场，古地磁记录、地磁场与星际物质相互作用过程为研究热点，发表论文数量最多。极区天文、空间领域的研究对南极和北极的区分不明显，多为同时讨论两极的共同情况而不是单一讨论南极或者北极。由于地磁场是偶极场，研究地磁场时往往需要考虑其对称结构，如研究地磁场时，关于古地磁倒转是从南极和北极地磁记录中提取的信息，单一的南极或者北极信息不够全面。而有关地磁场产生的发电机理论，则更需要同时考虑地球两极的情况。文献中的极区中高层大气现象、太阳对地球作用也是同时讨论了南极和北极。极区大气中层顶有云区，与陨石以及水汽输送情况有关。极区这些自然现象基本上都是同时出现在地球南北两极的。

将文章数量按所署第一单位所在国家进行排列(图9.2)，美国发表的文章数最多，高达9篇，其次为英国，有6篇，日本和法国各有2篇，丹麦和澳大利亚各1篇。

美国和英国在极区天文、空间方面发表的 *Nature*，*Science* 文章数量最多，美国在南极、北极也建立了专门研究天文、空间的观测站，可见其在南极、北极开展的科学研究工作较为深入。欧洲和美国还发射了很多极区观测卫星，如AIM卫星专门观测地球两极中高层大气密度成分和波动。日本也在南极建立了天文、空间的考察站，近来也有了很多科研成果。丹麦和澳大利亚本身接近极区，开展极地研究工作比较容易，发表关于极地的高水平文章不足为奇。中国的子午工程南极站点还在不断完善之中，但离世界领先水平还有很大差距，暂无可用的南极、北极观测数据。

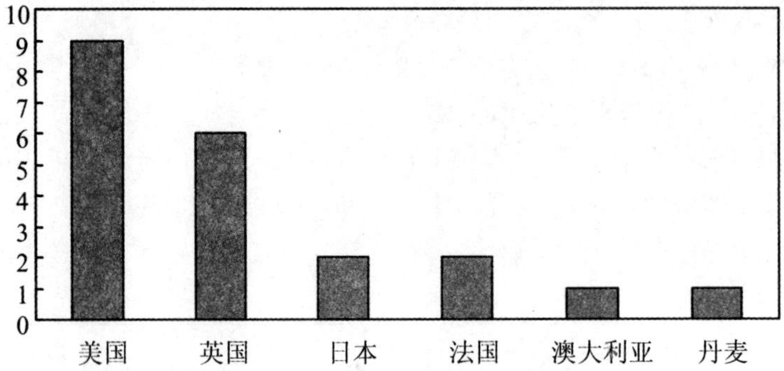

图 9.2 第一作者所在国家发表的文章数量

9.2 摘要翻译

1. Benoit P H, Sears D W G. The breakup of a meteorite parent body and the delivery of meteorites to Earth[J]. Science, 1992, 255(5052): 1685-1687.

The breakup of a meteorite parent body and the delivery of meteorites to Earth

Benoit P. H., Sears D. W. G.

(Cosmochemistry Group, Department of Chemistry and Biochemistry, University of Arkansas, Fayetteville, AR 72701)

Abstract: Whether many of the 10000 meteorites collected in the Antarctic are unlike those falling elsewhere is contentious. The Antarctic H chondrites, one of the major classes of stony meteorites, include a number of individuals with higher induced thermoluminescence peak temperatures than observed among non-Antarctic H chondrites. The proportion of such individuals decreases with the mean terrestrial age of the meteorites at the various ice fields. These H chondrites have cosmic-ray exposure ages of about 8 million years, experienced little cosmic-ray shielding, and suffered rapid postmetamorphic cooling. Breakup of the H chondrite parent body, 8 million years ago, may have produced two types of material with different size distributions and thermal histories. The smaller objects reached Earth more rapidly through more rapid orbital evolution.

一个陨石母体的解体及到达地球

Benoit P. H., Sears D. W. G.

(美国阿肯色州立大学化学与生物化学系宇宙化学组)

摘要:在南极收集到的10000多颗陨石中,与其他地方下沉的陨石不一样是有争论的。在南极石质陨石的主要类型之一H球粒陨石,包含了一些比非南极的H球粒陨石更高的致热释光峰。这些个体的比例随着它们在不同冰原上暴露的年龄而逐步减少。这些H球粒陨石拥有约800万年宇宙射线暴露年龄,没有经历多少宇宙射线屏蔽,并遭受快速后期变质作用冷却。破碎的H球粒陨石母体,在800万年前,可能产生了两种具有不同尺寸分布和热历史的材料。更小的物体因快速的轨道旋转可更快地到达地球。

2. Hoffman K A. Dipolar reversal states of the geomagnetic-field and core mantle dynamics[J]. Nature, 1992, 359(6398): 789-794.

Dipolar reversal states of the geomagnetic-field and core mantle dynamics

Hoffman K. A.

(Physics Department, California Polytechnic State University, San Luis Obispo California, USA)

Abstract: Palaeomagnetic records from lavas suggest that at least two specific inclined dipolar field configurations have dominated the reversal process for the past ten million years. These long-lived states provide a way to explain directional rebounds, aborted reversals and the recording in sediments of what appear to be preferred longitudinal paths of the virtual geomagnetic pole. The polar orientations correlate with near-radial flux concentrations recognizable when today's field is stripped of its axial dipole, and with lower-mantle seismic anomalies, suggesting a tie to deep-Earth dynamics.

地磁场两极反转和核幔对流

Hoffman K. A.

(美国加州波利技术州立大学)

摘要:从熔岩古地磁资料显示,在过去1000万年的地磁场两极反转过程中,至少有两个特定倾斜偶极场组态起主导作用。这些长期状态提供了一种方法来解释方向反弹、中止逆转,以及在沉积物中的可能是虚地磁极的纵向路径记录。极地取向与将辐射状偶极从当今磁场拿走后可识别的近辐射状通量浓度相关良好,也与低地幔地震异常相关,表明与地球深部的动力学相关联。

3. Valet J P, Tucholka P, Courtillot V, et al. Paleomagnetic constraints on the geometry of the geomagnetic-field during[J]. Nature,1992,356 (6368):400-407.

Paleomagnetic constraints on the geometry of the geomagnetic-field during

Valet J. P. , Tucholka P. , Courtillot V. , Meynadier L.

(Laboratoire de Géophysique, Université Paris-Sud, Bâtiment 504, 91405 Orsay, France)

Abstract: Palaeomagnetic records of the path of the pole during reversals of the Earth's magnetic field provide a test of the hypothesis that dipolar or low-order axisymmetric components of the field dominate during reversals. Multiple records of reversals during the past 12 Myr show no simple or consistent geographical pattern. Although a more robust analysis of the transitional field awaits a greater number of well-distributed sampling sites, the present data are not inconsistent with the simplest models, in which a field reminiscent of the non-dipole component of the present-day field becomes dominant.

古地磁在地磁场倒转时的几何约束条件

Valet J. P. , Tucholka P. , Courtillot V. , Meynadier L.

(法国地球物理研究所)

摘要:地磁场的极倒转路径记录对这个假设:偶极或低位轴对称场在地磁反转过程中是否占主导地位,提供了一个检验。过去1200万年的多个逆转记录没有显示简单或一致的地理模式。尽管对过渡场的更强有力分析需要大量分布的取样点,现代的数据并不与这些最简单的模型相矛盾,在这些简单模型中,当今磁场的非偶极组分也可以占主导地位。

4. Gubbins D,Coe R S. Longitudinally confined geomagnetic reversal paths from Nondipolar transition fields[J]. Nature,1993,362(6415):51-53.

Longitudinally confined geomagnetic reversal paths from Nondipolar transition fields

Gubbins D. , Coe R. S.

(Department of Earth Sciences, Leeds University, Leeds LS2 9JT, UK)

Abstract: It has long been thought that conditions at the boundary between the core and mantle influence the Earth's magnetic field, but the supporting evidence is rather indirect. Recent palaeomagnetic results, suggesting that there are persistent preferred longitudinal paths for the virtual geomagnetic pole (VGP) during reversals, would provide the first direct evidence of the solid mantle's influence on the core, although their statistical significance has been disputed. The results are potentially exciting because the preferred paths lie close to the Pacific rim, where the present geomagnetic secular variation changes character. Here we present a simple model, based on an extension of a previous theory, that produces reversals with VGP paths confined within relatively narrow longitude bands despite the transition field having a substantially non-dipolar structure. Thus, although longitude bias of the VGP paths is definitive evidence for core-mantle interaction, simple VGP paths are not evidence of near-dipolar transition fields.

来自 Nondipolar 过渡区域的经向封闭地磁倒转

Gubbins D. , Coe R. S.

(英国利兹大学地球科学系)

摘要:长期以来,科学家一直认为,在核心和地幔之间的边界条件影响地球的磁场,但是支持的证据是很间接的。最近的古地磁结果表明,在地磁倒转时,虚拟地磁极(VGP)有持续优先的经向通道,这为固体地幔对核心的影响提供了直接的证据,尽管其统计学显著性有争论。结果是令人兴奋的,因为首选路径多集中在环太平洋地区,目前的地磁变化在那里变更。这里,基于对以前理论的延伸,我们提出了一个简单的模型,当 VGP 路径限制在相对狭窄的范围内时,尽管过渡区域有显著的非偶极场,也能产生地磁倒转。因此,VGP 路径的经度偏差是核幔相互作用的确切证据,简单 VGP 路径不是近偶极核幔相互作用的证据,也不是近偶极过渡区域的证据。

5. Gubbins D,Kelly P. Persistent patterns in the geomagnetic-field over the past 2.5 Myr[J]. Nature, 1993,365(6449):829-832.

Persistent patterns in the geomagnetic-field over the past 2.5 Myr

Gubbins D., Kelly P.

(Department of Earth Sciences, Leeds University, Leeds LS2 9JT, UK)

Abstract: Historical geomagnetic measurements covering the past 400 years reveal a symmetrical pattern of four relatively stationary flux concentrations (lobes) at the surface of the liquid core and regions of rapid change extending from the Atlantic to the Indian Ocean. Palaeomagnetic data define a time-average over several thousand years which might reflect the stationary parts of the present field, but unfortunately the historical record is too short to provide a satisfactory average. Here we model palaeomagnetic directions from the past 2.5 Myr using the same methods as for modern data, and find the two northern lobes in the same position as today, over Arctic Canada and Siberia. In southern regions, by contrast, the field appears to have been smoothed out, as might be expected from the current rapid secular variation. We propose that the present geomagnetic field morphology and pattern of secular variation have persisted for several million years, as would occur if the solid mantle controls flow at the top of the core.

过去250万年的地磁场的持续模式

Gubbins D., Kelly P.

(英国利兹大学地球科学系)

摘要:过去400年的地磁测量揭示了4个在液体的核心快速从大西洋延伸到印度洋区域相对静态的流量浓度的对称格式。古地磁数据定义在几千年的时间的平均值,这可能反映了当前磁场的静态部分,但不幸的是历史记录太短,无法提供令人满意的均值。在这里,我们使用现代数据方法,对过去的250万年古地磁方向进行模拟,在今天的加拿大北极地区和西伯利亚相同的位点,找到两个北方的流量浓度场。相比之下在南部地区,因当前快速的变化,磁场似乎已经被同化了。目前的地磁场长期变化的形态和格局已经持续了数百万年,固体地幔控制其在地核上的流通。

6. Laxon S,Mcadoo D. Arctic Ocean gravity field derived from Ers-1 satellite altimetry[J]. Science, 1994,265(5172):621-624.

Arctic Ocean gravity field derived from Ers-1 satellite altimetry

Laxon S., Mcadoo D.

(Mullard Space Science Laboratory, Department of Space and Climate Physics, University College London, Holmbury Saint Mary, Dorking, Surrey RH5 6NT, UK)

Abstract: The derivation of a marine gravity field from satellite altimetry over permanently ice-covered regions of the Arctic Ocean provides much new geophysical information about the structure and development of the Arctic sea floor. The Arctic Ocean, because of its remote location and perpetual ice cover, remains from a tectonic point of view the most poorly understood ocean basin on Earth. A gravity field has been derived with data from the ERs-1 radar altimeter, including permanently ice-covered regions. The gravity field described here clearly delineates sections of the Arctic Basin margin along with the tips of the Lomonosov and Arctic mid-ocean ridges. Several important tectonic features of the Amerasia Basin are clearly expressed in this gravity field. These include the Mendeleev Ridge; the Northwind Ridge; details of the Chukchi Borderland; and a north-south trending, linear feature in the middle of the Canada Basin that apparently represents an extinct spreading center that "died" in the Mesozoic. Some tectonic models of the Canada Basin have proposed such a failed spreading center, but its actual existence and location were heretofore unknown.

卫星测量得到的北冰洋重力场

Laxon S. , Mcadoo D.

(英国伦敦大学学院空间和气候物理系 Mullard 太空科学实验室)

摘要:北冰洋的永久冰雪覆盖的地区,一个由卫星测高仪得到的海洋重力场,为北极海底的结构和开发,提供了许多新的地球物理信息。北冰洋,由于位置偏远和为永久冰层覆盖,从板块的角度来看,仍是地球上了解最少的海洋盆地。在包括永久冰雪覆盖的地区,从 ERs-1 号雷达高度计数据,可导出其重力场。这里所描述的重力场,包括北极盆地随着北极的罗蒙诺索夫和洋中脊的尖端边缘部分。这个地球重力场明确地显示了美亚盆地的几个重要板块构造。这些包括门捷列夫洋脊、Northwind 洋脊、楚科奇边缘海和在加拿大盆地中部南北向的线性特征,显然代表了一个在中生代已死亡的传播中心。一些加拿大海盆的构造模式提出这样的传播中心,但其是否实际存在和位置迄今不明。

7. Cande S C,Raymond C A,Stock J,et al. Geophysics of the pitman fracture zone and Pacific-Antarctic plate motions during the cenozoic[J]. Science,1995,270(5238):947-953.

Geophysics of the pitman fracture zone and Pacific-Antarctic plate motions during the cenozoic

Cande S. C. , Raymond C. A. , Stock J. , Haxby W. F.

(Scripps Institution of Oceanography , La Jolla, California 92093, USA)

Abstract: Multibeam bathymetry and magnetometer data from the Pitman fracture zone (FZ) permit construction of a plate motion history for the South Pacific over the past 65 million years. Reconstructions show that motion between the Antarctic and Bellingshausen plates was smaller than previously hypothesized and ended earlier, at chron C27 (61 million years ago). The fixed hot-spot hypothesis and published paleomagnetic data require additional motion elsewhere during the early Tertiary, either between East Antarctica and West Antarctica or between the North and South Pacific. A plate reorganization at chron C27 initiated the Pitman FZ and may have been responsible for the other right-stepping fracture zones along the ridge. An abrupt (8 degrees) clockwise rotation in the abyssal hill fabric along the Pitman flowline near the young end of chron C3a (5. 9 million years ago) dates the major change in Pacific-Antarctic relative motion in the late Neogene.

新生代皮特曼断裂区及太平洋-南极板块运动的地球物理学

Cande S. C., Raymond C. A., Stock J., Haxby W. F.

(美国加州大学圣地亚哥分校斯克里普斯海洋研究所)

摘要:从皮特曼断裂带(FZ)多波束测深和磁强计数据可以重建过去 6500 万年南太平洋板块运动的历史。重建表明,别林斯高普与南极板块运动小于原先的假设,并在 6100 万年前结束(chron C27)。固定热点假说和已有的古地磁数据表明在第三纪期间,无论是东西南极洲之间,或南北太平洋之间,都存在额外的运动。在 C27 发生的板块重组引发了皮特曼 FZ,并可能导致沿洋脊的断裂带。在 C3a 结束期的早期(590 万年前),沿皮特曼流线的深海山突然发生的顺时针转动(8°)标志在太平洋-南极新近纪末期相对运动的重大变革。

8. Moore T E, Chappell C R, Chandler M O, et al. High-altitude observations of the polar wind[J]. Science, 1997, 277(5324): 349-351.

High-altitude observations of the polar wind

Moore T. E., Chappell C. R., Chandler M. O., Craven P. D., Giles B. L., Pollock C. J., Burch J. L., Young D. T., Waite J. H., Nordholt J. E., Thomsen M. F., McComas D. J., Berthelier J. J., Williamson W. S., Robson R., Mozer F. S.

(NASA Marshall Space Flight Center, Huntsville, AL, USA)

Abstract: Plasma outflows, escaping from Earth through the high-altitude polar caps into the tail of the magnetosphere, have been observed with a xenon plasma source instrument to reduce the floating potential of the Polar spacecraft. The largest component of H^+ flow, along the local magnetic field (30 to 60 kilometers per second), is faster than predicted by theory. The flows contain more O^{2+} than predicted by theories of thermal polar wind but also have elevated ion temperatures. These plasma outflows contribute to the plasmas energized in the elongated nightside tail of the magnetosphere, creating auroras, substorms, and storms. They also constitute an appreciable loss of terrestrial water dissociation products into space.

高纬观测的极地风

Moore T. E., Chappell C. R., Chandler M. O., Craven P. D., Giles B. L., Pollock C. J., Burch J. L., Young D. T., Waite J. H., Nordholt J. E., Thomsen M. F., McComas D. J., Berthelier J. J., Williamson W. S., Robson R., Mozer F. S.

(美国航空和航天局 Marshall 太空飞行中心)

摘要:等离子外流,从地球高纬度极帽逃逸到磁层尾,已为氙离子源仪器观测到,以减少极地飞船的潜在浮力。沿当地磁场氢离子流速的最大的组分(每秒 30～60 km),比理论预计的要快。氢离子流中含有的氧离子比热极地风理论预期的多,其氧离子温度也升高。这些等离子流出在磁层狭长的阴面尾部,激发其等离子形成极光、亚暴和磁暴。它们也造成了相当可观的地球上水分解产物流失至太空。

9. Hartley D E, Villarin J T, Black R X, et al. A new perspective on the dynamical link between the stratosphere and troposphere[J]. Nature, 1998, 391(6666): 471-474.

A new perspective on the dynamical link between the stratosphere and troposphere

Hartley D. E., Villarin J. T., Black R. X., Davis C. A.

(Georgia Institute of Technology, Atlanta, Georgia 30332-0340, USA)

Abstract: Atmospheric processes of tropospheric origin can perturb the stratosphere, but direct feedback in the opposite direction is usually assumed to be negligible, despite the troposphere's sensitivity to changes in the release of wave activity into the stratosphere. Here, however, we present evidence that such a feedback exists and can be significant. We find that if the wintertime Arctic polar stratospheric vortex is distorted, either by waves propagating upward from the troposphere or by eastward-travelling stratospheric waves, then there is a concomitant redistribution of stratospheric potential vorticity which induces perturbations in key meteorological fields in the upper troposphere. The feedback is large despite the much greater mass of the troposphere: it can account for up to half of the geopotential height anomaly at the tropopause. Although the relative strength of the feedback is partly due to a cancellation between contributions to these anomalies from lower altitudes, our results imply that stratospheric dynamics and its feedback on the troposphere are more significant for climate modelling and data assimilation than was previously assumed.

对平流层和对流层动力连接的一个新看法

Hartley D. E., Villarin J. T., Black R. X., Davis C. A.

(美国佐治亚技术研究院)

摘要:来自对流层的大气过程可以干扰平流层,但在相反方向的直接反馈通常被认为是微不足道的,尽管对流层对重力波释放至平流层的变化敏感。在这里,我们的证据表明,这种反馈的存在,可能会显著。我们发现,如果冬季北极平流层旋涡,被对流层上升的波或被在平流层向东传播的波所扭曲后,会伴随着平流层位涡再分配,并诱发在对流层上层的关键气象场的扰动。尽管对流层的物质量要大得多,但是这种反馈仍然是很大的:在对流层顶它可以占有位势高度异常的一半。尽管反馈的相对强度部分原因是由于低海拔地区异常之间的抵消作用造成的,我们的研究结果意味着,平流层动力学对对流层反馈对气候模拟以及数据分析比以前认为的更重要。

10. Pepin R O. Isotopic evidence for a solar argon component in the Earths mantle[J]. Nature, 1998, 394 (6694): 664-667.

Isotopic evidence for a solar argon component in the Earths mantle

Pepin R. O.

(School of Physics and Astronomy, University of Minnesota, Minneapolis, Minnesota 55455, USA)

Abstract: Determining the presence of solar argon, krypton and xenon in the Earth's mantle is important for understanding the source, incorporation mechanism and transport of noble gases in the Earth, as well as the evolutionary history of the Earth's atmosphere. There are strong indications in the mid-ocean ridge basalt database that solar helium and neon are indeed present, and modelling exercises indicate that the compositions of all five noble gases in the Earth's primordial inventory were solar-like. But solar isotopic signatures of the heavier noble gases argon and xenon, which differ significantly from atmospheric compositions, have appeared only subtly if at all in analyses of mantle-derived samples-their non-radiogenic isotope ratios are generally found to be indistinguishable or only slightly different from those in the atmosphere. The first promising isotopic evidence for a solar-like argon component in the Earth's mantle appeared in a recent analysis of basalt glasses from the Hawaiian Loihi seamount. Here I show that recent measurements of neon and argon isotopes in a suite of mid-ocean ridge basalt samples from the southern East Pacific Rise greatly strengthen the case for the presence of solar argon, and by inference krypton and xenon, in the Earth's mantle.

同位素证据显示地幔中存在来自太阳的氩

Pepin R. O.

(美国明尼苏达大学天体物理学院)

摘要:确定太阳氩、氪和氙气在地球地幔的存在对了解其来源、渗入机理和惰性气体在地球中的传输,以及地球大气层的进化史是重要的。在大洋中脊玄武岩数据库中有强的迹象表明,来自太阳的氦、氖确实存在,模拟实验表明,在地球的原始库存中的所有 5 个稀有气体成分与太阳类似。但太阳来源重惰性气体氩气和氙气,其同位素标志与来自大气的差异显著,在地幔源样品的分析中仅出现了微妙的相似,其非放射同位素比值与大气来源的没有区别或只是略有不同。在最近对夏威夷 Loihi 海山的玄武岩分析中,找到了在地球的地幔中存在太阳来源氩的第一个有希望的同位素证据。在这里,我最近分析了来自南东太平洋海隆,洋中脊玄武岩一组样品,对氖、氩同位素的测量大大强化了太阳氩气存在的证据,同样可推断太阳氪气和氙气的存在。

11. Gee J S, Cande S C, Hildebrand J A, et al. Geomagnetic intensity variations over the past 780 kyr obtained from near-seafloor magnetic anomalies[J]. Nature, 2000, 408(6814): 827-832.

Geomagnetic intensity variations over the past 780 kyr obtained from near-seafloor magnetic anomalies

Gee J. S., Cande S. C., Hildebrand J. A., Donnelly K., Parker R. L.

(Scripps Institution of Oceanography, La Jolla, California 92093, USA)

Abstract: Knowledge of past variations in the intensity of the Earth's magnetic field provides an important constraint on models of the geodynamo. A record of absolute palaeointensity for the past 50 kyr has been compiled from archaeomagnetic and volcanic materials, and relative palaeointensities over the past 800 kyr have been obtained from sedimentary sequences. But a long-term record of geomagnetic intensity should also be carried by the thermoremanence of the oceanic crust. Here we show that near-seafloor magnetic anomalies recorded over the southern East Pacific Rise are well correlated with independent estimates of geomagnetic intensity during the past 780 kyr. Moreover, the pattern of absolute palaeointensity of seafloor glass samples from the same area agrees with the well-documented dipole intensity pattern for the past 50 kyr. A comparison of palaeointensities derived from seafloor glass samples with global intensity variations thus allows us to estimate the ages of surficial lava flows in this region. The record of geomagnetic intensity preserved in the oceanic crust should provide a higher-time-resolution record of crustal accretion processes at mid-ocean ridges than has previously been obtainable.

从近海底磁异常得到的在过去 78 万年的地磁强度变化

Gee J. S., Cande S. C., Hildebrand J. A., Donnelly K., Parker R. L.

(美国加州大学圣地亚哥分校斯克里普斯海洋研究所)

摘要:关于过去地球磁场强度变化的知识是地球磁场发电机模型的重要制约因素。从古地磁和火山物质,已编制出过去 5 万年里地球磁场的绝对记录,从沉积序列也获得过去 80 万年的相对古地磁强度记录。但是,地磁强度长期记录也应保存在大洋地壳的热剩磁里。在这里,我们显示,在南部的东太平洋海隆,海底磁异常在过去 78 万年与独立的磁场强度的估计吻合良好。此外,在过去 5 万年,海底玻璃样品的绝对古地磁强度的模式与来自同一地区的偶极强度模式相关。比较从海底玻璃样品得到古地磁强度与全球的地磁强度变化,使我们可以估算该地区的地表熔岩流的年龄。保存在大洋地壳中的地磁强度,为大洋中脊的地壳增生过程提供一个高时间分辨率的记录。

12. Pearce J A, Leat P T, Barker P F, et al. Geochemical tracing of Pacific-to-Atlantic upper-mantle flow through the Drake passage[J]. Nature, 2001, 410(6827): 457-461.

Geochemical tracing of Pacific-to-Atlantic upper-mantle flow through the Drake passage

Pearce J. A., Leat P. T., Barker P. F., Millar I. L.

(Department of Earth Sciences, Cardiff University, Cardiff, CF10 3YE, UK)

Abstract: The Earth's convecting upper mantle can be viewed as comprising three main reservoirs, beneath the Pacific, Atlantic and Indian oceans. Because of the uneven global distribution and migration of ridges and subduction zones, the surface area of the Pacific reservoir is at present contracting at about 0.6 $km^2 \cdot yr^{-1}$, while the Atlantic and Indian reservoirs are growing at about 0.45 $km^2 \cdot yr^{-1}$ and 0.15 $km^2 \cdot yr^{-1}$, respectively. Garfunkel and others have argued that there must accordingly be net mantle flow from the Pacific to the Atlantic and Indian reservoirs (in order to maintain mass balance), and Alvarez further predicted that this flow should be restricted to the few parts of the Pacific rim (here termed "gateways") where there are no continental roots or subduction zones that might act as barriers to shallow mantle flow. The main Pacific gateways are, according to Alvarez, the southeast Indian Ocean, the Caribbean Sea and the Drake passage. Here we report geochemical data which confirm that there has been some outflow of Pacific mantle into the Drake passage-but probably in response to regional tectonic constraints, rather than global mass-balance requirements. We also show that a mantle domain boundary, equivalent to the Australian-Antarctic discordance, must lie between the Drake passage and the east Scotia Sea.

地球化学追踪通过德雷克海峡从太平洋到大西洋的上地幔流

Pearce J. A., Leat P. T., Barker P. F., Millar I. L.

(英国威尔士加迪夫大学地球科学系)

摘要:地球的上地幔对流由太平洋、大西洋和印度洋下方三个主要库组成。由于洋脊和俯冲带在全球性不均匀分布和迁移的太平洋库的面积目前以每年 0.6 km^2 的速度收缩,而大西洋和印度洋库每年约增 0.45 km^2 和 0.15 km^2。加芬克尔等人认为,从太平洋库,地幔净流入大西洋和印度洋库(为了保持质量平衡),阿尔瓦雷斯进一步预测,这个幔流应限于少数太平洋边缘地区(这里称为"门户"),那里没有可能会障碍浅地幔流的大陆根部或俯冲。主要的太平洋出入口,根据阿尔瓦雷斯预测,有东南印度洋、加勒比海和德雷克海峡。在这里,我们报告地球化学数据,证实太平洋地幔库确实经德雷克海峡流出,不过是因为区域板块限制,而不是全球质量平衡的要求。我们的报告还表明,地幔域边界,相当于澳大利亚南极不整合带,位于德雷克海峡和东斯科舍海之间。

13. Li J H, Sato T, Kageyama A. Repeated and sudden reversals of the dipole field generated by a spherical dynamo action[J]. Science, 2002, 295(5561): 1887-1890.

Repeated and sudden reversals of the dipole field generated by a spherical dynamo action

Li J. H., Sato T., Kageyama A.

(Department of Fusion Science, The Graduate University for Advanced Studies, Toki, 509-5292, Japan)

Abstract: Using long-duration, three-dimensional magnetohydrodynamic simulation, we found that the magnetic dipole field generated by a dynamo action in a rotating spherical shell repeatedly reverses its polarity at irregular intervals (that is, punctuated reversal). Although the total convection energy and magnetic energy alternate between a high-energy state and a low-energy state, the dipole polarity can reverse only at high-energy states where the north-south symmetry of the convection pattern is broken and the columnar vortex structure becomes vulnerable. Another attractive finding is that the quadrupole mode grows, exceeding the dipole mode before the reversal; this may help to explain how Earth's magnetic field reverses.

球形发电机产生的磁极反复而快速的反转

Li J. H., Sato T., Kageyama A.

（日本综合研究大学聚变科学系）

摘要：使用长时间，三维磁流体模拟，我们发现，在球壳中，发电机旋转所产生的磁偶极子场是不规则的，反复的逆转其极性（即间断逆转）。虽然总对流能量和磁场能量在高能量状态和低能量状态之间交替，偶极反转只发生在高能量状态，其对流格局的南北对称被打破，柱状旋涡结构变得脆弱。另一个吸引人的发现是，四极模式的增长，超过了两极模式的逆转，这可能有助于解释地球磁场的逆转。

14. Feldman W C, Boynton W V, Tokar R L, et al. Global distribution of neutrons from Mars: results from Mars Odyssey[J]. Science, 2002, 297(5578): 75-78.

Global distribution of neutrons from Mars: results from Mars Odyssey

Feldman W. C., Boynton W. V., Tokar R. L., Prettyman T. H., Gasnault O., Squyres S. W., Elphic R. C., Lawrence D. J., Lawson S. L., Maurice S., McKinney G. W., Moore K. R., Reedy R. C.

(Los Alamos National Laboratory, Los Alamos, NM 87545, USA)

Abstract: Global distributions of thermal, epithermal, and fast neutron fluxes have been mapped during late southern summer/northern winter using the Mars Odyssey Neutron Spectrometer. These fluxes are selectively sensitive to the vertical and lateral spatial distributions of H and CO_2 in the uppermost meter of the martian surface. Poleward of ±60 degrees latitude is terrain rich in hydrogen, probably H_2O ice buried beneath tens of centimeter-thick hydrogen-poor soil. The central portion of the north polar cap is covered by a thick CO_2 layer, as is the residual south polar cap. Portions of the low to middle latitudes indicate subsurface deposits of chemically and/or physically bound H_2O and/or OH.

火星中子的全球性分布：来自火星探测器奥德赛的结果

Feldman W. C., Boynton W. V., Tokar R. L., Prettyman T. H., Gasnault O., Squyres S. W., Elphic R. C., Lawrence D. J., Lawson S. L., Maurice S., McKinney G. W., Moore K. R., Reedy R. C.

（美国洛杉矶 Los Alamos 国家实验室）

摘要：使用火星奥德赛中子光谱仪测量南方夏季/北方冬季的热、超热、快速中子的通量分布。这些通量对于火星表面最上层的氢和二氧化碳的横向和纵向分布选择性敏感。极地的±60°纬度地区富含氢，可能在几十厘米厚的氢贫瘠土壤之下埋藏着水。北方的极盖的中央部分，如南极极盖一样，为厚厚的二氧化碳覆盖。低到中纬度地区地表下存在化学或物理形式结合的 H_2O 或 OH。

15. Hulot G, Eymin C, Langlais B, et al. Small-scale structure of the geodynamo inferred from Oersted

and Magsat satellite data[J]. Nature,2002,416(6881):620-623.

Small-scale structure of the geodynamo inferred from Oersted and Magsat satellite data

Hulot G., Eymin C., Langlais B., Mandea M., Olsen N.

(Département de Géomagnétisme et Paléomagnétisme, CNRS UMR 7577, Institut de Physique du Globe de Paris, 4 Place Jussieu, B89, Tour 24, 75252 Paris cedex 05, France)

Abstract: The "geodynamo" in the Earth's liquid outer core produces a magnetic field that dominates the large and medium length scales of the magnetic field observed at the Earth's surface. Here we use data from the currently operating Danish Oersted satellite, and from the US Magsat satellite that operated in 1979/80, to identify and interpret variations in the magnetic field over the past 20 years, down to length scales previously inaccessible. Projected down to the surface of the Earth's core, we found these variations to be small below the Pacific Ocean, and large at polar latitudes and in a region centred below southern Africa. The flow pattern at the surface of the core that we calculate to account for these changes is characterized by a westward flow concentrated in retrograde polar vortices and an asymmetric ring where prograde vortices are correlated with highs (and retrograde vortices with lows) in the historical (400 year average) magnetic field. This pattern is analogous to those seen in a large class of numerical dynamo simulations, except for its longitudinal asymmetry. If this asymmetric state was reached often in the past, it might account for several persistent patterns observed in the palaeomagnetic field. We postulate that it might also be a state in which the geodynamo operates before reversing.

卫星资料推测的地球发电机的小尺度结构

Hulot G., Eymin C., Langlais B., Mandea M., Olsen N.

(法国国家科学院地球物理研究所)

摘要:"地球发电机"在地球的液态外核产生的磁场主宰了在地球表面观察到的大中尺度的磁场。在这里,我们使用目前在丹麦经营的奥斯特卫星数据和美国在1979/1980年运行的Magsat卫星数据,来识别和解释在过去20年的、尺度小到以前无法触及的磁场变化。预计到地球的核心表面,我们发现这些变化在太平洋下幅度小,极地以下幅度大,中心在南部非洲地区。我们对核心表面流进行一些计算,来解释这些变化,所得到的流动模式,其特点是有一个西向径流集中于极涡环和一个非对称环,在该环处,在顺行的涡环与历史(400年的平均值)上的磁场高点(和有低点的逆行涡)相关。这种模式与发电机的数值模拟过程所见到的类似,主要的差别表现在其纵向不对称上。如果这种不对称状态,在过去常常形成,它可以解释古地磁数据观测到的几个持续的模式。我们推测,这可能是磁场逆转之前地球发电机的一个运行状态。

16. Leitch E M, Kovac J M, Pryke C, et al. Measurement of polarization with the Degree Angular Scale Interferometer[J]. Nature, 2002, 420(6917): 763-771.

Measurement of polarization with the Degree Angular Scale Interferometer

Leitch E. M., Kovac J. M., Pryke C., Carlstrom J. E., Halverson N. W., Holzapfel W. L., Dragovan M., Reddall B., Sandberg E. S.

(Department of Astronomy & Astrophysics, University of Chicago, 5640 South Ellis Avenue, Chicago, Illinois 60637, USA)

Abstract: Measurements of the cosmic microwave background (CMB) radiation can reveal with remarkable precision the conditions of the Universe when it was, 400000 years old. The three most fundamental properties of the CMB are its frequency spectrum (which determines the temperature), and the fluctuations in both the temperature and polarization across a range of angular scales. The frequency spectrum has been well determined, and considerable progress has been made in measuring the power spectrum of the temperature fluctuations. But despite many efforts to measure the polarization, detection of this property of the CMB has hitherto been beyond the reach of even the most sensitive observations. Here we describe the Degree Angular Scale Interferometer (DASI), an array of radio telescopes, which for the past two years has conducted polarization-sensitive observations of the CMB from the Amundsen-Scott South Pole research station.

用度角刻度干涉仪测量极化偏振

Leitch E. M., Kovac J. M., Pryke C., Carlstrom J. E., Halverson N. W., Holzapfel W. L., Dragovan M., Reddall B., Sandberg E. S.

(美国芝加哥大学天文与天体理物理系)

摘要:宇宙微波背景辐射(CMB)的测量可以显著精确地揭示宇宙在形成后 400000 年的情况。CMB 的这三个最根本特性是它的频谱(它决定了温度)、温度波动和极化偏振波动。频谱已得到很完善的研究,对于温度波动的频谱测量已经取得了相当重要的研究进展。但是,尽管经过了许多努力,CMB 的极化偏振探测迄今为止仍然超出了最灵敏观测范围。在这里,我们描述了度角规模干涉仪(DASI),这是射电望远镜阵列,过去两年在阿蒙森和斯科特南极研究站用于 CMB 极化偏振观测。

17. Tomeoka K, Kiriyama K, Nakamura K, et al. Interplanetary dust from the explosive dispersal of hydratedasteroids by impacts[J]. Nature, 2003, 423(6935): 60-62.

Interplanetary dust from the explosive dispersal of hydratedasteroids by impacts

Tomeoka K., Kiriyama K., Nakamura K., Yamahana Y., Sekine T.

(Department of Earth and Planetary Sciences, Faculty of Science, Kobe University, Nada, Kobe 657-8501, Japan)

Abstract: The Earth accretes about 30000 tons of dust particles per year, with sizes in the range of

20-400 mμm. Those particles collected at the Earth's surface-termed micrometeorites-are similar in chemistry and mineralogy to hydrated, porous meteorites, but such meteorites comprise only 2.8% of recovered falls. This large difference in relative abundances has been attributed to "filtering" by the Earth's atmosphere, that is, the porous meteorites are considered to be so friable that they do not survive the impact with the atmosphere. Here we report shock-recovery experiments on two porous meteorites, one of which is hydrated and the other is anhydrous. The application of shock to the hydrated meteorite reduces it to minute particles and explosive expansion results upon release of the pressure, through a much broader range of pressures than for the anhydrous meteorite. Our results indicate that hydrated asteroids will produce dust particles during collisions at a much higher rate than anhydrous asteroids, which explains the different relative abundances of the hydrated material in micrometeorites and meteorites: the abundances are established before contact with the Earth's atmosphere.

水合小行星碰撞所产生的星际尘埃

Tomeoka K., Kiriyama K., Nakamura K., Yamahana Y., Sekine T.

(日本神户大学地球与行星科学系)

摘要：每年地球吸收约30000吨粉尘颗粒，大小范围为20～400 mμm。这些被收集的粒子在地球表面被称为微小陨石，这些陨石在水合化学和矿物学上与多孔陨石相似，但这样的陨石仅占2.8%。这种相对丰度的大幅差异归因于地球大气的过滤作用，即多孔陨石被认为非常易碎以至于它们不能在大气层的冲击下保存。这里，我们报告两个多孔陨石的冲击恢复实验，其中一个是水合的，另一个是无水的。对于水合陨石来说，冲击可以转变为微小颗粒，并且随着压力的释放，爆炸性的膨胀；其压力的范围要比无水合陨石大。我们的研究结果表明，水合陨石在相互碰撞过程中产生尘埃的速率远高于非水合小行星，这解释了水合陨石在微型陨石和陨石中的不同丰度；在它们到达地球大气之前，丰度就已经确定。

18. Jackson A. Intense equatorial flux spots on the surface of the Earth's core[J]. Nature, 2003, 424 (6950): 760-763.

Intense equatorial flux spots on the surface of the Earth's core

Jackson A.

(School of Earth Sciences, University of Leeds, Leeds LS2 9JT, UK)

Abstract: A large number of high-accuracy vector measurements of the Earth's magnetic field have recently become available from the satellite Oersted, complementing previous vector data from the satellite Magsat, which operated in 1979/80. These data can be used to infer the morphology of the magnetic field at the surface of the fluid core, similar to 2900 km below the Earth's surface. Here I apply a new methodology to these data to calculate maps of the magnetic field at the core surface which show intense flux spots in equatorial regions. The intensity of these features is unusually large-some have intensities comparable to high-latitude flux patches near the poles, previously identified as the major component of the dynamo field. The tendency for pairing of some of these spots to the north and south of the geographical equator suggests they might be associated with the tops of equatorially symmetric columnar structures in the fluid, or their antisymmetric equivalents. The drift of the equatorial features may represent material flow or could represent wave motion; discrimination of these two effects based on future data could provide new information on the strength of the hidden toroidal magnetic field of the Earth.

地心表层赤道位置强烈的通量点

Jackson A.

（英国利兹大学地球科学系）

摘要：从奥斯特卫星已获得大量的对地球磁场的高精度矢量测量数据，补充了在 1979/1980 年从 Magsat 卫星所得的矢量数据。这些数据可用于推断在液核表面 2900 km 以下的磁场形态。在这里，我应用一个新的方法来分析这些数据，计算磁场图，结果表明地核表面在赤道地区存在强通量点。这些点的强度非常大，接近于极地附近一些高纬度地区的强度，这些高纬度地区以前被认为是地球发电机场的主要组成部分。这些热点趋向于赤道南北的区域偶联，表明它们可能与流体中沿赤道是对称的，或是反对称的，柱状结构的顶部相关。这些赤道附近结构的漂移可能代表物质流或波动；基于未来数据区别这两种效应，可提供有关隐藏环形地球磁场强度的新信息。

19. Lawrence J S, Ashley M C B, Tokovinin A, et al. Exceptional astronomical seeing conditions above Dome C in Antarctica[J]. Nature, 2004, 431(7006): 278-281.

Exceptional astronomical seeing conditions above Dome C in Antarctica

Lawrence J. S., Ashley M. C. B., Tokovinin A., Travouillon T.

(School of Physics, University of New South Wales, New South Wales 2052, Australia)

Abstract: One of the most important considerations when planning the next generation of ground-based optical astronomical telescopes is to choose a site that has excellent "seeing"—the jitter in the apparent position of a star that is caused by light bending as it passes through regions of differing refractive index in the Earth's atmosphere. The best mid-latitude sites have a median seeing ranging from 0.5 to 1.0 arcsec. Sites on the Antarctic plateau have unique atmospheric properties that make them worth investigating as potential observatory locations. Previous testing at the US Amundsen-Scott South Pole Station has, however, demonstrated poor seeing, averaging 1.8 arcsec. Here we report observations of the wintertime seeing from Dome C, a high point on the Antarctic plateau at a latitude of 75 degrees S. The results are remarkable: the median seeing is 0.27 arcsec, and below 0.15 arcsec 25 per cent of the time. A telescope placed at Dome C would compete with one that is 2 to 3 times larger at the best mid-latitude observatories, and an interferometer based at this site could work on projects that would otherwise require a space mission.

南极冰穹 C 位置卓越的天文观测条件

Lawrence J. S. , Ashley M. C. B. , Tokovinin A. , Travouillon T.

（澳大利亚新威尔士大学物理学院）

摘要：在规划地面光学天文望远镜的下一代时最重要的考虑是选择了一个地点，具有好的观视性，也就是当光线通过具不同折射率的地球大气层时，其弯曲引起恒星视位置的抖动尽可能小。最好在中纬度地区，观视性范围从 0.5 到 1.0 角秒。南极高原上的站点有独特的大气性质，可成为潜在的天文观测点，因此值得研究。先前在美国阿蒙森·斯科特南极站前的测试，得到的观视性较差，平均为 1.8 角秒。本文中我们报道了来自南极高原南纬 75°高点——Dome C 的冬季的观测。观测结果非常显著：中位数为 0.27 角秒；其 25%的时间低于 0.15 角秒。Dome C 处放置的望远镜可与在中纬度最好地点放置的比其大 2 至 3 倍的望远镜相媲美，而在此放置的干涉仪可用于有太空使命的项目。

20. Plane J M C，Murray B J，Chu X Z，et al. Removal of meteoric iron on polar mesospheric clouds[J]. Science，2004，304(5669)：426-428.

Removal of meteoric iron on polar mesospheric clouds

Plane J. M. C. , Murray B. J. , Chu X. Z. , Gardner C. S.

(School of Environmental Sciences, University of East Anglia, Norwich NR4 7TJ, UK)

Abstract: Polar mesospheric clouds are thin layers of nanometer-sized ice particles that occur at altitudes between 82 and 87 kilometers in the high-latitude summer mesosphere. These clouds overlap in altitude with the layer of iron (Fe) atoms that is produced by the ablation of meteoroids entering the atmosphere. Simultaneous observations of the Fe layer and the clouds, made by lidar during midsummer at the South Pole, demonstrate that essentially complete removal of Fe atoms can occur inside the clouds. Laboratory experiments and atmospheric modeling show that this phenomenon is explained by the efficient uptake of Fe on the ice particle surface.

极地中层云中陨石铁的去除

Plane J. M. C. , Murray B. J. , Chu X. Z. , Gardner C. S.

（英国东英吉利大学环境科学学院）

摘要：极区中层云是一种毫微米级的冰粒子薄层，通常发生在高纬度夏季 82～87 km 高度。这些云与铁原子层重叠，铁原子层是陨石进入大气层烧蚀产生的。激光雷达夏季在南极对铁层和云的同时观测证明，云内的铁原子可以完全被排除。实验室实验和大气模拟表明，这种现象是由于冰颗粒表面上对铁的高效吸收形成的。

21. Maule C F，Purucker M E，Olsen N，et al. Heat flux anomalies in Antarctica revealed by satellite magnetic data[J]. Science，2005，309(5733)：464-467.

Heat flux anomalies in Antarctica revealed by satellite magnetic data

Maule C. F., Purucker M. E., Olsen N., Mosegaard K.

(Center for Planetary Science, Juliane Maries vej 30, 2100 Copenhagen Oe, Denmark)

Abstract: The geothermal heat flux is an important factor in the dynamics of ice sheets; it affects the occurrence of subglacial takes, the onset of ice streams, and mass tosses from the ice sheet base. Because direct heat flux measurements in ice-covered regions are difficult to obtain, we developed a method that uses satellite magnetic data to estimate the heat flux underneath the Antarctic ice sheet. We found that the heat flux underneath the ice sheet varies from 40 to 185 megawatts per square meter and that areas of high heat flux coincide with known current volcanism and some areas known to have ice streams.

卫星磁力数据揭示了在南极洲热通量的异常

Maule C. F., Purucker M. E., Olsen N., Mosegaard K.

(丹麦行星科学中心)

摘要:地热热流通量是冰盖动力学的一个重要因素,它会影响次冰川的形成、冰流的开始和冰盖基部的大规模丢失。因为在冰封地区很难直接测量热流密度,我们找到了一种方法,利用卫星磁力数据估计的热流密度,我们利用卫星磁数据,发展一个方法,估计南极冰盖下的热通量。我们发现南极冰层下面的热流密度在每平方公尺 40～185 兆瓦之间变化,高热通量的区域恰好与已知的火山活动区和一些已知有冰流的区域一致。